Applied Calculus

INTERPRETATIONS IN BUSINESS, LIFE, AND SOCIAL SCIENCES

About the Authors

Denny Burzynski

Denny Burzynski has taught mathematics at West Valley College since 1980. His office is 7' by 12' with no windows, but is close to the laserwriter. He enjoys listening to his daughter, eating Thai food for lunch, running (sort of), and summer vacation. Denny received his BA and MA in mathematics at the California State University, Long Beach. He has coauthored four other mathematics textbooks with his colleagues at West Valley. During the spring semester of 1992, he interned as a program director at the National Science Foundation in Washington, DC. Along with his colleagues Wade Ellis, Jr, and Ed Lodi, Denny has given numerous state and national workshops on technology in mathematics education. He, Wade, Ed, and Guy are committed to making mathematics more accessible, relevant, and understandable to all students. Denny is a member of the MAA and AMATYC, and a past-president of the California Mathematics Council, Community Colleges. He lives in San Jose, California but still wishes he lived at the beach again.

Guy D. Sanders

Guy D. Sanders has been a member of the mathematics department at West Valley College since 1983. He received his mathematics education at UCLA and has been teaching mathematics since 1972. Guy's strengths and interests lie in strategies to ensure that mathematics is more accessible, understandable, usable, and interesting to all students. He has participated in numerous workshops to that end. Guy lives in San Jose, California with his wife Patty and daughters Sheanna and Alana. He purports to golf, but has never been known to break 110.

Applied Calculus

INTERPRETATIONS IN BUSINESS, LIFE, AND SOCIAL SCIENCES

Denny Burzynski
Guy D. Sanders
West Valley College

PWS PUBLISHING COMPANY
I(T)P *An International Thomson Publishing Company*

BOSTON • ALBANY • BONN • CINCINNATI
DETROIT • LONDON • MADRID • MELBOURNE
MEXICO CITY • NEW YORK • PARIS • SAN FRANCISCO
SINGAPORE • TOKYO • TORONTO • WASHINGTON

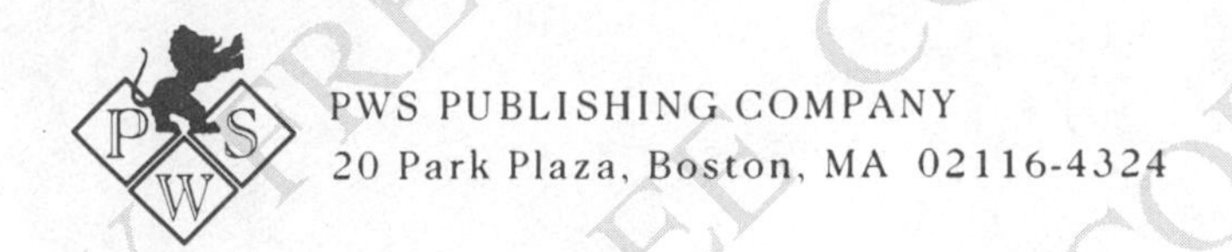

I(T)P™
International Thomson Publishing
The trademark ITP is used under license.

This book is printed on recycled, acid-free paper.

For more information, contact:
PWS Publishing Co.
20 Park Plaza
Boston, MA 02116

International Thomson Publishing Europe
Berkshire House I68-I73
High Holborn
London WC1V 7AA
England

Thomas Nelson Australia
102 Dodds Street
South Melbourne, 3205
Victoria, Australia

Nelson Canada
1120 Birchmont Road
Scarborough, Ontario
Canada M1K 5G4

International Thomson Editores
Campos Eliseos 385, Piso 7
Col. Polanco
11560 Mexico D.F., Mexico

International Thomson Publishing GmbH
Konigswinterer Strasse 418
53227 Bonn, Germany

International Thomson Publishing Asia
221 Henderson Road
#05-10 Henderson Building
Singapore 0315

International Thomson Publishing Japan
Hirakawacho Kyowa Building, 31
2-2-1 Hirakawacho
Chiyoda-ku, Tokyo 102
Japan

Library of Congress Cataloging-in-Publication Data

Burzynski, Denny.
Applied calculus : interpretations in business, life, and social sciences / Denny Burzynski & Guy D. Sanders.
p. cm.
Includes index.
ISBN 0-534-175988 (hardcover)
1. Calculus. I. Sanders, Guy D. II. Title
QA303.B947 1995 95-35155
515--dc20 CIP

Acquisitions Editor: *Steve Quigley*
Production Editor: *Monique A. Calello*
Editorial Assistant: *Anna Aleksandrowicz*
Manufacturing Coordinator: *Wendy Kilborn*
Marketing Manager: *Marianne Rutter*
Cover/Text Designer: *Julia Gecha*
Production: *Wade Ellis, Jr. / Art Ogawa*
Electronic Prepress Services: *Pure Imaging*
Cover image: *"Loma Amarillo," © 1989 Eyvind Earle*
Cover Printer: *New England Book Components, Inc.*
Text Printer: *Quebecor Printing/Martinsburg*

Printed and bound in the United States of America.
95 96 97 98 99---10 9 8 7 6 5 4 3 2 1

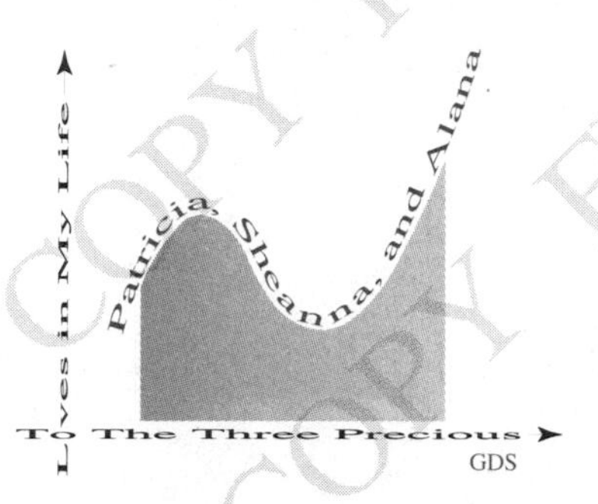

To my daughter, Sandi, who is still the star at the center of my universe, and to Rio, the star at the center of hers. To the memory of my father, an artist of great feeling and humor.
ILYF. . . DB

Contents

CHAPTER 7 Applications of Integration 377

CHAPTER 8 Calculus of Functions of Several Variables 450

APPENDIX Algebra Review 545

Preface

Applied Calculus: Interpretations in Business, Life, and Social Sciences, is the product of more than 15 years experience teaching the applied calculus course. The clearest of all our experience in teaching the course was that emphasis needed to be placed on interpreting the results obtained in calculus operations. We have written the book with precisely that in mind and have taken every opportunity to show students just how to make such interpretations. We have also offered an appreciation for the power of visualization. We believe that visualizing concepts helps one to know not only when and how to use a calculus operation, but helps one to know what the result of the operation means. We have also come to realize that technology has an important place in an applied course. We have many years of experience using technology in teaching mathematics and have included it in nearly every section of the book. We have done so in such a way that it can be incorportated by instructors and/or students if and when they choose do to so. It does not get in the way of students and instructors who wish not to use it. Little or no experience with graphing calculators is assumed, and by following the short directions provided in the sections, students and instructors can make powerful use of this important tool. The technology is intended to enhance the study of calculus in such a way that a greater variety of applications can be examined and hence, more interpretations made.

Goals and Philosophy

Some of our major goals in creating *Applied Calculus: Interpretations in Business, Life, and Social Sciences* are

1. To make applied calculus more accessible to students and to provide them with greater opportunitites for success with mathematics.
2. To provide students with a rich and broad mathematical experience that encourages independent thinking, mathematical exploration, and that builds mathematical confidence.
3. To connect mathematics to other disciplines as well as to present it as a developing human discipline.
4. To provide a comfortable and inviting environment in which to explore mathematical concepts and techniques.

To meet these goals, this book emphasizes several areas that the authors feel are most important to students of business, the life, and the social sciences. We have tried to develop a text that

1. Promotes a sincere understanding of the power and utility of the calculus.
2. Helps to develop in the student the ability to recognize when and the knowledge of how to use the methods of the calculus.
3. Provides the student with the ability to use technology effectively to solve problems that involve change or accumulation.
4. Is comfortable, non-intimidating, and pleasing to look at and read.

Major Themes

Interpretation The main emphasis of *Applied Calculus: Interpretations in Business, Life, and Social Sciences* is on interpretation. Applied calculus is an instrument used by the businessperson, life scientist, or social scientist to solve applied problems. Most often, then, a calculus operation results in a number. Numbers arrived at through calculus operations have contextual meanings attached to them. The book continually and consistently illustrates how to interpret numerical results in the context of the application and also develops the students' ability to do so.

Students often have trouble understanding mathematical rules and definitions. Therefore, we have included interpretations of rules and definitions when we feel they need clarification. The goal is to show students that rules and definitions have precise meanings and can be understood when read carefully. Another goal is to bring students to a point where they can correctly make their own interpretations of rules and definitions.

Functions Success in mathematics and in fields that apply mathematics depends on a solid understanding of the function. This text is about functions and their use in mathematics and in investigating applied and theoretical problem situations. The first chapter presents functions more thoroughly than presentations made in intermediate algebra, the recommended prerequisite for this course. The book encourages students to see functions as mathematical instruments than can be used to describe relationships between various quantities.

Functions are used to model a wide variety of problem situations. Here, we take advantage of the close relationship between a function defined by an expression and the representation of a function as a graph. Although some functions may be defined for all real numbers, the domains of most functions arising in problem situations are restricted in some natural way. These restrictions are emphasized and students are encouraged to think about the domain of applicability of the functions they create and use.

Multiple Representations of Functions Multiple representations of a function are used throughout this book. A function can be represented by an English sentence, a graph, a symbolic expression, or a table. Each of these representations has important and specific uses and the interrelationship between them can bring additional information about the behavior of a function. As students learn to work with each of these representations and to exploit their connections, they become more aware of the nature of functions and more able to use them as instruments in investigating both applied and theoretical problems.

Use of Technology

We believe that one of the most important outcomes in this course is the ability to describe the behavior of a function. We have been very careful to use technology so that it helps the student understand the behavior of the function they are studying. *Applied Calculus: Interpretations in Business, Life, and Social Sciences* provides the student with a solid foundation in the use of calculator technology, and to a lesser degree, computer algebra system technology (*Derive*). For the most part, we use examples (applications when possible) from the text to demonstrate how to effectively use technology to

- Evaluate functions of one and several variables.
- Graph functions of one variable.
- To approximate limits of functions using graphs, tables, lists, or sequences.
- To approximate the numerical derivative of a function.
- To approximate the relative extreme points of a function both graphically and by using the solve command to approximate the critical values.
- To approximate the inflection points of a function both graphically and by using the solve command to approximate the hypercritical values.
- To use *Derive* to find both the symbolic and numerical derivative of a function that is expressed implicitly.
- To summarize the behavior of a function.
- To approximate the definite integral of a function.
- To use *Derive* to find both the symbolic and numerical integral of a function.
- To use *Derive* to find both the symbolic and numerical solution to a differential equation.
- To approximate the partial derivative of a function.
- To approximate the total differential of a function.

We have at least 3 years of experience using technology in this course. We know how and when to use it this course, and apply it only in a way that enhances the student's understanding of calculus concepts. We did not just tack the technology on to a standard set of materials because it is the current thing to have. In all but 2 or 3 sections, a concise and to-the-point technology subsection appears as the last subsection, just before the exercise set. We present methods for solving calculus problems using both the graphing calculator **Using Your Calculator** and the computer algebra system *Derive* (**Using Derive**). The calculator technology material was written with the TI–82 in mind. It is generic enough that the entries can be readily converted to the corresponding entries of the other graphing calculators. The ideas behind the *Derive* commands should allow them to be converted to *Maple* commands.

Pedogogical Features

Writing Style Our experience has shown that most students taking this course are not entirely comfortable with symbolic definitions and explanations. For this reason we have offered written explanations of definitions and concepts. We have tried to unpack the information contained in mathematical sentences, expressions, and symbols. Our writing style speaks to the student, is somewhat informal, but mathematically accurate.

We have made an effort to reinforce the fact that differentiation is the language of change and that integration is the language of accumulation. Students need to know when to apply the calculus to a problem. By recognizing that a problem involves change, the student will think of the derivative. By recognizing that a problem involves accumulation, the student will think of the integral.

We use a dynamic approach that, at times is more verbal and uses many illustrations. By dynamic, we mean we present examples much the way they are presented on the board in class. The student is able to see the development of the problem as time passes. For example, when showing that the limit of a secant line is a tangent line, we use several diagrams to show the movement of the secant line toward the tangent line. This is the approach we use in class and it works better than one diagram with several secant lines and a tangent line. Such diagrams are static and the student sees only the result, not the process.

Writing Requirements We have been careful to include exercises that require the students to convey their answers in sentence or paragraph form. To that end, we have included many examples in which problem situations are interpreted in sentence and paragraph form.

Objectives Each section begins with a list of the topics that are examined in the section. Each topic is set off as a subsection to make reading, studying and reviewing easier.

Example Sets Example problems that are to be solved are set off from the main text in boxes that are called **Example Sets**. Example sets include one or more examples of similar types. They have been carefully chosen and developed to illustrate concepts, calculator skills, and problem solving techniques in the most instructive way.

Illuminator Sets Example problems that are meant to illuminate concepts rather than demonstrate a problem solving technique are set off from the main text in boxes called **Illuminator Sets**.

Section Exercises The exercises are grouped into five sections, and in many cases, are odd-even paired.

1. **A Understanding the Concepts** These exercises are generally noncomputational. They may ask the student to interpret the result of a computation, or to interpret the information contained in a graph or a table.
2. **B Skill Acquisition** These exercises consist of problems that help to develop the students pencil-and-paper computational skills. Each can also be solved using electronic technology. We have carefully created problems that develop the student's calculus skills as well as hone his/her algebra skills without introducing unruly algebraic manipulations that overshadow the point of the exercise and discourage the student.
3. **C Applying the Concepts** The exercises in this section are applied problems from the such areas as business, economics, life science, social science, and psychology. To solve these problems, the student must use his or her computational skills as well as interpretive skills. Each exercise in this section can be worked using pencil-and-paper technology as well as calculator or computer technology.
4. **D Describing your Thoughts** There are typically two, three, or four exerices in this group. These problems ask the student to answer a question that a student might ask an instructor or another student. We belive these exercises help develop the students critical thinking and communicatons skills.
5. **E Review** Each section ends with 5–10 review exercises chosen from previous sections throughout the text. The exercises we selected for review are from those areas we believe are most important for the student in terms of his or her future needs. Each exercise is marked with the section from which it is discussed.

Chapter Summaries Each chapter includes a chapter summary that we have written in an informal style, much the way we would hope a student would write the summary. Each summary touches on the main points of the chapter and serves as a good, quick overview of the chapter material.

Supplementary Materials

Applied Calculus: Interpretations in Business, Life, and Social Sciences comes with

1. A Complete Solutions Manual that contains the worked-out solution to all the exercises except the *Describe your Thoughts* exercises.
2. A Student Solutions Manual that contains worked-out solutions to selected odd numbered exercises.
3. A Graphing Calculator Manual.
4. A Test Bank that is composed of EXP Test for DOS and Windows.
5. A Derive Notebook.

Construction of the Textbook and Acknowledgments

Books are rarely constructed alone and this one is no exception. Many talented people worked on the construction of this book and we most gratefully acknowledge their contributions.

We thank the staff at PWS publishing for believing that we could produce a textbook using a production process that is significantly different from the standard process. Our sincere gratitude goes to our editors,

1. Barbara Holland (formerly of the Wadsworth/PWS family) for her vision and for believing that we could write and complete a book that we hope begins a new evolutionary path in applied calculus textbooks.
2. Steve Quigley for bringing the book through the review process where many good ideas were discovered.
3. Monique Calello for her wonderful attention to detail in content and design. We very much appreciate having a production editor who is open to our comments and suggestions, but who still sees the bigger picture, working on our book.

We also thank Anna Aleksandrowicz for her work on the ancillary materials.

We believe PWS has taken a giant step forward in textbook production and are most pleased that they took that step with us.

Typesetting This book was typeset by the authors and their colleagues Wade Ellis, Jr., and Ed Lodi, using Textures, a phototypesetting package developed by Donald Knuth of Stanford University. The design for the text was created by PWS and Wadsworth Publishing Companies. The Textures macro package that directed TEX to typeset the text to the design was initially created by Rachel Goldeen and Art Ogawa and then enhanced by Art Ogawa in the California Sierras. All of the book, with the exception of the answer section, was created by the authors using the Textures macro package and we are responsible for the finished product. The answers to the exerices were produced by Todd Timmons of Westark Community College, Fort Smith, Arkansas.

Proofreading Although we've read and reread the text what seems like a thousand times, we also assume credit for any errors that remain. Those proofreaders who read the text in its entirety are Amy Mayfield and John Torzilli of PWS, and Amy Miller of Green Bean Graphics. Sandi Wiedemann Sethna of UC Santa Cruz read parts of the book and made many useful comments and recommendations.

Graphics The graphics were produced by the authors, Amy Miler of Green Bean Graphics, Woodside, California, Wade Ellis, Jr. of West Valley College, and Ariadna Pierson of Santa Clara, California. The graphics were created using Adobe Illustator and PSMathGraphs II, a graphics package created by John Jacobs of the College of Marin and MaryAnn Software, San Rafael, California.

Cover We are grateful to Eyvind Earle, Monterey, California, for allowing us to use his image of Loma Amarillo on the cover of the book. We have long admired his work and have always experienced feelings of fascination, optimism, confidence, and inspiration when viewing it. We hope our writing reflects the qualities we feel from Eyvind Earle's art.

Supplementary Materials The Instructor's Manual and Student's Solution Manual was constructed by Todd Timmons of Westark Community College, Fort Smith, Arkansas. He worked extraordinarily hard to produce the solutions to the exercises we developed and politely corrected us when some of our exercises made no sense. We also thank Joe Kenstowicz of West Valley College, Saratoga, California, for spot checking the solutions.

Reviewers We are very grateful to the following mathematics educators for their comments on the manuscript and recommendations on how it could be improved. They provided numerous insights and suggestions that markedly improved the quality of *Applied Calculus: Interpretations in Business, Life, and Social Sciences* and their help was invaluable and appreciated.

- Richard Armstrong, *Florissant Valley Community College*
- Fred Brauer, *University of Wisconsin at Madison*
- Len Bruening, *Cleveland State University*
- David Burton, *Chabot College*
- Connie Carruthers, *Scottsdale Community College*
- Richard Fast, *Glendale Community College*
- Karl Ray Gentry, *University of North Carolina at Greensboro*
- Harvey Greenwald, *California Polytechnic State University*
- John Haverhals, *Bradley University*
- Yvette Hester, *Texas A&M University*
- Jaclyn LeFobore, *Illinois Central College*
- Arthur Lieberman, *Cleveland State University*
- Steve Marsden, *Glendale Community College*
- John Martin, *Santa Rosa Junior College*
- Kay Moneyhun, *Murray State University*
- Karla V. Neal, *Louisiana State University*
- Robert H. Pervine, *Murray State University*
- Claudia Pinter-Lucke, *California Polytechnic State University*
- Arthur Rosenthal, *Salem State College*
- Gordon Shilling, *University of Texas*
- Jane Sieberth, *Franklin University*
- Mark L. Tepley, *University of Wisconsin at Milwaukee*
- Todd Timmons, *Westark Community College*
- Richard Tondra, *Iowa State Universtiy*
- Melvin R. Woodard, *Indiana University of Pennsylvania*

Denny Burzynski
Guy D. Sanders

CHAPTER 1

Functions

1.1 An Intuitive Description of Functions

Introduction

The function is of fundamental interest in the calculus. In the calculus you will make a detailed study of the behavior of functions. You will study how it can be used to describe change and accumulation in physical and theoretical systems. To do so requires a genuine understanding of the concept of the function. This section is intended to provide you with an intuitive understanding of the function concept.

It is often convenient, and many times necessary, to describe relationships between quantities of physical or theoretical phenomena. These relationships can be modeled by mathematical instruments called functions. Functions can then be used to predict the future behavior of the system from which the phenomena develops.

We will illustrate how information can be described using mathematical symbols, show how function notation is quite descriptive, and demonstrate how to use the notation in computation.

Functions as Instruments for Describing Information

Functions: Describing Relationships

A **function** is a mathematical instrument used to describe relationships and present information about those relationships. You say: *My taxes are a function of my income, since the amount of my taxes depends on the amount of money I make.* This common usage of the word *function* conveys the idea that one thing depends on another. It is useful for you to think up some relationships yourself. For example, your interest in a lecture is a function of the quality of the lecture; your grade in a course is a function of your exam scores; your grade in a course is a function of your effort in the course; your telephone bill is a function of the number of long distance phone calls you make; the number of units of a product a company can produce is a function of the number of dollars it invests in labor. Try a few yourself.

Not all these relationships or functions are well defined or quantifiable, but they do indicate how one thing depends on another thing. Sometimes one thing can depend on more than one other thing. For example, the profit a company makes each year depends on the amount of money it spends on advertising *and* the number of salespeople it employs. When a definite, well-defined dependence of one thing on another can be developed, we may be able to express the relationship between those things mathematically (using mathematical symbols) and then extract information from that relationship.

Function Notation

It is cumbersome to continue to describe a relationship between quantities with words and sentences. We need to develop a descriptive notation. We will now develop the commonly used function notation.

If one thing depends on another, we have

one thing depends on another thing

How Functions Describe Relations

To distinguish the two things, we number them 1 and 2. Let's agree to say that the second thing is determined from the first, that is, that the second thing depends on the first. Then we have,

$thing_2$ depends on $thing_1$,

or

$thing_2$ is determined by $thing_1$

The idea is that once $thing_1$ is known, it is placed into the rule that describes the dependence, and $thing_2$ is produced. This is the *input–output* idea. Input $thing_1$ and output $thing_2$.

For example, putting names to $thing_2$ and $thing_1$, when something travels at a constant speed,

distance depends on *time*,

or

distance is determined by *time*

We can eliminate the words by using symbols. If we let d represent *distance* and t represent *time*, then we have

d depends on t,

or

d is determined by t

We can compact the notation even more by using parentheses "( )" to represent the words *depends on* or *is determined by*. Thus,

$d(t)$ means that *distance* depends on *time*,

or

$d(t)$ means that *distance* is determined by *time*

We read the notation $d(t)$ as *dee of tee*, or as *dee at tee*, and think of it as shown in Figure 1.1.

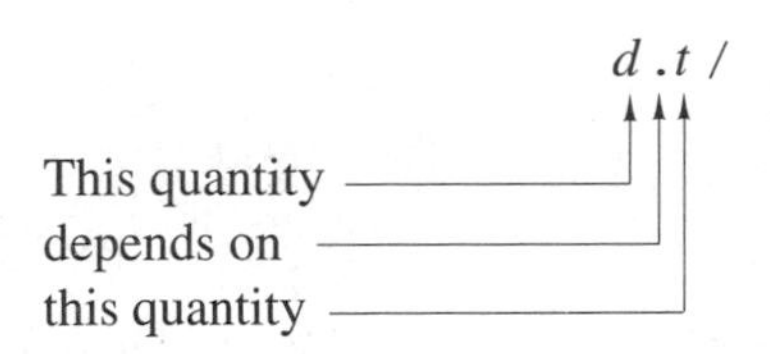

Figure 1.1

The notation is actually visually descriptive. The quantity inside the parentheses is the input quantity. (After all, it has been put into the parentheses.) The entire quantity $d(t)$ is the output quantity. *Rather than using only the quantity outside the parentheses to signify the output quantity, we include the parentheses to reinforce the fact that output depends on the noted input quantity.* In the case of our example, $d(t)$, t is the input quantity and $d(t)$ is the output quantity. When *time* is placed into the rule that describes the dependence, the *distance* traveled is output.

Table 1.1

t	$d(t)$
1	60
2	120
3	180
4	240
5	300
6	360

Suppose that for six hours we observe a car traveling at the constant speed of 60 miles per hour. Let's organize our observations into a table and look for a pattern. Careful observation of the table seems to indicate that the value of $d(t)$ is always 60 *times* the value of t. This observation helps us to establish the rule that describes the relationship between t and $d(t)$. It appears that t determines $d(t)$ by the rule $60 \cdot t$, that is, $60t$. We express the relationship by writing an "=" sign between the output notation and the rule expressing what to do with the input. In this case,

$$\underbrace{d(t)}_{\textit{output}} = \underbrace{60t}_{\textit{rule}}$$

Thus, when we read this notation the left side tells us that two particular quantities are related, and the right side tells us precisely how one (the output) is determined from the other (the input). In this example, the left side tells us that distance and time are related, and the right side tells us that distance is determined by multiplying travel time by 60.

We can now use the function to produce the output values for any acceptable input values. For example, if $t = 8$, then $d(8) = 60 \cdot 8 = 480$. That is, if the time spent traveling is 8 hours, then the distance traveled is 480 miles. A value of t such as -3 is not an acceptable value since time and distance are positive quantities and $d(-3) = 60(-3) = -180$ has no physical meaning.

Before we proceed to several examples, we remind you of some terminology associated with functions. The set of values that are used as inputs to a function

Domain and Range

is called the **domain** of the function. The corresponding set of output values is called the **range** of the function. In Section 1.2, we will define the domain and range of a function more rigorously.

We will illustrate the use of function notation with several examples.

ILLUMINATOR SET A

If we represent the radius of a circle by r and the circumference of the circle by C, then the relationship between the circumference and radius of a circle can be expressed as

$$C(r) = 2\pi r$$

The left side tells us that the circumference of a circle depends on the radius of the circle, and the right side tells us the circumference is determined by multiplying the radius by 2π.

We can build a table that displays the relationship visually by inputting several values for the radius and computing the corresponding values of the circumference. The table below shows several exact values of the circumference and several approximations (using $\pi \approx 3.14$). Complete the table by filling in the missing parts. You may wish to use your calculator to carry out the multiplications.

Table 1.2

r	Exact $C(r)$	Approximate $C(r)$
1	2π	6.28
2	4π	12.56
3	6π	18.84
4	8π	25.12
5		
6		

The table exhibits only seven of the infinitely many possible choices for input values and their corresponding output values. The table therefore represents only a partial description rather than a complete description of this function.

Judging by what we can compute, we might imagine that both the domain and the range of this function are the set of all real numbers. It is true that any real number, positive, negative, or zero, can be substituted for r and a value for $C(r)$ computed. However, this function is describing some physical or theoretical situation and therefore its domain must be chosen to reflect the reality of that situation. To exist, a circle must have a positive radius and circumference. Therefore, both the domain and the range must be at most the set of all positive real numbers. In practical situations, circles also have some limit as to how big they can be. For example, an architect using this function to construct circles that represent trees and shrubs on the land surrounding a housing development might set the domain as $0 < r \leq 60$, where r is in millimeters. The point is that *the domain of a function is chosen by the user of the function to describe a particular situation.* When a particular situation is not known to us, we will select the biggest domain that reflects the reality of the situation the function describes. In this case we would say that the domain is some subset of the set of positive real numbers.

ILLUMINATOR SET B

If we represent the length of the side of a square by s and the area of the square by A, then the relationship between the length of a side and the area of a square can be expressed as

$$A(s) = s^2$$

Table 1.3

s	$A(s)$
1	1
2	4
3	9
4	16
6	
10	

The left side tells us that the area of a square depends on the length of a side, and the right side tells us the area is determined by squaring the length of the side.

We can build a table that displays the relationship visually by inputting several values for the length of the side and computing the corresponding values of the area. Complete the table by filling in the blank part of the output column. You may wish to use your calculator to carry out the multiplications.

For much the same reasoning as in the previous example, the domain and range of this function are some subsets of the set of positive real numbers. The table exhibits only seven input values and four output values and therefore presents only a partial description of this function.

ILLUMINATOR SET C

Let A represent the amount of money accumulated, P represent the amount of money invested, i represent the interest rate per period, and t represent the time (in years) of an investment. Then the amount of money accumulated in an investment is related to the amount of money invested, the interest rate, and the length of time for which the money is invested by the function

$$A(P, i, t) = P(1 + i)^t$$

This function illustrates how one quantity depends on three other quantities.

Write down the meaning of both the left and right sides of the equal sign. The left side tells us that ______________________. The right side tells us that ______________________. Then build a table that displays three or four values of A for some values of P, i, and t that you choose. One set of input values and the corresponding output value have been entered to get you started. (Use your graphing calculator to check that the three input values actually do generate the specified output value.)

As in the previous examples, the table, Table 1.4, presents only a partial description of the function. The domain and range of this function are somewhat difficult to specify. Any real number, positive, negative, or zero, could be substituted for P, i, or t, and a value of A computed. But input values cannot be negative or zero, for in that case the function gives no meaningful information. They can only be certain positive numbers, and there must be some upper limit as to what they can be. Nor can they be irrational (you can't invest $\sqrt{6}$ dollars), so they must be rational. What upper bounds can be set on P, i, and t? Again, *the domain of a function is chosen by the user of the function to describe a particular situation.* For example, an investor who can invest at most $10,000 with a conventional bank and wants to collect his or her earnings sometime within the next 20 years would set the following domain: all ordered triples (P, i, t), where $0 < P \leq 10{,}000$, $0 < i \leq 100$, and $0 < t \leq 20$, and where the numbers in each of these sets are rational numbers.

Table 1.4

P	i	t	A
15,000	0.08	4	20,407.33

Table 1.5

x	$f(x)$
-2	-4
0	0
5	10
20	40
a	$2a$
$a+h$	$2\cdot(a+h)$
-3	
$x+3$	

ILLUMINATOR SET D

The function $f(x) = 2x$ can be thought of as a *doubling* function since it acts to double the value of the quantity that is input into it. Complete Table 1.5.

We might denote this function as $Double(x)$ rather than $f(x)$ since $Double(x)$ tells us what to do with the input quantity to get the output quantity—we double it. Since this function is not associated with any particular physical or theoretical situation, we describe the domain and range in the most general manner. The domain and range of this function is the set of all real numbers. As in the previous examples, the table presents only a partial description of the function.

Functions and Technology

Using Your Calculator You can use your graphing calculator to create the table representation of a function from the defining expression form. The following entries show the creation of a table using the function of Exercise 6 of this section. The function is

$$N(t) = 38 + \frac{32t^2}{t^2 + 18}$$

where $N(t)$ approximates the number of words a community college court reporting student can record after t hours of instruction.

In your functions menu, enter

```
Y1 = 38 + 32X^2/(X^2 + 18)
```

· Access the Table Set feature and enter the appropriate quantities. In this case, the minimum value for X is 0 since X represents the number of hours of instruction. The change in the table value is reasonably 1. Now create the table by accessing the Table feature.

You should be able to read and interpret the ordered pairs. For example, $(0, 38)$ means that with 0 hours of instruction, a court reporting student should be able to record about 38 words each minute. The pair $(22, 68.853)$ means that with 22 hours of instruction, a court reporting student should be able to record about 69 words each minute.

Calculator Exercises Use your graphing calculator to create the table representation of each function from the defining expression form.

1. A state's forestry service is attempting to establish a herd of deer as part of a forestry restoration program. Their plan is to initially place 150 deer in various locations in the forest so that the population of deer t years from now will be approximated by the function

$$N(t) = \frac{150 + 38t}{1 + 0.037t}$$

Create a table that displays the time and the population for the first 30 years in 3-year increments. Can you see a trend in the growth of the population? Does the increase in population seem to be speeding up or slowing down? How can you tell?

2. Healthy human skin cells reproduce themselves. In an experiment in which a tissue sample of skin was taken from a healthy person, a biologist found that the number of cells N in the tissues at time t could be approximated by the function

$$N(t) = -0.82x^4 + 4.4x^3 + 5.1x^2 + 3.6$$

where N is in thousands and t is in hours from the beginning of the experiment. Create a table that displays the time and the number of cells for the first 8 hours in 1/2-hour increments. Using the table, write, in sentence form, a description of the behavior of this function. Your description should include an estimate of the number of cells initially present, the maximum number of cells present and at what time that maximum appears, the minimum number of cells present and at what time that minimum appears, and whether or not the number of cells is increasing or decreasing between these times.

EXERCISE SET 1.1

For Exercises 1–11, write an English sentence description of the relationship that is being expressed by the function notation on the left side of the "= " sign. For example, the population $P(t)$ of a town t months from now is approximated by the function

$$P(t) = 3625 + 36t - 0.45t^2$$

Your response should be: *The population depends on the number of months from now.*

1. ***Economics: City Population*** The number of people $N(t)$ living in a midwestern town t months from now is approximated by the function

$$N(t) = 5201 + 28t - 0.42t^2$$

2. ***Business: Revenue*** The revenue $R(x)$ realized by a distributor of hair care products on the sale of x units of the product is approximated by the function

$$R(x) = 62x - 0.95x^2$$

3. ***Business: Worker Efficiency*** A worker's efficiency $E(t)$ is approximated by the function

$$E(t) = 35 + 15t - 3t^2$$

where t represents the number of hours after 7:00 A.M.

4. ***Manufacturing: Cost of Production*** The cost $C(x)$ to a manufacturer for producing x units of a product is approximated by the function

$$C(x) = 525 + 55x - 0.3x^2$$

5. ***Economics: Air Pollution*** Environmentalists suggest that t years from now, if the current laws are left unchanged, the level of pollution $P(t)$ in the air will be approximated by

$$P(t) = \frac{0.55\sqrt{9t^2 + 10.2t + 55}}{(t+1)^2}$$

parts per million.

6. ***Education: Skill Enhancement*** The administrators of a community college court reporting program advertise that the typical student enrolled in their program can record

$$N(t) = 38 + \frac{32t^2}{t^2 + 18}$$

words per minute after t hours of course instruction.

7. ***Business: Quality Assurance*** Engineers in the quality control department of a company say their test data indicates that the percentage $P(t)$ of units of a new mechanical device that fail after t hours of use is approximated by

$$P(t) = 1 - e^{-0.2t}$$

8. ***Medicine: Body Temperature and Illness*** A person's temperature $T(t)$ during a particular type of illness t days from the onset of the illness is approximated by the function

$$T(t) = 100.3^\circ + 2.8 \sin\left(\frac{\pi}{7}\right), \quad 0 \leq t \leq 12$$

9. ***Economics: Profit from Advertising*** The marketing department of a company suggests that the company's profit $P(x, y)$ on a new product is given by the function

$$P(x, y) = 622(15x + 4.6y + 18xy - x^2)$$

where x represents the number of dollars spent on radio advertisements and y represents the number of dollars spent on newspaper advertisements.

10. ***Economics: Stock Dividends*** The amount of money $A(d, p)$ yielded by a stock is given by the function

$$A(d, p) = \frac{d}{p}$$

where d is the dividends per share, and p is the price per share.

11. ***Psychology: Reading Ability*** The *reading ease* $E(w, s)$ of a passage of words is given by the function

$$E(w, s) = 206.835 - 0.846w - 1.015s$$

where w represents the number of syllables in a 100-word section, and s represents the average number of words per sentence.

For Exercises 12–23, write, using symbols, the defining function expression for each situation. As best you can, specify the domain of your function.

12. ***Mathematics: Volume*** The volume of a cube is a function of the length of an edge. Use V to represent volume and s to represent the length of an edge.

13. ***Mathematics: Volume*** The volume of a cylinder with base 5 square inches is a function of its height. Use V to represent volume and h to represent height.

14. ***Mathematics: Area*** The area of a circle is a function of its radius.

15. ***Mathematics: Volume*** The volume of a sphere is a function of its radius.

16. ***Business: Cost of Production*** The total cost of some items, each of which costs $0.20, is a function of the number of items that are produced and sold.

17. ***Mathematics: Perimeter*** The perimeter of a square is a function of the length of a side of the square.

18. ***Business: Cost or Purchase*** The total cost of n papers (if each paper costs $0.25) is a function of the number of papers purchased.

19. ***Mathematics: Area*** The area of a triangle with base 3 inches is a function of its altitude.

20. ***Mathematics: Area*** The area of a rectangle with base 3 inches is a function of its height.

21. ***Mathematics: Area*** The area of a rectangle is a function of its base and height. (Remember the parentheses.)

22. ***Mathematics: Area*** The area of a triangle is a function of its base and height.

23. ***Business: Cost of Purchase*** The total cost of n items is a function of price and the number n of items purchased.

For Exercises 24–31, find what is asked for.

24. ***Manufacturing: Automobile Production*** The number of cars produced at an automobile factory is a function of time and the rate r at which the cars are produced. Determine appropriate units of time and rate.

25. *Manufacturing: Button Production* The number of items produced at a button factory is a function of time and the rate r at which the buttons are produced. Determine appropriate units of time and rate. These units should not be the same as the units in Exercise 24.

26. *Application: Your Creation* Think of two quantities that are related, choose letters to represent each of the quantities, and use function notation to express that relationship. Specify, as best you can, the domain of your function.

27. *Application: Your Creation* Think of two quantities that are related, choose letters to represent each of the quantities, and use function notation to express that relationship. Specify, as best you can, the domain of your function.

28. *Business: Cost* Suppose

$$Cost(\#units) = 3(\#units)^2 - 2(\#units)$$

Complete the following table.

Table 1.6

Input	Output
2	
3	
a	
x	
$x+3$	
$x+h$	

Does this table present a complete or partial description of the function?

29. *Business: Revenue* Suppose

$$Revenue(\#units) = (\#units)^2 - 4(\#units) - 2$$

Complete the following table.

Table 1.7

Input	Output
3	
4	
5	
b	
x	
$x+3$	
$x+h$	

30. *Data Summary: Function Creation* Given the following table, determine a function definition that summarizes the information in the table.

Table 1.8

Input	Output
1.2	1.44
1.3	1.69
1.4	1.96
1.5	2.25

31. *Data Summary: Function Creation* Given the following table, determine a function definition that summarizes the information in the table.

Table 1.9

Input	Output
2	5
3	10
4	17
5	26

1.2 Relations and Functions

Introduction

In Section 1.1 we considered functions from an intuitive point of view. In this section we will take a more rigorous look at them. We will

- present two alternate mathematical definitions of a function and its more general concept, the relation,
- introduce the various parts of a function,
- present four different ways of describing a function,
- examine the vertical line test, a test that indicates if a graph does or does not represent a function, and
- discuss how to determine the domain of a function.

Relations

Relation

Suppose an assembly machine uses pressure to produce a particular shape of a piece of equipment. Consider the following relationship between the set of pressures and the set of defective pieces of equipment. (See Table 1.10.) With each element in the set of pressures, the table associates an element from a set of defectives. This constitutes what is called a **relation**.

Table 1.10

Pressure	Number of Defectives
30	0
35	1
40	2
45	2
50	3

Relation, Domain, and Range

A *relation* is a process or method, such as a rule or procedure, of associating with each element of a first set, called the *domain*, one or more elements of a second set, called the *range*.

In the pressure/defective example, if we call the relation *Pressures to Defectives*, R, then

$$\text{Domain of } R = \{30, 35, 40, 45, 50\} \quad \text{Range of } R = \{0, 1, 2, 3\}$$

Notice that each domain element in the relation R is directed to precisely one range element.

$30 \longrightarrow 0$ (and only 0)

$35 \longrightarrow 1$ (and only 1)

$40 \longrightarrow 2$ (and only 2)

$45 \longrightarrow 2$ (and only 2)

$50 \longrightarrow 3$ (and only 3)

Although both 40 and 45 are directed to 2, they each go only to 2 and nowhere else.

On the other hand, if we call the relation *Defectives to Pressures*, S, then

$$\text{Domain of } S = \{0, 1, 2, 3\} \quad \text{Range of } S = \{30, 35, 40, 45, 50\}$$

The relation S does not have the property that each domain element is directed to precisely one range element. There is a domain element that is associated with more than one range element—2 is associated with both 40 and 45. The following list illustrates this visually.

$0 \longrightarrow 30$ (and only 30)

$1 \longrightarrow 35$ (and only 35)

$2 \longrightarrow 40$ (but also)

$2 \longrightarrow 45$ (a second value)

$3 \longrightarrow 50$ (and only 50)

Functions

In Section 1.1 we defined a function intuitively as a way to describe relationships and present information about those relationships. We now define function more precisely.

Function

A *function* is a relation in which each domain element is associated with precisely one range element.

A more practical, or working definition of a function is as follows.

Function

A *function* is two sets, called the domain and range, and a rule that prescribes how each domain element is assigned to exactly one range element.

The Parts of a Function

Notice that a function has three parts, a domain, a range, and a rule. To completely describe the function, all three must be precisely specified. If one part is not specified, the function is not described completely. The rule of a function need not be algebraic, and some nonalgebraic rules will be illustrated in the examples that follow. It is quite common when working with physical phenomena to describe a corresponding function only partially. We will illustrate how functions are described later in the section.

In Section 1.1, we thought in terms of *input* and *output*. The domain of a function is composed of the input elements and the range, the output elements. So, functions consist of three parts, the domain (the set of input values), the range (the set of output values), and the rule (or assignment maker) that assigns each input value to precisely one output value.

It is important to understand that the rule of a function assigns one input value to precisely one output value. Illuminator Set A illustrates this point.

ILLUMINATOR SET A

1. The relation R that assigns pressures to defectives is a function. Each pressure is assigned to one and only one number of defectives. The domain elements 40 and 45 are both assigned to the range element 2, but this does not violate the definition of function since each of the numbers, 40 and 45, are *individually* assigned to precisely one number. R assigns 40 to 2 and to no other number,

and 45 to 2 and to no other number. Thus, $(40, 2)$ and $(45, 2)$ are both part of this function. For each input, R assigns one and only one output.

2. The relation S that assigns a number of defectives to a pressure is not a function. There is one domain element that is assigned to more than one range element. S assigns the number 2 to both the numbers 40 and 45. The relation S is such that for each input, there can be more than one output. The ordered pairs $(2, 40)$ and $(2, 45)$ are part of a relation, but not part of a function.

Describing Relations and Functions

Some of the more common ways of describing a function are:

Four Ways to Describe a Function

1. tables,
2. ordered pairs,
3. symbolic or verbal rules, and
4. graphs.

These various methods are illustrated with the example that follows. It should be clear, after observing these methods, that it can be a difficult task to describe a function completely. Of the three parts of a function, the domain is usually the easiest to specify. The range, which is generally more difficult to specify, is commonly omitted. Also, when working with physical phenomena, it may be difficult or even impossible to develop a rule that assigns input values to output values. In such cases, the description of the function is only partial because the rule is omitted. The following is a standard assumption regarding domains.

Assumption Regarding Domains

If the rule of a function is specified but the domain is not, assume the domain is the set of all real numbers that, when acted upon by the function rule, result in a real number.

The following example illustrates these four methods of description.

ILLUMINATOR SET B

Quantum Tunneling

Physicists have observed that as the temperature of helium approaches absolute zero ($-273°$ Celsius), it exhibits properties that are contrary to many of the conventional laws of physics. One such property is that solid helium allows a proportion of another solid to pass through it. Physicists call the phenomenon of one solid passing through another *quantum tunneling*. Suppose we call the relation that assigns temperature to proportion Q. At each temperature of the helium, a particular proportion of another solid is allowed passage. The relation Q is therefore a function and can be described in any of the ways we noted above.

Table 1.11

Temp	%
−265	50.2
−270	76.9
−272	87.6
−272.9	92.9

1. **Description by Tables**
The function Q can be described, at least partially, by a table. (See Table 1.11.) The domain values are listed in the left column of the table and the range values in the right column. The rule is the nonalgebraic rule, *assign the element in the left column of a particular row to the element in the right column of that same row*. For example, −270°C is assigned to 76.9%. Notice that the description is only partial since not *every* domain and range value has been listed.

2. **Description by Ordered Pairs**
The function Q can be described, at least partially, using ordered pairs (x, y). The domain values are listed as the first element in the ordered pair and the range values as the second element. The rule is the nonalgebraic rule, *assign the first element in an ordered pair to the second element of that same ordered pair*.

Letting the first element of an ordered pair represent temperature and the second element represent proportion, we can represent the function Q partially as the set

$$\{(-265, 50.2), (-270, 76.9), (-272, 87.6), (-272.9, 92.9)\}$$

For example, −272 is assigned to 87.6.

Notice that the description is only partial since not *every* domain and range value has been listed.

Description of a relation or function by table or ordered pairs has the disadvantage that if the domain consists of a great number of elements, constructing the table or listing the ordered pairs can become tedious if not impossible. What's worse, the relationship may be very difficult to visualize if the list is big.

3. **Description by Rule**

Complete Description

A rule, if one can be found, that describes how an output value is obtained from an input value has the advantage that it may be concise and, when applied, can generate the output value associated with any chosen input value. The rule is commonly given in the symbolic form presented in Section 1.1. For example, for quantum tunneling with x representing temperature and $f(x)$ representing proportion, the rule of the function can be expressed in the form $f(x) = -0.0533x - 13.622$. To describe the function completely, both the domain and the range must be specified. Helium exhibits quantum tunneling effects only when its temperature is near absolute zero (−273°C). A scientist working with the quantum tunneling effects of helium who has his equipment adjusted to produce temperatures between −255°C and −273°C has set the function's domain as $-255 \leq x \leq -273$. The function's range is $0 \leq f(x) \leq 100$, since anywhere from 0% to 100% of a solid can pass through the helium. Thus, for this scientist's purposes, the function is completely described as

$$f(x) = -0.0533x - 13.622, \quad \underbrace{-255 \leq x \leq -273}_{\textit{input values}}, \quad \underbrace{0 \leq f(x) \leq 100}_{\textit{output values}}$$

(Remember, $f(x)$ represents an output value.)

Notice two things. The function rule is symbolic (algebraic), and the function would be different if another scientist had adjusted the equipment so that it produced temperatures between $-265°C$ and $-270°C$. In this case, the function would be completely described as

$$f(x) = -0.0533x - 13.622, \quad -265 \leq x \leq -270, \quad 0 \leq f(x) \leq 100$$

4. Description by Graph

Graph of a Function

The geometric representation (picture) of the input/output pairs of a function is called the **graph of the function**. The graphical description has an advantage over the other types of descriptions in that the graph may reveal information (such as trends) that may not be evident from the table, the list of ordered pairs, or the rule alone. The local behavior of a function is a consideration of the graph over a small portion of its domain while the global behavior considers the entire domain of the function. We will begin by defining graphical completeness.

Complete Graph

The graph of a function is *complete* if it suggests all of the possible input/output combinations and all the local and global behavior of the function.

ILLUMINATOR SET C

Describe the function $f = \{(x, f(x)) | f(x) = x^3 + 1\}$ graphically.

Solution:

We first note that although the domain of f is not specified, it is the set of all real numbers. Table 1.12 shows some ordered pairs that satisfy this function.

Table 1.12

x	$y = x^3 + 1$
-2	-7
-1	0
0	1
1	2
2	9

We plot these five points and connect them with a smooth curve (see Figure 1.2).

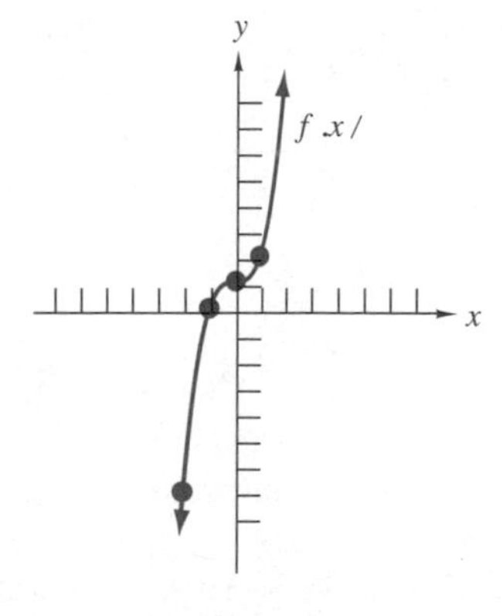

Figure 1.2

Note that f is truly a function since each input value is assigned to precisely one output value. Notice also a trend that may not be quite as evident in the observation of the rule alone. As the input values increase, the output values increase. This is an example of an *increasing* function.

For a function that describes a physical situation, the rule, taken alone, may have a larger domain than is warranted by the situation. In such cases, the graph of the function may not accurately describe the situation.

Situation Graph

A *situation* graph is a subset of a complete graph and displays only the input/output combinations and behavior of the function that are pertinent to the situation.

ILLUMINATOR SET D

The function $f(x) = -0.0533x - 13.622$ is the rule used in quantum tunneling and its graph appears in Figure 1.3.

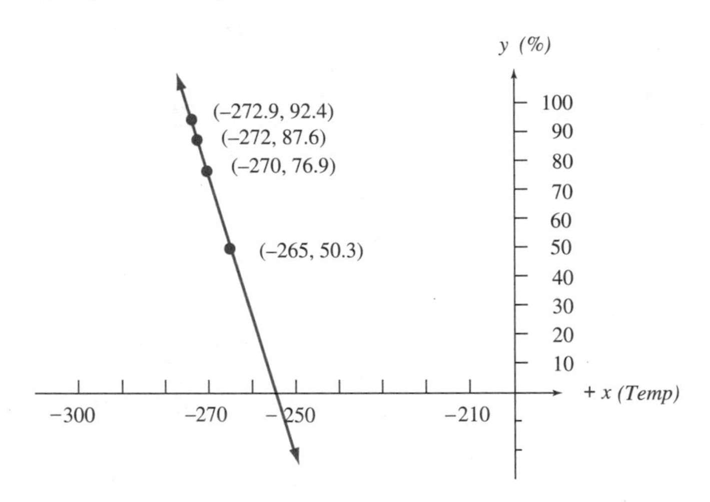

Figure 1.3

The points were plotted using the information provided in the table presented earlier in this section. The graph consists of only four points. Since both temperature and proportion are continuous, it is reasonable to assume that the points can be connected to produce a continuous, straight line.

The complete graph is not the appropriate graph to describe quantum tunneling. It displays information that is not relevant to the situation. It displays input values that are not in the domain and output values that are not in the range. For example, it indicates that percentage can be negative, and that temperature can be less than absolute zero. The appropriate graph is a situation graph. Figure 1.4 shows the quantum tunneling situation graph.

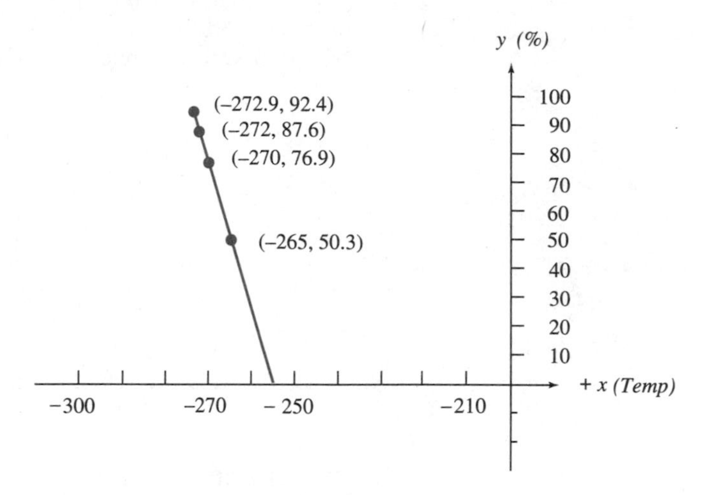

Figure 1.4

If the function is specified in such a way that the domain and range correspond to the physical situation, the complete graph and the situation graph will be the same.

Determining the Domain of a Function

The problems of the next example set illustrate how the domain and range of a function can be determined from a function rule that is given symbolically. The domain of a function can usually be found by algebraic methods. The range can be rather difficult to determine algebraically but can be estimated from the graph of the function.

EXAMPLE SET A

Find the domain of each function from the function rule.

1. $f(x) = 2x - 11$.

Solution:
Since real values of $f(x)$ result when any real number is substituted for x, the domain is the set of all real numbers.

2. $f(x) = \dfrac{5}{x-7}$.

Solution:
If $x = 7$, the expression $\dfrac{5}{x-7}$ is not defined and $f(7)$ is therefore not a real number. The domain is $\{x | x \neq 7\}$.

3. $f(x) = \sqrt{2x+9}$.

Solution:
$f(x)$ will be a real number only when $2x + 9 \geq 0$. Solve for x.

$$2x + 9 \geq 0$$
$$2x \geq -9$$
$$x \geq \frac{-9}{2}$$

The domain is $\{x | x \geq \frac{-9}{2}\}$.

Functions and Technology

Using Your Calculator You can use your graphing calculator to construct the graphs of functions from their defining expression. The following entries show how the graph of Exercise 6 of Section 1.1 is constructed. The function is

$$N(t) = 38 + \frac{32t^2}{t^2 + 18}$$

where $N(t)$ approximates the number of words a community college court reporting student can record after t hours of instruction.

In your functions menu, enter

Y1 = 38 + 32X^2 / (X^2 + 18)

Set your graphing window so as to obtain a complete graph. Since the input variable is t (X on the calculator), and t represents the number of hours in a court reporting course, reasonable values might be

Xmin= 0 and Xmax= 60

(a typical semester course). Also, since $N(t)$ represents the number of words that are recorded, reasonable values might be

Ymin= 0 and Ymax= 100.

Use your TRACE key to trace along the curve to view the ordered pairs of the functions and to see the trend of the function.

Note One of the most difficult features of setting up the graphing window is setting the Ymax value. Try to judge a reasonable value from the context of the application. If you're still not sure, set it to be *any* value you like, like 10 or 100. Construct the graph and activate the TRACE cursor. The cursor will appear on the graph and its coordinates will appear at the bottom of the viewing screen. Checking the Y coordinate will give you an idea about the y values of the function and you can make a judgment about Ymax using that value.

Calculator Exercises Use your graphing calculator to construct the graph of each function. In each case, label any minimum values and maximum values that appear on the graph. Specify the intervals upon which the graph increases (the curve goes up as you look from left to right) and the intervals upon which the graph decreases (the curve goes down as you look from left to right).

1. $f(x) = |x - 16|$

2. $f(x) = x^2 e^{-0.5x}$

3. $f(x) = \dfrac{50}{x + 25}$

4. $f(x) = (x - 7)(x - 3)(x + 2)$

5. A manufacturer of a particular product has determined that her cost, C, in dollars, for producing the product depends on the number of cases, x, that are produced. The cost function is

$$C(x) = -0.01x^2 + 4.7x + 155$$

where $0 \leq x \leq 200$. Sketch the graph of this function and, in sentence form, discuss its meaning relative to the problem situation. (As the number of cases produced increases (looking left to right), is the cost increasing or decreasing? Is the increase speeding up or slowing down?)

6. A drug company claims that t seconds after its new antitoxin medication is administered for bee or wasp stings, the amount, A, in parts per million (ppm), of toxin in the bloodstream can be approximated by the function

$$A(t) = \frac{0.3t + 1950}{1.5t + 2.2}$$

Sketch the graph of this function and, in sentence form, describe its meaning as it relates to this problem situation. (Is there a maximum amount of toxin in the bloodstream? If so, at what time does it occur? What happens as time increases? Is the toxin ever finally removed from the bloodstream?)

EXERCISE SET 1.2

For Exercises 1–15, determine if the given relation is or is not a function.

1. $R = \{(9, 1), (9, 2)\}$

2. $T = \{(6, 0), (4, 1), (8, 3)\}$

3. $M = \{5, 2), (6, 2), (7, 2), (8, 2)\}$

4. $S = \{(4, 1), (8, 3), (11, 0), (8, 10)\}$

5. $A = \{(a, 0), (7, 1), (0, a)\}$

6. $Q = \{(m, m), (n, m), (4, m), (0, 0)\}$

7. See the following table.

x	y
3	8
1	4
5	4
2	7

8. See the following table.

x	y
1	2
2	1
3	4
4	1
5	1

9. See the following table.

x	y
2	2/3
3	1/8
4	1/4
3	5/7
0	4/5

10. See the following table.

x	y
8	4
8	1
8	10
8	12

11. When a fair die is rolled once, the probability $1/6$ is assigned to each of the numbers 1, 2, 3, 4, 5, and 6.

12. ***Business: Auto Insurance*** For determining the base premium of auto insurance, states are divided into territories according to accident rate experience. Suppose that for a particular territory in a particular state, the following table associates the base premium with the amount of bodily injury insurance.

Base Premium	Amt of Coverage
\$65	\$10,000 – \$20,000
\$70	\$15,000 – \$30,000
\$75	\$25,000 – \$50,000
\$80	\$50,000 – \$100,000

13. ***Economics: Government Bonds*** Series E bonds issued by the Federal government can be purchased for 75% of their maturity value. At maturity, the value will be paid to the holder. The following table illustrates the assignment of a maturity value to a purchase price.

Maturity Value	Purchase Price
\$25	\$18.75
\$50	\$37.50
\$75	\$56.25
\$100	\$75.00
\$200	\$150.00
\$500	\$375.00
\$1000	\$750.00

14. ***Manufacturing: Air Pollution*** The amount of carbon monoxide $C(t)$ (in cubic centimeters) produced by a particular Chevrolet engine is related to the number of seconds t the engine idles. The equation describing the relationship is $C = 1000t$.

15. ***Science: Shadows*** At a particular time of the day, the equation $L(h) = 0.3h$ relates the height of a person to the length of the shadow he or she casts.

For Exercises 16–24, determine the domain of each function.

16. $f(x) = 2x + 8$

17. $f(x) = \dfrac{7}{x - 8}$

18. $f(x) = \dfrac{x + 1}{2x + 3}$

19. $f(x) = -4x^2 + x - 1$

20. $f(x) = \sqrt{3x - 6}$

21. $f(x) = \dfrac{\sqrt{x + 2}}{x - 2}$

22. $f(x) = \dfrac{\sqrt{x + 1}}{x + 3}$

23. $f(x) = \dfrac{1}{\sqrt{x + 4}}$

24. $f(x) = \dfrac{\sqrt{x - 6}}{\sqrt{x + 5}}$

Exercises 25–30 require the use of a graphing calculator. Use your graphing calculator to construct the complete graph of the function. Then, using the complete graph, estimate the range of the problem situation.

25. ***Business: Demand for a Product*** The marketing department of a plastic container manufacturer has estimated that the monthly demand $D(x)$ (in thousands of units) is related to the price x (in dollars) of a unit by the function $D(x) = \dfrac{48.26}{\sqrt{x}}$. Because of competition, labor, and capital, the company has to price each item at no less than \$1 and no more than \$9.

26. ***Medicine: Drug Concentration*** The concentration $C(t)$ (in milligrams per liter) of a particular drug in a person's bloodstream t hours after the drug has been injected into the bloodstream is approximated by the function $C(t) = \dfrac{80.06t}{25 + 0.08t^3}$. After 36 hours, the drug has essentially no effect on the patient.

27. ***Medicine: Weight of Infants*** Physicians use the function $W(t) = 0.033t^2 - 0.3974t + 7.3032$ to estimate the average weight of infants for their first 14 days.

28. ***Manufacturing: Costs of Ordering and Transportation*** Manufacturers have costs due to ordering and transportation. For a particular manufacturer, the cost $C(x)$ (in thousands of dollars) is related to the size of the order x (in hundreds of units) by the function $C(x) = \dfrac{8.06}{x} + \dfrac{8.06x}{x + 2.54}$. The manufacturer must always order at least 1 unit, but, due to space restrictions, no more than 10 units.

29. ***Business: Temperature of Food*** A restaurant estimates that the temperature $T(t)$ of food placed in its main freezer is related to the time t (in hours) that the food has been in the freezer by the function

$$T(t) = \frac{580}{t^2 + 10.3t + 33.665}$$

30. ***Economics: Investment*** When \$10,000 is invested in an account paying $r\%$ interest compounded monthly, the amount of money accumulated at the end of 5 years is given by the function

$$A(r) = 10{,}000\left(1 + \frac{r}{1200}\right)^{60}$$

$0 \le r \le 5$.

1.3 Evaluating Functions

Introduction

Evaluating Functions

Introduction

We know that functions are instruments for describing information. When an input value is known, it can be placed into the function, and the corresponding output value determined through computation. The process of computing the output value for a specified input value is called *evaluating a function*. In fact, the phrase *evaluate a function* means to compute the output value that corresponds to a specified input value. The input and output values can be numerical or symbolic. In this section we will present a detailed look at the process of evaluating functions. The process is important because one of the duties of a function is to produce an output value for a specified input value for not only physical phenomena, but for theoretical purposes as well. In fact, one of your first important evaluations in calculus is theoretical and is demonstrated in Example Set B.

Evaluating Functions

Because the characteristic action of a function is to assign one output value to each input value, we can (and should) think of a function as a rule that provides us with instructions for assigning an output value to a chosen input value. If the input value is represented by x, the output value can be represented by $f(x)$,

$g(x)$, $h(x)$, $\phi(x)$, or some letter or symbol other than f, g, h, or ϕ. Figure 1.5 and Figure 1.6 may be helpful in illustrating this concept.

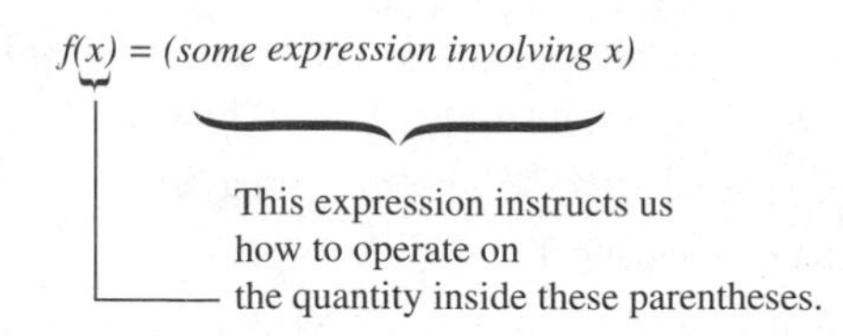

Figure 1.5

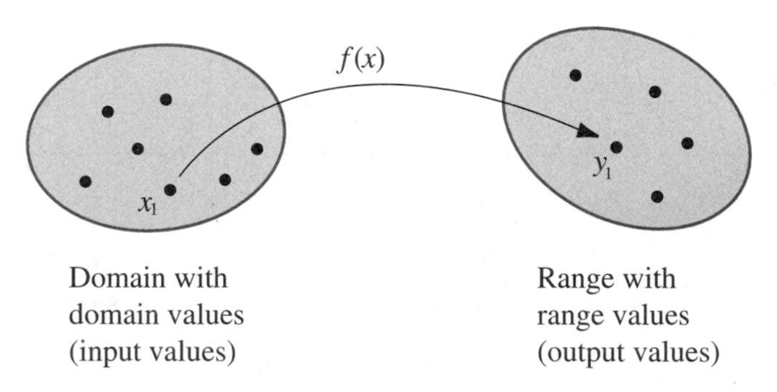

Figure 1.6

For example, the function $f(x) = x^3 + 8$ instructs us to take the input quantity (the quantity that appears inside the parentheses) and

$$\text{cube it} \quad \cdots \quad x^3, \quad \text{and}$$

$$\text{add } 8 \quad \cdots \quad x^3 + 8.$$

The variable inside the parentheses is just a placeholder for an input value that is to be inserted later. The input value may be represented by a number, a letter, a symbol, or even an expression such as $a + h$.

Evaluation of functions plays a big part in the study of much of mathematics from the calculus level up. So as to make the notation and computations as clear as possible, we will illustrate the evaluation process with six examples. This technique is important. Study it carefully.

EXAMPLE SET A

1. For the function $f(x) = x^3 - x + 5$, find $f(2)$.

Solution:

This rule instructs us to take the input quantity (the quantity that appears inside the parentheses), 2, and

$$\text{cube it} \quad \cdots \quad 2^3 \quad \text{and}$$

$$\text{subtract } 2 \quad \cdots \quad 2^3 - 2 \quad \text{and}$$

$$\text{add } 5 \quad \cdots \quad 2^3 - 2 + 5$$

So,

$$f(2) = 2^3 - 2 + 5$$

$$f(2) = 8 - 2 + 5$$

$$f(2) = 11$$

Thus, when the input is 2, the output is 11.

2. For the function $f(x) = -2x + 10$, find $f(6)$.

Solution:
This rule instructs us to take the input quantity (the quantity that appears inside the parentheses), 6, and

$$\text{take the opposite of twice it} \quad \cdots \quad -2(6) \quad \text{and}$$

$$\text{add } 10 \quad \cdots \quad -2(6) + 10$$

So,

$$f(6) = -2(6) + 10$$

$$f(6) = -12 + 10$$

$$f(6) = -2$$

Thus, when the input is 6, the output is -2.

3. For the function $g(x) = \dfrac{-x + 2a}{a}$, find $g(a)$, where $a \neq 0$.

Solution:
This rule instructs us to take the input quantity, a, and

$$\text{take the negative of it} \quad \cdots \quad -a \quad \text{and}$$

$$\text{add } 2a \quad \cdots \quad -a + 2a \quad \text{and}$$

$$\text{divide this result by } a \quad \cdots \quad \frac{-a + 2a}{a}$$

So,

$$g(a) = \frac{-a + 2a}{a}$$

$$g(a) = \frac{a}{a}$$

$$g(a) = 1$$

Thus, when the input is a, the output is 1.

The next three examples illustrate an evaluation that occurs frequently in calculus and more advanced mathematics.

EXAMPLE SET B

1. For the function $f(x) = 5x - 7$, find $f(a + h)$.

Solution:
This rule instructs us to take the input quantity (the quantity that appears inside the parentheses), $a + h$, and

multiply it by 5 $\cdots$ $5(a + h)$ and

subtract 7 $\cdots$ $5(a + h) - 7$

So,

$$f(a + h) = 5(a + h) - 7$$

$$f(a + h) = 5a + 5h - 7$$

Thus, when the input is $a + h$, the output is $5a + 5h - 7$.

2. For the function $f(x) = x^2 + 4$, find $f(x + h) - f(x)$.

Solution:
We shall compute $f(x + h)$ first. This rule instructs us to first take the input quantity, $x + h$, and

square it $\cdots$ $(x + h)^2$ and

add 4 $\cdots$ $(x + h)^2 + 4$

So,

$$f(x + h) = (x + h)^2 + 4$$

Next, we perform the subtraction $f(x + h) - f(x)$.

$$f(x+h)-f(x) = \overbrace{(x+h)^2+4}^{f(x+h)} \overbrace{-}^{-} \overbrace{(x^2+4)}^{f(x)} \quad \leftarrow \text{Use ( ) here.}$$

$$= x^2+2hx+h^2+4-x^2-4$$

$$= 2hx+h^2$$

Thus, $f(x+h)-f(x) = 2hx+h^2$.
Notice that if this difference is computed correctly, all terms not involving h will drop out.

3. For the function $f(x) = 5x^2 - 8$, find $\dfrac{f(x+h)-f(x)}{h}$, $h \neq 0$.

Solution:
We shall compute $f(x+h)$ first. This rule instructs us to first take the input quantity, $x+h$, and

$$\begin{array}{rcl} \text{square it} & \cdots & (x+h)^2 \quad \text{and} \\ \text{multiply it by 5} & \cdots & 5(x+h)^2 \quad \text{and} \\ \text{subtract 8} & \cdots & 5(x+h)^2 - 8 \end{array}$$

Thus, $f(x+h) = 5(x+h)^2 - 8$.
Next, we compute the difference-quotient $\dfrac{f(x+h)-f(x)}{h}$.

$$\frac{f(x+h)-f(x)}{h} = \frac{\overbrace{5(x+h)^2-8}^{f(x+h)} \overbrace{-}^{-} \overbrace{(5x^2-8)}^{f(x)}}{h}$$

$$= \frac{5(x^2+2hx+h^2)-8-5x^2+8}{h}$$

$$= \frac{5x^2+10hx+5h^2-8-5x^2+8}{h}$$

$$= \frac{10hx+5h^2}{h}$$

$$= \frac{5h(2x+h)}{h}$$

$$= \frac{5\not{h}(2x+h)}{\not{h}} \quad \text{since } h \neq 0$$

$$= 5(2x+h)$$

Thus, $\dfrac{f(x+h)-f(x)}{h} = 5(2x+h)$.

Notice that the difference is computed correctly and all terms not involving h dropped out.

EXERCISE SET 1.3

For Exercises 1–22, evaluate each function.

1. For $f(x) = 8x + 7$, find $f(3)$.
2. For $f(x) = -2x - 11$, find $f(8)$.
3. For $f(x) = -4x + 1$, find $f(-6)$.
4. For $f(r) = -5r^2 - r - 1$, find $f(1)$.
5. For $h(n) = n^2 + 3n + 8$, find $h(-4)$.
6. For $f(s) = s^3 - 27$, find $f(3)$.
7. For $f(z) = \dfrac{z^3 + 64}{z+4}$, find $f(0)$.
8. For $P(h) = 4n^3 - 6n^2 + 5n - 8$, find $P(kh)$.
9. For $f(x) = 4x^2 + 3x - 1$, find $f(h)$.
10. For $f(x) = 5x - 8$, find $f(x+h)$.
11. For $f(x) = x^2 + x - 5$, find $f(x+h)$.
12. For $f(x) = 7x^2 - 3x - 1$, find $f(x+h)$.
13. For $f(x) = 3x - 7$, find $f(x)$.
14. For $f(x) = 10x^2 - 11x + 4$, find $f(x)$.
15. For $f(x) = 5x + 6$, find $-f(x)$.
16. For $f(x) = 8x - 9$, find $-f(x)$.
17. For $f(x) = 4x + 1$, find $f(x+h) - f(x)$.
18. For $f(x) = 2x - 5$, find $f(x+h) - f(x)$.
19. For $f(x) = x^2 + x + 7$, find $f(x+h) - f(x)$.
20. For $f(x) = x^2 + 5x + 10$, find $f(x+h) - f(x)$.
21. For $f(x) = 8x^2 - 3x - 7$, find $f(x+h) - f(x)$.
22. For $f(x) = 7x^2 - 8x + 12$, find $f(x+h) - f(x)$.

For Exercises 23–26, write in paragraph form the steps you take to find $\dfrac{f(x+h)-f(x)}{h}$ for each of the following functions. Then carry out those steps.

23. For $f(x) = 6x + 1$, find $\dfrac{f(x+h)-f(x)}{h}$.
24. For $f(x) = 4x + 7$, find $\dfrac{f(x+h)-f(x)}{h}$.
25. For $f(x) = 3x^2 + 7x + 1$, find $\dfrac{f(x+h)-f(x)}{h}$.
26. For $f(x) = 5x^2 + 9x - 4$, find $\dfrac{f(x+h)-f(x)}{h}$.

For Exercises 27–30, use a graphing calculator.

27. ***Medicine: Drug Concentration*** The percent P of concentration of a particular drug in the bloodstream t hours after the drug has been injected is given by the function

$$P(t) = \frac{4.06t}{0.64t^2 + 8.1}.$$

Use your graphing calculator to graph this function and then from the graph determine (a) how long it will take for the drug to reach its maximum percentage level and specify that level, and (b) the number of hours after injection that the percentage level of the drug is approximately 50%. (Hint: Use the TRACE key. Use the ZOOM key for part (b).)

28. ***Business: Profit on CDs*** The profit $P(x)$ (in thousands of dollars) a company realizes for selling x (in hundreds) compact disc players is given by the function

$$P(x) = 0.92x^2(x - 8.06)^2, \quad \text{for} 0 < x \le 8.$$

Use your graphing calculator to graph this function and then from the graph (a) approximate the number of compact disc players the company must sell to maximize its profit, (b) specify that profit, and (c) approximate the number of compact disc players the company must sell to make a profit of about \$135,000.

29. ***Agriculture: Yield of an Orchard*** The owner of an avocado orchard has determined that the function

$$Y(x) = -0.2x^2 + 59.73x$$

relates the number of trees, x, planted per acre and the yield of avocados, $Y(x)$. Use your graphing calculator to graph this function. (a) Approximate the number of trees per acre that will produce the maximum yield, (b) specify that yield (do you really need the graph for this?), (c) approximate the number of trees per acre that will produce about 3500 avocados, and (d) approximate the number of trees per acre that will result in no avocado yield at all.

30. ***Economics: Equilibrium Point*** When a company produces x number of units of a particular commodity, the supply of the commodity is given by the supply function

$$S(x) = 12 + 0.16x^2$$

and the demand for the commodity is given by the demand function

$$D(x) = \frac{70}{1 + 0.088x}$$

Use your graphing calculator to graph each of these functions on the same coordinate system and then find the coordinates of the point of intersection. The point of intersection is called the *point of equilibrium* and the x coordinate of the point is called the *equilibrium price*. It is the price at which supply equals demand. The y coordinate is the level of supply and demand.

1.4 Composite Functions

Introduction

Just as with numbers, it is possible to operate on functions. In this section you will see how to use one function as the input quantity to another function.

Composite Functions

Composite Function

A **composite function** is a function that results from a chain-type process of composing one function with another. One function, say $g(x)$, is used as the input to another function, say $f(x)$. For example, consider the two functions $f(x) = x^2 + 4$ and $g(x) = 5x - 11$. Evaluate $g(x)$ at $x = 2$. That is, evaluate $g(2)$.

$$\begin{aligned} g(2) &= 5(2) - 11 \\ &= 10 - 11 \\ &= -1 \end{aligned}$$

Now, evaluate $f(x)$ at this result. That is, evaluate $f(-1)$.

$$\begin{aligned} f(-1) &= (-1)^2 + 4 \\ &= 1 + 4 \\ &= 5 \end{aligned}$$

Thus, $g(2) = -1$ and $f(-1) = 5$. Figure 1.7 shows this chain of events that leads from 2 to 5.

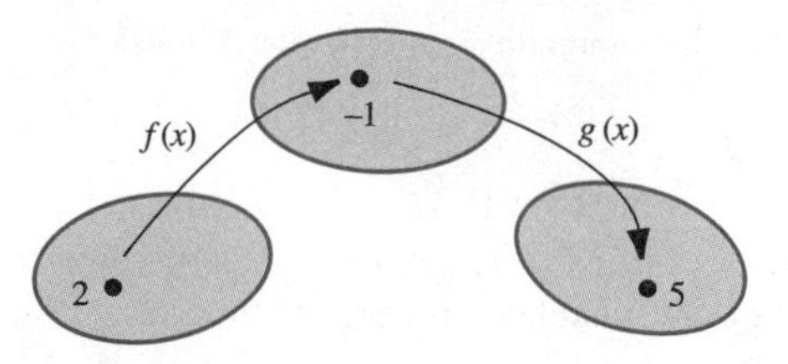

Figure 1.7

The output 5 is obtained by first applying the rule $g(x)$ at $x = 2$, getting the intermediate result -1, and then applying the rule $f(x)$ to $x = -1$. In other words, 5 is obtained by composing (to make something by combining other things) the functions $f(x)$ and $g(x)$, and evaluating the composition at 2. It is possible, by composing the functions $f(x)$ and $g(x)$ at the outset, to produce 5 directly from 2, bypassing the intermediate value -1.

To see how, we first develop a logical notation to describe the composition of two functions. Consider

$5 = f[-1]$. But $-1 = g(2)$. Replace -1 with $g(2)$. So
$5 = f[-1] = f[g(2)]$

This is the notation we are looking for, and we restate it as follows.

Composition of Functions

The *composition of the functions* $f(x)$ and $g(x)$ is denoted by $(f \circ g)(x)$ and is given by $(f \circ g)(x) = f[g(x)]$. $(f \circ g)(x)$ is read as *f circle g, of x* or *f composite g, of x.*

The composite function $f[g(x)]$ is evaluated by first evaluating $g(x)$ for some particular x value, and then evaluating $f(x)$ at that resulting value. Also, although it is not entirely accurate, it is common to denote $(f \circ g)(x)$ by $f \circ g$. In fact, it is common to relax the output notation by writing simply f rather than the more accurate description $f(x)$. Mathematicians have traded some accuracy in notation for simplicity of notation. Figure 1.8 shows that the composite of $f(x)$ and $g(x)$ bypasses the intermediate value -1.

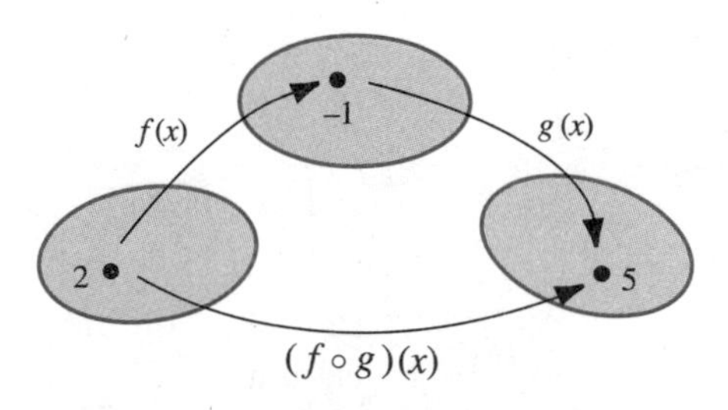

Figure 1.8

EXAMPLE SET A

If $f(x) = \sqrt{x-3}$ and $g(x) = 2x+1$, find the algebraic expression for

1. $(f \circ g)(x)$,

2. $(g \circ f)(x)$.

Solution:

1. $(f \circ g)(x)$ indicates we should input $g(x)$ into $f(x)$.

$$
\begin{aligned}
(f \circ g)(x) &= f[g(x)] && \text{(definition of } f \circ g\text{)} \\
&= f(2x+1) && \text{(definition of } g\text{)} \\
&= \sqrt{(2x+1)-3} && \text{(definition of } f\text{)} \\
&= \sqrt{2x-2}
\end{aligned}
$$

2. $(g \circ f)(x)$ indicates we should input $f(x)$ into $g(x)$.

$$
\begin{aligned}
(g \circ f)(x) &= g[f(x)] && \text{(definition of } g \circ f\text{)} \\
&= g(\sqrt{x-3}) && \text{(definition of } f\text{)} \\
&= 2\sqrt{x-3}+1 && \text{(definition of } g\text{)}
\end{aligned}
$$

It is important to note that, in general, $f \circ g \neq g \circ f$. The problems of Example Set A illustrate the fact that the composition operation $\circ$ is not commutative.

The Domain of a Composite Function

A natural question is how the domain of a composite function is related to the domains of the functions involved in the composition. To help answer this question, consider further the functions $f(x) = \sqrt{x-3}$ and $g(x) = 2x+1$ as defined in Example Set A.

$$f(x) = \sqrt{x-3} \text{ with domain } \{x|x \geq 3\} \text{ and range } \{f(x)|f(x) \geq 0\}$$

$$g(x) = 2x+1 \text{ with domain } \{x|x \in R\} \text{ and range } \{g(x)|g(x) \in R\}$$

You need to give extra concentration to the idea of determining the domain of the composition of two functions. This idea is a bit more involved than the algebraic combinations of functions. The idea may be more understandable by considering the following. Since you are looking at $f(g(x))$, you know that any replacement you make for $g(x)$ has to be an allowable domain value for $f(x)$. The domain of $f(x)$ is $\{x|x \geq 3\}$ so that the only values of $g(x)$ we can use must satisfy $g(x) \geq 3$. Now, what domain values of $g(x)$ produce range values satisfying

$g(x) \geq 3$? This brings the problem to a more manageable position. We simply set $g(x) \geq 3$ as follows:

$$g(x) \geq 3$$
$$2x + 1 \geq 3$$
$$2x \geq 2$$
$$x \geq 1$$

Since domain values of $g(x)$ that are ≥ 1 produce range values of $g(x) \geq 3$ that are the allowable domain values of $f(x)$, the set $\{x|x \geq 1\}$ is the domain of the composite function $f \circ g$.

This discussion helps suggest the following definition regarding the domain of a composite function.

The Domain of a Composite Function

The domain of the composite function $(f \circ g)(x)$ is the subset of numbers in the domain of $g(x)$ that produce values in the domain of $f(x)$.

We omit the proof of this theorem but illustrate its use in Example Set B.

EXAMPLE SET B

1. For the functions $f(x) = \frac{1}{x}$ and $g(x) = \left(x - \frac{1}{3}\right)(x+1)(x+2)$, find an algebraic expression for $(f \circ g)(x)$ and then find the domain of $(f \circ g)(x)$.

Solution:
By its definition, the function $f(x)$ produces the reciprocal of the input quantity. Thus,

$$\begin{aligned}(f \circ g)(x) &= f[g(x)] \\ &= f\left[\left(x - \frac{1}{3}\right)(x+1)(x+2)\right] \\ &= \frac{1}{\left(x - \frac{1}{3}\right)(x+1)(x+2)}\end{aligned}$$

The function $g(x)$ has as its domain the set of all real numbers (R) while the function $f(x)$ is undefined for $x = 0$. The numbers $\frac{1}{3}$, -1, and -2 in the domain of $g(x)$ all produce 0 as outputs, and therefore cannot be in the domain of $f(x)$. Hence,

$$\text{The domain of } (f \circ g)(x) = \{x|x \in R \text{ and } x \neq \tfrac{1}{3}, -1, -2\}.$$

2. For the functions $f(x) = \sqrt{4 - x}$ and $g(x) = \sqrt{x}$, find an algebraic expression for $(f \circ g)(x)$ and find the domain of $(f \circ g)(x)$.

Solution:
We begin by computing $(f \circ g)(x)$.

$$\begin{aligned}(f \circ g)(x) &= f[g(x)] \\ &= f[\sqrt{x}] \\ &= \sqrt{4 - \sqrt{x}}\end{aligned}$$

The domain of $g(x) = \{x|x \geq 0\}$ and the domain of $f(x) = \{x|x \leq 4\}$. The numbers in the domain of $g(x)$ for which $g(x) \leq 4$ are all those numbers x such that $\sqrt{x} \leq 4$. Since $x \geq 0$, this implies that $0 \leq x \leq 16$. Hence,

$$\text{The domain of } (f \circ g)(x) = \{x|x \in R \text{ and } 0 \leq x \leq 16\}$$

Example Set C illustrates why we cannot simply examine $(f \circ g)(x)$ to determine the domain of $f \circ g$.

EXAMPLE SET C

For the functions $f(x) = x^2$ and $g(x) = \sqrt{x}$, find an expression for $(f \circ g)(x)$ and determine the domain of $(f \circ g)(x)$.

Solution:

$$\begin{aligned}(f \circ g)(x) &= f[g(x)] \\ &= f[\sqrt{x}] \\ &= [\sqrt{x}]^2 \\ &= x\end{aligned}$$

We would conclude, from looking only at $(f \circ g)(x) = x$, that the domain of $(f \circ g)(x)$ is the set of all real numbers. However, the domain of $g(x)$ is restricted to the set $\{x|x \geq 0\}$. The domain of $f[g(x)]$ cannot be any bigger than the domain of $g(x)$. In addition, all of the values in the domain of $g(x)$ have output in the domain of $f(x)$. We conclude that the domain of $(f \circ g)(x) = \{x|x \geq 0\}$.

Composite Functions and Technology

Using Your Calculator You can use your graphing calculator to evaluate a composite function. The following entries show how to evaluate $(f \circ g)(2)$ for $f(x) = \sqrt{x-3}$ and $g(x) = 2x + 1$. Enter

Y1 = √(x − 3)

Y2 = 2x + 1

In the computation window, enter

Y1(Y2(2))

The calculator responds with the value 2.

When $(g \circ f)(2)$ is evaluated, the calculator responds with an error. Can you get your calculator to respond with an error? Why does it do so?

Calculator Exercises Use your graphing calculator to evaluate each composite function.

1. If $f(x) = 5x^2 - x$ and $g(x) = \sqrt{x}$, evaluate both $f[g(16)]$ and $g[f(16)]$.
2. A study of a northwestern community indicates that the average daily level C of carbon monoxide in the air, in parts per million (ppm), is approximated by the function $C(p) = \sqrt{0.18p^2 + 9.6}$, $0 < p \leq 20$, where p represents the population of the community in thousands of people. (Enter this function using $Y1$.) The population of the community depends on the time t in years from now and is approximated by the function $p(t) = 1.8 + 0.04t^2$, $0 < t \leq 20$. (Enter this function using $Y2$.) Compute the level of carbon monoxide in this community's air 15 years from now in two ways. First, compute $p(15)$ and then substitute this result into $C(p)$. Second, compute directly using the method given above $C[p(15)]$. Compare the two results. If they are equal, write your conclusion in sentence form. If they are unequal, try again.
3. For the previous exercise, make a single change in the defining expression of $Y1$ so that when you turn off the graphing capability of $Y2$, the graph of the composite function $C[p(t)]$ appears in the window $0 < t \leq 25$ and $0 \leq C \leq 25$. Sketch your graph.

EXERCISE SET 1.4

For Exercises 1–13, find the domain and rule for $(f \circ g)(x)$, and the domain and rule for $(g \circ f)(x)$.

1. $f(x) = 2x + 5$ and $g(x) = x^2 - x$
2. $f(x) = -5x + 1$ and $g(x) = x^2 - 10$
3. $f(x) = x^2 + x - 4$ and $g(x) = 2x^2 - 5$
4. $f(x) = -(x+1)^3$ and $g(x) = 4x + 2$
5. $f(x) = \dfrac{3}{5x}$ and $g(x) = x - 4$
6. $f(x) = \dfrac{x+1}{x-3}$ and $g(x) = 3x + 6$
7. $f(x) = \sqrt{x}$ and $g(x) = x + 1$
8. $f(x) = \dfrac{3}{5x}$ and $g(x) = \sqrt{x+5}$
9. $f(x) = \dfrac{x^2-1}{x+1}$ and $g(x) = \dfrac{5}{x}$
10. $f(x) = \dfrac{x+3}{x-4}$ and $g(x) = \dfrac{x+2}{x-4}$
11. $f(x) = \dfrac{2x-1}{x+2}$ and $g(x) = \dfrac{3x+22}{2x-6}$

For exercises 12 and 13, find the domain for $f \circ g)(x)$.

12. $f = \{(2,5), (3,8), (4,11), (5,12)\}$ and $g = \{(1,2), (6,3), (11,4), (15,5)\}$
13. $f = \{(2,0), (1,-2), (3,2), (4,4)\}$ and $g = \{(-5,4), (-1,3), (0,2), (4,1)\}$

For Exercises 14–20, find what is asked for.

14. If $f(x) = 3x - 4$, $g(x) = 2x + 11$, and $h(x) = 5x + 6$, find the rule for $(f \circ g \circ h)(x)$ and then evaluate this composition at $x = 2$.

15. If $f(x) = x^2 - 1$, $g(x) = -5$, and $h(x) = x - 2$, find the rule for $(f \circ g \circ h)(x)$ and then evaluate this composition at $x = 0$.

16. If $f(x) = 2x + 7$ and $g(x) = \frac{x-7}{2}$, find the rules for both $(f \circ g)(x)$ and $(g \circ f)(x)$.

17. If $f(x) = 5x + 4$ and $g(x) = \frac{x-4}{5}$, find the rules for both $(f \circ g)(x)$ and $(g \circ f)(x)$.

18. If $f(x) = 2x + 9$, find both the rule and the domain for $(f \circ f)(x)$.

19. If $f(x) = \frac{x}{x-1}$, find both the domain and the rule for $(f \circ f)(x)$.

20. If $f(x) = \frac{-x}{1-x}$, find both the domain and the rule for $(f \circ f)(x)$.

21. *Manufacturing: Cost of Production* A manufacturer believes that the function $C(x) = 0.16x^2 + 0.92x + 850$ closely approximates the cost of a daily production run of x units. The manufacturer also knows that the number of units produced depends on the time of day t (in hours since the start of a shift) and is given by the function $x(t) = 0.95t^2 + 90t$. Find the approximate cost of producing x units (a) 5 hours after the start of a shift, and (b) 6 hours after the start of the shift. (c) About how many hours after the beginning of a shift will the cost of a production run be \$20,000?

22. *Medicine: Size of Tumors* The volume of a spherical cancer tumor is related to the diameter of the tumor by the function $V(x) = \frac{\pi x^3}{6} + 1.5$, where x is in millimeters. However, the diameter of the tumor is related to the time t (in days) since detection of the tumor by the function $x(t) = 0.00006t^2 - 0.002t + 0.005$, $0 < t \leq 210$. Find the volume of the tumor (a) 60 days after detection, and (b) 90 days after detection. (c) About how many days after detection will the volume of the tumor be 20 cubic millimeters?

23. *Economics: Consumer Imports* A government study estimates that consumers will buy a quantity $Q(p)$ of imported spices when the price per pound is p dollars. The price per pound depends on the time of year and is given by the function $p(t) = 0.03t^2 + .09t + 5.8$, $0 < t \leq 26$, where t is given in weeks from the beginning of the year. The function $Q(p) = \frac{52860}{p^{1.8}}$ represents the quantity, in pounds, of spice that people are willing to buy at the price p. Estimate the number of pounds of spice that is bought (a) 10 weeks after the beginning of the year, and (b) 20 weeks after the beginning of the year. (c) Estimate the number of weeks after the beginning of the year that people will be buying 9000 pounds of spice.

Summary

The main focus of Chapter 1 was the function. The function is of fundamental importance in calculus and it is with respect to the function that all calculus operations are undertaken. It was the intent of this chapter to provide you with a clear understanding of the concept of the function and a working knowledge of its notation and use for describing information.

An Intuitive Description of Functions

The first section of the text presented an intuitive description of the function concept. This section began by illustrating how a function can be used as an instrument for describing relationships between quantities and then presented information about those relationships. This section then carefully developed the function notation. For example, a function such as $f(x) = x^2 - 3x + 100$ should tell you that the value of f depends only on the value of x and that the value of f can be obtained by applying the function rule $x^2 - 3x + 100$ to the chosen x

value. Similarly, the function $C(A, D) = \dfrac{A^2}{A + 1.25D}$ should tell you that the net audience of an advertisement, C, depends on both the total audience of all outlets that are used, A, and the sum of the duplicated audiences for all pairs of outlets used, D.

Relations and Functions

Section 1.2 presented functions in a more rigorous way than did Section 1.1. Functions were discussed as particular types of relations, relations for which one input value produces precisely one output value. This section pointed out that for a function to be completely described, the domain, the range, and the defining rule must all be specified. The section then illustrated four equivalent ways of describing functions: tables, ordered pairs, symbolic or verbal rules, and graphs. This section also discussed the concepts of complete graph and situation graph of a function. Essentially, a graph is a complete graph if it suggests all of the possible input/output combinations and all the local and global behavior of the function. A situation graph is a subset of a complete graph and displays only the information pertinent to the problem situation.

Evaluating Functions

Section 1.3 illustrated how to evaluate a function. You can think of a function as a rule that provides you instructions for assigning an output value to a chosen input value. This section showed several examples, the most important being the evaluation of the difference quotient $\dfrac{f(x+h) - f(x)}{h}$. You should pay particular attention to this evaluation as it is a fundamental evaluation in calculus. You will see its geometric significance in Section 2.5, when the instantaneous rate of change of a function is examined. Having a good understanding of the evaluation of the difference quotient will provide you with a richer understanding of how results of calculus evaluations are interpreted.

Composite Functions

Section 1.4 discussed how functions can be composed. You should pay some attention to how the domain of a function that is the composition of two or more functions is related to the domain of the original functions. You have many occasions to work with composite functions in your study of calculus. A composite function is a function that results from a chain-type process of composing one function with another. One function, say $g(x)$, is used as the input quantity to another function, say, $f(x)$. The composition, $f[g(x)]$, is denoted by $(f \circ g)(x)$.

Supplementary Exercises

For Exercises 1–8, write, using symbols, the defining function expression and rule for each situation. As best you can, specify the domain of your function.

1. The volume of a cube is a function of the length of an edge cubed.

2. The volume of a cylinder with base 10 square inches is a function of its height multiplied by $\frac{\pi}{3}$.
3. The area of a circle is a function of its radius squared multiplied by π.
4. The volume of a sphere is a function of its radius cubed multiplied by $\frac{4\pi}{3}$.
5. The perimeter of a square is a function of the length of a side of the square multiplied by 4.
6. The total cost of n papers (if each paper costs \$0.35) is a function of the number of papers purchased multiplied by \$0.35.
7. The area of a triangle with base 5 inches is a function of its altitude multiplied by $\frac{5}{2}$.
8. The area of a rectangle with base 7 inches is a function of its height multiplied by 7.

For Exercises 9–14, determine if the relation is or is not a function.

9. $L = \{(19, 3), (19, 7)\}$
10. $H = \{(57, 0), (14, 7), (34, 17)\}$
11. $W = \{12, 4), (15, 4), (33, 4), (18, 7)\}$
12. $P = \{(5, 1), (55, 3), (55, 78), (55, 190)\}$
13. $B = \{(b, 6), (3, 8), (6, b), (32, b)\}$
14. $M = \{(m, 2m), (2n, m), (14, 2m), (23, m^2)\}$

For Exercises 15–23, determine the domain of each function.

15. $f(x) = 24x - 34$
16. $f(x) = \dfrac{23}{3x - 4}$
17. $f(x) = \dfrac{3x - 9}{5x + 4}$
18. $f(x) = 3x^2 + 4x - 31$
19. $f(x) = \sqrt{-4x - 7}$
20. $f(x) = \dfrac{\sqrt{x + 4}}{x - 4}$
21. $f(x) = \dfrac{\sqrt{2x + 1}}{3x + 1}$
22. $f(x) = \dfrac{5x}{\sqrt{5x - 3}}$
23. $f(x) = \dfrac{\sqrt{12x - 6}}{\sqrt{-7x + 6}}$

For Exercises 24–48, find the requested information.

24. For $h(x) = 4x + 9$, find $h(67)$.
25. For $f(x) = 2x - 13$, find $f(0)$.
26. For $j(y) = 12 - 6y$, find $j(-53)$.
27. For $f(x) = x^3 + 11$, find $f(-z)$.
28. For $s(x) = 75x - 2$, find $s(r^4)$.
29. For $H(q) = -2q^4 - 2q + 9$, find $H(-1)$.
30. For $W(n) = 3n^2 + n + 1$, find $W(-4)$.
31. For $P(c) = c^2 - 7$, find $P(3 + c)$.
32. For $Q(q) = \dfrac{q + 4}{2q + 3}$, find $Q(2r)$.
33. For $E(b) = b^3 - 2b^2 + 4b - 12$, find $E(a^3)$.
34. For $f(x) = 2x - 3$, find $f(x + h)$.
35. For $f(x) = 3x^2 + 4x - 2$, find $f(x + h)$.
36. For $f(x) = x^2 - 8x - 13$, find $f(x + h)$.
37. For $f(x) = 5x + 34$, find $f(x - h)$.
38. For $f(x) = x^3 + 2$, find $-f(x)$.
39. For $f(x) = x^2 - x + 2$, find $-f(x)$.
40. For $f(x) = 3x + 5$, find $f(x + h) - f(x)$.
41. For $f(x) = 4x - 15$, find $f(x + h) - f(x)$.
42. For $f(x) = 2x^2 + 3x + 1$, find $f(x + h) - f(x)$.
43. For $f(x) = \dfrac{x + 5}{x + 10}$, find $f(x + h) - f(x)$.
44. For $f(x) = -6x - 2$, find $\dfrac{f(x + h) - f(x)}{h}$.
45. For $f(x) = 2x^2 - 5x$, find $\dfrac{f(x + h) - f(x)}{h}$.
46. For $f(x) = -3x^2 + x - 2$, find $\dfrac{f(x + h) - f(x)}{h}$.
47. For $f(x) = 2x^2 + 5x + 3$, find $\dfrac{f(x + h) - f(x)}{h}$.
48. For $f(x) = -x^2 + 7x + 8$, find $\dfrac{f(x + h) - f(x)}{h}$.

For Exercises 49–58, find the domain and rule for $(f \circ g)(x)$, and the domain and rule for $(g \circ f)(x)$.

49. $f(x) = x + 25$ and $g(x) = 7x^2 - 2x$
50. $f(x) = -x - 3$ and $g(x) = x^2 + 2$
51. $f(x) = 2x^2 - 4$ and $g(x) = x^2 + 3x - 5$
52. $f(x) = (x + 1)^2$ and $g(x) = x + 12$
53. $f(x) = \dfrac{1}{3x}$ and $g(x) = 3x - 7$
54. $f(x) = \dfrac{x - 2}{x - 1}$ and $g(x) = -2x + 9$
55. $f(x) = \sqrt{3x - 1}$ and $g(x) = 2x + 1$
56. $f(x) = \dfrac{x^2 - 9}{x - 3}$ and $g(x) = \dfrac{-2}{3x}$
57. $f(x) = \dfrac{x + 1}{x - 3}$ and $g(x) = \dfrac{x + 1}{x - 5}$

58. $f(x) = \frac{3x-2}{3x+2}$ and $g(x) = \frac{x+2}{4x-1}$

For Exercises 59–65, find what is asked for.

59. If $f(x) = 5x + 3$, $g(x) = x + 1$, and $h(x) = 2x$, find the rule for $(f \circ g \circ h)(x)$ and then evaluate this composition at $x = 7$.

60. If $f(x) = x^2 - 16$, $g(x) = -7$, and $h(x) = 3x - 1$, find the rule for $(f \circ g \circ h)(x)$ and then evaluate this composition at $x = \frac{1}{3}$.

61. If $f(x) = 6x + 9$ and $g(x) = \frac{2x-1}{5}$, find the rules for both $(f \circ g)(x)$ and $(g \circ f)(x)$.

62. If $f(x) = \sqrt{3x+14}$ and $g(x) = x^2 - 4$, find the rules for both $(f \circ g)(x)$ and $(g \circ f)(x)$.

63. If $f(x) = -x+5$, find the rule and the domain for $(f \circ f)(x)$.

64. If $f(x) = \frac{1}{x-1}$, find the domain and the rule for $(f \circ f)(x)$.

65. If $f(x) = \frac{-2x}{3-x}$, find the domain and the rule for $(f \circ f)(x)$.

For Exercises 66–69, compute, then interpret the slope of the line passing through the given pair of points. For example, the slope of the line passing through the points $(6, 2)$ and $(10, 9)$ is $\frac{7}{4}$. This means that as the input value increases by 4 units, the output value increases by 7 units. Remember, the slope of a line is denoted by the letter m where

$$m = \frac{\text{Change in output}}{\text{Change in input}} = \frac{y_2 - y_1}{x_2 - x_1}$$

The slope measures the change in the output for a given change in the input.

66. $(4, 2)$ and $(4, 1)$

67. $(-3, -1)$ and $(-1, 3)$

68. $(12, 38)$ and $(22, -61)$

69. $(-3, -10)$ and $(10, 0)$

For Exercises 70–76, construct the graph of each function.

70. $f(x) = \frac{1}{2}x + 3$

71. $f(x) = \frac{-1}{5}x - 8$

72. $f(x) = \frac{-3}{4}x + 13$

73. $f(x) = \frac{3}{12}x + 3$

74. $f(x) = -4x + 12$

75. $f(x) = 3x - 1$

76. $f(x) = \frac{-1}{2}x$

CHAPTER 2

Limits

2.1 Limits Viewed Intuitively

Introduction

Calculus is an effective tool for examining the behavior of functions. Functions can display varying behavior at different points in their domain. Many functions behave quite predictably at most points in their domain, but some exhibit strange and interesting behavior at points near the limits of their domain.

An Intuitive View of the Limit of a Function

Suppose that, while waiting on the fifth floor of the mathematics building for your college calculus class to start, you see a big crow fly up and land on a ledge just outside the window. To investigate the behavior of the crow you would not spring out of your seat, jump through the window, and pounce on top of the crow, but rather you would move carefully toward the window observing the crow's behavior as you approach its location. It is not necessary to get to the precise location of the crow, only near it. So it is with investigating the behavior of a function at interesting points in its domain. To examine the behavior of a function near an interesting point, we need only *approach* that point, investigating associated function values along the way.

In Illuminator Set A, we will examine the behavior of an entire function so that we may illustrate what is meant by the *behavior* of a function. In the examples that follow Illuminator Set A, however, we will investigate the behavior of the given function only at its interesting points.

ILLUMINATOR SET A

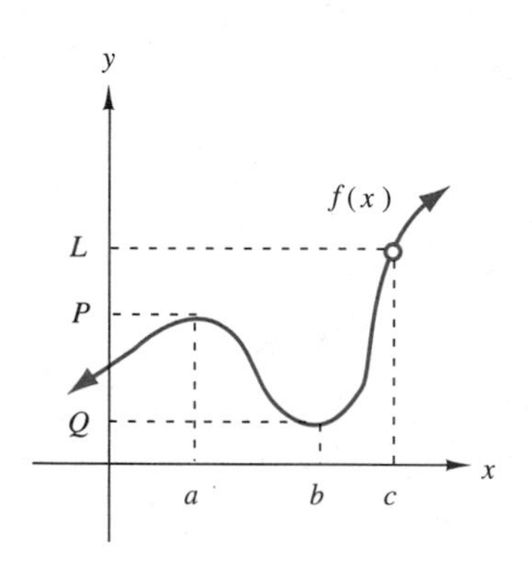

Figure 2.1

Consider the function represented graphically in Figure 2.1. We first notice that the x value c is not in the domain of the function since there is no point on the graph at $x = c$. Although there is a y value, namely L, on the vertical axis at the same level as the hole in the curve, there is no x value that will produce it. The x value c cannot be used and, hence, the y value L is never realized. That is, $f(c)$ is not defined, and therefore cannot be equal to L.

We can investigate the behavior of the function by directly observing its graph. We observe the following behavior.

1. The function increases as x increases through the interval $(-\infty, a)$.
2. The function attains a relatively high point (a maximum) at $x = a$, in fact, $f(a) = P$.
3. The function decreases as x increases through the interval (a, b).
4. The function attains a relatively low point (a minimum) at $x = b$, in fact, $f(b) = Q$.
5. The function increases as x increases through the interval (b, c).
6. The function is not defined (and therefore has no behavior at all) at $x = c$.
7. The function increases as x increases through the interval $(c, +\infty)$. Since the x value c is not in the domain of the function $f(x)$, the function has no behavior at all at $x - c$. However, we are interested in the behavior of the function at x values near $x = c$. We can investigate the function's behavior by approaching $x = c$. In fact, we can approach $x = c$ from both the left and the right sides, that is, through x values less than c and through x values greater than c. We observe the following behavior of the function near $x = c$.
8. As we *approach* $x = c$ from the left side, the function values, as measured by their heights, approach $y = L$ from below. We also observe that as we *approach* $x = c$ from the left side only, the function values, although climbing toward $y = L$, are limited by $y = L$ and cannot go beyond it. The y value L acts as a limit to the increasing y values.
9. As we *approach* $x = c$ from the right side, the function values, as measured by their heights, approach $y = L$ from above. We also observe that as we *approach* $x = c$ from the right side only, the function values, although descending toward $y = L$, are limited by $y = L$ and cannot go beyond it. The y value L acts as a limit to the decreasing y values.

   Thus, regardless from which side we approach $x = c$, the function values approach $y = L$. We point out that we are observing the behavior of the function *near* the x value c, and not at c itself. Remember, the function is not defined at $x = c$ and, therefore, has no behavior there at all.

In observations 8 and 9 of the example above, we noted that the y value L acted as a limit to the ascending and descending y values. The values of $f(x)$ could be made very close to L by choosing x values that are very close to c. The closer x gets to c, the closer y gets to L, and if x approaches c without going beyond it, y will approach L without going beyond it. It is helpful to introduce some notation at this point that alerts us to the existence of a limiting y value, say L, as the x values *approach* some x value, say c.

The Limit Notation

The statement "The limit of $f(x)$, as x approaches c, is L" is described by the notation

$$\lim_{x \to c} f(x) = L$$

lim* Reminds Us of the Word *approach

When you see the symbol "lim," you should think that the word reminds us of the word *approach*. Since we are interested in the behavior of the function only at values near $x = c$, and not at $x = c$ itself, we are interested in the behavior of the function only as we *approach* $x = c$.

Thus, $\lim_{x \to c} f(x) = L$ is read as "The limit of $f(x)$, as x approaches c, is L." $\lim_{x \to c} f(x) = L$ might be interpreted in any one of three ways.

1. As x approaches c, $f(x)$ approaches L.
2. As the x values get close to c, the $f(x)$ values get close to L.
3. As the input values get close to c, the output values get close to L.

The examples of Illuminator Set B illustrate this point.

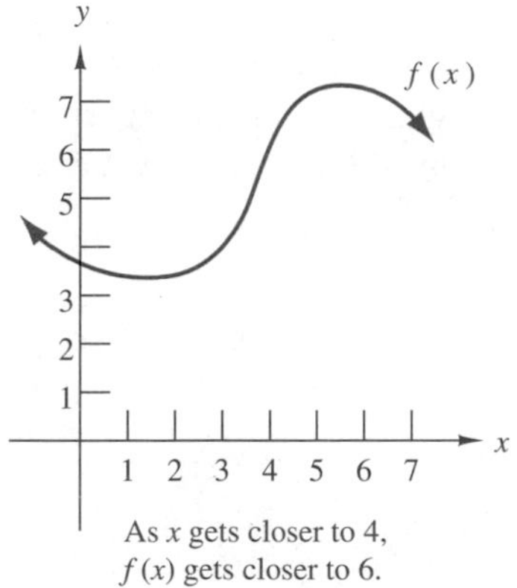

As x gets closer to 4, $f(x)$ gets closer to 6.

Figure 2.2

ILLUMINATOR SET B

1. $\lim_{x \to 4} f(x) = 6$ is read as "The limit of $f(x)$, as x approaches 4, is 6." It might be interpreted in any one of three ways.
   a. As x approaches 4, $f(x)$ approaches 6.
   b. As the x values get close to 4, the $f(x)$ values get close to 6.
   c. As the input values get close to 4, the output values get close to 6.

   Figure 2.2 shows a function $f(x)$ for which 6 is a limit as x approaches 4.

2. $\lim_{x \to 3} \dfrac{x^2 + x - 12}{x^2 - x - 6} = \dfrac{7}{5}$ is read as "The limit of $\dfrac{x^2 + x - 12}{x^2 - x - 6}$, as x approaches 3, is $\dfrac{7}{5}$."
   It might be interpreted in any one of three ways.
   a. As x approaches 3, $\dfrac{x^2 + x - 12}{x^2 - x - 6}$ approaches $\dfrac{7}{5}$.

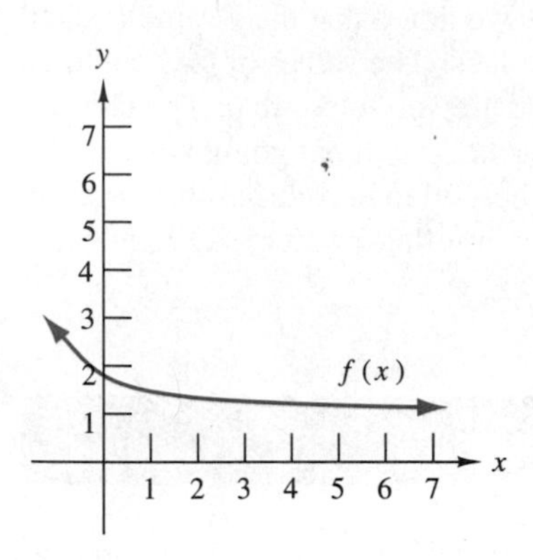

Figure 2.3

b. As the x values get close to 3, the values of $\dfrac{x^2+x-12}{x^2-x-6}$ get close to $\dfrac{7}{5}$.

c. As the input values get close to 3, the output values get close to $\dfrac{7}{5}$.

Figure 2.3 illustrates this fact.

3. Figure 2.4 shows a function for which the limit does not exist as x approaches some number c. The closer x gets to c, the larger y gets. The output value y does not approach any particular number as x approaches c, but rather keeps getting larger and larger.

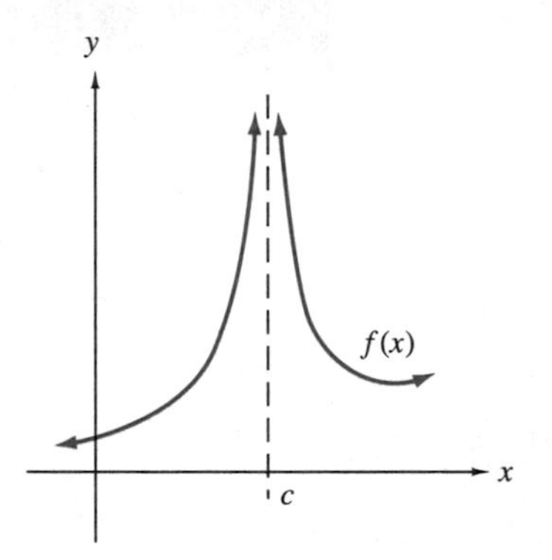

Figure 2.4

We'll now make an important observation and then expand the limit notation. In Figure 2.5, observe that as x approaches c from the left side, $f(x)$ approaches L and will not go beyond it.

In Figure 2.6, observe that as x approaches c from the right side, $f(x)$ approaches L and will not go beyond it. The important observation we make here is that regardless of the side, the left or the right, from which x approaches the number c, $f(x)$ approaches the unique, limiting value L. That is, the limit

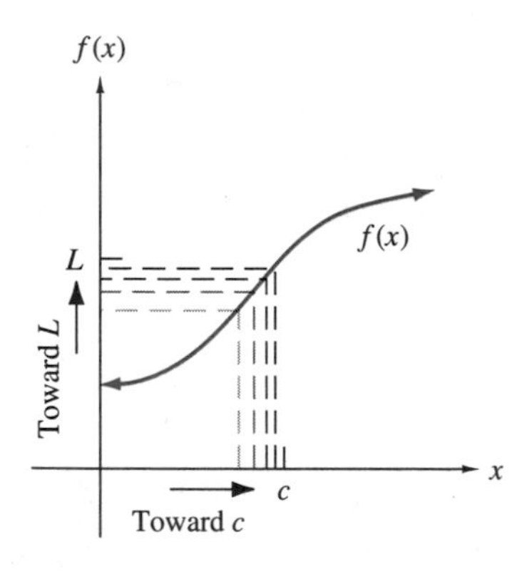

Figure 2.5

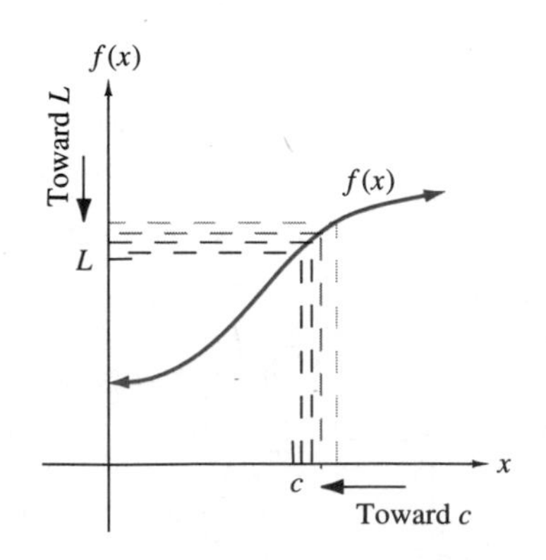

Figure 2.6

existed because the limiting value obtained as we approached from the left side was the same limiting value obtained as we approached from the right side.

It is now helpful to expand the limit notation to describe the approach to an x value c from only one side.

One-Sided Limits

One-Sided Limits

The statement "The limit of $f(x)$, as x approaches c from the left side, is L" is described by the notation

$$\lim_{x \to c^-} f(x) = L$$

The statement "The limit of $f(x)$, as x approaches c from the right side, is L" is described by the notation

$$\lim_{x \to c^+} f(x) = L$$

The important observation we made above about the existence of the limit can now be stated using the limit notation.

The Existence of the Limit

The limit of a function, $f(x)$, if it exists, is always a unique real number, and it will exist at an x value c provided that

$$\lim_{x \to c^-} f(x) = \lim_{x \to c^+} f(x)$$

Furthermore, if the two one-sided limits equal the same, unique number L, the number L is called the limit of the function. This means that

$$\text{If } \lim_{x \to c^-} f(x) = \lim_{x \to c^+} f(x), \text{ then } \lim_{x \to c} f(x) = L$$

and

$$\text{If } \lim_{x \to c} f(x) = L, \text{ then } \lim_{x \to c^-} f(x) = \lim_{x \to c^+} f(x)$$

The following examples visually illustrate this important concept.

ILLUMINATOR SET C

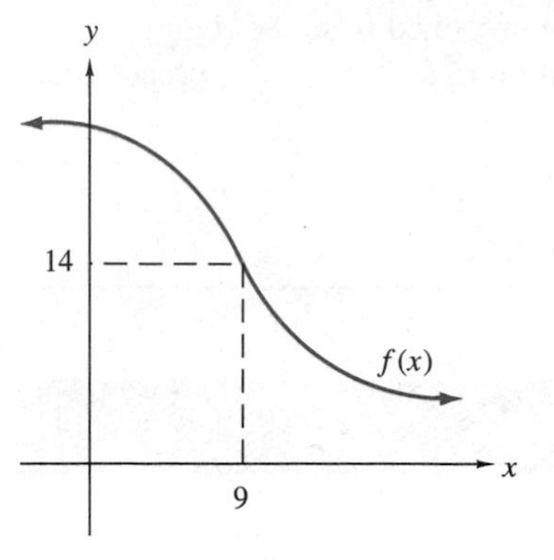

Figure 2.7

Suppose the graph pictured in Figure 2.7 displays the relationship between the quantities represented by x and y. As x approaches 9 from the left, the function values descend toward and approach the y value 14. Figure 2.8 shows both these movements. Figure 2.9 shows the movement of x from the right. As x approaches

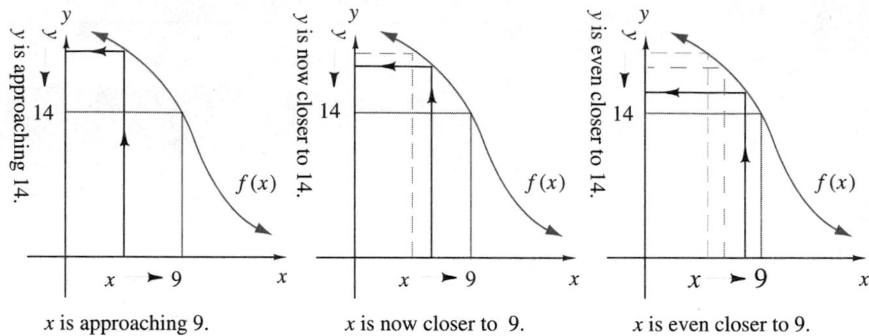

Figure 2.8

9 from the right, the function values ascend toward and approach the y value 14. We see that regardless of the side, the left or the right, from which x approaches

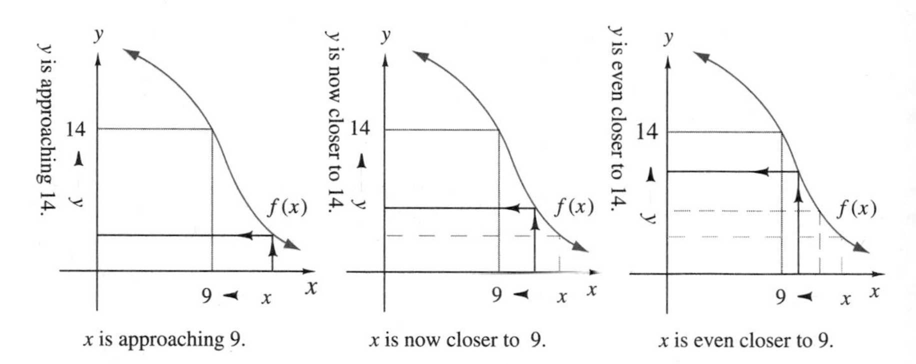

Figure 2.9

9, $f(x)$ approaches 14. That is, we see that

$$\lim_{x\to 9^-} f(x) = 14 \text{ and } \lim_{x\to 9^+} f(x) = 14$$

Since each one-sided limit equals the same unique value, 14, they are equal to each other. That is,

$$\lim_{x\to 9^-} f(x) = \lim_{x\to 9^+} f(x)$$

This equality guarantees us that the limit exists and in fact is equal to 14. This fact is expressed as

$$\lim_{x\to 9} f(x) = 14$$

Interpretation: As the x values get close to 9, the y values get close to 14.

ILLUMINATOR SET D

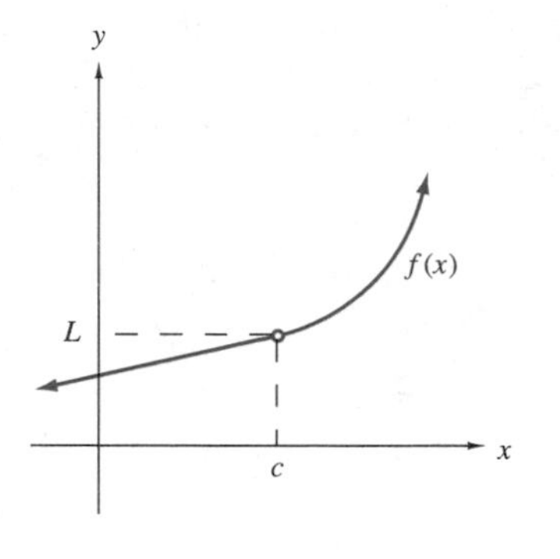

Figure 2.10

Suppose the graph pictured in Figure 2.10 displays the relationship between the quantities represented by x and y. As x approaches c from the left, the function values ascend toward and approach the y value L. Figure 2.11 shows both these movements. Figure 2.12 shows the movement of x from the right. As

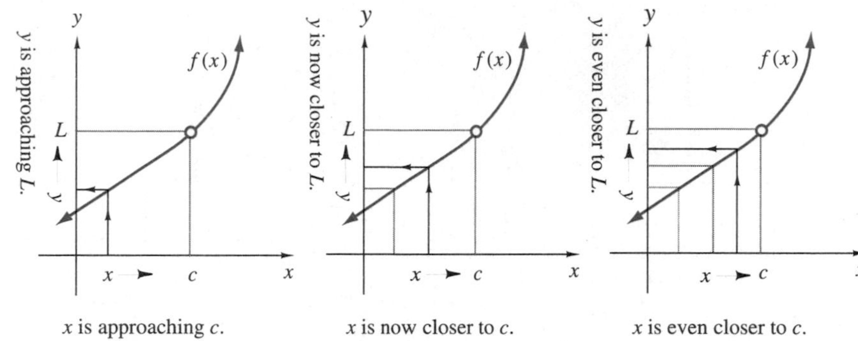

Figure 2.11

x approaches c from the right, the function values descend toward and approach the y value L. We see that regardless of the side, the left or the right, from which

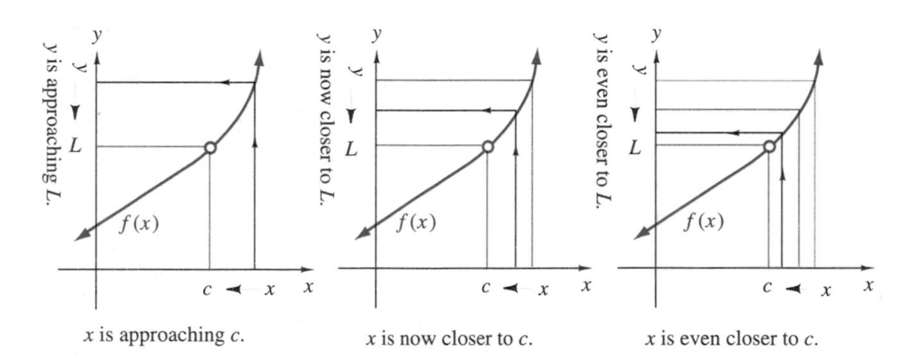

Figure 2.12

x approaches c, $f(x)$ approaches L. That is, we see that

$$\lim_{x \to c^-} f(x) = L \text{ and } \lim_{x \to c^+} f(x) = L$$

Since each one-sided limit equals the same unique value, L, they are equal to each other. That is,

$$\lim_{x \to c^-} f(x) = \lim_{x \to c^+} f(x)$$

This equality guarantees us that the limit exists and in fact is equal to L. This fact is expressed as

$$\lim_{x \to c} f(x) = L$$

ILLUMINATOR SET E

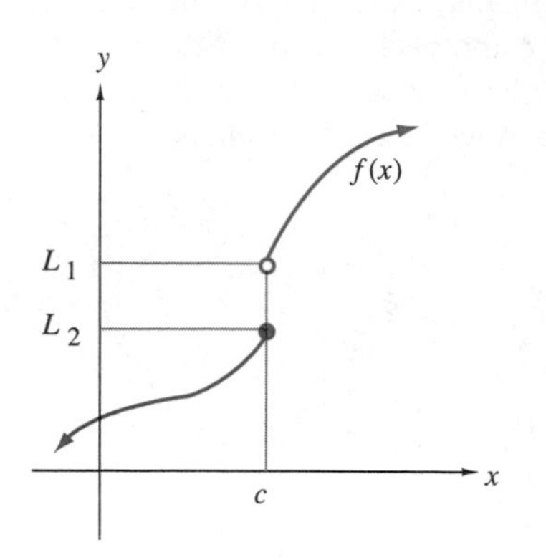

Figure 2.13

Suppose the graph pictured in Figure 2.13 displays the relationship between the quantities represented by x and y. As x approaches c from the left, the function values ascend toward and approach the y value L_1. (The subscript 1 on the L just identifies this L as the *first* L.) Figure 2.14 shows both these movements. We see

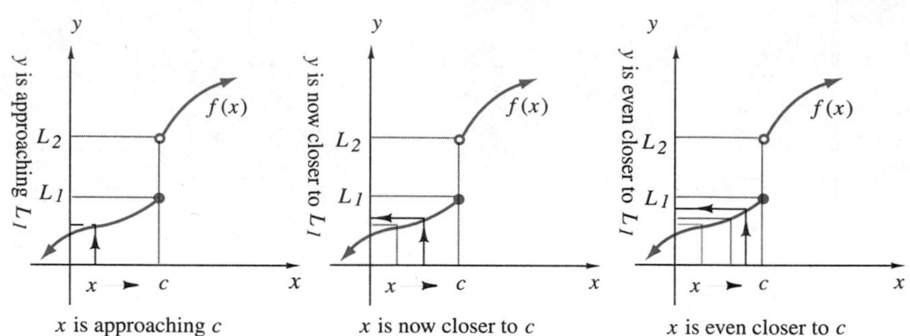

Figure 2.14

that as x approaches c from the left side, $f(x)$ approaches the y value L_1. That is, we see that

$$\lim_{x \to c^-} f(x) = L_1$$

As x approaches c from the right, the function values descend toward and approach the y value L_2. (The subscript 2 on the L just identifies this L as the *second* L.) Figure 2.15 shows both these movements. We see that as x approaches

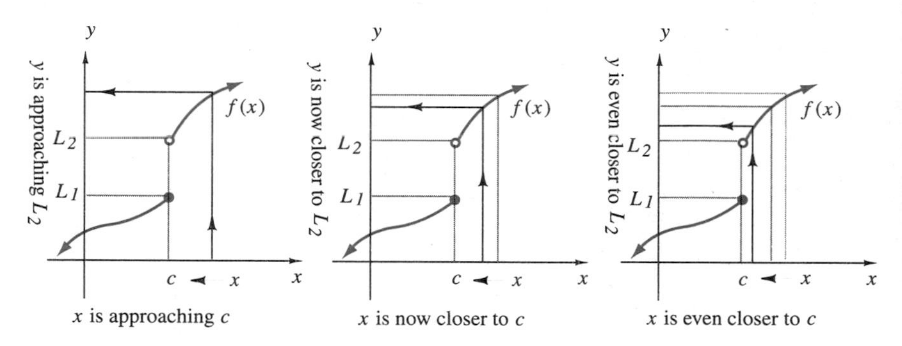

Figure 2.15

c from the right side, $f(x)$ approaches the y value L_2. That is, we see that

$$\lim_{x \to c^+} f(x) = L_2$$

But we notice that L_1 and L_2 are two distinct values. That is, we notice that

$$\lim_{x \to c^-} f(x) \neq \lim_{x \to c^+} f(x)$$

Since these one-sided limits are not the same unique value, we conclude that the limit of this function, as x approaches c, does not exist.

Computational Methods and Apparent Limits

In the next two examples we will illustrate the limit concept using actual numerical values and specific functions. The method used in these two examples is computational and is quite useful for complicated functions. The computational method, however, only helps us to *suggest* a limit. Calculators or computers are used to compute several function values using x values near the x value to be approached. We never know if we have computed enough function values to actually have the limit or if we have stopped short and some peculiar behavior takes place closer in. So, computational methods produce only *apparent limits*. Algebraic methods and pencil and paper can be used to determine limits for simpler functions. We will investigate some of these algebraic methods in Section 2.2.

Table 2.1

x	$\dfrac{3x^2-2x-8}{x-2}$
1	7
1.5	8.5
1.7	9.1
1.9	9.7
1.99	9.97
1.999	9.997
1.9999	9.9997
1.99999	9.99997

EXAMPLE SET A

Find the limit, if it exists, as x approaches 2, of the function

$$f(x) = \frac{3x^2-2x-8}{x-2}$$

Solution:
We can see that $f(x)$ is defined for all numbers x except 2. At $x = 2$, the denominator is 0. In fact, $f(2) = \frac{0}{0}$ which is indeterminate. The interesting points in the domain of this function are those points near $x = 2$. (This is why the problem directs us to find the limit as x approaches 2. Remember, we are not trying to determine the behavior at 2, but rather at points close to 2. Since $x = 2$ produces an indeterminate form, there is no behavior at $x = 2$.) We will investigate the behavior of $f(x)$ as x approaches 2 from both the left side and the right side by constructing a table of values consisting of several x values close to 2 and their associated function values. We first examine $\lim_{x\to 2^-} f(x)$ by letting x approach 2 through values slightly less than 2, and by observing the associated function values. (See Table 2.1.)

Note: Your calculator may produce the number 10 for the last value entered due to the fact that it will automatically round off. This type of round-off error restricts our ability to use tables of numbers to find limits. That is why algebraic techniques, using a pencil and paper, are the preferred method for finding limits.

Table 2.2

x	$\dfrac{3x^2-2x-8}{x-2}$
3	13
2.5	11.5
2.3	10.9
2.1	10.3
2.01	10.03
2.001	10.003
2.0001	10.0003
2.00001	10.00003

According to our table, it appears that as the x values get close to 2 from the left, the $f(x)$ values get close to 10. We conclude that, apparently, $\lim_{x\to 2^-} f(x) = 10$.

We next examine $\lim_{x\to 2^+} f(x)$ by letting x approach 2 through values slightly greater than 2 and observing the associated function values.(See Table 2.2.) (Your calculator may have produced exactly 10 for the $f(x)$ value on the last line due to round-off error.)

It appears that as the x values get close to 2 from the right, the $f(x)$ values get close to 10. We conclude that, apparently, $\lim_{x\to 2^+} f(x) = 10$.

Then, since $\lim_{x\to 2^-} f(x) = 10$ and $\lim_{x\to 2^+} f(x) = 10$, $\lim_{x\to 2^-} f(x) = \lim_{x\to 2^-} f(x)$, and, apparently,

$$\lim_{x\to 2} f(x) = 10$$

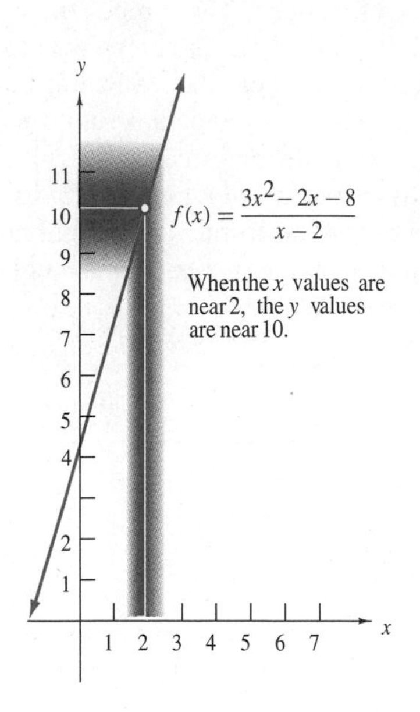

Figure 2.16

Interpretation: The graph of $f(x) = \dfrac{3x^2 - 2x - 8}{x - 2}$ is pictured in Figure 2.16. Observation of the graph helps convince us that our conclusion, $\lim\limits_{x \to 2} f(x) = 10$, is correct.

It is worthwhile to note that

$$f(x) = \frac{3x^2 - 2x - 8}{x - 2} = \frac{(3x + 4)(x - 2)}{x - 2} = 3x + 4, \quad x \neq 2,$$

so that the graph of $f(x)$ is the graph of the straight line $y = 3x + 4$ with a hole located at $x = 2$.

EXAMPLE SET B

Find the limit, if it exists, as x approaches 4, of the piecewise function

$$f(x) = \begin{cases} \frac{1}{2}x + 3, & \text{if } x < 4 \\ -x^2 + 8x - 13, & \text{if } x \geq 4 \end{cases}$$

Solution:

We will examine the behavior of the function by examining the function values first as x approaches 4 from the left side and then as x approaches 4 from the right side.

(Your calculator may have produced exactly 5 on the last line due to round-off error.)

It appears that as the x values get close to 4 from the left, the $f(x)$ values get close to 5. We conclude that, apparently, $\lim\limits_{x \to 4^-} f(x) = 5$. (See Table 2.3.)

Table 2.3

x	$f(x)$
3	4.5
3.5	4.75
3.7	4.85
3.9	4.95
3.99	4.995
3.999	4.9995
3.9999	4.99995
3.99999	4.999995

(Your calculator may have produced exactly 3 on the last line due to round-off error.)

It appears that as the x values get close to 4 from the right, the $f(x)$ values get close to 3. We conclude that, apparently, $\lim\limits_{x \to 4^+} f(x) = 3$. (See Table 2.4.)

Table 2.4	
x	$f(x)$
5	2
4.5	2.75
4.3	2.91
4.1	2.99
4.01	2.9999
4.001	2.999999
4.0001	2.99999999
4.00001	2.9999999999

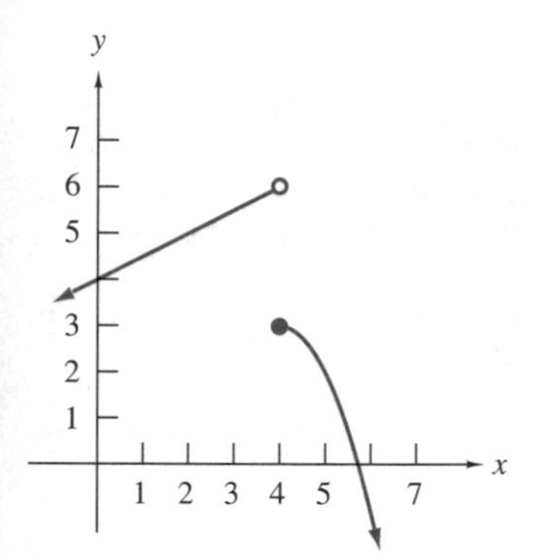

Figure 2.17

Since $\lim_{x\to 4^-} f(x) = 5$ and $\lim_{x\to 4^+} f(x) = 3$, then $\lim_{x\to 4^-} f(x) \neq \lim_{x\to 4^+} f(x)$, and $\lim_{x\to 4} f(x)$ does not exist. (Remember, for the limit to exist, both the left-hand limit and the right-hand limit must equal the same unique value. In this case, the one-sided limits are two different values.)

Interpretation: As x approaches 4, $f(x)$ does not approach a unique value, and apparently, $\lim_{x\to 4} f(x)$ does not exist.

The graph of $f(x)$ is pictured in Figure 2.17. Observation of the graph helps convince us of the correctness of our conclusion that $\lim_{x\to 4} f(x)$ does not exist.

Limits and Technology

Using Your Calculator You can use your graphing calculator to estimate the limit of a function. Following are three different approaches you could take. The following entries show the computation of $\lim_{x\to 2} \dfrac{x^2+4x-12}{x^2-5x+6}$.

1. Using the Graph feature, enter the function, set the viewing window, and graph the function. Activate the Trace cursor and move it along the curve toward $x = 2$. Observe the values of the y–coordinate.
2. Using the Table feature, enter the function as Y1 and select Table Set-Up. To observe the behavior of the function as $x \to 2^-$, set the minimum value, TblMin, near 2, say at 1.9, and the table increment at your desired decimal value, say 0.01. Choose Indpnt: Auto, and Depend: Auto. Now observe a table of values by selecting the Table feature. You should be able to see that as the x values approach 2 from the left, the function values approach -8.
3. Using the Seq (sequence) and List features, in the computation window, enter

   Seq(2+.1^N,N,1,8,1)→ L1

   Enter Y1(L1)→L2. View the results in lists 1 and 2.

Calculator Exercises Use the Trace feature of your graphing calculator to find each limit, if it exists. Draw a sketch of each function and, if the limit exists, indicate it on your sketch.

1. $\lim_{x\to 4}(3x-5)$

2. $\lim_{x\to 2}\frac{x^2-4}{x+2}$

3. $\lim_{t\to 15}\frac{4t+45,000}{t^2+45}$

4. $\lim_{x\to 6}\frac{500}{x-6}$

5. University botanists conducting experiments on a particular type of plant have found that it is very sensitive to temperature. They have constructed the function $S(T) = 4500 - (T-60)^2$ to approximate the number of plants that survive S at temperature T. Find and interpret $\lim_{x\to 24.5^-} S(T)$. Use the graph of $S(t)$ to describe, in sentence form, the behavior of this function.

EXERCISE SET 2.1

A UNDERSTANDING THE CONCEPTS and B SKILL ACQUISITION

For Exercises 1–9, use the examples of Illuminator Set A as models to describe the behavior of the function. Use the limit notation (one-sided, if necessary) to describe all appropriate limits.

1.

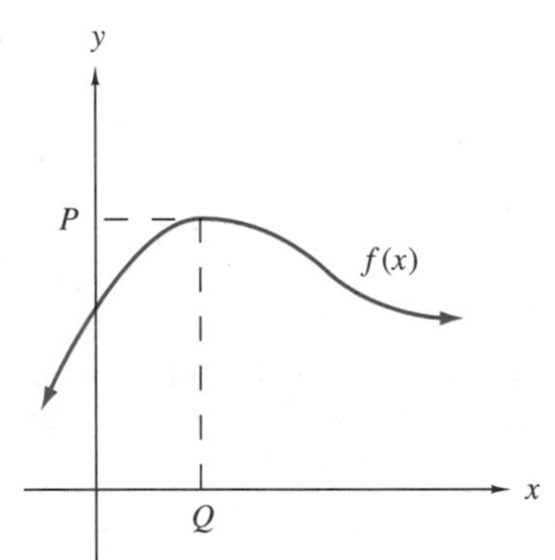

2.

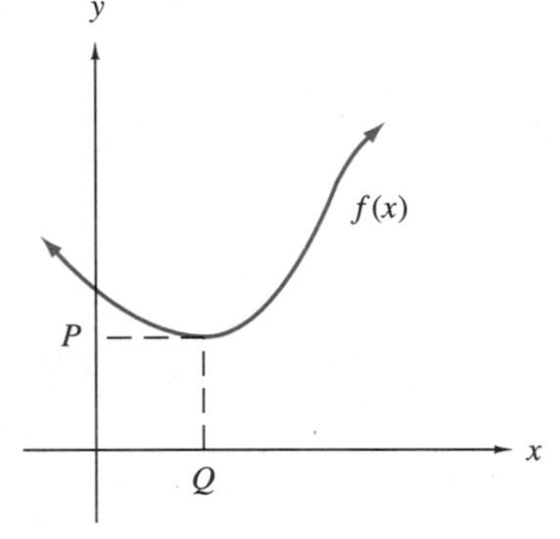

3.

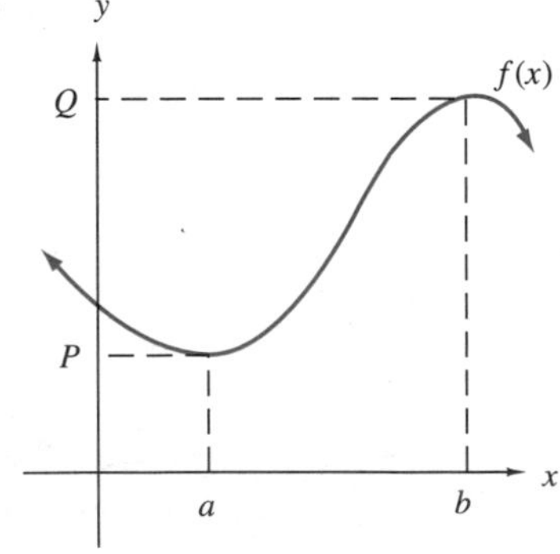

4.

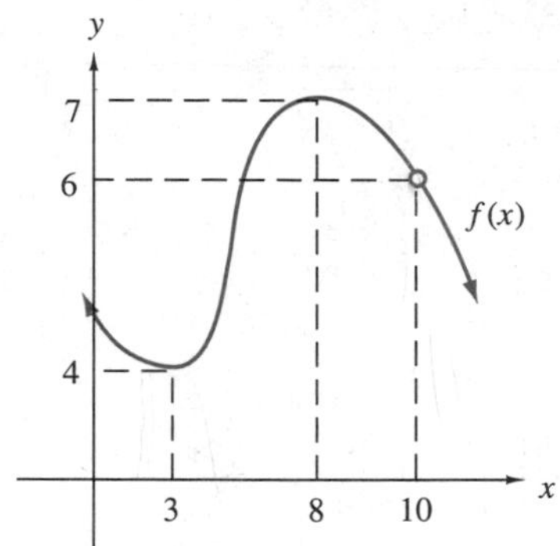

5.

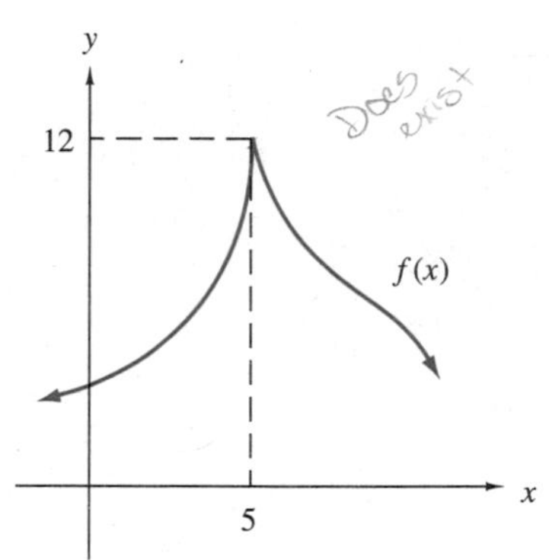

6.

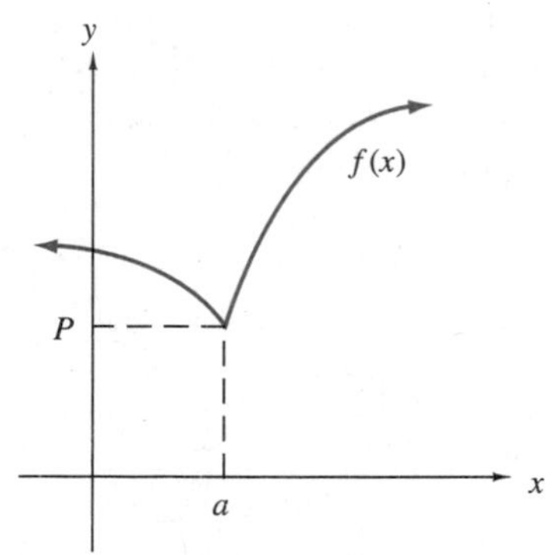

7.

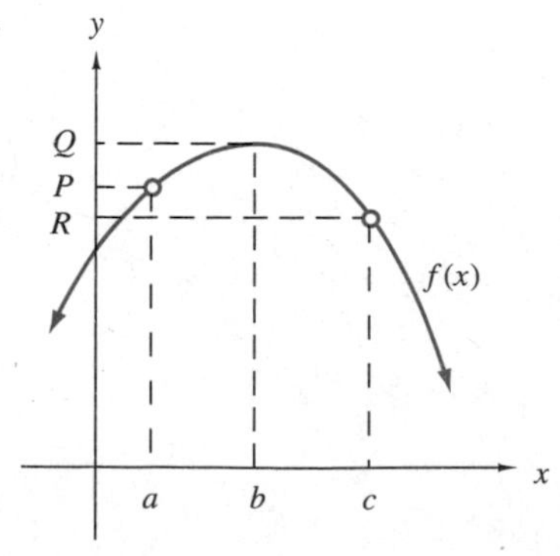

8.

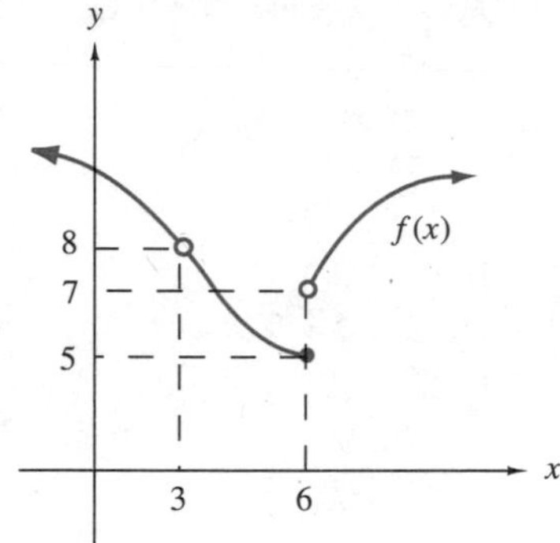

9.

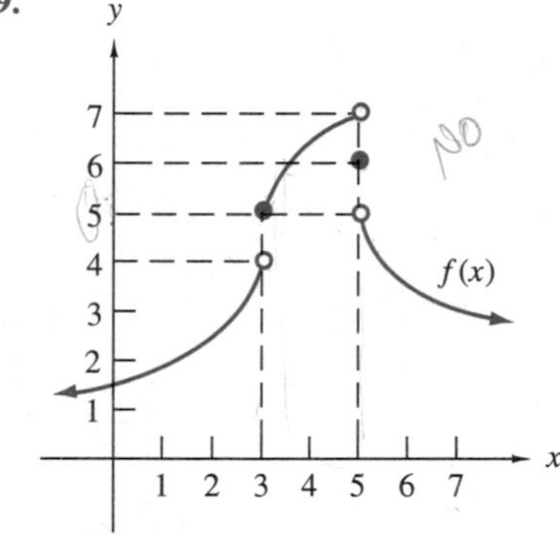

In Exercises 10–13, use observations from the table to specify the apparent limit, if it exists.

10. These tables are for $f(x)$.

x (from the left)	$f(x)$
7	12.1
7.5	12.3
7.7	12.95
7.9	12.995
7.99	12.9995
7.999	12.99995

x (from the right)	$f(x)$
9	13.2
8.5	13.1
8.3	13.01
8.1	13.001
8.01	13.0001
8.001	13.00001

11. These tables are for $f(x)$.

x (from the left)	$f(x)$
4	−6
4.5	−5.5
4.7	−5.1
4.9	−5.01
4.99	−5.001
4.999	−5.0001

x (from the right)	$f(x)$
6	−4.87
5.5	−4.8
5.3	−4.9
5.1	−4.99
5.01	−4.999
5.001	−4.9999

12. These tables are for $f(x)$.

x (from the left)	$f(x)$
0.5	1.3
0.7	1.8
0.9	1.9
0.99	1.99
0.999	1.999
0.9999	1.9999

x (from the right)	$f(x)$
1.5	3.5
1.3	3.3
1.1	3.1
1.01	3.01
1.001	3.001
1.0001	3.0001

13. These tables are for $f(x)$.

x (from the left)	$f(x)$
−8.5	6.8
−8.3	6.2
−8.1	6.1
−8.01	6.01
−8.001	6.001
−8.0001	6.0001

x (from the right)	$f(x)$
−7.5	4.2
−7.7	4.7
−7.9	4.9
−7.99	4.99
−7.999	4.999
−7.9999	4.9999

For Exercises 14–20, from the graph and the specified value of c, determine the $\lim_{x \to c} f(x)$, if it exists. If the limit does not exist, then indicate so. If the limit does exist, specify its value and its meaning.

14.

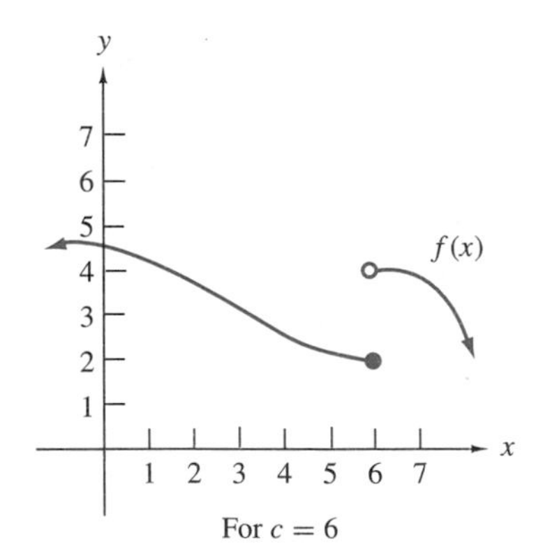

For $c = 6$

15.

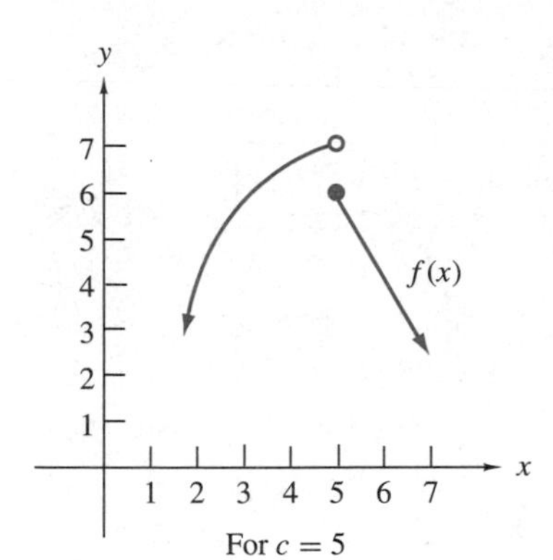

For $c = 5$

18.

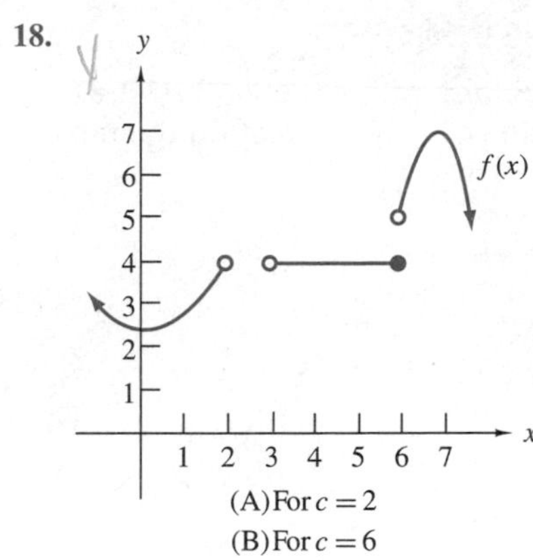

(A) For $c = 2$
(B) For $c = 6$

16.

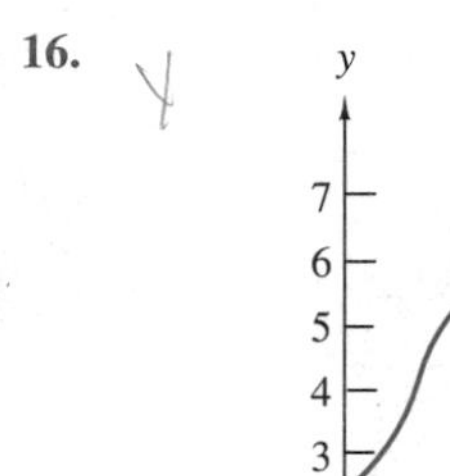

For $c = 3$

19.

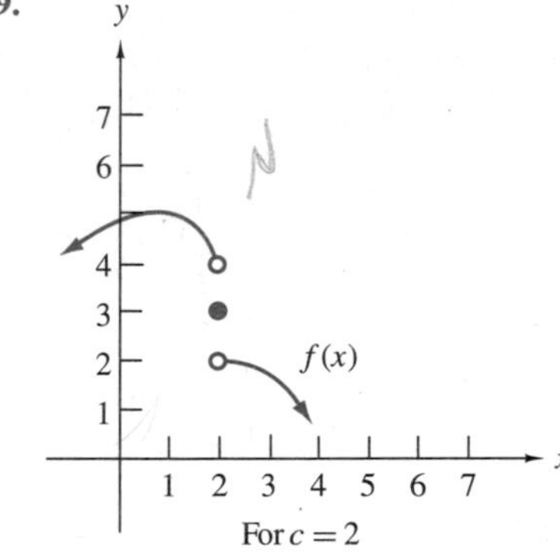

For $c = 2$

17.

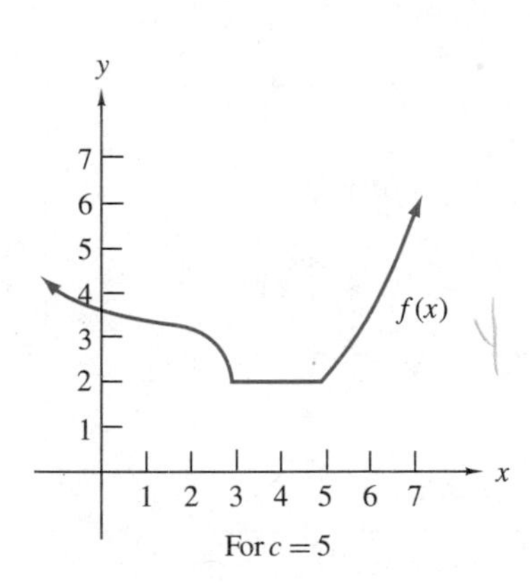

For $c = 5$

20.

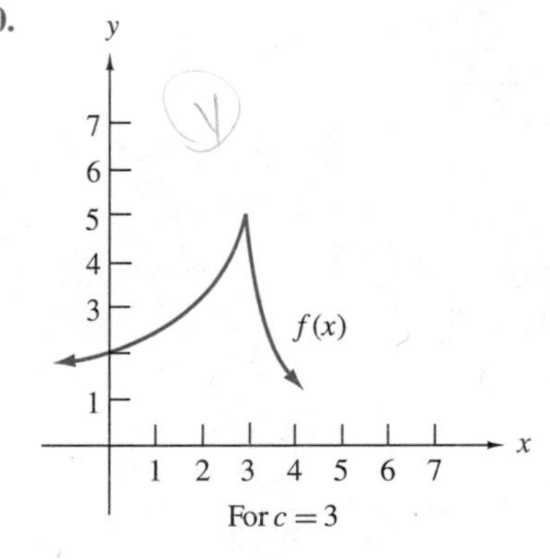

For $c = 3$

C APPLYING THE CONCEPTS

For Exercises 21–26, determine the limit, if it exists, at the specified value of x by completing the tables.

21. $f(x) = x^2 + 4x - 8$ at $x = 1$

x (from the left)	$f(x)$
0.5	
0.7	
0.9	
0.99	
0.999	

x (from the right)	$f(x)$
1.5	
1.3	
1.1	
1.01	
1.001	

22. $f(x) = -5x^2 + 2x + 7$ at $x = -3$

x (from the left)	$f(x)$
−3.5	
−3.3	
−3.1	
−3.01	
−3.001	

x (from the right)	$f(x)$
−2.5	
−2.7	
−2.9	
−2.99	
−2.999	

23. $f(x) = \begin{cases} -3x + 14, & \text{if } x < 3 \\ x^2 - 6x + 14, & \text{if } x > 3 \end{cases}$ at $x = 3$

x (from the left)	$f(x)$
2.5	
2.7	
2.9	
2.99	
2.999	

x (from the right)	$f(x)$
3.5	
3.3	
3.1	
3.01	
3.001	

24. $f(x) = \begin{cases} |x - 2| + 4, & \text{if } x < 5 \\ x^2 - 10x + 32, & \text{if } x > 5 \end{cases}$ at $x = 5$

x (from the left)	$f(x)$
4.5	
4.7	
4.9	
4.99	
4.999	

x (from the right)	$f(x)$
5.5	
5.3	
5.1	
5.01	
5.001	

25. $f(x) = \begin{cases} x^2 - 8x + 20, & \text{if } x < 4 \\ 3, & \text{if } x = 4 \\ 5x - 18, & \text{if } x > 4 \end{cases}$ at $x = 4$

x (from the left)	$f(x)$
3.5	
3.7	
3.9	
3.99	
3.999	

x (from the right)	$f(x)$
4.5	
4.3	
4.1	
4.01	
4.001	

26. $f(x) = \begin{cases} -x^2 + 14x - 46, & \text{if } x < 7 \\ 2, & \text{if } x = 7 \\ x^2 - 14x + 50, & \text{if } x > 7 \end{cases}$ at $x = 7$.

x (from the left)	$f(x)$
6.5	
6.7	
6.9	
6.99	
6.999	

x (from the right)	$f(x)$
7.5	
7.3	
7.1	
7.01	
7.001	

For Exercises 27–32, construct the graph of each function and determine if the limit exists at the specified value of x. If the limit exists, specify its value.

27. $f(x) = \begin{cases} \frac{1}{2}x + 1, & \text{if } x < 4 \\ x^2 - 8x + 17, & \text{if } x > 4 \end{cases}$ at $x = 4$

28. $f(x) = \begin{cases} 5, & \text{if } x < 3 \\ -x^2 + 6x - 6, & \text{if } x > 3 \end{cases}$ at $x = 3$

29. $f(x) = \begin{cases} \frac{2x^2 - 3x - 2}{x - 2}, & \text{if } x < 2 \\ 5, & \text{if } x = 2 \\ -3x + 11, & \text{if } x > 2 \end{cases}$ at $x = 2$

30. $f(x) = \begin{cases} \frac{x^2 - 4x + 3}{x - 3}, & \text{if } x < 2 \\ 2, & \text{if } x = 2 \\ |x - 3| + 2, & \text{if } x > 2 \end{cases}$ at $x = 2$

31. $f(x) = \dfrac{x^2 - 2x - 3}{x + 1}$ at $x = -1$.

32. $f(x) = \dfrac{x^2 - x - 2}{x - 2}$ at $x = 2$.

For Exercises 33–36, sketch a graph of a function having the specified properties. Since many different functions may be possible for each of these exercises, answers may vary. (Sketches need not be accurate graphs, but need only present the basic structure.)

33. $f(2)$ is not defined, but $\lim_{x \to 2} f(x) = 3$

34. $f(1) = 3$, $f(0) = 1$, $f(2) = 2$ and $\lim_{x \to 1} f(x) = 2$

35. $f(1) = 1$, $f(0) = 3$, $f(2) = 0$, $\lim_{x \to 1^-} f(x) = 4$, $\lim_{x \to 1^+} f(x) = 3$

36. $f(0) = 1$, $f(3) = 2$, $f(5) = 0$, $\lim_{x \to 3^-} f(x) = 1$, $\lim_{x \to 3^+} f(x) = 3$

D DESCRIBE YOUR THOUGHTS

37. A friend that does not know much mathematics asks you to describe to him what is meant by the limit of a function. Try to do so.

38. Describe the difference between an apparent and an actual limit of a function.

39. Explain why it is possible for $\lim_{x \to 3} f(x) = 7$, but for $f(3)$ to not exist.

E REVIEW

40. **(1.2)** Construct the graph of the function

$$f(x) = \frac{2}{3}x - 2.$$

41. **(1.3)** For $f(x) = x^2 + 3x + 5$, find $\dfrac{f(x+h) - f(x)}{h}$.

42. **(1.4)** Let $f(x) = 5x - 6$ and $g(x) = x^2 + 4$. Find $f \circ g$.

2.2 Limits Viewed Algebraically

Introduction

Determining Limits Using the Properties of Limits

Determining Limits when the Properties Do Not Directly Apply

Limits at the Endpoints of an Open Interval

Summary of the Properties of Limits

Limits and Technology

Introduction

Computational or graphical methods for determining limits, as illustrated in Section 2.1, are neither efficient nor accurate to use. They provide only the *apparent* limit of the function. In this section we examine two preferred methods that are not only more efficient than computing or graphing, but that provide the *actual* limit of the function rather than the *apparent* limit. The first method uses the *properties of limits*, and the second uses algebraic manipulations (the most common of which involve factoring polynomials and rationalizing radical expressions).

Determining Limits Using the Properties of Limits

Properties of Limits

Limits of functions can often be determined directly and precisely using the properties of limits. We will state, interpret, and illustrate eight of the properties of limits. We will not prove any of the properties since the proofs rely on the theoretical definition of the limit and we are concerned with only the applications of calculus.

Limit of a Constant

If k is a constant, then

$$\lim_{x \to c} k = k$$

The limit property of constants states that the limit of a constant will always exist and is precisely the constant indicated. This makes sense when we observe that changing values of x have no effect on the value of a constant (after all, a constant is a value that never changes). Figure 2.18 shows the graph of the constant function $f(x) = k$. The graph is the horizontal line through k on the y axis. As the values of x get near c, the y values are always equal to k.

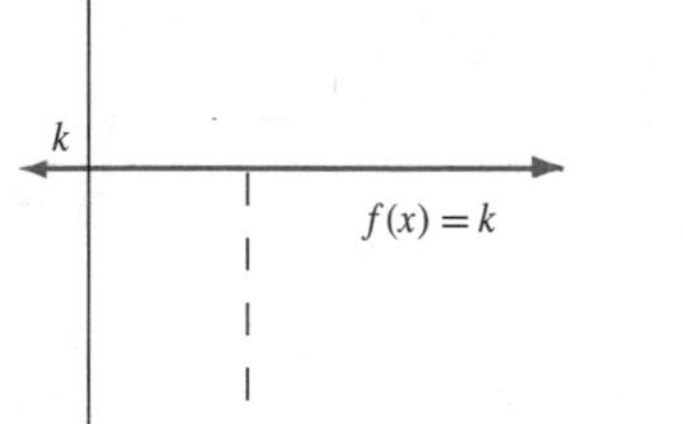

Figure 2.18

ILLUMINATOR SET A

If $f(x) = 4$, then

$$\lim_{x\to 7} f(x) = \lim_{x\to 7} 4$$
$$= 4$$

Limit of a Polynomial Function

If $P(x) = a_n x^n + a_{n-1}x^{n-1} + \cdots + a_1 x + a_0$ is a polynomial function, then

$$\begin{aligned}\lim_{x\to c} P(x) &= P(c)\\ &= a_n c^n + a_{n-1}c^{n-1} + \cdots + a_1 c + a_0\end{aligned}$$

The limit property of polynomial functions states that the limit of a polynomial will always exist and can be obtained by direct substitution. The graphs of polynomial functions are always smooth, unbroken curves. Figure 2.19 shows the graph of a polynomial function. As the values of x get close to c, the values of y get close to $f(c)$.

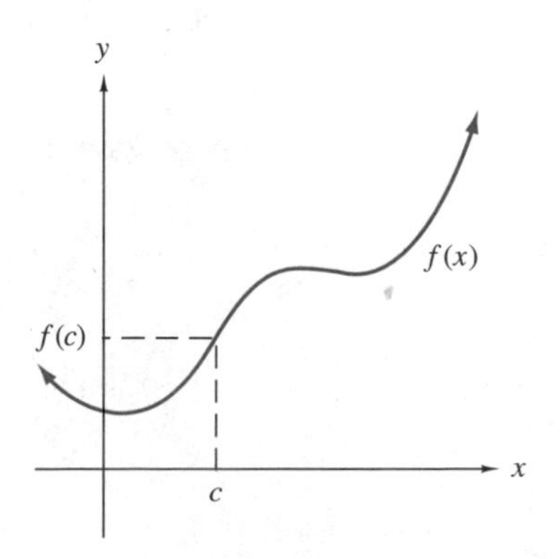

Figure 2.19

ILLUMINATOR SET B

If $f(x) = 3x^2 - 5x + 1$, then

$$\begin{aligned}\lim_{x\to 2} f(x) &= \lim_{x\to 2}(3x^2 - 5x + 1)\\ &= 3(2)^2 - 5(2) + 1\\ &= 3\cdot 4 - 10 + 1\\ &= 3\end{aligned}$$

Table 2.5

x	$f(x)$
1.9	2.33
1.99	2.9303
1.999	2.993003
2.001	3.007003
2.01	3.0703
2.1	3.73

As you did in Section 2.1, you could construct a table of values near $x = 2$ to help you intuitively suggest that the limit is 3. Such a table is illustrated in Table 2.5.

Limit of a Rational Function

If $R(x) = \dfrac{P(x)}{Q(x)}$, where $P(x)$ and $Q(x)$ are polynomial functions, then

$$\begin{aligned}\lim_{x\to c} R(x) &= \lim_{x\to c}\frac{P(x)}{Q(x)}\\ &= \frac{P(c)}{Q(c)}\\ &= R(c), \quad \text{if } Q(c) \neq 0\end{aligned}$$

The limit property of rational functions states the limit of a rational function can be determined by direct substitution only if that substitution does not produce 0 in the denominator. (We will see in Example Set I that substitution may produce 0 in the denominator and the limit still exist. If this should happen, this theorem cannot be used and other techniques must be applied to produce the limit.)

ILLUMINATOR SET C

If $f(x) = \dfrac{x^2 - 3}{x + 5}$, then

$$\begin{aligned}\lim_{x\to -2} f(x) &= \lim_{x\to -2}\frac{x^2 - 3}{x + 5} = \frac{(-2)^2 - 3}{-2 + 5}\\ &= \frac{4 - 3}{3} = \frac{1}{3}, \quad \text{since } x + 5 \neq 0 \text{ at } x = -2\end{aligned}$$

Limit of a Sum or Difference

If $f(x)$ and $g(x)$ are functions and $\lim_{x\to c} f(x)$ and $\lim_{x\to c} g(x)$ both exist, then

$$\lim_{x\to c} \left[f(x) \pm g(x)\right] = \lim_{x\to c} f(x) \pm \lim_{x\to c} g(x)$$

The limit property of sums or differences states that the limit of a sum or a difference of two functions can be obtained from direct substitution provided the limits of each of the individual functions exist. The limit is obtained by determining the limit of each individual function, then adding or subtracting those limit values.

EXAMPLE SET A

1. Find $\lim_{x\to 4} (x-3)$.

Solution:

$$\begin{aligned}\lim_{x\to 4} (x-3) &= \lim_{x\to 4} x - \lim_{x\to 4} 3\\ &= 4-3\\ &= 1\end{aligned}$$

(Since $f(x) = x - 3$ is a polynomial function, we could have used the limit property of polynomials to complete this evaluation.)

2. Find $\lim_{x\to -1} (4x^2 - 2x + \frac{5}{x})$.

Solution:

$$\begin{aligned}\lim_{x\to -1} (4x^2 - 2x + \frac{5}{x}) &= \lim_{x\to -1} 4x^2 - \lim_{x\to -1} 2x + \lim_{x\to -1} \frac{5}{x}\\ &= 4(-1)^2 - 2(-1) + \frac{5}{-1}\\ &= 4 + 2 - 5\\ &= 1\end{aligned}$$

Limit of a Product of Functions

If $f(x)$ and $g(x)$ are functions and $\lim\limits_{x \to c} f(x)$ and $\lim\limits_{x \to c} g(x)$ both exist, then

$$\lim_{x \to c} [f(x) \cdot g(x)] = \left[\lim_{x \to c} f(x)\right] \cdot \left[\lim_{x \to c} g(x)\right]$$

The limit property of products states that the limit of a product of two functions can be determined by direct substitution provided the limits of the two individual functions exist. It is obtained by determining the limit of each individual function, then multiplying those limit values together.

EXAMPLE SET B

Find $\lim\limits_{x \to 1} [(2x - 3)(x^2 + 1)]$.

Solution:

$$\begin{aligned}
\lim_{x \to 1} [(2x - 3)(x^2 + 1)] &= \left[\lim_{x \to 1} (2x - 3)\right] \cdot \left[\lim_{x \to 1} (x^2 + 1)\right] \\
&= [2(1) - 3][1^2 + 1] \\
&= [2 - 3][1 + 1] \\
&= [-1][2] \\
&= -2
\end{aligned}$$

Limit of a Quotient of Functions

If $f(x)$ and $g(x)$ are functions and $\lim\limits_{x \to c} f(x)$ and $\lim\limits_{x \to c} g(x)$ both exist and $\lim\limits_{x \to c} g(x) \neq 0$, then

$$\lim_{x \to c} \frac{f(x)}{g(x)} = \frac{\lim\limits_{x \to c} f(x)}{\lim\limits_{x \to c} g(x)}$$

The limit property of quotients states that the limit of a quotient of two functions can be obtained by direct substitution provided the limits of the two individual functions exist and the limit of the denominator is not 0. It is obtained by determining the limit of each individual function, then finding the indicated quotient.

EXAMPLE SET C

Find $\lim_{x \to -2} \frac{3x - 2}{x + 1}$.

Solution:

$$\lim_{x \to -2} \frac{3x - 2}{x + 1} = \frac{\lim_{x \to -2} (3x - 2)}{\lim_{x \to -2} (x + 1)}$$

$$= \frac{3(-2) - 2}{-2 + 1}$$

$$= \frac{-6 - 2}{-1}$$

$$= \frac{-8}{-1}$$

$$= 8$$

Limit of a Power of a Function

If $f(x)$ is a function and $\lim_{x \to c} f(x)$ exists, then for any integer n,

$$\lim_{x \to c} [f(x)]^n = [\lim_{x \to c} f(x)]^n$$

The limit property of a power states that the limit of an integer power of a function can be determined by direct substitution provided that the limit of the function exists, and it is obtained by determining the limit of the function, and raising that value to the n th power.

EXAMPLE SET D

Find $\lim_{x \to 1/5} (5x - 3)^5$.

Solution:

$$\lim_{x \to 1/5} (5x - 3)^5 = \left[\lim_{x \to 1/5} (5x - 3)\right]^5$$

$$= (5 \cdot \frac{1}{5} - 3)^5$$

$$= (-2)^5$$
$$= -32$$

Limit of a Radical

If $f(x)$ is a function, $\lim_{x \to c} f(x)$ exists, n is a positive integer, and the nth root of this limit is defined, then

$$\lim_{x \to c} \sqrt[n]{f(x)} = \sqrt[n]{\lim_{x \to c} f(x)}$$

The limit property of a radical states that the limit of the nth root of a function under a radical can be determined by direct substitution provided that the limit of the function exists. The limit is obtained by determining the limit of the function, and taking the nth root of that value, if it is defined.

EXAMPLE SET E

Find $\lim_{x \to 8} \sqrt[3]{x^2}$.

Solution:

$$\lim_{x \to 8} \sqrt[3]{x^2} = \sqrt[3]{\lim_{x \to 8} x^2}$$
$$= \sqrt[3]{8^2}$$
$$= \sqrt[3]{64}$$
$$= 4$$

Determining Limits when the Properties Do Not Directly Apply

All the limits we have examined so far were determined by direct application of the limit properties and direct substitution. Quite often, however, direct substitution fails. We will examine several such possibilities now.

The Form $k/0$

Limits of rational functions, $\lim_{x \to c} \frac{P(x)}{Q(x)}$, that produce upon direct substitution the form $\frac{k}{0}$, where k is a nonzero constant, will never approach a specific value. Instead, as the values of x approach some x value c, the values of the numerator approach some value k, and the values of the denominator approach 0. This means the numerator of the fraction is becoming k, and the denominator is getting smaller and smaller. This, in turn, means that the fraction itself is getting bigger and bigger. It gets bigger without bound and, therefore, approaches no specific value. When a rational expression produces the form $\frac{k}{0}$, we say that *the limit does not exist.* Figure 2.20 shows a typical rational function for which the limit as x approaches c produces the form $\frac{k}{0}$.

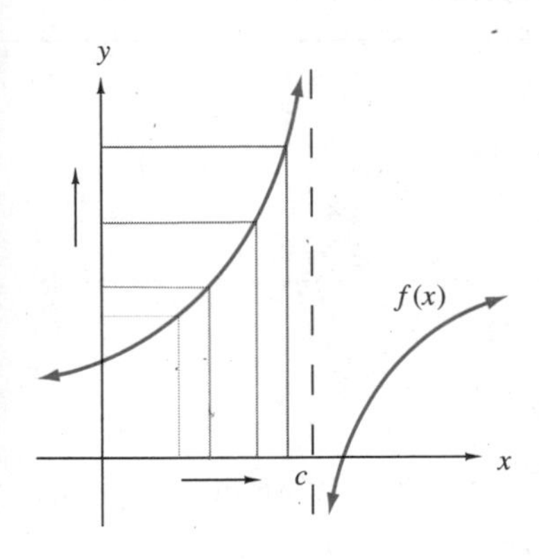

Figure 2.20

As the figure illustrates, as x gets close to c, the y values grow without bound. Since they never approach any unique value, we say that the limit does not exist.

EXAMPLE SET F

Find $\lim_{x \to -1} \frac{x^2 - x - 3}{x + 1}$.

Solution:

$$\lim_{x \to -1} \frac{x^2 - x - 3}{x + 1} = \frac{(-1)^2 - (-1) - 3}{-1 + 1}$$

$$= \frac{1 + 1 - 3}{0}$$

$$= \frac{-1}{0}, \quad \text{which is undefined.}$$

Thus, $\lim_{x \to -1} \frac{x^2 - x - 3}{x + 1}$ does not exist.

Interpretation: As the values of x approach -1, the values of $\frac{x^2 - x - 3}{x + 1}$ do not approach any specific value, but instead get smaller and smaller without bound.

Rational Fractions and the Indeterminate Form 0/0

Limits of rational expressions that produce upon direct substitution the *indeterminate* form $\frac{0}{0}$ when computing the limit need further work. Both the numerator and denominator of the fraction are, upon direct substitution, producing 0. This will occur when the numerator and denominator have a common factor that is approaching 0 while x is approaching some value c. The problem can often be eliminated by factoring and dividing out the problem-causing common factor. It is important to note that the factors that are approaching zero can be divided out because they are only close to zero in value and not equal to zero. Remember, x is approaching c and is not actually equal to c.

The indeterminate form $\frac{0}{0}$ is unlike the form $\frac{k}{0}$, where $k \neq 0$. A function producing the form $\frac{k}{0}$ upon direct substitution of c for x has a graph like that pictured in Figure 2.20. It is broken at $x = c$, and the limit does not exist. The graph of a function having the form $\frac{0}{0}$ upon direct substitution of c for x may have a hole in it at $x = c$, and may have a limit. Example Set G will illustrate this point. When the indeterminate form $\frac{0}{0}$ occurs from the direct substitution

CAUTION

of c for x, be very careful not to mentally predetermine the limit. Analyze the behavior of the function by factoring and looking for the factor that is common to the numerator and denominator. Divide it out and then try direct substitution again.

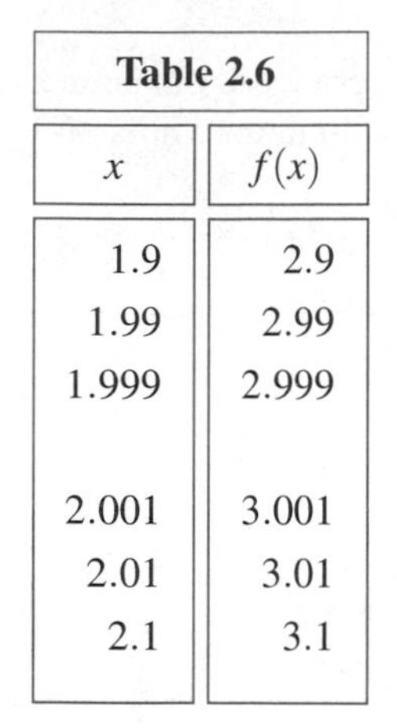

Table 2.6

x	$f(x)$
1.9	2.9
1.99	2.99
1.999	2.999
2.001	3.001
2.01	3.01
2.1	3.1

EXAMPLE SET G

Find $\lim_{x \to 2} \dfrac{x^2 - x - 2}{x - 2}$.

Solution:

Direct substitution produces $\dfrac{2^2 - 2 - 2}{2 - 2} = \dfrac{0}{0}$, which is indeterminate. Since the limit of the denominator is 0, the quotient rule for limits doesn't apply, and further work is needed.

The indeterminate form $\dfrac{0}{0}$ along with a rational function alerts us to the fact that there must be a factor common to the numerator and denominator that is approaching 0 while x is approaching 2. We'll try to factor and divide out the problem-causing factors.

$$\lim_{x \to 2} \frac{x^2 - x - 2}{x - 2} = \lim_{x \to 2} \frac{(x-2)(x+1)}{x-2} \qquad \text{Divide out the common factor } x - 2.$$

$$= \lim_{x \to 2} (x + 1)$$

$$= 2 + 1$$

$$= 3$$

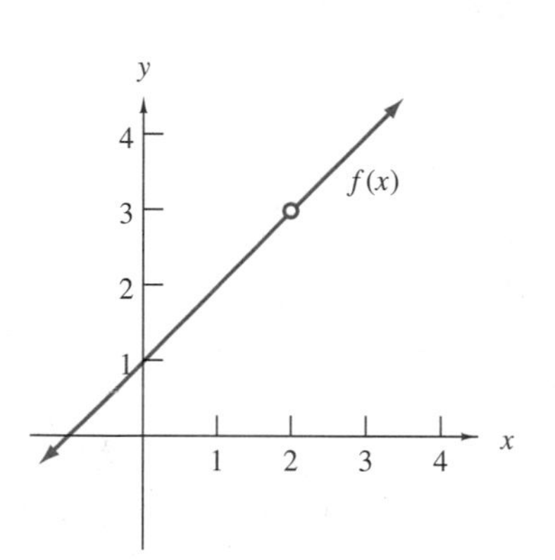

Figure 2.21

So, the limit actually does exist! This is a case where the function has a hole at $x = 2$. Figure 2.21 visually verifies this limit and Table 2.6 intuitively suggests it.

Radical Fractions and the Indeterminate Form 0/0

As we saw in Example Set G, the appearance of the indeterminate form $\dfrac{0}{0}$ along with a rational expression alerts us to the technique of factoring and dividing out the problem-causing factors. Similarly, the appearance of the indeterminate form $\dfrac{0}{0}$ along with a radical fractional expression alerts us to the technique of rationalizing the numerator or denominator and then dividing out the problem-causing factors.

EXAMPLE SET H

Find $\lim_{x \to 6} \dfrac{\sqrt{x+3} - 3}{x - 6}$.

Solution:
Direct substitution produces

$$\frac{\sqrt{6+3}-3}{6-6} = \frac{\sqrt{9}-3}{0} = \frac{3-3}{0} = \frac{0}{0}$$

which is indeterminate. Since the limit of the denominator is 0, the quotient rule for limits doesn't apply, and further work is needed. The indeterminate form $\frac{0}{0}$ along with a radical fraction alerts us to the fact that there must be a factor common to the numerator and denominator that is approaching 0 while x is approaching 6. We will try to rationalize the numerator and then factor and divide out the problem-causing factors.

$$\lim_{x\to 6}\frac{\sqrt{x+3}-3}{x-6} = \lim_{x\to 6}\frac{\sqrt{x+3}-3}{x-6}\cdot\frac{\sqrt{x+3}+3}{\sqrt{x+3}+3}$$

$$= \lim_{x\to 6}\frac{(x+3)-9}{(x-6)(\sqrt{x+3}+3)} \quad \text{Simplify the numerator only.}$$

$$= \lim_{x\to 6}\frac{x-6}{(x-6)(\sqrt{x+3}+3)} \quad \text{Divide out the common factor } x-6.$$

$$= \lim_{x\to 6}\frac{1}{\sqrt{x+3}+3}$$

$$= \frac{1}{\sqrt{6+3}+3}$$

$$= \frac{1}{\sqrt{9}+3}$$

$$= \frac{1}{3+3}$$

$$= \frac{1}{6}$$

EXAMPLE SET I

1. Find, if it exists, $\lim_{x\to 4} f(x)$ for the piecewise function

$$f(x) = \begin{cases} 3x-1, & \text{if } x < 2 \\ 2, & \text{if } x = 2 \\ -x^2+9, & \text{if } x > 2 \end{cases}$$

Solution:
To find the limit as x approaches 4, we need to approach 4 from both the left and the right. Since approaching 4 from both the left and the right places us in the interval $x > 2$, we use only the bottom piece of the function, $-x^2 + 9$. Thus,

$$\begin{aligned}\lim_{x\to 4} f(x) &= \lim_{x\to 4}(-x^2 + 9)\\ &= -4^2 + 9\\ &= -16 + 9\\ &= -7\end{aligned}$$

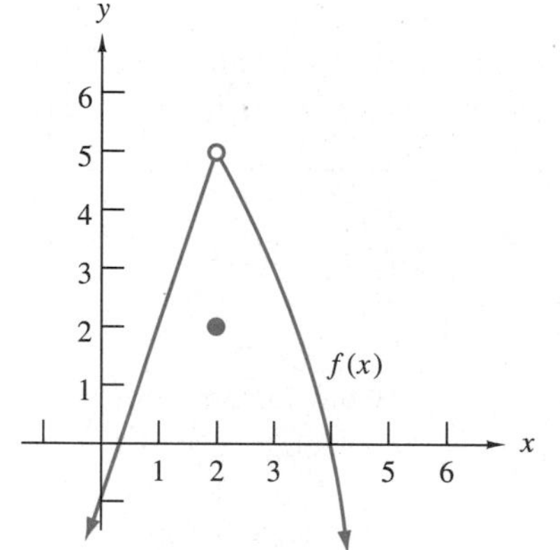

Figure 2.22

2. Find, if it exists, $\lim_{x\to 2} f(x)$ for the piecewise function

$$f(x) = \begin{cases} 3x - 1, & \text{if } x < 2 \\ 2, & \text{if } x = 2 \\ -x^2 + 9, & \text{if } x > 2 \end{cases}$$

Solution:
To find the limit as x approaches 2, we need to approach 2 from both the left and the right. As x approaches 2 from the left, $x < 2$, and we need to use the top piece of the function, $3x - 1$.

$$\begin{aligned}\lim_{x\to 2^-} f(x) &= \lim_{x\to 2^-}(3x - 1)\\ &= 3\cdot 2 - 1\\ &= 5\end{aligned}$$

As x approaches 2 from the right, $x > 2$, and we need to use the bottom piece of the function, $-x^2 + 9$.

$$\begin{aligned}\lim_{x\to 2^+} f(x) &= \lim_{x\to 2^+}(-x^2 + 9)\\ &= -2^2 + 9\\ &= 5\end{aligned}$$

Table 2.7

x	$f(x)$
1.9	4.7
1.99	4.97
1.999	4.997
2.001	4.995999
2.01	4.9599
2.1	4.59

Then, since $\lim_{x\to 2^-} f(x) = \lim_{x\to 2^+} f(x)$, $\lim_{x\to 2} f(x)$ exists. In fact, $\lim_{x\to 2} f(x) = 5$. Notice the limit has no relation to the value of $f(x)$ at $x = 2$. The graph of $f(x)$ is pictured in Figure 2.22, which shows that there is a hole in the graph at the point $(2, 5)$. When $x = 2$, $f(x) = 2$, and the point $(2, 2)$ exists and is part of the function, but it is not part of the smooth curve that we see in the sketch. Table 2.7 (which you can construct on your calculator) intuitively suggests that the limit is 5.

EXAMPLE SET J

For the function $f(x) = 5x^2 - 3x + 7$, find $\lim\limits_{h \to 0} \dfrac{f(x+h) - f(x)}{h}$.

Solution:

This limit will play an important role in differential calculus and we will consider it in detail in Section 2.5. Notice that the quotient has two variables, x and h, but we are only concerned with h approaching 0.

Initial substitution of $h = 0$ into the quotient results in the indeterminate form $\dfrac{0}{0}$.

$$\lim_{h \to 0} \frac{f(x+h) - f(x)}{h} = \lim_{h \to 0} \frac{5(x+h)^2 - 3(x+h) + 7 - (5x^2 - 3x + 7)}{h}$$

$$= \lim_{h \to 0} \frac{5(x + 2hx + h^2) - 3x - 3h + 7 - 5x^2 + 3x - 7}{h}$$

$$= \lim_{h \to 0} \frac{5x^2 + 10hx + 5h^2 - 3x - 3h + 7 - 5x^2 + 3x - 7}{h}$$

$$= \lim_{h \to 0} \frac{10hx + 5h^2 - 3h}{h} \qquad \text{Factor out the } h.$$

$$= \lim_{h \to 0} \frac{h(10x + 5h - 3)}{h} \qquad \text{Divide out the } h\text{s.}$$

$$= \lim_{h \to 0} (10x + 5h - 3) \qquad \text{Use direct substitution.}$$

$$= 10x - 3$$

Thus, $\lim\limits_{h \to 0} \dfrac{f(x+h) - f(x)}{h} = 10x - 3$.

Interpretation: As the values of h approach zero, the values of $\dfrac{f(x+h) - f(x)}{h}$ approach $10x - 3$. That is, as h gets close to zero, the quotient $\dfrac{f(x+h) - f(x)}{h}$ resembles the expression $10x - 3$.

Keep in mind that these last several examples considered the indeterminate form $\dfrac{0}{0}$. If the indeterminate form does not arise from the initial direct substitution, the limit can be determined using the limit properties.

Limits at the Endpoints of an Open Interval

Sometimes it is necessary or convenient to consider a function defined over an open interval (a, b). For example, political scientists have determined that in a city experiencing severe drug problems, the number N, in thousands, of people voting each year will increase according to the function $N(t) = 16 + 7t^2 - t^3$, where t is years, and $0 < t < 6$. Political campaign managers might be interested in the number of potential voters near year 6. That is, political campaign managers might be interested in $\lim_{x \to 6}(16 - 7t^2 - t^3)$. For a limit to exist, we know that the limit from the left must equal the limit from the right. But this function is defined only on the left of 6 and not at all on the right. In order to give existence to limits of functions over open intervals at their endpoints, we make the following two definitions.

The limit at the Left Endpoint of an Interval

If $f(x)$ is a function defined over the open interval (a, b), then the limit of the $f(x)$, as x approaches the left endpoint a, is defined as

$$\lim_{x \to a} f(x) = \lim_{x \to a^+} f(x)$$

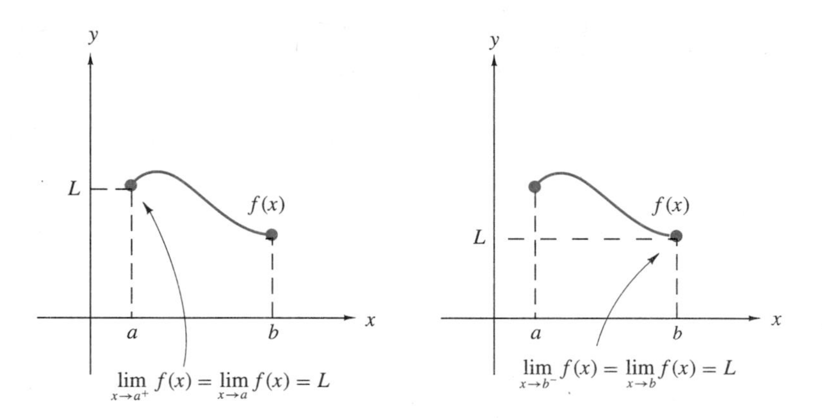

Figure 2.23

The definition states that the limit of the function at the left endpoint is defined to be the value of the right-sided limit of the function. The leftmost figure in Figure 2.23 illustrates this definition.

The limit at the Right Endpoint of an Interval

If $f(x)$ is a function defined over the open interval (a, b), then the limit of the $f(x)$, as x approaches the right endpoint b, is defined as

$$\lim_{x \to b} f(x) = \lim_{x \to b^-} f(x)$$

The definition states that the limit of the function at the right endpoint is defined to be the value of the left-sided limit of the function. The rightmost figure in Figure 2.23 illustrates this definition.

EXAMPLE SET K

The function $N(t) = 16 + 7t^2 - t^3$, defined on $(0, 6)$, describes the number of people voting in a city experiencing severe drug problems. $N(t)$ represents the number of voters (in thousands), and t is the number of years from the present. Find $\lim_{x \to 6} f(x)$, and interpret the result.

Solution:
Since 6 is the right endpoint of the interval on which $f(x)$ is defined, $\lim_{x \to 6} f(x) = \lim_{x \to 6^-} f(x)$. Thus,

$$\begin{aligned} \lim_{x \to 6} f(x) &= \lim_{x \to 6^-} (16 + 7t^2 - t^3) \\ &= 16 + 7 \cdot 6^2 - 6^3 \\ &= 16 + 252 - 216 \\ &= 52 \end{aligned}$$

Interpretation: As the 6th year is approached, the number of people voting approaches 52,000.

Summary of the Properties of Limits

1. If k is a constant, then $\lim_{x \to c} k = k$.
2. If $P(x)$ is a polynomial function, then $\lim_{x \to c} P(x) = P(c)$.
3. If $R(x) = \dfrac{P(x)}{Q(x)}$ is a rational function, then

$$\lim_{x \to c} R(x) = \lim_{x \to c} \frac{P(x)}{Q(x)} = \frac{P(c)}{Q(c)} = R(c)$$

   where $Q(c) \neq 0$.
4. If $f(x)$ and $g(x)$ are functions and $\lim_{x \to c} f(x)$ and $\lim_{x \to c} g(x)$ both exist, then

$$\lim_{x \to c} \left[f(x) \pm g(x)\right] = \lim_{x \to c} f(x) \pm \lim_{x \to c} g(x)$$

5. If $f(x)$ and $g(x)$ are functions and $\lim_{x \to c} f(x)$ and $\lim_{x \to c} g(x)$ both exist, then

$$\lim_{x \to c} \left[f(x) \cdot g(x)\right] = \left[\lim_{x \to c} f(x)\right] \cdot \left[\lim_{x \to c} g(x)\right]$$

6. If $f(x)$ and $g(x)$ are functions and $\lim_{x \to c} f(x)$ and $\lim_{x \to c} g(x)$ both exist and $\lim_{x \to c} g(x) \neq 0$, then

$$\lim_{x \to c} \frac{f(x)}{g(x)} = \frac{\lim_{x \to c} f(x)}{\lim_{x \to c} g(x)}$$

7. If $f(x)$ is a function and $\lim_{x \to c} f(x)$ exists, then for any integer n,

$$\lim_{x \to c} [f(x)]^n = [\lim_{x \to c} f(x)]^n.$$

8. If $f(x)$ is a function, $\lim_{x \to c} f(x)$ exists, n is a positive integer, and the nth root of this limit is defined, then

$$\lim_{x \to c} \sqrt[n]{f(x)} = \sqrt[n]{\lim_{x \to c} f(x)}$$

Limits and Technology

Using Derive You can use Derive to compute the limit of a function. The following entries show the computation of $\lim_{x \to 2} \frac{x^2 - x - 2}{x - 2}$. You need to press Enter at the end of each line. The **Calculus** command in the third line activates Derive's built-in calculus functions, one of which is the **Limit** command you see in the fourth line.

Author
(x^2 - x - 2)/(x - 2)
Calculus
Limit (*Press* Enter *twice*)
2
Simplify
The limit, 3, is displayed.

You can graph this expression to visually verify that 3 is the limit as x approaches 2.

Options
Display
Graphics (*Press* Enter *for high resolution*)
Use the up arrow key to highlight $\frac{x^2 - x - 2}{x - 2}$.
Plot
Overlay
Plot

The graph is displayed. The **O**verlay command overlays a coordinate system on the screen.

Derive Exercises Use Derive to compute each limit.

1. $\lim_{x \to 3} 6x - 4$

2. $\lim_{x \to 2} \frac{x + 5}{x^2 - 2x - 8}$

3. $\lim_{x \to -3} \dfrac{x+3}{2x^2 - 18}$

4. $\lim_{x \to 1} \dfrac{\sqrt{x} - 1}{x - 1}$

EXERCISE SET 2.2

A UNDERSTANDING THE CONCEPTS

1. If $\lim_{x \to 2} f(x) = 5$ and $\lim_{x \to 2} g(x) = 6$, find
   a. $\lim_{x \to 2} [f(x) + g(x)]$ b. $\lim_{x \to 2} [f(x) - g(x)]$
   c. $\lim_{x \to 2} [f(x) \cdot g(x)]$ d. $\lim_{x \to 2} \dfrac{f(x)}{g(x)}$
2. If $\lim_{x \to 4} [h(x) - g(x)] = -4$ and $\lim_{x \to 4} g(x) = 8$, find
   a. $\lim_{x \to 4} h(x)$ b. $\lim_{x \to 4} \sqrt{h(x)}$
   c. $\lim_{x \to 4} \left[[h(x)]^2 + [g(x)]^2 \right]$ d. $\lim_{x \to 4} \dfrac{h(x)}{h(x) - g(x)}$
3. If $\lim_{x \to 3^-} f(x) = 2$, $\lim_{x \to 3^+} f(x) = 2$, and $f(3) = 4$, find $\lim_{x \to 3} f(x)$, if it exists, and explain your answer.
4. If $\lim_{x \to 1^-} f(x) = 5$, $\lim_{x \to 1^+} f(x) = 2$, and $f(1) = 5$, find $\lim_{x \to 1} f(x)$, if it exists, and explain your answer.
5. In the following figure, $f(x)$ is defined for $-3 \le x \le 5$. State whether the following statements are true or false.
   a. $\lim_{x \to 4} f(x) = 3$ b. $\lim_{x \to 3^-} f(x) = \lim_{x \to 3^+} f(x)$
   c. $\lim_{x \to -1} f(x) = f(-1)$ d. $\lim_{x \to 1} f(x)$ does not exist.
   e. $\lim_{x \to -3^-} f(x) = 1$ f. $\lim_{x \to -3} f(x) = 1$
   g. $\lim_{x \to 5} f(x) = \lim_{x \to 5^-} f(x)$
   h. $\lim_{x \to c} f(x)$ exists for all c in the interval $-3 \le c \le 1$.

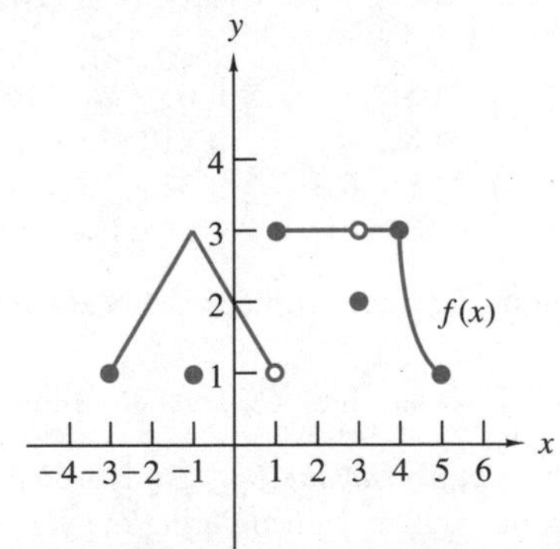

6. In the following figure, $f(x)$ is defined for $-3 \le x \le 5$. State whether the following statements are true or false.
   a. $\lim_{x \to 4} f(x) = -1$ b. $\lim_{x \to 2} f(x) = f(2)$
   c. $\lim_{x \to -1^-} f(x) = 1$ d. $\lim_{x \to -1^+} f(x) = -1$
   e. $\lim_{x \to -3^+} f(x) = \lim_{x \to -3} f(x)$ f. $\lim_{x \to 5^-} f(x) = \lim_{x \to 5} f(x)$
   g. $\lim_{x \to -3} f(x) = -2$
   h. $\lim_{x \to c} f(x)$ exists for all c in the interval $-1 \le c \le 5$.

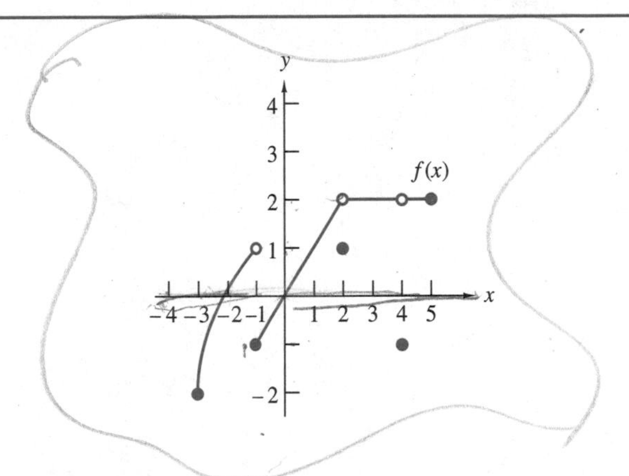

B SKILL ACQUISITION

For Exercises 7–38, find each limit, if it exists.

7. $\lim_{x \to 3} 5$
8. $\lim_{x \to 2} -4$
9. $\lim_{x \to 2} x$
10. $\lim_{x \to -4} x$
11. $\lim_{x \to -1} 3x$
12. $\lim_{x \to 5} 4x$
13. $\lim_{x \to -2} (5x - 1)$
14. $\lim_{x \to -1} (3x + 5)$
15. $\lim_{x \to 0} (2x^3 - 3x + 1)$
16. $\lim_{x \to 0} (5x^3 - 2x^2 + 5x - 3)$
17. $\lim_{x \to 2} \dfrac{x - 3}{x + 1}$
18. $\lim_{x \to -1} \dfrac{3x - 1}{2 - x}$
19. $\lim_{x \to 4} \dfrac{x^2 - 16}{x + 4}$
20. $\lim_{x \to 1} \dfrac{x^2 - x}{x - 2}$
21. $\lim_{x \to 0} \dfrac{3}{x}$
22. $\lim_{x \to -2} \dfrac{5}{x + 2}$

23. $\lim_{x\to 1} \dfrac{x^2-3x+1}{x^2+x+1}$

24. $\lim_{x\to 2} \dfrac{2x^2-5x-2}{2-x}$

25. $\lim_{x\to -3} \dfrac{x^2-9}{x+3}$

26. $\lim_{x\to 4} \dfrac{16-x^2}{4-x}$

27. $\lim_{x\to 0} \dfrac{x^3-4x^2}{2x^3-4x^2}$

28. $\lim_{x\to 0} \dfrac{x^3-3x}{x}$

29. $\lim_{x\to -3} \dfrac{x+3}{2x^2-18}$

30. $\lim_{x\to -3} \dfrac{2x^2-18}{x+3}$

31. $\lim_{x\to 4} \dfrac{\sqrt{x}}{x^2-1}$

32. $\lim_{x\to 2} \dfrac{3(x+2)}{\sqrt{x^2+1}}$

33. $\lim_{x\to 1} \dfrac{\sqrt{x}-1}{x-1}$

34. $\lim_{x\to 3} \dfrac{\sqrt{x^2+1}}{x-3}$

35. $\lim_{x\to 1} \dfrac{\sqrt{1-x^2}}{1-x}$

36. $\lim_{x\to 3} \dfrac{\sqrt{x^2-9}}{x+3}$

37. $\lim_{h\to 0} \dfrac{2(x+h)^2-2x}{h}$

38. $\lim_{h\to 0} \dfrac{\sqrt{x+h}-\sqrt{x}}{h}$

For Exercises 39–44, find $\lim_{h\to 0} \dfrac{f(x+h)-f(x)}{h}$ for the given function.

39. $f(x) = 5x - 7$

40. $f(x) = 8x + 3$

41. $f(x) = 6x^2 - 4x + 7$

42. $f(x) = 3x^2 + 7x - 4$

43. $f(x) = \dfrac{6}{x+4}$

44. $f(x) = \dfrac{8}{x-9}$

45. Find the following limits, if they exist, for

$$f(x) = \begin{cases} 2x-1, & \text{if } x < -3 \\ x-4, & \text{if } x \geq -3 \end{cases}$$

a. $\lim_{x\to -3^-} f(x)$ b. $\lim_{x\to -3^+} f(x)$ c. $\lim_{x\to -3} f(x)$

46. Find the following limits, if they exist, for

$$f(x) = \begin{cases} x^2-1, & \text{if } x < 2 \\ 4, & \text{if } x = 2 \\ x+1, & \text{if } x > 2 \end{cases}$$

a. $\lim_{x\to 2^-} f(x)$ b. $\lim_{x\to 2^+} f(x)$ c. $\lim_{x\to 2} f(x)$

47. Find the following limits, if they exist, for

$$f(x) = \begin{cases} \sqrt{x-4}, & \text{if } 4 \leq x < 13 \\ 3, & \text{if } x = 13 \\ \dfrac{15-x}{2}, & \text{if } x > 13 \end{cases}$$

a. $\lim_{x\to 13^-} f(x)$ b. $\lim_{x\to 13^+} f(x)$ c. $\lim_{x\to 13} f(x)$

48. Find the following limits, if they exist, for

$$f(x) = \begin{cases} \dfrac{3}{x}, & \text{if } x \leq \dfrac{1}{2} \\ 2x+4, & \text{if } x > \dfrac{1}{2} \end{cases}$$

a. $\lim_{x\to 1/2^-} f(x)$ b. $\lim_{x\to 1/2^+} f(x)$ c. $\lim_{x\to 1/2} f(x)$

C APPLYING THE CONCEPTS

49. ***Economics: City Water*** The monthly charge, in dollars, for water in a particular city is given by the function

$$C(x) = \begin{cases} 0.3x, & \text{if } 0 \leq x \leq 15 \\ 0.9x-9, & \text{if } x > 15 \end{cases}$$

where x is the number of cubic feet of water used. Find $\lim_{x\to 15} C(x)$, if it exists, and interpret the result.

50. ***Economics: Residential Electricity*** The monthly charge, in dollars, for x kilowatt hours of electricity used by a residential customer in a particular city is given by the function

$$C(x) = \begin{cases} 0.05x, & \text{if } 0 \leq x < 210 \\ 0.08x, & \text{if } 210 \leq x < 340 \\ 0.1x-6.8, & \text{if } 340 \leq x < 450 \\ 0.12x-15.8, & \text{if } x \geq 450 \end{cases}$$

Find the following limits, if they exist, and interpret the results.

a. $\lim_{x\to 210} C(x)$ b. $\lim_{x\to 340} C(x)$ c. $\lim_{x\to 450} C(x)$

51. ***Economics: County Income Tax*** In a particular county, income tax in x dollars earned during the year is determined by the function

$$T(x) = \begin{cases} 0, & \text{if } 0 \leq x < 10{,}000 \\ 0.15x, & \text{if } 10{,}000 \leq x < 25{,}000 \\ 0.2x-1{,}250, & \text{if } 25{,}000 \leq x < 40{,}000 \\ 0.25x-3{,}000, & \text{if } 40{,}000 \leq x < 60{,}000 \\ 0.30x-6{,}000, & \text{if } x \geq 60{,}000 \end{cases}$$

Find the following limits, if they exist, and interpret the results.

a. $\lim\limits_{x \to 10{,}000} T(x)$ b. $\lim\limits_{x \to 25{,}000} T(x)$

c. $\lim\limits_{x \to 40{,}000} T(x)$ d. $\lim\limits_{x \to 60{,}000} T(x)$

52. ***Manufacturing: Production*** Suppose the production of N items of a new line of products is given by the function

$$N(t) = \begin{cases} 300\left[t + 10 - \dfrac{600}{(t+60)}\right], & \text{if } 0 \le t < 20 \\ 500\left[t - 7 + \dfrac{1}{(t-18)}\right], & \text{if } 20 \le t \le 52 \end{cases}$$

where t is the number of weeks the new line has been in production. Find the following limits, if they exist, and interpret the results.

a. $\lim\limits_{x \to 19} N(t)$ b. $\lim\limits_{x \to 20} N(t)$ c. $\lim\limits_{x \to 21} N(t)$

D DESCRIBE YOUR THOUGHTS

53. $\lim\limits_{x \to 4} \dfrac{3x+5}{x-4}$ does not exist. Explain why this is so in terms of the behavior of the function near $x = 4$.

54. Upon the direct substitution of 6 for x in $\lim\limits_{x \to 6} \dfrac{x^2 - 3x - 18}{x^2 - 4x - 12}$, the expression $\dfrac{0}{0}$ results. Explain what this result indicates and what should be done at this point to find the limit, if it exists.

55. As if you were explaining it to a friend, explain how you would determine the limit as x approaches a of a piecewise function of the form

$$f(x) = \begin{cases} g(x), & \text{if } x < a \\ h(x), & \text{if } x = a \\ k(x), & \text{if } x > a \end{cases}$$

56. Near the end of the section we discussed limits on open intervals. Explain how you would find the $\lim\limits_{x \to b} f(x)$ on the closed interval $[a, b]$, assuming the function is defined over $[a, b]$.

E REVIEW

57. (1.2) Use your calculator to construct the graph of the function $f(x) = x^2 - x - 2$.

58. (1.3) For $f(x) = \dfrac{3}{x}$, find $\dfrac{f(x+h) - f(x)}{h}$.

59. (1.3) If

$$f(x) = \begin{cases} 3x - 2, & \text{if } x \le 1 \\ x^2 + 1, & \text{if } x > 1 \end{cases}$$

find: a. $f(-2)$ b. $f(1)$ c. $f(3)$.

60. (1.4) Let $f(x) = \dfrac{2}{x-1}$ and $g(x) = \dfrac{3}{x+2}$. Find $f \circ g$.

61. (1.5) Let $f(x) = \sqrt{x+3}$ and $g(x) = \dfrac{2}{x-1}$. Find $(g \circ f)(6)$.

2.3 Limits Involving Infinity

Introduction

Limits as the Independent Variable Approaches Infinity

Asymptotes

Asymptotes and Technology

Introduction

It often happens that, when working with particular problems involving relationships between two or more quantities, people in the life, social, or biological sciences and in business are interested in what happens in the long term. In such situations, they would be asking, "What is happening to the values of the dependent variable as the values of the independent variable grow larger and larger?"

Limits as the Independent Variable Approaches Infinity

Consider the function $A(x) = 200 + \dfrac{100{,}000}{x}$, and ask, "What is happening to A as x grows larger and larger?" You could construct a table of values as in Table 2.8. The table suggests that A is approaching 200 as x gets larger and larger. Rather than looking at specific values of $A(x)$ as you did in the table, you can get a quick and accurate picture by considering the limit as x grows large without bound. We express this as the limit as x approaches infinity and symbolize it as $\lim_{x\to\infty} A(x)$. (The symbol ∞ does not name a real number. It indicates boundlessness.)

Table 2.8

x	$A(x)$
1,000	300
100,000	201
1,000,000	200.1

Although we would never expect x to get infinitely large values in a practical situation, this limit demonstrates *trends* in the function as x grows larger.

Let's suppose that $A(x) = 200 + \dfrac{100{,}000}{x}$ represents the average cost $A(x)$ (in dollars) of producing x bicycles. If we are interested in what will happen to the average cost per bicycle as the number of bicycles produced increases, we can see any *trends* by considering the limit of $A(x)$ as x approaches infinity. To do so, we first note that as the values of x get larger and larger, the values of $\dfrac{1}{x}$ get smaller and smaller and approach 0.

$$\begin{aligned}
\lim_{x\to\infty} A(x) &= \lim_{x\to\infty}\left(200 + \frac{100{,}000}{x}\right)\\
&= \lim_{x\to\infty}(200) + \lim_{x\to\infty}\frac{100{,}000}{x}\\
&= 200 + 100{,}000 \lim_{x\to\infty}\frac{1}{x}\\
&= 200 + 100{,}000 \cdot 0 \qquad \text{since } \frac{1}{x} \to 0\\
&= 200 + 0\\
&= 200
\end{aligned}$$

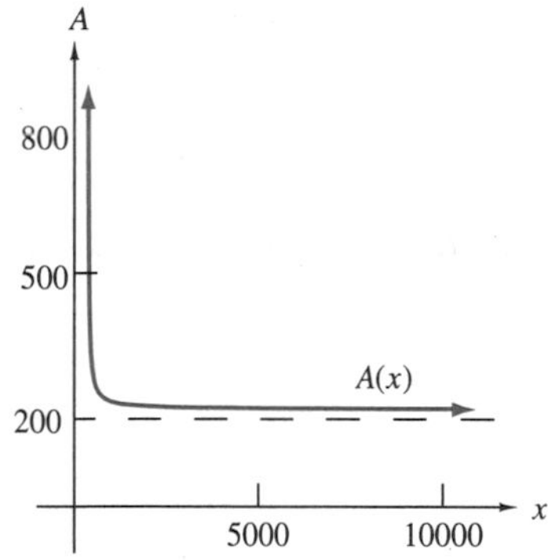

Figure 2.24

Consequently, we observe that the average cost of each bicycle tends toward \$200 as more and more bicycles are built. Note that the average cost of producing one bicycle, $A(1)$, is \$100,200. The cost is so high because of the initial production set-up costs. As the production equipment is used, the cost per bicycle decreases. These facts are illustrated in Figure 2.24. Although this function is defined only for discrete values of x, that is, $x = 1, 2, 3, \ldots$, we draw the graph as a smooth curve for the simplicity of the sketch and the ease of viewing. Notice that the average cost of each bicycle drops dramatically as the number of bicycles produced increases from 1 to 2000, but drops gradually after that.

To evaluate limits of this type quickly and accurately, let's consider the following two limit properties. In doing so, we will take the general concept of infinity (∞) and break it into two specific types, positive infinity ($+\infty$), in which x grows in the positive direction, and negative infinity ($-\infty$), in which x grows in the negative direction. The symbol ∞ without an affixed $+$ or $-$ sign represents the general concept of infinity, that is, x growing in either direction.

Property 1

$$\lim_{x\to\infty} (x^n) = \infty \quad \text{(or } \textit{does not exist}\text{)}, \quad \text{where } n > 0$$

This limit property tells us that the positive powers of x will grow without bound as x grows larger and larger.

Recall that there are two reasons for a limit not to exist. (1) As the values of the independent variable approach a particular value, the values of the dependent variable do not approach a unique constant, but rather grow larger or smaller without bound. (2) The left- and right-hand limits are not the same constant. In the first case, as an alternative to saying *the limit does not exist*, we state that *the limit equals infinity*. In the second case, we always state that *the limit does not exist*.

Property 2

$$\lim_{x\to\infty} \left(\frac{1}{x^n}\right) = 0, \qquad \text{where } n > 0$$

Property 1 tells us that as x grows larger and larger, x^n also grows larger and larger. Property 2 tells us that as x grows larger and larger, the reciprocal of x^n, $\frac{1}{x^n}$, shrinks toward 0.

EXAMPLE SET A

Find and interpret $\lim_{x\to\infty} \left(3 + \frac{2}{x} - \frac{1}{x^2}\right)$.

Solution:

$$\begin{aligned}
\lim_{x\to\infty} \left(3 + \frac{2}{x} - \frac{1}{x^2}\right) &= \lim_{x\to\infty} (3) + \lim_{x\to\infty} \left(\frac{2}{x}\right) - \lim_{x\to\infty} \left(\frac{1}{x^2}\right) \\
&= 3 + 2 \lim_{x\to\infty} \left(\frac{1}{x}\right) - 0 \\
&= 3 + 2 \cdot 0 - 0 \\
&= 3
\end{aligned}$$

Interpretation: As the x values get larger and larger, the values of $3 + \frac{2}{x} - \frac{1}{x^2}$ get closer and closer to 3.

Often we are confronted with limits of rational functions such as

$$\lim_{x\to\infty} \frac{5x^2 + 7x - 2}{8x^2 - 3x + 1}$$

Notice that

$$\lim_{x\to\infty} \frac{5x^2 + 7x - 2}{8x^2 - 3x + 1} = \frac{\lim_{x\to\infty} (5x^2 + 7x - 2)}{\lim_{x\to\infty} (8x^2 - 3x + 1)}$$

$$= \frac{\infty}{\infty}$$

The Indeterminate Form ∞/∞

Like $\frac{0}{0}$, $\frac{\infty}{\infty}$ is an *indeterminate form.* Thus, when we encounter it, we must look further to find the limit, if it exists. A helpful technique is to identify *dominant terms.*

Dominant Terms

In an expression, when x is approaching ∞, a term is **dominant** if, in the expression, it is the term with the highest degree.

In our example, the dominant term in the numerator is $5x^2$, and the dominant term in the denominator is $8x^2$. Limits in which the independent variable approaches infinity can be found by considering only the dominant terms. The process works by multiplying the numerator and denominator of the rational expression by 1 in the form of the reciprocal of the highest power of x that is present. The value of the expression does not change because we are multiplying by 1. The following example illustrates the technique.

The highest degree of x occurring in the expression $\frac{5x^2 + 7x - 2}{8x^2 - 3x + 1}$ is 2, so the highest power of x is 2. The reciprocal of x^2 is $\frac{1}{x^2}$. Thus, we multiply the expression by 1 in the form $\frac{1/x^2}{1/x^2}$.

$$\lim_{x\to\infty} \frac{5x^2 + 7x - 2}{8x^2 - 3x + 1} = \lim_{x\to\infty} \frac{5x^2 + 7x - 2}{8x^2 - 3x + 1} \cdot \frac{1/x^2}{1/x^2}$$

$$= \lim_{x\to\infty} \frac{\dfrac{5x^2}{x^2} + \dfrac{7x}{x^2} - \dfrac{2}{x^2}}{\dfrac{8x^2}{x^2} - \dfrac{3x}{x^2} + \dfrac{1}{x^2}}$$

$$= \lim_{x\to\infty} \frac{5 + \dfrac{7}{x} - \dfrac{2}{x^2}}{8 - \dfrac{3}{x} + \dfrac{2}{x^2}}$$

$$= \frac{\lim_{x\to\infty}\left(5 + \frac{7}{x} - \frac{2}{x^2}\right)}{\lim_{x\to\infty}\left(8 - \frac{3}{x} + \frac{1}{x^2}\right)} \quad \text{by the quotient property of limits}$$

$$= \frac{5+0-0}{8-0+0}$$

$$= \frac{5}{8}$$

Notice that the 5 comes from the term $5x^2$, the dominant term in the numerator, and that the 8 comes from the term $8x^2$, the dominant term in the denominator. Thus, the limit can also be found by considering only the dominant terms.

$$\lim_{x\to\infty} \frac{5x^2+7x-2}{8x^2-3x+1} = \lim_{x\to\infty} \frac{5x^2}{8x^2}$$

$$= \lim_{x\to\infty} \frac{5}{8}$$

$$= \frac{5}{8}$$

EXAMPLE SET B

Find $\lim_{x\to\infty} \frac{2-3x+5x^2-4x^3}{2x^3-13x+7}$.

Solution:

Initially (from direct substitution), $\lim_{x\to\infty} \frac{2-3x+5x^2-4x^3}{2x^3-13x+7} = \frac{-\infty}{\infty}$. We will look further for the limit by considering the dominant terms.

$$\lim_{x\to\infty} \frac{2-3x+5x^2-4x^3}{2x^3-13x+7} = \lim_{x\to\infty} \frac{-4x^3}{2x^3}$$

$$= \lim_{x\to\infty} \frac{-4}{2}$$

$$= \lim_{x\to\infty} -2$$

$$= -2$$

Notice that the degrees of the dominant terms in the numerator and denominator are the same (both are 3).

EXAMPLE SET C

Find $\lim_{x\to\infty} \frac{8x^3+7x-2}{5x^2+3x+11}$.

Solution:

Initially (from direct substitution), $\lim_{x\to\infty} \frac{8x^3 + 7x - 2}{5x^2 + 3x + 11} = \frac{\infty}{\infty}$. We will look further for the limit by considering the dominant terms.

$$\begin{aligned}\lim_{x\to\infty} \frac{8x^3 + 7x - 2}{5x^2 + 3x + 11} &= \lim_{x\to\infty} \frac{8x^3}{5x^2} \\ &= \lim_{x\to\infty} \frac{8x}{5} \\ &= \infty\end{aligned}$$

and we conclude that the limit does not exist.

Notice that the degree of the dominant term in the numerator is greater than the degree of the dominant term in the denominator.

EXAMPLE SET D

Find $\lim_{x\to\infty} \frac{7x - 3}{2x^2 + 5}$.

Solution:

Initially (from direct substitution), $\lim_{x\to\infty} \frac{7x - 3}{2x^2 + 5} = \frac{\infty}{\infty}$. We'll look further for the limit by considering the dominant terms.

$$\begin{aligned}\lim_{x\to\infty} \frac{7x - 3}{2x^2 + 5} &= \lim_{x\to\infty} \frac{7x}{2x^2} \\ &= \lim_{x\to\infty} \frac{7}{2x} \\ &= \frac{7}{2} \lim_{x\to\infty} \frac{1}{x} \\ &= \frac{7}{2} \cdot 0 \\ &= 0\end{aligned}$$

Notice that the degree of the dominant term in the numerator is less than the degree of the dominant term in the denominator.

The last Example Sets, B, C, and D, demonstrate the three possible types of results that can occur when evaluating the limits of a rational function as the independent variable approaches infinity. Notice that the degrees of the numerator and denominator in Example Sets B, C, and D, respectively, had the following properties: (1) the degrees were the same, (2) the degree of the numerator was larger than the degree of the denominator, and (3) the degree of the numerator was smaller than the degree of the denominator. We summarize these three possibilities letting $N(x)$ and $D(x)$ represent the numerator and denominator, respectively, of the rational function.

Case (1): (Degree of $N(x)$) $<$ (Degree of $D(x)$)

If, initially (from direct substitution), $\lim_{x\to\infty} \frac{N(x)}{D(x)} = \frac{\infty}{\infty}$, and the (degree of the numerator) $<$ (degree of the denominator), then

$$\lim_{x\to\infty} \frac{N(x)}{D(x)} = 0$$

In case (1), the denominator grows significantly faster than the numerator, and the ratio tends toward 0.

Case (2): (Degree of $N(x)$) $=$ (Degree of $D(x)$)

If, initially (from direct substitution), $\lim_{x\to\infty} \frac{N(x)}{D(x)} = \frac{\infty}{\infty}$, and the (degree of the numerator) $=$ (degree of the denominator), then

$$\lim_{x\to\infty} \frac{N(x)}{D(x)} = \text{the ratio of the coefficients of the dominant terms.}$$

In case (2), both the numerator and the denominator grow at about the same rate. Thus, the numbers multiplying the larger powers become important.

Case (3): (Degree of $N(x)$) $>$ (Degree of $N(x)$)

If, initially (from direct substitution), $\lim_{x\to\infty} \frac{N(x)}{D(x)} = \frac{\infty}{\infty}$, and the (degree of the numerator) $>$ (degree of the denominator), then

$$\lim_{x\to\infty} \frac{N(x)}{D(x)} = \infty \textit{ or does not exist.}$$

In case (3), the numerator grows significantly faster than the denominator, and the ratio tends toward ∞.

Each of these results can be quickly determined by paying attention to only the *dominating terms*. But keep in mind that this process works only when the indeterminate form $\frac{\infty}{\infty}$ occurs upon initial substitution.

EXAMPLE SET E

Find $\lim_{x\to\infty} \frac{3x}{x^2+1}$.

Solution:

Initially (from direct substitution), $\lim_{x\to\infty} \frac{3x}{x^2+1} = \frac{\infty}{\infty}$. We will determine the limit, if it exists, by considering only the dominant terms. Notice that the degree of the dominant term in the numerator is less than the degree of the dominant term in the denominator. Hence,

$$\lim_{x\to\infty} \frac{3x}{x^2+1} = \lim_{x\to\infty} \frac{3x}{x^2} = \lim_{x\to\infty} \frac{3}{x}$$
$$= 0$$

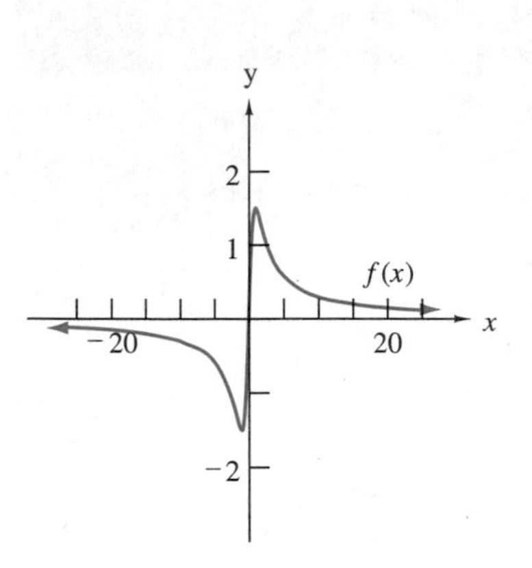

Figure 2.25

This is an example of case (1). The degree of the numerator is less than the degree of the denominator. The graph of $f(x) = \frac{3x}{x^2+1}$ is displayed in Figure 2.25. Notice that as x grows large (or small), $f(x)$ approaches 0.

EXAMPLE SET F

Find $\lim_{x\to\infty} \frac{2x+1}{x-3}$.

Solution:

Initially (from direct substitution), $\lim_{x\to\infty} \frac{2x+1}{x-3} = \frac{\infty}{\infty}$. We will find the limit, if it exists, by considering only the dominant terms. Notice that the degree of the dominant term in the numerator is equal to the degree of the dominant term in the denominator. Hence,

$$\lim_{x\to\infty} \frac{2x+1}{x-3} = \lim_{x\to\infty} \frac{2x}{x} = \lim_{x\to\infty} 2$$
$$= 2$$

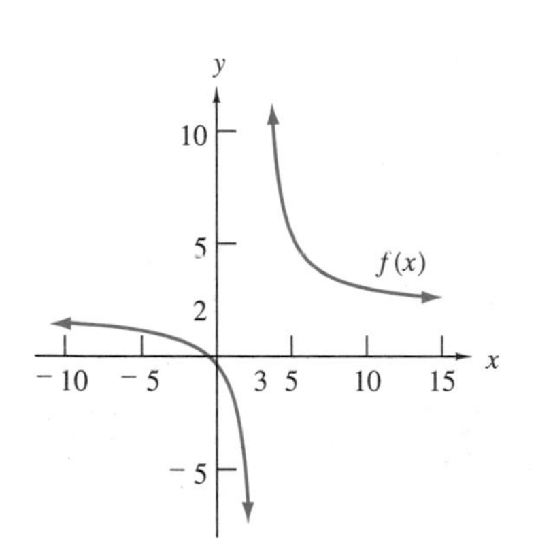

Figure 2.26

This is an example of case (2). The degree of the numerator is equal to the degree of the denominator. The graph of the $f(x) = \frac{2x+1}{x-3}$ is displayed in Figure 2.26.

EXAMPLE SET G

Find $\lim_{x\to\infty} \frac{x^2}{2x-4}$.

Solution:

Initially, $\lim_{x\to\infty} \frac{x^2}{2x-4} = \frac{\infty}{\infty}$. We will find the limit, if it exists, by considering only the dominant terms. Notice that the degree of the dominant term in the numerator is greater than the degree of the dominant term in the denominator. Hence,

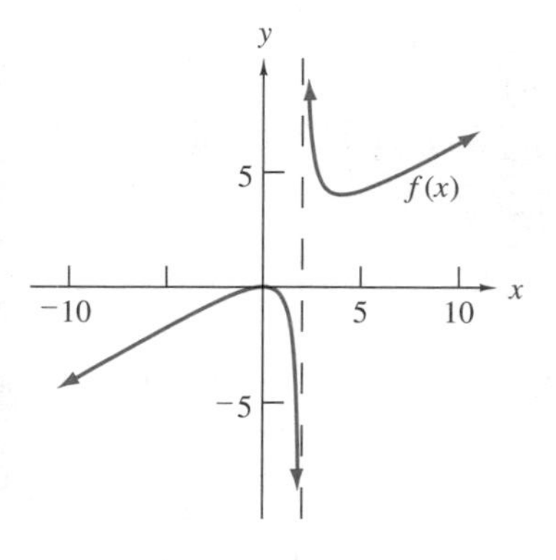

Figure 2.27

$$\lim_{x\to\infty} \frac{x^2}{2x-4} = \lim_{x\to\infty} \frac{x^2}{2x} = \lim_{x\to\infty} \frac{x}{2}$$

$$= \infty$$

This is an example of case (3). The degree of the numerator is greater than the degree of the denominator. The graph of the $f(x) = \dfrac{x^2}{2x-4}$ is displayed in Figure 2.27.

Asymptotes

In Example Sets E and F, the limits existed and resulted in constants, and the curves approached a horizontal line representing those constants. (In Example Set E, the horizontal line was the x axis, and in Example Set F, it was the line $y = 2$.)

Horizontal Asymptote

A horizontal line that is approached by a curve as the independent variable tends toward infinity is called a **horizontal asymptote**.

Dashed Lines as Asymptotes

A dashed line is commonly used on the graph to represent an asymptote. The horizontal asymptotes illustrated in Example Sets E and F are approached as x approaches both $+\infty$ and $-\infty$ (or simply ∞). In most applied situations, our concern will be to only approach $+\infty$. Of course, in Example Set G, no horizontal asymptote exists since the limit, as x approaches infinity, does not exist.

Example Set F can illustrate another type of asymptote since there is a restriction on the domain of the function. At $x = 3$, division by zero occurs. A quick and accurate method for determining what is occurring as we approach $x = 3$ from both sides is to consider

$$\lim_{x\to 3} \frac{2x+1}{x-3} = \frac{7}{0}$$

which is undefined, so that the limit does not exist. (This is the $\dfrac{k}{0}$ case in Section 2.2.)

As we noted in Section 2.2, the limit does not exist because the function is without bound. Using our new infinity notation, we can express this fact as

$$\lim_{x\to 3} \frac{2x+1}{x-3} = \infty$$

or, more specifically, as

$$\lim_{x \to 3^-} \frac{2x+1}{x-3} = -\infty$$

(since, for $x \to 3^-$, the numerator is always positive while the denominator is always negative), and

$$\lim_{x \to 3^+} \frac{2x+1}{x-3} = +\infty$$

(since, for $x \to 3^+$, the numerator is always positive and the denominator is now always positive also).

Geometrically, this situation determines a **vertical asymptote**.

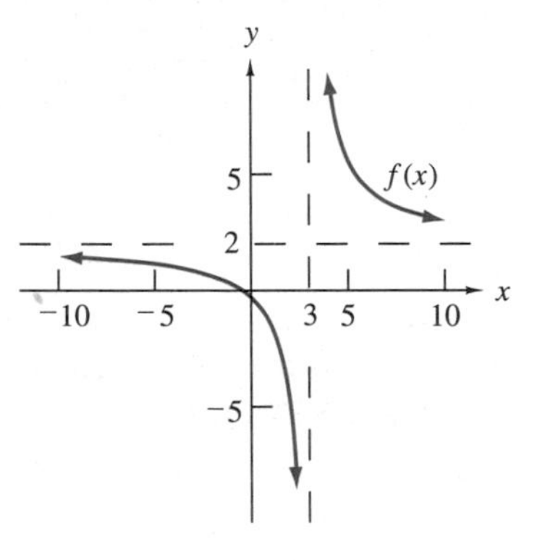

Figure 2.28

Vertical Asymptote

If $\lim_{x \to c} f(x) = \infty$, then $x = c$ is a **vertical asymptote**.

Thus, in Example Set F, there is a vertical asymptote at $x = 3$ and a horizontal asymptote at $y = 2$. Using dashed lines to represent asymptotes, the graph of $f(x) = \frac{2x+1}{x-3}$ can now be represented as in Figure 2.28. In Example Set G there is a vertical asymptote at $x = 2$ since $\lim_{x \to 2} \frac{x^2}{2x-4} = \infty$.

EXAMPLE SET H

Find the horizontal and vertical asymptotes, if they exist, for the function $f(x) = \frac{2x-3}{x+2}$.

Solution:

Horizontal asymptotes:

$$\begin{aligned}
\lim_{x \to \infty} f(x) &= \lim_{x \to \infty} \frac{2x-3}{x+2} && \text{(which is, initially, } \tfrac{\infty}{\infty}\text{)} \\
&= \lim_{x \to \infty} \frac{2x}{x} && \text{(considering only dominant terms)} \\
&= \lim_{x \to \infty} 2 \\
&= 2
\end{aligned}$$

Therefore, there is a horizontal asymptote at $y = 2$.

Vertical asymptotes:

Since the domain of the function excludes $x = -2$,

$$\lim_{x \to -2} f(x) = \lim_{x \to -2} \frac{2x - 3}{x + 2}$$
$$= \infty$$

Therefore, there is a vertical asymptote at $x = -2$.

The graph of $f(x) = \dfrac{2x - 3}{x + 2}$, with asymptotes indicated, is illustrated in Figure 2.29.

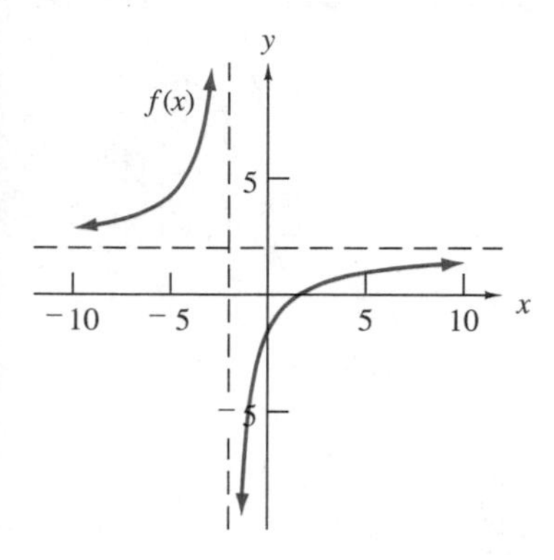

Figure 2.29

Asymptotes and Technology

Using Your Calculator You can use your graphing calculator to construct the graph of a function. If that function has vertical asymptotes, however, you will need to be careful when you describe its behavior. For example, the function

$$f(x) = \frac{2x + 1}{x - 3}$$

has a vertical asymptote at $x = 3$. Your calculator may construct the graph so that it appears that the asymptote is part of the curve by connecting it to the branches of the curve. This connection happens when your calculator is set in the Connected mode. The calculator is being instructed to connect the *turn-on* pixels. The problem can be fixed by placing your calculator in Dot mode. However, the Dot mode gives the curve a very discontinuous appearance. It is best if you can determine the existence of vertical asymptotes by analyzing the function.

Calculator Exercises Use your graphing calculator to construct the graph of each function. Indicate on your graph, by drawing dashed lines, all vertical and all horizontal asymptotes.

1. $f(x) = \dfrac{3}{x + 5}$

2. $f(x) = \dfrac{5}{x^2 - 4x - 12}$

3. $f(x) = \dfrac{5x^2 - 4}{2x^2 - 32}$

4. The expenditure by a manufacturer of x millions of dollars brings in $S(x)$ millions of dollars in sales according to the function,

$$S(x) = \frac{4.3x^2 - 8.8x + 6.4}{1.3x^2 - 1.5x + 1.6}$$

If this trend in expenditures continues, use the graph to determine the largest amount of money this manufacturer can expect from sales. Using the graph, describe the behavior of this function.

EXERCISE SET 2.3

A UNDERSTANDING THE CONCEPTS

1. In the following figure, find $\lim_{x\to\infty} f(x)$

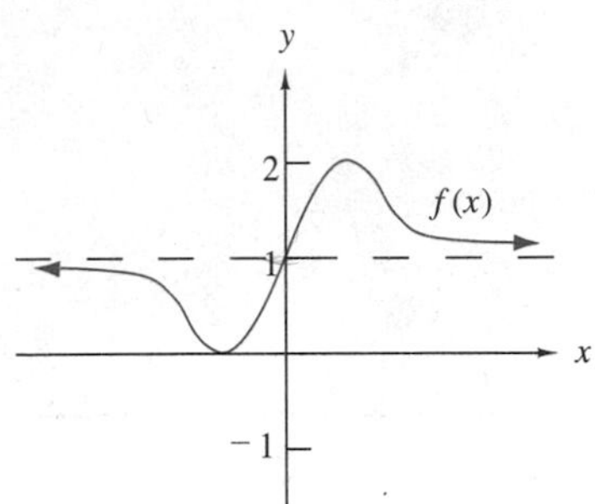

2. In the following figure, find
   a. $\lim_{x\to\infty} f(x)$ b. $\lim_{x\to-\infty} f(x)$ c. $\lim_{x\to-1} f(x)$

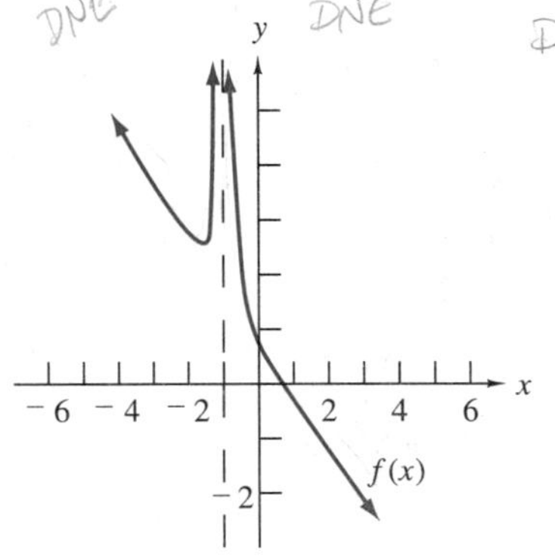

3. In the following figure, find
   a. $\lim_{x\to\infty} f(x)$ b. $\lim_{x\to 2} f(x)$

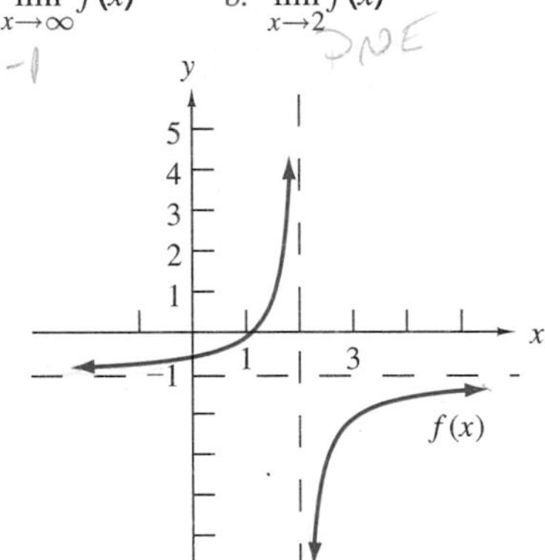

4. In the following figure, find
   a. $\lim_{x\to\infty} f(x)$ b. $\lim_{x\to 2} f(x)$ c. $\lim_{x\to-2} f(x)$

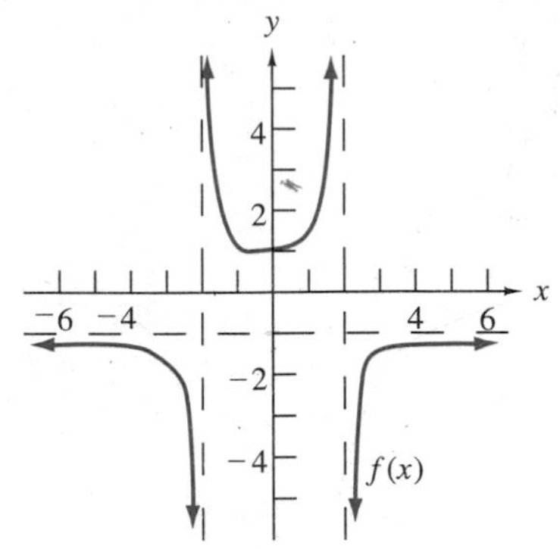

5. In the following figure, find
   a. $\lim_{x\to\infty} f(x)$ b. $\lim_{x\to-\infty} f(x)$ c. $\lim_{x\to 2} f(x)$

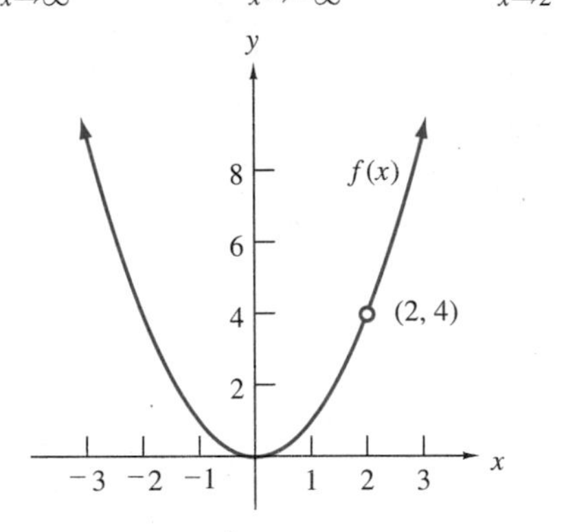

B SKILL ACQUISITION

For Exercises 6–30, find each limit, if it exists.

6. $\lim_{x\to\infty} \dfrac{3x}{x^2+2}$

7. $\lim_{x\to\infty} \dfrac{2x-5}{3x+1}$

8. $\lim_{x\to\infty} \dfrac{6x^2-3x+2}{5x^2+7x-3}$

9. $\lim_{x\to\infty} \dfrac{7x-3}{x^3+2}$

10. $\lim_{x\to\infty} \dfrac{5x^2-9}{8x+7}$

11. $\lim_{x\to\infty} \dfrac{16x^3-3x+7}{9x^2-5x}$

12. $\lim_{x\to\infty} \dfrac{(x+2)(x-3)}{2x^2}$

13. $\lim_{x\to\infty} \dfrac{(5x-3)(x+2)}{x^2-3}$

14. $\lim_{x\to\infty} \dfrac{9x-3}{(x-2)(3x-4)}$

15. $\lim_{x\to\infty} \dfrac{(5-3x)(3-4x)}{(6+x)(5-x)}$

16. $\lim_{x\to-\infty} \frac{3x^2-4}{5x+2}$

17. $\lim_{x\to-\infty} \frac{5-3x^2}{8-4x}$

18. $\lim_{x\to\infty} \frac{8x^3+2x^2-8x+7}{-2x^2+3x-4}$

19. $\lim_{x\to\infty} \frac{(2x+3)^2}{(x-2)^3}$

20. $\lim_{x\to 3} \frac{x^2-6x+9}{x-3}$

21. $\lim_{x\to -4} \frac{x-5}{x+4}$

22. $\lim_{x\to -1} \frac{x+7}{x+1}$

23. $\lim_{x\to 5^-} \frac{x-4}{x-5}$

24. $\lim_{x\to 9^+} \frac{2x+1}{81-x^2}$

25. $\lim_{x\to\infty} \frac{-5x^5-3x^3+7x-2}{x^4(-2x+3)}$

26. $\lim_{x\to-\infty} \frac{8x^2-5x+4}{4-3x-7x^2}$

27. $\lim_{x\to\infty} \frac{\sqrt{3x^2+2}}{5x-1}$

28. $\lim_{x\to\infty} \frac{\sqrt{5x-2}}{9x+1}$

29. $\lim_{x\to-\infty} \frac{\sqrt{8-3x}}{2-7x}$

30. $\lim_{x\to\infty} \sqrt{\frac{8x+3}{7x-1}}$

C APPLYING THE CONCEPTS

31. *Business: Average Cost* The average cost C (in dollars) per tube of toothpaste incurred by a company in producing x tubes is given by the average cost function

$$\overline{C}(x) = 0.89 + \frac{1500}{x}$$

Find and interpret $\lim_{x\to\infty} \overline{C}(x)$.

32. *Economics: Real Estate* The population P (in thousands) of a small valley t years from the beginning of 1990 is given by the function

$$P(t) = \frac{35t^2+105t+140}{t^2+7t+70}$$

a. What is the population of the valley in 1990?

b. What is the populaton of the valley at the beginning of the year 2000?

c. What is the expected population of the valley in the long term?

33. *Life Science: Fish Population* The number N of fish in a pond is related to p, the level of PCB (in parts per million) in the pond by the function

$$N(p) = \frac{500}{1+p}$$

If the level of PCB were allowed to increase without control, what would happen to the number of fish in the pond in the long run?

34. *Life Science: Anatomy* The focal length f (in millimeters) of the human eye is related to the distance d (in millimeters) from the lens of the eye to the object by the function

$$f(d) = \frac{25d}{25+d}$$

What is the focal length of the eye for an object very far away (such as 2 miles)?

35. *Manufacturing: Production Costs* The average weekly cost $\overline{C}(n)$ (in dollars) of a computer company to produce n computers is given by the function

$$\overline{C}(n) = \frac{15,000+240n+60n^{0.5}}{n}$$

What happens to the average cost in the long run, that is, as the number of computers increases without bound?

36. *Medicine: Drug Concentration* The concentration C (in milligrams per liter) of a drug in the human body is related to the time t that the drug has been in the body by the function

$$C(t) = \frac{36t}{t^2+12}$$

What will be the concentration of the drug in the body in the long run?

37. *Business: Pollution Clean-up* The amount $A(t)$ of pollution in a lake is related to time t in years by the function

$$A(t) = (t^{1/5}+5)^4$$

What is the expected amount of pollution in the lake in the long run?

D DESCRIBE YOUR THOUGHTS

38. Describe the strategy you would use to find the limit of a rational function.

39. Describe, as if you were explaining it to someone who knows some algebra, how to determine the vertical asymptotes of a rational function.

40. Describe, as if you were explaining it to someone who knows some algebra, how to determine the horizontal asymptotes of a rational function.

41. Describe how you could use a graphing calculator to help you suggest a horizontal asymptote.

E REVIEW

42. (1.2) Construct the graph of the function $f(x) = \frac{-1}{3}x + 4$.

43. (1.3) For $f(x) = 4x^2 - 4x + 5$, find $\frac{f(x+h) - f(x)}{h}$.

44. (1.4) Find $(f \circ g)(x)$ when $f(x) = x^2 - 3$ and $g(x) = 5x - 1$.

45. (2.2) Find, if it exists, $\lim_{x \to 6} \frac{x^2 - 8x + 12}{x^2 - 10x + 24}$.

46. (2.2) Find, if it exists, $\lim_{x \to 0} \frac{x^2 + 3x}{x}$.

2.4 Continuity of Functions

Introduction

Continuous and Discontinuous Functions

Continuity and Idealized Functions

Continuity on an Interval

Continuity, Piecewise Functions, and Technology

Introduction

Continuous and Discontinuous Functions

Continuous functions have an essential role in calculus. Nearly all of the rules for the calculus operations assume that the function to be operated on is continuous. In this section we examine the conditions necessary for a function to be continuous at a point and then how to construct an argument that proves or disproves continuity.

Continuous and Discontinuous Functions

Intuitively, a continuous function is a function whose graph runs continuously from one point to another point without any breaks or jumps. That is, the curve is completely unbroken and can be drawn, from start to finish, without lifting the pencil from the paper. Some functions, however, are not continuous. If a hole or a jump appears at some point, $x = a$, the function is discontinuous at that point. The leftmost illustration in Figure 2.30 shows a continuous function, and the rightmost illustration shows a discontinuous function (with two points of discontinuity).

By observing the behavior of a function that is continuous at a point $x = a$, we may describe the conditions that are needed for a function to be continuous at a point $x = a$. Consider the continuous function pictured in Figure 2.31. This function is continuous at the point $x = a$, and it meets the following conditions:

1. A point exists at $x = a$. That is, $f(a)$ is defined. In fact, $f(a) = L$.

2. The branches of the curve come nearer to each other as the x values get nearer to $x = a$. That is, $\lim_{x\to a^-} f(x) = \lim_{x\to a^+} f(x)$ implies that $\lim_{x\to a} f(x)$ exists. In fact, $\lim_{x\to a} f(x) = L$.
3. Not only is there a point at $x = a$, and not only do the branches of the curve come near each other near $x = a$, but the branches of the curve join together at $x = a$. That is, $\lim_{x\to a} f(x) = f(a)$, $(L = L)$.

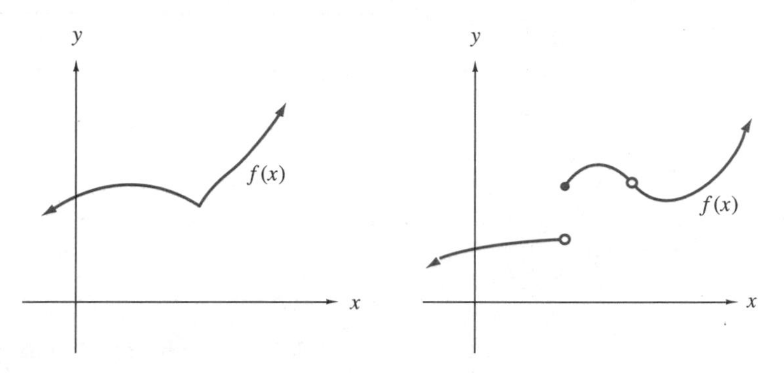

Figure 2.30

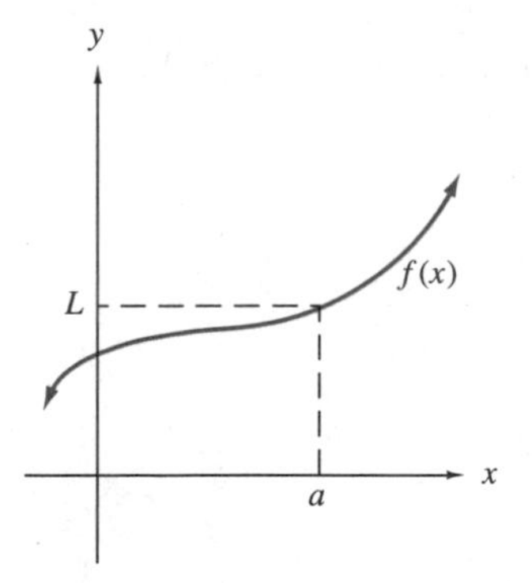

Figure 2.31

All of these conditions must be satisfied for a function to be continuous at a point $x = a$. We now state them symbolically.

Continuity Conditions

If $f(x)$ is a function, then $f(x)$ is continuous at a point $x = a$ if all three of the following conditions are satisfied.

1. $f(a)$ is defined.
2. $\lim_{x\to a} f(x)$ exists.
3. $\lim_{x\to a} f(x) = f(a)$

- Condition 1 means that a point exists at $x = a$.
- Condition 2 means that the branches of the curve come nearer to
- Condition 3 means that not only is there a point at $x = a$, and not only do the branches of the curve come near each other near $x = a$, but that the branches of the curve actually join together at $x = a$.

To prove that a function is continuous at a point $x = a$, it is necessary to show that all three continuity conditions are satisfied. If one of the conditions fails at $x = a$, the function is discontinuous at that point, and no further investigation is necessary. The following examples illustrate the procedure for determining whether a function is continuous at a point $x = a$. Each example uses the phrase *discuss the continuity of the function $f(x)$ at $x = a$.* This phrase directs us to *specify* whether or not $f(x)$ is continuous at $x = a$, and to substantiate our conclusion.

In Example Sets A–D that follow, we will discuss the continuity of a given function by writing an analysis of the continuity conditions and then a summary in which we will concisely state the results of the analysis.

EXAMPLE SET A

Discuss the continuity of the function $f(x) = \begin{cases} \dfrac{x^2 - 7x + 12}{x - 3}, & \text{if } x \neq 3 \\ -1, & \text{if } x = 3 \end{cases}$

at $x = 3$.

Solution:
The interesting point is $x = 3$ since some change in behavior appears to occur here. We need to examine the continuity conditions.

OUR ANALYSIS

1. Is $f(3)$ defined? Substitute 3 into the appropriate piece of the function and find out. Since $x = 3$, we use the bottom piece of the function.
   $f(3) = -1$. Yes, there is a point at $x = 3$.
2. Does $\lim\limits_{x \to 3} f(x)$ exist? Since we are only approaching $x = 3$, it is true that $x \neq 3$ so that we use the top piece of the function to evaluate the limit.

$$\begin{aligned} \lim_{x \to 3} f(x) &= \lim_{x \to 3} \frac{x^2 - 7x + 12}{x - 3} \\ &= \lim_{x \to 3} \frac{(x-4)(x-3)}{x - 3} \\ &= \lim_{x \to 3} (x - 4) \\ &= 3 - 4 \\ &= -1 \end{aligned}$$

   Yes, $\lim\limits_{x \to 3} f(x)$ exists. In fact, $\lim\limits_{x \to 3} f(x) = -1$.
3. Is $\lim\limits_{x \to 3} f(x) = f(3)$? Since $-1 = -1$ (from 1 and 2), we answer yes.
   Not only is there a point at $x = 3$ (condition 1), and not only do the branches of the curve come near each other near $x = 3$ (condition 2), but the branches of the curve actually join together at $x = 3$ (condition 3). All three continuity conditions are satisfied.

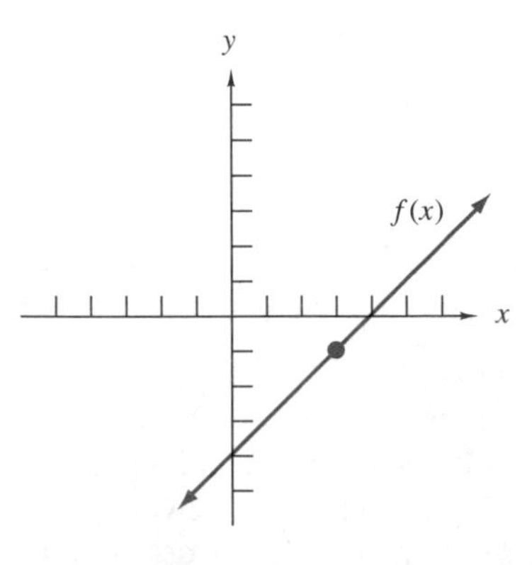

Figure 2.32

SUMMARY

1. $f(3)$ is defined. In fact, $f(3) = -1$.
2. $\lim\limits_{x \to 3} f(x)$ exists. In fact, $\lim\limits_{x \to 3} f(x) = -1$.
3. $\lim\limits_{x \to 3} f(x) = f(3)$.

Since all three continuity conditions are satisfied, we conclude that this function is continuous at $x = 3$. The graph of $f(x)$ is pictured in Figure 2.32.

EXAMPLE SET B

Discuss the continuity of the function $f(x) = \dfrac{x^2 - 5x + 6}{x - 2}$ at $x = 2$.

Solution:
The interesting point is $x = 2$ since the denominator is undefined at $x = 2$. We need to examine the continuity conditions.

OUR ANALYSIS

1. Is $f(2)$ defined? Substitute 2 into $f(x)$ and find out.

$$f(2) = \frac{2^2 - 5 \cdot 2 + 6}{2 - 2}$$

$$= \frac{0}{0} \quad \text{which has no value assigned to it}$$

Thus, $f(2)$ is not defined and, hence, there is no point for $x = 2$. Continuity condition 1 fails.

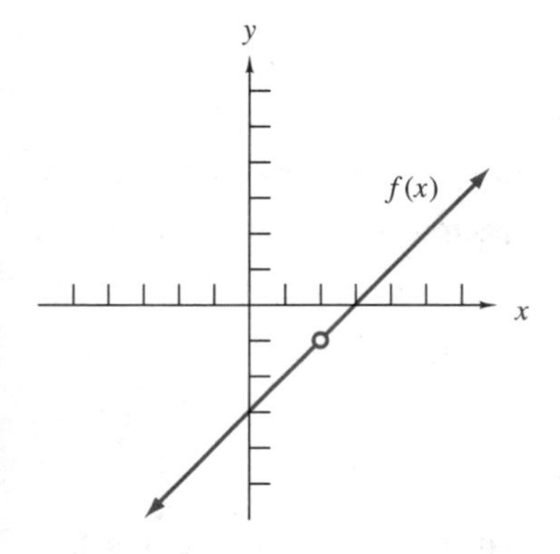

Figure 2.33

SUMMARY

1. $f(2)$ is not defined.

Since one of the continuity conditions is not satisfied, we conclude that this function is discontinuous at $x = 2$. The graph of $f(x)$ is pictured in Figure 2.33. Notice the hole at $x = 2$.

EXAMPLE SET C

Discuss the continuity of the function $f(x) = \begin{cases} x^2 - 2x - 2, & \text{if } x \leq 3 \\ \frac{1}{3}x - \frac{7}{3}, & \text{if } x > 3 \end{cases}$ at $x = 3$.

Solution:
The interesting point is $x = 3$ since some change in behavior appears to occur here. We need to examine the continuity conditions.

OUR ANALYSIS

1. Is $f(3)$ defined? Substitute 3 into the appropriate piece of $f(x)$ and find out. Since $x = 3$, we use the top piece of the function.

$$f(3) = 3^2 - 2 \cdot 3 - 2$$

$$= 1$$

Yes, there is a point at $x = 3$.

2. Does $\lim_{x \to 3} f(x)$ exist? Check the equality of the left- and right-hand limits.

- Check $\lim_{x \to 3^-} f(x)$. Since $x \to 3^-$, $x < 3$, and we use the top piece of the function.

$$\begin{aligned}\lim_{x \to 3^-} f(x) &= \lim_{x \to 3^-} (x^2 - 2x - 2) \\ &= 3^2 - 2 \cdot 3 - 1 \\ &= 1\end{aligned}$$

- Check $\lim_{x \to 3^+} f(x)$. Since $x \to 3^+$, $x > 3$, and we use the bottom piece of the function.

$$\begin{aligned}\lim_{x \to 3^+} f(x) &= \lim_{x \to 3^+} (\frac{1}{3}x - \frac{7}{3}) \\ &= \frac{1}{3}(3) - \frac{7}{3} \\ &= -\frac{4}{3}\end{aligned}$$

Then, $\lim_{x \to 3^-} f(x) \neq \lim_{x \to 3^+} f(x)$ $\quad (1 \neq -\frac{4}{3})$, and $\lim_{x \to 3} f(x)$ does not exist. The branches of the curve do not come nearer to each other as the x values come nearer to 3. Continuity condition 2 fails.

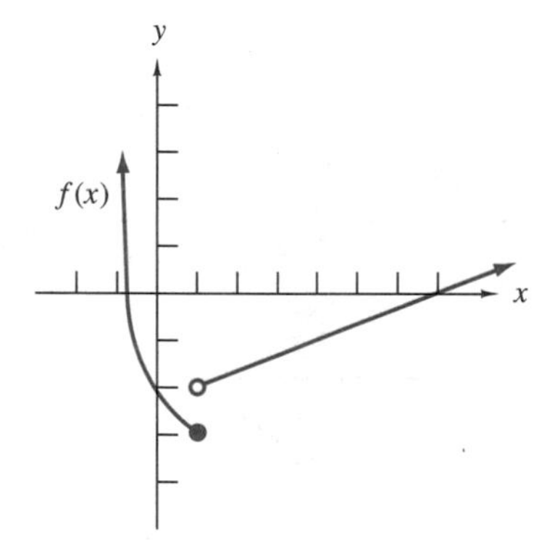

Figure 2.34

SUMMARY

1. $f(3)$ is defined. In fact, $f(3) = 1$.

2. $\lim_{x \to 3} f(x)$ does not exist.

Since one of the continuity conditions is not satisfied, we conclude that this function is discontinuous at $x = 3$. The graph of $f(x)$ appears in Figure 2.34. Notice the jump at $x = 3$.

EXAMPLE SET D

Discuss the continuity of the function $f(x) = \begin{cases} x^2 + 4x + 6, & \text{if } x < -2 \\ 3, & \text{if } x = -2 \\ -x, & \text{if } x > -2 \end{cases}$

at $x = -2$.

Solution:

The interesting point is $x = -2$ since some change in behavior appears to occur here. We need to examine the continuity conditions.

OUR ANALYSIS

1. Is $f(-2)$ defined? Substitute -2 into the appropriate piece of $f(x)$ and find out. Since $x = -2$, we use the middle piece of the function.

$f(-2) = 3$, and the function is defined. Yes, there is a point at $x = -2$.

2. Does $\lim_{x \to -2} f(x)$ exist? Check the equality of the left- and right-hand limits.

   - Check $\lim_{x \to -2^-} f(x)$. Since $x \to -2^-$, $x < -2$, and we use the top piece of the function.

$$\begin{aligned} \lim_{x \to -2^-} f(x) &= \lim_{x \to -2^-} (x^2 + 4x + 6) \\ &= (-2)^2 + 4(-2) + 6 \\ &= 2 \end{aligned}$$

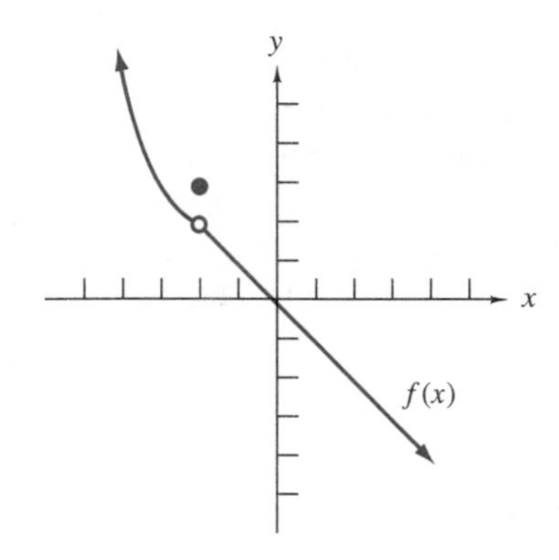

Figure 2.35

   - Check $\lim_{x \to -2^+} f(x)$. Since $x \to -2^+$, $x > -2$, and we use the bottom piece of the function.

$$\begin{aligned} \lim_{x \to -2^+} f(x) &= \lim_{x \to -2^+} (-x) \\ &= -(-2) \\ &= 2 \end{aligned}$$

Then, since $\lim_{x \to -2^-} f(x) = \lim_{x \to -2^+} f(x)$ $(2 = 2)$, $\lim_{x \to -2} f(x)$ exists. The branches of the curve come nearer to each other as the x values come nearer to -2.

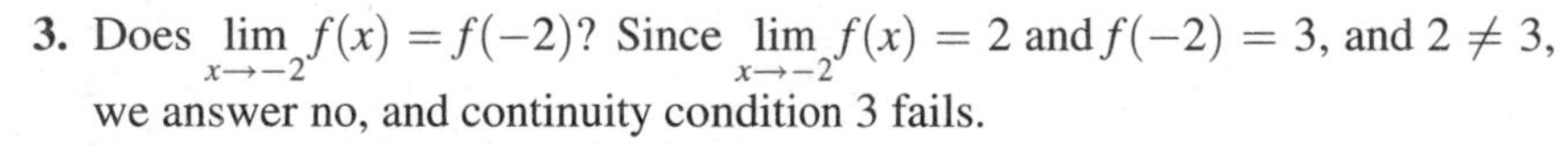

3. Does $\lim_{x \to -2} f(x) = f(-2)$? Since $\lim_{x \to -2} f(x) = 2$ and $f(-2) = 3$, and $2 \neq 3$, we answer no, and continuity condition 3 fails.

Although there is a point at $x = -2$ and the branches of the curve come near each other near $x = -2$, the branches of the curve are not joined at $x = -2$.

SUMMARY

1. $f(-2)$ is defined. In fact, $f(-2) = 3$.
2. $\lim_{x \to -2} f(x)$ exists. In fact, $\lim_{x \to -2} f(x) = 2$.
3. $\lim_{x \to -2} f(x) \neq f(-2)$

Since one of the continuity conditions is not satisfied, we conclude that this function is discontinuous at $x = -2$. The graph of $f(x)$ appears in Figure 2.35. Notice the hole at $x = -2$.

ILLUMINATOR SET A

Some people, when placed in extremely stressful situations, exhibit explosive bursts of anger. Such behavior might be described by the function pictured in Figure 2.36.

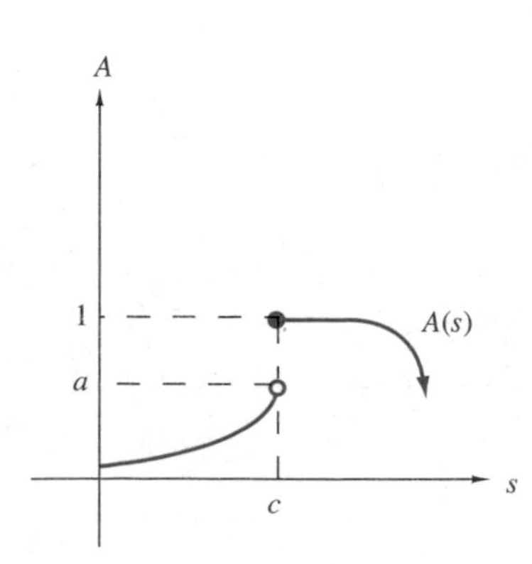

Figure 2.36

The horizontal axis records the amount of stress and the vertical axis, the amount of anger. Make a statement about the continuity of the function $A(s)$ as s approaches c and give a possible interpretation to the point $s = c$.

This function is discontinuous at $s = c$ since continuity condition 2 fails ($\lim_{s \to c} A(s)$ does not exist). However, $\lim_{s \to c^-} A(s)$ does exist, and in fact, $\lim_{s \to c^-} A(s) = a$. We might say that as this person's stress level approaches c, that is, as $s \to c^-$, he is losing his patience and as the stress level gets very close to c, that he is *reaching his limit* of patience. Finally, at $s = c$, he loses his patience completely and exhibits explosive behavior.

Continuity and Idealized Functions

Idealized Functions

Often in business, in the social sciences, and in the life sciences, the domain of the function of interest consists of integers in some interval rather than all real numbers in that interval. So that problems can be analyzed, it is necessary to *idealize* functions so that they are continuous and, therefore, useful. The domain of such idealized functions is then all real numbers in some interval rather than just the integers in that interval.

ILLUMINATOR SET B

A hotel might describe the relationship between its occupancy rate and the week of the year by the continuous function pictured as the leftmost illustration of Figure 2.37. The function is idealized as continuous. In reality, it is not continuous. The actual function may look more like the function pictured in the rightmost illustration in Figure 2.37. Calculus may be applied to the *idealized* continuous function and not the *actual* discontinuous function.

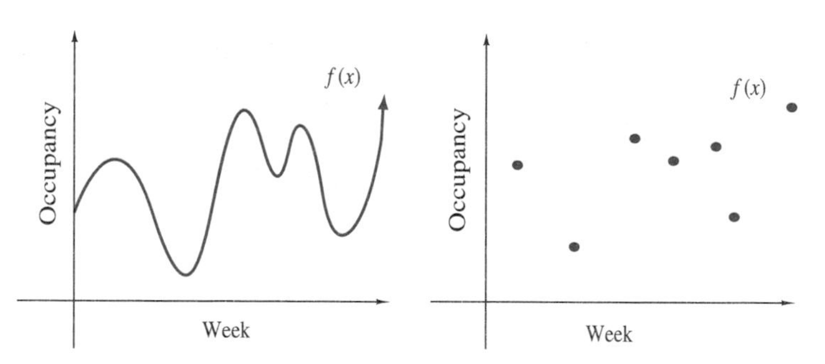

Figure 2.37

Continuity on an Interval

Quite often we will need to discuss the continuity of a function *on an interval* rather than at just one point. As we defined one-sided limits (Section 2.2), we can define one-sided continuity. One-sided continuity is defined in the same way as regular continuity except that the limits for one-sided continuity are one-sided limits.

One-Sided Continuity

If $[a, b]$ is a closed interval and the function $f(x)$ is defined over it, then

a. $f(x)$ is continuous at the right endpoint $x = b$ if

1. $f(b)$ is defined.

2. $\lim_{x \to b^-} f(x)$ exists.

3. $\lim_{x \to b^-} f(x) = f(b)$.

b. $f(x)$ is continuous at the left endpoint $x = a$ if

1. $f(a)$ is defined.

2. $\lim_{x \to a^-} f(x)$ exists.

3. $\lim_{x \to a^-} f(x) = f(a)$.

We now define continuity on an open interval (a, b) and on a closed interval $[a, b]$.

Continuity on an Interval

1. A function $f(x)$ is *continuous on the open interval* (a, b) if $f(x)$ is continuous at every point in (a, b); that is, $f(x)$ is continuous at every x value between, but not including, a and b.
2. A function $f(x)$ is *continuous on the closed interval* $[a, b]$ if $f(x)$ is continuous at every point in $[a, b]$; that is, $f(x)$ is continuous at every x value between *and* including a and b.

We conclude by stating two important continuity properties.

Continuity of Polynomials and Rational Expressions

1. Every polynomial is continuous on every open interval (a, b) and on every closed interval $[a, b]$.
2. Every rational expression is continuous on every open interval (a, b) and on every closed interval $[a, b]$ except for the points in the interval that make the denominator 0.

These important continuity properties tell us that polynomials will always graph as unbroken curves, and that rational expressions will be broken only at the x values for which the denominator is 0.

Continuity, Piecewise Functions, and Technology

Using Your Calculator The function $f(x) = \dfrac{x^2 - 7x + 12}{x - 3}$ is discontinuous at $x = 3$ because continuity condition 1 fails; that is, $f(3)$ is undefined. However, the graph constructed by the graphing calculator shows a graph that appears continuous at $x = 3$. This apparent continuity occurs when the graph is constructed in Connected mode rather than the Dot mode. In Connected mode, the calculator is directed to connect each *turn-on* pixel.

Your calculator can construct graphs of functions that are defined piecewise. The following entry shows how to construct the graph of

$$f(x) = \begin{cases} x^2 - 2x - 2, & \text{if } x \leq 3 \\ \dfrac{1}{3}x - \dfrac{7}{3}, & \text{if } x > 3 \end{cases}$$

```
Y1=(x^2−2x−2)(x ≤ 3)+(x/3−7/3)(x > 3)
```

Calculator Exercises Use your graphing calculator to construct the graphs of each piecewise defined function.

1. $f(x) = \begin{cases} 3 - x^2, & \text{if } x \leq 1 \\ x^3 - 4x, & \text{if } x > 1 \end{cases}$

2. $f(x) = \begin{cases} x^2 - 0.9x + 3.5, & \text{if } x \leq 3 \\ 2.8x - 4.5, & \text{if } x > 3 \end{cases}$

3. $f(x) = \begin{cases} 2.6 - 1.1x^2, & \text{if } |x| < 6 \\ 2.6, & \text{if } |x| > 6 \end{cases}$

4. $f(x) = \begin{cases} x^3 - 8x, & \text{if } -3 \leq x < 3 \\ 10, & \text{if } 3 \leq x < 7 \\ \dfrac{10}{x - 7}, & \text{if } x \geq 7 \end{cases}$

EXERCISE SET 2.4

A UNDERSTANDING THE CONCEPTS

For Exercises 1–12, determine if each illustrated function is continuous or discontinuous at the point $x = a$. If the function is continuous, so state. If the function is discontinuous, specify which of the three continuity condition is the first to fail.

1.

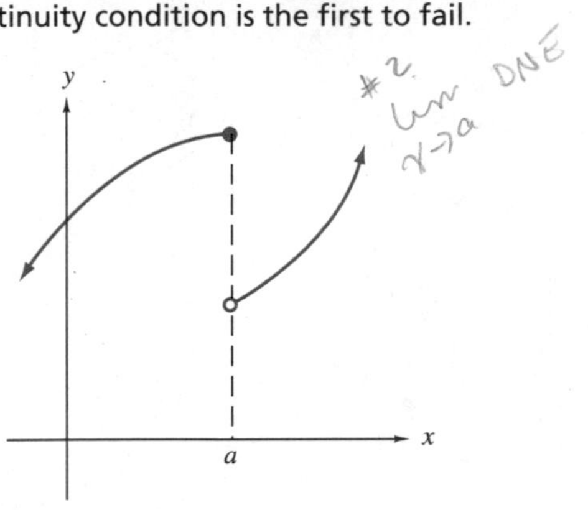

2.

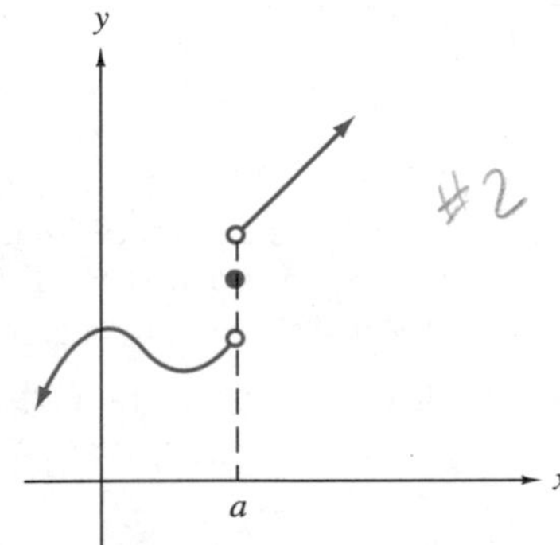

3.

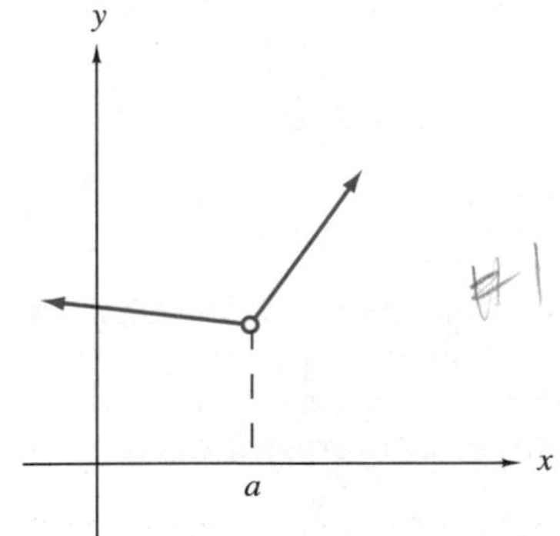

4.

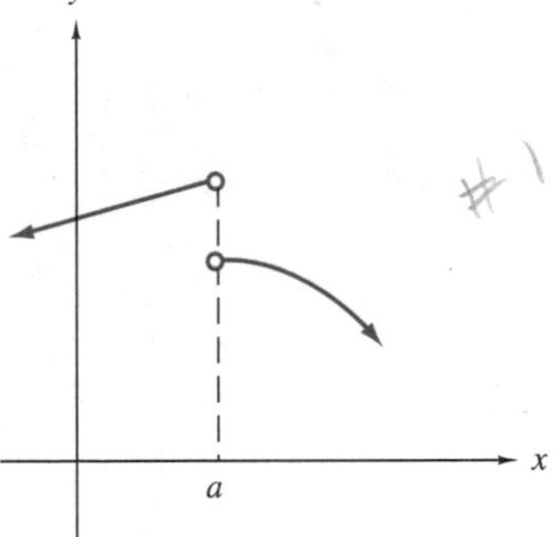

5.

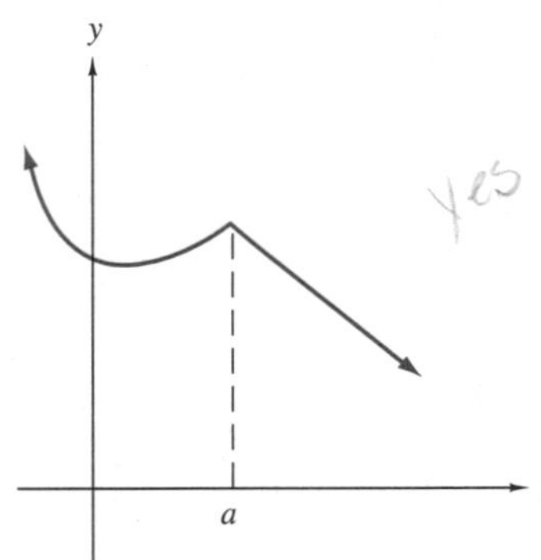

6.

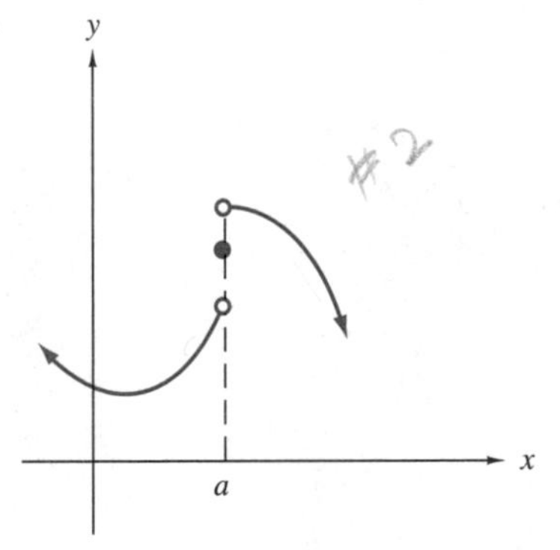

7.

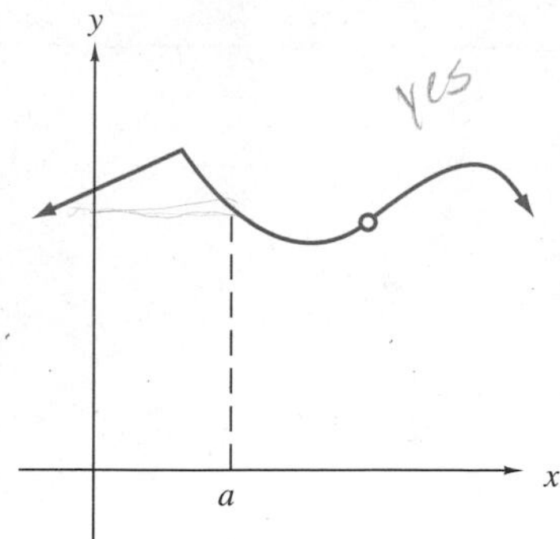

8.

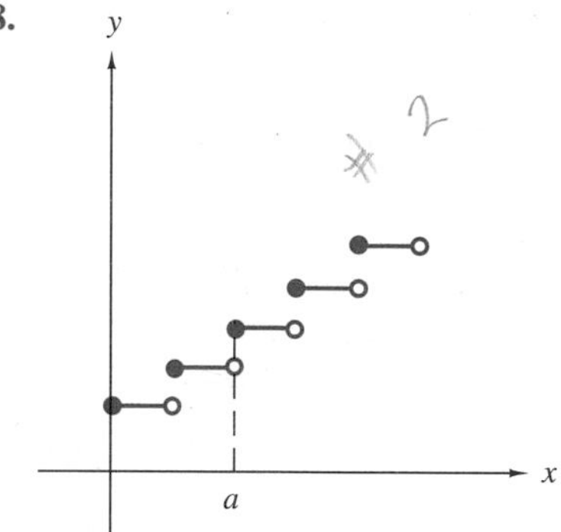

9.

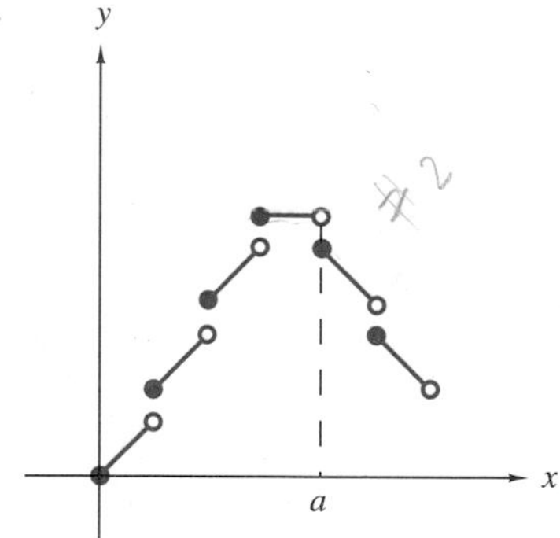

10.

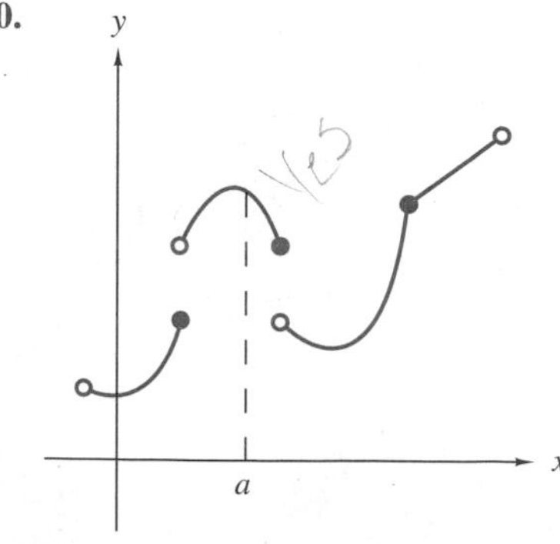

11.

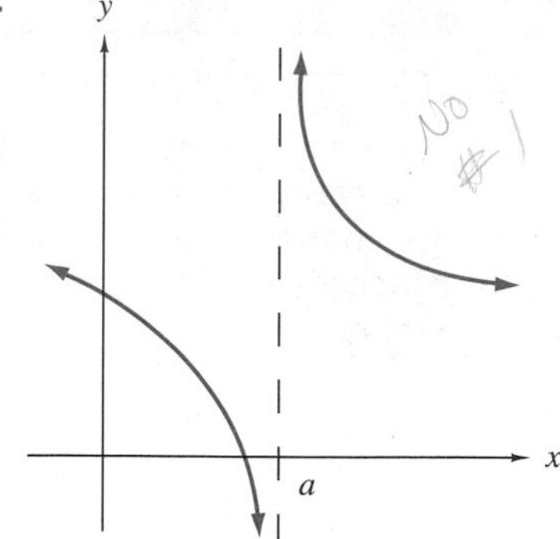

12.

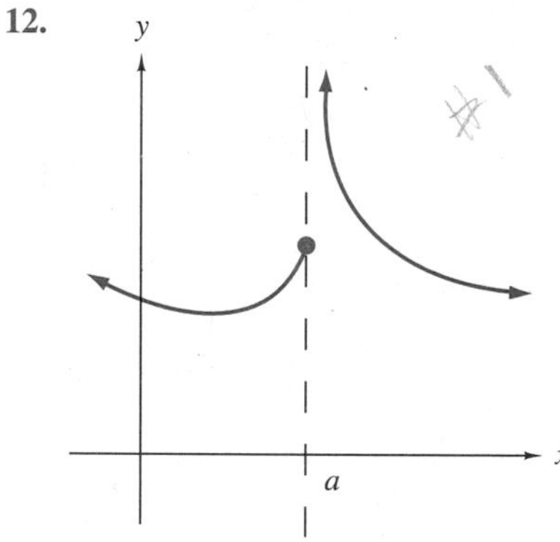

B SKILL ACQUISITION

For Exercises 13–18, construct the sketch of a curve that is discontinuous at $x = a$ and meets all the specified conditions. (Answers may vary.)

13. $a = 3$ and $f(3)$ is not defined.

14. $a = -1$ and $f(-1)$ is not defined.

15. $a = 2, \quad f(2) = 3,$
$\lim_{x \to 2^-} f(x) = 3, \quad \lim_{x \to 2^+} f(x) = 4.$

16. $a = 3, \quad f(3) = 1,$
$\lim_{x \to 3^-} f(x) = 4, \quad \lim_{x \to 3^+} f(x) = 4.$

17. $a = 3, \quad f(3) = 2, \quad \lim_{x \to 3} f(x) = 1.$

18. $a = 4, \quad f(4)$ is not defined, $\quad \lim_{x \to 4} f(x) = 3.$

For Exercises 19–31, discuss the continuity of each function at the specified value of x. (You are encouraged to write an analysis as we did in each example, but you need only present the summary.)

19. $f(x) = 3x^2 - 7x + 4$ at $x = -1$

20. $f(x) = -5x^3 + x^2 + 2$ at $x = 2$

21. $f(x) = \dfrac{x^2 - 5x + 6}{x - 2}$ at $x = 3$

22. $f(x) = \dfrac{x^2 - 9x + 8}{x - 1}$ at $x = -2$

23. $f(x) = \dfrac{x^2 - 5x + 6}{x - 2}$ at $x = 2$

24. $f(x) = \dfrac{x^2 - 9x + 8}{x - 1}$ at $x = 1$

25. $f(x) = \begin{cases} 3x + 1, & \text{if } x < 1 \\ x^2 - 2x + 5, & \text{if } x > 1 \end{cases}$ at $x = 1$

26. $f(x) = \begin{cases} x^2 - 4, & \text{if } x \le 0 \\ \dfrac{3}{4}x - 3, & \text{if } x > 0 \end{cases}$ at $x = 1$

27. $f(x) = \begin{cases} |x - 3|, & \text{if } x < 3 \\ -x^2 + 6x - 8, & \text{if } x \ge 3 \end{cases}$
at $x = 0$

28. $f(x) = \begin{cases} -x^2 - 2x, & \text{if } x < -1 \\ 3, & \text{if } x = -1; \\ x^2 + 2x + 2, & \text{if } x > -1 \end{cases}$
at $x = -1$

29. $f(x) = \begin{cases} x^2 + 8x + 13, & \text{if } x < -4 \\ 0, & \text{if } x = -4 \\ -3, & \text{if } x > -4 \end{cases}$
at $x = -4$

30. $f(x) = \begin{cases} \dfrac{x^2 + 7x + 10}{x + 2}, & \text{if } x < -2 \\ 3, & \text{if } x = -2 \\ -x + 1, & \text{if } x > -2 \end{cases}$
at $x = -2$

31. $f(x) = \begin{cases} \dfrac{3x^2 - 7x + 4}{x - 1}, & \text{if } x < 1 \\ -1, & \text{if } x = 1 \\ \dfrac{x^2 - 3x + 2}{x - 1}, & \text{if } x > 1 \end{cases}$
at $x = 12$

C APPLYING THE CONCEPTS

For Exercises 32–37, sketch the graph of a discontinuous function for which only the specified continuity condition is not satisfied. (Answers may vary.)

32. $f(2) = 2$ and $f(x)$ is discontinuous at $x = 3$ because continuity condition 2 is not satisfied.

33. $f(-3) = 0$ and $f(x)$ is discontinuous at $x = -3$ because continuity condition 2 is not satisfied.

34. $f(4) = 3$ and $f(x)$ is discontinuous at $x = 4$ because continuity condition 3 is not satisfied.

35. $f(1) = -3$ and $f(x)$ is discontinuous at $x = 1$ because continuity condition 3 is not satisfied.

36. $f(1) = 1$ and $f(x)$ is discontinuous at $x = 1$ because continuity conditions 2 and 3 are not satisfied.

37. $f(-2) = 3$ and $f(x)$ is discontinuous at $x = -2$ because continuity conditions 2 and 3 are not satisfied.

38. ***Psychology: Learning Curve*** Psychologists have established that many types of learning follow a well-established "learning curve." Such a curve is pictured below. The horizontal axis records the amount of time spent studying and the vertical axis records the amount of understanding of a particular concept or technique.

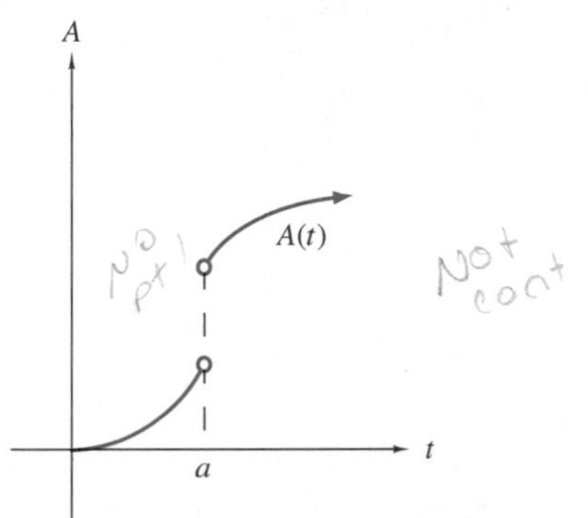

Make a statement about the continuity of the function $A(t)$ at $t = a$, and give a possible interpretation to the point $t = a$.

39. ***Economics: Population Growth*** When observations are made over long enough periods of time, the growth of many populations can be viewed as a smooth, continuous curve. However, some growth can be disrupted and produce a curve such as that pictured below. The horizontal axis records the amount of time the population has existed and the vertical axis records the number of members in the population.

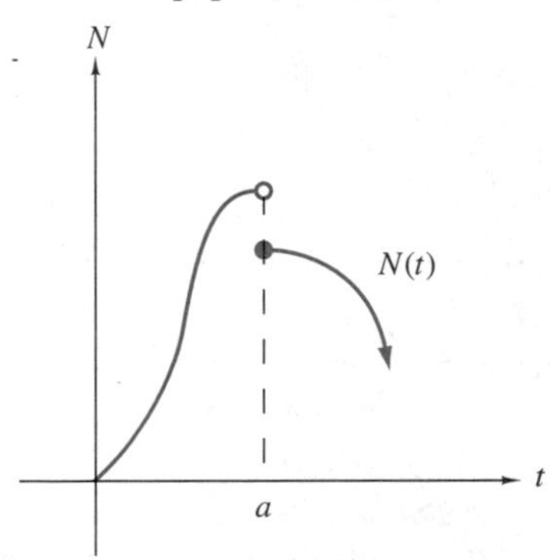

Make a statement about the continuity of the function $N(t)$ at $t = a$ and give a possible interpretation to the point $t = a$.

40. ***Economics: Profit*** When observations are made over long enough periods of time, the curve that represents the profits of a Wall Street financial brokerage house can be viewed as a smooth, continuous curve. However, certain conditions, market or otherwise, can produce a curve such as that pictured below.

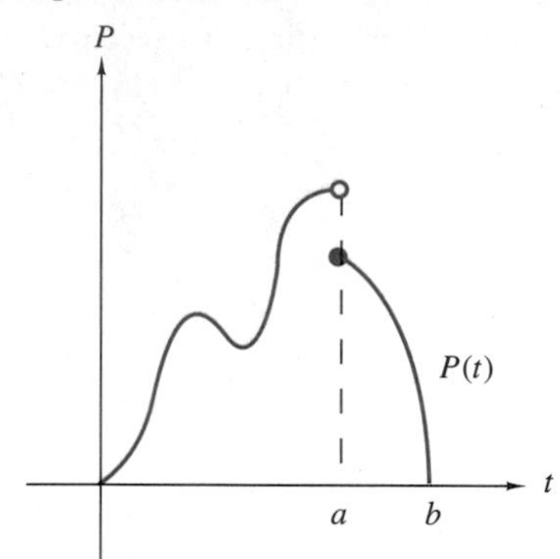

Make a statement about the continuity of the function $P(t)$ at $t = a$ and give possible interpretations to the points $t = a$ and $t = b$.

41. ***Economics: Demand*** When idealized, the function relating time (in months) and the demand for personal computers can be viewed as a smooth, continuous curve. However, certain market conditions can produce a curve such as that pictured below.

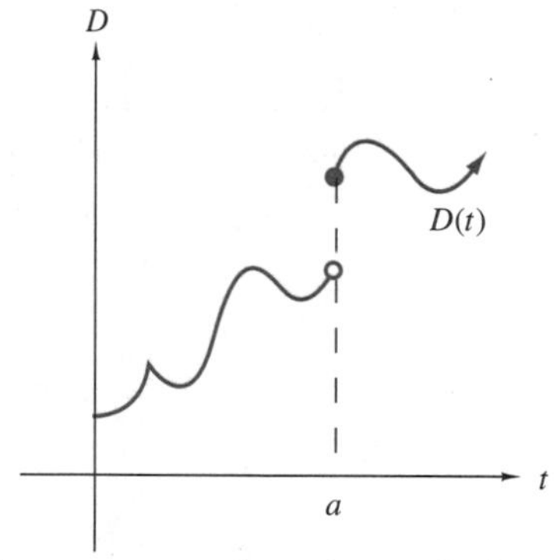

Make a statement about the continuity of the function $D(t)$ at $t = a$ and give a possible interpretation to the point $t = a$.

42. ***Ecology: Oil Pollution*** When idealized, the function relating time (in months) to the amount (in thousands of gallons) of crude oil polluting a particular stretch of a coastal wildlife preserve can be viewed as a smooth, continuous curve. However, certain phenomena can produce a curve such as that pictured in the following illustration.

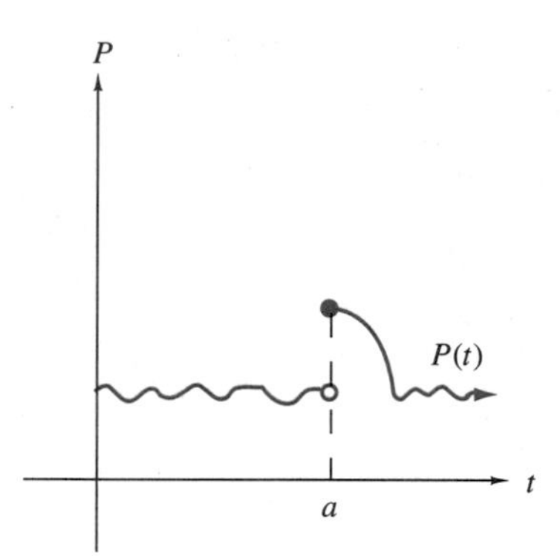

Make a statement about the continuity of the function $P(t)$ at $t = a$ and give a possible interpretation to the point $t = a$.

43. ***Business: Parking Fees*** When idealized, the function relating time (in semesters or quarters) and a college's or university's parking fee can be viewed as the broken curve pictured below.

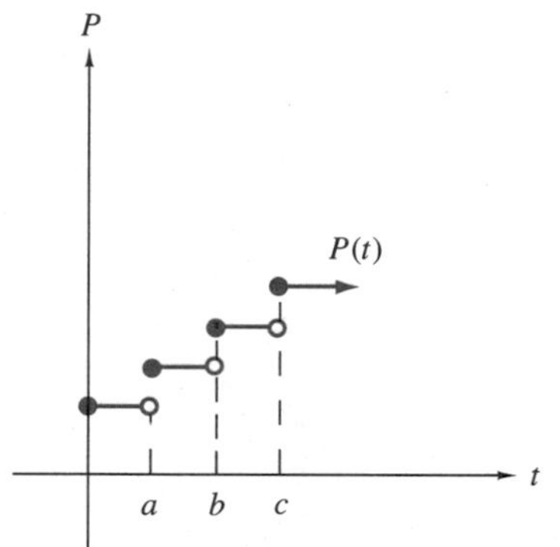

Give a possible interpretation for the discontinuities at the points $t = a$, $t = b$, and $t = c$.

D DESCRIBE YOUR THOUGHTS

44. Describe the intuitive difference between continuous and discontinuous functions.

45. Explain why, for a function to be continuous at $x = a$, it is not enough for $f(a)$ and $\lim_{x \to a} f(x)$ to exist, but it also must be that $\lim_{x \to a} f(x) = f(a)$.

46. When determining $\lim_{x \to 3} f(x)$, where

$$\begin{cases} x^2 - 2x - 2, & \text{if } x \leq 3 \\ \frac{1}{3}x - \frac{7}{3}, & \text{if } x > 3 \end{cases}$$

it is necessary to consider both one-sided limits. However, when determining $\lim_{x\to 3} f(x)$, where

$$\begin{cases} \dfrac{x^2 - 7x + 12}{x - 3}, & \text{if } x \neq 3 \\ -1, & \text{if } x = 3 \end{cases}$$

it is not necessary to analyze the one-sided limits. Explain the difference between these two cases.

E REVIEW

47. **(1.4)** If $f(x) = 3x^2 + 5$ and $g(x) = 4x + 7$, find an expression for $(f \circ g)(x)$.

48. **(2.1)** Sketch the graph of a function having the specified properties.

$$f(1) = 2, \quad \lim_{x\to 1^-} f(x) = 2, \quad \lim_{x\to 1^+} f(x) = 4$$

49. **(2.2)** Find $\lim_{x\to -4} \dfrac{x^2 - 3x - 28}{x + 4}$.

50. **(2.2)** Find $\lim_{x\to 1} \dfrac{x + 6}{x - 1}$.

51. **(2.3)** Find $\lim_{x\to \infty} \dfrac{5x^2 + 3x - 1}{2x^3 - 4}$.

2.5 Average and Instantaneous Rates of Change

Introduction

Average Rate of Change of a Function

Instantaneous Rate of Change

Interpreting the Instantaneous Rate of Change

Rates of Change and Technology

Introduction

Differential Calculus as the Language of Change

In most situations in which one quantity is related to another, a change in the value of one will produce a change in the value of the other. It is often important to know the rate at which such changes are taking place. For example, it may be important to know that a one-unit change in the value of one quantity (such as the sales price of an item) may produce a three-tenths unit change in the value of another (such as the profit realized on an item). It is precisely to this idea of change that differential calculus speaks. **Differential calculus** is the language of change and its methods provide a way of analyzing the rate at which one quantity changes relative to another. We have begun the transition from algebra to calculus by looking at the limit of a function. We now continue that transition by considering first the average rate of change of a function and then the instantaneous rate of change of a function.

Average Rate of Change of a Function

Dependent and Independent Variables

From our study of functions, we know that when the value of one quantity, say, y, depends on the value of another quantity, say, x, the quantity y is called the *dependent variable* and that the quantity x is called the *independent variable*. Let's consider an example.

Suppose that the price of an item depends on time in such a way that over a period of time, the price changes from \$15 to \$35. In this case, price is the dependent variable and time is the independent variable. If the time period is short, say, 3 months, the rate at which the price is changing can be viewed as dramatic. If the time period is longer, say, 48 months, the rate at which the price is changing is probably not so remarkable. A reasonable measure of the rate at which the dependent quantity changes relative to the independent quantity is the **average rate of change**.

$$\text{Average rate of change} = \frac{\text{change in the dependent quantity}}{\text{change in the independent quantity}}$$

All changes in a quantity are found by computing

$$(\text{final value}) - (\text{initial value})$$

If the dependent quantity is represented by y, and y_1 = initial value and y_2 = final value, then

$$\text{Change in the dependent quantity} = y_2 - y_1$$

If the independent quantity is represented by x, and x_1 = initial value and x_2 = final value, then

$$\text{Change in the independent quantity} = x_2 - x_1$$

Then, the average rate of change can be found using the following formula.

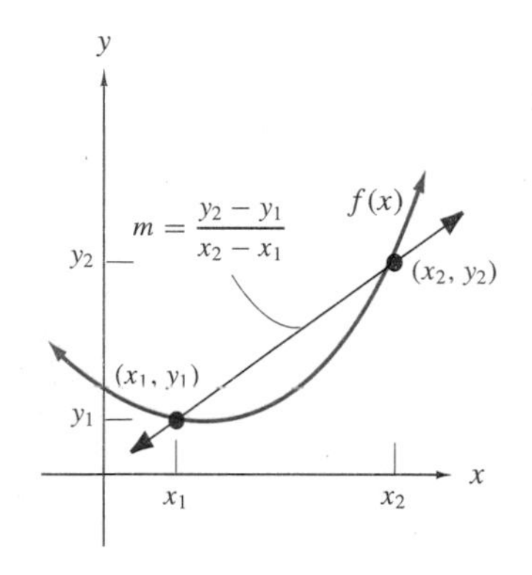

Figure 2.38

Average Rate of Change of a Function

$$\text{Average rate of change} = \frac{\text{change in dependent quantity}}{\text{change in independent quantity}} = \frac{y_2 - y_1}{x_2 - x_1}$$

You might recognize the average rate of change as the slope of a line. In fact, it is precisely that, and we will make use of this fact as we proceed. Figure 2.38 illustrates the fact that the slope of the line through the points (x_1, y_1) and (x_2, y_2) is the average rate of change of the function as the input values change from x_1 to x_2.

Interpreting the Value

Since the average rate of change of a function is defined as a fraction, interpreting its computed value is more easily accomplished if the value is expressed in fractional form. The problem in Example Set A will illustrate this point.

EXAMPLE SET A

Find the average rate of change of the price of an item with respect to time if from the beginning of April to the beginning of August of the same year, the price of the item changes from \$15 to \$35.

Solution:
If y represents price, then $y_1 = 15$ and $y_2 = 35$.
If x represents time, then $x_1 = 4$ (the fourth month) and $x_2 = 8$.

$$\begin{aligned}\text{Average rate of change of price with respect to time} &= \frac{\text{change in price}}{\text{change in time}} \\ &= \frac{\text{final price} - \text{initial price}}{\text{final month} - \text{initial month}} \\ &= \frac{35 - 15}{8 - 4} \\ &= \frac{20}{4} \\ &= 5\end{aligned}$$

Interpreting this value is easier if it is in fractional form.

$$\text{Average rate of change} = 5 = \frac{5}{1} = \frac{\$5 \leftarrow \text{change in price}}{1 \text{ month} \leftarrow \text{change in time}}$$

Now the interpretation of this result is more evident.

Interpretation: From the beginning of April to the beginning of August of the same year, the price of this item has increased, on the average, $5 each month.

If the relationship between the dependent variable y and the independent variable x is given in equation form, that is, $y = f(x)$, then

$$y_1 = f(x_1) \quad \text{and} \quad y_2 = f(x_2)$$

and the formula for the average rate of change can be expressed completely in terms of x.

Average Rate of Change

$$\text{Average rate of change} = \frac{f(x_2) - f(x_1)}{x_2 - x_1}$$

EXAMPLE SET B

A college bookstore sells mechanical pencils for which it pays $1 each. The manager of the bookstore feels that if the selling price is $x each, she will sell

$12 - x$ pencils each week. The bookstore has a fixed cost of \$12. The function relating profit P and selling price x is

$$P(x) = -x^2 + 11x - 12, \quad 2 \leq x \leq 12$$

Find the average rate of change of the profit with respect to the selling price if the selling price changes from \$2 to \$5.

Solution:

$$\begin{aligned}\text{Average rate of change} &= \frac{\text{change in profit}}{\text{change in price}} \\ &= \frac{P(5) - P(2)}{5 - 2} \\ &= \frac{18 - 6}{3} \\ &= \frac{12}{3} \\ &= 4 \quad \left(= \frac{4 \leftarrow \text{change in profit}}{1 \leftarrow \text{change in price}} \right)\end{aligned}$$

Interpretation: For selling prices between \$2 and \$5, each \$1 increase in the selling price increases the bookstore's weekly profit, on the average, by \$4. We take this to mean that

1. As the selling price increases from \$2 to \$3, the bookstore's weekly profit will increase by about \$4.
2. As the selling price increases from \$3 to \$4, the bookstore's weekly profit will increase by about \$4.
3. As the selling price increases from \$4 to \$5, the bookstore's weekly profit will increase by about \$4.

Since the average is only an expected value, we don't expect each \$1 increase in the selling price to increase the weekly profit by exactly \$4, but rather by about \$4.

In fact, by computing the average rate of change on the intervals \$2 to \$3, \$3 to \$4, and \$4 to \$5, we can get better estimates for the changes in weekly profit.

$$\frac{P(3) - P(2)}{3 - 2} = \frac{12 - 6}{1} = 6$$

$$\frac{P(4) - P(3)}{4 - 3} = \frac{16 - 12}{1} = 4$$

$$\frac{P(5) - P(4)}{5 - 4} = \frac{18 - 16}{1} = 2$$

Thus,

1. Rather than estimating a weekly profit increase of \$4 as the selling price increases from \$2 to \$3, the shorter interval allows us to improve the estimate as \$6.

2. The weekly profit increase averages \$4 as the selling price increases from \$3 to \$4.
3. Instead of the weekly profit increasing by \$4 as the selling price increases from \$4 to \$5, the shorter interval allows us to improve the estimate as \$2.

It appears that as the price of mechanical pencils increases by units of \$1, the weekly profit increase realized on their sale is decreasing.

This example illustrates how the average rate of change acts only as an expected change. More accurate information is obtained by making a smaller change in the value of the independent variable (such as a 1-unit change rather than a 3-unit change).

We can write the expression $\dfrac{f(x_2) - f(x_1)}{x_2 - x_1}$ for the average rate of change in a more useful way. Rather than depending on both the initial value x_1 and the final value x_2, we can express the final value in terms of the initial value. Suppose that x represents the initial value of the independent variable. If we change this value by h units, $x + h$ will represent the new value. If h is positive, then $x + h$ is the final value and represents an *increase* in the initial value. If h is negative, then $x + h$ is the final value and represents a decrease in the initial value. (If h is negative, $x + h$ is smaller than x.) We can now express the average rate of change formula in a more general and useful way.

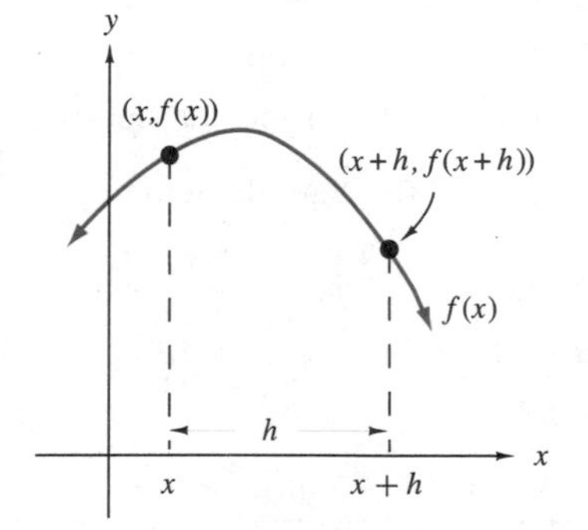

Figure 2.39

$$\frac{\text{final dependent value} - \text{initial dependent value}}{\text{final independent value} - \text{initial independent value}} = \frac{f(x+h) - f(x)}{(x+h) - x} = \frac{f(x+h) - f(x)}{h}$$

Figure 2.39 illustrates this notation graphically. This form of the average rate of change formula is commonly called the **difference quotient**.

Difference Quotient

$$\text{Average rate of change} = \frac{f(x+h) - f(x)}{h}$$

This form of the difference quotient is convenient because it allows us easy access to the amount of change in the independent variable. We can adjust the change in the independent variable, making it smaller or larger, simply by adjusting the value of h.

EXAMPLE SET C

1. For the function $f(x) = 5x + 8$, find the difference quotient $\dfrac{f(x+h) - f(x)}{h}$.

Solution:

$$\begin{aligned}\frac{f(x+h) - f(x)}{h} &= \frac{5(x+h) + 8 - (5x + 8)}{h} \\ &= \frac{5x + 5h + 8 - 5x - 8}{h} \\ &= \frac{5h}{h} \\ &= \frac{5\cancel{h}}{\cancel{h}}, \quad \text{if } h \neq 0 \\ &= 5 \quad \left(= \frac{5 \leftarrow \text{change in } f(x)}{1 \leftarrow \text{change in } x}\right)\end{aligned}$$

Interpretation: The average rate of change of the function $f(x) = 5x + 8$ with respect to x is 5. Notice that this rate of change is constant; it does not depend on x. Regardless of the initial value of x, a 1-unit change in its value produces a 5-unit change in the value of the dependent variable.

2. For the function $f(x) = -x^2 + 11x - 12$, compute the difference quotient $\dfrac{f(x+h) - f(x)}{h}$. (This is the function we examined in Example Set B.)

Solution:

$$\begin{aligned}\frac{f(x+h) - f(x)}{h} &= \frac{-(x+h)^2 + 11(x+h) - 12 - (-x^2 + 11x - 12)}{h} \\ &= \frac{-(x^2 + 2hx + h^2) + 11x + 11h - 12 + x^2 - 11x + 12}{h} \\ &= \frac{-x^2 - 2hx - h^2 + 11x + 11h - 12 + x^2 - 11x + 12}{h} \\ &= \frac{-2hx - h^2 + 11h}{h} \\ &= \frac{h(-2x - h + 11)}{h} \\ &= \frac{\cancel{h}(-2x - h + 11)}{\cancel{h}} \quad \text{if } h \neq 0 \\ &= -2x - h + 11\end{aligned}$$

Interpretation: The average rate of change of $f(x) = -x^2 + 11x - 12$ with respect to x is given by $-2x - h + 11$. Unlike the rate of change of example 1 of Example Set C, this rate of change depends on both the initial value of x and the amount of change, h. For example, recalling that x represents the initial dollar selling price of the mechanical pencils and h represents the dollar change in the selling price,

1. When $x = 2$ and $h = 3$,

$$\begin{aligned} -2x - h + 11 &= -2(2) - 3 + 11 \\ &= 4 \end{aligned}$$

2. When $x = 2$ and $h = 1$,

$$\begin{aligned} -2x - h + 11 &= -2(2) - 1 + 11 \\ &= 6 \end{aligned}$$

Both these results agree with the results we obtained in Example Set B.

Interpretation:

1. For $x = 2$ and $h = 3$: When the initial value of x is 2, a 3-unit change will produce a $4 \cdot 3 = 12$-unit change in the function value.

2. For $x = 2$ and $h = 1$: When the initial value of x is 2, a 1-unit change will produce a $6 \cdot 1 = 6$-unit change in the function value.

Secant Line

Since the difference quotient is so important in calculus, viewing it geometrically will be helpful. Figure 2.40 exhibits the graph of a function $f(x)$ along with an initial x value (producing the point $(x, f(x))$), and an increase of h units in x (producing the point $(x + h, f(x + h))$). The line passing through the points $(x, f(x))$ and $(x + h, f(x + h))$ is called a **secant line**.

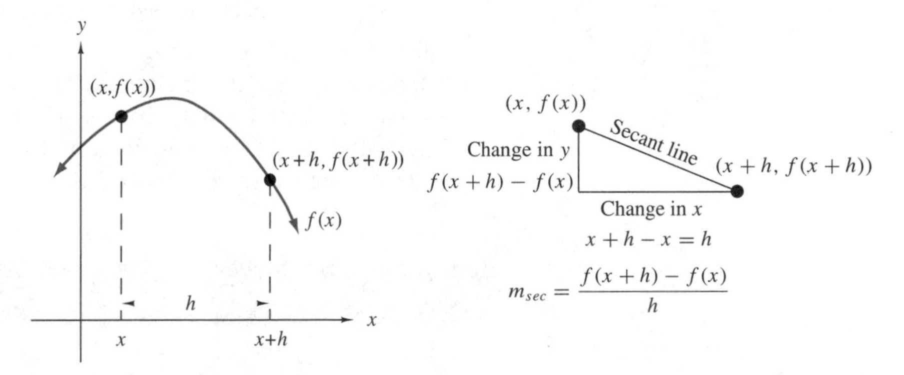

Figure 2.40

Using the formula for the slope of a line passing through two points, the slope of the secant line is

$$\frac{\text{(final } y \text{ value)} - \text{(initial } y \text{ value)}}{\text{(final } x \text{ value)} - \text{(initial } x \text{ value)}} = \frac{f(x+h) - f(x)}{x+h-x}$$

$$= \frac{f(x+h) - f(x)}{h}$$

But, this is precisely the difference quotient!

Difference Quotient and the Slope of the Secant Line

$$\text{Average rate of change} = \text{slope of the secant line}$$
$$= m_{sec}$$

where m_{sec} is a notation for the *slope of the secant line*.

Instantaneous Rate of Change

We have seen that the average rate of change gives only an expected value for the rate at which the dependent variable, y, is changing with respect to changes in the independent variable, x. The average rate of change describes the change in y in the long term. To see how y is changing in the short term, or even at a particular instant, the change in x should be made small. The smaller the change in x, the more accurately the average rate of change describes the short term change in y. The difference quotient $\lim_{h\to 0} \frac{f(x+h) - f(x)}{h}$ allows us to easily adjust the change in x. Recall that h represents a change in x. By letting h get smaller and smaller, the average rate of change, as computed by the difference quotient $\lim_{h\to 0} \frac{f(x+h) - f(x)}{h}$, better reflects the short-term change in y. If h is very close to 0, the average rate of change very closely approximates the immediate, or instantaneous, change in y. The idea of examining a function for changing x values reminds us of the limit process. The **instantaneous rate of change** of a function is found by finding the limit of the difference quotient of the function as h gets smaller and smaller, that is, as h approaches 0.

Instantaneous Rate of Change

$$\text{Instantaneous rate of change of } f(x) = \lim_{h\to 0} \frac{f(x+h) - f(x)}{h}$$

if this limit exists.

Notice that although there are two variables, x and h, in this limit expression, we are interested only in changes in h. Evaluation of this limit by direct substitution will always result in the indeterminate form $\frac{0}{0}$ (verify this by direct substitution). Therefore, when evaluating this limit, we will always proceed immediately with algebraic manipulations.

EXAMPLE SET D

Find the instantaneous rate of change of the function $f(x) = 3x^2 + 7x - 2$ at **1.** $x = 1$, **2.** $x = 2$, and **3.** $x = -4$.

Solution:

Since the instantaneous rate of change $= \lim_{h\to 0} \frac{f(x+h) - f(x)}{h}$, we need to compute this limit, if it exists.

$$\begin{aligned}
\lim_{h\to 0} \frac{f(x+h)-f(x)}{h} &= \lim_{h\to 0} \frac{\overbrace{3(x+h)^2 + 7(x+h) - 2}^{f(x+h)} - \overbrace{(3x^2 + 7x - 2)}^{f(x)}}{h} \\
&= \lim_{h\to 0} \frac{3(x^2 + 2hx + h^2) + 7x + 7h - 2 - 3x^2 - 7x + 2}{h} \\
&= \lim_{h\to 0} \frac{3x^2 + 6hx + 3h^2 + 7x + 7h - 2 - 3x^2 - 7x + 2}{h} \\
&= \lim_{h\to 0} \frac{6hx + 3h^2 + 7h}{h} \\
&= \lim_{h\to 0} \frac{h(6x + 3h + 7)}{h} \\
&= \lim_{h\to 0} \frac{\not{h}(6x + 3h + 7)}{\not{h}} \qquad \text{since } h \neq 0 \\
&= \lim_{h\to 0} (6x + 3h + 7) \qquad \text{Take the limit by substitution.} \\
&= 6x + 7
\end{aligned}$$

Thus, the instantaneous rate of change is $6x + 7$. Since the expression involves the variable x, the instantaneous rate of change is itself variable and depends on the initial value selected for x. For example,

1. When $x = 1$, the instantaneous rate of change of $y = 6(1) + 7 = 6 + 7 = 13$.

2. When $x = 2$, the instantaneous rate of change of $y = 6(2) + 7 = 12 + 7 = 19$.

3. When $x = -4$, the instantaneous rate of change of $y = 6(-4) + 7 = -24 + 7 = -17$.

We are not quite ready to interpret these values. A short discussion of the geometry associated with the instantaneous rate of change will put us in the right position.

Interpreting the Instantaneous Rate of Change

We saw that, geometrically, the average rate of change is the slope of a secant line. The instantaneous rate of change has its geometric counterpart. By making h smaller and smaller in the difference quotient $\lim_{h\to 0} \frac{f(x+h)-f(x)}{h}$, we know we get more accurate information about the changes in y. But,

$$\lim_{h\to 0} \frac{f(x+h)-f(x)}{h} = \lim_{h\to 0} m_{sec}$$

To find the instantaneous rate of change, we let h approach 0. We want to know what happens to the slope of the secant line as h approaches 0. Geometrically, this means we need to examine $\lim_{h\to 0} m_{sec}$. The graphs in Figure 2.41 illustrate this examination and help us to suggest a meaning for $\lim_{h\to 0} m_{sec}$.

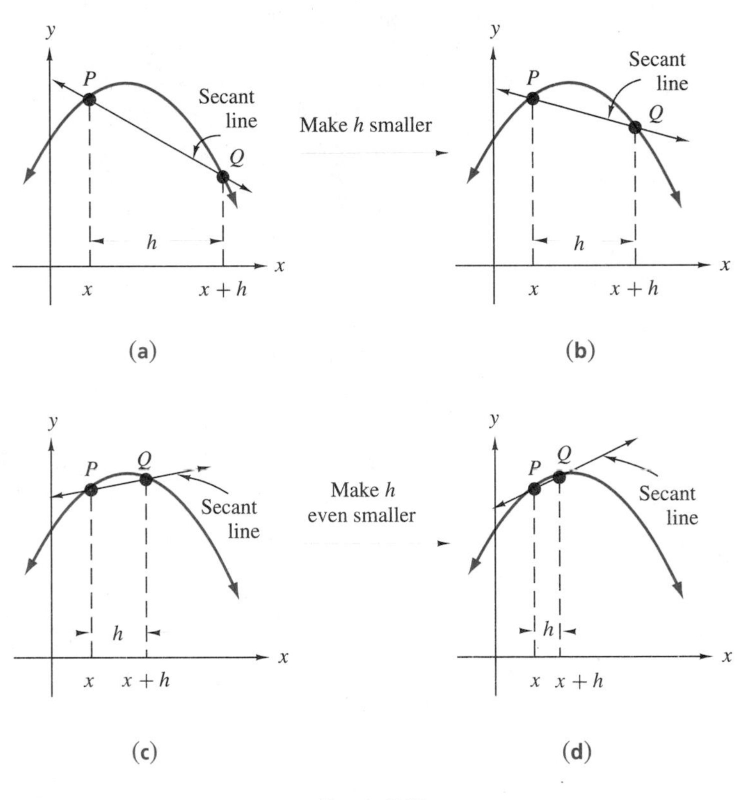

Figure 2.41

As h approaches 0, the point Q moves along the curve toward point P, and the secant line has changing slope. As h gets very close to 0 (or as Q gets very close to P), the changes in the secant lines' slopes are very small. In fact, as h approaches 0, the secant line approaches a fixed line that passes through point P, and the slope of the secant line approaches the slope of this fixed line. Figure 2.42 illustrates this fact.

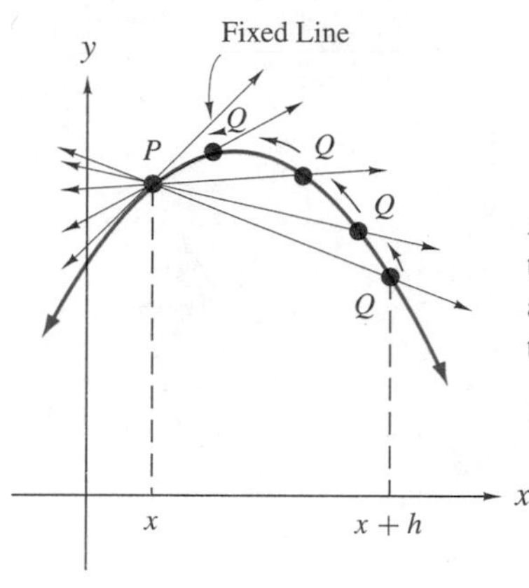

As Q moves along the curve toward P, h approaches zero and the secant line approaches the fixed line.

Figure 2.42

The fixed line passing through the point P is called the **tangent line** to the curve at the point P. The slope of this tangent line is the limit, as h approaches 0, of the slope of the secant line. We conclude from our geometric examination that $\lim_{h \to 0} m_{sec}$ does in fact have a meaning. Specifically, $m_{tan} = \lim_{h \to 0} m_{sec}$, where m_{tan} symbolizes the *slope of the tangent line*.

Tangent Line

We now know that the slope of the tangent line of a function is the instantaneous rate of the change of the function.

The Slope of a Tangent Line

For a function $f(x)$, the instantaneous rate of change of $f(x)$ at x is

$$\lim_{h \to 0} \frac{f(x+h) - f(x)}{h} = m_{tan}$$

That is, the instantaneous rate of change of a function is the slope of the tangent line to the graph of the function.

Interpretation: We now present two interpretations for the instantaneous rate of change of a function.

Interpreting the Instantaneous Rate of Change

Suppose the instantaneous rate of change of a function $f(x)$ at a point x is computed to be the number m. Then, expressed as a fraction,

$$m = \frac{m \leftarrow \text{change in } f(x)}{1 \leftarrow \text{change in } x}$$

which can be interpreted in the following ways.

1. Observe Figure 2.43. Suppose that between the points x and $x+1$, the function is a straight line. Then the tangent line to the function is exactly the function itself, and its slope remains constant between x and $x+1$. Since the slope of the tangent line is the instantaneous rate of change, it follows that between x

and $x+1$, the instantaneous rate of change remains constant. We can now state an interpretation.

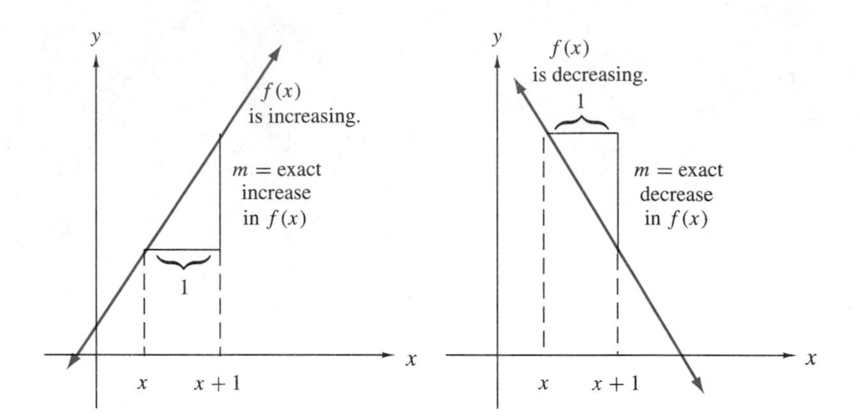

Figure 2.43

If the instantaneous rate of change were to *remain constant* (so that the tangent line is exactly the function itself) as x increases 1 unit in value, the value of the function would change by *exactly* m units. (a) If m is a positive number, a 1-unit increase in x will produce an m-unit increase in $f(x)$. (b) If m is a negative number, a 1-unit increase in x will produce an $|m|$-unit decrease in $f(x)$.

2. Observe Figure 2.44. Suppose that between the points x and $x+1$, the function is a curved line. Then the tangent line to the function is an approximation to the function, and has various values of slope between x and $x+1$. Since the slope of the tangent line is the instantaneous rate of change, it follows that between x and $x+1$, the instantaneous rate of change varies in value. We can now state an interpretation.

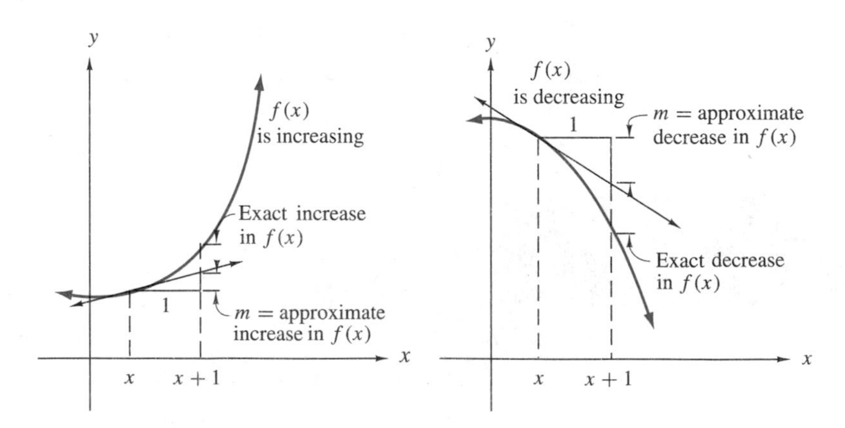

Figure 2.44

If the rate of change *varies* (so that the tangent line approximates the function) as x increases 1 unit in value, the value of the function would change by *approximately* m units. (a) If m is a positive number, a 1-unit increase in x will produce, approximately, an m-unit increase in $f(x)$. (b) If m is a negative number, a 1-unit increase in x will produce, approximately, an $|m|$-unit decrease in $f(x)$.

EXAMPLE SET E

1. For the function $f(x) = 2x^2 - 7x + 5$, $\lim_{h \to 0} \frac{f(x+h) - f(x)}{h} = 4x - 7$. Interpret this result for (a) $x = 5$, (b) $x = 8$. Compare the results of parts (a) and (b).

Solution:

a. Since $\lim_{h \to 0} \frac{f(x+h) - f(x)}{h}$ is the instantaneous rate of change of $f(x)$ with respect to x, the instantaneous rate of change of $f(x)$ at $x = 5$ is

$$\begin{aligned} 4x - 7 &= 4(5) - 7 \\ &= 13 \end{aligned}$$

Since $4x - 7$ contains a variable, the rate of change varies. Thus, we make the following interpretation.

Interpretation: As x increases 1 unit, from 5 to 6, the value of $f(x)$ increases by *approximately* 13 units. That is, a 1-unit increase in the x value produces, approximately, a 13-unit increase in the y value.

b. If $x = 8$, then $4x - 7 = 4(8) - 7 = 25$.

Interpretation: As x increases 1 unit, from 8 to 9, the value of $f(x)$ increases by *approximately* 25 units. That is, a 1-unit increase in the x value produces, approximately, a 25-unit increase in the y value.

Comparing parts (a) and (b), we can see that the instantaneous rate of change has changed. We expect this since the instantaneous rate of change is variable ($4x - 7$ involves a variable).

2. For the function $f(x) = \frac{x+3}{x-1}$, $\lim_{h \to 0} \frac{f(x+h) - f(x)}{h} = \frac{-4}{(x-1)^2}$. Interpret this result for $x = 4$.

Solution:

$$\begin{aligned} \frac{-4}{(x-1)^2} &= 7\frac{-4}{(4-1)^2} \\ &= \frac{-4}{3^2} \\ &= \frac{-4}{9} \quad \left(= \frac{-4/9}{1}\right) \end{aligned}$$

Since $\frac{-4}{(x-1)^2}$ contains a variable, the rate of change varies.

Interpretation: As x increases 1 unit, from 4 to 5, the value of $f(x)$ decreases by approximately $\frac{4}{9}$ units. That is, a 1-unit increase in the x value produces, approximately, a $\frac{4}{9}$-unit decrease in the y value.

3. For the function $f(x) = -8x + 3$, $\lim_{h \to 0} \frac{f(x+h) - f(x)}{h} = -8$. Interpret this result for $x = 15$.

Solution:
Since $\lim_{h \to 0} \frac{f(x+h) - f(x)}{h} = -8$, and -8 contains no variables, the instantaneous rate of change of this function with respect to x is constant. For any x value, a 1-unit increase will produce, exactly, an 8-unit decrease (since -8 is negative) in the corresponding $f(x)$ value. Specifically, for $x = 15$, we make the following interpretation.

Interpretation: As x increases 1 unit, from 15 to 16, the value of $f(x)$ decreases by exactly 8 units. That is, a 1-unit increase in the x value produces, exactly, an 8-unit decrease in the y value.

EXAMPLE SET F

A magazine publisher has determined that the cost C (in dollars) associated with x number of half-page articles is given by the function

$$C(x) = -5x^2 + 200x, \quad 0 \le x \le 20$$

1. If the magazine is currently running 12 half-page articles, what is the expected increase in costs if it were to run 13?

Solution:
Since we are interested in the change in cost with respect to a 1-unit change in the number of half-page articles, the instantaneous rate of change is appropriate.

$$\lim_{h \to 0} \frac{f(x+h) - f(x)}{h} = \lim_{h \to 0} \frac{-5(x+h)^2 + 200(x+h) - (-5x^2 + 200x)}{h}$$

$$= \lim_{h \to 0} \frac{-5(x^2 + 2hx + h^2) + 200x + 200h + 5x^2 - 200x}{h}$$

$$= \lim_{h \to 0} \frac{-5x^2 - 10hx - 5h^2 + 200x + 200h + 5x^2 - 200x}{h}$$

$$= \lim_{h \to 0} \frac{-10hx - 5h^2 + 200h}{h}$$

$$= \lim_{h \to 0} \frac{h(-10x - 5h + 200)}{h}$$

$$= \lim_{h \to 0} \frac{\not{h}(-10x - 5h + 200)}{\not{h}} \quad \text{if } h \neq 0$$

$$= \lim_{h \to 0} (-10x - 5h + 200) \quad \text{Take the limit by substitution.}$$

$$= -10x + 200$$

Thus, the instantaneous rate of change $= -10x + 200$. It involves a variable and therefore will produce an approximation as x increases by 1 unit. If $x = 12$,

$$\begin{aligned} -10x + 200 &= -10(12) + 200 \\ &= -120 + 200 \\ &= 80 \end{aligned}$$

Interpretation: If the number of half-page articles is increased by 1, from 12 to 13, the cost will increase by approximately \$80.

The exact increase is given by $C(13) - C(12) = \$75$.

2. If the magazine is currently running 15 half-page articles, what is the expected increase in cost if it were to run 16?

Solution:
If $x = 15$,

$$\begin{aligned} -10x + 200 &= -10(15) + 200 \\ &= -150 + 200 \\ &= 50 \end{aligned}$$

Interpretation: If the number of half-page articles is increased by 1, from 15 to 16, the cost will increase by approximately \$50.

Comparing parts 1 and 2, it appears that rate of cost increase decreases as the number of half-page articles increases.

Rates of Change and Technology

Using Your Calculator You can use your graphing calculator to find the average rate of change of a function and to approximate the instantaneous rate of change of a function. The following entries illustrate one way of computing the average rate of change of the function $f(x) = (5x - 3)^{2/3}$ as x increases from 16 to 20, and the approximate instantaneous rate of change of $f(x)$ at $x = 16$.

```
Y1=(5x−3)^(2/3)
```

Now, in the computation window, enter

```
(Y1(20)−Y1(16))/(20−16)
```

The result is 0.7530075664.

To approximate the instantaneous rate of change of $f(x) = (5x - 3)^{2/3}$ at $x = 16$, you can use $h = 0.001$. Then, in the computation window, enter

```
(Y1(16.001)−Y1(16))/(16.001−16)
```

The result is 0.783508665.

Calculator Exercises For Exercises 1 and 2, use your calculator to find the average rate of change of the function over the given interval and to approximate the instantaneous rate of change of the function at the left endpoint of the given interval.

1. $f(x) = 4x - 5$ on $[1, 7]$

2. $f(x) = 0.85x^2 - 1.82x - 5$ on $[0, 6]$

3. A company that manufactures a particular product has determined that the cost of producing x units of the product is described by the function $C(x) = 0.035x^2 + 1855$. Find the average rate of change of the cost as the number of units produced increases from 200 to 500. Find the instantaneous rate of change of the cost when the level of production is 200 units. Draw the graph of this function, so that it includes the tangent line (you may have to refer to your calculator manual) and label the slope with m_{tan} = (the number you compute). Interpret both the values you just computed.

EXERCISE SET 2.5

A UNDERSTANDING THE CONCEPTS

1. The quantities x and y are related in such a way that when the value of x is increased from 18 to 21, the value of y is increased from 5 to 17. Find and interpret the average rate of change of y with respect to x.

2. The quantities x and y are related in such a way that when the value of x is increased from 3 to 10, the value of y is increased from 28 to 49. Find and interpret the average rate of change of y with respect to x.

3. The quantities x and y are related in such a way that when the value of x is increased from 25 to 40, the value of y is decreased from 95 to 65. Find and interpret the average rate of change of y with respect to x.

4. The quantities x and y are related in such a way that when the value of x is increased from 42 to 44, the value of y is decreased from 70 to 68. Find and interpret the average rate of change of y with respect to x.

5. The quantities x and y are related in such a way that when the value of x is increased from 12 to 16, the value of y is decreased from 40 to 37. Find and interpret the average rate of change of y with respect to x.

6. The quantities x and y are related in such a way that when the value of x is increased from 35 to 42, the value of y is increased from 16 to 17. Find and interpret the average rate of change of y with respect to x.

7. ***Medicine: Drug Concentration*** The concentration C (in mg/cc) of a particular drug in a person's bloodstream is related to the time t (in hours) after injection in such a way that as t changes from 0 to 8, C changes from 0.8 to 0.1. Find and interpret the average rate of change of C with respect to t.

8. ***Economics: City Redevelopment*** The population P (in thousands of people) of a city involved in redevelopment is related to time t (in years) after the beginning of redevelopment in such a way that as t changes from 2 to 10, P changes from 65 to 82. Find and interpret the average rate of change of P with respect to t.

9. ***Business: Retail Profit*** When a retailer increases the selling price of an item from 75 cents to 80 cents, his weekly profit in sales of that item changes 300 cents. With these changes, the average rate of change of profit with respect to selling price is 60. Interpret the average rate of change of profit with respect to selling price.

10. ***Psychology: Reaction Time*** A psychologist has determined that for 20-year-old women, the reaction time t (in seconds) to a particular stimulus is 2 seconds, whereas for 23-year-old women, the reaction time to the same stimulus is 8 seconds. The average rate of change of reaction time with respect to age is 2. Interpret the average rate of change of time with respect to age.

11. The relationship between the quantities x and y is expressed by the function $y = f(x)$. The instantaneous rate of change of y with respect to x is $8x - 1$. Find and interpret the instantaneous rate of change of y with respect to x when $x = 4$.

12. The relationship between the quantities x and y is expressed by the function $y = f(x)$. The instantaneous rate of change of y with respect to x is $-3x^2 + 5x - 3$. Find and interpret the instantaneous rate of change of y with respect to x when $x = 1$.

13. The relationship between the quantities x and y is expressed by the function $y = f(x)$. The instantaneous rate of change of y with respect to x is -2. Find and interpret the instantaneous rate of change of y with respect to x when $x = 0$.

14. The relationship between the quantities x and y is expressed by the function $y = f(x)$. The instantaneous rate of change of y with respect to x is -10. Find and interpret the instantaneous rate of change of y with respect to x when $x = 7$.

For Exercises 15–18, determine from the graph of $f(x)$ if the instantaneous rate of change of $f(x)$ measures the approximate or the exact rate of change as x increases one unit from $x = a$ to $x = a + 1$.

15.

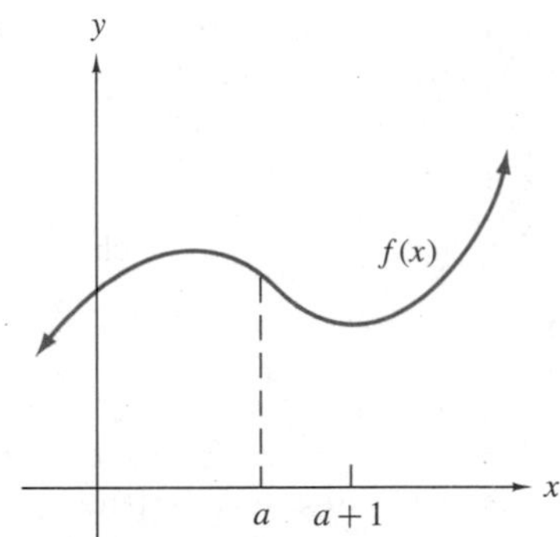

16.

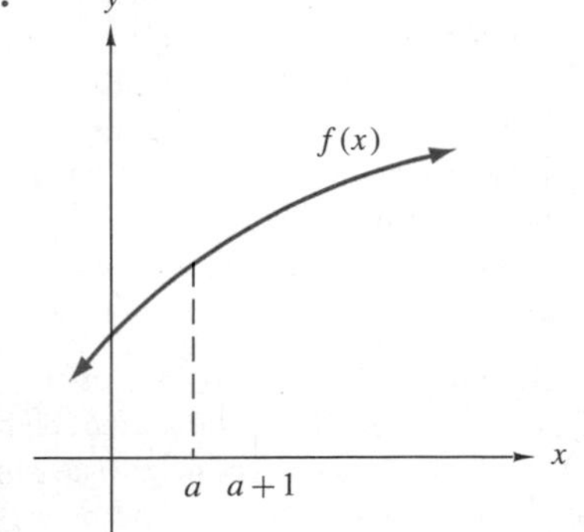

17.

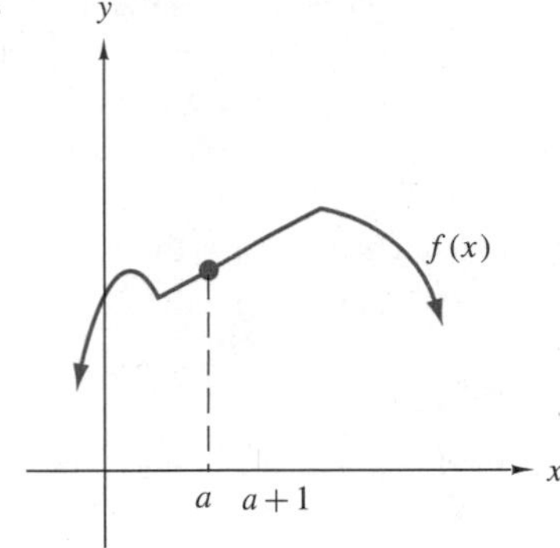

18.

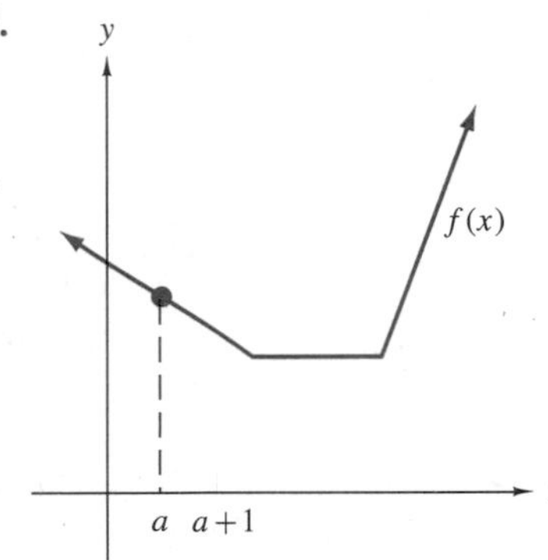

For each of Exercises 19–23, y is a function of x. Determine from the illustrated computation if the average rate of change of y or the instantaneous rate of change of y is being computed. If the instantaneous rate of change of y is being computed, specify if it measures the exact or the approximate rate of change.

19. $\lim_{h \to 0} \dfrac{f(x+h) - f(x)}{h} = 5x^2 + 2x + 1$

20. $\dfrac{f(x+h) - f(x)}{h} = 3x + 2h - 1$

21. $\dfrac{f(x+h) - f(x)}{h} = 9x - h$

22. $\lim_{h \to 0} \dfrac{f(x+h) - f(x)}{h} = -6$

23. $\lim_{h \to 0} \dfrac{f(x+h) - f(x)}{h} = \dfrac{x+4}{x-7}$

For Exercises 24–28, sketch the graph of a function so that the instantaneous rate of change of y at $x = a$ satisfies the given condition. (It may be helpful to examine Figure 2.43 and Figure 2.44 again.)

24. The instantaneous rate of change is approximate; the actual rate of change is greater than the instantaneous rate of change.

25. The instantaneous rate of change is approximate; the actual rate of change is less than the instantaneous rate of change.
26. The instantaneous rate of change equals the actual rate of change.
27. The instantaneous rate of change is negative.
28. The instantaneous rate of change is positive.

B SKILL ACQUISITION

For Exercises 29–34, find the average rate of change of $f(x)$ with respect to x on each interval.

29. $f(x) = x^2 - 2;\quad [1,4]$
30. $f(x) = x^2 + 3x - 11;\quad [6,10]$
31. $f(x) = 4x - 5;\quad [-1,5]$
32. $f(x) = 2\sqrt{x} - 1;\quad [4,9]$
33. $f(x) = \dfrac{x^2 - 25}{x - 5};\quad [6,15]$
34. $f(x) = \begin{cases} 2x^2 + x - 4, & \text{if } 1 \le x < 3 \\ x - 7, & \text{if } 3 \le x \le 6 \end{cases}\quad [1,6]$

For Exercises 35–42, find the instantaneous rate of change of y of each function at the specified value of x.

35. $f(x) = x^2 + 4;\quad x = 3$
36. $f(x) = x^2 - 3x + 1;\quad x = 6$
37. $f(x) = 5x^2 + 4x + 2;\quad x = 0$
38. $f(x) = -6x^2 - 3x + 11;\quad x = 1$
39. $f(x) = 4x + 4;\quad x = -2$
40. $f(x) = -6x + 5;\quad x = 5$
41. $f(x) = \dfrac{x+2}{x-3};\quad x = 2$
42. $f(x) = \dfrac{3x+2}{x+5};\quad x = 1$

C APPLYING THE CONCEPTS

For Exercises 43–44, use the information presented in Table 2.9. The information was excerpted from *Workforce 2000* by William B. Johnston and Arnold H. Packer, published by Hudson Institute. It presents the current number of jobs, in thousands, as of 1984 in an occupation and the anticipated number of jobs in that occupation in the year 2000.

Table 2.9

Occupation	1984	2000
Market/Sales	10,656	14,806
Health	2,478	3,862

43. Find and interpret the average rate of change with respect to time of the number of jobs in the Marketing and Sales occupations. Round your result to the nearest whole number before making your interpretation.

44. Find and interpret the average rate of change with respect to time of the number of jobs in the Health Diagnosing and Treating occupations. Round your result to the nearest whole number before making your interpretation.

45. ***Business: Profit*** The buyer for a retail store located in a shopping mall has convinced himself that if he prices small personal fans at $\$x$ apiece, the weekly profit made by the store on the sale of the fans will be given by the profit function $P(x) = -x^2 + 38x - 240$.
    a. If the store is currently selling the small personal fans for \$10 each, what is the expected change in the weekly profit if it increases the selling price of each fan to \$11?
    b. If the store is currently selling the small personal fans for \$20 each, what is the expected change in the weekly profit if it increases the selling price of each fan to \$21?

46. ***Business: Profit*** The buyer for a retail store located in a shopping mall has convinced herself that if she prices leather wallets at $\$x$ apiece, the weekly profit made by the store on the sale of the wallets will be given by the profit function $P(x) = -x^2 + 100x - 875$.
    a. If the store is currently selling the leather wallets for \$40 each, what is the expected change in the weekly profit if it increases the selling price of each wallet to \$41?
    b. If the store is currently selling the wallets for \$55 each, what is the expected change in the weekly profit if it increases the selling price of each wallet to \$56?

47. ***Social Science: Marriage*** Studies show that there is a relationship between the median age A (in years) of a woman at her first marriage and the time t (in years since 1950). The relationship between A and t is described by the function $A(t) = 0.08t + 19.7$. By how many years would we expect the median age of a woman at her first marriage to change between the years
    a. 1975 to 1976? (Hint: $t = 1975 - 1950 = 25$.)
    b. 1990 to 1991?

48. ***Anthropology: Anatomy*** Anthropologists often use the function $h(x) = 2.75x + 71.48$ to estimate the height h (in centimeters) of a human female. The height h depends on the length x (in centimeters) of the humerus bone (the bone from the shoulder to the elbow). By how much do we expect the projected height of a 400-year-old fossil female to change if more detailed research shows that the length of the remains of her humerus bone is closer to 41 centimeters than an earlier estimate of 40 centimeters?

D DESCRIBE YOUR THOUGHTS

49. Describe the difference between the average rate of change and the instantaneous rate of change of a function.

50. Explain why we must keep the lim symbol affixed to each expression except the last when computing the instantaneous rate of change of a function using the difference quotient.

51. Explain why the word *approximately* needs to be used when interpreting the instantaneous rate of change of a function. (A picture may help your explanation.)

E REVIEW

52. (2.1) For the function pictured below, determine if $\lim_{x \to 4} f(x)$ exists. If it exists, specify its value.

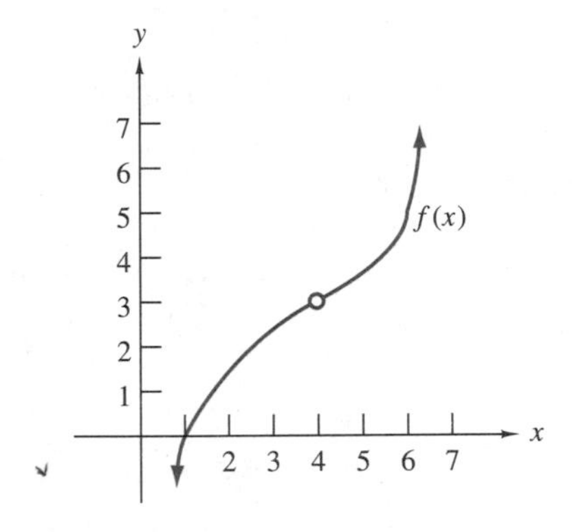

53. (2.2) Find, if it exists, $\lim_{x \to -3} (2x^2 + 4)$.

54. (2.2) Find, if it exists, $\lim_{x \to 6} \frac{x^2 - 8x + 12}{x - 6}$.

55. (2.3) Find, if it exists, $\lim_{x \to \infty} \frac{5x^2 - 3x + 1}{4x^2 + 2}$.

56. (2.4) Discuss the continuity of

$$f(x) = \begin{cases} 1, & \text{if } x \le 2 \\ x^2 - 4x + 7, & \text{if } x > 2 \end{cases} \quad \text{at } x = 2.$$

Summary

This chapter represented a transition from algebra to calculus. Its central theme was the limit of a function. As Chapter 1 began with an intuitive view of functions, Chapter 2 began with an intuitive view of limits of functions. Limits are defined more rigorously in Section 2.2 in terms of one-sided limits. This chapter also discussed continuity of functions and made a point of examining and interpreting the average and instantaneous rates of change of a function.

Limits Viewed Intuitively

The first section presented an intuitive description of the limit concept. The word *limit* should make you think of the word *approach*. The notation $\lim_{x \to c} f(x) = L$ represents the idea that *as the values of x get closer and closer to the number c, the values of $f(x)$ get closer and closer to the number L.* Keep in mind that the input values can approach a particular value from one side or the other. The notations $\lim_{x \to c^-} f(x) = L$ and $\lim_{x \to c^+} f(x) = L$ mean, respectively, that *as the values of x get closer and closer to the number c from the left, the values of $f(x)$ get* *closer and closer to the number L,* and *as the values of x get closer and closer to the number c from the right, the values of f(x) get closer and closer to the number L.*

Also keep in mind that if the limit of a function exists, it is unique. Furthermore, the limit of a function exists only when both the left- and right-sided limits exist. That is, the limit of a function, $f(x)$, if it exists, is always a unique real number, and it will exist at an x value c provided that

$$\lim_{x \to c^-} f(x) = \lim_{x \to c^+} f(x)$$

Additionally, if the two one-sided limits equal the same unique number L, the number L is called the limit of the function. This means that

$$\text{If } \lim_{x \to c^-} f(x) = \lim_{x \to c^+} f(x), \text{ then } \lim_{x \to c} f(x) = L$$

and

$$\text{If } \lim_{x \to c} f(x) = L, \text{ then } \lim_{x \to c^-} f(x) = \lim_{x \to c^+} f(x)$$

You can use a calculator or computer to compute several function values using x values near the x value to be approached. However, when you do, you never know if you have computed enough function values to actually have the limit or if you have stopped short and some peculiar behavior takes place closer in. So, computational methods produce only *apparent limits.* Algebraic methods and pencil and paper can be used to determine limits for simpler functions.

Limits Viewed Algebraically

Section 2.2 presented limits in a more rigorous way. Limits of functions can often be determined directly and precisely using properties of limits. This section presented and illustrated eight useful limit properties. These properties can be used only under certain circumstances. This section pointed out how to determine limits when the properties do not directly apply. The forms $\frac{k}{0}$ and $\frac{0}{0}$ were discussed and illustrated.

Limits Involving Infinity

Section 2.3 presented limits that involve the concept of infinity. The section focused on the meaning of $\lim_{x \to \infty} x = \infty$ and illustrated what is meant by: as the values of x grow larger and larger, so do the values of $f(x)$. The section ended with the introduction to asymptotes to a function. A horizontal line that is approached by a curve as the independent variable tends toward infinity is called a *horizontal asymptote*. If $\lim_{x \to c} f(x) = \infty$, then $x = c$ is a *vertical asymptote*. A vertical asymptote is a vertical line that the curve approaches, but does not intercept.

Continuity of Functions

Section 2.4 discussed the continuity of functions. Intuitively, a continuous function is a function whose graph runs continuously from one point to another point without any breaks or jumps. That is, the curve is completely unbroken and can be drawn, from start to finish, without lifting the pencil from the paper. A function is continuous at the point $x = a$ if it meets the following conditions:

1. A point exists at $x = a$. That is, $f(a)$ is defined. In fact, $f(a) = L$.
2. The branches of the curve come nearer to each other as the x values get nearer to $x = a$. That is, $\lim_{x \to a^-} f(x) = \lim_{x \to a^+} f(x) \longrightarrow \lim_{x \to a} f(a)$ exists. In fact, $\lim_{x \to a} f(x) = L$.
3. Not only is there a point at $x = a$, and not only do the branches of the curve come near each other near $x = a$, but the branches of the curve join together at $x = a$. That is, $\lim_{x \to a} f(x) = f(a) \quad (L = L)$.

This section showed you how to determine continuity by first analyzing the function relative to the three continuity conditions, then summarizing your findings.

Average and Instantaneous Rates of Change

Section 2.5 presented a detailed and illustrated description of the concepts of average and instantaneous rates of change. This section was first to note that differential calculus is the language of change and its methods provide a way of analyzing the rate at which one quantity changes relative to another. In the study of functions, we know that when the value of one quantity, say, y, depends on the value of another quantity, say, x, then the quantity y is called the *dependent variable* and the quantity x is called the *independent variable*. The independent variable is the input quantity and the dependent variable is the output quantity.

Two reasonable measures of the rate at which the dependent quantity changes relative to the independent quantity are the *average rate of change* and the *instantaneous rate of change*. A *secant line* is a line that intercepts a curve in more than one point. The slope of a secant line to a function corresponds to the average rate of change of the function.

$$\begin{aligned} \text{Average rate of change} &= \text{slope of the secant line} \\ &= m_{sec} \end{aligned}$$

where m_{sec} is a notation for the *slope of the secant line*. The instantaneous rate of change of a function $f(x)$ is defined to be the limiting value of the slope of the secant line to the function. That is,

$$\text{Instantaneous rate of change} = \lim_{h \to 0} \frac{f(x+h) - f(x)}{h}$$

A *tangent line* to a curve is a line that intercepts the curve at exactly one point (at least in some neighborhood around the point). The instantaneous rate of change of a function can be interpreted as the slope of the tangent line of a function.

Supplementary Exercises

For Exercises 1–4, describe the behavior of each function. Use the limit notation (one-sided, if necessary) to describe all appropriate limits.

1.

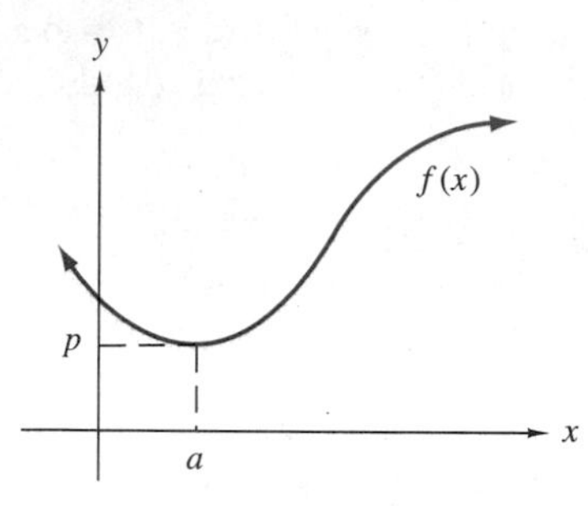

2.

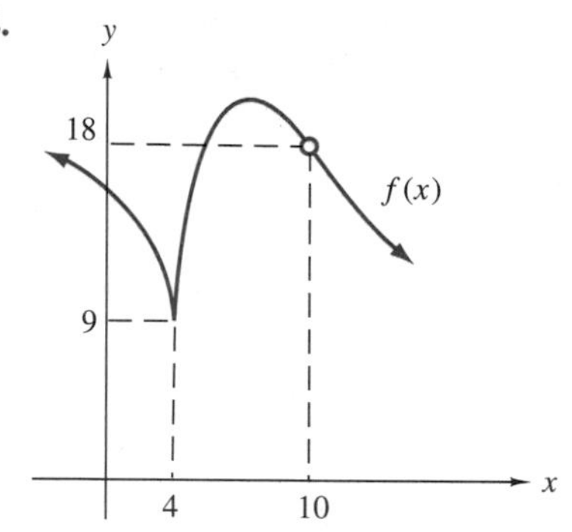

3.

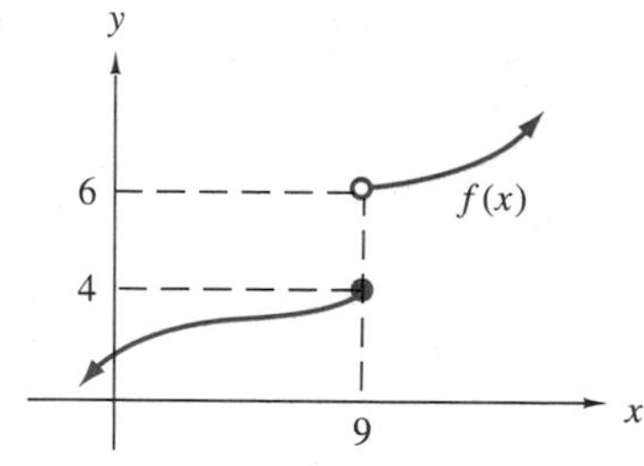

4.

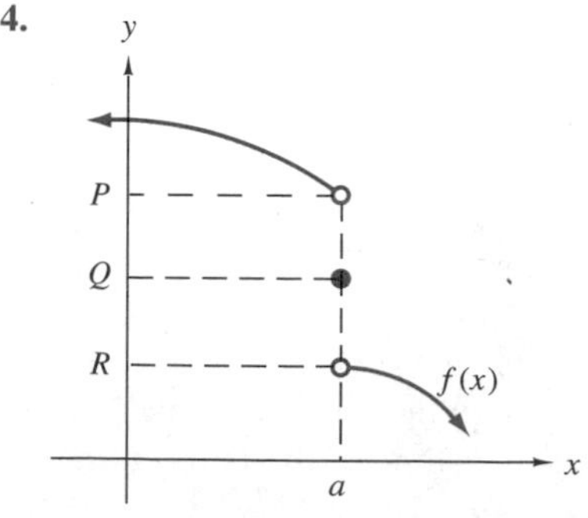

In Exercises 5–7, use observations from the tables to specify the apparent limit, if it exists.

5. This table is for $f(x)$.

x (from the left)	$f(x)$
12.5	1.7
12.7	1.9
12.9	1.99
12.99	1.999
12.999	1.9999

x (from the right)	$f(x)$
13.5	2.2
13.3	2.1
13.1	2.01
13.01	2.001
13.001	2.0001

6. This table is for $f(x)$.

x (from the left)	$f(x)$
5.1	8.42
5.2	8.425
5.3	8.4257
5.4	8.4259
5.49	8.42599
5.499	8.425999

x (from the right)	$f(x)$
5.8	8.49
5.7	8.45
5.6	8.427
5.51	8.4426
5.501	8.4261
5.5001	8.42601

7. This table is for $f(x)$.

x (from the left)	$f(x)$
1.80	6.04
1.85	6.02
1.90	6.01
1.99	6.001
1.999	6.0001
1.9999	6.00001

x (from the right)	$f(x)$
2.2	5.8
2.1	5.85
2.09	5.88
2.009	5.888
2.0009	5.8888
2.00009	5.88888

For Exercises 8–10, graph the function and determine if the limit at the specified point exists. If it exists, specify its value.

8. $f(x) = \begin{cases} \frac{2}{3}x + 3, & \text{if } x \le 0 \\ -x + 3, & \text{if } x > 0 \end{cases}$
at $x = 0$

9. $f(x) = \begin{cases} x^2 - 2x + 3, & \text{if } x < 1 \\ 4, & \text{if } x \ge 1 \end{cases}$
at $x = 1$

10. $f(x) = \begin{cases} \dfrac{x^2 - 5x + 6}{x - 3}, & \text{if } x \ne 3 \\ 2, & \text{if } x = 3 \end{cases}$
at $x = 3$

For Exercises 11–14, sketch the graph of a function having the specified properties. (Graphs may vary.)

11. $f(3)$ is not defined, but $\lim_{x\to3} f(x) = 2$.

12. $f(1) = 3$, $f(2) = 1$, $f(4) = 3$,
$\lim_{x\to2} f(x) = 1$

13. $f(-2) = 3$, $f(0) = 3$,
$\lim_{x\to-2^-} f(x) = 1$, $\lim_{x\to-2^+} f(x) = 4$

14. $f(0) = 1$, $f(2) = 5$, $f(4) = 0$,
$\lim_{x\to2} f(x) = 3$

For Exercises 15–30, find the indicated limit, if it exists.

15. $\lim_{x\to1} 6$

16. $\lim_{x\to-4} (x + 6)$

17. $\lim_{x\to-9} (x^2 - 5)$

18. $\lim_{x\to6} 8$

19. $\lim_{x\to0} (3x^2 - 2x + 7)$

20. $\lim_{x\to3} (-x^2 + 8x - 2)$

21. $\lim_{x\to2} \dfrac{x+6}{x-1}$

22. $\lim_{x\to4} \dfrac{x^2 - 2x - 8}{x+1}$

23. $\lim_{x\to0} \dfrac{x^2 - x + 5}{3x+1}$

24. $\lim_{x\to3} \dfrac{x^2 - 2x - 3}{x-3}$

25. $\lim_{x\to-2} \dfrac{x^2 - 4x - 12}{x+2}$

26. $\lim_{x\to5} \dfrac{5-x}{x-5}$

27. $\lim_{x\to0} \dfrac{4x^3 + x^2}{3x^3 + 5x^2}$

28. $\lim_{x\to0} \dfrac{2x^3 + 5x^2}{x^3 - 7x^2}$

29. $\lim_{x\to0} \dfrac{4x^2 - 3x}{x}$

30. $\lim_{x\to0} \dfrac{3x^2 - 8x}{2x}$

For Exercises 31–40, find $\lim_{h\to0} \dfrac{f(x+h) - f(x)}{h}$ for each function.

31. $f(x) = 2x + 5$

32. $f(x) = 6x - 10$

33. $f(x) = x^2 + 1$

34. $f(x) = 5x^2 + 6$

35. $f(x) = x^2 + 6x - 11$

36. $f(x) = 3x^2 + 7x - 4$

37. $f(x) = \dfrac{x+2}{x-3}$

38. $f(x) = \dfrac{4}{x-7}$

39. $f(x) = \dfrac{-2}{x+4}$

40. $f(x) = \sqrt{x}$

For Exercises 41–52, find the limit, if it exists.

41. $\lim\limits_{x\to\infty} \dfrac{7x^2+3x-1}{8x^2+x+3}$

42. $\lim\limits_{x\to\infty} \dfrac{9x^3-4}{5x^4+6}$

43. $\lim\limits_{x\to\infty} \dfrac{x^4+2x^3+x+1}{x^5+3x-9}$

44. $\lim\limits_{x\to\infty} \dfrac{2x^2+5x-7}{x+11}$

45. $\lim\limits_{x\to\infty} \dfrac{3x^2+7x-7}{4}$

46. $\lim\limits_{x\to\infty} 200+\dfrac{3}{x}$

47. $\lim\limits_{x\to\infty} \dfrac{30x+4}{x^2}+\dfrac{25}{x}+6$

48. $\lim\limits_{x\to\infty} \dfrac{x^2-2x-15}{x-5}$

49. $\lim\limits_{x\to\infty} \dfrac{x^2-10x+24}{x^2-4}$

50. $\lim\limits_{x\to\infty} \dfrac{6}{x+2}$

51. $\lim\limits_{x\to\infty} \dfrac{x^2+5x-14}{x-2}$

52. $\lim\limits_{x\to\infty} \dfrac{x^2+11x+10}{x+1}$

53. ***Business: Price Setting*** A parcel express company determines that the price $P(x)$ (in dollars) they charge a customer for mailing a parcel of weight w (in ounces) anywhere in the continental United States is given by the piecewise function

$$f(x) = \begin{cases} 8, & \text{if } 0 < w \le 8 \\ w, & \text{if } 8 < w \le 16 \\ 2w-14, & \text{if } 16 < w \le 32 \\ w^2-31w+18, & \text{if } w > 32 \end{cases}$$

Find

a. $\lim\limits_{w\to 8} P(w)$

b. $\lim\limits_{w\to 16} P(w)$

c. $\lim\limits_{w\to 32} P(w)$

d. Interpret the results of (a), (b), and (c).

54. ***Business: Demand for a Product*** The quantity demanded x (in hundreds) of a calculator per week is related to the unit price p (in dollars) of the calculator by the equation $x = \sqrt{900-30p}$, $0 \le p \le 30$.

Find

a. $\lim\limits_{p\to 0} (x)$

b. $\lim\limits_{p\to 15} (x)$

c. $\lim\limits_{p\to 30} (x)$

d. Interpret the results of (a), (b), and (c).

For Exercises 55–58, determine if each illustrated function is discontinuous or continuous at the point $x = a$. If the function is continuous, so state. If the function is discontinuous, specify the first continuity condition that is not satisfied.

55.

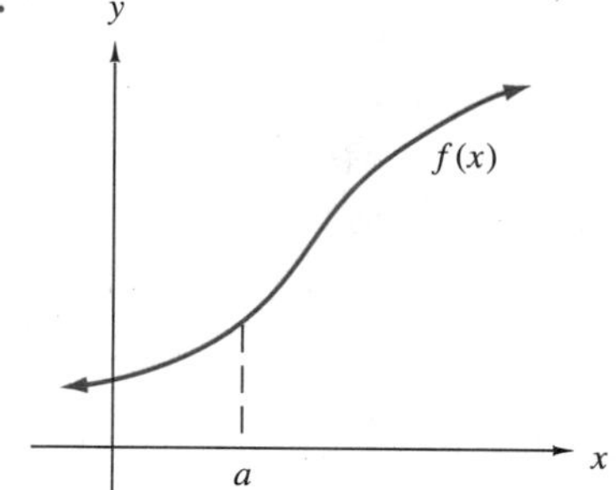

56.

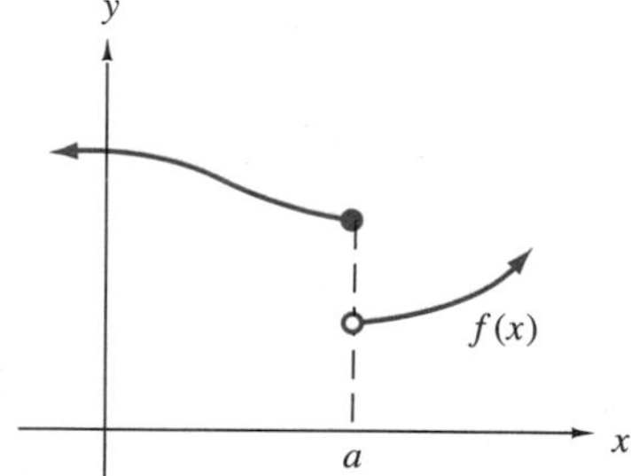

57.

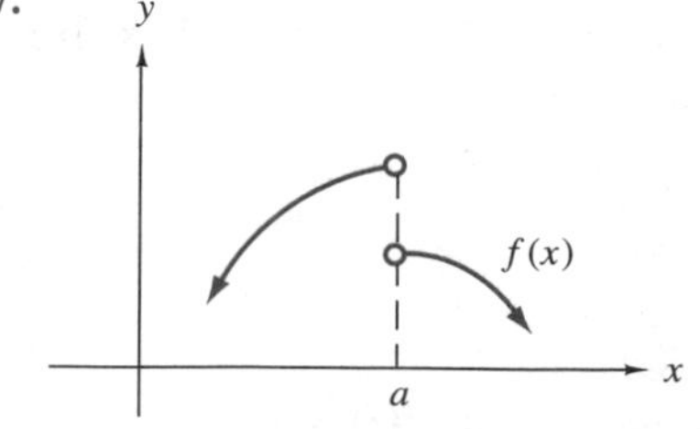

58.

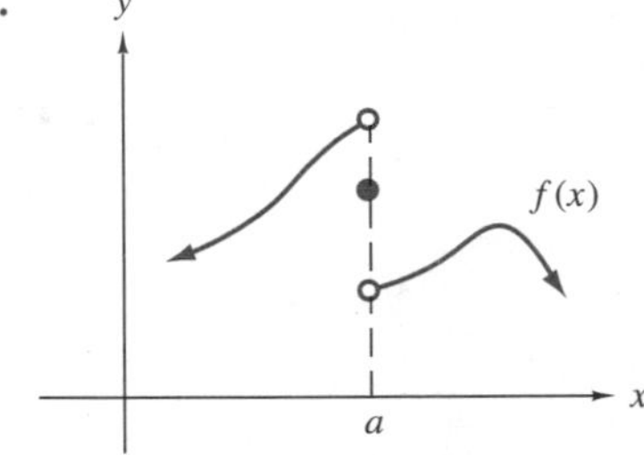

For Exercises 59–61, sketch the graph of a curve that is discontinuous at $x = a$ and meets all the specified conditions.

59. $a = 2$, and $f(2)$ is not defined.

60. $a = 4$, and $f(4) = 3$, $\lim_{x\to 4^-} f(x) = 3$, $\lim_{x\to 4^+} f(x) = 2$

61. $a = 3$, $f(3) = 2$, $\lim_{x\to 3} f(x) = 3$

For Exercises 62–68, discuss the continuity of each function at the specified value of x. (You are encouraged to write an analysis as we did in each example, but you need only present the summary.)

62. $f(x) = x^2 - 4x + 6$ at $x = 2$

63. $f(x) = 7x^2 - x + 2$ at $x = 1$

64. $f(x) = \dfrac{2x^2 + 5x - 1}{x + 4}$ at $x = 3$

65. $f(x) = \dfrac{x^2 - 5x - 24}{x + 3}$ at $x = 6$

66. $f(x) = \dfrac{x^2 - 8x + 12}{x - 6}$ at $x = 6$

67. $f(x) = \begin{cases} -x^2 + 4x, & \text{if } x \le 2 \\ -2x + 8, & \text{if } x > 2 \end{cases}$ at $x = 2$

68. $f(x) = \begin{cases} \dfrac{x^2 + x - 2}{x - 1}, & \text{if } x < 1 \\ 2, & \text{if } x = 1 \\ -x + 4, & \text{if } x > 1 \end{cases}$ at $x = 1$

For Exercises 69–72, sketch the graph of a discontinuous function for which only the specified continuity condition is not satisfied. (Answers may vary.)

69. $f(2) = 3$ and $f(x)$ is discontinuous at $x = 2$ because continuity condition 2 is not satisfied.

70. $\lim_{x\to 3} f(x) = 4$ and $f(x)$ is discontinuous at $x = 3$ because continuity condition 1 is not satisfied.

71. $f(3) = 2$, $\lim_{x\to 3} f(x) = 3$, and $f(x)$ is discontinuous at $x = 2$ because continuity condition 3 is not satisfied.

72. $f(1) = 4$, $\lim_{x\to 1} f(x) = 3$ and $f(x)$ is discontinuous at $x = 1$ because continuity condition 3 is not satisfied.

73. ***Economics: Number of Speeding Tickets*** When idealized, the function relating time t (in days) and the number N of speeding tickets given by a police officer using radar on a particular stretch of roadway can be viewed as a smooth, continuous curve. However, certain conditions can produce a curve such as that pictured below.

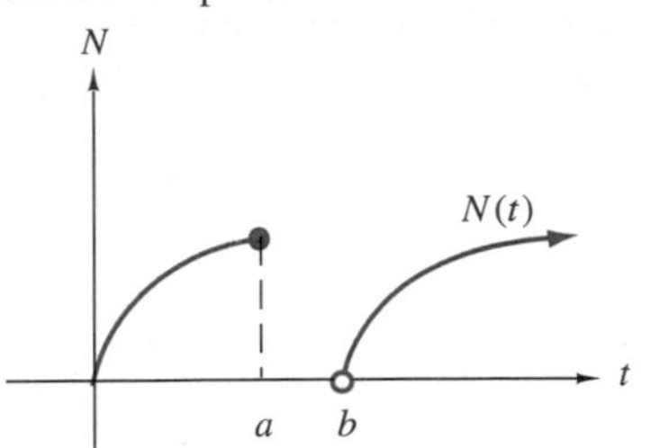

Make a statement about the continuity of the function $N(t)$ at $t = a$ and give a possible interpretation to the point $t = a$.

74. ***Medicine: Cholesterol*** When idealized, the function relating time t (in weeks) and the amount A (in mg/dl) of cholesterol in some person's blood can be viewed as a smooth, continuous curve. However, certain medical conditions can produce a curve such as that pictured below.

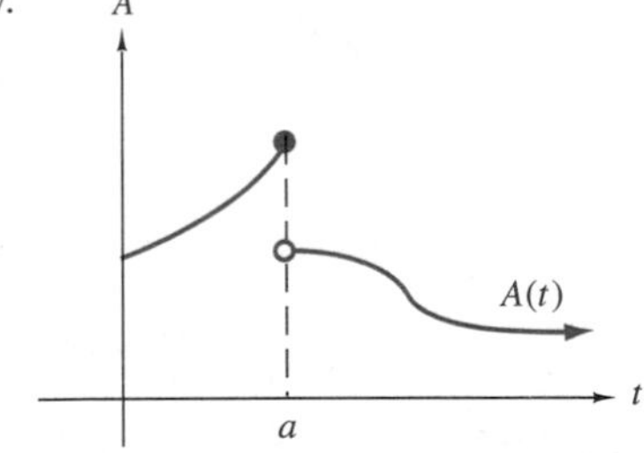

Make a statement about the continuity of the function $A(t)$ at $t = a$ and give a possible interpretation to the point $t = a$.

75. The quantities x and y are related in such a way that when the value of x is increased from 35 to 42, the value of y is increased from 107 to 142. Find and interpret the average rate of change of y with respect to x.

76. The quantities x and y are related in such a way that when the value of x is increased from 16 to 25, the value of y is decreased from 88 to 61. Find and interpret the average rate of change of y with respect to x.

77. ***Economics: Number of Speeding Tickets*** A traffic police officer notices that for the 6-month period of January through June (180) days, she and the other officers in her unit wrote 8280 traffic tickets. Find and interpret the average rate of change of the number of traffic tickets written with respect to time.

78. ***Business: Sale of a Product*** The number of taco salads sold at a college cafeteria during the months February through September (242 days) was 5566. Find and interpret the average rate of change of the number of taco salads sold with respect to time. Why does this number not reflect the true situation very well?

For Exercises 79–83, determine from the graph of $f(x)$ if the instantaneous rate of change of $f(x)$ with respect to x measures the approximate or the exact rate of change as x increases one unit from $x = a$ to $x = a + 1$.

79.

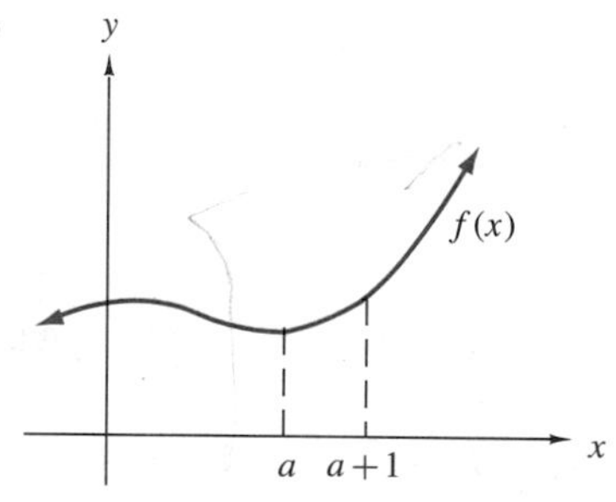

80.

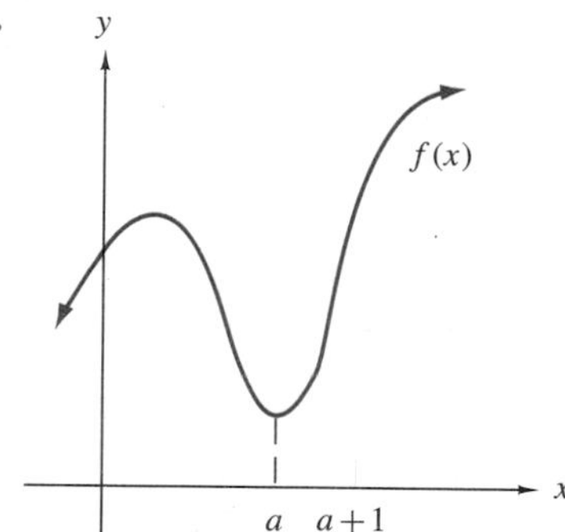

81.

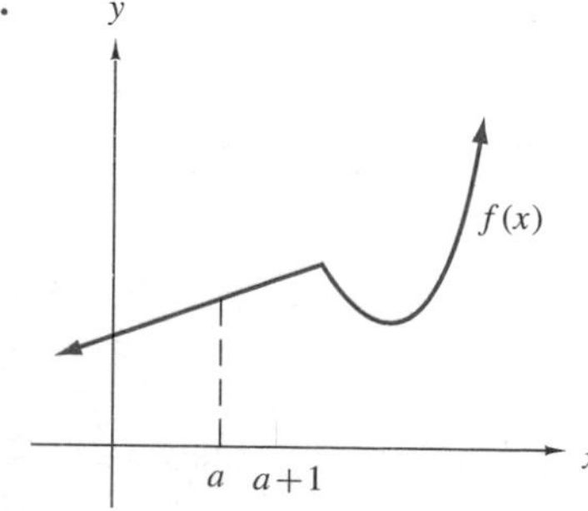

82.

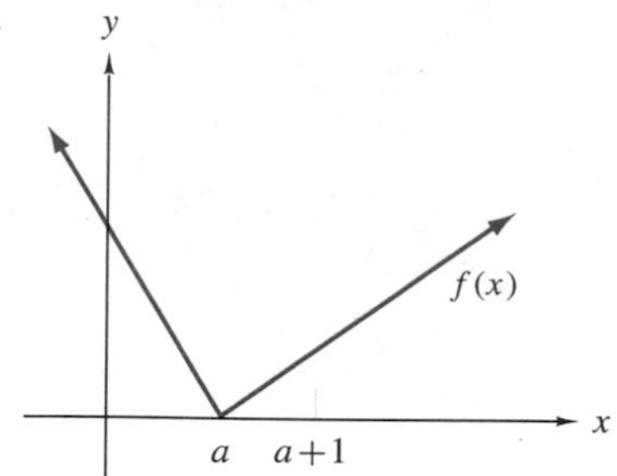

83.

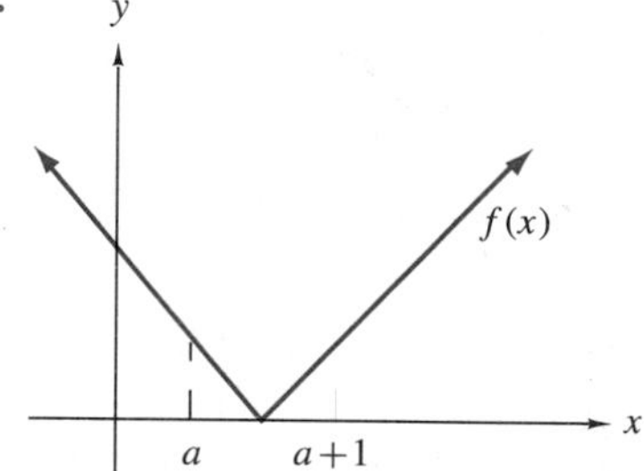

For Exercises 84–89, y is a function of x. Determine from the illustrated computation if the average rate of change of y or the instantaneous rate of change of y is being computed. If the instantaneous rate of change of y is being computed, specify if it measures the approximate or the exact rate of change.

84. $\lim_{h \to 0} \dfrac{f(x+h) - f(x)}{h} = 5x + 2$

85. $\lim_{h \to 0} \dfrac{f(x+h) - f(x)}{h} = 11$

86. $\dfrac{f(x+h) - f(x)}{h} = 6x + 4h + 5$

87. $\dfrac{f(x+h) - f(x)}{h} = 5x^2 + 3hx + 5h - 4$

88. $\lim_{h \to 0} \dfrac{f(x+h) - f(x)}{h} = x^2 - 8x - 152$

For Exercises 89–90, sketch the graph of a function so that the instantaneous rate of change of y at $x = a$ satisfies the given conditions.

89. The instantaneous rate of change of c is approximate; the actual rate of change is less than the instantaneous rate of change of c.

90. The instantaneous rate of change of c is approximate; the actual rate of change is greater than the instantaneous rate of change of c.

For Exercises 91–94, find the average rate of change of $f(x)$ with respect to x on each interval.

91. $f(x) = 6x + 8$; on $[3, 5]$

92. $f(x) = 4x^2 + x + 2$; on $[1, 3]$

93. $f(x) = \sqrt{x}$; on $[9, 16]$

94. $f(x) = \dfrac{x^2 - 4}{x + 2}$; on $[-1, 1]$

For Exercises 95–100, find the instantaneous rate of change of $f(x)$ of each function at the specified value of x.

95. $f(x) = x^2 + x - 4; \quad x = 2$

96. $f(x) = 3x^2 - 8x + 1; \quad x = 5$

97. $f(x) = -9x + 2; \quad x = 7$

98. $f(x) = 5x + 4; \quad x = 1$

99. $f(x) = \dfrac{x + 2}{x - 6}; \quad x = 5$

100. $f(x) = \dfrac{x - 1}{x + 4}; \quad x = 52$

101. ***Biology: Body and Heart Weight*** The function $y = 0.213x - 4.44$ expresses the relationship between the body weight y (in milligrams) and the heart weight x (in milligrams) of 10-month-old diabetic offspring of crossbred male mice.

a. Find and interpret the average rate of change of y of body weight with respect to heart weights for a change in heart weights from 195 mg to 235 mg.

b. Find and interpret the instantaneous rate of change of y of body weight with respect to heart weights for heart weights of 210 mg.

c. Find the actual rate of change of body weight with respect to heart weights for a change in heart weights from 210 mg to 211 mg.

102. ***Business: Reading Program*** The function $G = 26.68W - 7.44$ relates the approximate speed gain G (in words per minute) for a student enrolled in a particular speed reading program for W weeks.

a. Find and interpret the average rate of change of G with respect to the W for a change in W from 6 weeks to 8 weeks.

b. Find and interpret the instantaneous rate of change of G with respect to W at $W = 6$.

c. Find the actual rate of change of speed gain with respect to number of weeks in the program between 6 and 8 weeks.

103. ***Economics: City Population*** Analysts from a national real estate company have determined that t years from now the population P of a particular city can be found using the function $P(t) = t^2 - 14t + 12{,}040$.

a. Find and interpret the average rate of change of P of the city's population with respect to time for a change in time from 3 years to 7 years.

b. Find and interpret the instantaneous rate of change of $P(t)$ with respect to t at $t = 3$.

c. Find the actual rate of change of $P(t)$ with respect to t between $t = 3$ and $t = 4$.

104. ***Economics: City Air Quality*** An environmental group studying the air quality of a particular city has determined that t years from now, the amount A (in parts per million) of carbon monoxide in the city's air can be found using the function $A(t) = 0.3t^2 + 0.1t + 4$.

a. Find and interpret the average rate of change of $A(t)$ with respect to t for a change in t from 7 years to 10 years.

b. Find and interpret the instantaneous rate of change of $A(t)$ with respect to t at $t = 4$.

c. Find the actual change of the amount of carbon monoxide in the air between 4 and 5 years.

CHAPTER 3

Differentiation: The Language of Change

3.1 The Derivative of a Function and Two Interpretations

Introduction

This section presents the derivative, one of two main operations of calculus. The derivative is the calculus instrument used for examining change. This section presents two ways of interpreting the derivative of a function, one as the slope of the tangent line to a curve and the other as the instantaneous rate of change of the function.

The Derivative of a Function

We have discovered that for a function $f(x)$, the instantaneous rate of change of f with respect to x at $x = a$ is the slope of the tangent line to $f(x)$ at $x = a$, which is given by the function

$$m_{tan} = \lim_{h \to 0} \frac{f(x+h) - f(x)}{h}$$

For example, if $f(x) = x^2 + 6x - 2$, the instantaneous rate of change of f with respect to x is given by the function $m_{tan} = 2x + 6$. (Can you verify this? See Section 2.5 for a review of the meaning of this statement and the mechanics of carrying it out. It's important!)

Derived Function and Differentiation

This new function — the instantaneous rate of change function — that is *derived* from the original function $f(x)$, is called the **derivative of f(x)** because it is derived from $f(x)$ using the slope formula. A common notation for the derivative of the function $f(x)$ is $f'(x)$ (read *f prime of x*). The process of determining the derivative of a function is called **differentiation**. In a problem in which change is encountered, differentiation should come to mind because, mathematically, *differentiation is the language of change.*

Limit Definition of the Derivative

The *derivative* of the function $f(x)$, denoted $f'(x)$, is

$$f'(x) = \lim_{h \to 0} \frac{f(x+h) - f(x)}{h}$$

provided this limit exists. For all points where the limit exists, $f(x)$ is said to be *differentiable.*

Since $f'(x) = \lim_{h \to 0} \frac{f(x+h) - f(x)}{h}$ and $m_{tan} = \lim_{h \to 0} \frac{f(x+h) - f(x)}{h}$, we have $f'(x) = m_{tan}$, so that we can interpret the derivative in two ways.

Interpreting the Derivative

The derivative $f'(x)$ of the function $f(x)$

1. is the slope of the tangent line to $f(x)$.
2. is the instantaneous rate of change of f with respect to x.

As we saw in Section 2.5, the instantaneous rate of change of a function, when working with an interval of length 1, can produce the exact or approximate rate of change.

Exact and Approximate Rates of Change

1. If the rate of change remains constant as x increases one unit from $x = a$ to $x = a + 1$, the derivative, $f'(x)$, produces the exact rate of change. (The expression for $f'(x)$ will involve no variable.)
2. If the rate of change varies as x increases one unit from $x = a$ to $x = a + 1$, the derivative, $f'(x)$, produces the approximate rate of change. (The expression for $f'(x)$ will involve variables.)

Intervals of length 1 are very important in business applications. Their importance is described in the definition of the term *marginal*.

Marginal Change

The change in the output variable that is produced by a 1-unit change in the input variable is called **marginal** change in the output.

For example, a manufacturer that is currently producing and selling x units of a commodity realizes a revenue of $R(x)$ dollars from the sale of those x units. If the manufacturer increases the number of sales by 1, from x to $x + 1$, the marginal revenue is approximated by $R'(x)$ and is the amount of money realized from the sale of that $(x + 1)$th unit. It is important to understand that marginal change provides only an approximation to the actual change because it uses and interprets the derivative of a function over an interval of length 1 rather than at a single point.

There are several notations for the derivative of a function $f(x)$. Depending on the situation, one notation or description may be more useful than another.

Derivative Notations

If $y = f(x)$, then each of the following notations represents the derivative of $f(x)$.

$$f'(x) \qquad y' \qquad \frac{dy}{dx} \qquad \frac{df}{dx} \qquad \frac{d}{dx}f$$

The notation $\dfrac{dy}{dx}$ is read as *dee y dee x* and was developed by Godfried Leibniz. For now, $\dfrac{dy}{dx}$ is to be considered only a notation, not a fraction. It indicates the derivative of y with respect to x, not the operation dy divided by dx.

Derivative Evaluation Notations

Common notations indicating the value of the derivative at $x = a$ are

$$f'(a) \qquad \left.\frac{df}{dx}\right|_{x=a} \qquad \left.\frac{dy}{dx}\right|_{x=a} \qquad y'|_{x=a}$$

EXAMPLE SET A

Find and interpret

1. the derivative of $f(x) = x^2 + 6x - 2$.
2. $f'(5)$.

Solution:

1. Using the definition of the derivative, we have

$$\begin{aligned}
f'(x) &= \lim_{h\to 0}\frac{f(x+h)-f(x)}{h} \\
&= \lim_{h\to 0}\frac{\overbrace{(x+h)^2+6(x+h)-2}^{f(x+h)}-\overbrace{(x^2+6x-2)}^{f(x)}}{h} \\
&= \lim_{h\to 0}\frac{x^2+2hx+h^2+6x+6h-2-x^2-6x+2}{h} \\
&= \lim_{h\to 0}\frac{2hx+h^2+6h}{h} \\
&= \lim_{h\to 0}\frac{h(2x+h+6)}{h} \\
&= \lim_{h\to 0}\frac{\not{h}(2x+h+6)}{\not{h}} && \text{since } h \neq 0 \\
&= \lim_{h\to 0}(2x+h+6) && \text{Take the limit by substitution.} \\
&= 2x+6
\end{aligned}$$

Thus, $f'(x) = 2x + 6$.

Interpretation: $f'(x) = 2x+6$ gives the instantaneous rate of change of f with respect to x. If you are working with 1-unit changes in the input variable, then $f'(x) = 2x + 6$ provides an approximation to the actual change. (The change is approximate because $f'(x)$ involves a variable.) If this function came from a business application, we might say that $f'(x)$ produces the marginal change in the function.

2. To find $f'(5)$, we will substitute 5 for x in $2x + 6$.

$$\begin{aligned} f'(x) &= 2x + 6 \\ f'(5) &= 2(5) + 6 \\ &= 10 + 6 \\ &= 16 \quad \text{or} \quad \frac{16}{1} \end{aligned}$$

Interpretation: If x increases 1 unit in value, from 5 to 6, $f(x)$ will increase by approximately 16 units.

Differentials

In Example 2 of Example Set A, we noted that for $f'(x) = 2x + 6$, $f'(5) = 16 = \frac{16}{1}$ meant that as x increased by 1 unit, from 5 to 6, $f(x)$ increases by approximately 16 units. But what happens if x increases by more than 1 unit, or less than 1 unit, or even decreases? We can answer this question using the equality property of fractions.

Suppose that for the function $f(x) = 2x + 6$, x increases 3 units from 5 to 8. Then

$$\frac{16}{1} = \frac{16}{1} \cdot \frac{3}{3} = \frac{48}{3}$$

and we conclude that $f(x)$ increases by approximately 48 units.

If x increases by $\frac{1}{2}$ unit, then

$$\frac{16}{1} = \frac{16}{1} \cdot \frac{1/2}{1/2} = \frac{8}{1/2}$$

and we conclude that $f(x)$ increases by approximately 8 units.

If x decreases by 2 units, then

$$\frac{16}{1} = \frac{16}{1} \cdot \frac{-2}{-2} = \frac{-32}{-2}$$

and we conclude that $f(x)$ decreases by approximately 32 units. (Be careful with this interpretation. The statement that $f(x)$ decreases by approximately -32 units contains a double negative — decreases by a negative — and actually means $f(x)$ increased. Such a statement is incorrect.)

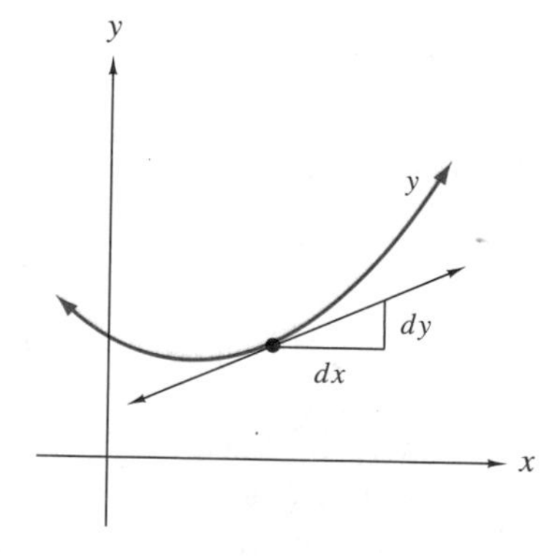

Figure 3.1

The $\frac{dy}{dx}$ notation is especially descriptive in these instances. The denominator, dx, can be taken to represent the change in the x value, and the numerator, dy, the approximate change in the function value. Figure 3.1 illustrates these changes.

Notice that $f'(x) = \frac{dy}{dx}$. That is, the slope of the tangent line, $f'(x)$, equals the rise, dy, over the run, dx.

Differential

The quantity dx is called the **differential of x**, and represents the change in the x value. The quantity dy is called the **differential of y**, and represents the *approximate* change in the function, or y value. The actual change in the y value as x changes from x_1 to x_2 is given by $\Delta y = f(x_2) - f(x_1)$. You can see in the figure that the smaller the change in x, the better dy approximates Δy.

Conditions for Nondifferentiability

A function $f(x)$ will not be differentiable at $x = a$ (that is, $f'(x)$ will not exist at $x = a$) if any of the following three conditions exists.

1. $f(x)$ is not continuous at $x = a$. See Figure 3.2.

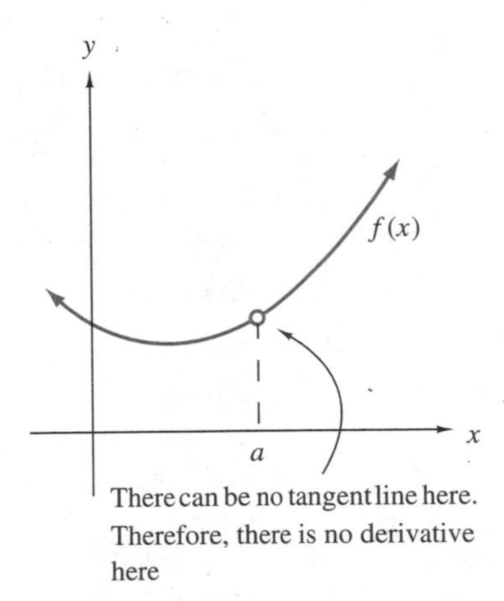

Figure 3.2

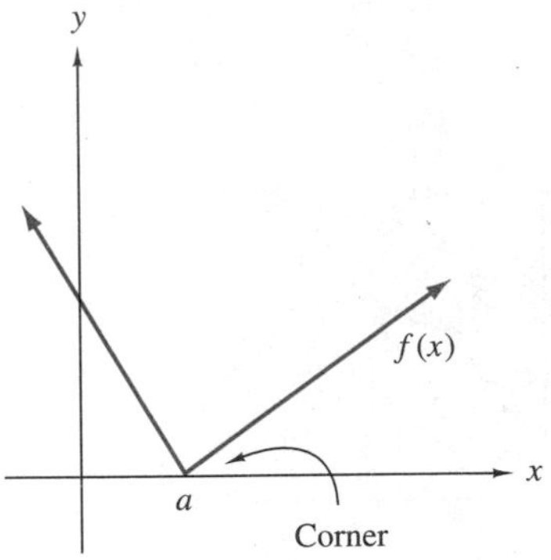

Condition 1 for Nondifferentiability

If $f(x)$ is discontinuous at $x = a$, then $f'(x)$ does not exist at $x = a$.

For example, $f(x) = \dfrac{3x}{x-4}$ is discontinuous at $x = 4$. Therefore, $f'(4)$ does not exist.

A function may be continuous at a point $x = a$, but may not be differentiable there. Continuity does not imply differentiability. Conditions 2 and 3 illustrate this point.

y

f(x)

x

a

Corner

Figure 3.3

2. $f(x)$ has a corner (Figure 3.3) or a cusp (Figure 3.4) at $x = a$.
To see why this is true, recall that

$$f'(x) = \text{the slope of the tangent line to } f(x)$$

$$= \lim_{h \to 0} \frac{f(x+h) - f(x)}{h}$$

which involves a limit. But the limit of a function is unique, and there is no unique tangent line at a corner. Even though there are two potential tangent lines—one as the corner is approached from the left, and one as the corner is approached from the right—there is no tangent line right at the corner. This is because the tangent line, just like the derivative, must be unique if it exists. (See Figure 3.4.)

Condition 2 for Nondifferentiability

If $f(x)$ has a corner or a cusp at $x = a$, $f'(x)$ does not exist at $x = a$.

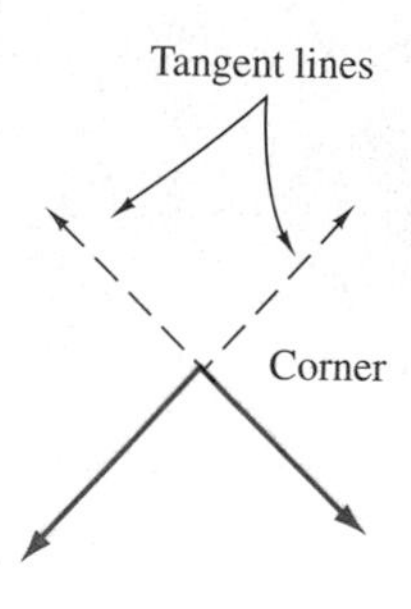

Figure 3.4

For example, $f(x) = |x - 2|$ has a corner at $x = 2$. Therefore, $f'(2)$ does not exist. Although $f(x)$ is continuous at $x = 2$, $f'(x) = \dfrac{-1}{(x-2)^2}$ is not defined at $x = 2$. This shows that the continuity of the function does not imply the existence of the derivative.

3. $f(x)$ has a vertical tangent line at $x = a$.

Remember, vertical tangent lines have undefined slopes. The graph on the left of Figure 3.5 shows a point, called a cusp, on a continuous curve where there is a vertical tangent line, and the graph on the right of Figure 3.5 shows another situation of a continuous curve where there is a vertical tangent line.

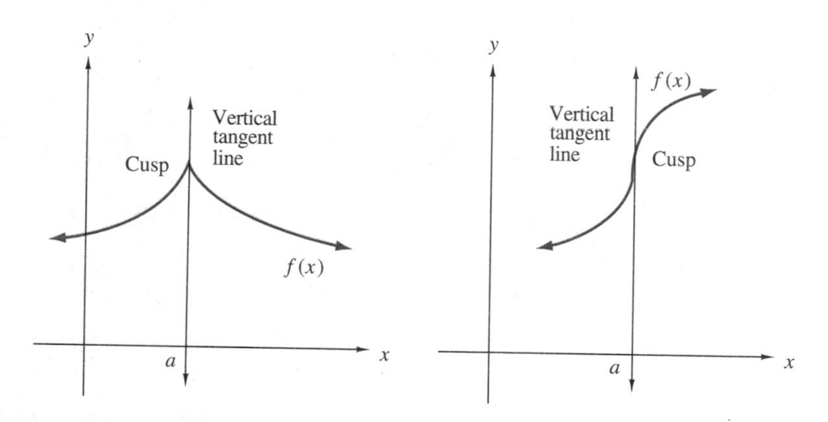

Figure 3.5

Condition 3 for Nondifferentiability

If $f(x)$ has a vertical tangent at $x = a$, $f'(x)$ does not exist at $x = a$.

For another example, $f(x) = \sqrt{x}$ has a vertical tangent at $x = 0$. (See Figure 3.6.)

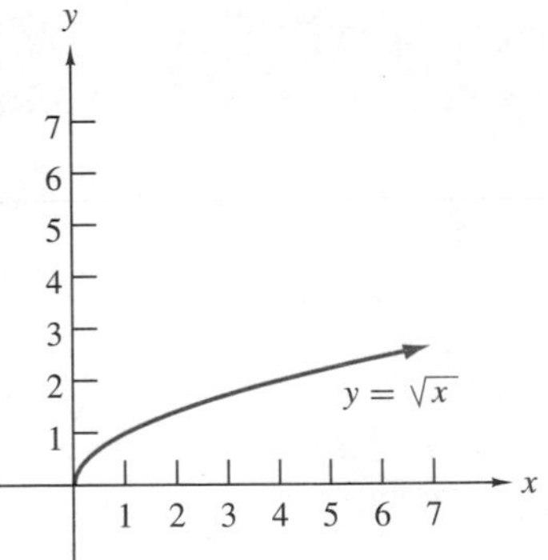

The y axis is a vertical asymptote.

Figure 3.6

Fundamental Differentiation Rules

Finding the derivative of a function $f(x)$ using the limit definition

$$\lim_{h \to 0} \frac{f(x+h) - f(x)}{h}$$

can be cumbersome and time consuming; it is not the most efficient use of our time and energy. By using the limit definition, however, we could prove differentiation rules that are both manageable and efficient. Since we are concerned more with the applications of calculus than the theory on which it is built, we will simply state these rules without proof.

Derivative of a Constant

If $y = c$, where c is any real number, then $\dfrac{dc}{dx} = 0$.

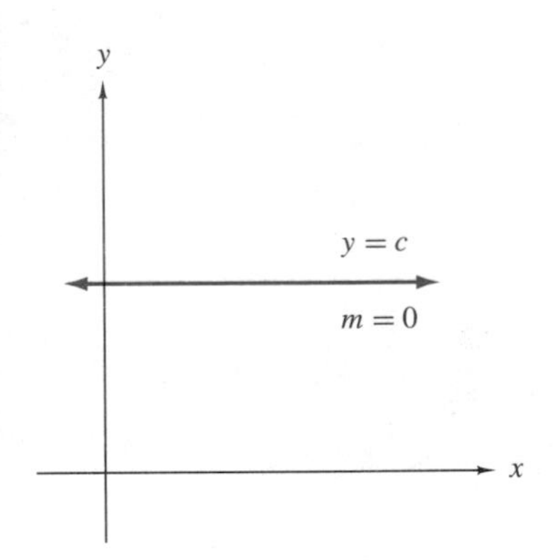

Figure 3.7

Alternatively, if the function is expressed as $f(x) = c$, then $f'(x) = 0$. This derivative rule states that the derivative of *any* constant is zero.

This result is understood geometrically by observing that the graph of $y = c$ is a horizontal line passing through c on the y axis, and that the slope of a horizontal line is 0. Also, an increase of 1 unit in any x value will produce no change in the y value. (See Figure 3.7.)

ILLUMINATOR SET A

1. If $f(x) = 4$, then $f'(x) = 0$.

   Interpretation: For any value of x, if x increases 1 unit in value to $x + 1$, then $f(x)$ will change by exactly 0 units in value. That is, a 1-unit increase in the x value produces no change at all in the $f(x)$ value.

2. If $f(x) = -16$, find $f'(2)$.
   Since $f(x) = -16$, a constant, $f'(x) = 0$ for every value of x. In particular, $f'(2) = 0$.

   Interpretation: If x increases 1 unit, from 2 to 3, $f(x)$ does not change at all in value; $f(2) = -16$ and $f(3) = -16$. In fact, $f(\text{any number}) = -16$.

Derivative of a Simple Power

If $y = x^n$, where n is any real number, then $\frac{dy}{dx} = n \cdot x^{n-1}$.

Alternatively, if the function were expressed as $f(x) = x^n$, then $f'(x) = n \cdot x^{n-1}$. This derivative rule states that the derivative of x^n is found by decreasing the exponent on x by 1, then multiplying x by the original exponent.

ILLUMINATOR SET B

1. If $y = x^2$, then $\frac{dy}{dx} = 2x^{2-1} = 2x^1 = 2x$.
2. If $f(x) = x^5$, then $f'(x) = 5x^{5-1} = 5x^4$.
3. If $f(x) = \frac{1}{x^4}$, find $f'(x)$.

   Solution:
   We must try to make $f(x)$ match one of our derivative forms. We can get $f(x) = \frac{1}{x^4}$ into the form $f(x) = x^n$ by using the rules of exponents. $f(x) = \frac{1}{x^4} = x^{-4}$. Now, we can differentiate $f(x)$ as follows.

$$\begin{aligned} f'(x) &= -4x^{-4-1} \\ &= -4x^{-5} \\ &= \frac{-4}{x^5} \end{aligned}$$

4. If $f(x) = x^{2/3}$, find $f'(x)$.

   Solution:

$$\begin{aligned} f'(x) &= \frac{2}{3}x^{(2/3)-1} \\ &= \frac{2}{3}x^{-1/3} \\ &= \frac{2}{3x^{1/3}} \end{aligned}$$

5. If $y = x = x^1$, then $\frac{dy}{dx} = 1x^{1-1} = 1x^0 = 1 \cdot 1 = 1$.

Result 5 above occurs so often that we make a special note of it.

The Derivative of $y = x$

If $y = x$, then $y' = 1$.

Derivative of a Constant Times a Function

If $f(x)$ is a differentiable function, then $\frac{d}{dx}cf(x) = c \cdot \frac{d}{dx} f(x)$.

This rule states that to differentiate a function that is multiplied by a constant, we differentiate the function then multiply it by the constant.

ILLUMINATOR SET C

1. If $y = 6x^4$, then

$$\begin{aligned}\frac{dy}{dx} &= 6 \cdot \frac{d}{dx}x^4 \\ &= 6 \cdot 4x^{4-1} \\ &= 6 \cdot 4x^3 \\ &= 24x^3\end{aligned}$$

Interpretation: If x increases 1 unit in value, from, say, $x = a$ to $x = a + 1$, then $f(x)$ will increase or decrease in value by approximately $24a^3$ units. $f(x)$ will increase if a is positive since a^3 will be positive, and decrease if a is negative since a^3 will be negative.

2. If $f(x) = \frac{4}{x^8}$, find $f'(x)$.

Solution:
We will write $f(x) = \frac{4}{x^8}$ in the more convenient form $f(x) = 4x^{-8}$.

$$\begin{aligned}f'(x) &= 4 \cdot (-8)x^{-8-1} \\ &= -32x^{-9} \\ &= \frac{-32}{x^9}\end{aligned}$$

3. If $y = cx$, where c is any real number, then

$$\frac{dy}{dx} = c \cdot \frac{d}{dx}x$$

$$= c \cdot 1$$

$$= c$$

Result number 3 in Illuminator Set C occurs so often that we make a special note of it.

The Derivative of $y = cx$

If $y = cx$, then $y' = c$.

ILLUMINATOR SET D

1. If $y = 8x$, then $y' = 8$.

2. If $f(x) = -14x$, then $f'(x) = -14$.

The Derivative of a Sum or a Difference

If $f(x)$ and $g(x)$ are both differentiable functions, then

$$\frac{d}{dx}[f(x) \pm g(x)] = \frac{d}{dx}f(x) \pm \frac{d}{dx}g(x)$$

This rule states that to differentiate the sum or difference of two functions (that are themselves differentiable), we differentiate each function individually, then add or subtract the resulting derivatives.

In all the examples that follow, we will eliminate some of the intermediate steps.

EXAMPLE SET B

1. Find $\dfrac{dy}{dx}$ for $y = 5x^3 + 4x^2 + 9x - 7$.

Solution:

$$\frac{dy}{dx} = \frac{d}{dx}(5x^3) + \frac{d}{dx}(4x^2) + \frac{d}{dx}(9x) - \frac{d}{dx}(7)$$

Differentiate each individual function.

$$\frac{dy}{dx} = 15x^2 + 8x + 9 - 0 = 15x^2 + 8x + 9$$

2. If $y = 9\sqrt[3]{x^2}$, find $\left.\frac{dy}{dx}\right|_{x=8}$.

Solution:
We will begin by writing y in the more convenient form $y = 9x^{2/3}$, recalling that the notation $\left.\frac{dy}{dx}\right|_{x=8}$ means to evaluate $\frac{dy}{dx}$ at $x = 8$. Then

$$\frac{dy}{dx} = 6x^{-1/3}$$

$$= \frac{6}{x^{1/3}} = \frac{6}{\sqrt[3]{x}}$$

Now we evaluate the derivative at $x = 8$.

$$\left.\frac{6}{\sqrt[3]{x}}\right|_{x=8} = \frac{6}{\sqrt[3]{8}}$$

$$= \frac{6}{2} = 3$$

Interpretation: If x increases 1 unit in value, from 8 to 9, $f(x)$ will increase approximately 3 units in value.
Similarly, if x increases by $\frac{1}{2}$ unit, then

$$f'(8) = 3 = \frac{3}{1} = \frac{3}{1} \cdot \frac{1/2}{1/2} = \frac{3/2}{1/2}$$

so that $f(x)$ increases by approximately $\frac{3}{2}$ units.

EXAMPLE SET C

Health officials, concerned with a flu epidemic that has hit the Atlanta area, have estimated that the number of people ill with flu symptoms can be approximated by the function $N(t) = 90t^2 - t^3, 0 \leq t \leq 70$, where t is measured in days since the beginning of the epidemic.

1. At what rate are flu symptoms spreading on day 15 of the epidemic, and how many have been taken ill with flu symptoms as of day 15?
2. At what rate are flu symptoms spreading on day 65 of the epidemic?
3. When are the flu symptoms spreading at the rate of 1500 people per day?

Solution:

1. The rate at which the flu symptoms are spreading is given by $N'(t)$, the derivative of $N(t)$. Using the sum rule,

$$N'(t) = 180t - 3t^2, \qquad 0 \leq t \leq 70$$

To find the rate of change at $t = 15$, we compute

$$\begin{aligned} N'(15) &= 180(15) - 3(15)^2 \\ &= 2025 \end{aligned}$$

On day 15, there are

$$\begin{aligned} N(15) &= 90 \cdot 15^2 - 15^3 \\ &= 16{,}875 \end{aligned}$$

people ill with flu symptoms.

Interpretation: On day 15 of the epidemic, there are 16,875 people ill with flu symptoms. If t increases by 1, from 15 to 16, $N(t)$ increases by approximately 2025. That is, from day 15 to day 16 of the epidemic, we expect to see an increase of approximately 2025 cases of flu symptoms. The epidemic is on the rise. So, on day 16 of the epidemic, we expect to see approximately 16,875 + 2025 = 18,900 people ill with flu symptoms.

2. The word *rate* implies *change*, and *differentiation is the language of change.* Hence, we need to differentiate. To find the rate of change at $t = 65$, we compute

$$\begin{aligned} N'(65) &= 180(65) - 3(65)^2 \\ &= -975 \end{aligned}$$

Interpretation: From day 65 to day 66 of the epidemic, we expect to see a decrease of approximately 975 cases of flu symptoms. The epidemic is on the decline.

3. Since we are asked the question *when*, we need to find t. Specifically, we need to find t when $N'(t) = 1500$. We set $N'(t) = 180t - 3t^2$ equal to 1500 and solve for t.

$$\begin{aligned} 1500 &= 180t - 3t^2 \\ 3t^2 - 180t + 1500 &= 0 \\ 3(t^2 - 60t + 500) &= 0 \\ 3(t - 10)(t - 50) &= 0 \end{aligned}$$

The values 10 and 50 are solutions to this equation.

Interpretation: At both 10 days and 50 days after the beginning of the epidemic, the rate at which people are taken ill with flu symptoms is 1500 people per day.

Derivatives and Technology

Using Your Calculator You can use your graphing calculator to compute the numerical derivative of a function. This numerical derivative will be an approximation to the exact numerical derivative when the derivative is computed symbolically. The reason is that the calculator approximates the slope for the tangent line with the slope of a secant line that is just slightly different from the tangent line. The following entries show how to compute the derivative of $f(x) = 90x^2 - x^3$ at $x = 15$. This is the function of Example Set C.

```
Y1 = 90X^2 − X^3
Y5 = nDeriv(Y1,X,X,0.001)
```

Now, in the computation window, enter

```
Y5(15)
```

The calculator responds with the value 2024.999999, which you can approximate as 2025. Thus, as x increases 1 unit, from 15 to 16, the function increases approximately 2025 units.

You could also compute the value of the derivative at 15 by storing 15 to X and computing Y5 as follows. In the computation window, enter

```
15 → X
Y5
```

This will produce the value $2024.999999 \approx 2025$. Note that the actual change in the function value can be computed as

```
Y1(16) − Y1(15)
```

The actual change is 2069.

Calculator Exercises

1. Compute $f(2.6)$ and $f'(2.6)$ for $f(x) = x^{3.6} - x^{-1.6} + 5$.
2. Compute $f(8)$ and $f'(8)$ for $f(x) = \dfrac{4}{x-8} - \dfrac{10}{5.5x + 1.1}$.
3. According to the Center for Education Statistics in Washington, D.C., the total amount A, in millions of dollars, of money spent by U.S. primary and secondary private educational institutions between 1975 and 1989 can be approximated by the function

$$A(t) = 894.6t + 7717.4$$

where t is the number of years from 1975. Find and interpret both $A(5)$ and $A'(5)$ and then $A(10)$ and $A'(10)$. Based on this information, would you say that the total amount of money spent by U.S. primary and secondary private educational institutions is increasing, decreasing, or has stabilized? Do you recognize the value of $A'(5)$ and $A'(10)$? If so, what is it?

4. According to the American Bar Association, the number N of practicing lawyers t years from 1960 through 1985 can be approximated by the function

$$N(t) = 471t^2 + 2225t + 273{,}063$$

Find and interpret both $N(15)$ and $N'(15)$ and then $N(20)$ and $N'(20)$. Based on this information, would you say that the number of practicing lawyers is increasing, decreasing, or has stabilized?

5. According to Alex Brown and Sons, the total revenue R realized by U.S. car rental agencies between 1984 and 1990 can be approximated by the function

$$R(t) = \frac{1}{0.015t^2 - 0.326t + 2.39}$$

where t is the number of years from 1984. Find and interpret both $R(2)$ and $R'(2)$ and then $R(4)$ and $R'(4)$. Based on this information, would you say that the total revenue realized by U.S. car rental agencies is increasing, decreasing, or has stabilized?

Using Derive Derive can compute the exact derivative of a function. The following entries show the computation of $N'(x)$ and the evaluation $N'(15)$ for $N(x) = 90x^2 - x^3$. As you saw in Section 2.2, the **C**alculus command accesses Derive's built-in calculus functions. In this case that built-in function is **Differentiate**. Remember to press Enter at the end of each line.

Author
Declare
Function
n
90x^2 - x^3
Calculus
Differentiate (*Press* Enter *three times to differentiate with respect to x.*)
Simplify
Author
n(15)
Simplify

The derivative at $x = 15$ is displayed. This set of commands declares and names a function, n, differentiates it with respect to x, then evaluates it at $x = 15$.

Derive Exercises Use Derive to find each derivative and to perform each evaluation.

1. $f(x)$, $f(25)$, and $f'(25)$ for $f(x) = x^2 + 6x + 1000$. Interpret both these values.
2. $f(x)$, $f(6.25)$, and $f'(6.25)$ for $f(x) = -4.5x^3 - 3.2x - 11.6$. Interpret both these values.
3. The cost C to a manufacturer for producing x units of a product is given by the cost function

$$C(x) = 1.2x^3 - 20.4x^2 + 15x + 12.5$$

Use Derive to compute $C'(x)$, then compute and interpret both $C(40)$ and $C'(40)$.

EXERCISE SET 3.1

A UNDERSTANDING THE CONCEPTS

For Exercises 1–10, each derivative involves a variable.

1. ***Business: Profit from Advertising*** Suppose, for a particular item, $P(x)$ relates the profit P (in thousands of dollars) to the amount x (in thousands of dollars) spent on advertising. Interpret both $P(60) = 40$ and $\left.\frac{dP}{dx}\right|_{x=60} = 2$.

2. ***Anatomy: Animal Weight and Tail Thickness*** Suppose that for a particular animal, the animal's weight w (in pounds) and tail thickness x (in millimeters) are related by the function $w(x)$. Interpret both $w(30) = 20$ and $w'(30) = 0.76$.

3. ***Psychology: Task Proficiency*** Psychologists have determined that as a person repeatedly performs a task, he gets more proficient at it. Suppose that t hours are necessary to build N objects. The function $N(t)$ relates t and N. Interpret both $N(150) = 1200$ and $N'(150) = 21.25$.

4. ***Business: Compact Disc Sales*** For short periods of time, the number N (in thousands) of compact discs sold by a particular rock group each week is related to the number x of times the group's music video is played on MTV. For a particular type of rock music, N and x are related by the function $N(x)$. Interpret both $N(30) = 74{,}000$ and $N'(30) = 7{,}000$.

5. ***Education: Exam Scores*** An analysis by a group of educational psychologists shows that the average score S (in points) attained by students at a particular high school on a standardized exam is strongly related to the number n of teachers having college degrees in the subject they teach. The research indicates that the function $S(n)$ relates S and n. Interpret both $S(44) = 87$ and $S'(44) = 1.5$.

6. ***Business: Magazine Circulation*** A computer magazine publishing company has estimated that its national circulation C (in thousands) depends on the number x of reviews of current computer software it places in each issue. The function $C(x)$ relates C and x. Interpret both $C(12) = 2{,}000$ and $C'(12) = 20$.

7. ***Ecology: Air Pollution*** In a particular community in which a large chemical producer is located, under low wind conditions the function $d = f(t)$ relates the distance d (in miles) a cloud of toxic gas travels and the time t (in hours) since the release of the cloud. Interpret both $f(2) = 7$ and $f'(2) = 3$.

8. ***Business: Job Applications*** The director of the personnel office of a large company has determined that the function $N(x)$ approximates the relation between the number N of applicants applying for an accounting position and the amount of dollars x (in hundreds) it advertises as benefits. Interpret both $N(42) = 340$ and $\left.\frac{dN}{dx}\right|_{x=42} = 66$.

9. ***Economics: Inflation*** An economist is studying the graph of inflation versus time. The function relating the amount of inflation I (in percentage points) and time t (in months from the beginning of the year) is $I(t)$. What computation does the economist need to make to determine the rate of inflation in June ($t = 6$)?

10. ***Sociology: Interest in a Subject*** A sociologist claims that the level of interest L (on a scale of 0 to 10, 0 being low and 10 being high) an American has in AIDS is related to the number x of public service messages concerning AIDS the person sees. The sociologist has constructed the function $L(x)$ to approximate the relationship. What computation should the sociologist make to predict the number of points a person's interest level will change if the person has seen 65 AIDS messages?

For Exercises 11–14, explain why $f'(x)$ does not exist at each specified value of x.

11.

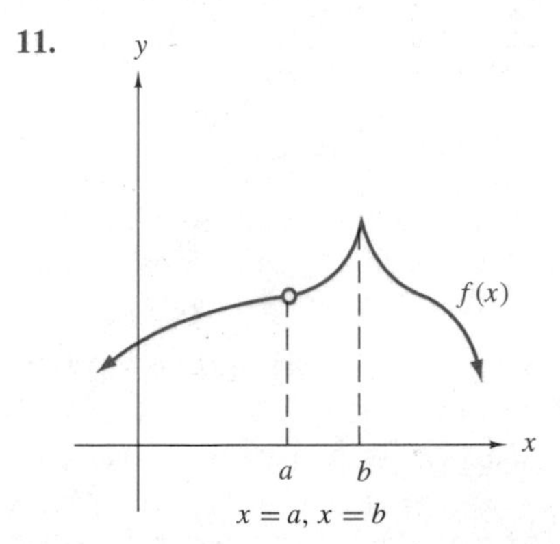

$x = a,\ x = b$

12.

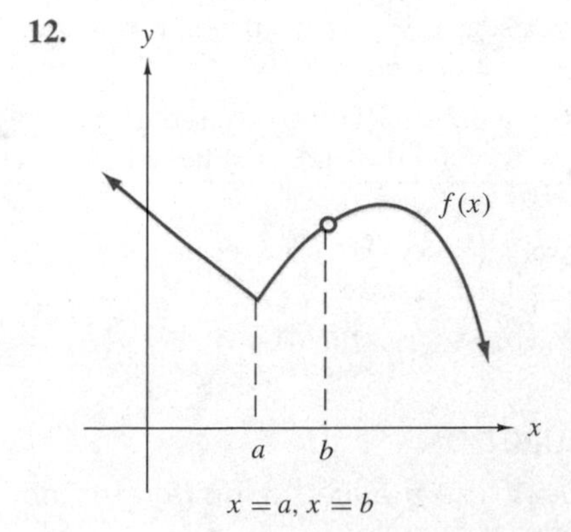

$x = a,\ x = b$

13.

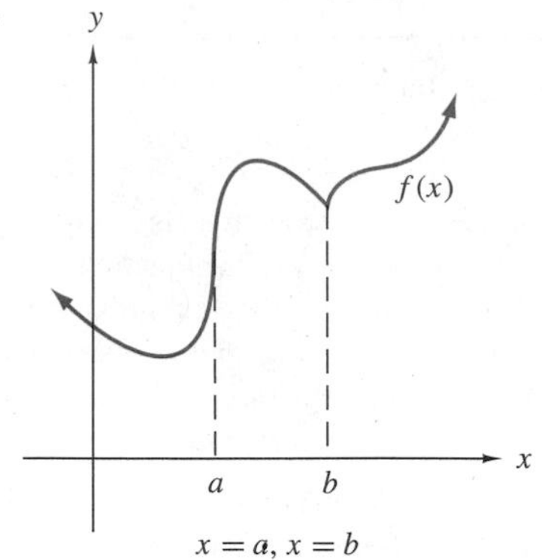

$x = a, x = b$

14.

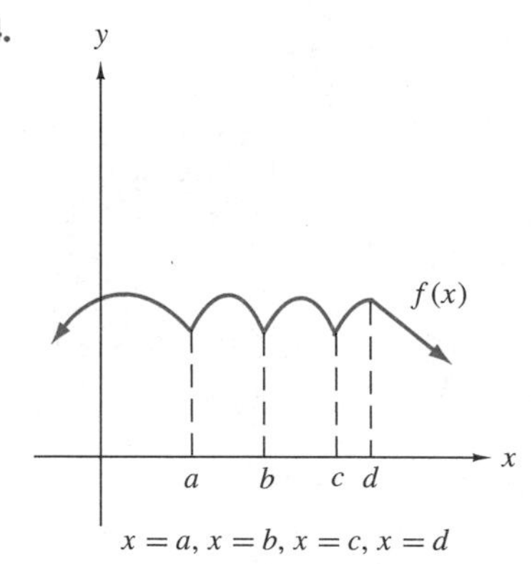

$x = a, x = b, x = c, x = d$

For Exercises 15–19, sketch the graph of a function that meets all of the specified conditions. (Since many curves are possible, answers may vary.)

15. $f(x)$ has a cusp at $(4, 6)$; $f'(7)$ does not exist because $f(x)$ is discontinuous at $x = 7$.

16. $f'(x)$ does not exist at $x = a$, but $f(x)$ is continuous at $x = a$; $f'(x)$ does not exist at $x = b$ because $f(x)$ is discontinuous at $x = b$; $f(x)$ is continuous at $x = c$, but $f'(x)$ does not exist at $x = c$ for a different reason than that $f'(x)$ does not exist at $x = a$.

17. $f'(a)$ does not exist because $f(x)$ has a vertical tangent at $x = a$; $f'(b) = 0$; $f'(c)$ does not exist because $f(x)$ is discontinuous at $x = c$.

18. $f'(a)$ is negative; $f'(b)$ does not exist because $f(x)$ has a cusp at $x = b$; $f'(c)$ is positive.

19. $f'(a)$ does not exist because $\lim_{x\to a^-} f(x) < \lim_{x\to a^+} f(x)$.

B SKILL ACQUISITION

For Exercises 20–23, use the limit definition to find the derivative of each function.

20. $f(x) = x^2 + 2x - 7$

21. $y = 5x^2 - x$

22. $y = 8x + 9$

23. $f(x) = \dfrac{3}{x}$

For Exercises 24–35, use the differentiation rules to find each derivative.

24. $f(x) = 10x^3 - 12x + 6$

25. $f(x) = 4x^4 + 2x^3 + 8x^2 - x - 10$

26. $f(x) = 8x^{5/2}$

27. $y = 9x^{2/3} + x + 1$

28. $y = 6x^{-3/2} + 4x^{1/2} + 8x^{3/2}$

29. $y = 10x^{-5/2} + 2x^{-7/2} - 4x^{-9/2}$

30. $y = \dfrac{1}{x}$

31. $y = \dfrac{5}{x^3} + \dfrac{1}{x} - \dfrac{1}{7}$

32. $f(x) = \dfrac{2}{x^3} + \dfrac{3}{x^2} + \dfrac{3}{x} - \dfrac{1}{4}$

33. $f(x) = \sqrt[3]{x} + \sqrt[4]{x^3} - x - 1$

34. $f(x) = \dfrac{-2}{\sqrt[5]{x^2}} + \dfrac{5x^{4/5}}{\sqrt[5]{x^3}}$

35. $y = 12$

For Exercises 36–41, find what is requested.

36. For $f(x) = x^2 + 6x - 4$, find $f(3)$, $f'(3)$, and $f(4) - f(3)$.

37. For $f(x) = 3x^2 + 11x + 1$, find $f(5)$, $f'(5)$, and $f(6) - f(5)$.

38. For $y = 2x^4 + 3x^3 + x + 16$, find $\left.\dfrac{dy}{dx}\right|_{x=0}$.

39. For $y = 8x^3 - 8x + 2$, find $\left.\dfrac{dy}{dx}\right|_{x=-1}$.

40. Find all the values of x where the slope of the tangent line to $f(x) = x^2 + 6x + 1$ is 0. (Hint: Find $f'(x)$, then set it equal to 0 and solve for x.)

41. Find all the values of x where the slope of the tangent line to $f(x) = 4x^2 + 2x + 3$ is 0.

C APPLYING THE CONCEPTS

42. ***Anatomy: Surface Area and Mass*** The function $A(m) = 0.11m^{2/3}$ approximates the relationship between the surface area A (in square meters) of a person and his mass m (in kilograms).

a. Find the approximate surface area of a man whose mass is 54 km; that is, find $A(54)$. (54 kg $\approx$ 120 pounds)

b. Find and interpret $A'(54)$.

43. ***Business: Winery Production*** A vat in a winery has wine flowing into it for fermentation. The amount A (in gallons) of wine in the vat after t minutes is given by the function $A(t) = 40t - 30\sqrt{t}$, $0 \le t \le 25$.

a. How much wine is in the vat after 16 minutes?

b. Approximately how many gallons of wine can be expected to flow into the vat between minutes 16 and 17? (Hint: Use $A'(t)$.)

44. ***Biology: Heart Rate*** Biologists have determined that the relationship between heart rate h (in beats per minute) and body weight w (in pounds) is approximated by the function $h(w) = \dfrac{250}{w^{1/4}}$.
    a. Find the heart rate for a 16-pound baby.
    b. Find the heart rate for a 256-pound man.
    c. Find and interpret $h'(w)$ for an 81-pound child. Express $h'(81)$ as a decimal rounded to two decimal places.

45. ***Medicine: Detection of a Tumor*** For a certain type of tumor, the weight w (in grams) of the tumor t months from the time of detection is given by the function $w(t) = 0.03t^3 + 0.006$. By how much can the weight of the tumor be expected to change in the next month if it has been 4 months since the detection of the tumor?

46. ***Life Science: Mound Occupation*** Mammologists (scientists who study mammals) have determined that the number N of animals occupying a mound is approximately related to the area a (in square meters) of the mound by the function $N(a) = 0.0071a + 0.60$. If the area of a mound is 110 square meters, how many more animals would be expected to occupy the mound
    a. if the area of the mound were increased to 111 square meters?
    b. if the area of the mound were decreased to 109 square meters?

47. ***Business: Advertising Revenue*** The function $R(x) = 85 + 7.130\sqrt{x}$, $0 \le x \le 15{,}000$, indicates that the monthly revenue R (in thousands of dollars) depends on the amount of dollars x spent on advertising each month.
    a. By how much would the monthly revenue be expected to change if the monthly expenditure on advertising were to be raised from its current level of \$6,000 to \$7,000?
    b. What is the revenue when the amount spent on advertising is \$6,000?

48. ***Medicine: Glucose Tolerance*** To test for hypoglycemia, physicians often have the patients undergo a glucose tolerance test. For one such test, the amount A (in milligrams) of glucose remaining in one cubic centimeter of blood t hours after ingestion of the glucose is given by the function $A(t) = 3.1 + \dfrac{4.8}{\sqrt{t}}$, $0 \le t \le 6$.
    a. Find the amount of glucose in the one cubic centimeter of blood 2 hours after ingestion of the glucose.
    b. Find the rate at which the concentration of glucose is changing 2 hours after ingestion of the glucose. Interpret this result.

49. ***Business: Inventory Costs*** The manager of a large department store has determined that the equation $C(x) = \dfrac{203{,}000}{x} + 2x$ relates the inventory cost C (in dollars) to the number of packages x of men's socks it stocks. The manager is currently ordering 317 packages of socks 18 times each year. What change in the inventory costs can the manager expect if she changes her order to 318 packages of socks 18 times each year?

D DESCRIBE YOUR THOUGHTS

50. As if you were explaining to a friend who has not yet taken a calculus course, explain what the derivative of a function is.

51. Describe the conditions under which the derivative of a function gives the exact change of the function or the approximate change of the function.

52. Explain why functions having cusps or corners do not have derivatives at those points.

53. Explain, with a geometric reference, why the derivative of a constant is zero.

E REVIEW

54. **(2.2)** Find, if it exists, $\lim_{x\to 2} \dfrac{x^2 + 4x - 12}{x - 2}$.

55. **(2.2)** Find, if it exists, $\lim_{x\to 3} \dfrac{x^2 - 3x - 4}{x + 1}$.

56. **(2.3)** Find, if it exists, $\lim_{x\to\infty} \dfrac{5x^2 + 2x + 1}{x + 6}$.

57. **(2.4)** Discuss the continuity of
$$f(x) = \begin{cases} \dfrac{x^2 + x - 12}{x - 3}, & \text{if } x \ne 3 \\ 7, & \text{if } x = 3 \end{cases} \quad \text{at } x = 3.$$

58. **(2.5)** Find and interpret the instantaneous rate of change of $f(x) = 3x^2 + 8x - 4$ at $x = 2$.

3.2 Differentiating Products and Quotients

Introduction
The Product Rule
The Quotient Rule
Products, Quotients, and Technology

Introduction

The differentiation rules we examined in Section 3.1 significantly increased our ability to differentiate functions. The product and quotient rules we will examine in this section will extend our differentiation ability to more sophisticated functions. The rules for the derivatives of products and quotients of functions are not as intuitive as are the rules for the derivatives of sums and differences.

The Product Rule

To differentiate the product of two differentiable functions, use the product rule.

Product Rule

If $u(x)$ and $v(x)$ are differentiable functions, then

$$\frac{d}{dx}[u(x)\cdot v(x)] = \frac{d}{dx}[u(x)]\cdot v(x) + u(x)\cdot\frac{d}{dx}[v(x)]$$

In the prime notation, the product rule is written as

$$\frac{d}{dx}[u(x)\cdot v(x)] = u'(x)\cdot v(x) + u(x)\cdot v'(x)$$

The product rule states that the derivative of a product is

(the derivative of the first factor) *times* (the second factor)

plus

(the first factor) *times* (the derivative of the second factor)

In symbols,

$$\frac{d}{dx}[u(x)v(x)] = \underbrace{u'(x)}_{(first)'}\cdot\underbrace{v(x)}_{(second)} + \underbrace{u(x)}_{(first)}\cdot\underbrace{v'(x)}_{(second)'}$$

EXAMPLE SET A

Find the derivative of $f(x) = (5x^2 + 4)(x^3 + 11)$.

Solution:
If we identify $u(x) = 5x^2 + 4$ and $v(x) = x^3 + 11$, we have the product $f(x) = u(x)v(x)$, and

$$\begin{aligned} f'(x) &= \underbrace{u'(x)}_{(\textit{first})'} \cdot \underbrace{v(x)}_{(\textit{second})} + \underbrace{u(x)}_{(\textit{first})} \cdot \underbrace{v'(x)}_{(\textit{second})'} \\ &= \underbrace{(10x)}_{(\textit{first})'} \cdot \underbrace{(x^3 + 11)}_{(\textit{second})} + \underbrace{(5x^2 + 4)}_{(\textit{first})} \cdot \underbrace{(3x^2)}_{(\textit{second})'} \\ &= 10x^4 + 110x + 15x^4 + 12x^2 \\ &= 25x^4 + 12x^2 + 110x \end{aligned}$$

Thus, $f'(x) = 25x^4 + 12x^2 + 110x$.

In addition to illustrating the product rule, Example Set A illustrates two important facts.

1. The derivative of a product could also be found by performing the multiplication, then differentiating term-by-term using the sum rule.

$$\begin{aligned} f(x) &= (5x^2 + 4)(x^3 + 11) \\ &= 5x^5 + 4x^3 + 55x^2 + 44 \end{aligned}$$

so that

$$f'(x) = 25x^4 + 12x^2 + 110x,$$

as before.
It is not always practical to perform the multiplication. For example, it is impractical to actually multiply out $(5x + 4)^7(x^3 + 11)^{-2/3}$.

2. A note of caution: the derivative of a product is *not* the product of the individual derivatives. That is,

$$\frac{d}{dx}[u(x)v(x)] \neq u'(x) \cdot v'(x)$$

For example,

$$f(x) = (5x^2 + 4)(x^3 + 11)$$

then

$$f'(x) \neq (10x)(3x^2)$$

The Quotient Rule

To differentiate the quotient of two differentiable functions, use the quotient rule.

Quotient Rule

If $u(x)$ and $v(x)$ are differentiable functions and $v(x) \neq 0$, then

$$\frac{d}{dx}\left[\frac{u(x)}{v(x)}\right] = \frac{v(x) \cdot \frac{d}{dx}[u(x)] - u(x) \cdot \frac{d}{dx}[v(x)]}{[v(x)]^2}$$

In the prime notation, the quotient rule is written as

$$\frac{d}{dx}\left[\frac{u(x)}{v(x)}\right] = \frac{v(x) \cdot u'(x) - u(x) \cdot v'(x)}{[v(x)]^2}$$

The quotient rule states that the derivative of a quotient is

$$\frac{d}{dx}\left[\frac{\text{numerator}}{\text{denominator}}\right] = \frac{(\text{denom.}) \cdot (\text{numer.})' - (\text{numer.}) \cdot (\text{denom.})'}{(\text{denom.})^2}$$

or (less technical, but easier to say),

$$\frac{d}{dx}\left[\frac{\text{top}}{\text{bottom}}\right] = \frac{(\text{bottom}) \cdot (\text{top})' - (\text{top}) \cdot (\text{bottom})'}{(\text{bottom})^2}$$

EXAMPLE SET B

Find $f'(2)$ for $f(x) = \dfrac{x^3 + 2x - 1}{x^2 - 1}$.

Solution:
We first recall that $f'(2)$ indicates that the derivative of $f(x)$ is to be evaluated at $x = 2$. Thus, initially, we need to find $f'(x)$. Since $f(x)$ is a quotient of functions, we will use the quotient rule. If we identify $u(x) = x^3 + 2x - 1$ and $v(x) = x^2 - 1$, we have the quotient $f(x) = \dfrac{u(x)}{v(x)}$.

$$f'(x) = \frac{v(x) \cdot u'(x) - u(x) \cdot v'(x)}{[v(x)]^2}$$

or

$$f'(x) = \frac{(\text{bottom}) \cdot (\text{top})' - (\text{top}) \cdot (\text{bottom})'}{(\text{bottom})^2}$$

$$= \frac{\overbrace{(x^2 - 1)}^{(bottom)}\overbrace{(3x^2 + 2)}^{(top)'} - \overbrace{(x^3 + 2x - 1)}^{(top)}\overbrace{(2x)}^{(bottom)'}}{(x^2 - 1)^2}$$

Now we simplify using the distributive property:

$$= \frac{3x^4 - x^2 - 2 - 2x^4 - 4x^2 + 2x}{(x^2 - 1)^2}$$

$$= \frac{x^4 - 5x^2 + 2x - 2}{(x^2 - 1)^2}$$

Thus,

$$f'(x) = \frac{x^4 - 5x^2 + 2x - 2}{(x^2 - 1)^2}$$

Now we can evaluate $f'(x)$ at $x = 2$.

$$f'(2) = \frac{2^4 - 5 \cdot 2^2 + 2 \cdot 2 - 2}{(2^2 - 1)^2}$$

$$= \frac{16 - 20 + 4 - 2}{9}$$

$$= \frac{-2}{9}$$

In addition to illustrating the quotient rule, Example Set B illustrates an important fact.

The derivative of a quotient is *not* the quotient of the individual derivatives of the function. That is,

$$\frac{d}{dx}\left[\frac{u(x)}{v(x)}\right] \neq \frac{u'(x)}{v'(x)}$$

For example, if

$$f(x) = \frac{x^3 + 2x - 2}{x^2 - 1}$$

then

$$f'(x) \neq \frac{3x^2 + 2}{2x}$$

EXAMPLE SET C

The campaign committee for a politician estimates that in an upcoming national election the politician can, by campaigning for x days is a particular county, secure $2.3x$ (in thousands) of votes. However, campaigning costs money and the committee estimates that the cost for campaigning x days will be $7.1x^2 + 210$ dollars. The function

$$f(x) = \frac{\text{number of votes}}{\text{cost for those votes}} = \frac{2.3x}{7.1x^2 + 210}$$

produces the number of votes per dollar spent. Find both $f'(4)$ and $f'(6)$.

Solution:
Since $f(x)$ is a quotient, we'll use the quotient rule. If we identify $u(x) = 2.3x$ and $v(x) = 7.1x^2 + 210$, we have

$$f'(x) = \frac{v(x) \cdot u'(x) - u(x) \cdot v'(x)}{[v(x)]^2}$$

or

$$f'(x) = \frac{(\text{bottom}) \cdot (\text{top})' - (\text{top}) \cdot (\text{bottom})'}{(\text{bottom})^2}$$

$$= \frac{\overbrace{(7.1x^2 + 210)}^{(bottom)}\overbrace{(2.3)}^{(top)'} - \overbrace{(2.3x)}^{(top)}\overbrace{(14.2x)}^{(bottom)'}}{(7.1x^2 + 210)^2}$$

Now we simplify:

$$= \frac{16.33x^2 + 483 - 32.66x^2}{(7.1x^2 + 210)^2}$$

$$= \frac{-16.33x^2 + 483}{(7.1x^2 + 210)^2}$$

Thus,

$$f'(x) = \frac{-16.33x^2 + 483}{(7.1x^2 + 210)^2}$$

Now, we can find $f'(4)$ and $f'(6)$.

$$f'(4) = \frac{-16.33(4)^2 + 483}{(7.1(4)^2 + 210)^2}$$

$$\approx 0.0021 \qquad \text{(2.1 votes per dollar spent per day)}$$

$$f'(6) = \frac{-16.33(6)^2 + 483}{(7.1(6)^2 + 210)^2}$$

$$\approx -0.0005 \qquad \text{(−0.5 votes per dollar spent per day)}$$

Interpretation: If the politician increases the number of campaigning days in the county by 1, from 4 days to 5 days, he will secure approximately 2.1 more votes for each dollar spent. If he increases the number of campaigning days in the county by 1, from 6 days to 7 days, he will lose approximately one-half vote for each dollar spent. His campaign committee must determine the optimal number of campaigning days. If he campaigns fewer days than that optimal number, he will not secure the greatest number of votes per dollar. If he campaigns more days than that optimal number, the efficiency of "buying" votes will go down. The committee must decide when the politician has had just the right amount of exposure. (Section 4.3 is concerned with the process of optimization.)

Products, Quotients, and Technology

Using Your Calculator You can use your graphing calculator to approximate the numerical derivative of a product or a quotient of functions. The following entries show the evaluation of $f'(4)$ for the function $f(x) = (5x^2 + 4)(x^3 + 1)$.

```
Y1 = (5x^2 + 4)(x^3 + 1)
Y5 = nDeriv(Y1,X,X,.001)
```

In the computation window, enter

```
Y5(4)
```

Calculator Exercises

1. Compute $f(3.5)$ and $f'(3.5)$ for $f(x) = (6.2x^{2.9} - 11)(1.4x^{-3.2} + 6)$.
2. Compute $f(4)$ and $f'(4)$ for $f(x) = \dfrac{5x^{-2.55} + 3x^{0.17}}{2x^{3/7} - 6x^{-5/9}}$.
3. The time t, in hours, it takes a private jet to fly the 1500 miles from City A to City B at the rate r miles per hour is approximated by the function
$$t(r) = \frac{1500}{1.2r + 18.2}$$
Compute and interpret both $t(350)$ and $t'(350)$.
4. When x liters of distilled water are added to 5 liters of a 30%-acid solution, the acid solution is diluted to an P%-acid solution according to the function
$$P(x) = \frac{1.5}{x + 5}$$
Compute and interpret both $P(3)$ and $P'(3)$.
5. The profit P, in dollars, made on the sale of x units of a product is given by the function
$$P(x) = (x - 3)\left(\frac{200}{x} + 5\right) - 200$$
Compute and interpret both $P(25)$ and $P'(25)$.
6. The average cost C, in dollars, to a company to produce x units of a product is approximated by the function
$$C(x) = \frac{0.21x^2 + 101}{3.2x + 2}$$
Compute and interpret both $C(1500)$ and $C'(1500)$.

Using Derive You can use Derive to compute the numerical derivative of a product or quotient of functions. The following entries show the evaluation of $f'(4)$ for the product $f(x) = (5x^2 + 4)(x^3 + 1)$. This set of commands uses Derive's numerical mathematics utility. It is not part of Derive's standard set of utilities, so you will need to load it. The first four commands do this.

Transfer
Load
Utility
numeric.mth
Declare
Function
f
(5x^2 + 4)/(x^3 + 1)
Author
dif_num(f(x),x,4,0.01)
approX

The derivative at $x = 4$ is displayed.

Derive Exercises Use Derive to find each derivative and to perform each evaluation.

1. $f'(x), f(3)$, and $f'(3)$ for $f(x) = \dfrac{1 - x^4}{x^3 + 1}$. Interpret both these values.
2. $f'(x)$, $f(2.5)$, and $f'(2.5)$ for $f(x) = \left(4\sqrt[4]{x^3} - 2.5\sqrt[3]{x^4}\right) \cdot \left(\dfrac{2.5}{x^{2.5}} - \dfrac{1.5}{\sqrt{x}}\right)$. Interpret both these values.
3. The percentage P of voters recognizing a candidate's name t months into an election campaign is approximated by the function
$$P(t) = \frac{14t}{t^{1.85} + 100}$$
Use Derive to find $P'(t)$ and compute and interpret both $P(8)$ and $P'(8)$.

EXERCISE SET 3.2

A UNDERSTANDING THE CONCEPTS

For Exercises 1–10, a function is given and the derivatives of its parts are given. Place the parts and the derivatives in their proper positions so as to express the derivative of the function. (You need not simplify.)

1. $f(x) = 3x^4(x^3 + 1)$;
$3x^4, \quad x^3 + 1,$
$\dfrac{d}{dx}(3x^4) = 12x^3, \quad \dfrac{d}{dx}(x^3 + 1) = 3x^2$

2. $f(x) = 5x(2x - 3)^4$;
$5x, \quad (2x - 3)^4,$
$\dfrac{d}{dx}(5x) = 5, \quad \dfrac{d}{dx}(2x - 3)^4 = 8(2x - 3)^3$

3. $f(x) = (x^5 + 2x + 6)(x^3 - 1)$;
$x^5 + 2x + 6, \quad x^3 - 1,$
$\dfrac{d}{dx}(x^5 + 2x + 6) = 5x^4 + 2, \quad \dfrac{d}{dx}(x^3 - 1) = 3x^2$

4. $f(x) = (3x + 7)\sqrt{5x + 2}$;
$3x + 7, \quad \sqrt{5x + 2},$
$\dfrac{d}{dx}(3x + 7) = 3, \quad \dfrac{d}{dx}\sqrt{5x + 2} = \dfrac{5}{2\sqrt{5x + 2}}$

5. $f(x) = \sqrt{x^2 + 6}\sqrt[3]{2x - 9}$;
$\sqrt{x^2 + 6}, \quad \sqrt[3]{2x - 9},$
$\dfrac{d}{dx}\sqrt{x^2 + 6} = \dfrac{x}{\sqrt{x^2 + 6}},$
$\dfrac{d}{dx}\sqrt[3]{2x - 9} = \dfrac{2}{3\sqrt[3]{(2x - 9)^2}}$

6. $f(x) = \dfrac{5x - 7}{8x + 1}$;
$5x - 7, \quad 8x + 1,$
$\dfrac{d}{dx}(5x - 7) = 5, \quad \dfrac{d}{dx}(8x + 1) = 8$

7. $f(x) = \dfrac{9x - 4}{2x - 5}$;

$9x-4,\quad 2x-5,$

$\frac{d}{dx}(9x-4)=9,\quad \frac{d}{dx}(2x-5)=2$

8. $f(x)=\frac{\sqrt{x^2+x}}{3x^2+5};$

$\sqrt{x^2+x},\quad 3x^2+5,$

$\frac{d}{dx}\sqrt{x^2+x}=\frac{1}{2}(x^2+x)^{-1/2}(2x+1),$

$\frac{d}{dx}(3x^2+5)=6x$

9. $f(x)=(x^2+3)^{1/4}(2x+7);$

$(x^2+3)^{1/4},\quad 2x+7,$

$\frac{d}{dx}(x^2+3)^{1/4}=\frac{1}{4}(x^2+3)^{-3/4}(2x),$

$\frac{d}{dx}(2x+7)=2$

10. $f(x)=x^{-2/3}(5x^2+x-4)^{1/6};$

$x^{-2/3},\quad (5x^2+x-4)^{1/6},$

$\frac{d}{dx}x^{-2/3}=\frac{-2}{3}x^{-5/3},$

$\frac{d}{dx}(5x^2+x-4)^{1/6}=\frac{1}{6}(5x^2+x-4)^{-5/6}(10x+1)$

B SKILL ACQUISITION

For Exercises 11–30, use the product and/or quotient rule to find each derivative.

11. $f(x)=(x^2+3x+5)(6x-7)$
12. $f(x)=(x^3+7x+2)(9x-1)$
13. $f(x)=(x-1)(x+3x^{2/3}+6)$
14. $f(x)=(3x+2)(x^2-5x^{2/5})$
15. $f(x)=\frac{3x+5}{2x-7}$
16. $f(x)=\frac{x-4}{8x+9}$
17. $f(x)=\frac{4x+3}{5x-8}$
18. $f(x)=\frac{11x+4}{x-6}$
19. $f(x)=\frac{x^2+6}{x^3-1}$
20. $f(x)=\frac{9x^2+4}{2x^2-3}$
21. $f(x)=\frac{3x^3-7x+1}{x^2-4}$
22. $f(x)=\frac{6x^2+8x-5}{2x^2+x+1}$
23. $f(x)=\frac{x^4-6}{x^3+1}$
24. $f(x)=\frac{2x^5+9}{x-7}$
25. $f(x)=\frac{(2x-1)(x^2+5)}{7x+2}$
26. $f(x)=\frac{(x-4)(9x+14)}{x^2+1}$
27. $f(x)=\frac{8}{x^2+3x+4}$
28. $f(x)=\frac{-6}{x^3+5}$
29. $f(x)=\frac{\sqrt{x}}{x-4}$
30. $f(x)=\frac{x^{2/3}}{x^{1/3}+x}$

For Exercises 31–34, find and interpret the derivative of the given function at the indicated value of the independent variable.

31. $f(x)=(9x+2)(x-6),\ x=4$
32. $f(x)=(x^2+x-1)(5x-7),\ x=2$
33. $g(t)=\frac{5t+4}{8t+6},\ t=10$
34. $h(s)=\frac{9s+1}{9s-4},\ s=1$

C APPLYING THE CONCEPTS

35. ***Psychology: Memorization*** The function

$$N(t)=\frac{80t}{105t-80}$$

is a form of a function developed by L. L. Thurston that relates the number N of facts that a person can remember t hours after memorizing them. Find and interpret $N'(1)$ and $N'(3)$.

36. ***Business: Company Takeovers*** During the buyout attempt of company X by company Y, the accountants of company Y developed a formula that relates the total assets A (in millions of dollars) company Y can expect from the acquisition of company X t years from the time of takeover. The function is

$$A(t)=\frac{t^{5/4}+1}{3t+6},\quad 0\le t\le 15$$

Find and interpret $A(4)$ and $A'(4)$.

37. ***Business: Cost of Production*** A producer and distributor of recycled paper notecards produces x boxes of cards per day at an average cost $C(x)$ (in dollars) of

$$C(x)=\frac{6x+8000}{2x+300}$$

If the producer is currently producing and distributing 280 boxes of cards each day, what change can she expect in the average daily cost of production if she increases production to 281 boxes?

38. *Medicine: Drug Concentration* The concentration $C(t)$ (in milligrams per cubic millimeter) of a particular drug in the human bloodstream t hours after ingestion is given by

$$C(t) = \frac{21t}{t^2 + t + 4}$$

What change in the concentration can the prescribing physician expect between the third and fourth hours after ingestion?

39. *Economics: Demand for a Product* Economists have established that the demand D for an item decreases as the price x increases. The daily number D of tubes of roofing tar that people are willing to buy at x cents each is given by

$$D(x) = \frac{85{,}000}{\sqrt{x} + 10}, \quad 200 \leq x \leq 700$$

If the current price per tube is $2.50, find and interpret the marginal demand for tubes of roofing tar.

40. *Business: Marginal Profit* The publisher of a fiction book expects a total profit from a recently published book that is expected to sell for many years to be P (in thousands of dollars). If t represents the number of years since the release of the book, P and t are related by the function

$$P(t) = \frac{850t^2}{t^2 + 20}$$

Find the publisher's marginal profit on this book at year 10 and then at year 11.

41. *Business: Marginal Revenue* A manufacturer of ceramic vases has determined that her weekly revenue and cost functions for the manufacture and sale of x vases are $R(x) = 95x - 0.08x^2$ dollars and $C(x) = 1200 + 35x - 0.04x^2$ dollars, respectively. Given that profit equals revenue minus cost,

   a. Find the marginal revenue, marginal cost, and marginal profit functions.

   b. Estimate the revenue realized by selling the 601st vase, the 900th vase. What seems to be happening?

   c. Estimate the cost incurred by producing the 601st vase, the 900th vase. What seems to be happening?

   d. Estimate the profit realized by selling the 601st vase, the 900th vase. What seems to be happening?

42. *Manufacturing: Productivity* A manufacturer's records over the last 96 months show that its productivity (in units produced each day) is $P(t) = 0.6t^2 - 400t + 9500$ in month number t, where $t = 0$ represents the first month of the eight-year period. Find and interpret the manufacturer's marginal productivity in (a) the 20th month, (b) the 60th month, and (c) the 75th month. What seems to be happening?

D DESCRIBE YOUR THOUGHTS

43. A person in your calculus class differentiates the function $f(x) = (5x^2 + 4)(x^3 + 11)$ as $f'(x) = 10x \cdot 3x^2 = 30x^3$. Explain to this person why his result is incorrect.

44. Explain how you would use the product rule to differentiate the function $f(x) = (x^2 + 2x)^2$.

45. Devise and explain a rule for differentiating a product that involves three factors. Generalize your rule to a product involving n factors.

E REVIEW

46. **(2.2)** Find, if it exists, $\lim_{x \to 4} \dfrac{x^2 - 9x + 20}{x^2 + 3x - 28}$.

47. **(2.3)** Find, if it exists, $\lim_{x \to \infty} \dfrac{3x^2 + 5x - 8}{x^3 + 27}$.

48. **(2.4)** Discuss the continuity of

$$f(x) = \begin{cases} \dfrac{x^2 - x - 6}{x - 3}, & \text{if } x \neq 3 \\ 5, & \text{if } x = 3 \end{cases} \quad \text{at } x = 3.$$

49. **(2.6)** Determine the average rate of change of $f(x)$ with respect to x of $f(x) = 2x^2 + 5$ as x changes in value from $x = 3$ to $x = 7$.

50. **(3.1)** Find $f'(x)$ for $f(x) = 5x^2 + 3x + 8$.

3.3 Higher-Order Derivatives

Introduction
Higher-Order Derivatives and Their Notations
Interpreting Higher-Order Derivatives
Higher-Order Derivatives and Technology

Introduction

The derivative $f'(x)$ gives information about the rate of change of the function $f(x)$. $f'(x)$ can tell us if $f(x)$ is increasing or decreasing over an interval. In this section we examine how to obtain information about $f'(x)$, that will tell us if $f(x)$ is increasing at an increasing or decreasing rate or decreasing at an increasing or decreasing rate.

Higher-Order Derivatives and Their Notations

From our experience with differentiation, we know (or may surmise) that the derivative f' of a function f is itself a function. Hence, it may be possible to differentiate f'. The derivative of f' is denoted by f'' and is called the *second derivative* of f. Again, f'' is a function and may be differentiated, giving the *third derivative*, f'''. Continuing this way, we get the fourth derivative of f, fifth derivative of f, sixth derivative of f, and all other **higher-order derivatives of f**.

For convenience, the notation changes slightly for the fourth and higher-order derivatives of f. Rather than using the *prime* notation, such as f', f'', and f''', the order of the derivative is enclosed in parentheses and is written as a superscript on f. For example, if f is a differentiable function, then

f' indicates the first derivative of f.
f'' indicates the second derivative of f.
f''' indicates the third derivative of f.
$f^{(4)}$ indicates the fourth derivative of f.
$f^{(5)}$ indicates the fifth derivative of f.
$f^{(6)}$ indicates the sixth derivative of f.
$\vdots$
$f^{(n)}$ indicates the nth derivative of f. The $\frac{d}{dx}$ notation is also used to indicate a higher-order derivative. If $y = f(x)$ is a differentiable function, then

$\frac{dy}{dx}$ and $\frac{df}{dx}$ both indicate the first derivative of the function.
$\frac{d^2y}{dx^2}$ and $\frac{d^2f}{dx^2}$ both indicate the second derivative of the function.
$\frac{d^3y}{dx^3}$ and $\frac{d^3f}{dx^3}$ both indicate the third derivative of the function.
$\frac{d^4y}{dx^4}$ and $\frac{d^4f}{dx^4}$ both indicate the fourth derivative of the function.

$\dfrac{d^5y}{dx^5}$ and $\dfrac{d^5f}{dx^5}$ both indicate the fifth derivative of the function.

$\vdots$

$\dfrac{d^ny}{dx^n}$ and $\dfrac{d^nf}{dx^n}$ both indicate the nth derivative of the function.

EXAMPLE SET A

Find $f'(x)$, $f''(x)$, $f'''(x)$, and $f^{(4)}$ for $f(x) = 3x^5 - 2x^4 + x^2 - 10x + 4$.

Solution:

$f'(x) = 15x^4 - 8x^3 + 2x - 10$. Now differentiate $f'(x)$ to get $f''(x)$.
$f''(x) = 60x^3 - 24x^2 + 2$. Now differentiate $f''(x)$ to get $f'''(x)$.
$f'''(x) = 180x^2 - 48x$. Now differentiate $f'''(x)$ to get $f^{(4)}(x)$.
$f^{(4)} = 360x - 24$.

Interpreting Higher-Order Derivatives

Just as the first derivative f' of a function f describes the rate of change of f, the second derivative f'' describes the rate of change of f'; that is, f'' provides the rate of change of the rate of change of f. To understand the meaning of *the rate of change of the rate of change*, consider the following example.

Every country experiences inflation, and inflation is related to time. At certain times, inflation can be higher or lower than it is at other times. That is, inflation potentially changes as time changes. The word *changes* makes us think of *differentiation* (since differentiation is the language of change). Let's represent the relation between inflation I and time t (in months from the beginning of the year) by the function $I(t)$. Then the derivative of $I(t)$ with respect to time $I'(t)$ represents the rate of change of inflation; that is, $I'(t)$ represents the rate of inflation. We wish to interpret $I''(t)$.

Suppose that $I'(4) = 0.005$. Rewriting this as a slope, we get

$$I'(4) = 0.005 = \frac{0.005 \quad \leftarrow \text{approximate change in inflation}}{1 \quad \leftarrow \text{change in time}}$$

We see that as time increases by 1 month, from month 4 to month 5, inflation increases by approximately 0.5%. Since the rate of inflation is positive, inflation is increasing. But is the increase accelerating or slowing? The second derivative answers this question.

Now suppose that $I''(4) = 0.2$. $I''(t)$ represents the rate of change of the rate of inflation; that is, $I''(t)$ indicates if the rate of inflation is accelerating, slowing, or remaining constant. Thus,

$$I''(4) = 0.2 = \frac{0.2 \quad \leftarrow \text{approximate change in the rate of inflation}}{1 \quad \leftarrow \text{change in time}}$$

means that as time increases by 1 month, from month 4 to 5, the rate of inflation will *increase* by approximately 20%. Thus, the rate of inflation is increasing, or accelerating.

Summarizing, we know that

1. A country experiences inflation.
2. Inflation is increasing at $t = 4$ (since $I'(4)$ is positive).
3. Not only is inflation increasing, but it is doing so at an increasing rate.

If the function relating inflation and time $I(t)$ in this example were described graphically, the part of the curve between 4 and 5 would look like the curve illustrated in Figure 3.8.

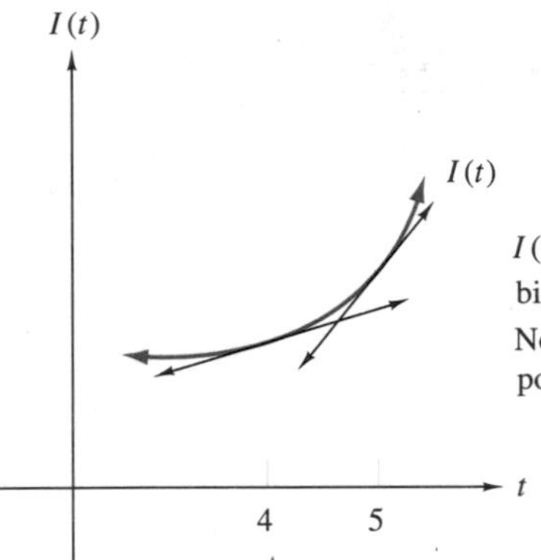

$I(t)$ increases at an increasing rate. $I(t)$ is getting bigger and, as $t \to 5^-$, $I(t)$ increases more quickly. Notice also, that the slopes of the tangent lines are positive and the positive trend in values is increasing.

Figure 3.8

The inflation example illustrated a function that was increasing at an increasing rate. The examples that follow illustrate the other three possibilities (see Figure 3.9, Figure 3.10, and Figure 3.11).

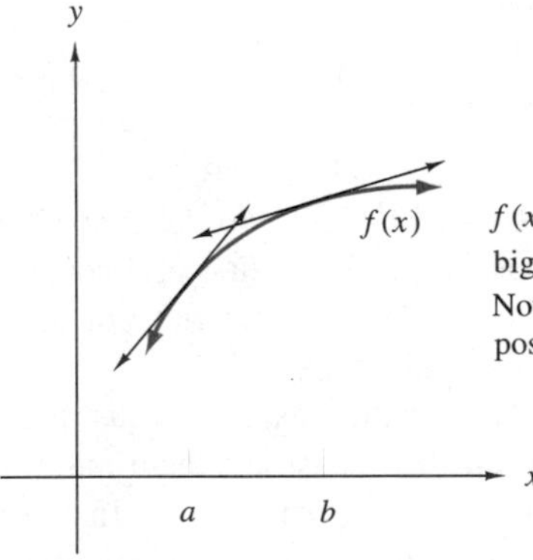

$f(x)$ increases at a decreasing rate. $f(x)$ is getting bigger but, as $x \to b^-$, $f(x)$ increases more slowly. Notice also, that the slopes of the tangent lines are positive but the positive trend in values is decreasing.

Figure 3.9

Each of the above two curves increase through the designated intervals. At every input value in those intervals, the first derivatives are positive. Each of the next two curves decrease through the designated intervals. At every input value in these intervals, the first derivatives are negative.

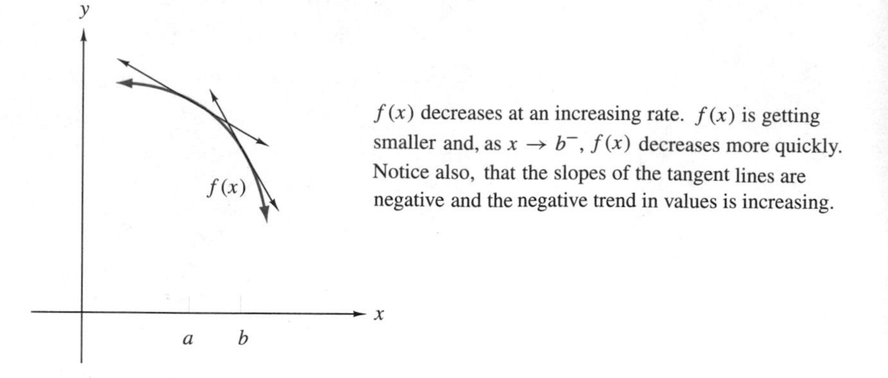

Figure 3.10

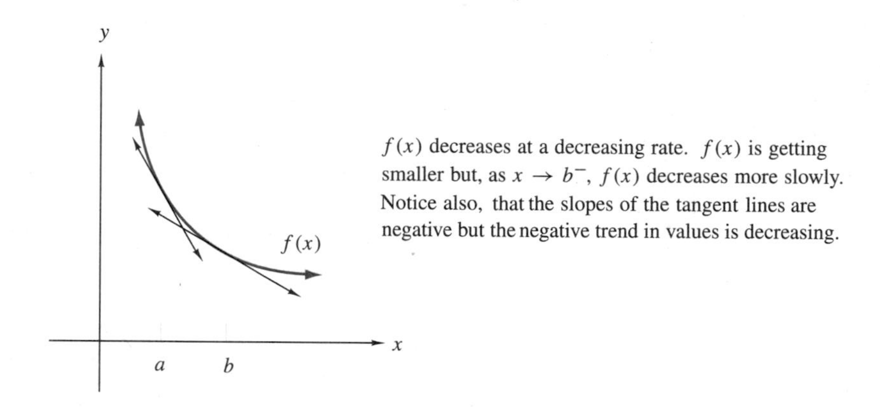

Figure 3.11

The third and higher-order derivatives can be useful in certain mathematical situations, but it is the first and second derivatives that are of primary use in applications. We will restrict our attention to interpreting only the first and second derivatives.

As the inflation example illustrates, the first and second derivative, f' and f'', of a function f describe the trend of the function over an interval and, hence, determine the shape of the curve. In the inflation example both the first and second derivatives were positive and the function increased at an increasing rate. The sign of the first and second derivatives also provides information about the other three possibilities, and we summarize them below. You can easily remember them by thinking of them in terms of the multiplication of signed numbers.

For multiplication:

- The product of two numbers with the *same sign* is positive (making us think of *increasing*).
- The product of two numbers with *opposite signs* is negative (making us think of *decreasing*).

For derivatives:

- Two derivatives, f' and f'' with the *same sign*, indicate that the function f changes at an *increasing rate*.
  $f' > 0,\ f'' > 0 \rightarrow f$ is increasing at an increasing rate.
  $f' < 0,\ f'' < 0 \rightarrow f$ is decreasing at an increasing rate.
- Two derivatives, f' and f'' with *opposite signs*, indicate that the function f changes at a *decreasing rate*.
  $f' > 0,\ f'' < 0 \rightarrow f$ is increasing at a decreasing rate.
  $f' < 0,\ f'' > 0 \rightarrow f$ is decreasing at a decreasing rate.

EXAMPLE SET B

1. Suppose the function $C(t)$ represents the amount of money (in thousands of dollars) a county spends on the local drug war t months from January, 1990. Interpret the information provided by $C'(15) = 1.6$ and $C''(15) < 0$.

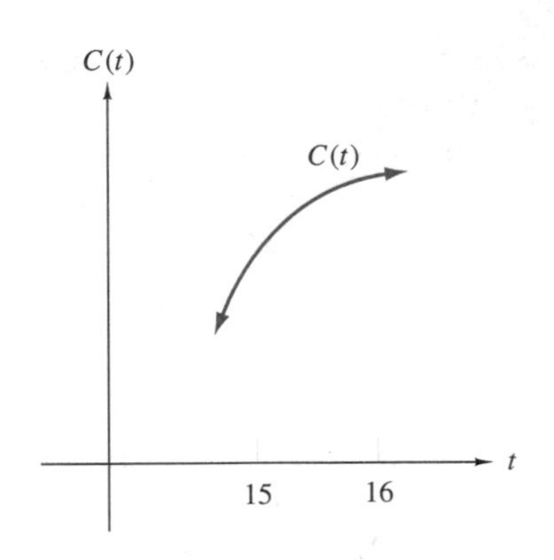

Figure 3.12

Solution:

Because the first derivative of C is positive, the cost of the drug war is increasing. Because the second derivative is negative (so that the derivatives have opposite signs), the function $C'(t)$ is decreasing and the function $C(t)$ itself is increasing at a decreasing rate. We make the following interpretation.

Interpretation: Fifteen months from January, 1990, although the cost of fighting the drug war is increasing, it is increasing at a decreasing rate. This means that the citizens of this county can expect the cost of fighting the drug war to increase each month, but they can expect each monthly increase to be less than the previous one.

Figure 3.12 illustrates this curve between $t = 15$ and $t = 16$. Notice that, as t increases, the curve rises, indicating that $C(t)$ is increasing. Notice also that as t increases, the slopes of the tangent lines decrease. This gives the curve its *opening downward* appearance, indicating that the rate of change in C is itself decreasing.

2. Suppose the function $W(x)$ represents the amount of money (in thousands of dollars) a county spends on welfare each month, where x is the number of people (in thousands) applying for and receiving welfare. Interpret the information provided by $W'(16) = -8$ and $W''(16) < 0$.

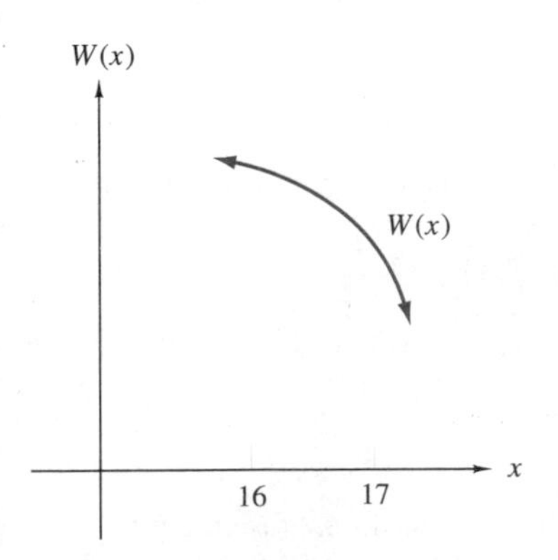

Figure 3.13

Solution:

Because $W'(16)$ is negative, $W(x)$ is decreasing when $x = 16$. Because the second derivative is negative (so that both derivatives have the same sign), $W(t)$ is decreasing at an increasing rate.

Interpretation: At the time when 16,000 people are applying for and receiving welfare, the citizens of this county can not only expect the amount of money spent each month on welfare to decrease, but they can expect each decrease to be greater than the previous decrease.

Figure 3.13 shows $W(x)$ for $x = 16$ to $x = 17$. Notice that as x increases, the curve falls, indicating that W is decreasing. Also notice that as x increases, the curve bends more sharply downward, indicating that not only is W decreasing, but it is doing so at an increasing rate.

3. Suppose the function $A(t)$ represents the amount (in square meters) of water hyacinth growing wild in a lake t months from the beginning of a university research project at the lake. Interpret the information provided by the first and second derivatives, $A'(25) = -14$ and $A''(25) > 0$.

Solution:
Because the first derivative is negative, the amount of water hyacinth growing wild in the lake 25 months after the beginning of the research project is decreasing. Because the second derivative is positive (so that the derivatives have opposite signs), $A(t)$ is decreasing at a decreasing rate.

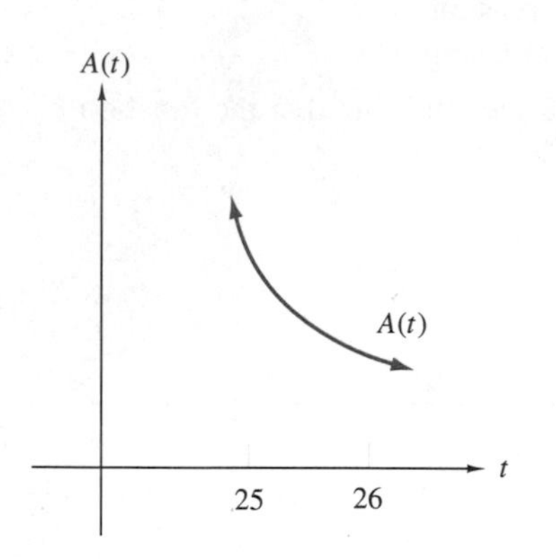

Figure 3.14

Interpretation: Twenty-five months after the beginning of the research project, the amount of water hyacinth in the lake is decreasing, but at a decreasing rate. This means that although the amount of water hyacinth in the lake is decreasing each month, each decrease will be less than the previous decrease.

Figure 3.14 shows $A(t)$ for $t = 25$ to $t = 26$. Notice that as t increases, the curve falls, indicating that W is decreasing. Also notice that as x increases, the curve is tending to bend upward, indicating that not only is W decreasing, but it is doing so at a decreasing rate.

Higher-Order Derivatives and Technology

Using Your Calculator You can use your graphing calculator to numerically approximate the second derivative of a function. The following entries show the computation of $f(3)$, $f'(3)$, and $f''(3)$ for the function $f(x) = -0.006x^4 - 0.05x^3 + 8$.

```
Y1 =-0.006x^4 − 0.05x^3 + 8
Y5 = nDeriv(Y1,X,X,.001)
Y6 = nDeriv(Y5,X,X,.001)
```

In the computation window, enter and compute

```
Y1(3)
Y5(3)
Y6(3)
```

Calculator Exercises

1. Compute $f(20)$, $f'(20)$, and $f''(20)$ for $f(x) = 30{,}000\left(2 + 0.3x + 0.8x^2\right)$.

2. Compute $f(200)$, $f'(200)$, and $f''(200)$ for $f(x) = \dfrac{14x^2 + 3}{7x - 2}$.

3. According to the Congressional Budget Office and the U.S. Bureau of the Census, the function

$$C(t) = 0.54t^2 + 12.64t + 107.1$$

approximates the total annual cost C, in billions of dollars, of Medicare t years from 1990, for the years 1990 through 1995. Compute and interpret $C(1)$, $C'(1)$, and $C''(1)$. Construct the graph of this function for $0 \leq C(t) \leq 200$ and use it to describe the function's behavior.

4. The cost C, in millions of dollars, for the government to seize x percent of a particular illegal drug as it is brought into the country is approximated by the function

$$C(x) = \frac{413x}{100 - x}$$

Compute and interpret $C(15)$, $C'(15)$, and $C''(15)$. Construct the graph of this function for $0 \leq C(t) \leq 1000$, and use it to describe its behavior.

Using Derive Derive can compute higher-order derivatives of functions. The following entries show the computations of $f(3)$, $f'(3)$, and $f''(3)$ for the function $f(x) = -0.006x^4 + 0.05x^3 + 8$. This set of commands uses Derive's numerical mathematics utility. It is not part of Derive's standard set of utilities, so you will need to load it. The first four commands do this.

Transfer
Load
Utility
numeric.mth
Declare
Function
f
-0.006x^4 + 0.05x^3 + 8
Author
f(3)
appro**X**
Author
dif__num(f(x), x, 3, 0.01)
appro**X**
Author
dif2__num(f(x), x, 3, 0.1)
appro**X**

The first and second derivatives at $x = 3$ are displayed.

The **dif__num** command computes the first derivative, and the **dif2__num** command computes the second derivative. Notice the difference in the forms of the first and second derivatives. The number 0.01 appears in the first derivative and the number 0.1 in the second derivative. Derive is approximating the slope of the tangent line at $x = 3$ with the slope of a secant line from $x = 3$ to $x = 3 + h$, where $h = 0.01$ and then $h = 0.1$. The value of h decreases so as not to increase the error of the approximation.

Can you devise another way of computing a second derivative using Derive?

Derive Exercises Use Derive to make each computation.

1. Compute $f(x)$, $f(7)$, $f'(7)$, and $f''(7)$ for $f(x) = \dfrac{x+6}{x-2}$.

2. Compute $f(x)$, $f(0)$, $f'(0)$, and $f''(0)$ for $f(x) = \dfrac{(x^3 - 4x)^{1/7}}{(3x+35)^{3/7}}$.

3. The number N of sales of a product t weeks after it is introduced into the market is given by the function

$$N(t) = \frac{185t}{(1.2t + 1.5)^{1.5}}, \quad 0 \leq t \leq 20$$

Use Derive to find $N'(t)$ and $N''(t)$ and compute and interpret $N(10)$, $N'(10)$, and $N''(10)$.

EXERCISE SET 3.3

A UNDERSTANDING THE CONCEPTS

1. ***Business: Production Costs*** Suppose that $C(x)$ represents the cost (in dollars) to a manufacturer for producing x decorative throw pillows. Interpret the information provided by $C'(4500) = 0.65$ and $C''(4500) > 0$.

2. ***Economics: Population*** Suppose that $A(t)$ represents the population (in thousands) of a particular geographic area in some state t years from 1980. Interpret the information provided by $P'(11) = 1.2$ and $P''(11) > 0$.

3. ***Management: Water Supply*** Suppose that $A(t)$ represents the amount of water (in thousands of gallons) flowing out of a reservoir each day into adjacent creeks t days from the beginning of the year. Interpret the information provided by $A'(230) = -685$ and $A''(230) < 0$.

4. ***Business: Sales Volume*** Suppose that $S(r)$ represents a company's monthly sales volume (in millions of dollars) when r dollars are spent on research and development. Interpret the information provided by $S'(1.6) = 5.8$ and $S''(1.6) < 0$.

5. ***Business: Marketing*** Suppose that $U(c)$ represents the useful life (in hours) of a toy of complexity index c (in a marketer's defined units). Interpret the information provided by $U'(7) = 30$ and $U''(7) < 0$.

6. ***Business: Capital Formation*** Suppose that $C(t)$ represents the capital formation (in thousands of dollars) of a revenue flow t years after the start-up of a company. Interpret the information provided by $C'(3) = 350$ and $C''(3) > 0$.

7. ***Law Enforcement: Drunk Driving*** Suppose that $N(t)$ represents the number of alcohol-related traffic accidents in a city t months after a strict new driving-under-the-influence law has taken effect. Interpret the information provided by $N'(15) = -15$ and $N''(15) < 0$.

8. ***Business: Advertising*** Suppose that $N(t)$ represents the number of people (in thousands) in a city who have heard of a new product being advertised on radio and television t days after the beginning of the advertisement. Interpret the information provided by $N'(21) = 0.6$ and $N''(21) < 0$.

9. ***Business: Sales Commission*** Suppose that $C(s)$ represents the percent of commission a salesperson makes on s sales each year. Interpret the information provided by $C'(60) = 0.008$ and $C''(60) > 0$.

10. ***Business: Cost of Commodities*** Suppose that $C(x)$ represents the cost (in dollars) of a particular commodity to a customer when the customer purchases x square feet of that commodity. Interpret the information provided by $C'(4000) = 150$ and $C''(4000) < 0$.

11. ***Business: Worker Efficiency*** Suppose that $E(t)$ represents the efficiency (in parts assembled) of an assembly line worker t hours after the worker begins his shift. Interpret the information provided by $E'(3.5) = 6$ and $E''(3.5) < 0$.

12. ***Economics: Educational Endowments*** Suppose that $A(n)$ represents the annual amount of endowments (in thousands of dollars) made to a university whose faculty publishes n papers in major journals each year. Interpret the information provided by the first and second derivatives, $A'(41) = 85$ and $A''(41) > 0$.

13. ***Botany: Photosynthesis*** Suppose that $A(x)$ represents the amount of photosynthesis that occurs in a particular type of plant that receives x intensity of sunlight. Suppose also that $A'(0.8) = 60$. Interpret $A''(0.8) > 0$.

14. ***Weather: Wind Chill*** Suppose that $W(v)$ represents the degree of wind chill when the velocity of wind is v. Interpret the information provided by $W'(20) = -0.4$ and $W''(20) > 0$.

15. ***Manufacturing: Ventilation*** In a research and development laboratory, a high-efficiency ventilation system begins to operate when the level of carbon monoxide reaches a certain critical level. The amount (in cubic centimeters) of carbon monoxide ventilated from the lab is related to the amount of time t (in minutes) that the system has been operating by the function $A(t)$. Interpret the information provided by $A'(3) = 4000$ and $A''(3) < 0$.

For Exercises 16–25, use the given information to make statements about the signs (positive/negative) of the first and second derivatives. (For example, for the inflation example in this section, we can state that $I'(t) > 0$ and $I''(t) > 0$.) Also, draw a rough sketch of the curve over this small interval.

16. ***Botany: Photosynthesis*** Suppose that $P(t)$ represents the amount of photosynthesis taking place in a plant t

hours after dawn. Suppose also that in the morning, photosynthesis increases and does so rapidly until early afternoon.

17. ***Economics: Gross National Product*** Suppose that $G(t)$ represents the gross national product of a country t years from the beginning of 1985. Suppose also that at a particular time, the gross national product is decreasing but is starting to improve.

18. ***Ecology: Water Consumption*** Suppose that $W(t)$ represents the amount of water consumed in a northern California county t months from the beginning of a drought year. Suppose also that because of public education to drought conditions and the need for water rationing over the summer months, the amount of water consumed in this county over the summer months is increasing, but is doing so more slowly each month.

19. ***Business: Cost of Contaminant Removal*** Suppose that $A(t)$ represents the percent of a contaminant removed from a tank of solvent t minutes after a decontamination process begins. Suppose also that as the process continues, more and more contaminant is removed, but it takes longer to remove the same percent of contaminant near the end of the process than it takes near the beginning of the process.

20. ***Business: Inventory*** Suppose that $I(t)$ represents a company's inventory of a particular product t weeks from the beginning of the year. Also, suppose that during the Christmas season, the inventory is being depleted and is being depleted rapidly.

21. ***Business: Profits in Publishing*** Suppose that $P(x)$ represents a publisher's profit on an applied calculus textbook. Suppose also that the profit rises when the text is first introduced on the market, but then slows as used books start replacing new books.

22. ***Business: Worker Efficiency*** Suppose that $E(t)$ represents the efficiency of a worker at a large company t months from the beginning of the year. Suppose also that because of possible layoffs, morale is low and worker efficiency is declining. But rumors of good times ahead for the company are helping to improve morale, and efficiency is beginning to increase.

23. ***Business: Agricultural Yield*** Suppose that $Y(t)$ represents the crop yield of corn in a state t weeks from the beginning of the harvesting season. Suppose also that because of severe and worsening drought conditions during the summer months, the crop yield is decreasing and is doing so more rapidly each week.

24. ***Economics: Stock Prices*** Suppose that $A(t)$ represents the average price of stocks on the New York Stock Exchange t minutes after trading begins on a particular day. During a crash, the average value of stocks tumbles quickly.

25. ***Politics: Nuclear Weapons*** Suppose that $N(t)$ represents the number of nuclear weapons possessed by the United States t months after the end of the Cold War. At the end of the Cold War and because of weapons treaties with the former Soviet Union, the United States began destroying many of its nuclear weapons. But as the number of weapons became smaller, the rate at which they were destroyed declined.

B SKILL ACQUISITION

For Exercises 26–31, find f', f'', f''', and $f^{(4)}$ for each function.

26. $f(x) = x^6 - 2x^5 + 6x - 4$

27. $f(x) = x^5 - x^3 + x + 6$

28. $f(x) = x^4 + 2x^3 + 11x^{-1}$

29. $f(x) = -3x^4 + x^3 - x^2 + 5x + 4x^{-1}$

30. $f(x) = x^3 + 6x^2 - 4\sqrt{x}$

31. $f(x) = 3x^2 + 8x + 10\sqrt{x}$

32. Find $f'(x)$ and $f''(x)$ for $f(x) = \dfrac{x-4}{x}$.

33. Find $f'(x)$ and $f''(x)$ for $f(x) = \dfrac{3x-5}{x}$.

34. Find $f'(x)$ and $f''(x)$ for $f(x) = \dfrac{2x+7}{x^3}$.

35. Find $f'(x)$ and $f''(x)$ for $f(x) = \dfrac{x^3+3}{x^2}$.

For Exercises 36–37, find and interpret $f'(1)$ and $f''(1)$.

36. $f(x) = x^2 + 5x - 4$

37. $f(x) = -5x^2 + x - 4$

C APPLYING THE CONCEPTS

For Exercises 38–40, determine the signs of the first and second derivatives and make a statement about the trend of the function as it applies to the given situation.

38. ***Business: Profits*** The function $P(x) = \dfrac{180x}{110 + x^2}$ relates the weekly profit P (in thousands of dollars) a company makes to the number x (in hundreds) of automobile sheepskin seat covers it sells each week. In this case,

$$P'(x) = \frac{180(110 - x^2)}{(110 + x^2)^2} \quad \text{and}$$

$$P''(x) = \frac{-360x(330 - x^2)}{(110 + x^2)^3}$$

What is the trend at $x = 10$?

39. ***Medicine: Drug Reaction*** The reaction function

$$S(x) = \frac{x}{2}\sqrt{260 - x}$$

measures the strength of the average person's reaction to x milligrams of a particular drug. In this case,

$$S'(x) = \frac{520 - x}{4\sqrt{260 - x}} \quad \text{and}$$

$$S''(x) = \frac{3x - 1040}{8\sqrt{(260 - x)^3}}$$

What is the trend at $x = 40$?

40. ***Biology: Bacterial Toxins*** The toxins function

$$P(t) = \frac{30t}{t^2 + 5}, \quad t > 1$$

relates the population P (in millions) of bacteria in a culture to the time t (in hours) after a particular toxin is introduced into the culture. In this case,

$$P'(t) = \frac{150 - 30t^2}{(t^2 + 5)^2} \quad \text{and}$$

$$P''(t) = \frac{60t(t^2 - 15)}{(t^2 + 5)^3}$$

What is the trend at $t = 4$?

For Exercises 41–43, find the first and second derivatives and make a statement about the trend of the function as it applies to the given situation.

41. ***Business: Sales Trend*** The number N of hand-held hair dryers a manufacturer believes it can sell each month at price p (in dollars) is given by the function

$$N(p) = \frac{1100}{p^2}$$

What is the trend when the price of a hair dryer is \$12?

42. ***Business: Advertising*** The sales department of a company believes that t days after the end of an advertising campaign, the number N of items it will sell each day is given by the function

$$N(t) = -3t^2 + 14t + 450$$

What is the trend 7 days after the end of the advertising campaign?

43. ***Education: Input/Output*** The faculty of a business school's word processing department believes that the number N of words a student can correctly input after t weeks of instruction is given by the function

$$N(t) = t^2 + 6t + 10, \quad 0 \le t \le 6$$

What is the trend at 4 weeks of instruction?

D DESCRIBE YOUR THOUGHTS

44. Describe the information the second derivative of a function can provide about the derivative of the function.

45. Describe the information the first and second derivative of a function can provide about the function.

46. Explain what could be happening to a function $f(x)$ for which $f'(x) > 0$ and $f''(x) = 0$. A picture may help your explanation (but a picture alone is not a sufficient explanation).

E REVIEW

47. **(1.2)** Find $f(3)$ for $f(x) = 3x^2 - x - 2$.

48. **(1.5)** If $f(x) = x^2 + 4$ and $g(x) = 5x - 3$, find $f \circ g$.

49. **(2.2)** Find, if it exists, $\lim_{x \to 3} \frac{x^2 - 7x + 12}{x^2 - 3x}$.

50. **(2.3)** Find, if it exists, $\lim_{x \to \infty} \frac{4x^2 - x - 1}{4x + 5}$.

51. **(3.3)** ***Business: Revenue*** Suppose that $R(t)$ represents the revenue realized by a manufacturer t weeks after the end of a large advertising campaign. Interpret the information provided by $A'(t) > 0$ and $A''(t) < 0$.

3.4 The Chain Rule and General Power Rule

Introduction

Composite Relationships

Relationships are often such that one quantity depends on another quantity, and that quantity in turn depends on yet another quantity. Such relationships are called **composite relationships** and are modeled by composite functions. In this section we examine the chain rule, a rule for differentiating composite functions.

The Chain Rule

Suppose that y is a function of u and that u, in turn, is a function of x. That is, the value of the variable y depends on the value of u and the value of u in turn depends on the value of x. A change in the value of x may produce a change in the value of u and a change in the value of u may produce a change in the value of y. Thus, through a chain of events, a change in the value of x may produce a change in the value of y. Since differentiation is the language of change, these changes can be described using derivative notation.

$\frac{dy}{du}$ represents the ratio of a change in y to a change in u.

$\frac{du}{dx}$ represents the ratio of a change in u to a change in x.

$\frac{dy}{dx}$ represents the ratio of a change in y to a change in x.

We can suggest an expression for $\frac{dy}{dx}$ in terms of $\frac{dy}{du}$ and $\frac{du}{dx}$ by making the following observation. Suppose $\frac{dy}{du} = 7$ and $\frac{du}{dx} = 4$. This means that y changes seven times as fast as u and that u changes four times as fast as x. Consequently, any change in x would produce a four times greater change in u which, in turn, would produce a seven times greater change in y. Since $\underbrace{7}_{dy/du} \cdot \underbrace{4}_{du/dx} = \underbrace{28,}_{dy/dx}$ we conclude that through the chain of changes, y changes 28 times as fast as x, and suggest the following formula.

$$\frac{dy}{du} \cdot \frac{du}{dx} = \frac{dy}{dx}$$

More conveniently,

$$\frac{dy}{dx} = \frac{dy}{du} \cdot \frac{du}{dx}$$

The formula indicates that since y is linked to u and u is linked to x, then y is linked to x and is subject to x's changes.

The Chain Rule

If y is a function of u and u is a function of x—that is, $y = f(u)$ and $u = g(x)$—then $y = f(u) = f[g(x)]$, and y is a function of x. Also,

$$\frac{dy}{dx} = \frac{dy}{du} \cdot \frac{du}{dx}$$

Alternatively,

$$\frac{dy}{dx} = f'[g(x)] \cdot g'(x)$$

EXAMPLE SET A

Suppose $y = u^2 + 3u$ and, in turn, $u = 2x - 9$. Find (1) $\frac{dy}{dx}$ and interpret this result, and (2) $\frac{dy}{dx}$ when $x = 5$ and interpret this result.

Solution:

1. Since y is a function of u and u is a function of x, y is a function of x and we can use the chain rule.

$$\begin{aligned}\frac{dy}{dx} &= \frac{dy}{du} \cdot \frac{du}{dx} \\ &= \frac{d}{du}(u^2 + 3u) \cdot \frac{d}{dx}(2x - 9) \\ &= (2u + 3) \cdot 2 \\ &= 4u + 6\end{aligned}$$

Thus, $\frac{dy}{dx} = 4u + 6$.

This statement does not precisely reflect what we wish to know. $\frac{dy}{dx}$ indicates that a relationship between y and x is to be illustrated. But $4u + 6$ involves u, not x. Since we know that $u = 2x - 9$, we replace all occurrences of u with $2x - 9$ and simplify.

$$\begin{aligned}\frac{dy}{dx} &= 4(2x - 9) + 6 \\ &= 8x - 30\end{aligned}$$

Thus, $\frac{dy}{dx} = 8x - 30$.

Interpretations:

a. The expression $8x - 30$ gives the slope of the tangent line at any point along the curve once the x value is specified.

b. The expression $8x - 30$ gives the instantaneous rate of change of y with respect to x.

2. When $x = 5$,

$$\frac{dy}{dx} = 8(5) - 30$$

$$= 40 - 30$$

$$= 10 \qquad \left(\frac{10}{1} \text{ as a slope}\right)$$

Interpretation: When the value of x increases by 1 unit, from 5 to 6, the value of y increases by approximately 10 units.

$$\left(\frac{dy}{dx} = 10 = \frac{10 \leftarrow \text{ change in } y}{1 \leftarrow \text{ change in } x}\right)$$

Notice that we could have found $\frac{dy}{dx}$ for the functions of Example Set A without using the chain rule by defining y as a function of x at the outset.

$$y = u^2 + 3u \quad \text{and} \quad u = 2x - 9$$

We then replace all occurrences of u in $y = u^2 + 3u$ with $2x - 9$ and simplify.

$$y = (2x - 9)^2 + 3(2x - 9)$$

$$= 4x^2 - 36x + 81 + 6x - 27$$

$$= 4x^2 - 30x + 54$$

Then, $\frac{dy}{dx} = 8x - 30$ as before.

Now, realize that if y was defined as $y = u^{15} + 3u$, this technique would involve an outrageous amount of work! It would take an extraordinary amount of time to expand and simplify $y = (2x - 9)^{15} - 3(2x - 9)$. (This function will be differentiated using the general power rule, a rule that we will examine later in this section.)

The function in Example Set B illustrates how the chain rule can be applied to a business application.

EXAMPLE SET B

For temperatures greater than 10°C, the average daily number of soft drinks sold, s, in a city depends on the temperature, t, as

$$s(t) = 1200 + 23t + 600(t - 10)^{1/3}$$

The number, in hundreds, of soft drink aluminum cans recycled, r, depends on the number of soft drinks sold, s, as

$$r(s) = s^{3/4}$$

How does the number of cans recycled change as the temperature changes from 27°C to 28°C?

Solution:

To answer this question, we need to find $\frac{dr}{dt}$. We know that r is a function of s and, in turn, s is a function of t. Hence, we can apply the chain rule.

$$\frac{dr}{dt} = \frac{dr}{ds} \cdot \frac{ds}{dt}$$

Since $\frac{dr}{ds} = \frac{3}{4}s^{-1/4}$ and $\frac{ds}{dt} = 23 + 200(t - 10)^{-2/3}$, we have

$$\begin{aligned}\frac{dr}{dt} &= \frac{dr}{ds} \cdot \frac{ds}{dt} \\ &= \frac{3}{4}s^{-1/4} \cdot \left[23 + 200(t - 10)^{-2/3}\right]\end{aligned}$$

But $s(t) = 1200 + 23t + 600(t - 10)^{1/3}$, so we replace all occurrences of s with $1200 + 23t + 600(t - 10)^{1/3}$.

$$\frac{dr}{dt} = \frac{3}{4}\left[1200 + 23t + 600(t - 10)^{1/3}\right]^{1/4} \cdot \left[23 + 200(t - 10)^{-2/3}\right]$$

At $t = 27$,

$$\begin{aligned}\frac{dr}{dt} &= \frac{3}{4}\left[1200 + 23(27) + 600(27 - 10)^{1/3}\right]^{-1/4} \cdot \left[23 + 200(27 - 10)^{-2/3}\right] \\ &\approx 5.244\end{aligned}$$

Interpretation: If the temperature in the city increases by 1°, from 27°C to 28°C, the number of aluminum cans recycled in this city will increase by approximately 524.

Global and Local View of a Function

Very often, several differentiation rules must be used to differentiate a function. It is *very* important to know in which order to use the differentiation rules. We can always determine the correct order by looking at the function globally and locally.

Global and Local View of a Function

To determine the *global* form of a function, we determine the form of the function as a power, a product, a quotient, or a sum or difference. This is the form in which the function appears as a whole. To determine the *local* forms of a function, we mean to determine the more elementary parts that compose the function. Functions are differentiated by first operating on the global form and then, if necessary, on the local forms.

ILLUMINATOR SET A

Examine each function globally and locally.

1. $f(x) = 7x^4$. This function is a power function.

2. $f(x) = (9x+1)(3x-4)$. Globally, this function is a product. It has the form

$$\underbrace{\left[\text{first factor}\right] \cdot \left[\text{second factor}\right]}_{\text{product}}$$

Locally, the first factor, $9x+1$, is a sum, and the second factor, $3x-4$, is a difference. We would begin differentiating this function using the product rule.

3. $f(x) = \left[(9x+1)(3x-4)\right]^8$. Globally, this function is a power. It has the form

$$\left[\text{expression}\right]^8$$

Locally, it is a product. We would begin differentiating this function using the general power rule (which we discuss momentarily) and then the product rule.

4. $f(x) = \left[\dfrac{(4x+5)^3(2x-1)^7}{(x+5)^3(6x+8)^5}\right]^4$. Globally, this function is a power. It has the form

$$\left[\text{expression}\right]^4$$

Locally, it is a quotient.

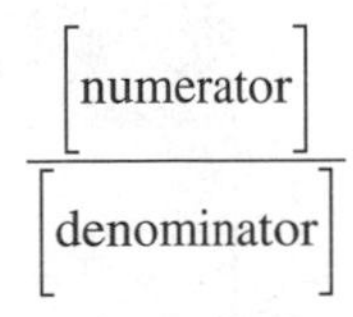

Even more locally, the quotient is composed of products, which in turn are powers. To differentiate this function, we would begin with the general power rule, then use the quotient rule. But when using the quotient rule, we would have to use the product rule and the general power rule.

The General Power Rule

Our list of differentiation rules includes the chain rule, which allows us to differentiate a function such as $y = (4x - 5)^{12}$. If we consider y as a composite function with $u = 4x - 5$ and $y = u^{12}$, we can apply the chain rule:

$$\begin{aligned}\frac{dy}{dx} &= \frac{dy}{du} \cdot \frac{du}{dx} \\ &= \frac{d}{du}(u^{12}) \cdot \frac{d}{dx}(4x - 5) \\ &= 12u^{11} \cdot 4\end{aligned}$$

Substituting $4x - 5$ for u, we get

$$\frac{dy}{dx} = 12(4x - 5)^{11} \cdot 4$$

Without simplifying, let's observe the result of this differentiation again, looking for a pattern.

$$\frac{dy}{dx} = \underbrace{12(4x - 5)^{11}}_{\textit{derivative of global part}} \quad \underbrace{\cdot}_{\textit{times}} \quad \underbrace{4}_{\textit{derivative of local part}}$$

The pattern we observe is in two parts. The global part follows the rule for powers:

$$\frac{d}{du}(u^n) = nu^{n-1} \rightarrow 12(4x - 5)^{11}$$

and the local part is $\dfrac{du}{dx} \rightarrow 4$.

Thus,

$$\frac{dy}{dx} = \underbrace{12(4x - 5)^{11}}_{nu^{n-1}} \cdot \underbrace{4}_{du/dx}$$

From this pattern, we can suggest the following differentiation rule for powers.

> **General Power Rule**
>
> If $y = [u(x)]^n$, then $\dfrac{dy}{dx} = n[u(x)]^{n-1} \cdot \dfrac{du}{dx}$, where n is any real number.

The general power rule might be remembered as: the derivative of a power is *the derivative of the global part times the derivative of the local part.* It may be that the derivative of the local part requires several differentiation rules. (See Example Set F.)

EXAMPLE SET C

Find the derivative of each of the following functions.

1. $f(x) = (4x^3 + 5x + 2)^6$

Solution:
From the general power rule, $f'(x) = n[u(x)]^{n-1} \cdot \frac{du}{dx}$, where $u = 4x^3 + 5x + 2$,

$$\begin{cases} \text{derivative of global part: } n[u(x)]^{n-1} = 6(4x^3 + 5x + 2)^5, & \text{and} \\ \text{derivative of local part: } \frac{du}{dx} = \frac{d}{dx}(4x^3 + 5x + 2) = 12x^2 + 5 & \end{cases}$$

$$f'(x) = 6(4x^3 + 5x + 2)^5(12x^2 + 5)$$

2. $f(x) = \dfrac{4}{(6 - 2x)^5}$

Solution:
Since this quotient has a constant in the numerator, we will rewrite it as

$$f(x) = 4(6 - 2x)^{-5}$$

From the general power rule, $f'(x) = n[u(x)]^{n-1} \cdot \frac{du}{dx}$, where $u = 6 - 2x$,

$$\begin{cases} \text{derivative of global part: } n[u(x)]^{n-1} = -20(6 - 2x)^{-6}, & \text{and} \\ \text{derivative of local part: } \frac{du}{dx} = \frac{d}{dx}(6 - 2x) = -2 & \end{cases}$$

$$f'(x) = -20(6 - 2x)^{-6}(-2)$$

$$= 40(6 - 2x)^{-6}$$ Eliminate the negative exponent.

$$= \frac{40}{(6 - 2x)^6}$$

3. $y = \sqrt[5]{(7x - 8)^3}$

Solution:
Since our rules of differentiation involve exponents and not radicals, we will begin by writing y in exponential form.

$$y = (7x - 8)^{3/5}$$

From the general power rule, $\frac{dy}{dx} = n[u(x)]^{n-1} \cdot \frac{du}{dx}$, where $u = 7x - 8$,

$$\begin{cases} \text{derivative of global part: } n[u(x)]^{n-1} = \frac{3}{5}(7x - 8)^{-2/5}, & \text{and} \\ \text{derivative of local part: } \frac{du}{dx} = \frac{d}{dx}(7x - 8) = 7 & \end{cases}$$

$$\frac{dy}{dx} = \frac{3}{5}(7x - 8)^{-2/5} \cdot 7$$

$$= \frac{21}{5}(7x - 8)^{-2/5}$$ Eliminate the negative exponent.

$$= \frac{21}{5(7x - 8)^{2/5}}$$ Convert back to radical notation.

$$= \frac{21}{5\sqrt[5]{(7x-8)^2}}$$

The problems of Example Sets D, E, and F illustrate how it is sometimes necessary to use several differentiation rules to differentiate a function.

EXAMPLE SET D

Find the derivative of the function $f(x) = \dfrac{(2x-1)^4}{3x+2}$, at $x = -1$, and interpret this result.

Solution:
Viewing this function globally, we see a quotient. Locally, we see a power. We will begin by using the quotient rule.

$$f'(x) = \frac{(3x+2)\cdot\frac{d}{dx}(2x-1)^4 - (2x-1)^4\cdot\frac{d}{dx}(3x+2)}{(3x+2)^2}$$

$$= \frac{(3x+2)\cdot 4(2x-1)^3\cdot 2 - (2x-1)^4\cdot 3}{(3x+2)^2} \qquad \text{Simplify.}$$

$$= \frac{8(3x+2)(2x-1)^3 - 3(2x-1)^4}{(3x+2)^2}$$

Now we need to factor the numerator. The smallest exponent on the common factor $2x-1$ is 3. Thus, the greatest common factor of the expression in the numerator is $(2x-1)^3$. Continuing to simplify the derivative we have

$$f'(x) = \frac{(2x-1)^3[8(3x+2) - 3(2x-1)]}{(3x+2)^2}$$

$$= \frac{(2x-1)^3[24x+16-6x+3]}{(3x+2)^2}$$

$$= \frac{(2x-1)^3[18x+19]}{(3x+2)^2}$$

When $x = -1$,

$$f'(-1) = \frac{(-27)(1)}{1}$$

$$= \frac{-27}{1} \quad \text{or} \quad -27$$

Interpretation: When the value of x increases by 1 unit, from -1 to 0, the value of y decreases by approximately 27 units. Similarly, if the value of x decreases by 2 units, from -1 to -3, then

$$f'(-1) = -27 = \frac{-27}{1} = \frac{-27}{1} \cdot \frac{-2}{-2} = \frac{54}{-2}$$

and we conclude that $f(x)$ increases by approximately 54 units.

EXAMPLE SET E

Find the derivative of the function $f(x) = x^5(4x - 1)^{1/4}$.

Solution:
Viewing this function globally, we see a product. Locally, we see powers. We will begin by using the product rule.

$$f'(x) = \frac{d}{dx}x^5 \cdot (4x-1)^{1/4} + x^5 \cdot \frac{d}{dx}(4x-1)^{1/4}$$

$$= 5x^4(4x+1)^{1/4} + x^5 \cdot \frac{1}{4} \cdot (4x+1)^{-3/4} \cdot 4 \qquad \text{Simplify.}$$

$$= 5x^4(4x+1)^{1/4} + x^5(4x+1)^{-3/4} \qquad \text{Factor.}$$

$$= x^4(4x+1)^{-3/4}[5(4x+1) + x]$$

$$= x^4(4x+1)^{-3/4}[20x + 5 + x] \qquad \text{Eliminate the negative exponent.}$$

$$= \frac{x^4(21x+5)}{(4x+1)^{3/4}}$$

EXAMPLE SET F

Find $f'(x)$ for the function $f(x) = \left(\frac{2x+5}{8x+7}\right)^4$.

Solution:
We see this function globally as a power. Locally, we see a quotient. To differentiate this function we will use both the power rule and the quotient rule.

$$f'(x) = 4\left(\frac{2x+5}{8x+7}\right)^3 \cdot \frac{d}{dx}\left(\frac{2x+5}{8x+7}\right)$$

$$= 4\left(\frac{2x+5}{8x+7}\right)^3 \cdot \frac{(8x+7)(2) - (2x+5)(8)}{(8x+7)^2}$$

$$= 4\frac{(2x+5)^3}{(8x+7)^3} \cdot \frac{16x + 14 - 16x - 40}{(8x+7)^2}$$

$$= \frac{4(2x+5)^3(-26)}{(8x+7)^5}$$

$$= \frac{-104(2x+5)^3}{(8x+7)^5}$$

The Chain Rule and Technology

Using Your Calculator You can use your graphing calculator to find the positive x value for which $f'(x) = 0$, where $f(x) = \frac{(2x-1)^4}{3x+2}$. The following entries show how this can be done.

```
Y1 = (2x − 1)^4/(3x + 2)
Y5 = nDeriv(Y1,X,X,.001)
Y7 = solve(Y5,X,G)
```

The function Y7 states that the derivative function, Y5, is the function to be set equal to zero and solved for x. The value G represents your *guess* at the zero-producing value.

You can make an educated guess at where Y5 will be 0 by looking at the graph of Y5. For example, the graph of Y5 in the standard $-10 \le x \le 10$ and $-10 \le y \le 10$ window shows that $f'(x) \approx 0$ when $x \approx 0.53$. Now, activate Y7 at G $= 0.53$ to solve $f'(x) = 0$ by entering

```
Y7(0.53)
```

in the computation window and computing. The result is .5000002143, and you can conclude that $f'(x) \approx 0$ when $x \approx 0.5$. (You might try any guess for G like 1, 10, or even 100. Your calculator is still likely to give you a zero-producing value. If you make such a guess however, you may not get the zero-producing value you want.)

Calculator Exercises

1. For the function $f(x) = \sqrt{\frac{x^3-20}{x^3+30}}$, find $f(3)$, $f'(3)$, and $f''(3)$.

2. For the function $f(x) = (1.5x^{3/4} - 10)^{4/3}\frac{(x^{1.8}-7)^{2/3}}{(x^{1.4}+5)^{3/2}}$, find $f(3)$, $f'(3)$, and $f''(3)$.

3. Sketch the graph of the function

$$f(x) = \frac{\sqrt{20.5-(x-5.2)^2}}{1+x^{0.8}}$$

and estimate the x value for which $f'(x) = 0$. In the same window, construct the graphs of $f(x)$ and $f'(x)$. You have estimated the value for which $f'(x) = 0$. Does this value appear to be the point at which the graph of $f(x)$ intercepts the x axis? What is the original function value at this x value? What can you say about the slope of the tangent line to the original function at this x value?

4. Conservationists estimate that if certain safeguards are put into effect and stay in effect over the next 15 years, t years from now there will be N hundred wild snorter pigs in the southwest United States, where

$$N(t) = \frac{(0.88t + 1.3)^3 + 400}{t + 1}$$

Sketch the graph of the function $N(t)$ and estimate the t value for which $N'(t) = 0$. If you denote the t value for which $N'(t) = 0$ by the letter c, then compute and interpret both $f(c)$ and $f'(c)$. Compute $f(c + 1)$ and compare it to the value $f(c) + f'(c)$. What are these values?

Using Derive You can use Derive to find the values of x for which $f'(x) = 0$. The following entries show how this can be done for the function $f(x) = \frac{(2x - 1)^4}{3x + 2}$. Press Enter at the end of each line. This set of commands uses Derive's numerical mathematics utility. It is not part of Derive's standard set of utilities, so you will need to load it. The first four commands do this.

Transfer
Load
Utility
numeric.mth
Author
(2x - 1)^4/(3x + 2)
Calculus
Differentiate (*Press* Enter *three times to differentiate with respect to x and get the first derivative.*)
Simplify
so**L**ve (*Press* Enter *three times to get the values of x for which the first derivative equals 0.*)

The soLve command sets this expression equal to 0 and solves for the variable.

Derive Exercises Use Derive to find $f'(x)$ and then the values of x for which $f'(x) = 0$.

1. $f(x) = \frac{1}{3}x^3 - x^2 - 3x + 5$

2. $f(x) = \frac{1}{4}x^4 - \frac{1}{3}x^3 - 3x^2 + 1$

3. $f(x) = \frac{2}{3\sqrt[3]{(x + 2)^2}}$

4. $f(x) = x^{2/3}(x - 4)$

5. The revenue R, in millions of dollars, realized by a company on a product t weeks after its release is given by the function

$$R(t) = \frac{44t}{t^2 + 31}$$

Find $R'(t)$ and the value(s) of t for which $R(t) = 0$.

EXERCISE SET 3.4

A UNDERSTANDING THE CONCEPT

For Exercises 1–8, state whether or not the chain rule is the appropriate differentiation technique for finding the specified derivative.

1. $y = 8u + 6$ and $u = 2x - 7$; $\frac{dy}{dx}$
2. $y = u^2 + 3u - 10$ and $u = 12x^2 + 2$; $\frac{dy}{dx}$
3. $m = 2x + 16$ and $x = n^2 + 6n + 1$; $\frac{dm}{dn}$
4. $r = 5s^2 + 2s$ and $s = (3t + 1)^4$; $\frac{dr}{dt}$
5. $y = 4a + 1$ and $a = 3b + 6$; $\frac{dy}{da}$
6. $y = 3s + 7$ and $s = (4t^2 + t - 6)^2$; $\frac{dy}{ds}$
7. $y = 15t^3 + 2t^2 - 8$ and $x = 9t - 1$; $\frac{dy}{dt}$
8. $s = 4t + 5$ and $t = 6$; $\frac{ds}{dt}$

For Exercises 9–10, suppose that $y = f(u)$, $u = g(x)$, and that $\frac{dy}{dx}$ can be determined using the chain rule.

9. Interpret the statement $\frac{dy}{dx} = 4$ when $x = 2$.
10. Interpret the statement $\frac{dy}{dx} = -6$ when $x = 15$.
11. *Business: Jobs* A population researcher has found that the population P (in thousands) of a city in a particular year is a function of the number N (in thousands) of the jobs available in that year and that the number of jobs available in that year is, in turn, a function of the economy E (in billions of dollars) of the state in which that city is located. The researcher has determined that for a particular year, $E = 204$, $\frac{dP}{dN} = 8$, and $\frac{dN}{dE} = 0.6$. Find $\frac{dP}{dE}$ and interpret its value for $E = 204$.
12. *Economics: Educational Endowments* A researcher has found that the number N of endowments made to a university in a particular year is a function of the size P (in hundreds of full-time students) of the university and that the size of the university is a function of the number J of papers published in major journals by the faculty. The researcher has determined that for a particular year, $J = 78$, $\frac{dN}{dP} = 4$, and $\frac{dP}{dJ} = 0.7$. Find $\frac{dN}{dJ}$ and interpret its value for $J = 78$.

For Exercises 13–16, specify the global form (power, product, or quotient) of the given function.

13. $f(x) = \frac{(x+2)^3}{(5x-8)^4}$
14. $f(x) = \left[(x-6)^2(x+4)^4\right]^8$
15. $f(x) = \left[\frac{(x+1)^3}{(x-1)^4}\right]^5$
16. $f(x) = \left[(x+2)^2(x+3)^2\right]^4\left[(x-1)^3(x+1)^5\right]^5$

B SKILL ACQUISITION

For Exercises 17–30, find $\frac{dy}{dx}$.

17. $y = 12u - 7$ and $u = 3x + 1$
18. $y = 16u^2 + u - 3$ and $u = 8x - 5$
19. $y = 10u + 1$ and $u = -x^2 + 3$
20. $y = 8u - 6$ and $u = -x^4 + 3x^2 + x$
21. $y = 3u^2 + 2$ and $u = 5x^2 + 4$
22. $y = 4u^2 + u - 3$ and $u = 3x^2 - 8$
23. $y = u^3 + 6u$ and $u = 5x^2 - 2$
24. $y = u^3 + 8u$ and $u = 3x^2 - 4$
25. $y = (5x^2 + 2x - 6)^3$
26. $y = (6x^3 + x)^4$
27. $y = \sqrt[5]{(x^2+7)^3}$
28. $y = \sqrt[6]{(4x^2+2x-1)^5}$
29. $y = (x^2 + 3)^3(2x - 1)^5$
30. $y = (4x + 1)^6(2x^2 + 3x - 4)^8$

For Exercises 31–46, find $\frac{dy}{dx}$ without using the quotient rule; rather, rewrite each function by using a negative exponent and then use the general power rule to find the derivative.

31. $y = \frac{1}{x+3}$
32. $y = \frac{1}{x-5}$
33. $y = \frac{8}{(x-4)^3}$
34. $y = \frac{5}{(x+3)^2}$
35. $y = \frac{3}{(2x+7)^4}$
36. $y = \frac{4}{(5x-11)^6}$

37. $y = \dfrac{(x^2+6)^7}{x-2}$

38. $y = \dfrac{(x^2-5)^4}{x-8}$

39. $y = \dfrac{3x^2+1}{(x+2)^3}$

40. $y = \dfrac{3x+4}{(x^2+1)^5}$

41. $y = x^4(5x+2)^{1/5}$

42. $y = 3x^2(6x-8)^{1/6}$

43. $y = 2x^7(4x+1)^{3/4}$

44. $y = 5x^3(8x^2+6)^{5/16}$

45. $y = \left(\dfrac{9x-4}{3x+2}\right)^3$

46. $y = \left(\dfrac{5x-4}{8x-1}\right)^5$

C APPLYING THE CONCEPTS

47. *Economics: Government Spending* The rate of government spending (in billions of dollars) in a country is related to the country's unemployment as

$$g(u) = 193 + u^{6/5}$$

where g is the level of government spending and u is the percentage of unemployment rate.

The percentage unemployment rate, in turn, is related to the inflation rate as

$$u(i) = 80(i+10)^{-1}$$

where i is the inflation rate. What change in spending can the government expect if the inflation rate rises from 7% to 8%?

48. *Business: Sale of Computers* An empirical relationship between the number of computers sold in a city and the median age of the people living in the city is given by the function

$$N(a) = -a^2 + 60a + 160$$

where N is the number of computers sold and a is the median age of the people.

The value of the software packages sold in the city depends on the number of computers sold as

$$V(N) = 100N + 500N^{3/4}$$

where V is the value of the software packages sold.

a. Write a symbolic description that shows how the value of the software packages sold changes with the median age of the people in the city.

b. How will the value change if the median age increases from 30 years to 31 years?

49. *Business: Demand for a Product* Demand for a product depends on the price of the product as

$$d(p) = 420p^{-2}$$

where d is the number of units demanded and p is the price of the product in dollars.

The cost per unit sold depends on the number produced as

$$c(d) = 20 + 10d^{-1/3}$$

where c is the cost per unit in dollars.

a. Write a symbolic description that shows how the cost of producing a unit changes with the sales price.

b. How does the cost of production change as the price increases from \$50 to \$51 per unit?

50. *Economics: Money Supply* The money supply is a function of interest rates as

$$m(i) = 360 - 84i^{3/5}$$

where m is the money supply and i is the interest rate.

The general price level is a function of the money supply as

$$p(m) = 10\sqrt{m}$$

where p is the general price level.

a. Write a symbolic description of how the general price level changes with the interest rate.

b. What change in prices can be expected if interest rates rise from 8% to 9%?

51. *Business: Production* The distribution manager of a national supplier of paper note pads knows that the number N (in thousands) of pads to be packaged is related to the price P (in dollars) of a pad. That relationship is expressed by the function $N(P) = -P^2 + 4P + 12$, $2 \le P \le 6$. The price of each pad is, in turn, related to the cost C (in dollars) of a unit of paper. That relationship is expressed by the function $P(C) = 1.5C + .5$. The cost of paper is currently \$1.70 per unit but is expected to increase by one cent next month. The distribution manager needs to prepare the packaging facility for a change in the number of pads to be packaged. What approximate change in the number of pads to be packaged should the manager expect?

52. *Ecology: Lake Toxicity* The amount A (in hundreds of gallons) of toxic material entering a lake is related to the time t (in months) an industrial factory near one of the lake's inlet rivers has been operating. The relationship is expressed by the function

$$(A(t) = (0.8t^{1/5} + 2)^4$$

At what rate is the toxic material in the lake changing after 32 months?

53. ***Business: Jobs*** The marketing department of a pool supply company has determined that the demand D (in thousands of tablets) for its pool chlorination tablets is related to the price x (in cents) by the function

$$D(x) = 400\left(\frac{8}{2x - 25}\right)^{1/4}$$

$15 \leq x \leq 100$. Currently, the price of a tablet is \$0.40. What can the marketing department expect of the demand for the tablet if the price of a tablet is raised to \$0.41?

54. ***Psychology: Memorization*** A psychologist has determined that the number N (in units) of nonsense letters a subject can memorize is related to the amount of time t (in seconds) the subject has to memorize them. The relationship is expressed by the function

$$N(t) = \frac{1}{2}\sqrt{4t^2 + 9} - \frac{3}{2}$$

$0 \leq t \leq 20$.

a. If the experimenter increases the amount of time available to a subject from 2 seconds to 3 seconds, approximately how many more letters should he expect the subject to memorize?
b. If the experimenter increases the amount of time available to a subject from 2 seconds to 7 seconds, approximately how many more letters should he expect the subject to memorize? (Hint: Look at the value of $N'(2)$ and interpret it as a slope.)

D DESCRIBE YOUR THOUGHTS

55. Explain why it is important to be able to look at a function globally and recognize it as a sum, difference, product, quotient, or power.

56. Describe how the function $f(x) = (5x - 4)^3$ can be thought of as a composite function.

57. Describe how you would differentiate the function $f(x) = (3x^2 + 4)^2(6x - 4)^5$. Do not perform the differentiation, just describe the strategy you would use.

E REVIEW

58. **(2.2)** Determine $\lim_{x \to 4} \frac{x^2 - x - 12}{x^2 - 2x - 8}$.

59. **(2.3)** Determine $\lim_{x \to \infty} \frac{5x^3 - 2x^2 + x - 1}{6x^4 - x + 6}$.

60. **(2.4)** Discuss the continuity of the following function

$$f(x) = \begin{cases} x^2 - 8x + 18, & \text{if } x \leq 4 \\ \frac{1}{2}x + 1, & \text{if } x > 4 \end{cases}$$

at $x = 4$.

61. **(2.6)** For the function $f(x) = x^3 + 6x - 12$, find
a. the average rate of change of $f(x)$ as x changes from -1 to 2.
b. the instantaneous rate of change of $f(x)$ at $x = 0$.

62. **(3.2)** Use the differential to find the approximate change in y as x changes from 8 to 8.03 for the function $y = 2x^3 - x - 4$.

3.5 Implicit Differentiation

Introduction
Explicit and Implicit Functions
Implicit Differentiation
Related Rates
Implicit Differentiation and Technology

Introduction

The functions we have differentiated so far have been of the form $f(x) =$ (some expression in x). That is, all instances of the input variable are on one and only one side of the equal sign. In this section we examine implicit differentiation, a technique for finding the derivative of a function when the input and output variables appear on the same side of the equal sign.

Explicit and Implicit Functions

Explicit Function

Equations that are expressed as $y = f(x)$ directly convey the relationship between y and x. Since y is expressed explicitly as a function of x, the form $y = f(x)$ conveys y as an **explicit function** of x. All the functions we have worked with so far have been explicit functions. Some examples are

$$y = 5x - 6, \qquad y = 8x^2 + 7x - 11,$$

$$y = \frac{2x - 5}{x^2 - 6}, \qquad y = (x - 4)^3(3x + 5)^2$$

Implicit Function

Some equations do not directly convey the relationship between y and x, but only imply it. Equations such as

$$xy = 3, \qquad x^2 + y^2 = 4,$$

$$y^2 = x, \qquad x^2 + 3xy + y^2 = 5$$

indirectly convey the relationship between y and x. Equations that only imply that y is a function of x are said to convey y as an **implicit function** of x.

Each of the above implicit functions can be converted to explicit functions by solving them for y. For example, for $x^2 + y^2 = 4$,

$$x^2 + y^2 = 4$$

$$y^2 = 4 - x^2$$

$$y = \pm\sqrt{4 - x^2}$$

so that $y = \sqrt{4 - x^2}$ and $y = -\sqrt{4 - x^2}$. We could now use the general power rule to find $\frac{dy}{dx}$.

Quite often, however, it is impractical, or even impossible, to convert an implicit function of x to an explicit function of x by solving for y. For example, it is impractical (or maybe even impossible) to solve the function

$$x^5 + 3x^2y^3 + y^5 = 2$$

for y. But we can find $\frac{dy}{dx}$ (or y') for such a function using a method called implicit differentiation.

Implicit Differentiation

Implicit Differentiation

Implicit differentiation is the method of finding $\frac{dy}{dx}$ (or y') when y is an implicit function of x. It is a direct application of the general power rule, and we can gain insight to it by studying the following example.

Suppose y is an implicit function of x. Since y is not conveyed directly as a function of x, its form, in terms of x, is unknown. Let $f(x)$ represent this unknown form. Then, with $y = f(x)$, $y^n = [f(x)]^n$, and

$$\frac{d}{dx}y^n = \frac{d}{dx}[f(x)]^n$$

$$= n \cdot [f(x)]^{n-1} \cdot f'(x) \qquad \text{by the general power rule}$$

$$= ny^{n-1}y' \qquad \text{since } f'(x) = y'$$

Derivative of an Implicit Function

If y is an implicit function of x, then

$$\frac{d}{dx}y^n = ny^{n-1}y'$$

CAUTION

It is very common for students to find the derivative of y^n, with respect to x, as ny^{n-1}. This is *not* correct. The expression ny^{n-1} represents $\frac{d}{dy}y^n$, the derivative with respect to y, not $\frac{d}{dx}y^n$, the derivative with respect to x.

ILLUMINATOR SET A

Suppose $y^3 - 5x^2 = 8$. We say that y is an implicit function of x. Since y is not conveyed directly in terms of x, its form, in terms of x, is unknown. Let $f(x)$ represent this unknown form. Then, with $y = f(x)$, we have that

$$y^3 - 5x^2 = 8 \text{ is equivalent to } [f(x)]^3 - 5x^2 = 8$$

Now differentiate each side with respect to x, using the general power rule.

$$\frac{d}{dx}[f(x)]^3 - \frac{d}{dx}[5x^2] = \frac{d}{dx}[8]$$

$$3 \cdot [f(x)]^2 \cdot f'(x) - 10x = 0$$

Now replace $f(x)$ with y and $f'(x)$ with y', and solve for y'.

$$3y^2y' - 10x = 0$$

$$3y^2y' = 10x$$

$$y' = \frac{10x}{3y^2}$$

Notice that y' depends on both x and y.

In the examples that follow, we will omit the step of replacing y with its unknown form $f(x)$, and directly use the fact that

$$\frac{d}{dx}[y^n] = ny^{n-1}y'$$

EXAMPLE SET A

Find and interpret y' for $3x^5 + 2y^4 + y = 37$ at the point $(1, 2)$.

Solution:
Since y is an implicit function of x, we need to differentiate implicitly.

$$\frac{d}{dx}[3x^5] + \frac{d}{dx}[2y^4] + \frac{d}{dx}y = \frac{d}{dx}[37]$$

$$15x^4 + 8y^3y' + y' = 0 \qquad \text{Isolate } y'.$$

$$8y^3y' + y' = -15x^4 \qquad \text{Factor out } y'.$$

$$y'(8y^3 + 1) = -15x^4 \qquad \text{Divide by } 8y^3 + 1.$$

$$y' = \frac{-15x^4}{8y^3 + 1}$$

Now evaluate y' at $x = 1$ and $y = 2$.

$$\begin{aligned} y'\big|_{x=1,y=2} &= \frac{-15x^4}{8y^3 + 1}\bigg|_{x=1,y=2} \\ &= \frac{-15(1)^4}{8(2)^3 + 1} \\ &= \frac{-15(1)}{64 + 1} \\ &= \frac{-15}{65} \\ &= \frac{-3}{13} \end{aligned}$$

Related Rates

Suppose the dependent variable y is related to the independent variable x by some function $y = f(x)$. Suppose also that y and x are both related to a third variable t (usually time in applications). Then $\frac{dy}{dt}$ and $\frac{dx}{dt}$, respectively, express the rates of change of y and x with respect to t. If the equation relating y and x is differentiated implicitly with respect to t, an equation involving $\frac{dy}{dt}$ and $\frac{dx}{dt}$ will result, and that equation will relate the rates of change $\frac{dy}{dt}$ and $\frac{dx}{dt}$. Problems in which rates of change are related to other rates of change are called **related rate problems**.

Related rate problems are solved using implicit differentiation. Example Sets B, C, and D illustrate the process.

EXAMPLE SET B

Suppose both y and x are differentiable functions of t and that the relationship between y and x is expressed by the equation $4x^3 + 3y^5 = 960$. Find and interpret $\frac{dy}{dt}$ when $\frac{dx}{dt} = 4$, $x = 6$, and $y = 2$.

Solution:

Since both y and x are functions of t, we need to differentiate them using implicit differentiation.

$$\frac{d}{dt}[4x^3] = 12x^2\frac{dx}{dt}, \quad \frac{d}{dt}[3y^5] = 15y^4\frac{dy}{dt}, \quad \frac{d}{dt}[960] = 0$$

Then, differentiating both sides of $4x^3 + 3y^5 = 960$, we get

$$\frac{d}{dt}(4x^3 + 3y^5) = \frac{d}{dt}960$$

$$\frac{d}{dt}[4x^3] + \frac{d}{dt}[3y^5] = \frac{d}{dt}[960]$$

$$12x^2\frac{dx}{dt} + 15y^4\frac{dy}{dt} = 0$$

Now we solve for $\frac{dy}{dt}$.

$$15y^4\frac{dy}{dt} = -12x^2\frac{dx}{dt}$$

$$\frac{dy}{dt} = \frac{-12x^2 dx/dt}{15y^4}$$

$$= \frac{-4x^2 dx/dt}{5y^4}$$

Now we can evaluate $\frac{dy}{dt}$ when $\frac{dx}{dt} = 4, x = 6$, and $y = 2$.

$$\begin{aligned}\frac{dy}{dt} &= \frac{-4(6)^2(4)}{5(2)^4} \\ &= \frac{-576}{240} \\ &= -7.2\end{aligned}$$

Interpretation: If t is increased by 1 unit, when $dx/dt = 4$, $x = 6$, and $y = 2$, then the value of y can be expected to decrease by approximately 7.2 units.

EXAMPLE SET C

Oil is leaking in a circular shape from a tanker in such a way that the radius of the circular spill is increasing at the rate of 2 feet per hour. To manage the spill, the oil company needs to know the rate at which the area covered by oil is increasing. Find the rate at which the area is increasing at the time that the radius of the spill is 600 feet. (Approximate π by 3.14.)

Solution:

When it is possible, drawings are very helpful in related rate problems. They make relationships more visual and promote better access to the equation, as shown in Figure 3.15. When we see the word *rate*, we think *derivative*. We wish to find $\frac{dA}{dt}$ when $\frac{dr}{dt} = 2$ and $r = 600$. (Note: $\frac{dr}{dt} = 2$, since the radius of the spill is increasing at the rate [which means *derivative*] of 2 feet per hour.)

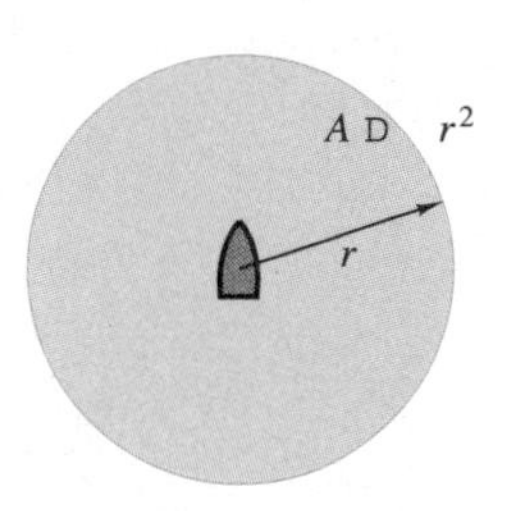

Figure 3.15

The equation relating A and r is $A = \pi r^2$ (the formula for the area of a circle). Since we are interested in the rate of change with respect to time t, we differentiate each side of the equation with respect to t.

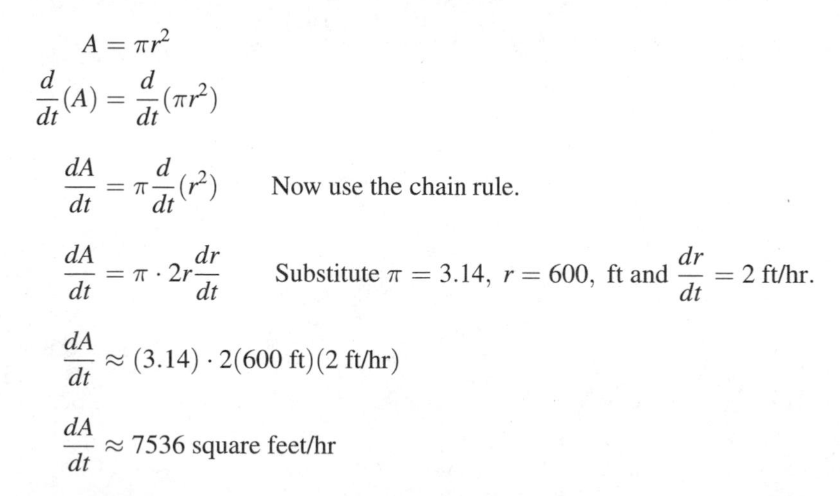

$$A = \pi r^2$$

$$\frac{d}{dt}(A) = \frac{d}{dt}(\pi r^2)$$

$$\frac{dA}{dt} = \pi \frac{d}{dt}(r^2) \qquad \text{Now use the chain rule.}$$

$$\frac{dA}{dt} = \pi \cdot 2r\frac{dr}{dt} \qquad \text{Substitute } \pi = 3.14,\ r = 600, \text{ ft and } \frac{dr}{dt} = 2 \text{ ft/hr.}$$

$$\frac{dA}{dt} \approx (3.14) \cdot 2(600 \text{ ft})(2 \text{ ft/hr})$$

$$\frac{dA}{dt} \approx 7536 \text{ square feet/hr}$$

Interpretation: In the next hour, the area of the oil spill can be expected to increase by approximately 7536 square feet.

EXAMPLE SET D

A manufacturer of precision instruments has determined that when the price of an instrument is p (in hundreds of dollars), it will sell x of those instruments each month. The demand function relating p and x is

$$x = \frac{20{,}000}{\sqrt[3]{2p^2 - 5}} + 350$$

Due to inflation and changing labor costs, both p and x depend on time t (in months). Find the rate at which the number of instruments sold is changing with respect to time, when the price of an instrument is \$400 and is changing at the rate of one dollar per month.

Solution:

We see the word *rate* and think *derivative*. We need to find $\frac{dx}{dt}$ when $\frac{dp}{dt} = 0.01$ and $p = 4$. (Remember, p is in hundreds so \$400 means that $p = 4$ and \$1 means that $\frac{dp}{dt} = \frac{1}{100} = 0.01$.)

We differentiate each side of $x = \frac{20,000}{\sqrt[3]{2p^2 - 5}} + 350$ with respect to t. Since the numerator is a constant, we will rewrite the equation in the more convenient form $x = 20,000(2p^2 - 5)^{-1/3} + 350$. Then

$$\begin{aligned}
\frac{dx}{dt} &= 20{,}000 \cdot \frac{d}{dt}(2p^2 - 5)^{-1/3} + \frac{d}{dt}(350) \\
&= 20{,}000\left[\frac{-1}{3}(2p^2 - 5)^{-4/3} \cdot 4p \cdot \frac{dp}{dt}\right] + 0 \\
&= \frac{-80{,}000(2p^2 - 5)^{-4/3} p}{3} \frac{dp}{dt} \\
&= \frac{-80{,}000p}{3(2p^2 - 5)^{4/3}} \frac{dp}{dt} \\
&= \frac{-80{,}000p}{3\sqrt[3]{(2p^2 - 5)^4}} \frac{dp}{dt}
\end{aligned}$$

Then, when $p = 4$ and $\frac{dp}{dt} = 0.01$,

$$\begin{aligned}
\frac{dx}{dt} &= \frac{-80{,}000(4)}{3\sqrt[3]{(2 \cdot 4^2 - 5)^4}} \cdot (0.01) \\
&= \frac{-3200}{3\left(\sqrt[3]{27}\right)^4} \\
&\approx -13.17 \\
&\approx -13
\end{aligned}$$

Interpretation: At the time when the price of a precision instrument is \$400 and increasing at the rate of \$1 per month, the company can expect to sell about 13 fewer instruments in the next month.

Implicit Differentiation and Technology

Using Derive Derive can produce the symbolic derivative of an implicitly defined function. Begin by replacing y with $f(x)$, just as we did in the development of the differentiation technique. The following entries show the differentiation of the function $3x^2 - x^2y^3 + 4y = 12$. This set of commands uses Derive's numerical mathematics utility. It is not part of Derive's standard set of utilities, so you will need to load it. The first four commands do this.

Transfer
Load
Utility
dif_apps
Author
imp_dif(3x^2 - x^2y^3 + 4y - 12, x, y)
Simplify

The **Transfer**, **Load**, **Utility**, and **dif_apps** commands load the special utilities needed for implicit differentiation. These utilities are not normally part of the running package. The **imp_dif** command performs the implicit differentiation. The x, y following the expression indicates that x is the input variable and that y is the output variable. You might interchange these variables and observe the difference in results.

Derive Exercises Use Derive to differentiate each function.

1. $-5x^3 + 2x^2y^4 + 6y^2 = 10$
2. $\frac{3}{x^2} + \frac{5}{y^3} - xy^2 = 1$
3. $x^3 + y^2 = x^2y^2$
4. $f(x) = x^{2/3}(x - 4)$

EXERCISE SET 3.5

A UNDERSTANDING THE CONCEPTS

For Exercises 1–2, determine if the equation as it is written expresses y as an explicit function of x or as an implicit function of x.

1. The equations are

   a. $3y + 5x^2 = 7$ b. $8x^2 + 3xy + 2y^3 = 4$

   c. $4y = 9x + 6$ d. $y = -6x^3 + 2x + 15$

   e. $x^2 + 3x - 1 = y$

2. The equations are

   a. $5y^4 - 2y^3 + 5x^2y^2 - y = 0$ b. $y = \frac{8x^3 + 3}{x - 1}$

   c. $y^2 = \frac{3x}{4 - 3x}$ d. $x = y^2 + \frac{2y + 6}{y - 1}$

   e. $\sqrt[4]{(x - 1)^3} = y$

In Exercises 3–10, y is expressed implicitly as a function of x. The function is then differentiated and solved for y'. However, each differentiation process contains an error. Specify the first step at which the error occurs and correct it. (The step numbers are shown in parentheses.)

3. $4y^3 - 8x^2 + y = 0$

$$12y^2 - 16x + y' = 0 \quad \ldots (1)$$
$$y' = 16x - 12y^2 \quad \ldots (2)$$

4. $-6y^4 + 3y^2 + 5x^2 + 3y = 9$

$$-24y^3 + 6y + 10x + 3y' = 0 \quad \ldots (1)$$
$$3y' = 24y^3 - 6y - 10x \quad \ldots (2)$$
$$y' = \frac{24y^3 - 6y - 10x}{3} \quad \ldots (3)$$

5. $(2y^3 - 5x^2)^4 + 6y = 0$

$$4(2y^3 - 5x^2)^3 + 5y' = 0 \quad \ldots (1)$$
$$5y' = -4(2y^3 - 5x^2)^3 \quad \ldots (2)$$
$$y' = \frac{-4(2y^3 - 5x^2)^3}{5} \quad \ldots (3)$$

6. $y - 8x = (3x + 2y)^2$

$$y' - 8 = 2(3x + 2y)(3 + 2) \quad \ldots (1)$$
$$y' - 8 = 10(3x + 2y) \quad \ldots (2)$$
$$y' = 10(3x + 2y) + 8 \quad \ldots (3)$$

7. $\sqrt[4]{(3x^4 - 2y^3)} = 12y + 7$

$$(3x^4 - 2y^3)^{1/4} = 12y + 7 \quad \ldots (1)$$
$$\frac{1}{4}(3x^4 - 2y^3)^{-3/4}(12x^3 - 6y^2) = 12y' \ldots (2)$$
$$\frac{1}{4}(3x^4 - 2y^3)^{-3/4} \cdot 6(2x^3 - y^2) = 12y' \ldots (3)$$
$$\frac{3}{2}(3x^4 - 2y^3)^{-3/4}(2x^3 - y^2) = 12y' \ldots (4)$$
$$\frac{3}{24}(3x^4 - 2y^3)^{-3/4}(2x^3 - y^2) = y' \quad \ldots (5)$$
$$\frac{2x^3 - y^2}{8(3x^4 - 2y^3)^{3/4}} = y' \quad \ldots (6)$$

8. $5x^3 + 3x^2y^4 - y = 7$

$$15x^2 + 6xy^4 + 3x^2 \cdot 4y^3 - y' = 0 \quad \ldots (1)$$
$$5x^2 + 6xy^4 + 12x^2y^3 - y' = 0 \quad \ldots (2)$$
$$15x^2 + 6xy^4 + 12x^2y^3 = y' \quad \ldots (3)$$

9. $6y^3 + x^5y^5 = 2x + 7$

$$18xy^2y' + 5x^4 \cdot 5y^4 = 2 \quad \ldots (1)$$
$$18y^2y' + 25x^4y^4 = 2 \quad \ldots (2)$$
$$18y^2y' = 2 - 25x^4y^4 \quad \ldots (3)$$
$$y' = \frac{2 - 25x^4y^4}{18y^2} \quad \ldots (4)$$

10. $\frac{3x}{2y} = 3y^5$

$$2y \cdot 3 - 3x \cdot 2y' = 15y^4y' \quad \ldots (1)$$
$$6y - 6xy = 15y^4y' \quad \ldots (2)$$
$$6y = 6xy' + 15y^4y' \quad \ldots (3)$$
$$6y = y'(6x + 15y^4) \quad \ldots (4)$$
$$y' = \frac{6y}{6x + 15y^4} \quad \ldots (5)$$
$$y' = \frac{6}{6x + 15y^3} \quad \ldots (6)$$
$$y' = \frac{2}{2x + 5y^3} \quad \ldots (7)$$

B SKILL ACQUISITION

For Exercises 11–25, find y' for each function.

11. $3y^4 + 2x^3 = 4$

12. $5y^2 - 11x^2 = 9$

13. $3y^3 - 2y^2 + 5x^4 - x = 1$

14. $3x^4 + 6y^4 - 5y^3 = 5x^4 - 2x$

15. $(y + 6)^8 = 4x^2 + x - 4$

16. $(3y^2 + 4)^4 - 3x^5 + 5 = 0$

17. $(2y^3 + 5x^2)^3 = 6x^2 + 11$

18. $5xy^2 + y^3 = 4$

19. $6x^2 + 4x^2y^4 = 7x - 1$

20. $6x - 4x^5y^3 + y^5 = 1$

21. $5x^2 - 3x^2y^2 - y^3 = 2x$

22. $\sqrt[4]{2y + 3} = 9x + 4$

23. $\sqrt[5]{(y^1+1)^2} = 1 - 4x$

24. $\dfrac{3x+1}{y^5-2} = 10x$

25. $\dfrac{2y-7}{y^3+1} = 8$

For Exercises 26–30, find and interpret y' at the specified point.

26. $y^2 + y^3 = 12x + 12$, at $(2, 3)$

27. $4y^3 + 3x^4 + y = -31$, at $(-2, 1)$

28. $\dfrac{1}{x+y} = \dfrac{1}{5}$, at $(6, -1)$

29. $x^{-3} + y^{-3} = -\dfrac{35}{216}$, at $(-2, -3)$

30. $(xy)^{2/3} = 9$ at $(3, 9)$

C APPLYING THE CONCEPTS

31. *Business: Price and Demand* The number x (in thousands) of cat flea collars demanded each year when the price of a collar is p dollars is expressed by the function $x^3 + 250p^2 = 18{,}000$. The collars are currently selling for \$4 each and the annual number of sales is 24,101. Find the approximate decrease in sales of the collar if the price of each collar is raised by \$1.

32. *Business: Price and Demand* Suppose that for a particular item the equation $8p + 2xp + 5x^2 = 400$ relates the price p (in dollars) and the demand x (in thousands of units) for the item. Find the approximate change in sales (demand) if the selling price is increased one dollar from \$11 to \$12.

Exercises 33–38 are related rate problems.

33. *Manufacturing: Computer Costs* The cost C (in dollars) of manufacturing x number of high-quality computer laser printers is

$$C(x) = 15x^{4/3} + 54x^{2/3} + 600{,}000$$

Currently, the level of production is 1728 printers and that level is increasing at the rate of 350 printers each month. Find the rate at which the cost is increasing each month.

34. *Management: Air Quality and Automobiles* The air quality monitoring department of a city has established a relationship between the quality of air and the number of automobiles driven in the city. The air quality is measured in pollution index units, I, $0 \leq I \leq 100$, and the number of automobiles driven in the city x is measured in thousands. The relationship is expressed by the function

$$I(x) = 30 + (5x - 277)^{3/5}.$$

Currently, there are 75,000 automobiles being driven in the city ($x = 75$) and that number is increasing at the rate of 4,500 per year. By how many points can the people at the air quality monitoring department expect the pollution index to increase in the next year?

35. *Business: Revenue* The monthly revenue R (in dollars) of a telephone polling service is related to the number x of completed responses by the function

$$R(x) = -12{,}000 + 25\sqrt{3.5x^2 + 25x}$$

$0 \leq x \leq 1500$. If the number of completed responses is increasing at the rate of 10 forms per month, find the rate at which the monthly revenue is changing when $x = 750$.

36. *Economics: Foreign Aid* The amount of economic aid E (in millions of dollars) a country receives from the United States is related to the amount of military aid x (in millions of dollars) it receives by the function

$$E(x) = 10 + \sqrt{5x^3 + x^2}$$

$0 \leq x \leq 25$. If military aid is decreasing at the rate of 4 million dollars per year, find the rate at which the economic aid is changing when the military aid is \$18 million a year.

37. *Business: Revenue* The monthly revenue R (in dollars) realized by a chain of nursery stores is related to the number x of house plants it can secure and sell each month by the function

$$R(x) = 35{,}000 + \sqrt{\frac{x^4 + 800}{3x + 650}}$$

$0 \leq x \leq 5000$. If the number of house plants being secured and sold is increasing at the rate of 120 plants each month, at what rate is the revenue changing when the number of plants the chain is securing and selling is 1500?

38. *Manufacturing: Cost of Production* The manufacturer of reinforced cardboard packing boxes has determined that its monthly cost C (in dollars) of producing such boxes is related to the price p (in dollars) it has to pay for a unit of cardboard by the function

$$C(p) = 45{,}000 + 18{,}000\left(\frac{250p + 20}{80p + 90}\right)^{3/4}.$$

If the current price of a unit of cardboard is \$0.40 but is increasing at the rate of \$0.02 per month, at what rate is the cost of producing the boxes changing?

D DESCRIBE YOUR THOUGHTS

39. Explain the difference between an explicit and implicit function of x.

40. The derivative of $8x^3y^4 = 5$ with respect to x is $24x^2y^4 + 32x^3y^3y'$. Explain the occurrence of the factor y'.

41. Explain what is meant by the term *related rate.*

E REVIEW

42. **(2.2)** Find $\lim_{x \to 2} \dfrac{x^2 - 7x + 10}{x - 2}$.

43. **(3.2)** Find and interpret $f'(x)$ for $f(x) = \dfrac{5x + 2}{x - 1}$.

44. **(3.3)** Find $f'(x)$ for $f(x) = (5x^2 + 8x)^3$.

45. **(3.3)** Find $f'(x)$ for $f(x) = (3x + 3)^3(2x - 1)^2$.

46. **(3.4)** Find $f'(x)$, $f''(x)$, and $f'''(x)$ for $f(x) = 20x^2 + 8x + 4$.

Summary

This chapter started your study of calculus. The chapter was devoted to the derivative of a function.

The Derivative of a Function and Two Interpretations

This first section presented the definition of the derivative of a function, and spoke of differentiation as the language of change. The *derivative* of the function $f(x)$, denoted $f'(x)$, is

$$f'(x) = \lim_{h \to 0} \frac{f(x+h) - f(x)}{h}$$

It is important that you understand the difference between $f(x)$ and $f'(x)$. $f(x)$ represents the output value for a given input value, while $f'(x)$ represents the rate at which the output value is changing. It is also important that when you are computing the derivative of a function using the definition, you write the lim at each step but the last. Remember that $\lim_{h \to 0} \dfrac{f(x+h) - f(x)}{h}$ is different from $\dfrac{f(x+h) - f(x)}{h}$. The first represents the *instantaneous* rate of change of the function $f(x)$ and the second the *average* rate.

Another important concept discussed was the two interpretations of the derivative of a function and when to use them. The derivative $f'(x)$ of the function $f(x)$ is (1) the slope of the tangent line to $f(x)$. Use this interpretation when you are thinking or visualizing geometrically (such as looking at the graph of the function); (2) the rate at which the function is changing. Use this interpretation when you are interpreting in applications.

This section also introduced the concept of the *marginal* change. It is the change in the output variable that is produced by a 1-unit change in the input variable and is called **the marginal** change in the output.

Also introduced were various notations for the derivative, and three conditions for nondifferentiability. You should note that a function is not differentiable at a point $x = a$ if (1) the function is discontinuous at $x = a$; (2) it has a corner at $x = a$; or (3) the function has a vertical tangent line at $x = a$.

This section ended by examining the fundamental rules of differentiating functions. The rules follow directly from the definition and are examined here.

1. If $y = c$, where c is any real number, then $\frac{dc}{dx} = 0$.
2. If $y = x^n$, where n is any real number, then $\frac{dy}{dx} = n \cdot x^{n-1}$.
3. An important consequence of rule 2 is that if $y = x$, then $y' = 1$.
4. If $f(x)$ is a differentiable function, then $\frac{d}{dx}cf(x) = c \cdot \frac{d}{dx}f(x)$.
5. An important consequence of rule 4 is that if $y = cx$, then $y' = c$.
6. If $f(x)$ and $g(x)$ are both differentiable functions, then $\frac{d}{dx}[f(x) \pm g(x)] = \frac{d}{dx}f(x) \pm \frac{d}{dx}g(x)$.

Differentiating Products and Quotients

Section 3.2 focused on differentiating products and quotients. The rules are:

1. If $u(x)$ and $v(x)$ are differentiable functions, then

$$\frac{d}{dx}[u(x)v(x)] = \underbrace{u'(x)}_{(\mathit{first})'} \cdot \underbrace{v(x)}_{(\mathit{second})} + \underbrace{u(x)}_{(\mathit{first})} \cdot \underbrace{v'(x)}_{(\mathit{second})'}$$

It is important to notice that the derivative of a product is not the product of the derivatives. That is, $\frac{d}{dx}[u(x)v(x)] \neq u'(x) \cdot v'(x)$.

2. If $u(x)$ and $v(x)$ are differentiable functions and $v(x) \neq 0$, then

$$\frac{d}{dx}\left[\frac{u(x)}{v(x)}\right] = \frac{v(x) \cdot u'(x) - u(x) \cdot v'(x)}{[v(x)]^2}$$

As with the product rule, it is important to keep in mind that the derivative of a quotient is not the quotient of the derivatives. That is, $\frac{d}{dx}\left[\frac{u(x)}{v(x)}\right] \neq \frac{u'(x)}{v'(x)}$.

Higher-Order Derivatives

Section 3.3 centered on higher-order derivatives. In this section we examined rates of change of rates of change. Of primary concern was the rate of change of the first derivative. This section discussed how the first and second derivatives of a function can tell us if the function is increasing at an increasing rate, increasing at a decreasing rate, decreasing at an increasing rate, or decreasing at a decreasing rate. For a function $f(x)$, two derivatives f' and f'' with the *same sign* indicate that the function f changes at an *increasing rate*.

$f' > 0,\ f'' > 0 \rightarrow f$ is increasing at an increasing rate.
$f' < 0,\ f'' < 0 \rightarrow f$ is decreasing at an increasing rate.

Two derivatives f' and f'' with *opposite signs* indicate that the function f changes at a *decreasing rate*.

$f' > 0,\ f'' < 0 \rightarrow f$ is increasing at a decreasing rate.
$f' < 0,\ f'' > 0 \rightarrow f$ is decreasing at a decreasing rate.

The Chain Rule and the General Power Rule

Section 3.4 examined the chain rule and the general power rule. These two rules are extraordinarily powerful in discussing changes in physical and theoretical phenomena. The chain rule states that if y is a function of u and u is a function of x, that is, $y = f(u)$ and $u = g(x)$, then $y = f(u) = f[g(x)]$ and y is a function of x, and

$$\frac{dy}{dx} = \frac{dy}{du} \cdot \frac{du}{dx}.$$

The general power rule is actually an application of the chain rule. If $y = [u(x)]^n$, then $\frac{dy}{dx} = n[u(x)]^{n-1} \cdot \frac{du}{dx}$, where n is any real number. The general power rule might be remembered as: the derivative of a power is *the derivative of the global part times the derivative of the local part*, and the derivative of the local part may require several differentiation rules. A common mistake when differentiating a general power is to neglect to differentiate the local part. Be careful to get the entire derivative. Remember, the chain rule involves a product; therefore, when differentiating a general power, you will get a product.

Implicit Differentiation

Section 3.5 introduced implicit differentiation. Some functions are explicitly defined. Some examples are $y = 3x^2 - 6x + 4$ and $f(x) = \frac{x^2+1}{x^2-4}$. Explicit functions directly express the relationship between the input and output quantities. Some functions, however, are defined implicitly. An example is $5x^2 + 8xy^3 - 3y^2 = 10$. In this function, the relationship between x and y is conveyed indirectly. The derivatives of implicitly defined functions can be found using the technique of implicit differentiation. If y is an implicit function of x, then

$$\frac{d}{dx}y^n = \frac{d(y^n)}{dy} \cdot \frac{dy}{dx} = ny^{n-1}y'$$

This section also examined problems in which a rate of change of one quantity is related to rates of change of other quantities and are called *related rate* problems. Related rate problems are solved using implicit differentiation.

Supplementary Exercises

For Exercises 1–3, each derivative involves a variable.

1. ***Business: Advertising and Profits*** Suppose that for a particular item, $p = f(x)$ relates the profit p (in thousands of dollars) to the amount x (in thousands of dollars) spent on advertising the item. Interpret $f(35) = 18$ and $f'(35) = 2.4$.

2. ***Manufacturing: Container Strength*** A manufacturer of plastic drinking cups has established that the function $d = f(t)$ relates the breaking strength (in pounds per square inch) of a plastic cup to the thickness t (in millimeters) of the cup. Interpret $f(18) = 10$ and $f'(18) = 0.8$.

3. ***Medicine: Time of Death*** Coroners have established that there is a relationship between the temperature T (in degrees Fahrenheit) of a dead body and the time t (in hours) since death occurred. The relationship depends on the temperature of the surrounding air. Suppose that for a particular constant surrounding air temperature, the function $T = f(t)$ expresses the relationship between the temperature of a dead body and the time of death. Interpret $f(4) = 92$ and $f'(4) = -1.8$.

For Exercises 4–6, explain why $f'(x)$ does not exist at each specified value of x.

4.

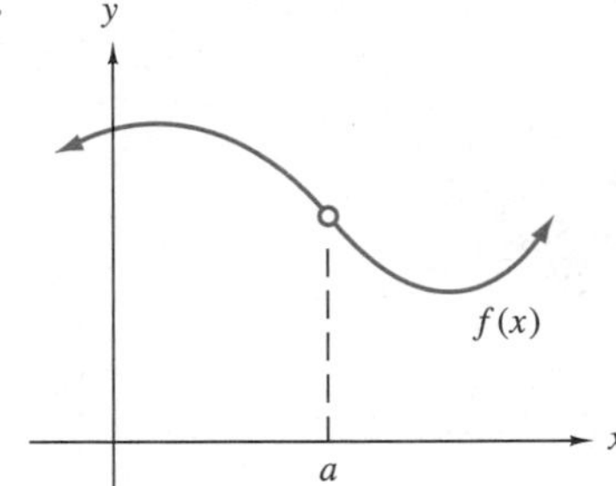

5.

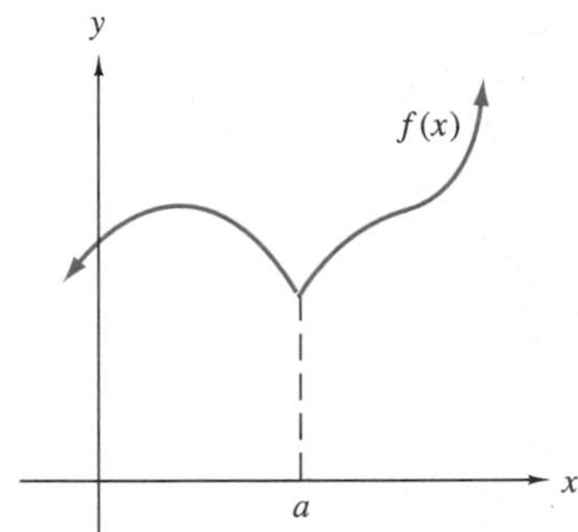

6.

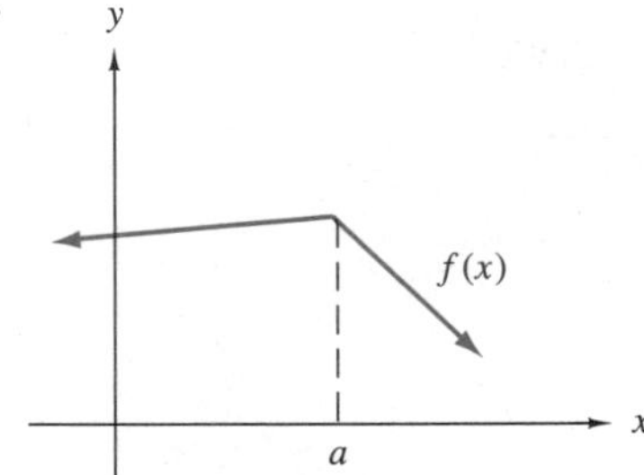

For Exercises 7–10, sketch the graph of a function that meets all the specified conditions. (Since many curves are possible, sketches may vary.)

7. $f(0) = 2$; $f'(2)$ does not exist because the function $f(x)$ is discontinuous at $x = 2$; $\lim_{x \to 2} f(x) = 3$; $f(x)$ has a cusp at $(5, 1)$.

8. $f(0) = 3$; $f(2) = 5$; $f(3) = 3$; $f(6) = 0$, $f'(0)$, $f'(2)$, and $f'(3)$ do not exist because $f(x)$ has corners at those points; $f'(5)$ does not exist because $f(x)$ is discontinuous at $x = 5$; $\lim_{x \to 5} f(x) = 5$.

9. $f(0) = 1$; $f(2) = 3$; $f(4) = 0$; $\lim_{x \to 2+} f(x) = 5$; $f'(4)$ does not exist because $f(x)$ has a cusp at $x = 4$.

10. $f(2) = 1$; $f(4) = 2$; $f(6) = 3$; for every value of x in the interval $0 < x \leq 2$, $f(x) = 0$; for every value of x in the interval $2 < x \leq 4$, $f(x) = 0$; for every value of x in the interval $4 < x \leq 6$, $f(x) = 0$; $f'(2)$, $f'(4)$, $f'(6)$ do not exist because $f(x)$ is discontinuous at those points.

For Exercises 11–14, suppose that $y = f(u)$, $u = g(x)$, and that dy/dx can be determined using the chain rule.

11. Interpret the statement $\dfrac{dy}{dx} = 5$ when $x = 9$.

12. Interpret the statement $\dfrac{dy}{dx} = 2$ when $x = 6$.

13. Interpret the statement $\dfrac{dy}{dx} = -2$ when $x = 4$.

14. Interpret the statement $\dfrac{dy}{dx} = -8$ when $x = 21$.

For Exercises 15–22, determine if the given equation expresses y as an explicit function of x or as an implicit function of x.

15. $2x^3 + y^2 + 4 = 0$

16. $y + 8x^3 = 4$

17. $5x^3 + 7xy - 2y^2 = 0$

18. $y = 4x + 1$

19. $4y + x^2 - 6$

20. $y^2 = x - 1$

21. $y = 5x^2 + 2x - 2$

22. $(x^2y)^{2/5} = 1$

For Exercises 23–27, use the graph to specify the differentials dx and dy and the actual change in y, Δy.

23.

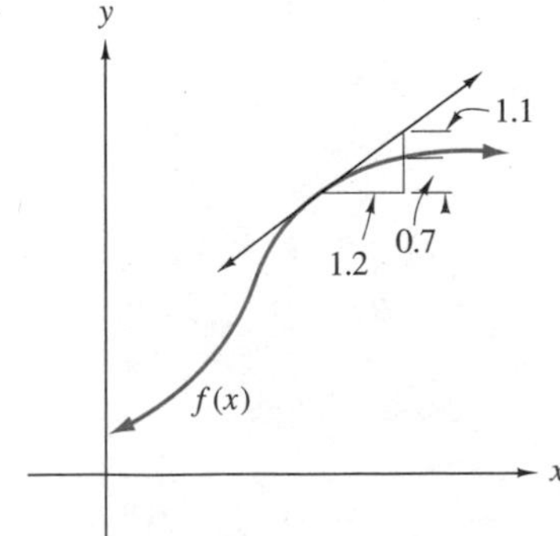

24.

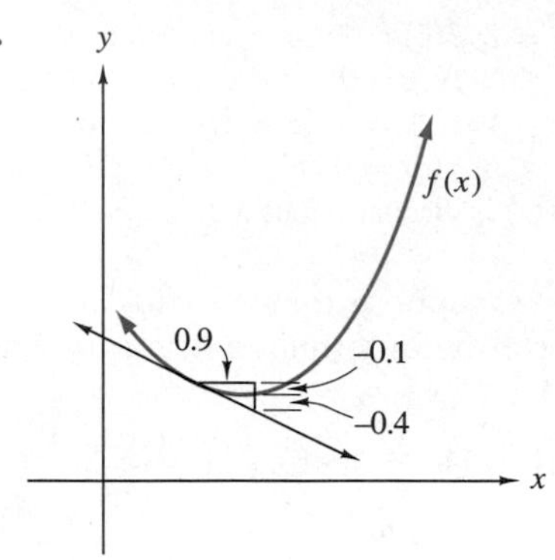

25.

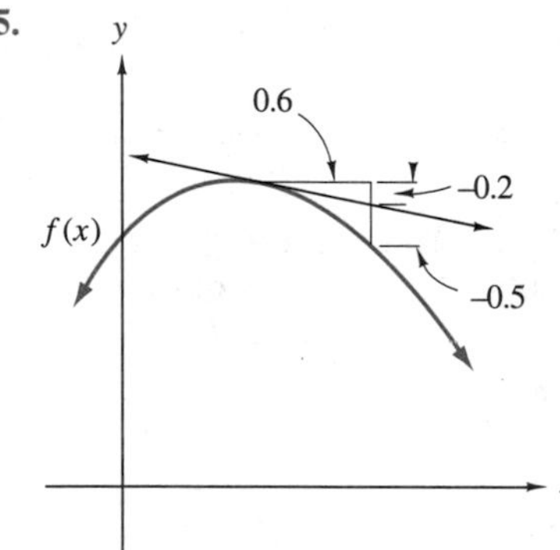

26.

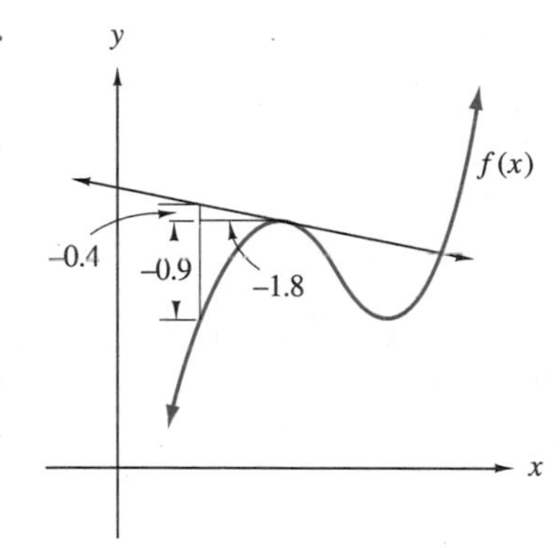

27.

For Exercises 28–32, specify the global form (power, product, quotient) of the given function.

28. $f(x) = \dfrac{(3x+1)^2}{(x-5)^2}$

29. $f(x) = \left[(x+1)^3(5x-2)^2\right]^2$

30. $f(x) = \left[\dfrac{(x^2+x-1)^3}{(x-4)^5}\right]^3$

31. $f(x) = \left[\dfrac{(x-1)^3(x+1)^2}{(x+5)^2(x+2)^3}\right]^4$

32. $f(x) = (x+4)^2(x-5)^3$

33. For $f(x) = (5x+4)^2, f'(x) = 10(5x+4)$. Find and interpret $f'(1)$.

34. For $f(x) = (x+3)^3(x-8)^4$, $f'(x) = (x+3)^2(x-8)^3(7x-12)$. Find and interpret $f'(0)$.

35. For $f(x) = \dfrac{\sqrt{x^2-1}}{\sqrt{x^2+1}}$,

$$f'(x) = \frac{2x}{(x^2+1)^{3/2}(x^2-1)^{1/2}}.$$

Find and interpret $f'(0)$.

36. For $f(x) = \dfrac{1}{(6x^3-x)^4}$,

$$f'(x) = \frac{-4(18x^2-1)}{(6x^3-x)^5}.$$

Find and interpret $f'(1)$.

For Exercises 37–76, find the derivative of each function.

37. $f(x) = x^3 + 5x + 4$

38. $f(x) = 8$

39. $f(x) = 5 - 3x$

40. $f(x) = -6x^3$

41. $f(x) = x - x^2$

42. $f(x) = 7x^2 + 3x$

43. $f(x) = 3x - 4x^2 + x^3$

44. $f(x) = x - 1$

45. $f(x) = 4 - x$

46. $f(x) = (2x+1)(3x-8)$

47. $f(x) = (5x-4)(x+6)$

48. $f(x) = (x+3)(8x-4)$

49. $f(x) = x^3(x^2+1)$

50. $f(x) = x^4(2x+5)$

51. $f(x) = 3x^2(5x+4)^3$

52. $f(x) = 5x^3(2x - 7)^2$
53. $f(x) = 3x(x + 6)^2$
54. $f(x) = 5x(3x - 2)^3$
55. $f(x) = (x - 4)^2(x + 1)^3$
56. $f(x) = (x + 6)^4(x + 7)^2$
57. $f(x) = \dfrac{1}{x - 4}$
58. $f(x) = \dfrac{1}{x + 8}$
59. $f(x) = \dfrac{3}{x + 6}$
60. $f(x) = \dfrac{5}{x - 1}$
61. $f(x) = \dfrac{2}{3x + 4}$
62. $f(x) = \dfrac{6}{5x - 7}$
63. $f(x) = \dfrac{-2}{2x + 3}$
64. $f(x) = \dfrac{-3}{4x + 1}$
65. $f(x) = \dfrac{x + 1}{x + 3}$
66. $f(x) = \dfrac{x - 7}{x - 1}$
67. $f(x) = \dfrac{3x + 8}{x - 1}$
68. $f(x) = \dfrac{4x + 3}{x + 3}$
69. $f(x) = \dfrac{x + 7}{2x + 3}$
70. $f(x) = \dfrac{x + 2}{10x - 3}$
71. $f(x) = \dfrac{(x + 2)^3}{(x - 4)^2}$
72. $f(x) = \dfrac{(x - 3)^2}{(x + 5)^3}$
73. $f(x) = \sqrt{6x + 1}$
74. $f(x) = \sqrt{8x - 3}$
75. $f(x) = \sqrt[3]{3x + 1}$
76. $f(x) = \sqrt[4]{8x - 3}$

For Exercises 77–80, find $\dfrac{dy}{dx}$.

77. $y = 3u + 4,\ u = x + 1$
78. $y = 2u - 1,\ u = 5x - 1$
79. $y = u^2 + 3,\ u = 2x + 3$
80. $y = 3u^2 + u + 1,\ u = x^2 + 8$

For Exercises 81–85, find y' for each function.

81. $x^2 + 3y^3 + 4 = 0$
82. $2x^3 + 4y^2 + x + 2 = 0$
83. $x^2 - y^2 + x + y = 5$
84. $5x - y^3 + x^2y^2 = 8$
85. $y^2 + 5xy + 1 = 0$

For Exercises 86–89, find the specified derivatives.

86. For $f(x) = 4x^3 + 3x^2 - x$, find $f'(x)$, $f''(x)$, $f'''(x)$.
87. For $f(x) = 2x^5 + x^3 - 7x^2 + x + 1$, find the first four derivatives.
88. For $f(x) = (x - 3)^3$, find $f'(x)$, $f''(x)$.
89. For $f(x) = (x + 4)^5$, find $f'(x)$, $f''(x)$.

For Exercises 90–96, find and interpret $f'(x)$ for the specified value of x.

90. $f(x) = 3x^2 - x, \quad x = 2$
91. $f(x) = x^3 - 2x + 1, \quad x = 50$
92. $f(x) = x^2(3x - 1), \quad x = 20$
93. $f(x) = \dfrac{1}{x + 7}, \quad x = 15$
94. $f(x) = \dfrac{4}{x + 2}, \quad x = 38$
95. $f(x) = \dfrac{x - 4}{x + 2}, \quad x = 3$
96. $f(x) = \dfrac{x + 3}{x - 4}, \quad x = 75$

CHAPTER 4

Applying the Derivative

4.1 Derivatives and the Behaviors of Functions

Introduction

Intervals on Which a Function Is Increasing or Decreasing

Relative Maxima and Minima

Concavity and Points of Inflection

Absolute Maxima and Minima

Describing Behavior and Technology

Introduction

Functions are important because they chronicle the behavior of physical or theoretical phenomena. They allow us to analyze the past, current, and future behavior of the phenomena. Of the various forms of a function (tables of values, sets of ordered pairs, equations, and graphs), the graphical form is particularly appealing because it can reveal information that may not be evident from the other forms. Graphs of functions can be generated by computers or some of the newer calculators by inputting the ordered pairs or the equation of the function.

Realistically, when information is needed about the behavior of a function, the function is graphed using a computer or calculator, and the needed information is drawn from the graph. Very rarely are graphs constructed using paper and pencil. Although we subscribe to the philosophy of graphing using machines, it is important, at least at the beginning of our study, to construct several graphs by hand. By constructing and analyzing in detail graphs of functions, we can gain expertise in the ability to interpret and chronicle the information that the picture is providing.

When examined from left to right, the graph of a function can exhibit properties that allow us to describe the behavior of the function. For example, the graph of the function can be (1) rising, falling, or staying constant; (2) opening upward or downward; (3) attaining a highest value or a lowest value relative to nearby values; (4) attaining a highest value or lowest value relative to the entire domain of the function; (5) approaching asymptotes; or (6) intersecting one or both of the coordinate axes.

Intervals on Which a Function Is Increasing or Decreasing

Increasing/Decreasing Function

A function $f(x)$ is **increasing** on an interval (a, b) if the graph of the function rises through (a, b). If a curve rises through (a, b), its tangent lines will also rise and will, therefore, have positive slopes. The graphs of Figure 4.1 illustrate functions that increase through an interval (a, b). A function $f(x)$ is **decreasing**

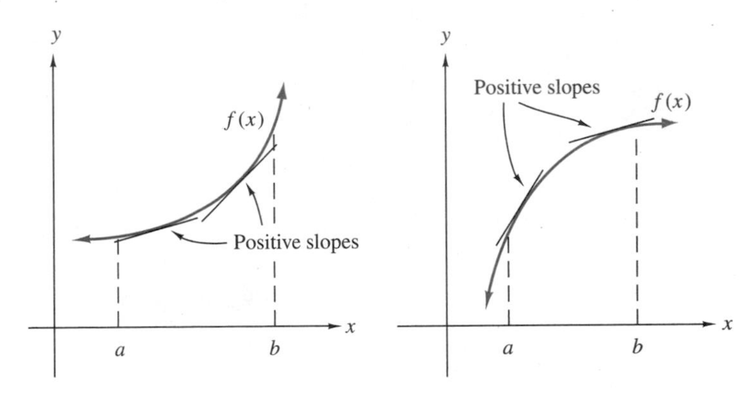

Figure 4.1

on an interval (a, b) if the graph of the function falls through (a, b). If a curve falls through (a, b), its tangent lines will also fall and will, therefore, have negative slopes. The graphs of Figure 4.2 illustrate functions that decrease through an interval (a, b). Since the slope of a tangent line to a curve is the derivative of the

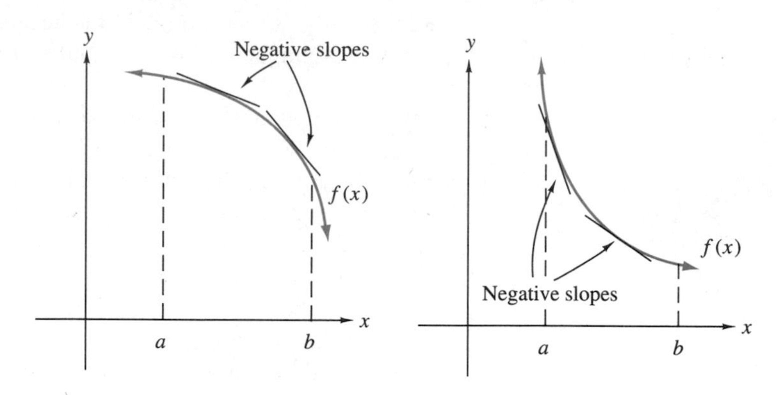

Figure 4.2

function corresponding to the curve, the increasing and decreasing behavior of the function can be described in terms of the signs of the first derivative of the function.

The First Derivative and the Increasing/Decreasing Behavior of a Function

Suppose that $f(x)$ is a continuous, differentiable function defined on the interval (a, b). Then

1. If $f'(x) > 0$ for every value of x in (a, b), then $f(x)$ is increasing on (a, b). Conversely, if $f(x)$ is increasing on (a, b), then $f'(x) > 0$ for every value of x in (a, b).
2. If $f'(x) < 0$ for every value of x in (a, b), then $f(x)$ is decreasing on (a, b). Conversely, if $f(x)$ is decreasing on (a, b), then $f'(x) < 0$ for every value of x in (a, b).
3. If $f'(x) = 0$ for every value of x in (a, b), then $f(x)$ is neither increasing nor decreasing on (a, b), but is constant on (a, b). Conversely, if $f(x)$ is constant on (a, b), then $f'(x) = 0$ for every value of x in (a, b).

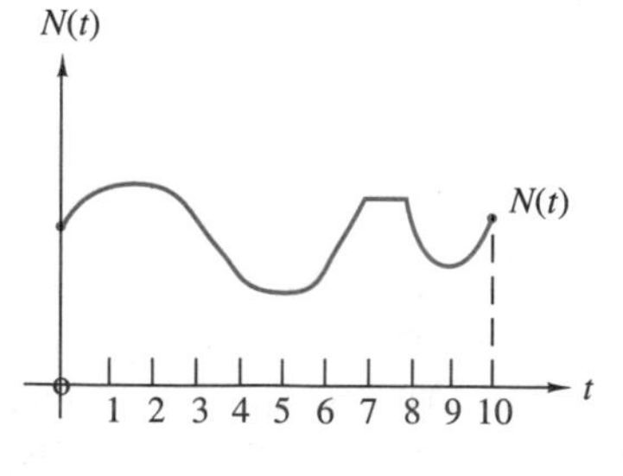

Figure 4.3

ILLUMINATOR SET A

The graph illustrated in Figure 4.3 displays the relationship between time t (in months from the beginning of 1980) and the number N of townhouse sales in the southeastern part of the United States in the period from 1980 to 1990. We can, in terms of increasing and decreasing, discuss the behavior of $N(t)$.

- $N(t)$ rises through $(0, 2)$, $(5, 7)$, and $(9, 10)$ so that $N(t)$ is increasing and $N'(t) > 0$ through each of these intervals.
- $N(t)$ falls through $(2, 5)$ and $(8, 9)$ so that $N(t)$ is decreasing and $N'(t) < 0$ through each of these intervals.

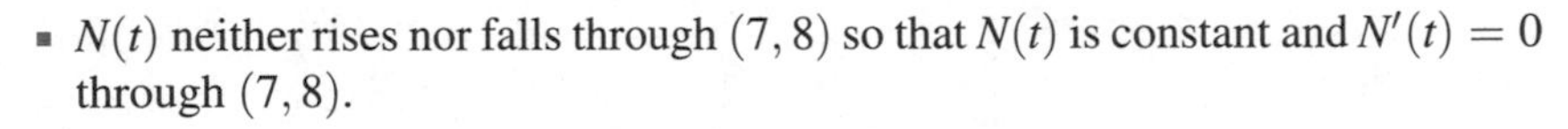

- $N(t)$ neither rises nor falls through $(7, 8)$ so that $N(t)$ is constant and $N'(t) = 0$ through $(7, 8)$.

Thus, we can conclude that the number of townhouse sales was increasing through the years 1980–1982, 1985–1987, and 1989–1990, decreasing through the years 1982–1985 and 1988–1989, and constant through the years 1987–1988.

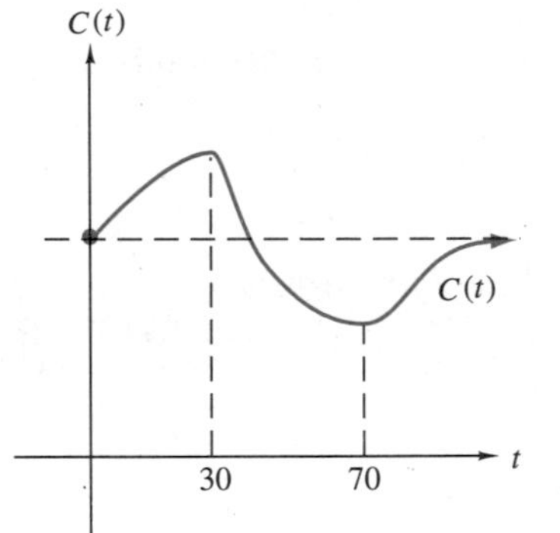

Figure 4.4

Relative Maxima and Minima

Relative Maximum and Critical Value

The function illustrated graphically in Figure 4.4 describes the relationship between time t (in minutes) and the concentration C (in milligrams per deciliter) of sugar in a person's blood when a highly concentrated sugar solution is ingested and then, 30 minutes later, an experimental drug that sharply reduces the concentration of sugar is injected. Notice that for values in a small interval around $t = 30$, it is $t = 30$ that produces the largest function value (the highest sugar concentration). That is, $t = 30$ produces a maximum value relative to the other t values near $t = 30$. Thus, the point $(30, C(30))$ is the **relative maximum** because, *relative* to points on the curve near $(30, C(30))$, the function has the largest value at that point. The leftmost graph of Figure 4.5 illustrates this idea. The t value 30 is called a **critical value** since it is critical to the accurate de-

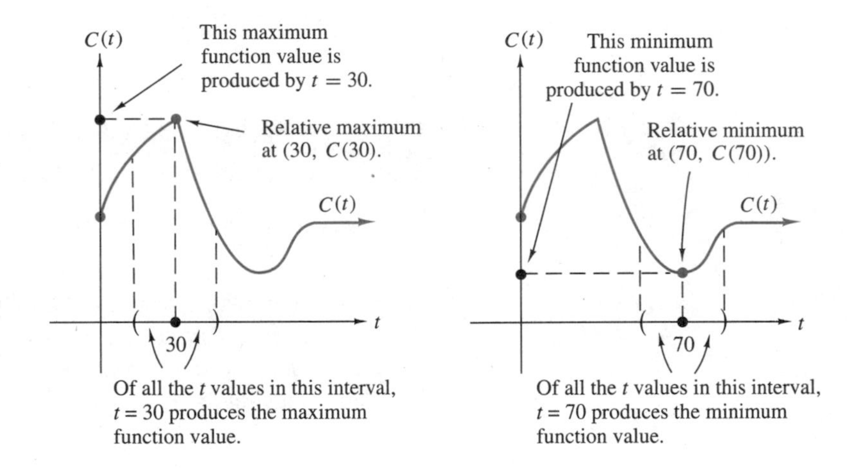

Figure 4.5

scription of the behavior of the function. Notice that at this critical value, the derivative $f'(30)$ is undefined (since the function has a corner here).

Relative Minimum

Notice that for values in a small interval around $t = 70$, it is $t = 70$ that produces the smallest function value (the lowest sugar concentration). That is, $t = 70$ produces a minimum value relative to the other t values near $t = 70$. Thus, the point $(70, C(70))$ is the **relative minimum** because, *relative* to points on the curve near $(70, C(70))$, the function has the smallest value at that point. The rightmost graph of Figure 4.5 illustrates this idea.

As before, the t value 70 is called a critical value since it is critical to the accurate description of the behavior of the function. Notice that at this critical value, the derivative $f'(x)$ equals zero.

These observations lead us to the following conclusions.

Relative Extrema and Critical Values

If a function $f(x)$ is defined at $x = c$, and either $f'(c) = 0$ or $f'(c)$ is undefined, then c is a **critical value** of $f(x)$. Conversely, if c is a critical value of $f(x)$, then either $f'(c) = 0$ or $f'(c)$ is undefined. Also, critical values can produce relative maxima and relative minima. The term **relative extrema** is used to describe points that can be relative maxima or minima.

Critical values for which the first derivative is zero are potential smooth relative maximum or minimum points. Critical values for which the first derivative is undefined are potential cusp or corner relative maximum or minimum points.

The next observation is known as the **first derivative test** for relative extrema.

First Derivative Test

If c is a critical value of the function $f(x)$, and

1. $f(x)$ is increasing immediately to the left of c and decreasing immediately to the right of c, then c produces a relative maximum. In terms of the first derivative, if $f'(x) > 0$ to the immediate left of c and $f'(x) < 0$ to the immediate right of c, then c produces a relative maximum.
2. $f(x)$ is decreasing immediately to the left of c and increasing immediately to the right of c, then c produces a relative minimum. In terms of the first derivative, if $f'(x) < 0$ to the immediate left of c and $f'(x) > 0$ to the immediate right of c, then c produces a relative minimum.

Figure 4.6 visually summarizes the first derivative test.

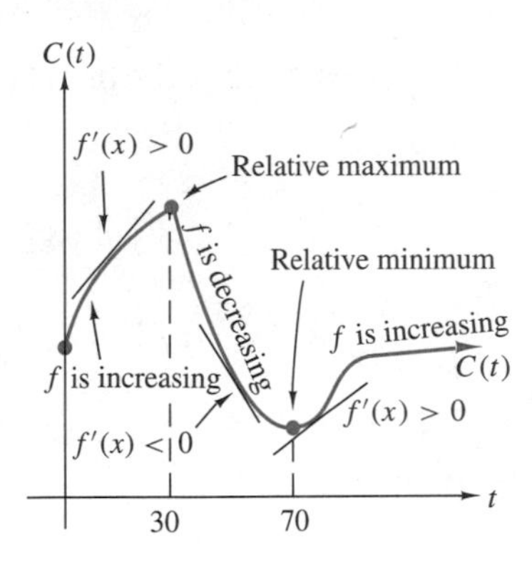

Figure 4.6

Concavity and Points of Inflection

Concavity

If the graph of a function opens upward at $x = c$— that is, the curve is above the tangent line at $x = c$— the function is said to be **concave upward** at $x = c$. For the curve to open upward, the slopes of the tangent lines must get larger. The rate at which the slopes change is described by the second derivative; thus, if $f''(x) > 0$, then the curve is concave upward. If the graph of a function opens downward at $x = c$— that is, the curve is below the tangent line at $x = c$— the function is said to be **concave downward** at $x = c$. For the curve to open downward, the slopes of the tangent lines must get smaller; thus, $f''(x) < 0$. Figure 4.7 illustrates a function that is concave upward at $x = 4$ and concave downward at $x = 8$. Notice that both $x = 4$ and $x = 8$ are critical values since $f'(4) = 0$, and $f'(8) = 0$.

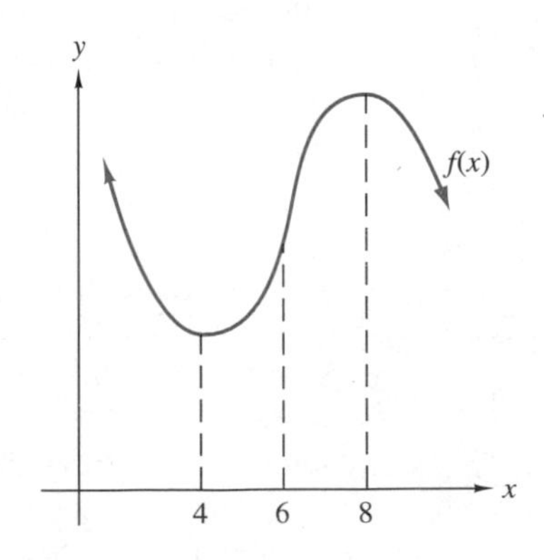

Figure 4.7

The Second Derivative and the Concavity of a Function

Suppose that $f(x)$ is a continuous, twice-differentiable function defined on the interval (a, b). Then

1. If $f''(x) > 0$ for every value of x in (a, b), then $f(x)$ is concave upward on (a, b). Conversely, if $f(x)$ is concave upward on (a, b), then $f''(x) > 0$ for every value of x in (a, b).
2. If $f''(x) < 0$ for every value of x in (a, b), then $f(x)$ is concave downward on (a, b). Conversely, if $f(x)$ is concave downward on (a, b), then $f''(x) < 0$ for every value of x in (a, b).
3. If $f''(x) = 0$ or is undefined for every value of x in (a, b), then the concavity of $f(x)$ cannot be determined at that value without further investigation.

Points of Inflection and Hypercritical Values

At $x = 6$ in Figure 4.7, there is a change in concavity; the graph of $f(x)$ changes from concave upward to concave downward. Points at which there is a change of concavity are called **points of inflection**. Since the function changes concavity at a point of inflection, the second derivative must change signs at that point (from positive to negative or from negative to positive). This means that at a point of inflection, the second derivative must either equal zero or be undefined; that is, if $x = c$ produces a point of inflection, then either $f''(c) = 0$, or $f''(c)$ is undefined. Values that produce possible points of inflection are called **hypercritical values**. In Figure 4.7, $x = 6$ is a hypercritical value and $(6, f(6))$ is a point of inflection.

Point of Diminishing Returns

A point of inflection where the concavity changes from upward to downward is often called the **point of diminishing returns.** Figure 4.8 illustrates a point of diminishing returns for a function $P(a)$ where a is the amount of money (in thousands of dollars) spent on advertising and P is the profit (in thousands of dollars) realized from the advertised product. As a increases from \$0 to \$90,000, the profit increases at an increasing rate. After \$90,000 has been spent on advertising, the profit still increases but progressively more slowly. It now takes more advertising money to realize the same increase in profit as it previously did.

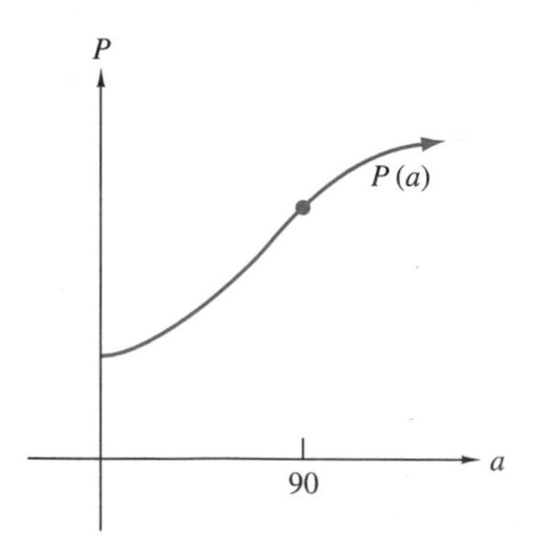

Figure 4.8

Absolute Maxima and Minima

Often when working with functions that describe physical or theoretical phenomena, the primary objective is to find the absolute maximum or the absolute minimum value of the function. For example, the function graphed in Figure 4.9, shows that on the interval $[a, b]$, there are two relative maxima, one at $x = r$ and another at $x = t$; one relative minimum at $x = s$; an absolute maximum at $x = t$;

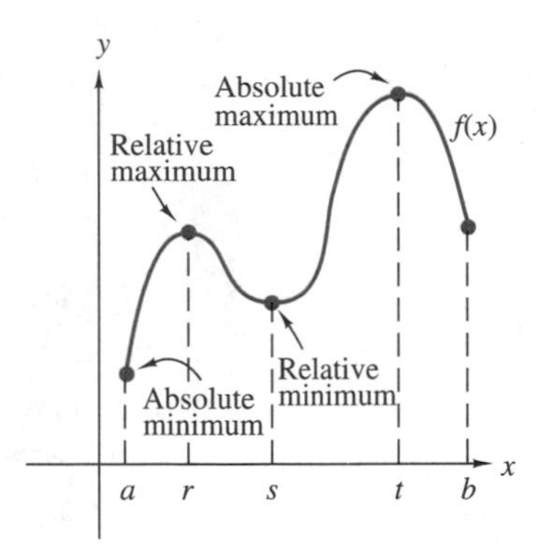

Figure 4.9

Absolute Maximum and Minimum

and an absolute minimum at the endpoint $x = a$. Notice that an absolute extreme point can occur at an endpoint or at a relative extreme point. The **absolute maximum** of a function is the y value of the highest point on the graph, and the **absolute minimum** is the y value of the lowest point on the graph. Not all continuous functions have an absolute maximum or absolute minimum. However, if a function is continuous on the closed interval $[a, b]$, it will have both an absolute maximum and an absolute minimum. (Try to draw a graph of a continuous function over a closed interval $[a, b]$ that does not have one or the

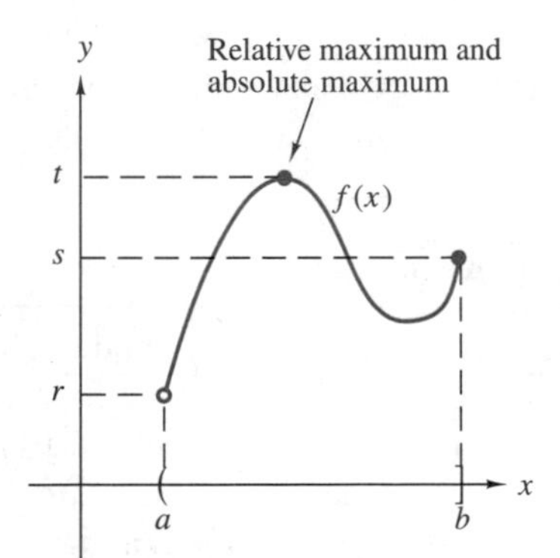

There is no smallest y value. As x approaches a from the right, y only approaches r from above. As x gets closer to a, y gets closer to r. Thus, there is no smallest y value and, hence, no absolute minimum.

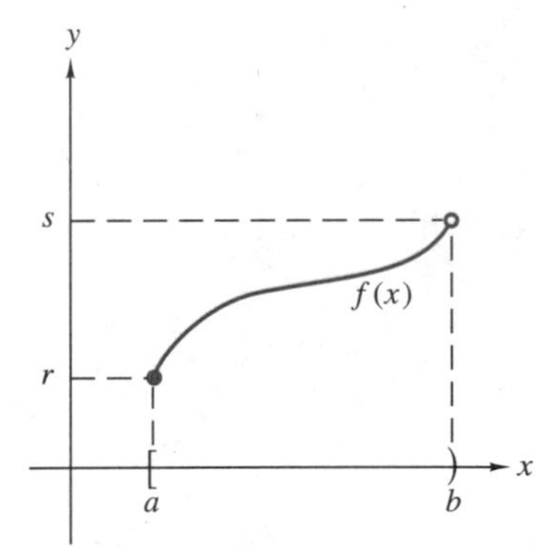

There is no largest y value. As x approaches b from the left, y only approaches s from below. As x gets closer to b, y gets closer to s. Thus, there is no largest y value and, hence, no absolute maximum.

Figure 4.10

other.) Figure 4.10 and Figure 4.11 show some of the possibilities. A continuous function that is defined over an interval that is not closed, such as $(a, b]$, $[a, b)$, or (a, b), may or may not have an absolute maximum or absolute minimum (or both). A function will not have an absolute maximum (or an absolute minimum) if the largest value (or smallest value) occurs adjacent to an open endpoint since we are not able to identify the value. The rightmost illustration of Figure 4.11

shows a function that is defined on an open interval (a, b) that has both types of absolute extrema. Absolute extrema are found by finding the y values of all the

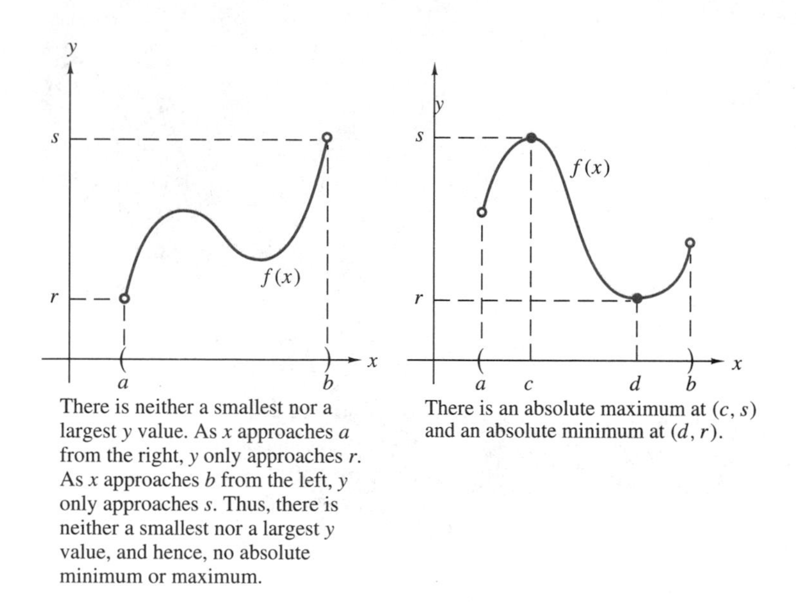

Figure 4.11

relative extrema and the y values of the endpoints, then comparing to determine which is the largest and which is the smallest.

Now, we assimilate all these properties with an example.

ILLUMINATOR SET B

We will construct a sign chart and a table that summarizes the behavior of the function illustrated in Figure 4.12. We notice first that $f(x)$ is defined on the interval $[2, 10)$ and that the line $x = 10$ is a vertical asymptote. Second, we see that $x = 4$, $x = 6$, and $x = 8$ are critical values and that $x = 7$ is a hypercritical value. The critical values $x = 4$ and $x = 8$ produce relative maxima at $(4, 4)$ and $(8, 9)$, and the critical value $x = 6$ produces a relative minimum at $(6, 1)$. The hypercritical value $x = 7$ produces the point of inflection at $(7, 5)$. We use this information to construct the sign chart shown in Figure 4.13. Now, $f(x)$ attains

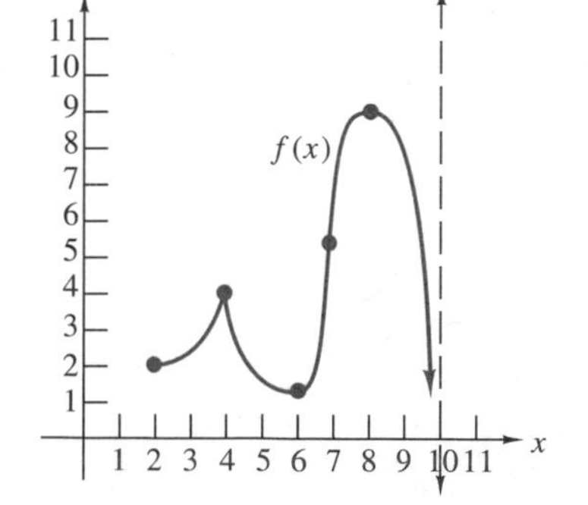

Figure 4.12

Test point	3		5		$6\frac{1}{2}$		$7\frac{1}{2}$		9
$f'(x)$	(+)		(–)		(+)		(+)		(–)
x		4		6		7		8	
$f''(x)$	(+)		(+)		(+)		(–)		(–)

Figure 4.13

an absolute maximum at $(8, 9)$ and attains no absolute minimum. We use these observations to construct the following table, which we will fill in after a more detailed analysis of the graph.

We will use the following abbreviations in all subsequent tables.

Table 4.1

Number	**Interval /Value**	f'	f''	$f(x)$	**Behavior of** $f(x)$
1	2				
2	$[2, 4)$				
3	4				
4	$(4, 6)$				
5	6				
6	$(6, 7)$				
7	7				
8	$(7, 8)$				
9	8				
10	$(8, 10)$				
11	10				

- Interval /Value ⟶ Int / Val
- undefined ⟶ und
- increasing ⟶ incr
- decreasing ⟶ decr
- concave upward ⟶ conc up (cu)
- concave downward ⟶ conc down (cd)
- vertical asymptote ⟶ VA
- horizontal asymptote ⟶ HA
- relative maximum ⟶ rel max
- relative minimum ⟶ rel min
- point of inflection ⟶ pt of infl

We now analyze the behavior of $f(x)$ at the endpoints and on the intervals $[2, 4)$, $(4, 6)$, $(6, 7)$, $(7, 8)$, and $(8, 10)$ using the number as listed in Table 4.1.

1 At the left-hand endpoint, $x = 2, f(x) = 2$.

2 On ¤minterval $[2, 4)$, $f'(x) > 0$, and $f''(x) > 0$, so $f(x)$ is increasing and concave upward.

4 On the interval $(4, 6)$, $f'(x) < 0$, and $f''(x) > 0$, so $f(x)$ is decreasing and concave upward.

6 On the interval $(6, 7)$, $f'(x) > 0$, and $f''(x) > 0$, so $f(x)$ is increasing and concave upward.

8 On the interval $(7, 8)$, $f'(x) < 0$, and $f''(x) < 0$, so $f(x)$ is increasing and concave downward.

10 On the interval $(8, 10)$, $f'(x) < 0$, and $f''(x) < 0$, so $f(x)$ is decreasing and concave downward.

11 Although there is no right-hand endpoint, $x = 10$ is a vertical asymptote.

We can now fill in items 1, 2, 4, 6, 8, 10, and 11 of the summary table. See Table 4.2.

Table 4.2

Number	Int / Val	f'	f''	$f(x)$	Behavior of $f(x)$
1	2			$(2, 2)$	
2	$[2, 4)$	$+$	$+$	incr, concave up	incr at an incr rate
3	4				
4	$(4, 6)$	$-$	$+$	decr, concave up	decr at an decr rate
5	6				
6	$(6, 7)$	$+$	$+$	incr, concave up	incr at an incr rate
7	7				
8	$(7, 8)$	$+$	$-$	incr, concave down	incr at an decr rate
9	8				
10	$(8, 10)$	$-$	$-$	decr, concave down	decr at an incr rate
11	10			VA	

We can now draw conclusions about the critical and hypercritical values, items 3, 5, 7, and 9, and fill in the rest of the table. (See Table 4.3.)

Table 4.3

Number	Int / Val	f'	f''	$f(x)$	Behavior of $f(x)$
1	2			$(2, 2)$	
2	$[2, 4)$	$+$	$+$	incr, concave up	incr at an incr rate
3	4	und		$(4, 4)$	rel max at $(4, 4)$
4	$(4, 6)$	$-$	$+$	decr, concave up	decr at a decr rate
5	6	0		$(6, 1)$	rel min at $(6, 1)$
6	$(6, 7)$	$+$	$+$	incr, concave up	incr at an incr rate
7	7		0	$(7, 5)$	pt of infl at $(7, 5)$
8	$(7, 8)$	$+$	$-$	incr, concave down	incr at a decr rate
9	8	0		$(8, 9)$	rel max at $(8, 9)$
10	$(8, 10)$	$-$	$-$	decr, concave down	decr at an incr rate
11	10			VA	

3 Using the first derivative test, $x = 4$ produces a relative maximum since $f(x)$ is increasing to its left and decreasing to its right.

5 Using the first derivative test, $x = 6$ produces a relative minimum since $f(x)$ is decreasing to its left and increasing to its right.

7 Using the first derivative test, $x = 7$ produces neither a relative maximum nor a relative minimum since $f(x)$ is increasing to its left and increasing to its right. However, since there is a change in concavity at $x = 7$, $x = 7$ produces a point of inflection.

9 Using the first derivative test, $x = 8$ produces a relative maximum since $f(x)$ is increasing to its left and decreasing to its right.

Describing Behavior and Technology

Using Your Calculator You can use your graphing calculator to describe the behavior of and construct a summary table for a function. The following entries show one way to describe, approximately, the behavior of the function $f(x) = 0.03x^3 - 2x + 3$.

```
Y1= .03x^3−2x+3
Y2=fMax(Y1,X,A,B)
Y3=fMin(Y1,X,A,B)
```

The parameters A and B are, respectively, the lower and upper bounds of the input values that produce the relative maximum and minimum points you see on the graph of $f(x)$. You can estimate these values directly from the graph. For particular extreme points, store your lower estimate as A, and your upper estimate as B. For $Y1$, the graph shows a relative maximum between $x = -10$ and $x = -2$, and a relative minimum between $x = 3$ and $x = 7$. Using these values as estimates, you obtain the critical values -4.714042912 and 4.714046928 as the relative maximum and relative minimum, respectively. Then, upon each of these results, the function values are computed as $Y1(ans) = 9.285393611$, and $Y1(ans) = -3.2853936111$.

Points of inflection are produced by hypercritical values and can be approximated using the SOLVE feature.

```
Y1= .03x^3−2x+3
Y5=nDeriv(Y1,X,X,.001)
Y6=nDeriv(Y5,X,X,.001)
Y7=solve(Y6,X,G)
```

The function $Y7$ produces the zeros of the function $Y6$, the second derivative of $Y1$. The parameter G in $Y7$ represents a *guess* at the hypercritical value. You can obtain a guess by observing the graph of the function. A guess of 0 produces $Y7 = 3.125\text{E} - 8 \approx 0$. Then $Y1(ans) = 2.999 \approx 3$, and you can conclude there is a point of inflection at about $(0, 3)$.

You now have enough information to construct a summary table for the behavior of this function.

Calculator Exercises

1. Construct a summary table that describes the behavior of the function $f(x) = x^3 - 4x + 5$. Use the window $-5 \leq x \leq 5$ and $-10 \leq y \leq 10$.

2. Construct a summary table that describes the behavior of the function $f(x) = x^4 - 4x^3 + 2$. Use the window $-5 \leq x \leq 5$ and $-25 \leq y \leq 25$.

3. A book publisher can print up to and including 20,000 books. The profit P, in dollars, realized by the publisher on the sale of the books depends on the number, n, of books sold and is given by the function

$$P(n) = 85(-n^3 + 14n^2 + 55n)$$

Construct a summary table that describes the behavior of this function.

4. The efficiency, E, of a worker who is learning a new construction technique is related to the number of hours, t, since learning began by the function

$$E(t) = 0.45t^3 - 3.2t^2 + 20t$$

Construct a summary table that describes the behavior of this function.

5. A manufacturer believes that a trainee's skill level, L, is related to the number of days, t, spent training by the function

$$L(t) = 1 - \frac{51.5}{t^2 + 51.5}$$

Construct a summary table that describes the behavior of this function.

EXERCISE SET 4.1

A UNDERSTANDING THE CONCEPTS

For Exercises 1–6, specify all the critical values, all the intervals on which the function $f(x)$ is increasing, and all the intervals on which the function $f(x)$ is decreasing.

1. $f(x)$ as illustrated below.

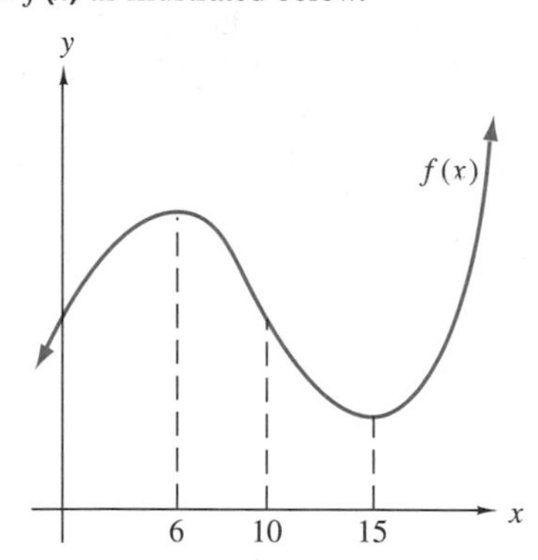

2. $f(x)$ as illustrated below.

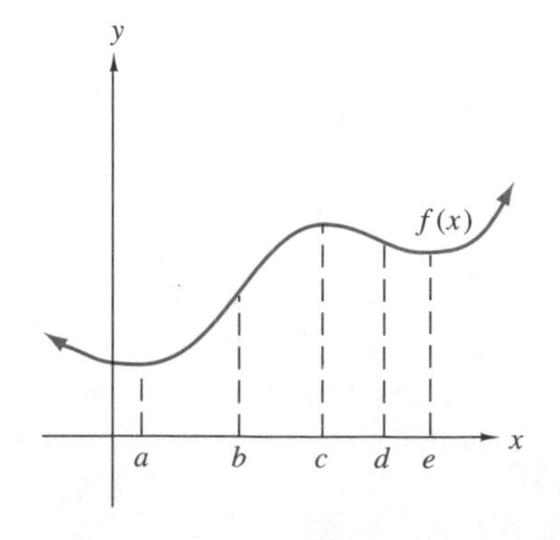

3. $f(x)$ as illustrated below.

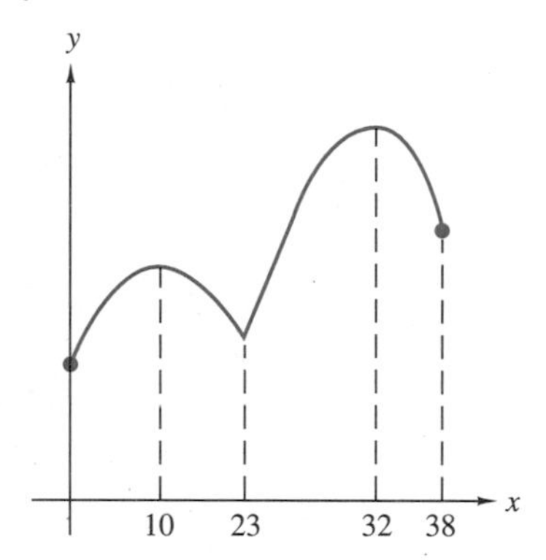

4. $f(x)$ as illustrated below.

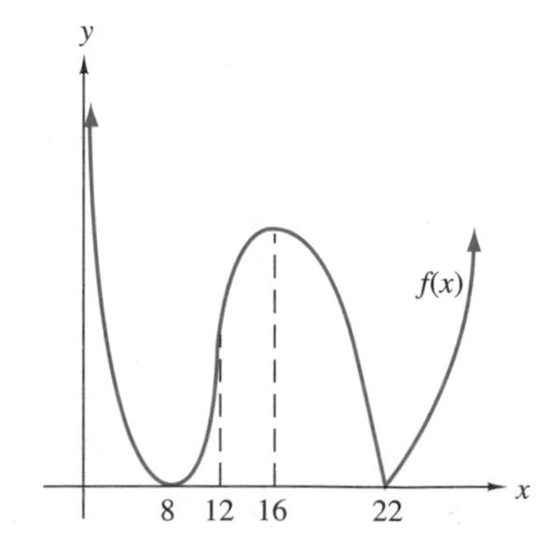

5. $f(x)$ as illustrated below.

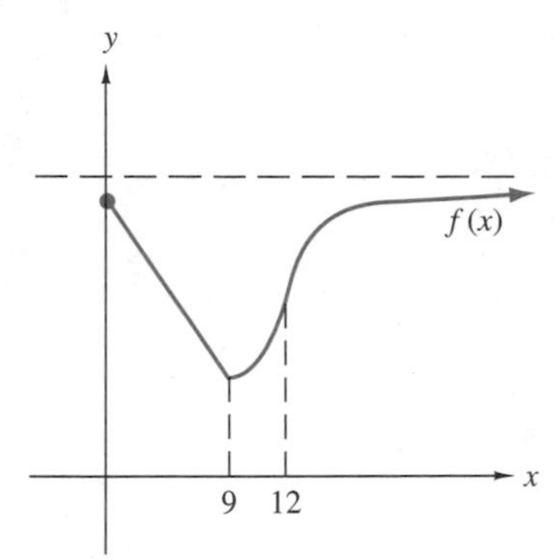

6. $f(x)$ as illustrated below.

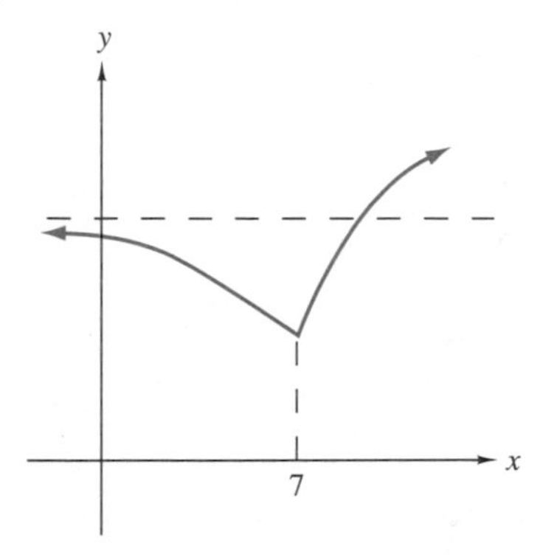

For Exercises 7–12, specify all the hypercritical values, all the intervals on which the function $f(x)$ is concave upward, and all the intervals on which the function $f(x)$ is concave downward.

7. $f(x)$ as illustrated in Exercise 1.
8. $f(x)$ as illustrated in Exercise 2.
9. $f(x)$ as illustrated in Exercise 3.
10. $f(x)$ as illustrated in Exercise 4.
11. $f(x)$ as illustrated in Exercise 5.
12. $f(x)$ as illustrated in Exercise 6.

For Exercises 13–18, specify all the critical values and hypercritical values and the relative and absolute extrema.

13. $f(x)$ as illustrated in Exercise 1.
14. $f(x)$ as illustrated in Exercise 2.
15. $f(x)$ as illustrated in Exercise 3.
16. $f(x)$ as illustrated in Exercise 4.
17. $f(x)$ as illustrated in Exercise 5.
18. $f(x)$ as illustrated in Exercise 6.

For Exercises 19–24, classify each curve as concave upward or concave downward.

19. A curve that is increasing at an increasing rate.
20. A curve that is decreasing at a decreasing rate.
21. A curve that is decreasing at an increasing rate.
22. A curve that is increasing at a decreasing rate.
23. A curve that increases at a decreasing rate, then attains a maximum value, then decreases at an increasing rate.
24. A curve that decreases at a decreasing rate, then attains a minimum value, then increases at an increasing rate.

25. Is it possible for a continuous function to have two relative minima but no relative maxima? If so, sketch such a function.
26. Is it possible for a continuous function to have a relative minimum but always be concave downward? If so, sketch such a function.
27. Is it possible for a continuous function to have two consecutive relative extrema, a relative minimum, and a relative maximum without having a point of inflection between them? If so, sketch such a function.
28. Is it possible to have a continuous function on a closed interval $[a, b]$ so that the absolute maximum and absolute minimum have the same value? If so, sketch such a function.

B SKILL ACQUISITION

For Exercises 29–32, construct a sign chart for each graph.

29. $f(x)$ as illustrated below.

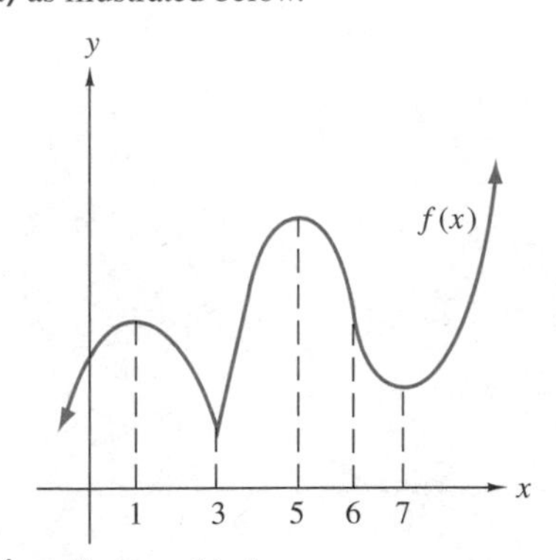

30. $f(x)$ as illustrated below.

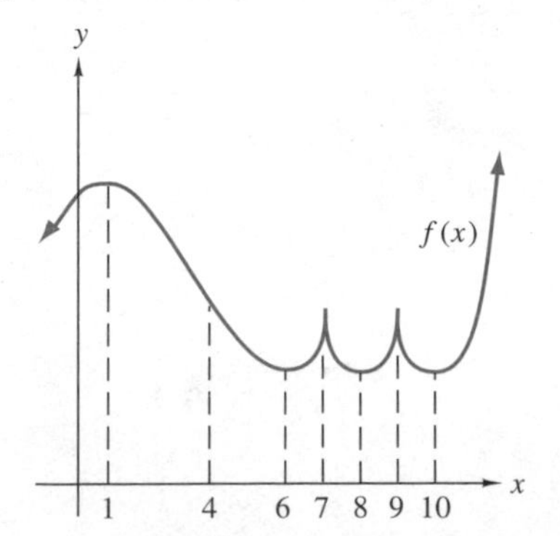

31. $f(x)$ as illustrated below.

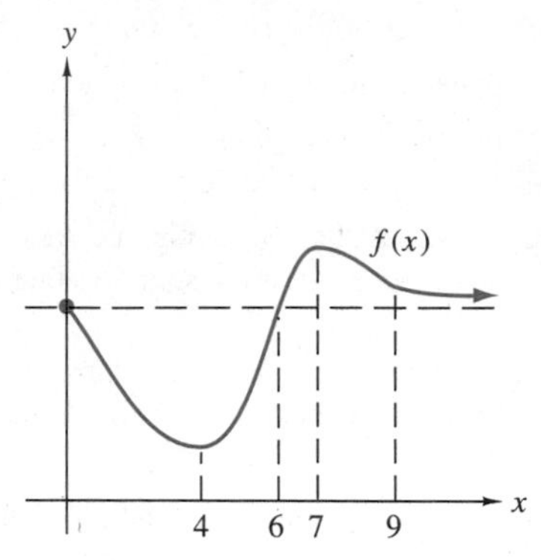

32. $f(x)$ as illustrated below.

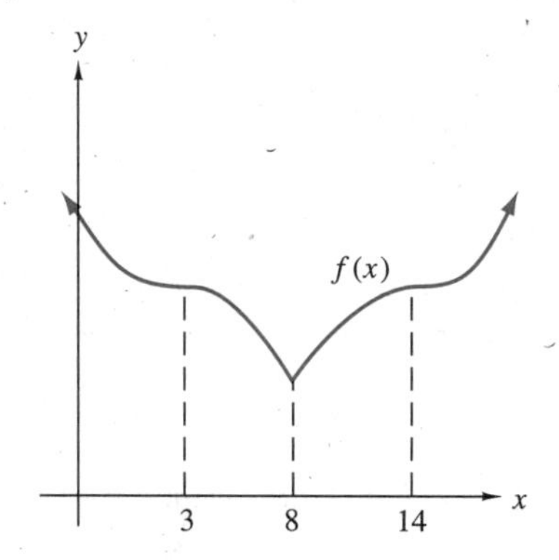

For Exercises 33–36, construct a summary table for each graph.

33. $f(x)$ as illustrated below.

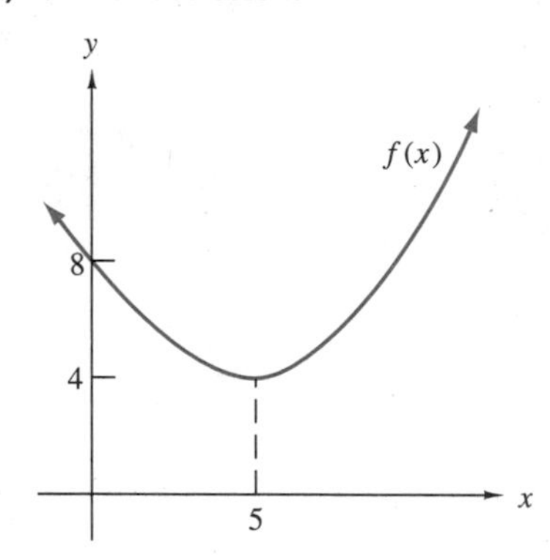

34. $f(x)$ as illustrated below.

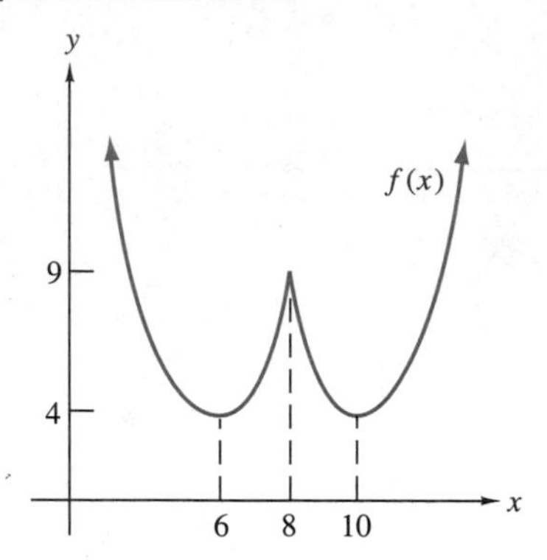

35. $f(x)$ as illustrated below.

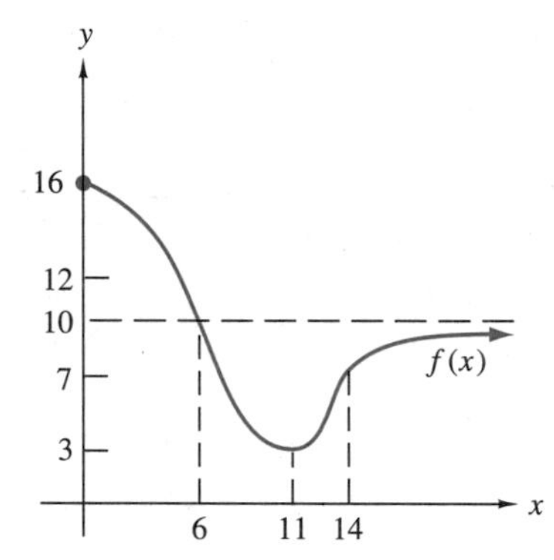

36. $f(x)$ as illustrated below.

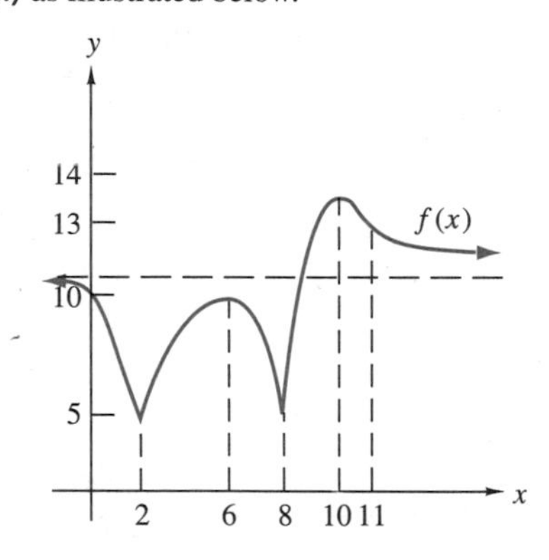

C APPLYING THE CONCEPTS

For Exercises 37–45, a situation is given along with the equation form and graph form of a function that describes the situation. As best you can (meaning you may have to approximate), use the graph to determine the critical value that maximizes or minimizes the function and then discuss the behavior of the function for values on both sides of the critical value.

37. ***Business: Manufacturing Costs*** A manufacturer of lightweight, durable containers is contracted to make right cylindrical containers that hold 20π in^3 (cubic inches) of liquid. The material used to construct the side of the can costs \$3 per in^2, and the material used to construct the top and bottom of the can costs \$4 per in^2. What radius will minimize the cost of the can? The function relating the cost C (in dollars) and the radius r (in inches) is $C(r) = 8\pi r^2 + \frac{377}{r}$, and its graph, for $r > 0$, appears below.

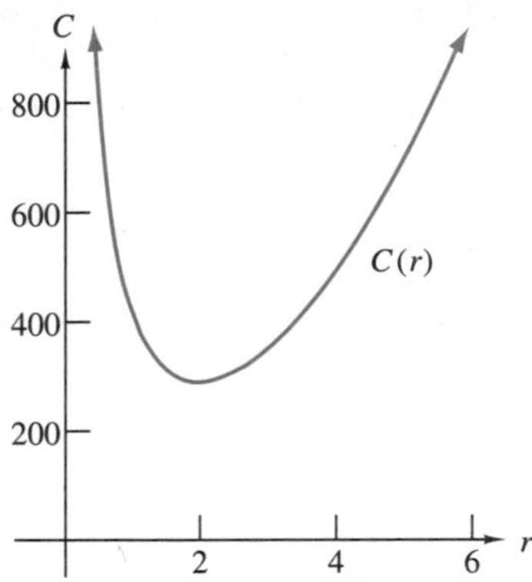

38. ***Business: Manufacturing*** A box with a square base and an open top is to be made from a square piece of cardboard by cutting out four squares of equal size from the corners and folding up the sides. What size should the corner cuts be so that the volume of the box is as large as possible? The function relating the volume V (in cubic inches) and the size x (in inches) of the corner cuts is $V(x) = 4x^3 - 60x^2 + 225x$, and its graph appears in below.

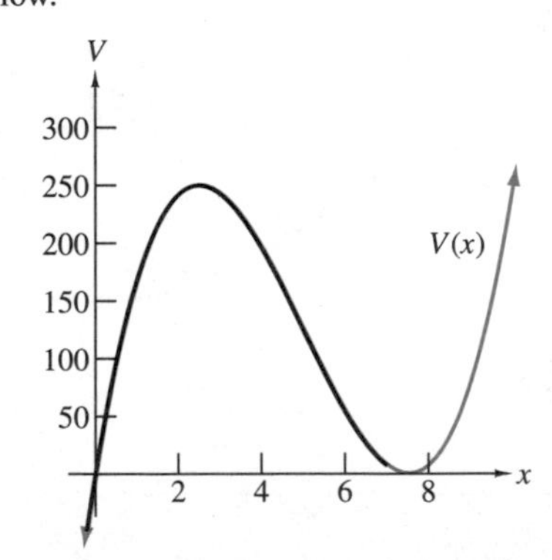

39. ***Business: Manufacturing*** A manufacturing company that uses toxic chemicals for cleaning precision parts plans to use 600 feet of fencing to enclose a rectangular region to store the chemical containers. One of the boundaries of the region will be a wall of a building and will, therefore, not need to be fenced. The function relating the area A (in square feet) to the width x (in feet) of the rectangular region is $A(x) = 600x - 2x^2$, and its graph appears below. What width will maximize the area?

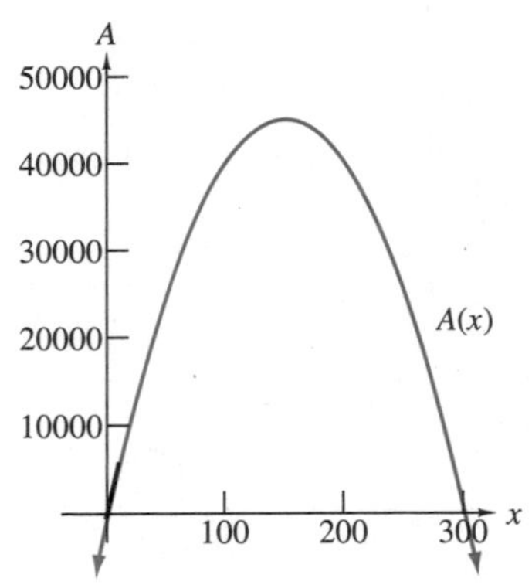

40. ***Business/Medicine: Drug Effectiveness*** A drug manufacturer claims that t hours after a particular drug is administered to a female weighing between 90 and 120 pounds, the concentration C (in milligrams per liter) is given by the function $C(t) = \frac{7t}{5t^2 + 4}$. When is the drug concentration maximized? The graph of $C(t)$ appears below.

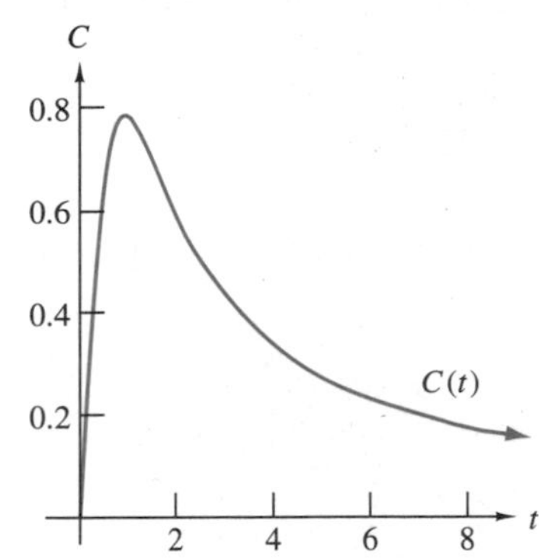

41. ***Medicine: Flu Epidemic*** A city's health officials believe that the number N of people contracting the flu t days from the beginning of an epidemic is predicted by the function $N(t) = -2t^2 + 180t + 40$. When is the number of people contracting the flu greatest? The graph of $N(t)$ appears below.

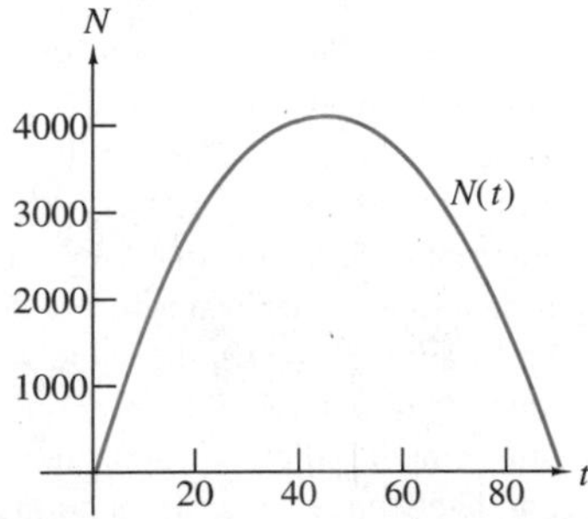

42. ***Business: Manufacturing Costs*** A container manufacturer plans to make cans that have a volume of 54 cubic inches. The cost of the material used for sides of the cans is 5 cents per square inch and the cost of the material used for the tops and bottoms of the cans is 3 cents per

square inch. The function relating the cost C (in dollars) of a can to its radius r (in inches) is $C(r) = 6\pi r^2 + \frac{540}{r}$. At what radius is the cost of a can minimized? The graph of $C(r)$ appears below.

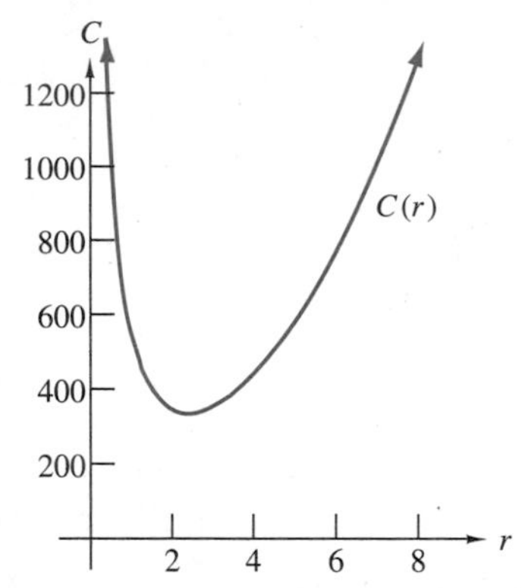

43. ***Business: Publishing Costs*** A publishing company has directed its printer to print pages that have one-half-inch margins on the sides and three-fourth-inch margins on the top and bottom and that contain 70 square inches of print. The function that relates the area A (in square inches) of a page and the width w (in inches) of a page is $A(w) = (w - 1)\left(\frac{140 - 3w}{2w}\right)$. At what width is the area of the page maximized? The graph of $A(w)$ appears below.

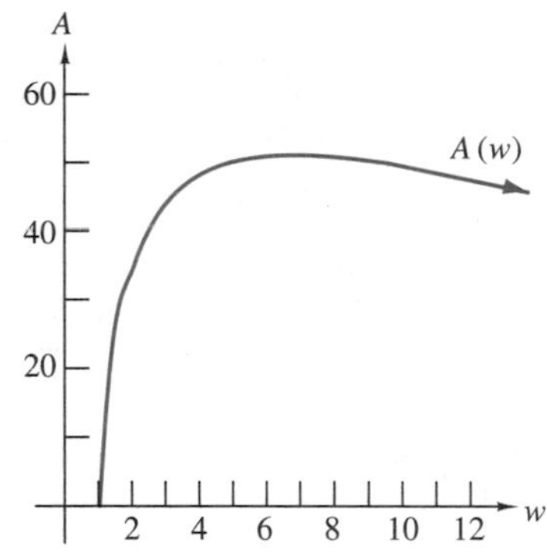

44. ***Business: Construction Costs*** In the following figure, a power station is located on a straight coastline at location B. A cable is to be laid that will connect the power station to a laboratory located at P on an island 2000 feet from point A. The cost of laying the cable on land is \$2 per foot, and under water the cost is \$7 per foot. In the first figure, if $x = 0$, the cable will be laid from point P directly to point A, then from point A along the coastline to point B. If $x = 7500$, then the cable will be laid completely under water from point P to point B. If x is between 0 and 7500, the cable will laid from B along the coastline to point C and then at an angle to point P. What value of x will minimize the cost of laying the cable?

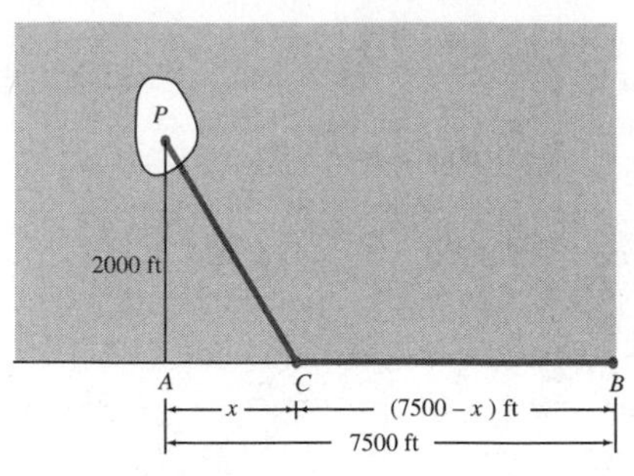

The function relating the cost C (in dollars) to the distance x (in feet) is $C(x) = 15{,}000 - 2x + 7\sqrt{x^2 + 2000^2}$. The graph of $C(x)$ appears below.

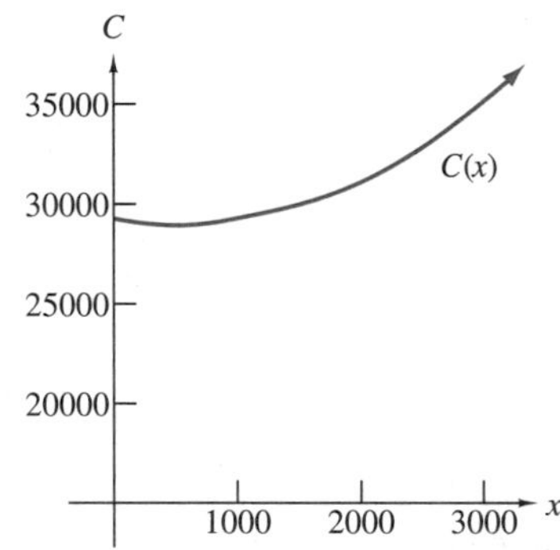

45. ***Politics: Candidate Recognition*** A political candidate's campaign officials believe that t weeks after the beginning of a campaign, the percentage P of the electorate favorably recognizing the candidate's name is given by the function $P(t) = \frac{46t}{t^2 + 15} + 0.39$. Find the value of t that maximizes $P(t)$. The graph of $P(t)$ appears below.

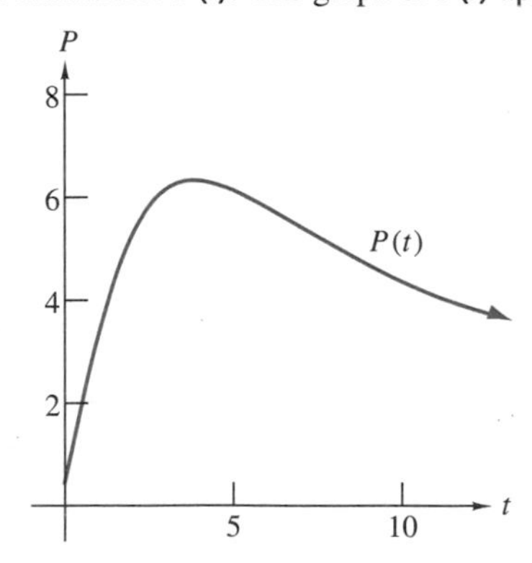

D DESCRIBING YOUR THOUGHTS

46. Suppose a function $f(x)$ is defined on the interval $[a, b]$. Describe the procedure you would use to determine the subintervals of $[a, b]$ upon which $f(x)$ is increasing/decreasing.

47. Describe the difference between a critical value and a hypercritical value.

48. Critical values of a function have been described in the text as potential relative maximum and relative minimum points of the function. Describe the different types of relative maximum and relative minimum that critical values can produce and discuss why they are only *potential* relative maximum and relative minimum points. A picture may be helpful to your explanation.

49. Describe the procedure you would use to construct a summary table for a function.

E REVIEW

50. (2.2) Find, if it exists, $\lim_{x\to 6} \frac{x^2 - 3x - 18}{x^2 - 4x - 12}$.

51. (2.3) Find, if it exists, $\lim_{x\to\infty} \frac{4x^3 + 3x + 4}{x^2 + x - 6}$.

52. (2.4) The function

$$f(x) = \begin{cases} -x^2 - 2x + 2, & x < 1 \\ x^2 + x - 6, & x \geq 1 \end{cases}$$

is discontinuous at $x = 1$. Which of the three continuity conditions is the first not to be satisfied?

53. (2.6) Find the average rate of change of the function $f(x) = \frac{x+1}{x+3}$ as x changes from $x = 2$ to $x = 5$.

54. (3.2) Find $f'(x)$ for $f(x) = \frac{3x - 4}{2x + 5}$.

55. (3.3) Suppose that for a function $f(x)$, $f'(3) > 0$ and $f''(3) < 0$. Make a statement about the behavior of the function near $x = 3$.

56. (3.4) Find $f'(x)$ for $f(x) = (5x^2 + 2)^3$.

57. (3.5) Find $\frac{dy}{dx}$ for $3x^2 + 5xy^2 + y^3 = 4$.

4.2 Describing the Behavior of a Function and Elementary Curve Sketching

Introduction

Examples

Describing Behavior and Technology

Introduction

Section 4.1 introduced us to many terms and conditions necessary to describe the behavior of a function. Fundamental to the description of the behavior of a function are the *critical values* and *hypercritical values* of the function. Recall that critical values occur where either $f'(x) = 0$ or $f'(x)$ is undefined, and hypercritical values occur where either $f''(x) = 0$ or $f''(x)$ is undefined. Thus, we need only find $f'(x)$ and $f''(x)$ and

1. solve, if possible, the equations $f'(x) = 0$ and $f''(x) = 0$. (In textbooks, for convenience, this is commonly done by factoring. In other situations, more sophisticated techniques may need to be used.)
2. determine which, if any, values of x *from the domain* of $f(x)$ that make $f'(x)$ and $f''(x)$ undefined. (This is usually done by checking for values of x that produce zero in the denominator, should there be a denominator. If $f'(x)$ or $f''(x)$ has no denominator, then both will always be defined for all the domain values of $f(x)$.)

It is important to remember that to find critical and hypercritical values, *both* conditions 1 and 2 have to be checked.

CAUTION

When working with rational functions, students commonly make an error in identifying critical and hypercritical points as the restricted values from the denominator. Remember, critical and hypercritical points must be *points* on the curve, and, therefore, *must* be in the domain of the function. (See Example Set C.)

Examples

We will now illustrate this process with a variety of different functions. Please make note of the common systematic process that is used regardless of the function.

EXAMPLE SET A

Describe the behavior of the function $f(x) = x^{3/5} - 1$. Then construct a sketch of the function.

Solution:
First we note that the domain of $f(x)$ is all real numbers. Next, we find the first and second derivatives.

$$f'(x) = \frac{3}{5x^{2/5}} \quad \text{and} \quad f''(x) = \frac{-6}{25x^{7/5}}$$

Then, to find the critical and hypercritical values, we check both conditions.

1. To find where $f'(x) = 0$ we need to solve $\frac{3}{5x^{2/5}} = 0$ for x.
   But a fraction is 0 *only* when the numerator is zero, and $3 \neq 0$. Thus, there are no critical values from $f'(x) = 0$.
   Likewise, to find where $f''(x) = 0$ we need to solve $\frac{-6}{25x^{7/5}} = 0$ for x. Again, a fraction is 0 *only* when the numerator is zero and $-6 \neq 0$. Thus, there are no hypercritical values from $f''(x) = 0$.
2. To find where $f'(x)$ is undefined we note that a fraction is undefined *only* when the denominator is 0. Thus, we set the denominator equal to 0 and solve for x.

   $$5x^{2/5} = 0 \quad \text{so that} \quad x = 0$$

   $f'(x)$ is undefined when $x = 0$, and 0 is in the domain of $f(x)$. Thus, 0 is a hypercritical value of $f(x)$ and, therefore, a possible point of inflection.
   Where is $f''(x)$ undefined? Again, we set the denominator equal to 0 and solve for x.

   $$25x^{7/5} = 0 \quad \text{so that} \quad x = 0$$

   $f''(x)$ is undefined when $x = 0$, and 0 is in the domain of $f(x)$. Thus, 0 is a critical value of $f(x)$ and, therefore, a possible extreme point.

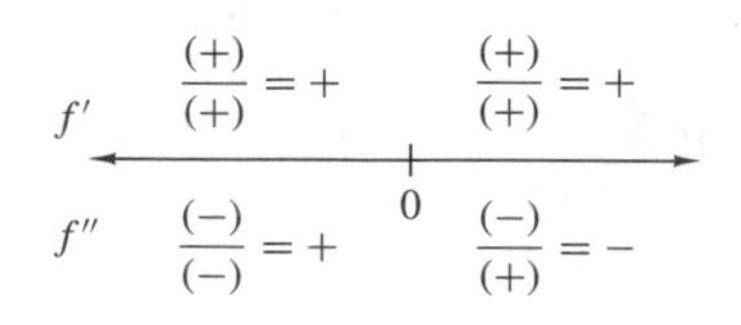

Figure 4.14

Thus, $x = 0$ is both a critical value and a hypercritical value. We can now construct a sign chart (illustrated in Figure 4.14) for $f(x)$. Using the sign chart, we can construct the following summary table. From the summary table (see Table 4.4) we can construct a sketch of $f(x)$ and describe its behavior (as shown in Figure 4.15).

Int / Val	f'	f''	$f(x)$	Behavior of $f(x)$
$(-\infty, 0)$	+	+	incr/cu	incr at an incr rate
0	und	und	-1	pt of infl at $(0, -1)$
$(0, +\infty)$	+	−	incr/cd	incr at a decr rate

Table 4.4

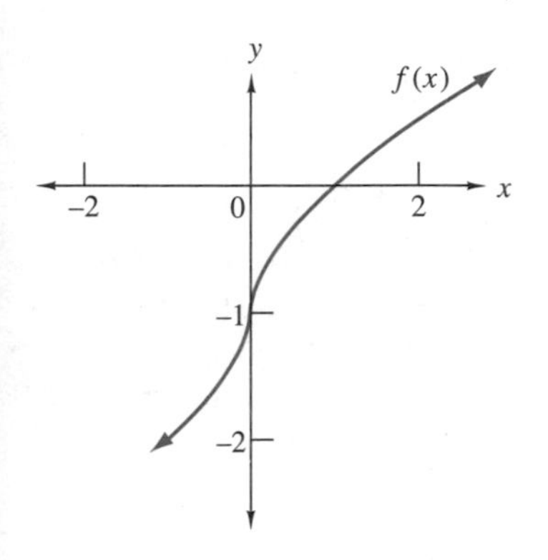

Figure 4.15

Summary of behavior: The function $f(x) = x^{3/5} - 1$ increases at an increasing rate when x is less than 0. When $x = 0, f(x) = -1$ and the function experiences a point of inflection (which alerts us to a coming change in concavity—that is, the speed of the increase is going to change). Then $f(x)$ increases at a decreasing rate when x is greater than 0.

EXAMPLE SET B

Summarize the behavior of $f(x) = x^3 - 3x^2 + 4$ and sketch its graph.

Solution:
Since $f(x)$ is a polynomial, its domain is all real numbers. The first and second derivatives are

$$f'(x) = 3x^2 - 6x \qquad \text{and} \qquad f''(x) = 6x - 6.$$

To find the critical and hypercritical values, we check both conditions.

1. Where is $f'(x) = 0$? We need to solve $3x^2 - 6x = 0$ for x. Factoring produces $3x(x - 2) = 0$. So, $x = 0,\ 2$.
   Where is $f''(x) = 0$? We need to solve $6x - 6 = 0$ for x. Factoring produces $6(x - 1) = 0$. So, $x = 1$.

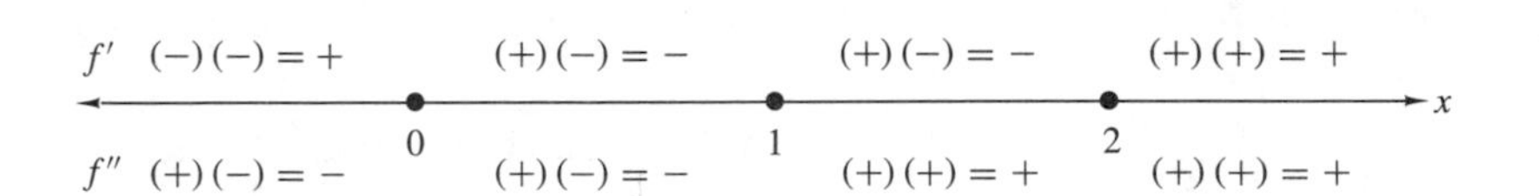

Figure 4.16

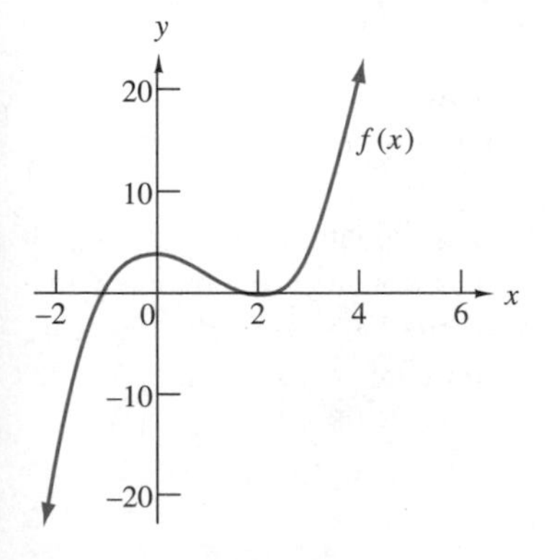

Figure 4.17

2. Where are $f'(x)$ and $f''(x)$ undefined? Since both $f'(x)$ and $f''(x)$ are polynomials, they are never undefined. Thus, $x = 0,\ 2$ are critical values and, therefore, possible relative extrema, and $x = 1$ is a hypercritical value and, therefore, a possible point of inflection. We construct the sign chart illustrated in Figure 4.16. Using the sign chart, we construct the summary table for $f(x)$.

Using the information and sketches from the summary table, we construct a sketch of $f(x)$ shown in Figure 4.17.

Summary of behavior: As x increases to 0, $f(x)$ increases but at a decreasing rate. At $x = 0, f(x) = 4$ and it attains a relative maximum. As x increases to $1, f(x)$ decreases at an increasing rate. At $x = 1$, the function experiences a

Table 4.5

Int / Val	$f'(x)$	$f''(x)$	$f(x)$	Behavior of $f(x)$
$(-\infty, 0)$	+	−	incr/cd	incr at a decr rate
0	0	−	$(0, 4)$	rel max at$(0, 4)$
$(0, 1)$	−	−	decr/cd	decr at an incr rate
1	−	0	$(1, 2)$	pt of infl at $(1, 2)$
$(1, 2)$	−	+	decr/cu	decr at a decr rate
2	0	+	$(2, 0)$	rel min at $(2, 0)$
$(2, +\infty)$	+	+	incr/cu	incr at an incr rate

point of inflection (which alerts us to a coming change in concavity—that is, the speed of the decrease is going to change). As x increases to 2, $f(x)$ decreases but does so now at a decreasing rate (the rate of decrease is slowing). At $x = 2$, $f(x) = 0$ and it attains a relative minimum. Then, as x increases without bound, $f(x)$ increases at an increasing rate.

EXAMPLE SET C

Summarize the behavior of $f(x) = \dfrac{5x+2}{3x-4}$, and sketch its graph.

Solution:

The rational expression $\dfrac{5x+2}{3x-4}$ is undefined when $3x - 4 = 0$; that is, when $x = 4/3$. But $4/3$ is not in the domain of $f(x)$ so it cannot be a critical or hypercritical value. The first and second derivatives are

$$f'(x) = \frac{-26}{(3x-4)^2} \qquad \text{and} \qquad f''(x) = \frac{156}{(3x-4)^3}$$

We search for critical and hypercritical values by checking the two conditions.

1. Since fractions are 0 only when their numerators are zero, $f'(x)$ and $f''(x)$ are never 0.
2. Since fractions are undefined only when their denominators are 0, $f'(x)$ and $f''(x)$ are undefined when $x = 4/3$. But $4/3$ is not in the domain of the original function $f(x)$ so it cannot be a hypercritical value. We will, however, plot a point on the sign chart at $x = 4/3$ because of its importance in the description of the behavior of $f(x)$.

f': $\frac{(-)}{(+)} = -$ $\qquad$ $\frac{(-)}{(+)} = -$

f'': $\frac{(+)}{(-)} = -$ $\quad \frac{4}{3} \quad$ $\frac{(+)}{(+)} = +$

Figure 4.18

Thus, $f(x)$ has no critical or hypercritical values and, therefore, no relative extreme points or points of inflection. The sign chart for $f(x)$ appears in Figure 4.18. Before constructing the summary table for $f(x)$, we notice that $\lim_{x\to\infty} \frac{5x+2}{3x-4} = \frac{5}{3}$. This means that $f(x)$ has a horizontal asymptote at $y = \frac{5}{3}$. (When working with fractions, we always check for horizontal asymptotes.)

Now, using the sign chart, we construct the summary table for $f(x)$. Using

Table 4.6

Int / Val	$f'(x)$	$f''(x)$	$f(x)$	**Behavior of** $f(x)$
$(-\infty, 4/3)$	$-$	$-$	decr/cd	decr at an incr rate
$x = 4/3$	und	und	und	VA
$(4/3, +\infty)$	$-$	$+$	decr/cu	decr at a decr rate
$y = 5/3$				HA

the information and sketches in the summary table, we construct the graph of $f(x)$ shown in Figure 4.19.

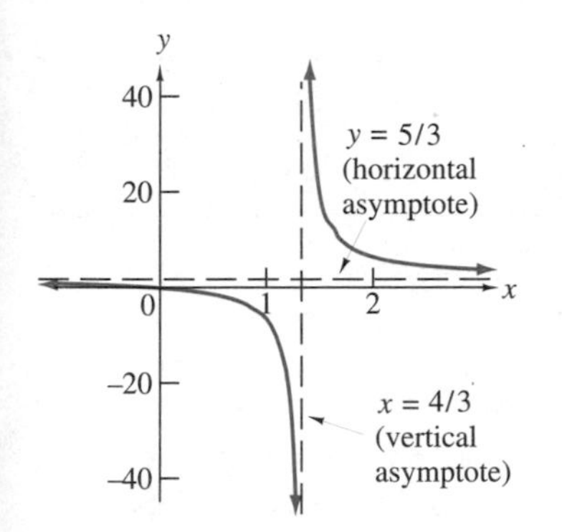

Figure 4.19

Summary of behavior: When x is very small (far to the left of 0), $f(x)$ is very near, but always less than 5/3. As x increases to 4/3, $f(x)$ decreases at an increasing rate. In fact, as x approaches 4/3, $f(x)$ gets unboundedly small. At $x = 4/3$, $f(x)$ is undefined. In fact, $x = 4/3$ is a vertical asymptote. When x is just beyond 4/3, $f(x)$ is unboundedly large. As x increases beyond 4/3, $f(x)$ decreases at a decreasing rate and approaches, but is always just greater than, 5/3.

EXAMPLE SET D

A function $A(t)$ exhibits the relationship between time t (in years) and the amount A (in thousands of dollars) a hair care products company spends annually on advertising. The year $t = 5$ corresponds to 1980, the year the company came under new management. Use the information in the following summary of behavior to construct a sketch of $A(t)$.

Summary of behavior: In 1975, under the previous management, the company spent \$35,000 ($35 \cdot \$1{,}000 = \$35{,}000$) annually on advertising. As time increased from 1975 to 1980, the amount spent on advertising decreased at an increasing rate. Then, in 1980 ($t = 5$) the amount spent on advertising attained a relative minimum of \$20,000. Also in 1980, the company came under new management. As time increased from 1980 to 1985, the amount spent on advertising increased at an increasing rate. Then, in 1985, the amount spent on advertising hit \$39,000 and the rate of increase began to decrease; the company hit the point of diminishing returns for advertising dollars. From 1985 through the present, the amount spent on advertising annually continued to grow toward, but was never more than, \$42,000. This growth was at a decreasing rate.

Solution:
We can extract the following information from the summary.

1. At $t = 0$ (1975), $A(t) = 35$ (\$35,000).
2. As t increases from 0 to 5 (1975 to 1980), $A(t)$ decreases at an increasing rate, so $A(t)$ must be decreasing and concave downward.
3. At $t = 5$ (1980), $A(t)$ attains a relative minimum of 20.
4. As t increases from 5 to 10 (1980 to 1985), $A(t)$ is increasing but at a decreasing rate. This means that the curve is increasing and concave upward.
5. We note the corner at $(5, 20)$, indicating that $A'(t)$ is undefined at this point.
6. At $t = 10$ (1985), $A(t) = \$39{,}000$, and the rate of spending begins to decrease; that is the point of diminishing returns. This means a change in concavity and that $A(t)$ must experience a point of inflection.
7. As t increases beyond 10, the rate of spending decreases so that the curve is still increasing but is now concave downward.
8. Spending is climbing toward, but never goes higher than, \$42,000. This means that 42 is a horizontal asymptote.

Using this information, we construct a sketch of $A(t)$ (as shown in Figure 4.20).

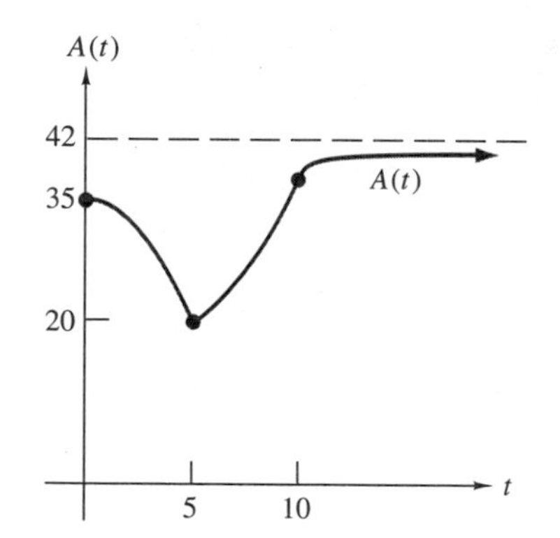

Figure 4.20

Describing Behavior and Technology

Using Your Calculator The behavior of a function can be described very well using the techniques described in the last section. Another interesting approach is to graph the first and second derivatives of the function to be summarized.

Construct the graph of $f'(x)$ for $f(x) = 0.03x^3 - 2x + 3$ on the standard window. (Turn off the graphing capability of all functions other than $f'(x)$.) You can see that on approximately $[-\infty, -4.9]$, $f'(x) > 0$ since the curve lies above the x axis. This means that $f(x)$ is increasing on this interval. On approximately $[-4.9, +4.9]$, $f'(x) < 0$ since the curve lies below the x axis. This means that $f(x)$ is decreasing on this interval. At $x \approx -4.9$ and $x \approx +4.9$, $f'(x) = 0$ since the curve intercepts the x axis at these values. This means that -4.9 and 4.9 are critical values and produce potential smooth relative maximums or minimums.

Construct the graph of $f''(x)$ for $f(x) = 0.03x^3 - 2x + 3$ on the standard window. (Turn off the graphing capability of all functions other than $f''(x)$.) You can see that on approximately $[-\infty, 0]$, $f''(x) < 0$ since the curve lies below the x axis. This means that $f'(x)$ is decreasing on this interval. On approximately $[0, +\infty]$, $f''(x) > 0$ since the curve lies above the x axis. This means that $f'(x)$ is increasing on this interval. At $x \approx 0$, $f''(x) = 0$ since the curve intercepts the

x axis at 0. This means that 0 is a hypercritical value and produces a potential point of inflection.

Calculator Exercises Construct the graphs of both the first and second derivatives of each of the following functions and use those graphs to describe the behavior of the functions.

1. A stockbroker who has studied the behavior of a stock over a long period of time believes that the cubic function

$$P(x) = 0.03x^3 - 0.75x^2 + 1.85x + 42.25$$

approximates the price, $P(n)$, per share n days from now over a period of 30 days.

2. A drug manufacturer claims that the effectiveness of a new drug is related to the time after the drug is introduced into the bloodstream by the function

$$E(t) = -0.04t^3 + 0.081t^2 + 0.39t$$

where $E(t)$ represents the effectiveness of the drug on a scale of 0 to 1, t represents the number of hours since introduction, and $0 \le t \le 5$.

3. Industrial psychologists use the function

$$N(t) = \frac{40t}{3.5t^{1.85} + 0.85}, \qquad 0 < t \le 5$$

to approximate the number N of technological concepts a worker can learn t hours after beginning a learning session.

4. The management of a supermarket checker training program claims that after t hours of training, its trainees can check N customers each hour, where N and t are related by the function

$$N(t) = 50 - \frac{35}{\sqrt{x + 0.5}}$$

5. A state highway engineer uses the function

$$v(\rho) = \frac{85}{\rho^{2.2} + 0.85}, \qquad 0 \le \rho \le 5$$

to relate the average velocity, v, in mph, of cars along a particular section of interstate highway, and the average density, ρ, in cars per 100 feet, of the cars along that section of highway.

EXERCISE SET 4.2

A UNDERSTANDING THE CONCEPTS

For Exercises 1–10, use the abbreviated summary table to construct a sketch of the corresponding function.

1. This table is for $f(x)$.

Int / Val	$f'(x)$	$f''(x)$	$f(x)$	Behavior of $f(x)$
$x = 0$				VA
$(0, 4)$	$-$	$+$		
4		0	$(4, 3)$	
$(4, 8)$	$-$	$+$		
$x = 8$				VA

2. This table is for $f(x)$.

Int / Val	$f'(x)$	$f''(x)$	$f(x)$	Behavior of $f(x)$
$(-\infty, 3)$	$-$	$+$		
3	0		$(3, 1)$	
$(3, 5)$	$+$	$+$		
5		0	$(5, 3)$	
$(5, 7)$	$+$	$-$		
7	0		$(7, 5)$	
$(7, 9)$	$-$	$-$		
9		0	$(9, 3)$	
$(9, 11)$	$-$	$+$		
11	0		$(11, 1)$	
$(11, +\infty)$	$+$	$+$		

3. This table is for $f(x)$.

Int / Val	$f'(x)$	$f''(x)$	$f(x)$	Behavior of $f(x)$
$(-\infty, 3)$	$-$			
3	und		$(3, 2)$	
$(3, 6)$	$+$			
6		0	$(6, 5)$	
$(6, +\infty)$	$+$	$+$		

4. This table is for $f(x)$.

Int / Val	$f'(x)$	$f''(x)$	$f(x)$	**Behavior of $f(x)$**
$(-\infty, -1)$	$-$	$+$		
-1	0		$(-1, -3)$	
$(-1, 0)$	$+$	$+$		
0		0	$(0, 0)$	
$(0, +\infty)$	$+$	$-$		

5. This table is for $f(x)$.

Int / Val	$f'(x)$	$f''(x)$	$f(x)$	**Behavior of $f(x)$**
$(-\infty, 0)$	$-$	$-$		
0	und		$(0, 0)$	
$(0, 2)$	$+$	$-$		
2	0	0	$(2, 5)$	
$(2, +\infty)$	$-$	$-$		

6. This table is for $f(x)$.

Int / Val	$f'(x)$	$f''(x)$	$f(x)$	**Behavior of $f(x)$**
$(-\infty, 0)$	$+$	$+$		
0		und	$(0, 0)$	
$(0, +\infty)$	$-$	$-$		

7. This table is for $f(x)$.

Int / Val	$f'(x)$	$f''(x)$	$f(x)$	**Behavior of $f(x)$**
$(-\infty, -2)$	$-$	$-$		
-2		0	$(-2, 2)$	
$(-2, 0)$	$-$	$+$		
0		0	$(0, 0)$	
$(0, 2)$	$+$	$+$		
2		0	$(2, 2)$	
$(2, +\infty)$	$+$	$-$		
$y = 4$				HA

8. This table is for $f(x)$.

Int / Val	$f'(x)$	$f''(x)$	$f(x)$	**Behavior of $f(x)$**
$x = 0$				VA
$(0, 3)$	$-$	$+$		
3	0	0	$(3, 1)$	
$(3, 6)$	$+$	$+$		
6	und	und	und	VA
$(6, +\infty)$	$-$	$+$		
$y = 0$				HA

9. This table is for $f(x)$.

Int / Val	$f'(x)$	$f''(x)$	$f(x)$	**Behavior of $f(x)$**
$(-\infty, 1)$	$-$	$-$		
1	und		$(1, 0)$	
$(1, 3)$	$+$	$-$		
3	0		$(3, 4)$	
$(3, 5)$	$-$	$-$		
5	und		$(5, 3)$	
$(5, 7)$	$+$	$-$		
7	0		$(7, 4)$	
$(7, 9)$	$-$	$-$		
9	und		$(9, 0)$	
$(9, +\infty)$	$+$	$+$		
$y = 5$				HA

10. This table is for $f(x)$.

Int / Val	$f'(x)$	$f''(x)$	$f(x)$	**Behavior of $f(x)$**
$(-\infty, 2)$	$+$	$+$		
2	und		$(2, 6)$	
$(2, 4)$	$-$	$+$		
4		0	$(4, 3)$	
$(4, 6)$	$-$	$-$		
6	und		$(6, -2)$	
$(6, +\infty)$	$+$	$-$		
$y = 0$				HA
$y = 4$				HA

For Exercises 11–18, use the graph shown in each figure to construct a summary table. Include asymptotes if applicable.

11.

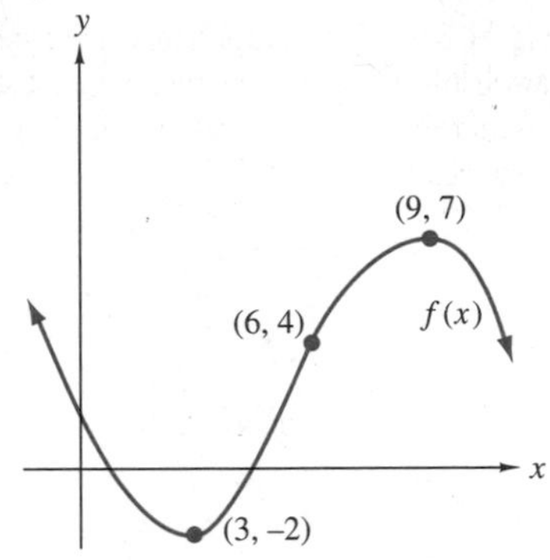

12.

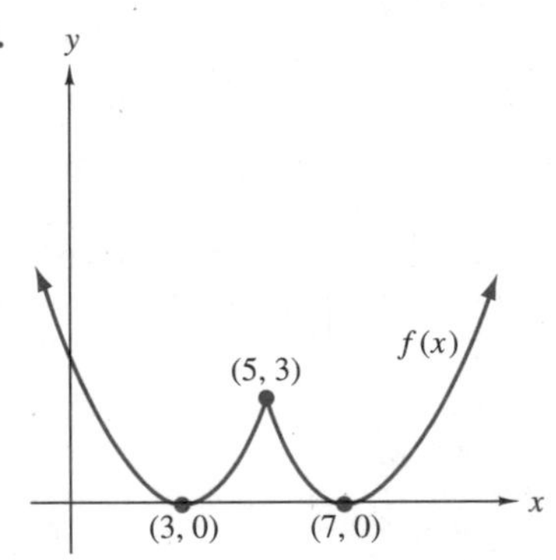

13.

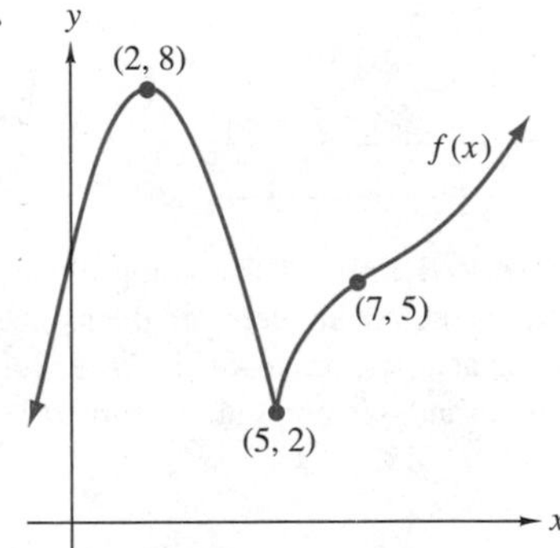

14.

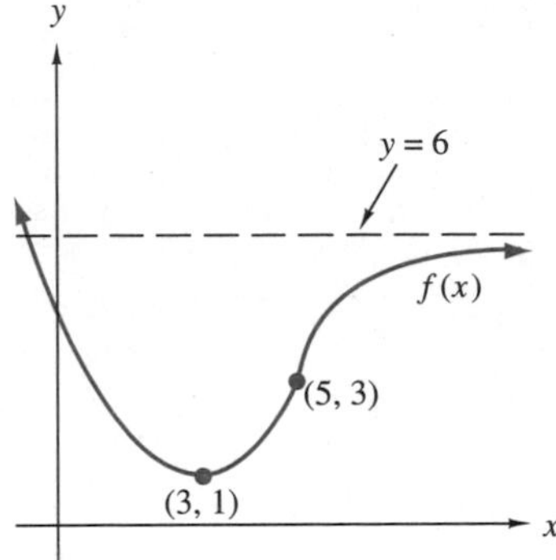

15.

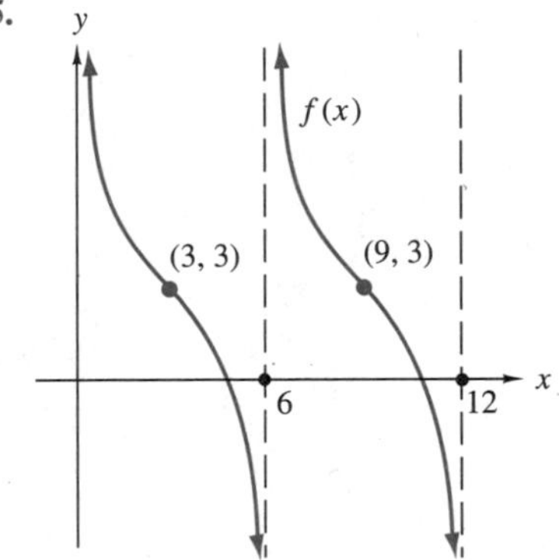

16.

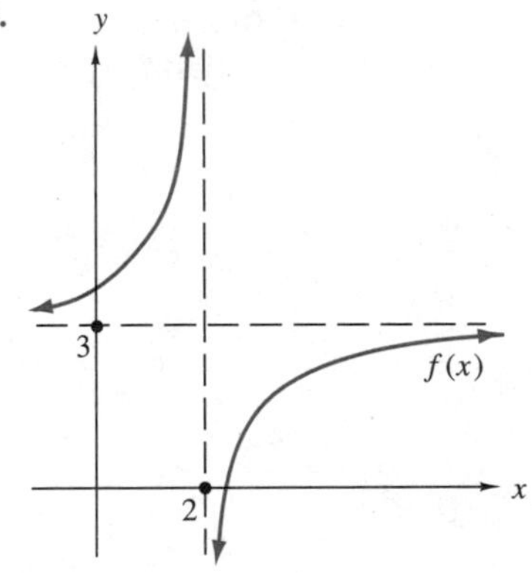

17.

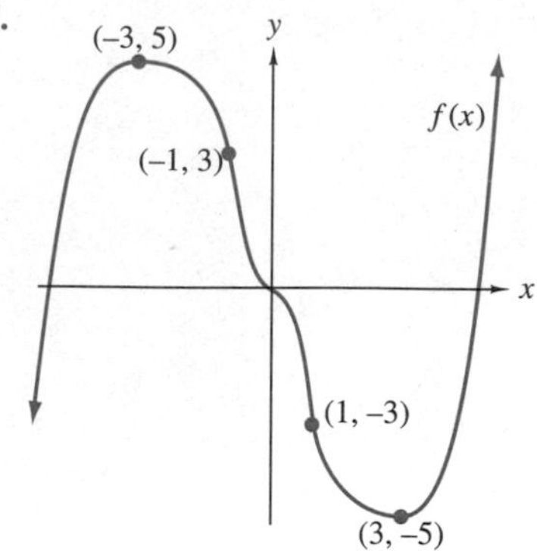

18.

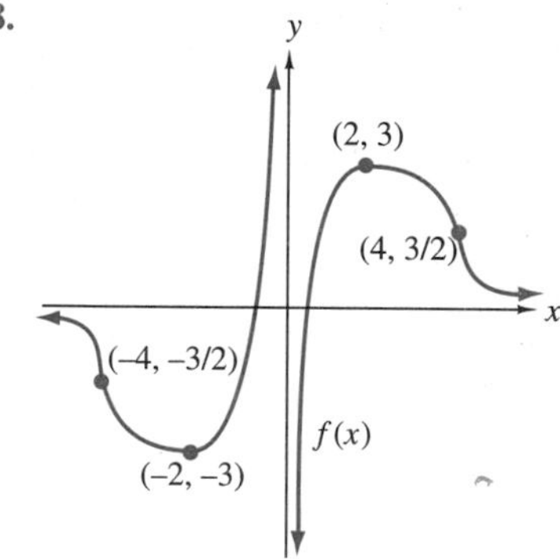

B SKILL ACQUISITION

For Exercises 19–30, sketch each function, designating asymptotes with dashed lines when they occur.

19. $f(x) = x^2 - 4x + 1$

20. $f(x) = -x^2 - 6x - 5$

21. $f(x) = x^3 - 3x^2 + 3$

22. $f(x) = x^3 + 3x - 2$

23. $f(x) = 3x^4 - 4x^3$

24. $f(x) = 6x^2 - 3x^4$

25. $f(x) = \dfrac{x^2}{x^2 + 3}$

26. $f(x) = x^{2/3} - 2$

27. $f(x) = 3x^{2/3} - 2x$

28. $f(x) = 2x^{5/3} - 5x^{4/3}$

29. $f(x) = \dfrac{2x + 2}{x}$

30. $f(x) = \dfrac{x}{x^2 - 4}$

C APPLYING THE CONCEPTS

For Exercises 31–34, describe the behavior of the given function. (You need not construct a summary table.)

31. ***Business: Average Cost*** The graph illustrated below displays the relationship between the average cost A (in dollars) of a particular commodity and the number q of the commodity produced.

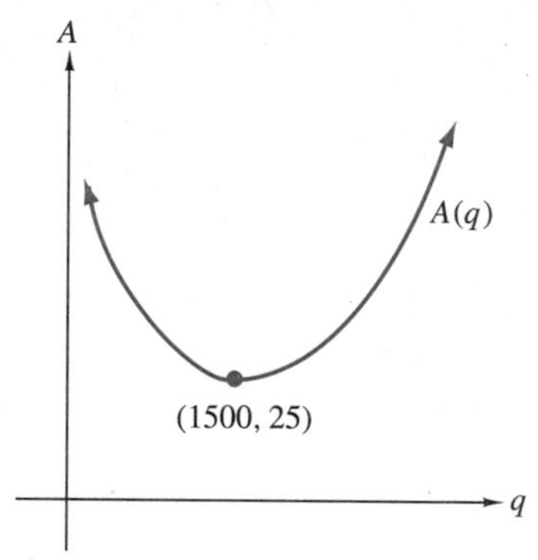

32. ***Psychology: Anger Level*** The graph illustrated below exhibits the relationship between the average anger level A (in appropriate units) and the amount of time t (in days) since the occurrence of a hate crime in the community.

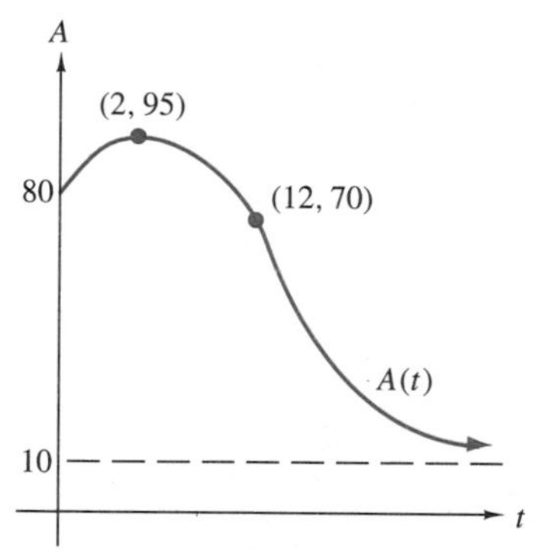

33. ***Economics: University Tuition*** The graph illustrated in below exhibits the relationship between the increase I (in percent) in tuition at a state university and the average enrollment E (in thousands) in community colleges near the state university.

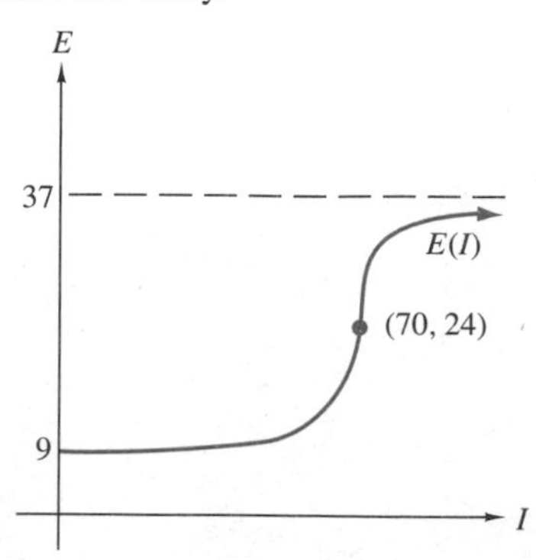

34. ***Economics: Use of Water*** The graph illustrated below displays the relationship between the amount A (in millions of gallons) of reclaimed water used by a county each year and the time t (in years) since 1940.

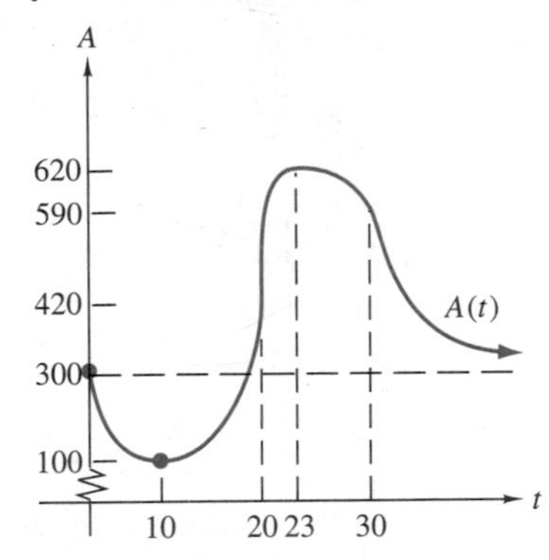

For Exercises 35–38, sketch the graph of a function that matches the summarized behavior.

35. ***Mathematics: Probability*** The probability of many types of events is described by a curve that increases at an increasing rate as x increases from negative infinity to -1. At $x = -1$, $f(x) \approx 0.24$ and the function experiences a point of inflection; in particular, this point of inflection is a point of diminishing returns. As x increases beyond -1 to 0, the function continues to increase but at a decreasing rate. When $x = 0$, $f(x) \approx 0.40$ and the function attains a relative maximum. As x then increases to $+1$, $f(x)$ decreases at an increasing rate. At $x = +1$, $f(x) \approx 0.24$ and the function experiences a point of inflection. As x increases from $+1$ to $+\infty$, the function continues to decrease but at a decreasing rate. The function is always positive and the x axis is a horizontal asymptote.

36. ***Economics: Demand for a Commodity*** The demand D (in single units) for a particular commodity is related to the price p (in dollars) of the commodity. For a particular commodity, $D = 0$ when $p = 0$. Then as the price increases from \$0 to \$250, the demand increases but at a decreasing rate. The demand attains a maximum of 1400 units when the price is \$250. As the price increases from \$250 to \$400, the demand for the commodity decreases at an increasing rate. When $p = 400$, $D = 350$ and experiences a point of inflection. When the price increases beyond \$400 per unit, the demand for it continues to decrease but now at a decreasing rate.

37. ***Manufacturing: Newton's Law of Cooling*** Newton's Law of Cooling describes how a heated object cools when placed into a cooler environment. A particular object has temperature $T = 98°$C when it is first taken from an oven. It then cools according to Newton's Law of Cooling. The temperature of the object decreases at a decreasing rate and approaches 0 as time goes on. In fact, the temperature of the object is only about 18°C after 60 minutes.

38. ***Business: Investment*** When an amount \$2000 is invested at 7% compounded continuously, the amount A accumulated grows at an increasing rate as time t (in years) goes by. In fact, after 30 years, \$16,332 will have accumulated.

D DESCRIBE YOUR THOUGHTS

39. Describe the steps you would take to summarize the behavior of a function.

40. Suppose that at $x = a$, the function $f(x)$ has a cusp. Describe the information you would expect to see, other than the word *cusp*, in the function's summary table that would help you conclude that there is a cusp at $x = a$.

41. Suppose that for a function $f(x)$, $f(a)$ is defined, $f'(a)$ is undefined, and $f''(a) = 0$. Describe the possible behaviors of $f(x)$ near $x = a$. A picture may enhance your description.

E REVIEW

42. **(2.2)** Find, if it exists, $\lim_{x \to 6} \frac{x^2 + 7x - 8}{x^2 - 3x + 2}$.

43. **(2.3)** Find, if it exists, $\lim_{x \to \infty} \frac{5x^2 + x + 7}{x^5 - 1}$.

44. **(2.4)** The function

$$f(x) = \begin{cases} -x^2 + 6x - 16, & x < 3 \\ x^2 - 6x + 11, & x \geq 3 \end{cases}$$

is discontinuous at $x = 3$. Which of the three continuity conditions is the first not to be satisfied?

45. **(2.6)** Find the average rate of change of the function $f(x) = x^2 + 2x + 6$ as x changes from $x = 5$ to $x = 7$.

46. **(3.2)** Find $f'(x)$ for $f(x) = \frac{x - 2}{8x + 2}$.

47. **(3.3)** Suppose that for a function $f(x)$, $f'(3) = 0$, and $f''(3) = 5$. Make a statement about the behavior of the function near $x = 3$.

48. **(3.4)** Find $f'(x)$ for $f(x) = (4x^3 + 2)^5$.

49. **(3.5)** Find $\frac{dy}{dx}$ for $4x^2 + 9x^4y^2 + 2y^2 = 1$.

50. **(4.1)** A point at which a curve changes concavity from upward to downward is called by a special name. Specify that name.

4.3 Applications of the Derivative-Optimization

Introduction

Optimization

The process of determining the maximum or minimum values of a function is called **optimization**. Using the differentiation techniques discussed in Sections 4.1 and 4.2, we will see how the derivative is applied to optimize functions that model applications from business, economics, medicine, and other fields.

A Strategy for Solving Optimization Problems

We will now consider how differential calculus can be used to solve optimization problems. These are problems in which we are interested in maximizing or minimizing a particular quantity. For example, we may be interested in knowing what sales price of an item will maximize the profit on that item, or how many days after an insecticide is applied will a population of insects be at a minimum.

A strategy that is helpful in solving optimization problems follows.

Optimization Strategy

1. Draw a figure when one is appropriate.
2. Assign a variable to each quantity mentioned in the problem.
3. Select the quantity that is to be optimized and construct a function that relates it to some or all of the other quantities.
4. Since all the rules of differentiation deal with functions of only one variable, the function must involve only one variable. If the function constructed in step 3 involves more than one variable, use the information contained in the problem to eliminate variables until you have a function of only one variable.
5. Find the critical values of the function and apply the test for absolute extrema.
6. Disregard all answers that are not relevant to the situation.

Applying the Strategy: Examples

We will illustrate the optimization strategy with five examples. We will show the first example with great detail and then include less detail in the remaining examples.

EXAMPLE SET A

A rectangular area along the bank of a straight section of a river is to be enclosed with 80 feet of fencing. Only three sides need to be fenced since the river will bound the other side. What are the dimensions that will produce the greatest enclosed area?

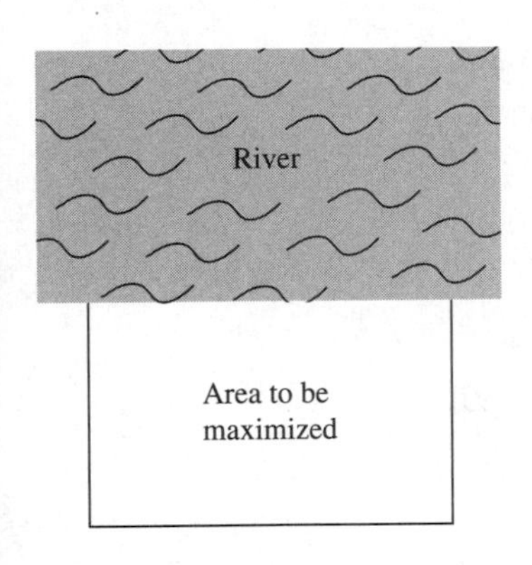

Figure 4.21

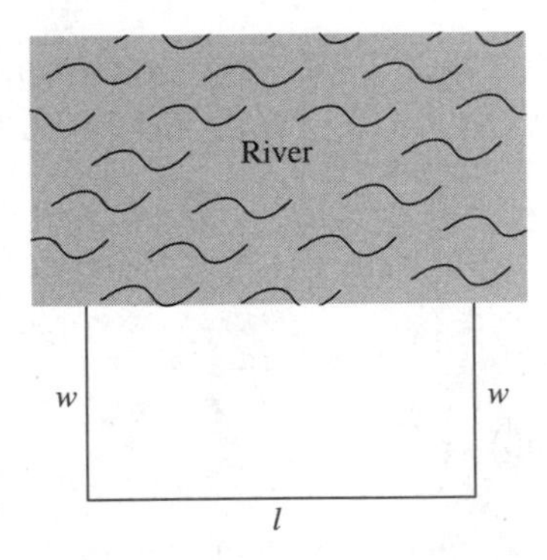

Figure 4.22

Solution:
We will apply the optimization strategy.

1. We will construct a figure since one is appropriate. (See Figure 4.21.)
2. The quantities mentioned in this problem are the area, the dimensions, and the amount of fencing available. We will label each with a variable and then affix the appropriate labels to our picture, as shown in Figure 4.22.
$A =$ the area of the rectangular region.
$l =$ the length of the rectangular region.
$w =$ the width of the rectangular region.
$P =$ the perimeter of the rectangular region (that is, the amount of fencing available.)
3. The quantity to be maximized is the area, so we need to construct an area function. Since the region is rectangular, we will use the formula for the area of a rectangle.

   Maximize: $A(l, w) = lw$.

   (Remember how functions are read: $A(l, w)$ means that the area A depends on both the length l and the width w.)
4. This function involves two variables, l and w; we need to eliminate one. To do so, we will use the information contained in the problem to express l in terms of w or w in terms of l. The problem restricts the amount of fencing to 80 ft, which means that the *perimeter*, P, of the fenced region is 80 ft. The perimeter of this shape is the sum of the lengths of the three fenced sides, so

$$P = w + l + w = 80$$

$$= l + 2w = 80$$

Thus, $l+2w = 80$, and we can solve for either l or w. (We can avoid fractions by solving for l. Avoiding fractions may not be possible in other problems.)

$$l + 2w = 80$$

$$l = 80 - 2w$$

Now, replacing l in the function $A(l, w) = lw$ with $80 - 2w$ gives us the function of the one variable w.

$$A(w) = (80 - 2w)w$$

$$A(w) = 80w - 2w^2$$

A negative area is not acceptable; thus, solving $A(w) \geq 0$ (by sign charting) gives us $0 \leq w \leq 40$. Therefore, $A(w)$ is defined over a closed interval $[0, 40]$.

5. To find the critical values and then the absolute extreme values, we first find the first derivative.

$$A'(w) = 80 - 4w$$

a. We then ask: Where is $A'(w) = 0$?

$$A'(w) = 0$$

$$80 - 4w = 0$$

$$-4w = -80$$

$$w = 20$$

So, $A'(w) = 0$ when $w = 20$.

b. We then ask: Where is $A'(w)$ undefined? It is never undefined since it is a polynomial function.

Thus, the only critical value is $w = 20$. Since we are only interested in the absolute extrema, we only need to evaluate $A(w)$ at 0, 20, and 40, to see which produces the greatest output (area).

$$A(0) = 0,$$

$$A(20) = 80(20) - 2(20)^2 = 800,$$

$$A(40) = 80(40) - 2(40)^2 = 0$$

Now, we conclude that 20 produces the absolute maximum value of the function. We can also use $l = 80 - 2w$ to find that $l = 80 - 2(20) = 40$.

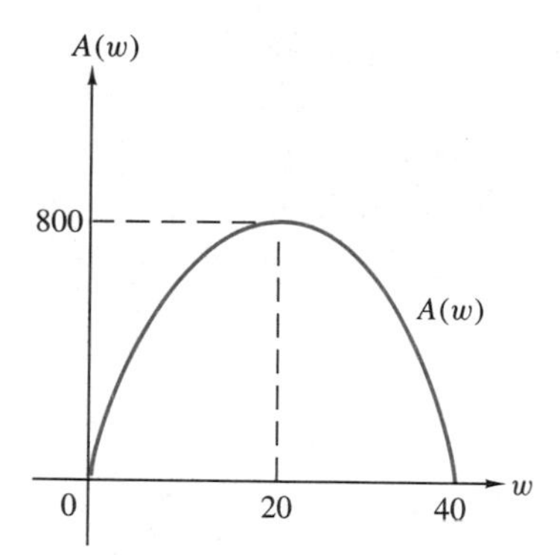

Figure 4.23

Interpretation: The dimensions that produce the maximum enclosed area for 80 feet of fencing are a length of 40 ft and a width of 20 ft. The graph of $A(w) = 80 - 2w^2$ appears in Figure 4.23 and reinforces the fact that the absolute maximum area occurs when the width is 20 feet.

EXAMPLE SET B

The concentration C of a drug in the bloodstream t hours after it has been administered is approximated by the function $C(t) = \dfrac{7t}{t^3 + 18}$, $t \geq 0$. When is the concentration the greatest?

Solution:

In this problem, we do not need to draw a picture, nor do we need to construct a function. From the outset, we have a function of one variable and we are ready to find the critical values.

$$C'(t) = \frac{(t^3 + 18) \cdot 7 - 7t(3t^2)}{(t^3 + 18)^2}$$

$$= \frac{7[(t^3 + 18) - 3t^3]}{(t^3 + 18)^2}$$

$$= \frac{7(18 - 2t^3)}{(t^3 + 18)^2}$$

$$= \frac{14(9 - t^3)}{(t^3 + 18)^2}$$

1. Where is the derivative 0? Recalling that a fraction is 0 only when the numerator is 0 gives us

$$14(9 - t^3) = 0$$

$$9 - t^3 = 0$$

$$t^3 = 9$$

$$t = \sqrt[3]{9}$$

2. Where is the derivative undefined? A fraction is undefined when the denominator is 0. But since the denominator is squared and $t \geq 0$, the denominator is never 0, so the fraction is never undefined.
   Thus, $C(t)$ has the single critical value $\sqrt[3]{9}$.

We need to evaluate $C(t)$ at 0 and $\sqrt[3]{9}$ to determine which produces the absolute maximum. Since $C(0) = 0$, $\sqrt[3]{9}$ must produce the absolute maximum of this function.

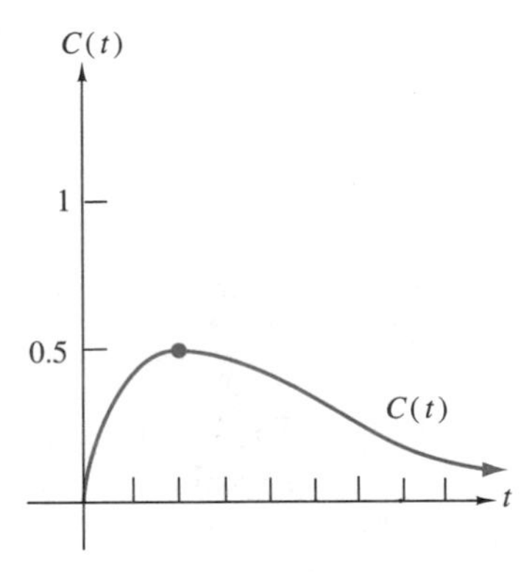

Figure 4.24

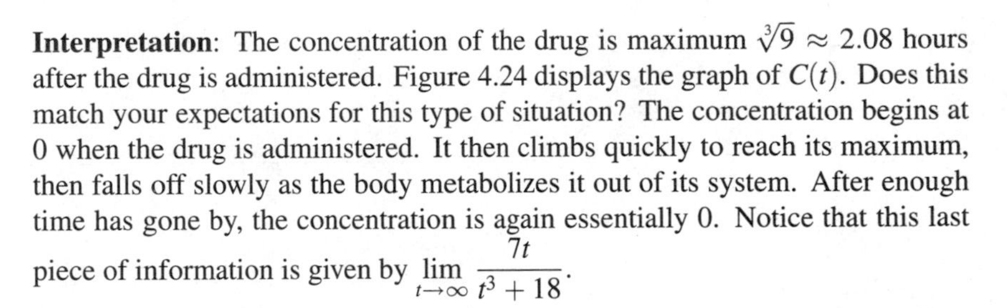
Interpretation: The concentration of the drug is maximum $\sqrt[3]{9} \approx 2.08$ hours after the drug is administered. Figure 4.24 displays the graph of $C(t)$. Does this match your expectations for this type of situation? The concentration begins at 0 when the drug is administered. It then climbs quickly to reach its maximum, then falls off slowly as the body metabolizes it out of its system. After enough time has gone by, the concentration is again essentially 0. Notice that this last piece of information is given by $\lim_{t\to\infty} \frac{7t}{t^3 + 18}$.

EXAMPLE SET C

A retail store can sell 20 hard-disk drives per week at a price of \$400 each. The manager estimates that for each \$10 reduction in price, she can sell two more drives per week. The drives cost the store \$200 each. Find the best price and the quantity that will maximize the store's profit. What is that profit?

Solution:

1. A picture is not appropriate here so we will not construct one.
2. We will introduce some variables to represent the quantities mentioned in the problem.
   r = the number of \$10 reductions in price.
   p = the price of each hard drive after a reduction.

$n =$ the number of hard drives purchased and sold.
$R =$ the revenue produced by all the sales of the hard drives.
$C =$ the cost to the store of n hard drives.
$P =$ the profit realized by the sale of n hard drives.

3. We are asked to find the price that maximizes the profit, so we need to construct a profit function. Since profit equals revenue minus cost, we have

$$P = R - C$$

But this is a function of two variables. Since revenue is price times the number sold, and cost is the individual cost times the number purchased,

$$R = \underbrace{(400 - 10r)}_{price} \cdot \underbrace{n}_{number\ sold} \quad \text{and} \quad C = \underbrace{200}_{cost} \cdot \underbrace{n}_{number\ purchased}$$

That is, $R = (400 - 10r)n$ and $C = 200n$. The expression for R comes from the fact that the sales price is \$400 minus the \$10 for every reduction; thus, \$400 − 10r. But also, the number of drives sold n is 20 plus 2 times the number of \$10 price reduction, that is, $n = 20 + 2r$. Hence,

$$P(r) = (400 - 10r)(20 + 2r) - 200(20 + 2r)$$

$$= 4000 + 200r - 20r^2$$

and the profit P depends on the number of \$10 price reductions.

4. $P'(r) = 200 - 40r$.

5. The critical value is $r = 5$, and the first derivative test shows this to produce a relative maximum.

Interpretation: The best price per drive is $\$400 - 10(5) = \$350$. This would produce sales of $n = 20 + 2(5) = 30$ drives per week, and a maximum profit of $P(5) = 4000 + 200(5) - 20(5)^2 = \$4500$ per week.

EXAMPLE SET D

A rectangular box with an open top is to be made from a 10-in.-by-16-in. piece of cardboard by removing small squares of equal size from the corners and folding up the remaining flaps. What should be the size of the squares cut from the corners so that the box will have the largest possible volume?

Mark the cardboard for cutting and folding.

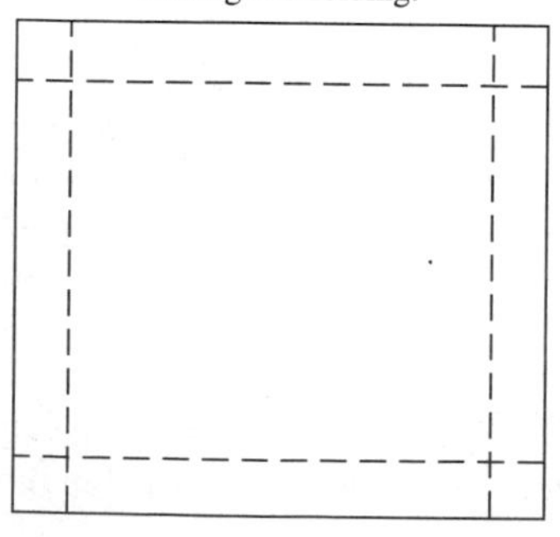

Figure 4.25

Solution:

Figure 4.25, Figure 4.26, and Figure 4.27 show the construction of the box from the flat piece of cardboard.

1. A figure is appropriate and very helpful. Figure 4.25 serves as a good picture of the situation.

2. The quantities mentioned or implied in this problem are volume, length, width, and height. We will label each with a variable and then affix the labels to our picture. Since the cut determines the height of the box which, in turn, determines both the length and width of the box, we will let
$h =$ the length of the side of the square that is cut from the flat piece of

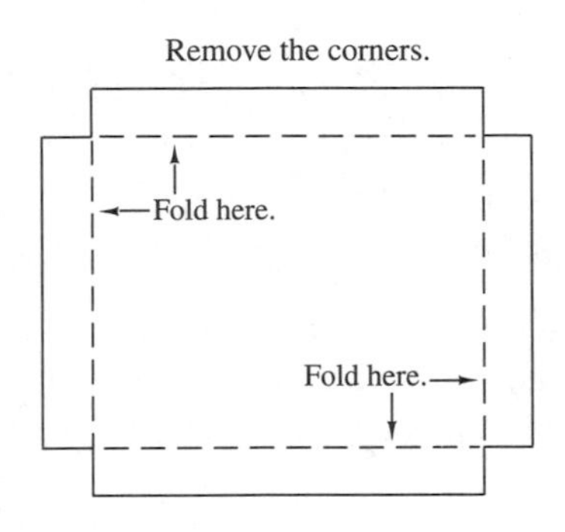

Figure 4.26

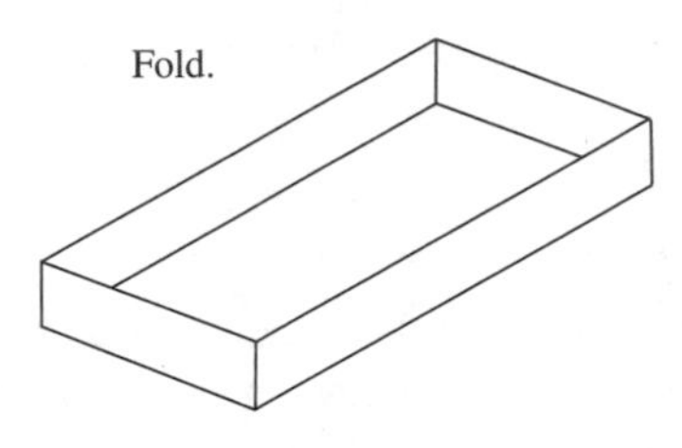

Figure 4.27

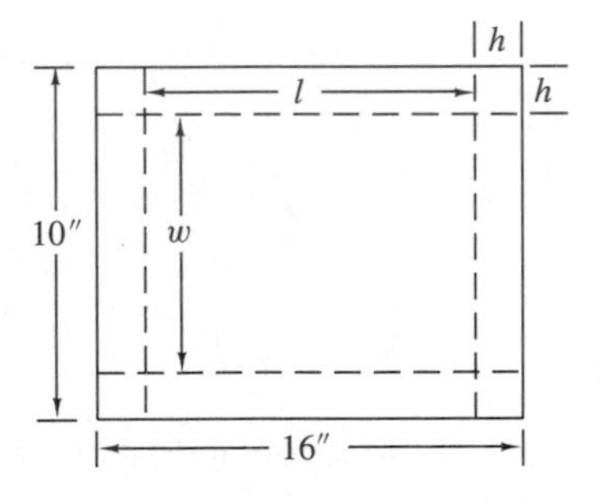

Figure 4.28

cardboard. We label h on the figure to help establish w and l.
$l =$ the length of the box $= 16 - 2h$.
$w =$ the width of the box $= 10 - 2h$.
$V =$ the volume of the box.

3. The quantity to be maximized is the volume, so we need to construct a volume function. Since the region is a rectangular box, we will use the formula for the volume of a rectangle box. Figure 4.28 shows the labeling. Maximize: $V(l, w, h) = lwh = (16 - 2h)(10 - 2h)h = 160h - 52h^2 + 4h^3$
4. $V'(h) = 160 - 104h + 12h^2$
   (1) Where is the derivative equal to 0?

$$160 - 104h + 12h^2 = 0$$

$$40 - 26h + 3h^2 = 0$$

$$(3h - 20)(h - 2) = 0$$

so that $h = \dfrac{20}{3}$ or 2.

(2) When is the derivative undefined? Never (since $V'(h)$ is a polynomial function).

Thus, the critical values are $h = \dfrac{20}{3}$ and 2. However, since $0 < h < 5$ (why?), only $h = 2$ needs to be tested. Figure 4.29 shows the sign chart for $V'(h)$ and indicates that 2 produces a relative maximum.

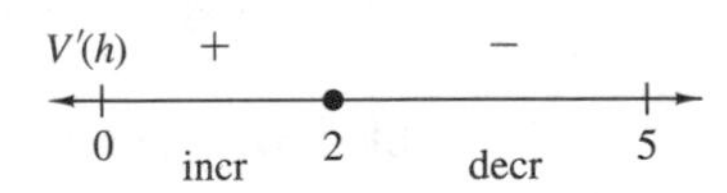

Figure 4.29

Interpretation: To maximize the volume of the box, the cutout squares should be 2 inches by 2 inches.

EXAMPLE SET E

A chemical company determines that the population P (in ten thousands) of a certain insect t days after its insecticide is applied to a one-acre field in which the insects reside is approximated by the function

$$P(t) = \frac{2[13 + (t - 5)^2]}{t + 1} + 15, \quad 0 \le t \le 20$$

A person wishes to determine the number of days after an application of the insecticide when the population of insects will be at a minimum, so that he will know when to reapply the insecticide for its greatest effectiveness.

Solution:

A figure is not appropriate here. Since we are given the function, we will proceed with differentiating it to find the critical values.

$$P'(t) = \frac{2(t+8)(t-6)}{(t+1)^2}$$

$P'(t) = 0$ at $t = -8$ and 6 and is undefined at $t = -1$. But $t = -8$ and -1 make no physical sense and are not in the domain of $P(t)$. The only critical value is $t = 6$. Figure 4.30 shows the sign chart for $P'(t)$ and indicates that 6 produces a relative minimum.

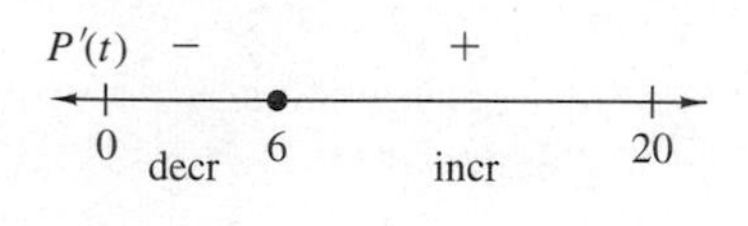

Figure 4.30

Interpretation: Six days after the insecticide is applied, the population of insects is at a minimum. This may be a good time to reapply the insecticide (assuming there will be no great harm to the soil).

Optimization and Technology

Using Your Calculator Many functions that model applied phenomena are data-generated functions and therefore well suited to the approximation techniques of the graphing calculator. The function of Example Set E can be optimized by entering the function

```
Y1=(2(13+(X-5)^2))/(X+1) + 15
```

activating the Trace cursor, and moving the cursor to the apparent minimum part of the curve. A minimum for this function is $(6, 19)$.

Calculator Exercises

1. Find the value of x that maximizes the function

$$f(x) = x^{5/3} - 5x^{2/3} + 3$$

Find also the maximum function value. Graph this function and label the maximum point with the appropriate ordered pair.

2. Find the value of x that maximizes the function

$$f(x) = -4x^4 + 4x^2 + 3$$

Find also the maximum function value. Graph this function and label the maximum point with the appropriate ordered pair.

3. Find both the maximum and minimum values of the profit function

$$P(x) = 350 - 35x(2x - 54)^{2/3}$$

where P represents the profit realized on the sale of x units of a product and $30 \le x \le 40$.

4. When learning a new subject, it is often the case that a person's interest level L increases near the beginning of learning but then declines after having spent some time with the material. Suppose for a 30-minute period, the function

$$L(t) = \frac{55t}{8.5 + 0.25t^2}$$

measures a person's interest level on a 0-to-25 scale. The variable t represents the number of minutes a person spends learning the new material. If a person begins studying new material at 3:00 in the afternoon, find the time at which her interest level is maximum. Sketch a picture of the interest curve and specify the maximum interest level.

EXERCISE SET 4.3

A UNDERSTANDING THE CONCEPTS

1. *Economics: Population* The following figure illustrates the relationship between time t and the number of people P in a given state over a 20-year period. During what year is the population at its maximum?

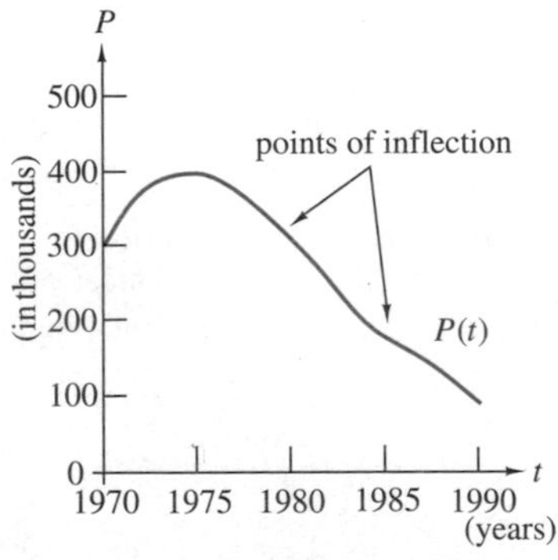

2. *Medicine: Drug Concentration* The following figure shows the concentration levels C of a certain drug in the bloodstream over the time t since consumption. What is the maximum concentration of the drug and when does it reach this level?

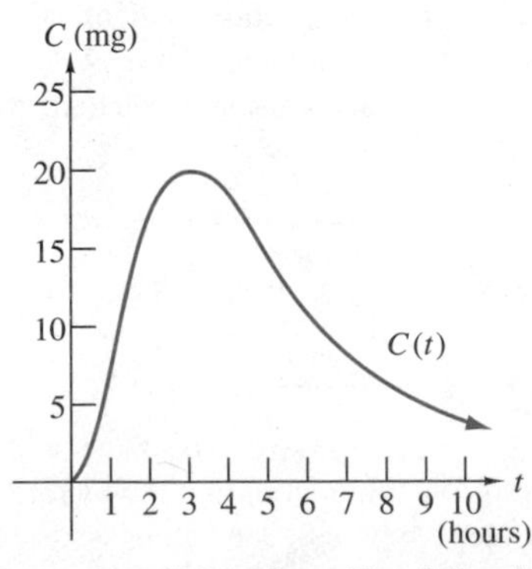

3. *Mathematics: Triangles* The following figure shows the relationship between the hypotenuse of a right triangle H and the length of one leg L when the sum of the two legs is always 20. What are the lengths of the two legs of the right triangle with a minimum hypotenuse length?

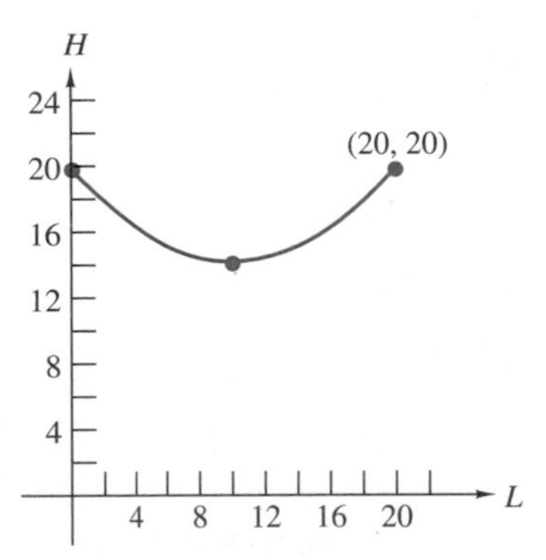

4. *Business: Operation Costs* The following figure illustrates the cost C, in dollars, of operating a truck at speeds V, in miles per hour. What is the most economical speed at which to operate the truck?

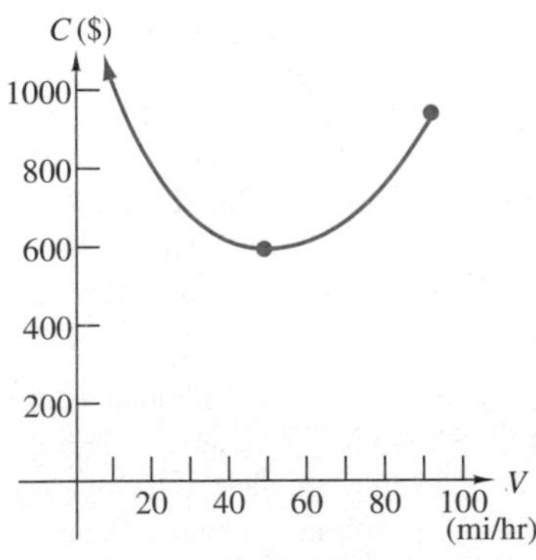

5. *Business: Profit* The following figure shows the profit P, in hundreds of dollars, a magazine publisher can make if it sells n, in hundreds, magazines. How many magazines should be sold to maximize profit?

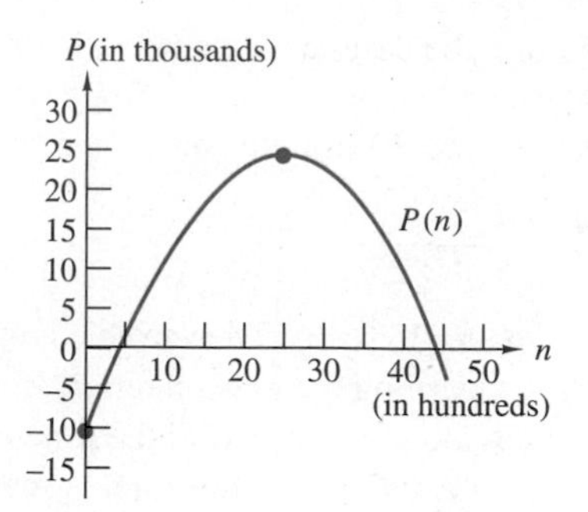

6. ***Economics: Taxes*** A state legislature is considering a bill that would impose a meals tax of t cents per dollar spent on food in restaurants. Economists estimate the tax will cause a decline in the amount of money $A(t)$, in millions of dollars, spent by the public on food in restaurants. This is given by

$$A(t) = 216 - 2t^2, \quad 0 \le t \le 8$$

Graph $A(t)$ and determine the value of t that maximizes $A(t)$.

B SKILL ACQUISITION

7. ***Medicine: Drug Concentration*** The concentration of a drug in the bloodstream $C(t)$ at any time t, in minutes, is described by the equation

$$C(t) = \frac{100t}{t^2 + 16}$$

where $t = 0$ corresponds to the time at which the drug was swallowed. Determine how long it takes the drug to reach its maximum concentration.

8. ***Medicine: Flu Outbreak*** According to a model developed by a public health group, the number of people $N(t)$, in hundreds, who will be ill with the Asian flu at any time t, in days, next flu season is described by the equation

$$N(t) = 90 + \frac{9}{4}t - \frac{1}{40}t^2, \quad 0 \le t \le 120$$

where $t = 0$ corresponds to the beginning of December. Find the date when the flu will have reached its peak and state the number of people who will have the flu on that date.

9. ***Business: Heating Costs*** A homeowner wishes to insulate his 1500-square-foot attic. The total cost $C(r)$ for r inches of insulation and the heating costs for that amount of insulation over the next 10 years is given by

$$C(r) = 120r + \frac{3410}{r}$$

How many inches of insulation should be placed in the attic if the total cost is to be minimized?

10. ***Business: Maximization of Space*** A rancher has 200 feet of fencing to enclose two adjacent rectangular corrals, as shown in the following figure. The equation describing the enclosed area is

$$A(x) = 2x\left(\frac{200}{3} - \frac{4}{3}x\right)$$

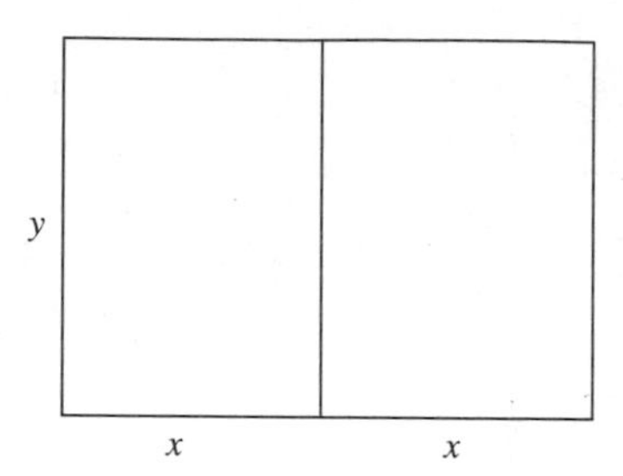

To enclose the maximum area, what should be the dimensions of each corral?

11. ***Business: Revenue*** Suppose a baby food company has determined that its total revenue R for its food is given by

$$R = -x^3 + 63x^2 + 1200x$$

where R is measured in dollars and x is the number of units (in thousands) produced. What production level will yield a maximum revenue?

12. ***Wildlife Management: Deer Population*** The game commission in a certain state introduces 500 deer into some newly acquired land. The population $N(t)$ of the herd is given by

$$N(t) = \frac{500 + 500t}{1 + 0.04t^2}$$

where t is time in years. Determine the number of years for the deer population to become a maximum. What is likely to happen to this deer population in the long run?

C APPLYING THE CONCEPTS

13. ***Business: Maximization of Space*** A farmer has 1200 feet of fence and wishes to build two identical rectangular enclosures, as in the following figure. What should be the dimensions of each enclosure if the total area is to be a maximum?

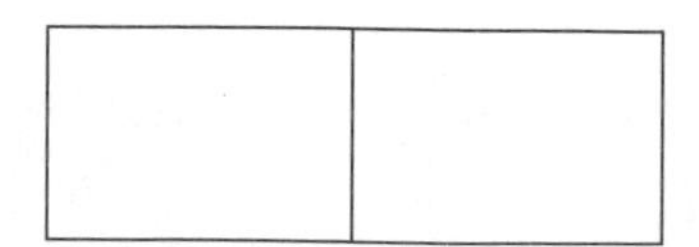

14. ***Business: Space Utilization*** The owner of a ranch has 3000 yards of fencing material with which to enclose a rectangular piece of grazing land along a straight portion of a river. If fencing is not required along the river, what are the dimensions of the pasture having the largest area? What is the largest area?

15. ***Business: Hotel Revenue*** A 210-room hotel is filled when the room rate is \$50 per day. For each \$1 increase in the rate, three fewer rooms are rented. Find the room rate that maximizes daily revenue.

16. ***Business: Airline Revenue*** A charter flight club charges its members \$250 per year. But for each new member above a membership of 65, the charge for all members is reduced by \$2 each. What number of members leads to a maximum revenue?

17. ***Business: Agricultural Yield*** An orchard presently has 30 trees per acre. Each tree yields 480 peaches. It is determined that for each additional tree that is planted per acre, the yield will be reduced by 12 peaches per tree. How many trees should be added per acre to maximize the yield?

18. ***Medicine: Spread of Disease*** One model for the spread of a disease assumes that the rate at which the disease spreads is proportional to the product of the number of people infected and the number not infected. If the size of the population, P, is constant, when is the disease spreading most rapidly?

19. ***Business: Demand and Revenue*** Demand for an electric fan is related to its selling price p by the equation

$$n = 2880 - 90p$$

where n is the number of fans that can be sold per month at a price p. Find the selling price that will maximize the revenue received.

20. ***Business: Packaging*** Postal regulations specify that a parcel sent by parcel post may have a combined length and girth (distance around) not exceeding 100 inches. If a rectangular package has a square cross section, find the dimensions of the package with the largest volume that may be sent through the mail.

21. ***Business: Packaging Costs*** The More Beef Company requires its corned beef hash containers to have a capacity of 64 cubic inches. The containers are in the shape of right circular cylinders. Find the radius and height of the container that can be made at a minimum cost, if the tin alloy for the side and bottom costs 4 cents per square inch and the aluminum for the pull-off lid costs 2 cents per square inch.

Bonus

22. ***Business: Maximum Revenue*** A tool company determines that it can achieve 500 daily rentals of jackhammers per year at a daily rental fee of \$30. For each \$1 increase in rental price, 10 fewer jackhammers will be rented. What rental price maximizes revenue?

23. ***Business: Motel Revenue*** A motel finds that it can rent 200 rooms per day if it charges \$80 per room. For each \$5 increase in rental rate, 10 fewer rooms will be rented per day. What room rate maximizes revenues?

24. ***Mathematics: Volume*** A block of ice is melting so that each edge is decreasing at the rate of 2 inches per hour. Find how fast the volume of the ice is decreasing at the instant when each edge is 8 inches long. (Assume the block is a perfect cube.)

Bonus

D DESCRIBE YOUR THOUGHTS

25. Describe the strategy you would use to solve Exercise 13. Do not solve the exercise; simply describe the strategy as if you were explaining to another person how to go about solving the problem.

26. Describe the strategy you would use to solve Exercise 17. Do not solve the exercise; simply describe the strategy as if you were explaining to another person how to go about solving the problem.

27. Describe the strategy you would use to solve Exercise 24. Do not solve the exercise; simply describe the strategy as if you were explaining to another person how to go about solving the problem.

E REVIEW

28. **(2.2)** Find, if it exists, $\lim_{x \to 4} \dfrac{x^2 + 2x + 5}{x^2 - 3x - 13}$.

29. **(2.3)** Find, if it exists, $\lim_{x \to \infty} \dfrac{5x^3 - 2x + 1}{x^4 + 5x - 7}$.

30. **(2.5)** Find the instantaneous rate of change of

$$f(x) = 2x^2 - 3x + 4 \text{ at } x = 5.$$

31. **(3.2)** Find $f'(5)$ if $f(x) = \dfrac{4x - 3}{2x + 3}$.

32. **(3.3)** Find $f''(x)$ for $f(x) = (2x^3 - 1)^3$.

33. **(4.1)** Find the points of inflection for

$$f(x) = 4x^3 - 12x^2 + 4.$$

4.4 Applications of the Derivative in Business and Economics

Introduction

In business and economic applications, effects of changes over wide intervals rather than effects of a change over an interval of length 1 are more frequently of interest. Analysis over wider intervals is defined as incremental analysis and it is best understood by having a basic understanding of marginal analysis. In this section we do not study incremental analysis. That subject is more appropriately studied in managerial courses. However, we lay the foundation for the study of incremental analysis by examining three marginal concepts: marginal revenue, marginal cost, and marginal profit. We also study profit maximization and cost minimization, inventory control, and elasticity.

Marginal Analysis

Marginal

In business and economics, the word *marginal* is equivalent to the phrase *rate of change*, or more precisely, the *instantaneous rate of change*. The marginal concept in business and economics is the derivative concept in mathematics.

1. **Marginal revenue** is the change in the revenue associated with a 1-unit change in the number of units sold. It is the derivative of the revenue function. Revenue functions are commonly denoted with the letter R. The marginal revenue is commonly represented with the letters MR. Symbolically, $R' = MR$.
2. **Marginal cost** is the change in the cost associated with a 1-unit change in the number of units produced. It is the derivative of the cost function. Cost functions are commonly denoted with the letter C. The marginal cost is commonly represented with the letters MC. Symbolically, $C' = MC$.
3. **Marginal profit** is the change in the profit associated with a 1-unit change in the number of units sold. It is the derivative of the profit function. Profit functions are commonly denoted with the letter P. The marginal profit is commonly represented with the letters MP. Symbolically, $P' = MP$.

EXAMPLE SET A

1. The revenue realized by a company on the sale of x units of its product is given by the revenue function $R(x) = x^3 + 4x^2 + 160x$. Compute and interpret both $R(70)$ and $R'(70)$.

   Solution:
   Since we are asked for information about $R'(x)$, we compute it first.

   $$R'(x) = 3x^2 + 8x + 160$$

   a. Substituting 70 for x in $R(x)$, we get $R(70) = 373{,}800$. This means that when the company sells 70 units of its product, it realizes a revenue of \$373,800.

   b. Substituting 70 for x in $R'(x)$, we get $R'(70) = 15{,}420$. This means that if the company increases the number of units it sells by 1, from 70 to 71, its revenue will increase by approximately \$15,420. (Notice that $R'(x)$ and $R'(70)$ could be denoted $MR(x)$ and $MR(70)$, respectively.)

2. The cost to a company to produce x units of a product is given by the cost function $C(x) = -0.035x^2 + 40x + 25$. Compute and interpret both $C(600)$ and $C'(600)$.

   Solution:
   Since we are asked for information about $C'(x)$, we compute it first.

   $$C'(x) = -0.070x + 40$$

   a. Substituting 600 for x in $C(x)$, we get $C(600) = 11{,}425$. This means that when the company produces 600 units of its product, it does so at a cost of \$11,425.

   b. Substituting 600 for x in $C'(x)$, we get $C'(60) = -2$. This means that if the company increases the number of units it produces by 1, from 600 to 601, the cost to do so will decrease by approximately \$2. (Notice that $C'(x)$ and $C'(600)$ could be denoted $MC(x)$ and $MC(600)$, respectively.)

3. The cost to a company to produce x units of a product is given by the cost function $C(x) = 0.02x^3 - 0.5x^2 + 10x + 150$. The revenue the company expects to realize from the sale of x units of the product is given by the revenue function $R(x) = -1.1x^2 + 41.5x$. If $P(x)$ represents the profit realized on the sale of x units of the product, compute and interpret both $P(20)$ and $P'(20)$.

   Solution:
   Since we are asked for information about $P'(x)$, we compute it first. To do so requires that we find the profit function $P(x)$. In business and economics, profit is defined to be the difference between the revenue and the cost. That is,

   $$P(x) = R(x) - C(x)$$

   In this case,

   $$\begin{aligned} P(x) &= -1.1x^2 + 41.5x - (0.02x^3 - 0.5x^2 + 10x + 150) \\ &= -1.1x^2 + 41.5x - 0.02x^3 + 0.5x^2 - 10x - 150 \\ &= -0.02x^3 - 0.6x^2 + 31.5x - 150 \end{aligned}$$

Differentiating, we get $P'(x) = -0.06x^2 - 1.2x + 31.5$.

a. Substituting 20 for x in $P(x)$, we get $P(20) = 80$. This means that when the company produces and sells 20 units of its product, it realizes a profit of \$80.

b. Substituting 20 for x in $P'(x)$, we get $P'(20) = -16.50$. This means that if the company increases the number of units it produces and sells by 1, from 20 to 21, the profit it realizes will decrease by approximately \$16.50. (Notice that $P'(x)$ and $P'(20)$ could be denoted $MP(x)$ and $MP(20)$, respectively.)

Profit, Revenue, and Cost Optimization

Businesses wish to optimize their profits, revenues, and costs. We can use the optimization techniques of Section 4.3 to compute these values and the input values that produce them.

EXAMPLE SET B

A company estimates that it can sell 2000 units each week of its product if it prices each unit at \$10. However, its weekly number of sales will increase by 50 units for each \$0.10 decrease in price. The company has fixed costs of \$800. The cost to make each unit is \$0.80. Find the level of production that maximizes the company's profit if the company must produce and sell between and including 2000 and 3500 units.

Solution:

We use the fact that profit = revenue − cost. To do so requires that we find both the revenue and cost functions. The revenue realized on the sale of a product is found by multiplying the number of units sold by the sales price. Letting x represent the number of units sold and p represent the sales price, $R(x) = x \cdot p$.

But the sales price need not be constant; it can change. As it decreases, the number of units sold increases above 2000. The sales price and the level of production are related. We need to find an expression that relates the number of units sold to the sales price. Let $p(x)$ represent the sales price at production level x.

The base production level is 2000 units and the base price is \$10. We can represent price decreases with the expression $10 - p(x)$. The number of units sold above 2000 is 50 times the number of \$0.10 price decreases. That number is found by finding the difference between the original \$10 price and the new price, $p(x)$, and then determining how many \$0.10 decreases are in that difference. Symbolically, this is

$$\frac{10 - p(x)}{0.10}$$

Then, the quantity sold is

$$x = 2000 + 50 \cdot \frac{10 - p(x)}{0.10}$$

Solving this equation for $p(x)$ gives us the sales price when the number of units sold is x.

$$p(x) = 14 - 0.002x$$

Thus, the revenue function is

$$\begin{aligned} R(x) &= x \cdot p(x) \\ &= x(14 - 0.002x) \\ &= 14x - 0.002x^2 \end{aligned}$$

Fixed Cost

The cost $C(x)$ of producing x units is the sum of the **fixed cost** \$800 and the **variable cost** of \$0.80 per unit, or \$0.80x. Thus,

$$C(x) = 800 + 0.80x$$

Variable Cost

Now, we can find the profit function, which is the revenue function minus the cost function.

$$\begin{aligned} P(x) &= R(x) - C(x) \\ &= 14x - 0.002x^2 - (800 + 0.80x) \\ &= 14x - 0.002x^2 - 800 - 0.80x \\ &= -0.002x^2 + 13.20x - 800 \end{aligned}$$

Table 4.7

x	$P(x)$
2000	17,600
3300	20,980
3500	20,900

To maximize this function, we compute the derivative, find the values of x for which $P'(x)$ is 0 or undefined, and test for an absolute maximum on the interval $[2000, 3500]$.

$$P'(x) = -0.004x + 13.2$$

$P'(x) = 0$ when $x = 3300$. Since $P(x)$ is a polynomial function, it is never undefined. Since 3300 is in the interval $[2000, 3500]$, it is the only critical point. We construct a table to find the absolute maximum. As the table shows, the maximum profit is \$20,980, and it occurs when the level of production is 3300 units each week.

Inventory Control

For retailers to do well, they must be careful about the number of units of a product they have in their inventory over some particular time period. The business can control its inventory in either of two ways. One way is to place one order that will cover the entire anticipated demand. This method is attractive since the company is likely to have enough units in stock to meet anticipated demand. The company will incur only one ordering cost, but it may also incur **carrying costs** such as storage and insurance costs as well as costs due to the possibility of obsolescence, spoilage, or breakage.

Carrying Costs

Another way is to place a series of smaller orders throughout the time period. This method is attractive since it will keep carrying costs down. However, this method will make the company susceptible to reordering costs as well the possibility of not having enough units in stock to meet demand.

Total Inventory Cost

The **total inventory cost** is the sum of the carrying costs and the order/reorder costs. The problem for a business, then, is to determine the **lot size**—that is, the number of units in an order—that will minimize its total inventory cost. This is a minimization problem for which we can use the optimization techniques of Section 4.3. To do so, we must get a total inventory cost function.

Lot Size

We will analyze this problem by making three assumptions: that the sales of the product are made at a relatively uniform rate over the time period under consideration; that the lot size, x, of each reorder is the same; and that as each inventory in stock falls to 0, another order immediately arrives. We will also make the following representations:

1. $C(x)$ = the total inventory cost for lot size x.
2. $K(x)$ = the carrying cost for lot size x.
3. $R(x)$ = the reordering cost for lot size x.

Since both carrying cost and reorder cost depend on the lot size x, they are functions of x. Therefore, the total inventory cost is a function of x. In symbols, the total inventory cost function is

$$C(x) = K(x) + R(x)$$

and our goal is to minimize $C(x)$.

Now, at any time, the largest inventory in stock is x. Since the units sell at a uniform rate, the average inventory during the time period is $x/2$. Figure 4.31 illustrates these ideas. To find explicit expressions for $K(x)$ and $R(x)$, we reason as follows.

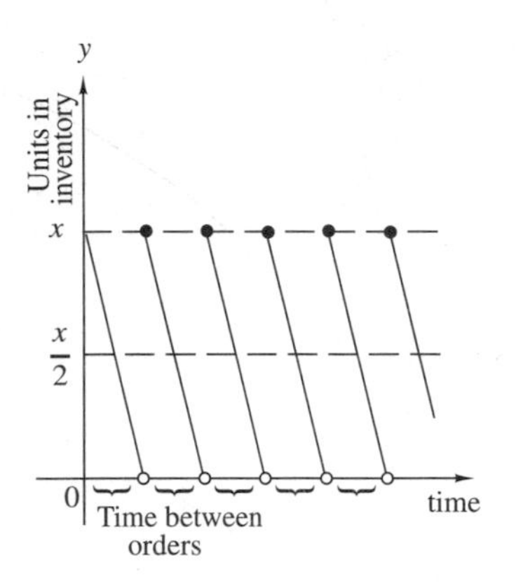

Figure 4.31

$$\text{Carrying Costs} = (\text{holding cost per unit}) \cdot (\text{average no. of units})$$

$$K(x) = (\text{holding cost per unit}) \cdot \frac{x}{2}$$

$$\text{Reordering Costs} = (\text{cost per order}) \cdot (\text{no. of orders})$$

$$R(x) = (\text{cost per order}) \cdot \left(\frac{\text{no. of units sold during time period}}{\text{lot size}}\right)$$

$$R(x) = (\text{cost per order}) \cdot \left(\frac{\text{no. of units sold during time period}}{x}\right)$$

EXAMPLE SET C

A company anticipates selling 1600 units of its product at a uniform rate over the next year. Each time the company places an order for x units, it is charged a flat fee of \$15. Carrying costs are \$4 per unit per year. How many times should the company reorder each year, and what should the lot size be to minimize inventory costs?

Solution:
Since the company anticipates selling 1600 units, it can place anywhere from 1 order for 1600 units to 1600 orders for 1 unit. We need to find the number of orders that will minimize inventory costs. To do so, we need an inventory cost function that is valid on the interval $[1, 1600]$. We know that

$$K(x) = (\text{holding cost per unit}) \cdot \frac{x}{2}$$

and that

$$R(x) = (\text{cost per order}) \cdot \left(\frac{\text{number of units sold during time period}}{x}\right)$$

In this case,

$$K(x) = (4) \cdot \frac{x}{2} = 2x$$

and

$$R(x) = (15) \cdot \left(\frac{1600}{x}\right) = \frac{24{,}000}{x}$$

Therefore,

$$C(x) = 2x + \frac{24{,}000}{x}, \quad 1 \le x \le 1600$$

Minimize $C(x)$ by computing $C'(x)$ and determining the x values for which it is 0 or undefined.

$$C'(x) = 2 - \frac{24{,}000}{x^2}$$

1. Where is $C'(x) = 0$?

$$2 - \frac{24{,}000}{x^2} = 0$$

$$2x^2 - 24{,}000 = 0$$

$$2x^2 = 24{,}000$$

$$x^2 = 12{,}000$$

$$x \approx \pm 109.54$$

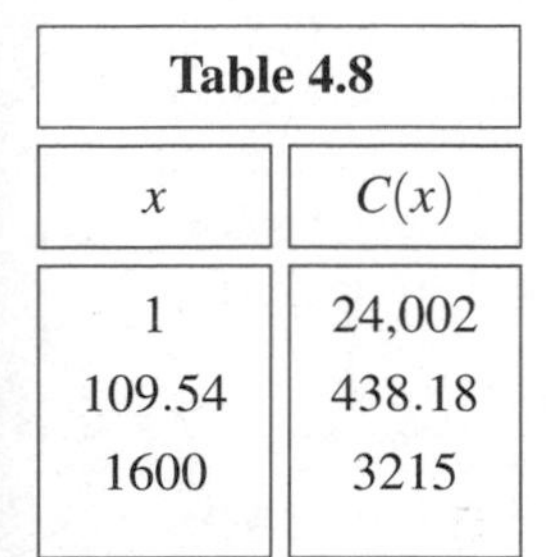

Table 4.8

x	$C(x)$
1	24,002
109.54	438.18
1600	3215

We can reject -109.54 since it is not inside the interval $[1, 1600]$.

2. Where is $C'(x)$ undefined? $C'(x)$ is undefined when $x = 0$, but since 0 is outside the interval $[1, 1600]$, we disregard it.

Thus, the only critical value is 109.54. We search for the absolute minimum by constructing a table (see Table 4.8) and observing the minimum inventory cost. The minimum cost is \$438.18, and it occurs when the lot size is 109.54 units. The number of reorders that minimizes the inventory cost is then

$$\text{number of orders} = \left(\frac{\text{number of units sold during time period}}{\text{lot size}}\right)$$

$$= \frac{1600}{109.54}$$

$$\approx 14.61$$

Since order sizes and lot sizes cannot be fractional, we need to round the lot size and the number of reorders to whole numbers that come as close to 1600 as possible. Consider the number of orders as either 14 or 15. For 14 orders per year, it would take 115 units as a lot size, giving a total of $14 \cdot 115 = 1610$ units per year. For 15 orders per year, it would take 107 units as a lot size, giving a total of $15 \cdot 107 = 1605$ units per year. The cost for each:

$$C(115) \approx \$438.70$$

$$C(107) \approx \$438.30$$

indicates that placing 15 orders per year with a lot size of 107 units is the most desirable.

Interpretation: The inventory costs can be minimized at about \$438.30 when 15 orders of 107 units each are placed during the year.

Elasticity of Demand

There are economic theories about the ways in which households and businesses are likely to respond to financial changes. If two variables, x and y, are related and a change in x causes almost no change in y, then y is not very sensitive to changes in x. If, however, a change in x causes a large change in y, then y is very sensitive to changes in x.

Elasticity

When a change in one variable causes a response in another variable, **elasticity** is a measure of the size of the response. Elasticity measures how sensitive the value of one variable is on the value of another. If a percent change in the input variable causes a relatively *large* change in the output variable, the relationship between the variables is called **elastic**. If a percent change in the input variable causes a relatively *small* change in the output variable, the relationship between the variables is called **inelastic**.

For example, if the price of a textbook is increased by 8% from \$42 to \$45.36, the percent of people who buy the textbook may decrease, but probably not by very much. In this case, the percent change in demand is not very sensitive to a percent change in price. However, if a college increases its 4-year total tuition by 8%, from $24,000 to $25,920, it may see a significant decline in the number of enrolling students. In this case, the percent change in college enrollment is very sensitive to the percent change in tuition costs. Since a price increase produces a demand decrease and a price decrease produces a demand increase, price and demand are inversely related. To measure how sensitive one variable is to another, economists use the concept of *elasticity*.

Elasticity of Demand

When the variables that are related are demand for a commodity and the price of the commodity, we speak of the **elasticity of demand**. Elasticity of demand is denoted with the lowercase Greek letter η (pronounced *eta*) and is computed by dividing the percent change in the quantity demanded by the percent change in the price. Since the relationship between price and quantity demanded is an inverse relationship, this ratio will always be negative. Symbolically, the percent change in q is represented by $\frac{\Delta q}{q}$, and the percent change in p is represented by $\frac{\Delta p}{p}$. Now, since η is computed by dividing the percent change in the quantity demanded by the percent change in the price,

$$\begin{aligned}\eta &= \frac{\Delta q}{q} \div \frac{\Delta p}{p} \\ &= \frac{\Delta q}{q} \cdot \frac{p}{\Delta p} \\ &= \frac{p}{q} \cdot \frac{\Delta q}{\Delta p}\end{aligned}$$

If the price p is a continuous function of the demand q, then

$$\lim_{\Delta p \to 0} \frac{\Delta q}{\Delta p} = \frac{dq}{dp}$$

Then, the **point elasticity of demand**, which is denoted ϵ (pronounced *epsilon*), is the instantaneous rate of change of the η. That is,

$$\epsilon = \lim_{\Delta p \to 0} \eta$$

$$\epsilon = \lim_{\Delta p \to 0} \frac{p}{q} \cdot \frac{\Delta q}{\Delta p}$$

$$\epsilon = \frac{p}{q} \cdot \lim_{\Delta p \to 0} \frac{\Delta q}{\Delta p}$$

$$\epsilon = \frac{p}{q} \cdot \frac{dq}{dp}$$

Point Elasticity of Demand

The **point elasticity of demand**, ϵ, is defined mathematically as

$$\epsilon = \frac{p}{q} \cdot \frac{dq}{dp}$$

where q is the original quantity demanded when the original price per unit is p. Point elasticity measures the elasticity at a particular point and is interpreted as the percent change in the demand for a 1% change in the price. Since price and demand are inversely related, ϵ will always be either negative or zero.

EXAMPLE SET D

A company estimates that the weekly sales q of its product is related to the product's price p by the function

$$q = \frac{4{,}500}{\sqrt[3]{p^2}}$$

where p is in dollars. Currently, each unit of the product is selling for \$27. Determine the point elasticity of demand of this product.

Solution:

We use the point elasticity of demand formula $\epsilon = \frac{p}{q} \cdot \frac{dq}{dp}$ with $p = 27$. To use the formula, however, we need the value of q and $\frac{dq}{dp}$.

First, when $p = 27$, $q = \frac{4500}{\sqrt[3]{27^2}} = 500$. Second,

$$\begin{aligned}
\frac{dq}{dp} &= \frac{d}{dp}\left[\frac{4500}{\sqrt[3]{p^2}}\right] \\
&= \frac{d}{dp}\left[\frac{4500}{p^{2/3}}\right] \\
&= \frac{d}{dp}\left[4500 \cdot p^{-2/3}\right] \\
&= -3000p^{-5/3} \\
&= \frac{-3000}{p^{5/3}}
\end{aligned}$$

When $p = 27$, $\dfrac{dq}{dp} = \dfrac{-3000}{27^{5/3}} \approx -12.35$, Now, we have

$$\begin{aligned}\epsilon &= \frac{p}{q} \cdot \frac{dq}{dp} \\ &\approx \frac{27}{500} \cdot (-12.35) \\ &\approx -0.67 \quad \left(\frac{-0.67}{1} = \frac{-0.67\%}{1\%}\right)\end{aligned}$$

Thus, $\epsilon \approx -0.67$.

Interpretation: If the price of each unit of this product is raised by 1%, from \$27 to \$27.27, the demand for the product will *decrease* by approximately 0.67%, from 500 units to approximately 497 units (a relatively small decrease).

Relating Point Elasticity to Expenditures and Revenues

Point elasticity, expenditures, and revenues are related to each other, and it is possible to derive and state that relationship symbolically. To do so, recall that

$$\text{revenue} = (\text{quantity sold}) \cdot (\text{price per unit})$$

and that

$$\text{expenditure} = (\text{quantity purchased}) \cdot (\text{price per unit})$$

Letting q represent either the quantity sold or the quantity purchased, and p the price per unit, and keeping in mind that the quantity sold or purchased is a function $q(p)$ of the price, we have

$$\begin{aligned}R(p) &= q \cdot p \\ &= q(p) \cdot p\end{aligned}$$

To observe how changes in the price can affect the revenue or the expenditure, we differentiate the product function $R(p) = q(p) \cdot p$ with respect to p.

$$\begin{aligned}R'(p) &= q'(p) \cdot p + q(p) \cdot 1 \\ &= q'(p) \cdot p + q(p) \\ &= q'(p) \cdot p + q\end{aligned}$$

We would like to write this derivative so that it involves ϵ, the point elasticity. We notice that the derivative involves p, as does the formula for ϵ. We will solve the formula relating p and ϵ for p. Since $\epsilon = \frac{p}{q} \cdot \frac{dq}{dp}$ and $\frac{dq}{dp} = q'(p)$, solving for p produces

$$p = \frac{\epsilon q}{q'(p)}$$

Now, we can rewrite $R'(p) = q'(p)p + q$ so that it involves ϵ.

$$R'(p) = q'(p) \cdot p + q$$

$$R'(p) = q'(p) \cdot \frac{\epsilon q}{q'(p)} + q$$

$$R'(p) = \epsilon q + q$$

$$R'(p) = q(1 + \epsilon)$$

Since q represents some number of units, it is always positive. Thus,

1. $R'(p)$ is negative when the quantity $(1 + \epsilon) < 0$. This will happen when $\epsilon < -1$.
2. $R'(p)$ is 0 when the quantity $(1 + \epsilon) = 0$. This will happen when $\epsilon = -1$.
3. $R'(p)$ is positive when the quantity $(1 + \epsilon) > 0$. Solving this inequality for ϵ gives $\epsilon > -1$. However, since ϵ is never positive, $-1 < \epsilon \leq 0$.

This leads us to three interesting cases involving values of ϵ.

Elastic, Unit Elastic, and Inelastic Demand

1. If $\epsilon < -1$, then $R'(p) < 0$ means that the total revenue falls with price increases and rises with price decreases. Demand is **elastic**.
2. If $\epsilon = -1$, then $R'(p) = 0$ means that the total revenue is unaffected by changes in price. Demand is **unit elastic**.
3. If $-1 < \epsilon < 0$, then $R'(p) > 0$ means that the total revenue rises with price increases and falls with price decreases. Demand is **inelastic**.

EXAMPLE SET E

1. In Example Set D, we computed ϵ to be -0.67. Since $-1 < \epsilon < 0$, demand is inelastic, meaning that a 1% increase in the price of the product, from \$27 to \$27 + \$27(0.01) = \$27.27, produces a 0.67% decrease in the demand, but an increase in revenue. Revenue rises because a price increase produces a relatively small decrease in demand. (A decrease from 500 units to $500 + 500(-0.0067) \approx 497$ units is relatively small.) In fact, revenue increases from

$$500 \text{ units} \cdot \frac{\$27}{\text{unit}} = \$13{,}500$$

to

$$497 \text{ units} \cdot \frac{\$27.27}{\text{unit}} = \$13{,}553.19$$

2. Suppose that in Example Set D the quantities q and p were related by the function

$$q = \frac{97{,}533}{\sqrt[5]{p^{24}}}$$

Now with an initial price of \$3 per unit, ϵ computes to -4.8, a number less than -1. In this case demand is elastic, meaning that a 1% increase in the price of the product, from \$3 to \$3.03, produces a 4.8% decrease in demand (from 500 units to $500+500(-0.048) = 476$ units) and a decrease in revenue from

$$500 \text{ units} \cdot \frac{\$3}{\text{unit}} = \$1500$$

to

$$476 \text{ units} \cdot \frac{\$3.03}{\text{unit}} = \$1442.28$$

Revenue falls because a price increase produces a relatively large decrease in demand. (A decrease from 500 units to 476 units is relatively large.)

3. Finally, suppose that in Example Set D the quantities q and p were related by the function

$$q = \frac{62{,}500}{p}$$

Now with an initial price of \$125 per unit, ϵ computes to -1. In this case demand is unit elastic, meaning that a 1% increase in the price of the product, from \$125 to \$126.25, produces no change in revenue. Revenue remains at \$62,500. Revenue is unaffected because a price increase is exactly offset by a decrease in demand: a \$1.25 increase in price is offset by a 5-unit decrease in demand.

EXERCISE SET 4.4

A UNDERSTANDING THE CONCEPTS

1. ***Business: Marginal Cost*** For a manufacturer, $C(x)$ represents the dollar cost of producing x units of product. Interpret both $C(250) = 12{,}000$ and $C'(250) = -85$.

2. ***Business: Marginal Revenue*** For a company, $R(x)$ represents the revenue, in thousands of dollars, realized from the expenditure of x thousands of dollars for advertising. Interpret both $R(6) = 125$ and $R'(6) = 16$.

3. ***Business: Marginal Profit*** For a company, $P(x)$ represents the profit, in thousands of dollars, realized from the sale of a product that sells for x dollars per unit. Interpret both $P(55) = 16$ and $P'(55) = -4$.

4. ***Business: Marginal Profit*** For a retail store, $P(x)$ represents the dollar profit realized from the sale of a product when it devotes x square feet to the display of the product. Interpret both $P(30) = 4500$ and $P'(30) = 150$.

5. ***Business: Inventory Costs*** The inventory cost C, in dollars, to a company is related to the lot size of inventory orders x by the function $C(x)$. Interpret both $C(200) = 1500$ and $C'(200) = 300$.

6. ***Business: Inventory Costs*** The inventory cost C, in dollars, to a company is related to the lot size of inventory orders x by the function $C(x)$. Interpret both $C(2600) = 3500$ and $C'(2600) = 700$.

7. ***Business: Elasticity of Demand*** A company determines that the point elasticity of demand for one of its products is -2.6. Is this demand elastic, unit elastic, or inelastic? If the price of the product is increased by 1%,

will the company's revenue increase, decrease, or remain the same?

8. *Business: Elasticity of Demand* A company determines that the point elasticity of demand for one of its products is -0.22. Is this demand elastic, unit elastic, or inelastic? If the price of the product is increased by 1%, will the company's revenue increase, decrease, or remain the same?

9. *Business: Elasticity of Demand* A company determines that the point elasticity of demand for one of its products is 1. Is this demand elastic, unit elastic, or inelastic? If the price of the product is increased by 1%, will the company's revenue increase, decrease, or remain the same?

B SKILL ACQUISITION and C APPLYING THE CONCEPTS

10. *Business: Marginal Cost* A manufacturer believes that the function

$$C(x) = -0.15x^2 + 18x + 260,$$

where $0 \leq x \leq 130$, approximates the dollar cost of producing x units of a product. Compute the marginal cost of producing 35 units. Interpret this value, as well as $C(35)$. What level of production will maximize the manufacturer's costs? What is that cost?

11. *Business: Marginal Cost* A manufacturer believes that the cost function

$$C(x) = \frac{3}{2}x^2 + 45x + 720$$

approximates the dollar cost of producing x units of a product. The manufacturer believes it cannot make a profit when the marginal cost goes beyond \$210. What is the most units the manufacturer can produce and still make a profit? What is the total cost at this level of production?

12. *Business: Marginal Revenue* The revenue function realized by a company by the sale of x units of a product is thought to be $R(x) = -0.0125x^2 + 112.5x$, where $0 \leq x \leq 9000$ dollars. Determine the interval(s) on which the marginal revenue is increasing/decreasing. At what level of production is the marginal revenue maximum? What is the total revenue at that level?

13. *Business: Marginal Profit* A retail company estimates that if it spends x thousands of dollars on advertising during the year, it will realize a profit of $P(x)$ dollars, where $P(x) = -0.25x^2 + 80x + 1400$, where $0 \leq x \leq 336$. What is the company's marginal profits at the \$100,000 and \$260,000 advertising levels? What advertising expenditure would you recommend to this company?

14. *Business: Marginal Profit* A company estimates that it can sell 5000 units each week of its product if it prices each unit at \$20. However, its weekly number of sales will increase by 100 units for each \$0.10 decrease in price. The company has fixed costs of \$1000. The cost to make each unit is \$1.60. Find the level of production that maximizes the company's profit if the company must produce and sell between and including 5000 and 20,000 units. What is the maximum profit?

15. *Business: Marginal Profit* A company estimates that it can sell 3000 units each month of its product if it prices each unit at \$75. However, its monthly number of sales will increase by 20 units for each \$0.25 decrease in price. The company has fixed costs of \$350. The cost to make each unit is \$4.20. Find the level of production that maximizes the company's profit if the company must produce and sell between and including 1000 and 7000 units. What is the maximum profit?

16. *Business: Inventory Costs* A company anticipates selling 8000 units of its product at a uniform rate over the next year. Each time the company places an order for x units, it is charged a flat fee of \$45. Carrying costs are \$6 per unit per year. How many times should the company reorder each year and what should the lot size be to minimize inventory costs?

17. *Business: Inventory Costs* A company anticipates selling 20,000 units of its product at a uniform rate over the next year. Each time the company places an order for x units, it is charged a flat fee of \$100. Carrying costs are \$2 per unit per year. How many times should the company reorder each year and what should the lot size be to minimize inventory costs?

18. *Business: Inventory Costs* A company figures it costs about \$10 to store a unit of a product for a year. To order a lot of size x, the cost is \$4 for each unit plus a flat fee of \$150. If the company anticipates selling 10,000 units at a uniform rate over the next year, what lot size will produce the smallest inventory cost?

19. *Business: Elasticity of Demand* A company estimates that the weekly sales q of its product is related to the product's price p by the function

$$q = \frac{2160}{\sqrt[5]{p^3}}$$

where p is in dollars. Currently, each unit of the product is selling for \$32. Determine the point elasticity of demand of this product, state whether the demand for this product is elastic, inelastic, or unit elastic, and interpret the meaning of this number in terms of the company's revenue. Determine the effect on revenue if the price were *decreased* by 1%. Determine the effect on revenue if the price were *decreased* by 5%.

20. *Business: Elasticity of Demand* A company estimates that the weekly sales q of its product is related to the product's price p by the function

$$q = \frac{14{,}625}{\sqrt[4]{p^{11}}}$$

where p is in dollars. Currently, each unit of the product is selling for \$4. Determine the point elasticity of demand of this product, state whether the demand for this product is elastic, inelastic, or unit elastic, and interpret the meaning of this number in terms of the company's revenue. Determine the effect on revenue if the price were *decreased* by 1%. Determine the effect on revenue if the price were *decreased* by 5%.

21. *Business: Elasticity of Demand* A company estimates that the weekly sales q of its product is related to the product's price p by the function

$$q = 15{,}000 - 0.65p$$

where p is in dollars. Currently, each unit of the product is selling for \$8,000. Determine the effect on revenue if the price were *increased* to \$9,200.

22. *Business: Elasticity of Demand* A company estimates that the weekly sales q of its product is related to the product's price p by the function

$$q = 1700\sqrt[5]{61{,}875 - p}, \quad 0 \le p \le 61{,}875$$

where p is in dollars. Currently, each unit of the product is selling for \$29,107. Determine the effect on revenue if the price were *decreased* to \$28,233.79.

23. *Business: Elasticity of Demand* Suppose that a company is currently selling 700 units of a product at \$16 per unit. Suppose also that the point elasticity of demand for this product is -2.6. By how much, if any, does the revenue increase or decrease?

24. *Business: Elasticity of Demand* Suppose that a company is currently selling 35,000 units of a product at \$160 per unit. Suppose also that the point elasticity of demand for this product is -8.8. By how much, if any, does the revenue increase or decrease?

25. *Business: Elasticity of Demand* Suppose that a company is currently selling 55,000 units of a product at \$38 per unit. Suppose also that the point elasticity of demand for this product is -0.39. By how much, if any, does the revenue increase or decrease?

26. *Business: Elasticity of Demand* Suppose that a company is currently selling x units of a product at p dollars per unit. Suppose also that the point elasticity of demand for this product is $-b, b > 0, b \neq 1$. Write an expression that gives the new revenue value.

27. *Business: Elasticity of Demand* Suppose that a company is currently selling x units of a product at p dollars per unit. Suppose also that the point elasticity of demand for this product is $-b, b > 0, b \neq 1$. Write an expression that gives the new revenue value if the price is increased by $a\%$, $a \neq 1$. (Hint: To convert a percent to a decimal, so that it may act as a multiplier, multiply it by 0.01.)

D DESCRIBE YOUR THOUGHTS

28. Elasticity of demand is always a negative number. Explain what this tells you about the behavior of the graph of the demand curve. Your explanation should be in terms of increasing/decreasing, max/min, concavity, and points of inflection.

29. Describe the relationship that exists between marginal revenue and marginal cost at the point where the marginal profit is zero.

30. As we analyzed the inventory control problem, we made an assumption that produced the value $\frac{x}{2}$, where x was the lot size. If that assumption were changed, could we still use $\frac{x}{2}$ in the computations?

E REVIEW

31. **(2.2)** Find, if it exists, $\lim_{x \to 8} \frac{x^2 - 11x + 24}{x - 8}$.

32. **(2.2)** Find, if it exists, $\lim_{x \to 3} \frac{x^2 - 3x - 4}{x + 1}$.

33. **(2.4)** Explain why a function with a cusp at the point $(a, f(a))$ cannot have a derivative at $x = a$.

34. **(2.5)** Find and interpret the instantaneous rate of change of $f(x) = 4x^2 - 2x - 6$ at $x = 3$.

35. **(3.2)** Find and interpret $f'(2)$ for

$$f(x) = \frac{7x + 8}{x + 4}.$$

36. **(3.3)** Find $f'(x)$ for $f(x) = (3x^2 - 4x)^3$.

37. **(3.3)** Find $f'(x)$ for $f(x) = (2x - 1)^4(2x + 1)^2$.

38. **(3.4)** Find $f'(x)$, $f''(x)$, and $f'''(x)$ for

$$f(x) = 4x^2 - 3x + 2.$$

39. **(3.5)** Find y' for $-2x^3y^2 + 5xy^3 + 3x - 4y = 10$.

40. **(4.3)** Find the relative extrema for the function

$$f(x) = \frac{1}{3}x^3 - \frac{5}{2}x^2 + 6x + 4.$$

Summary

This chapter first examined how the derivative of a function is used to describe the function's behavior and then how the derivative is used to solve problems that arise in business and economics.

Derivatives and the Behavior of Functions

Section 4.1 focused on describing the behavior of a function using the derivative of the function. The derivative of the function is used to determine the intervals on which a function is increasing or decreasing. A function $f(x)$ is increasing on an interval (a, b) if the graph of the function rises through (a, b). It is decreasing on an interval (a, b) if the graph of the function falls through (a, b). For continuous, differentiable functions $f(x)$,

1. If $f'(x) > 0$ for every value of x in (a, b), then $f(x)$ is increasing on (a, b). Conversely, if $f(x)$ is increasing on (a, b), then $f'(x) > 0$ for every value of x in (a, b).
2. If $f'(x) < 0$ for every value of x in (a, b), then $f(x)$ is decreasing on (a, b). Conversely, if $f(x)$ is decreasing on (a, b), then $f'(x) < 0$ for every value of x in (a, b).

Once we know the intervals on which a function is increasing/decreasing, we can use the first derivative test to determine the relative extrema of the function. To do so, we first identify the function's *critical points*. These are points for which $f(x)$ is defined, but also for which either $f'(x) = 0$ or $f'(x)$ is undefined. Once the critical points have been identified, we apply the *first derivative test* to locate relative maximum points and relative minimum points of the function $f(x)$. A point (a, b) is called a relative maximum of $f(x)$ if for all values of x in a small neighborhood near $x = a$, the output value $f(a) = b$ is the largest. Similarly, a point (a, b) is called a relative minimum of $f(x)$ if for all values of x in a small neighborhood near $x = a$, the output value $f(a) = b$ is the smallest. The first derivative test states that if c is a critical value of the function $f(x)$, and

1. $f(x)$ is increasing immediately to the left of c and decreasing immediately to the right of c, then c produces a relative maximum. In terms of the first derivative, if $f'(x) > 0$ to the immediate left of c and $f'(x) < 0$ to the immediate right of c, then c produces a relative maximum.
2. $f(x)$ is decreasing immediately to the left of c and increasing immediately to the right of c, then c produces a relative minimum. In terms of the first derivative, if $f'(x) < 0$ to the immediate left of c and $f'(x) > 0$ to the immediate right of c, then c produces a relative minimum.

This section continued with a discussion of concavity. A function is *concave upward* on an interval if the second derivative of the function is positive on that interval. A function is *concave downward* on an interval if the second derivative of the function is negative on that interval. Points at which the concavity changes are called *points of inflection*. Points of inflection occur at *hypercritical values*. These are input values for which either $f''(x) = 0$ or $f''(x)$ is undefined.

This section also discussed the *absolute extrema* of a function on an interval. Such points differ from relative extrema in that they are the largest/smallest values in the entire interval and not just in a small neighborhood around critical values. They are identified by evaluating the function at all critical values and at the endpoints of the interval. Continuous functions that are defined on closed intervals will always have absolute extrema. Continuous functions that are defined on open intervals may or may not have absolute extrema.

The section ended with examples of summary tables. These are tables that summarize the behavior of a function.

Describing the Behavior of a Function and Elementary Curve Sketching

Section 4.2 presented examples of how to construct the graph of a function from the function's summary table, as well as how to construct the summary table from the function's graph.

Applications of the Derivative: Optimization

Section 4.3 centered on demonstrating some applications of the derivative. This section brought together all the calculus material covered thus far and showed how to apply it to *optimization problems*. Optimization problems are problems that are solved by finding the absolute extrema of a function. The section began with an optimization strategy.

1. Draw a figure when one is appropriate.
2. Assign a variable to each quantity mentioned in the problem.
3. Select the quantity that is to be optimized and construct a function that relates it to some or all of the other quantities.
4. Since all the rules of differentiation deal with functions of only one variable, the function must involve only one variable. If the function constructed in step 3 involves more than one variable, use the information contained in the problem to eliminate variables until you have a function of only one variable.
5. Find the critical values of the function and apply the test for absolute extrema.
6. Disregard all answers that are not relevant to the situation.

Applications of the Derivative in Business and Economics

Section 4.4 focused on how the derivative can be used to produce important information in business and economics. This section began with the concept of the *marginal* change, which is the change in the output variable produced by a 1-unit change in the input variable. The term *marginal* is commonly used by business people and economists to mean *derivative*. Specifically introduced were the terms:

1. *Marginal revenue*, which is the change in the revenue associated with a 1-unit change in the number of units sold. It is the derivative of the revenue function.
2. *Marginal cost*, which is the change in the cost associated with a 1-unit change in the number of units produced. It is the derivative of the cost function.

3. *Marginal profit*, which is the change in the profit associated with a 1-unit change in the number of units sold. It is the derivative of the profit function.

The section then discussed optimization techniques. Using the optimization methods of Section 4.3, Example Set B illustrated how to maximize business profit.

Inventory control was discussed next. In making the analysis, three assumptions were made. They are: that the sales of the product are made at a relatively uniform rate over the time period under consideration; that the lot size, x, of each reorder is the same; and that as each inventory in stock falls to 0, another order immediately arrives.

The analysis demonstrated the procedure of constructing an inventory cost function. The optimization methods of Section 4.3 were then used in Example Set C to minimize an inventory cost and to determine the optimal number of reorders and reorder size a company should make.

This section ended with a discussion of *elasticity of demand*. Elasticity measures how sensitive the value of one variable is on the value of another. If a percent change in the input variable causes a relatively *large* change in the output variable, the relationship between the variables is called *elastic*. If a percent change in the input variable causes a relatively *small* change in the output variable, the relationship between the variables is called *inelastic*.

We developed a formula for measuring the point elasticity of demand. The *point elasticity of demand*, ϵ (pronounced *epsilon*), is defined mathematically as

$$\epsilon = \frac{p}{q} \cdot \frac{dq}{dp}$$

where q is the original quantity demanded when the original price per unit is p. Point elasticity measures the elasticity at a particular point and is interpreted as the percent change in the demand for a 1% change in the price.

There are three interesting cases involving values of ϵ.

1. If $\epsilon < -1$, then $R'(p) < 0$ means that the total revenue falls with price increases and rises with price decreases. Demand is *elastic*.
2. If $\epsilon = -1$, then $R'(p) = 0$ means that the total revenue is unaffected by changes in price. Demand is *unit elastic*.
3. If $-1 < \epsilon < 0$, then $R'(p) > 0$ means that the total revenue rises with price increases and falls with price decreases. Demand is *inelastic*.

Supplementary Exercises

For Exercises 1–6, specify all the critical values, all the intervals on which the function $f(x)$ is increasing, and all the intervals on which the function is decreasing.

1. $f(x)$ as illustrated below.

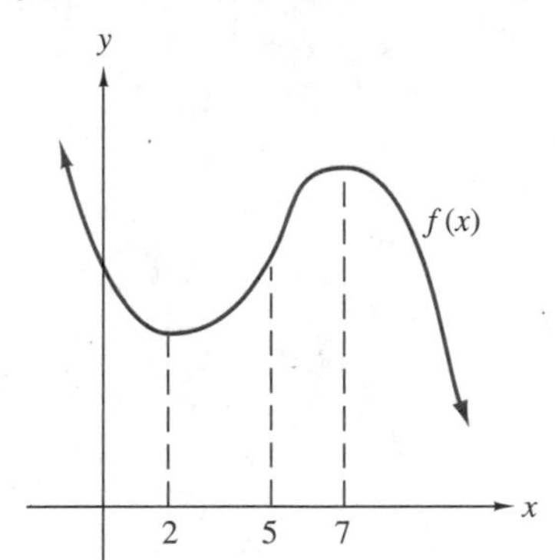

2. $f(x)$ as illustrated below.

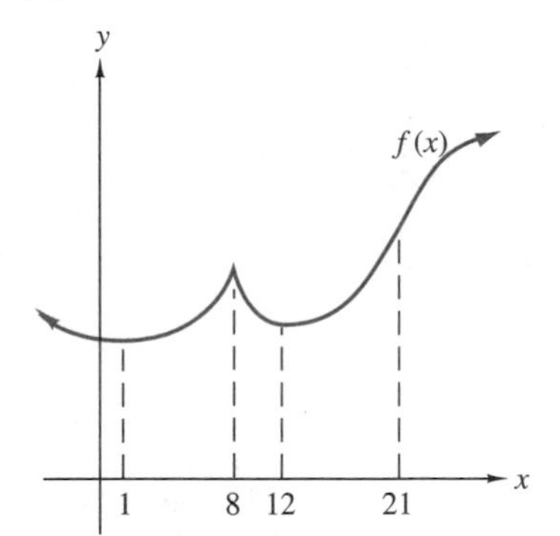

3. $f(x)$ as illustrated below.

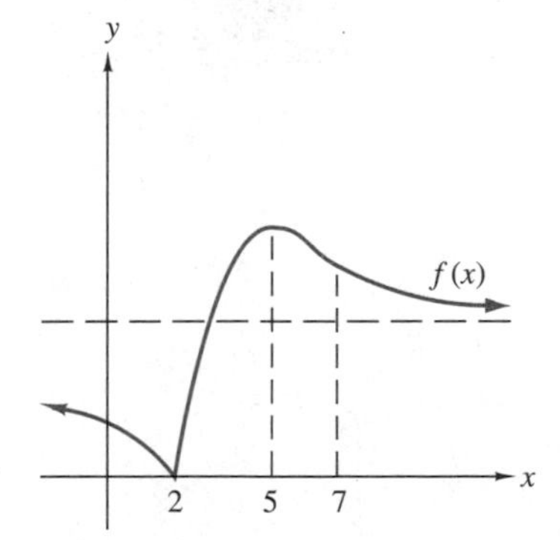

4. $f(x)$ as illustrated below.

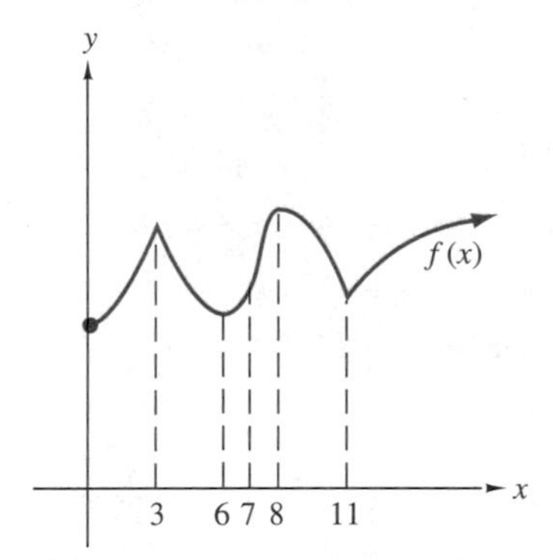

5. $f(x)$ as illustrated below.

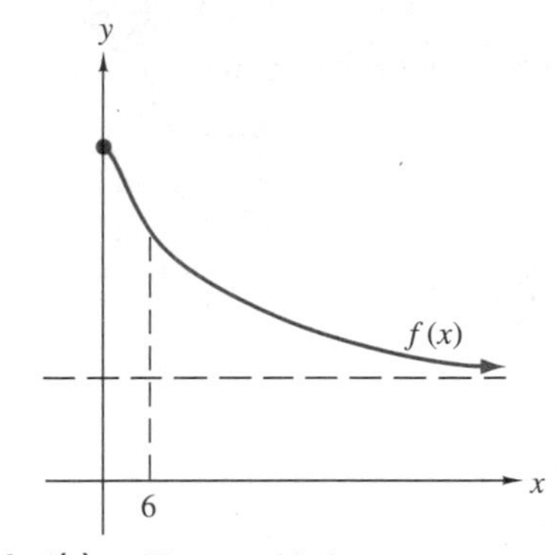

6. $f(x)$ as illustrated below.

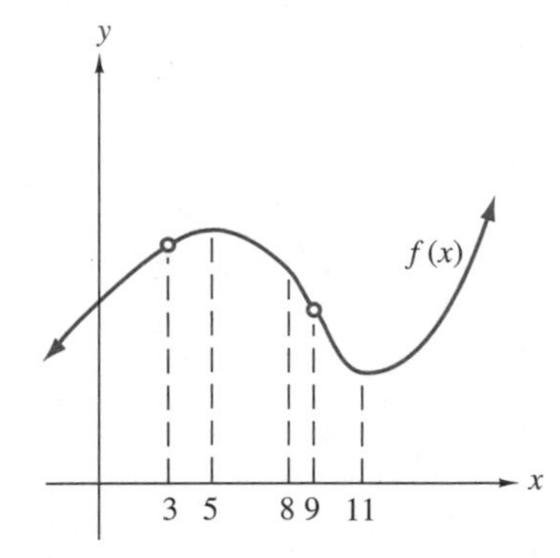

For Exercises 7–12, specify all the hypercritical values, all the intervals on which the function $f(x)$ is concave upward, and all the intervals on which the function is concave downward.

7. $f(x)$ as illustrated in the figure in Exercise 1.

8. $f(x)$ as illustrated in the figure in Exercise 2.

9. $f(x)$ as illustrated in the figure in Exercise 3.

10. $f(x)$ as illustrated in the figure in Exercise 4.

11. $f(x)$ as illustrated in the figure in Exercise 5.

12. $f(x)$ as illustrated in the figure in Exercise 6.

For Exercises 13–18, specify all the critical and hypercritical values and the relative and absolute extrema, if there are any, of the function $f(x)$.

13. $f(x)$ as illustrated in the figure in Exercise 1.

14. $f(x)$ as illustrated in the figure in Exercise 2.

15. $f(x)$ as illustrated in the figure in Exercise 3.

16. $f(x)$ as illustrated in the figure in Exercise 4.

17. $f(x)$ as illustrated in the figure in Exercise 5.

18. $f(x)$ as illustrated in the figure in Exercise 6.

For Exercises 19–22, classify each curve as concave upward or concave downward.

19. A curve that is increasing at a decreasing rate.

20. A curve that is increasing at an increasing rate.

21. A curve that is decreasing at a decreasing rate.

22. A curve that is decreasing at an increasing rate.

For Exercises 23–26, construct a sign chart for each graph.

23.

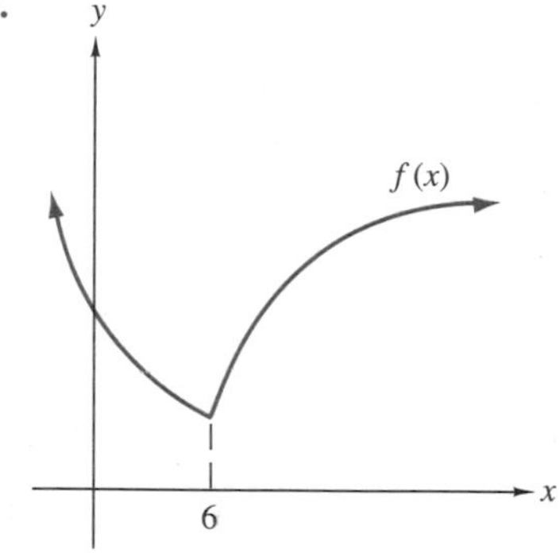

24.

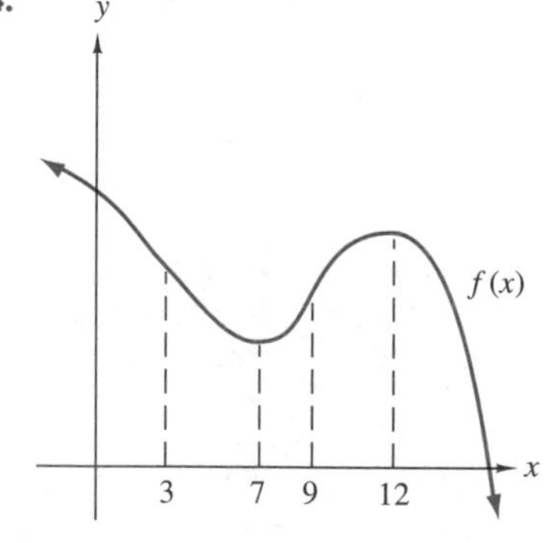

25.

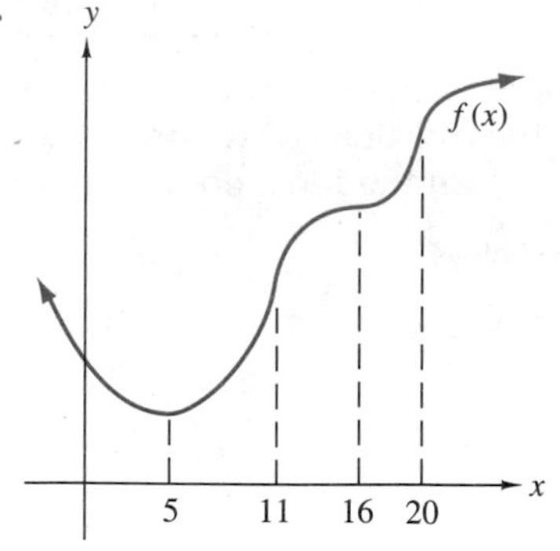

26.

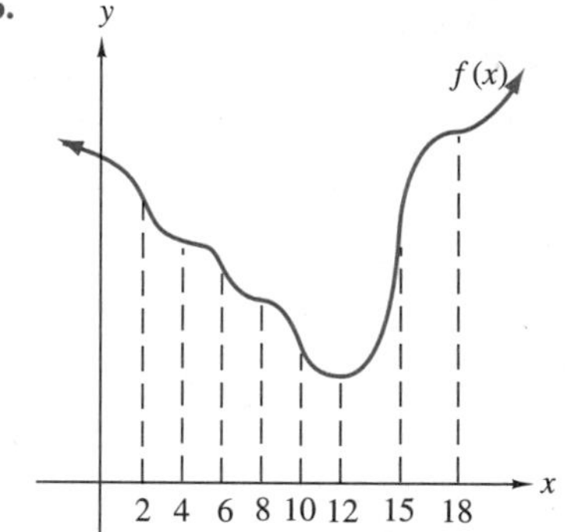

For Exercises 27–30, construct a summary table for each graph.

27.

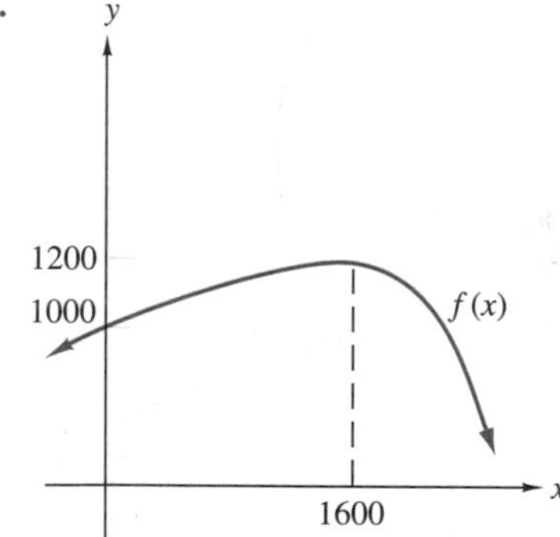

28.

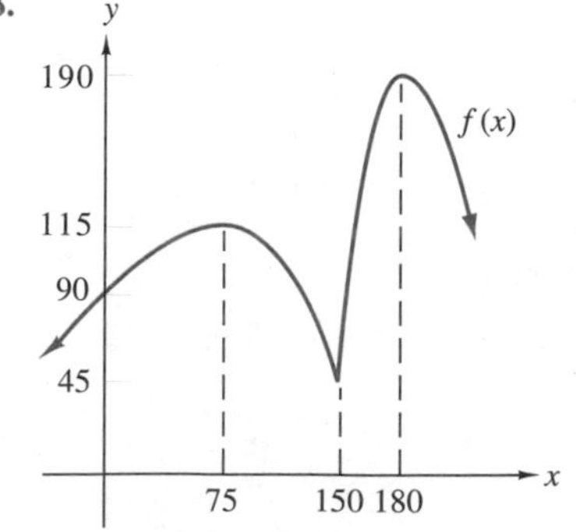

29.

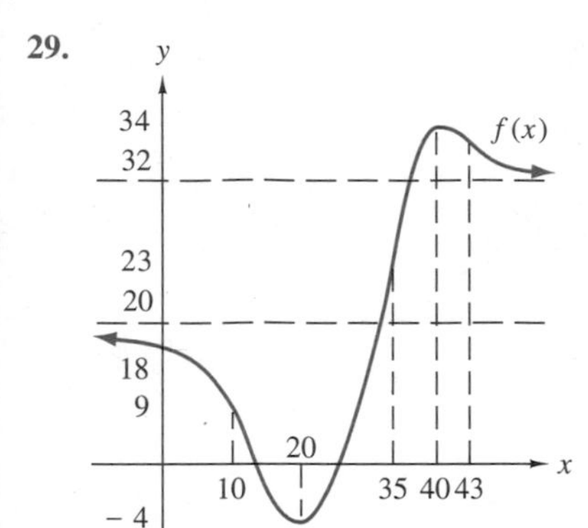

30.

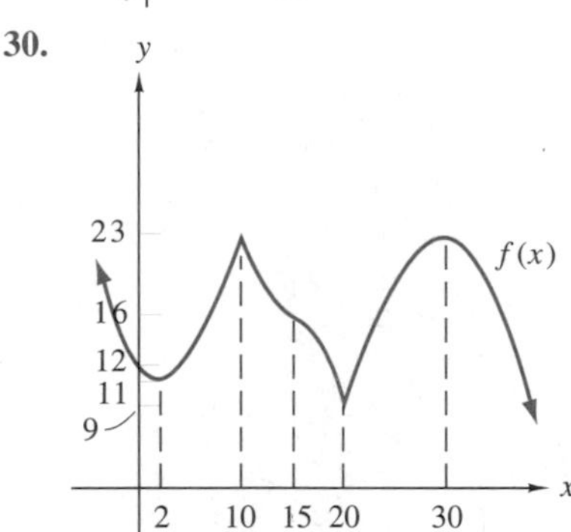

For Exercises 31–35, a situation is given along with the equation form and graph form of a function that describes the situation. As best you can (meaning you may have to approximate), use the graph to determine the critical value that maximizes or minimizes the function and then discuss the behavior of the function on both sides of the critical value.

31. ***Business: Packaging*** A box with no top is to be made from a square piece of cardboard by cutting out squares of equal size from each corner and folding up the sides. Approximately how long should the cut be so that the volume of the box is maximum? What is the approximate maximum volume? The function relating the length of the cut x to the volume of the box V is $V(x) = 625x - 100x^2 + 4x^3$. The graph of $V(x)$ appears in the following figure.

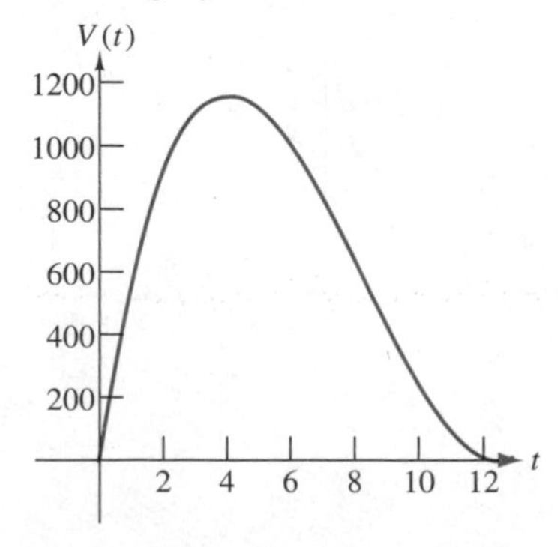

32. ***Business: Profit*** The profit P, in dollars, that a manufacturer realizes on the sale of x items is given by the function $P(x) = 35x - 0.12x^2 - 650$. What is the number of items that the company needs to sell to maximize its profits? What is the approximate maximum profit? The profit function appears in the following figure.

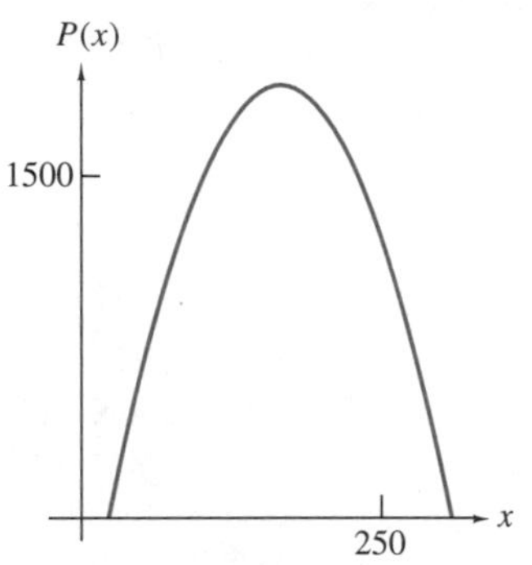

33. ***Business: Polling Costs*** A polling company has determined that the cost C of polling x people is given by the function $C(x) = 0.03x^2 - 55x + 52,000$. What is the approximate number of people the company should poll to minimize its costs? What is this minimum cost? The graph of this function appears in the following figure.

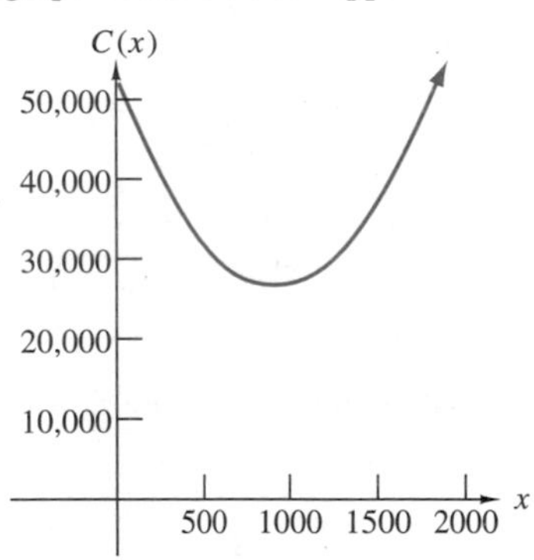

34. ***Business: Packaging*** The U.S. Post Office will accept for domestic delivery packages for which the sum of the length and girth (distance around) does not exceed 108 inches. For a box with a square base whose sides are of length x, the function $f(x) = 75x^2 - 10x^3$ relates the volume of the box to the length of a side of the base of the box. What is the maximum length of the side of the base of such a box? The function $f(x)$ is graphed in the following figure.

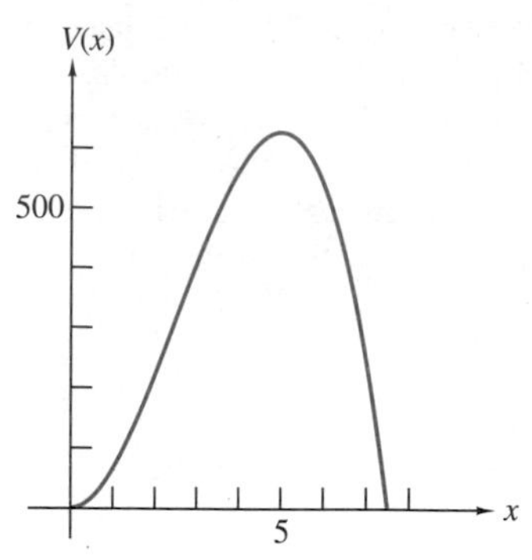

35. *Economics: Utility* An individual's satisfaction from the consumption of x units of a particular commodity during a specified unit of time is measured by *utility*. Every utility function $U(x)$ has associated with it a *saturation* quantity x_0, a number that maximizes the utility function. What is the saturation quantity for the utility function shown in the following figure?

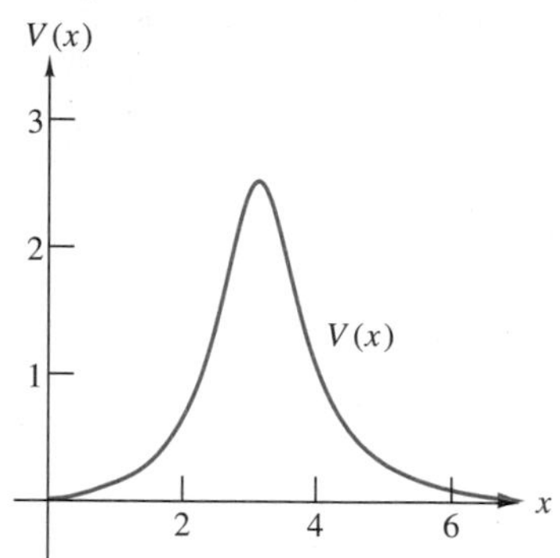

For Exercises 36–40, construct a summary table for each graph.

36.

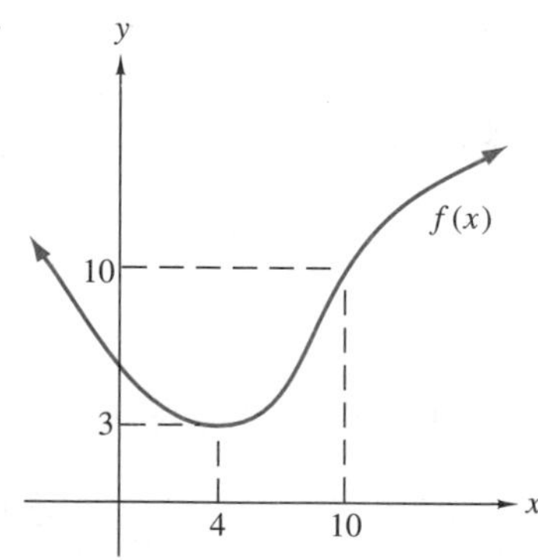

37.

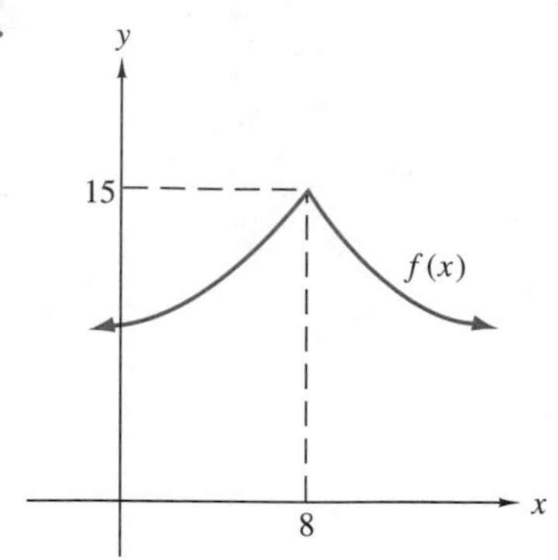

38.

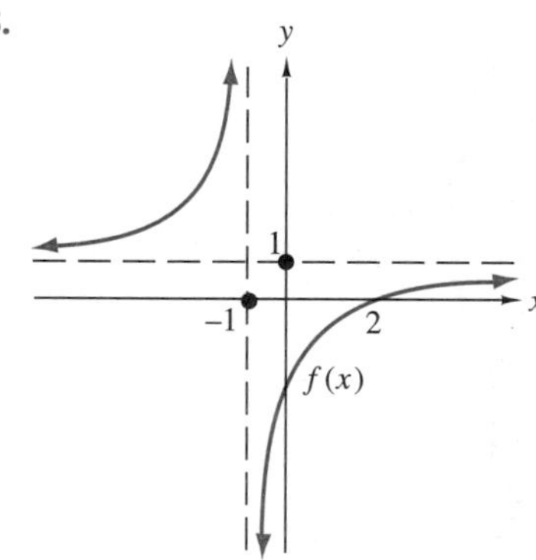

39.

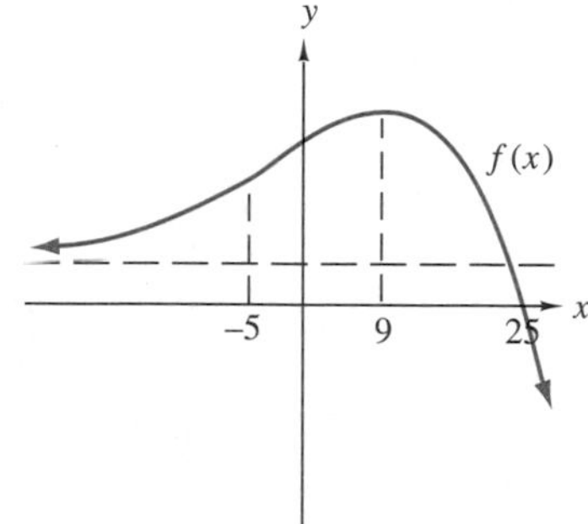

40.

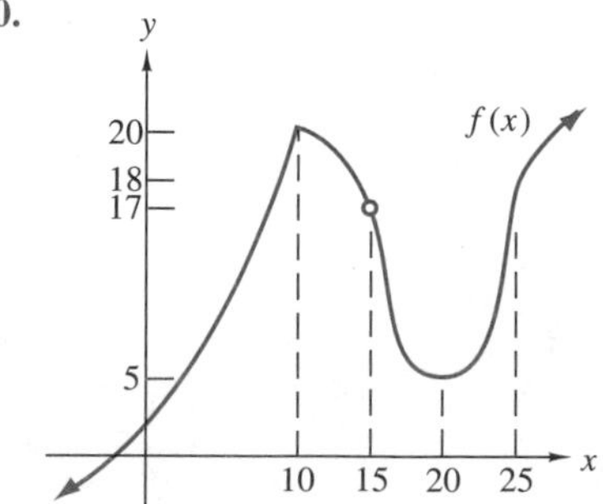

For Exercises 41–45, use the abbreviated summary table to construct a sketch of the corresponding function.

41. This table is for $f(x)$.

Int / Val	$f'(x)$	$f''(x)$	$f(x)$	Behavior of $f(x)$
$(-\infty, 5)$	$+$	$-$		
5	0		$(5, 8)$	
$(5, 9)$	$-$	$-$		
9	und		$(9, 6)$	
$(9, +\infty)$	$+$	$-$		

42. This table is for $f(x)$.

Int / Val	$f'(x)$	$f''(x)$	$f(x)$	Behavior of $f(x)$
$(-\infty, 2)$	$-$	$+$		
2	0		$(2, 2)$	
$(2, 5)$	$+$	$+$		
5		0	$(5, 4)$	
$(5, 7)$	$+$	$-$		
7	0		$(7, 7)$	
$7, +\infty)$	$-$	$-$		

43. This table is for $f(x)$.

Int / Val	$f'(x)$	$f''(x)$	$f(x)$	Behavior of $f(x)$
$(-\infty, 6)$	$+$	$+$		
6	und	und	und	
$(6, +\infty)$	$-$	$-$		
$x = 6$				VA
$y = 1$				HA

44. This table is for $f(x)$.

Int / Val	$f'(x)$	$f''(x)$	$f(x)$	Behavior of $f(x)$
$(-\infty, 0)$	$+$	$-$		
0	0		$(0, 4)$	
$(0, 4)$	$-$	$-$		
4	und	und	und	
$(4, 8)$	$+$	$-$		
8	0		$(8, 10)$	
$(8, 10)$	$-$	$-$		
10		0	$(10, 5)$	
$(10, +\infty)$	$-$	$+$		
$x = 4$				VA
$y = 2$				HA

45. This table is for $f(x)$.

Int / Val	$f'(x)$	$f''(x)$	$f(x)$	Behavior of $f(x)$
$(-\infty, -2)$	$-$	$-$		
-2	und		$(-2, 2)$	
$(-2, 2)$	$+$	$-$		
2		0	$(2, 4)$	
$(2, 4)$	$+$	$+$		
4	und	und	und	
$(4, 8)$	$+$	$-$		
8	0		$(8, 7)$	
$(8, 10)$	$-$	$-$		
10	und	und	und	
$(10, +\infty)$	$-$	$+$		
$x = 4$				VA
$x = 10$				VA

For Exercises 46–56, sketch each function, designating asymptotes with dashed lines when they occur.

46. $f(x) = 2x^2 - 4x + 3$

47. $f(x) = x^3 - 3x^2 + 2$

48. $f(x) = x^3 + 3x^2 - 9x - 11$

49. $f(x) = x^3 - 33x^2 + 216x$

50. $f(x) = x^4 - 32x + 48$

51. $f(x) = \dfrac{x - 1}{x - 2}$

52. $f(x) = \dfrac{1}{x^2 - 3x + 2}$

53. $f(x) = (x - 3)^{1/5}$

54. $f(x) = \dfrac{e^x}{x}$

55. $f(x) = x^{2/3}(x + 2)$

56. $f(x) = (x + 3)^{2/5}$

For Exercises 57–61, describe the behavior of the function in terms of the situation. (You need not construct a summary table.)

57. ***Economics: Investment*** The graph illustrated below displays the relationship between the amount in an account A when an initial investment of \$25,000 is compounded continuously and the number t of years for which the investment is made.

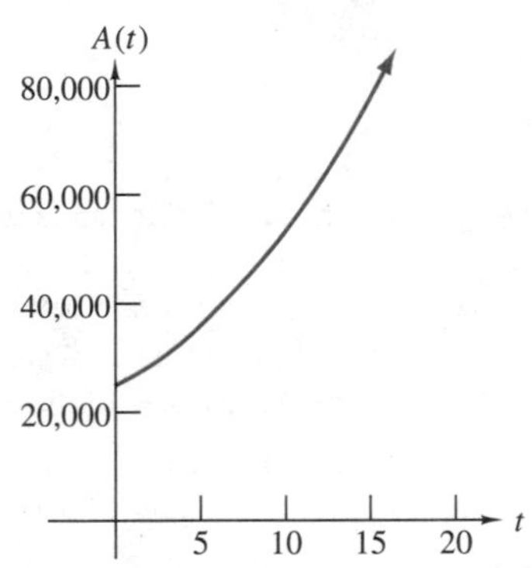

58. ***Business: Oil Production*** The graph illustrated below displays the relationship between time t, in months from now, and the number N of barrels of oil produced from an oil field.

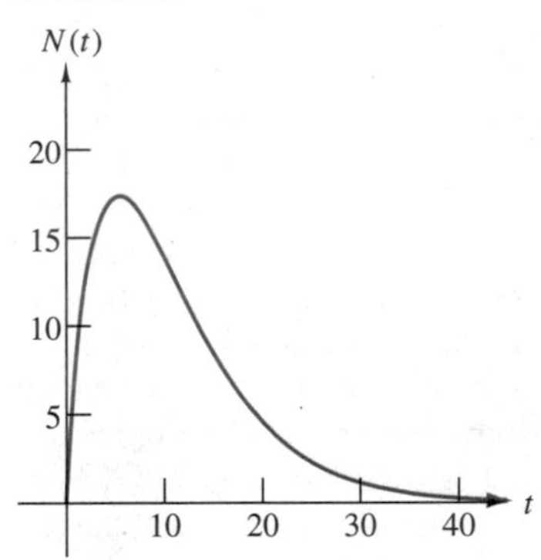

59. ***Medicine: Antibody Production*** The graph illustrated below displays the relationship between time t, in hours, after the injection of a medication into the body and the rate of production r, in units per hour, of antibodies.

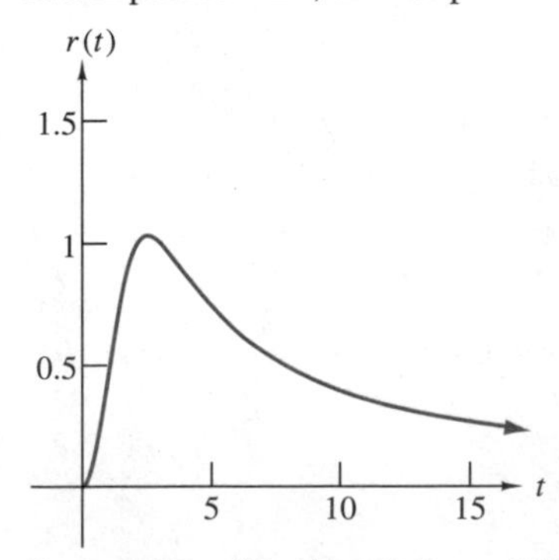

60. ***Economics: City Population*** The graph illustrated below displays the relationship between the population density D, as a percentage of the total population, of a city and the distance d, in miles, from the center of the city.

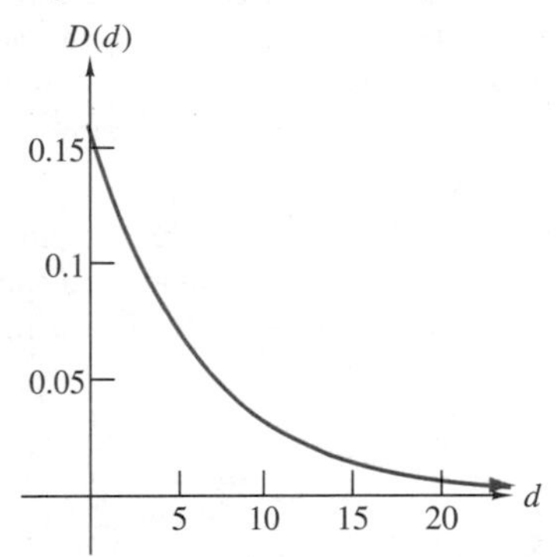

61. ***Business: Assembly time*** The graph illustrated below displays the relationship between the assembly time t, in hours, needed for a new product and the number n of those products being produced.

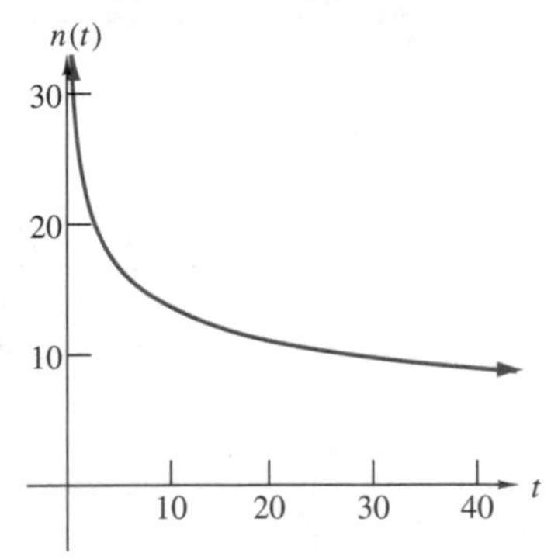

For Exercises 62–70, answer all of the indicated questions.

62. ***Business: Sales*** In some city locations, the size x, in thousands of square feet, of a retail store determines the number of monthly sales S, in dollars, the store will realize. Suppose that for a particular location, $S(x) = -60x^2 + 642x + 1200,\ 1.5 \le x \le 12$. Find the size of the store that maximizes the number of sales and specify the dollar value of that number of sales.

63. ***Medicine: Blood Pressure*** Some medications decrease blood pressure and the amount of decrease D depends on the amount x of the medication taken. Suppose that for a particular medication the relationship between D and x is described by $D(x) = \frac{1}{3}x^3(k - x),\ 0 \le x \le k$. Find the amount of the medication that will produce the maximum decrease in blood pressure.

64. ***Biology: Coughs*** When a person coughs, there is a decrease in the radius of that person's trachea (windpipe). If R represents the natural radius of a person's trachea and r represents the radius of the trachea during a cough, then R is considered a constant and r is a variable. The function $V(r) = kr^2(R - r)$, $\frac{R}{2} \leq r \leq R$ illustrates the relationship between the volume of air passing through the trachea and the radius of the trachea; k is a constant that depends on the units r is measured in. Find the radius of the trachea for which, during a cough, the volume of air passing through it is a maximum. (Your answer will be in terms of R and k.)

65. ***Business: Cost of Laying Pipe*** A freshwater pipeline is to be laid from point A on a straight shoreline to point B on an island 40 miles down off the coast. (See the figure below.)

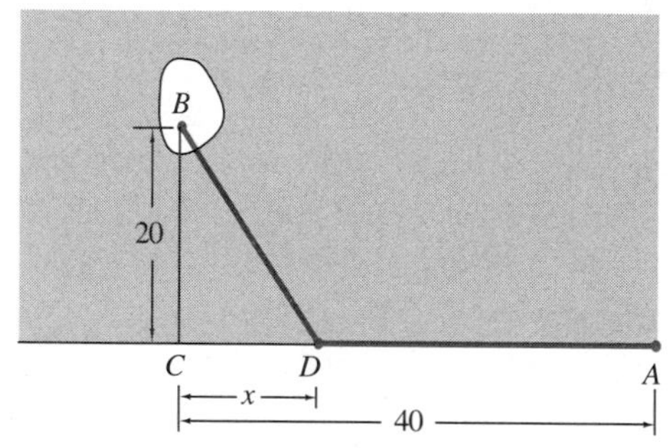

The pipeline may be run completely under the water directly from point A to point B, or it may be run along the shoreline to point C and then under the water to point B, or it may be run along the shoreline to some intermediate point D and then under the water to point B. Let x represent the distance from point C to point D. If $x = 0$, then points C and D are the same. If $x = 40$, then points A and D are the same. If the pipeline costs \$200 per mile on land and \$500 per mile under the water, find the distance x from point C that will minimize the cost of laying the pipeline.

66. ***Business: Construction*** Find the dimensions of a Norman window of perimeter 16 feet, that lets in the maximum amount of light. A Norman window is a window with a rectangular base and a semicircular top, as shown below. (Hint: Let $2x$ represent the length of the rectangular base so that the radius of the semicircular top is x.)

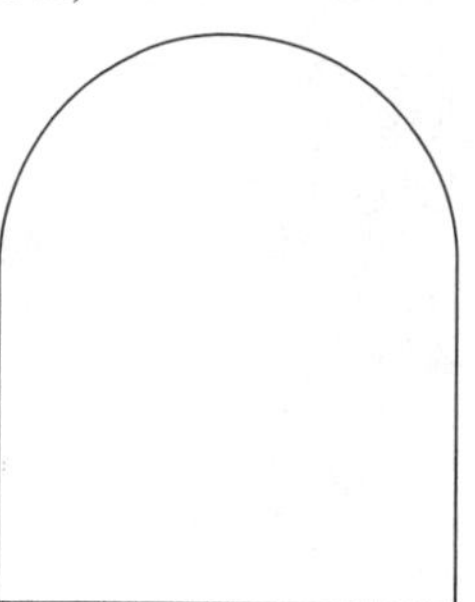

67. ***Business: Packaging*** A right cylindrical can is to be constructed so that it holds 16π cubic inches of a liquid. Find the radius of the top and bottom and the height of the side that will minimize the amount of material needed to be used. (Hint: Minimize the surface area of the can.)

68. ***Mathematics: Number Theory*** Find two numbers that add up to 20 but have a product that is as large as possible.

69. ***Mathematics: Geometry*** Show that of all the rectangles with a given perimeter P, the square is the one with the largest area.

70. ***Chemistry: Rate of Reaction*** In chemical reactions, a catalyst is a substance that controls the rate of the reaction but does not undergo any permanent change itself. An *autocatalytic* reaction is a reaction in which the product of the reaction is a catalyst for its own formation. Autocatalytic reactions can proceed at different rates. They can begin slowly, then speed up, then slow down again because the original substance gets used up. Sometimes, it is reasonable to assume that the rate r of the reaction is proportional to both the amount x of the original substance and the amount $a - x$ of the product. That is, we can sometimes assume that $r = kx(a-x)$. For what amount of the original substance will the reaction be taking place at the maximum rate?

CHAPTER 5

Derivatives of Exponential and Logarithmic Functions

5.1 The Exponential Functions

Introduction

Algebraic Functions

To this point in our study of calculus, we have worked with only **algebraic functions**; that is, functions that are created using the algebraic operations of addition, subtraction, multiplication, division, and powers. The functions

$$f(x) = 3x^2 + 5x - 2, \quad f(x) = \frac{x^2 + 5}{x^3 - 4}, \quad f(x) = \left(\frac{2x + 1}{x + 5}\right)^{2/3}$$

are examples of algebraic functions. In particular, the power function $f(x) = x^b$, where x is a variable and b is a rational number, is an algebraic function. In a power function, the base is a variable and the power is a constant.

We now turn our attention to another type of function that is extremely important in both pure and applied mathematics. This function is a nonalgebraic function called the **exponential function.**

Algebraic Functions Compared with Exponential Functions

We begin by defining the exponential function and we then distinguish it from the power function.

Exponential Function

The function

$$f(x) = b^x, \text{ where } b > 0,\ b \neq 1$$

and x is a variable, is the **exponential function**.

Note that the power function $f(x) = x^b$ and the exponential function $f(x) = b^x$ are two entirely distinct functions.

The power function $f(x) = x^b$ has the form: $(\text{variable})^{(\text{constant})}$.

The exponential function $f(x) = b^x$, has the form: $(\text{constant})^{(\text{variable})}$.

It is also important to note the restriction on the base b of the exponential function. The base cannot be just any constant. It must be greater than 0 but not equal to 1. (The reason for this restriction will be made evident at the end of this section.)

Graphs and Behavior of the Exponential Function

The exponential function is used to describe relationships in learning theory, population growth and decay, finance and economics, medicine, archaeology, ecology, psychology, sociology, chemistry, biology, physics, mathematics, and other fields. Because the exponential function is so useful and occurs so often, it is important to have a good "feel" for its behavior. To examine the behavior of the exponential function, we will examine the various forms of its graph. We will begin by looking at two specific examples, $f(x) = 2^x$ and $f(x) = (1/2)^x$.

To construct the graphs of $f(x) = 2^x$ and $f(x) = (1/2)^x$, we first build tables of values as in Table 5.1, plot the corresponding points, then connect those points with a smooth, continuous curve. To connect the points with a smooth, continuous curve, we will assume (correctly and without proof) that b^x is defined for all real number exponents and not just the rational exponents developed in elementary and intermediate algebra.

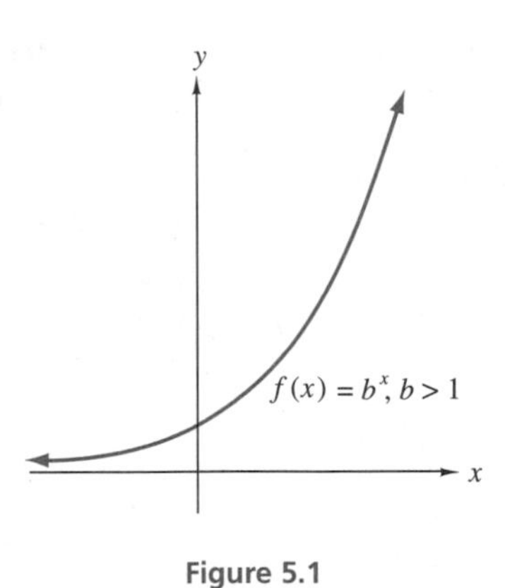

Figure 5.1

Table 5.1

x	$f(x) = 2^x$	$f(x) = \left(\frac{1}{2}\right)^x$
−3	1/8	8
−2	1/4	4
−1	1/2	2
0	0	1
1	2	1/2
2	4	1/4
3	8	1/8

In Figure 5.1, we see that the base is greater than 1 and that the function increases as x increases. In Figure 5.2, we see that the base is between 0 and 1 and that the function decreases as x increases. These graphs illustrate the general forms of the exponential function.

If the base is greater than 1 ($b > 1$), the output variable increases as the input variable increases, or geometrically, the curve rises as we look left to right.

Growth Function

Since the function $f(x) = b^x,\ b > 1$, increases as x increases, this function is referred to as a **growth** function. (See Figure 5.1.)

If the base is between 0 and 1 ($0 < b < 1$), the output variable decreases as the input variable increases, or geometrically, the curve falls as we look left to right.

Decay Function

Since the function $f(x) = b^x,\ 0 < b < 1$, decreases as x increases, this function is referred to as a **decay** function. (See Figure 5.2.)

Before examining growth and decay functions in more detail, we will next introduce a number that is very important in study of growth and decay processes.

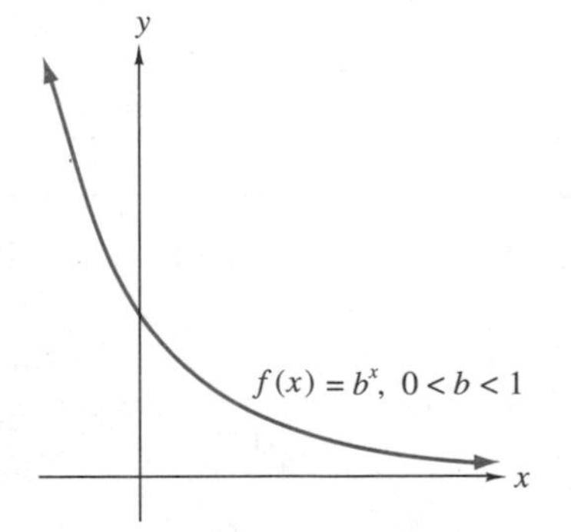

Figure 5.2

The Number *e*

When describing relationships using the exponential function $f(x) = b^x$, a base that occurs remarkably often is the number represented by the lowercase letter e. The number e is an irrational number and is defined precisely in the following way.

The Number *e*

$$e = \lim_{x \to \infty} \left(1 + \frac{1}{x}\right)^x$$

≈2.718

The expression $\left(1+\frac{1}{x}\right)^x$, or forms of it, often occurs when describing relationships in the fields mentioned earlier. We shall see it, or its various forms, in the applications we examine in this and the next few sections.

$e \approx 2.718$

We can calculate the value of e to any precision we want by computing the expression $\left(1+\frac{1}{x}\right)^x$ for large enough values of x. The value of e is 2.718281828459045.... Since the number e is irrational, it never ends and it contains no repeating block of digits. When calculating in applications with e, we commonly approximate its value to be 2.718 (although most calculators have the value built into them). The following table and Figure 5.3 will help convince us that as x grows larger and larger, the expression $\left(1+\frac{1}{x}\right)^x$ approaches e. (Notice that as x gets bigger and bigger, $\left(1+\frac{1}{x}\right)^x$ gets closer and closer to a limit that, to four significant digits, is 2.718.)

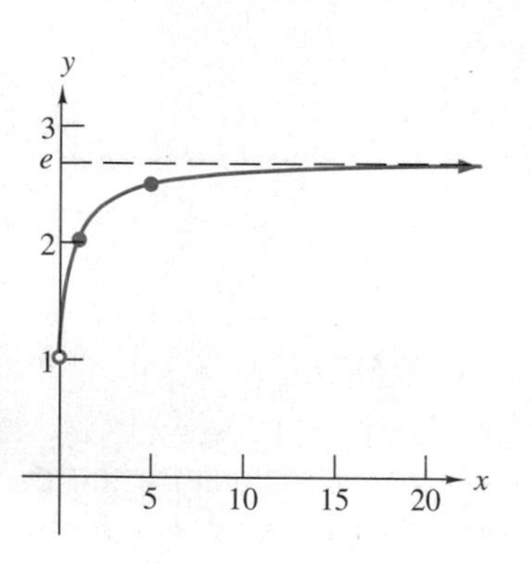

Figure 5.3

Table 5.2

x	$f(x)=\left(1+\frac{1}{x}\right)^x$
1,000	2.716923932...
10,000	2.718145927...
20,000	2.718213875...
30,000	2.718236498...
40,000	2.718247851...
50,000	2.718254646...
60,000	2.718259231...
70,000	2.718262358...
80,000	2.718264839...
90,000	2.718266666...
100,000	2.718268237...

Graphs of the General Exponential Function

General Exponential Function

With the number e as the base, we have the very important and useful general exponential function $f(x)=A_0e^{kx}$.

Exponential Growth Function

The forms commonly used to predict growth or decay in business and the life and social sciences use a positive number k as the coefficient of t. To predict growth, the form is

$$f(t)=A_0e^{kt}$$

Exponential Decay Function

and to predict decay, the form is

$$f(t)=A_0e^{-kt}$$

The number A_0 represents the amount of a quantity present at the initial observation; that is, at $t=0$, where t represents time. Notice that at $t=0$, $f(0)=A_0e^{k\cdot 0}=A_0e^0=A_0\cdot 1=A_0$. The number k is called the **growth constant** and represents the ratio of the rate at which a population is growing to the population itself. The constant k is always positive for a function undergoing exponential growth.

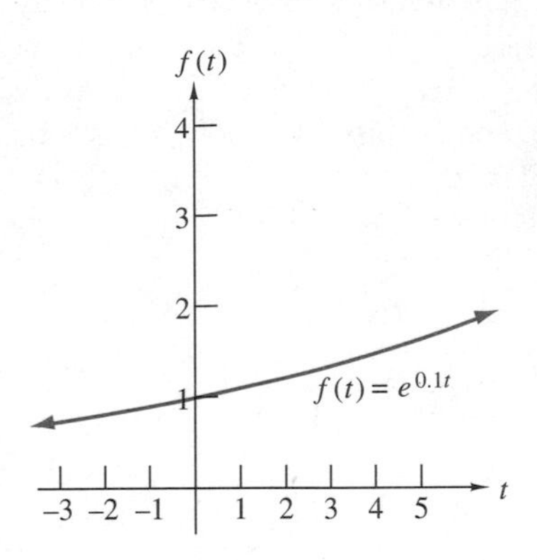

Figure 5.4

We can get a feel for exponential growth functions by examining the behavior of $f(t) = A_0e^{kt}$ for a specific value of A_0 and various values of the growth constant k. For convenience, we will take $A_0 = 1$. (Any other number would do. The only affect A_0 has on the graph is the vertical position of the curve. It does not affect the shape of the curve. Taking $A_0 = 1$ allows us to sketch the graph near the origin.) Figure 5.4, Figure 5.5, Figure 5.6, and Figure 5.7 show the graphs of $f(t) = e^{0.1t}, f(t) = e^{0.3t}, f(t) = e^{0.5t}, f(t) = e^{1t}, f(t) = e^{1.5t}$, and $f(t) = e^{2t}$, respectively.

We can see that each graph displays the recognizable upward trend of an exponential growth function. We also notice, looking at these graphs in order from Figure 5.4 to Figure 5.7, that as the value of the growth constant k increases, the curve bends more sharply upward. This indicates that for small values of k, growth is increasing relatively slowly and that for large values of k, growth is increasing relatively quickly.

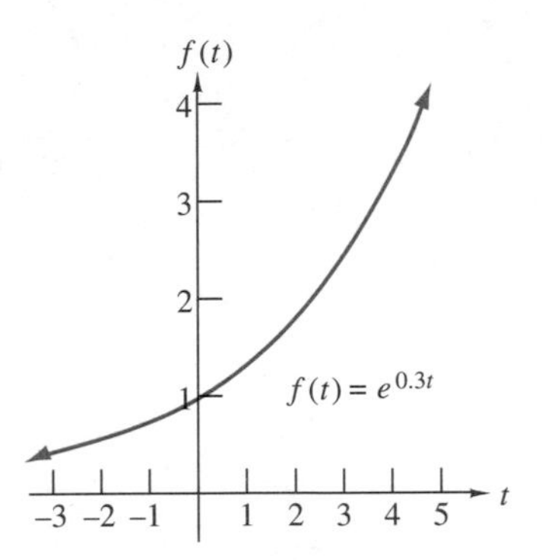

Figure 5.5

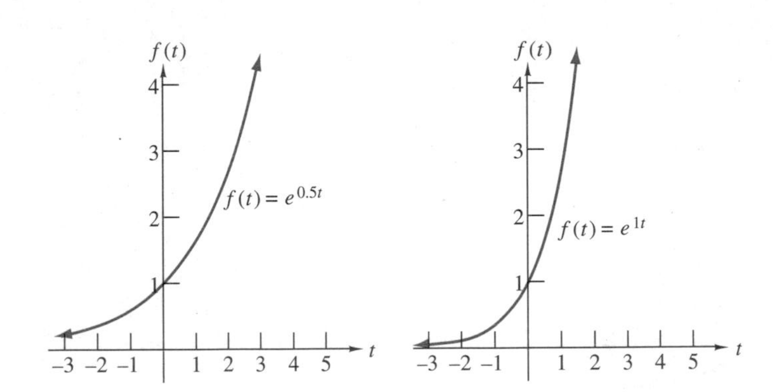

Figure 5.6

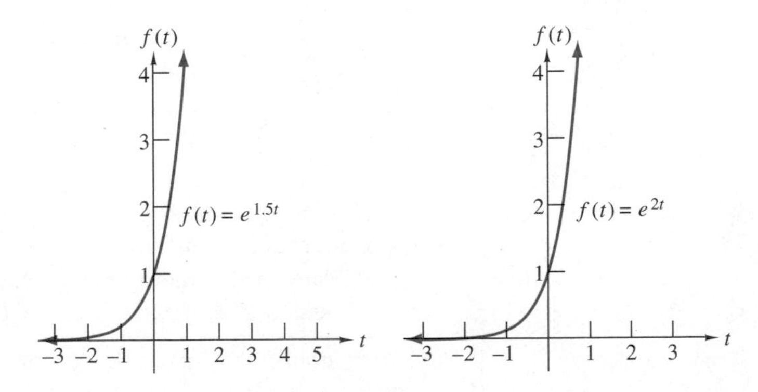

Figure 5.7

Next, to successfully examine the function $f(x) = A_0e^{-kx}$, we make an important observation. Since $e \approx 2.718 > 1$, it is true that $0 < \frac{1}{e} < 1$. Now, e^{-k} can be written as $e^{-1\cdot k}$. Then $e^{-k} = e^{-1\cdot k} = (e^{-1})^k = \left(\frac{1}{e}\right)^k < 1$. So, if e^{-k} is used as the base of an exponential function, it will describe a decay function.

Now let's examine the behavior of $f(t) = A_0 e^{-kt}$, an exponential decay function. Figure 5.8, Figure 5.9, Figure 5.10, and Figure 5.11 show the graphs of $f(t) = e^{-0.1t}$, $f(t) = e^{-0.3t}$, $f(t) = e^{-0.5t}$, $f(t) = e^{-1t}$, $f(t) = e^{-1.5t}$, and $f(t) = e^{-2t}$, respectively.

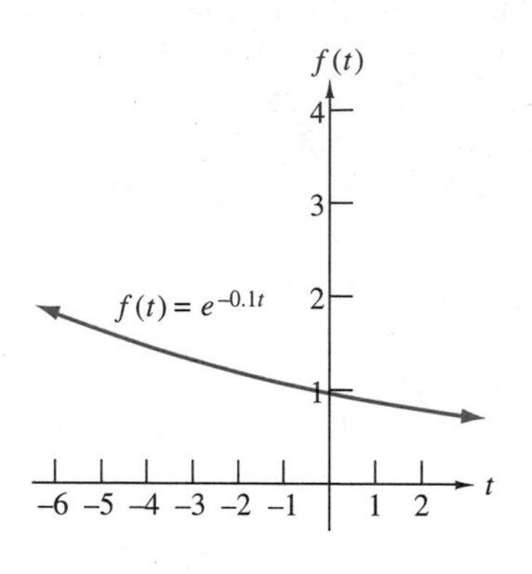

Figure 5.8

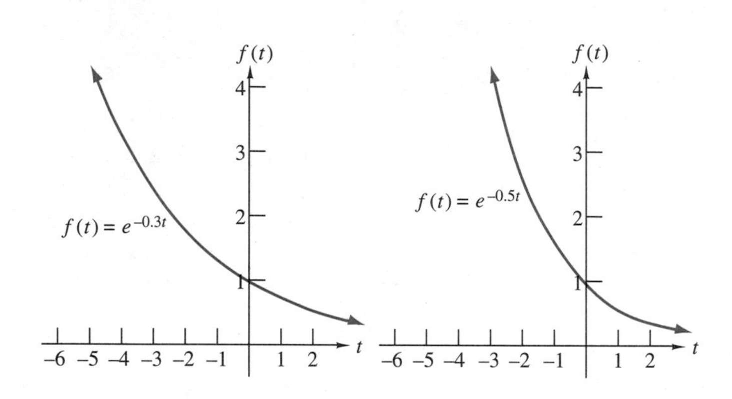

Figure 5.9

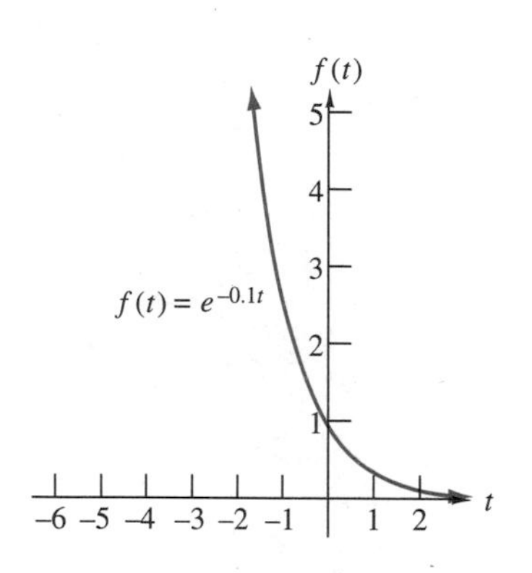

Figure 5.10

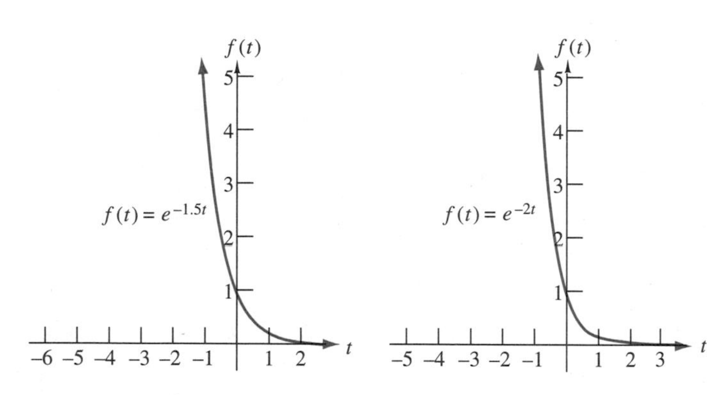

Figure 5.11

We can see that each graph displays the recognizable downward trend of an exponential decay function.

We now make the following general observations about exponential functions. We note that in each case, $k > 0$.

1. The domain of $f(x) = A_0 e^{kx}$ and $f(x) = A_0 e^{-kx}$ is all real numbers; that is, $f(x) = A_0 e^{kx}$ and $f(x) = A_0 e^{-kx}$ are computable for all values of x.
2. The range of $f(x) = A_0 e^{kx}$ and $f(x) = A_0 e^{-kx}$ is all positive numbers; that is, $f(x) = A_0 e^{kx}$ and $f(x) = A_0 e^{-kx}$ are never zero or negative.
3. The exponential functions $f(x) = A_0 e^{kx}$ and $f(x) = A_0 e^{-kx}$ are continuous for all values of x.
4. The exponential growth function, $f(x) = A_0 e^{kx}$, and exponential decay function, $f(x) = A_0 e^{-kx}$, both pass through the point $(0, A_0)$.

5. The growth function $f(x) = A_0 e^{kx}$ increases as x increases; that is, the graph of $f(x) = A_0 e^{kx}$ rises as we look at it left to right and always lies *above* the x axis (assuming that A_0 is positive).
6. The decay function $f(x) = A_0 e^{-kx}$ decreases as x increases; that is, the graph of $f(x) = A_0 e^{-kx}$ falls as we look at it left to right and always lies *above* the x axis (assuming that A_0 is positive).
7. For the growth function $f(x) = A_0 e^{kx}$, $\lim_{x \to \infty} A_0 e^{kx} = \infty$ and $\lim_{x \to -\infty} A_0 e^{kx} = 0$. Therefore, the growth function increases without bound as x increases without bound, and tends toward 0 as x decreases.
8. For the decay function $f(x) = A_0 e^{-kx}$, $\lim_{x \to \infty} A_0 e^{-kx} = 0$ and $\lim_{x \to -\infty} A_0 e^{-kx} = \infty$. Therefore, the decay function increases without bound as x decreases without bound, and tends toward 0 as x increases.

Applications of the Exponential Function

The following examples demonstrate how the exponential function can be used to obtain information about applied phenomena.

EXAMPLE SET A

The sociologists Stephan and Mischler found that the exponential function

$$N(p) = N_1 e^{-0.11(p-1)}, \quad 1 \leq p \leq 10$$

is produced when members of a discussion group of 10 people are ranked according to the number of times each participated. N_1 represents the number of times the first-ranked person participated, and $N(p)$ represents the number of times the pth-ranked person participated. If, in a discussion group of 10 people, the first-ranked person participated 35 times, how many times did the sixth-ranked person participate?

Solution:
We are asked to find the number of times the sixth-ranked person participated. This means that we need to find the value of $N(6)$. Since the first-ranked person participated 35 times, we substitute 35 for N_1. Since we are interested in the number of times the sixth-ranked person participated, we substitute 6 for p. This produces the computation

$$N(6) = 35e^{-0.11(6-1)}$$

Using a calculator, we find that

$$\begin{aligned} N(6) &= 35 \cdot e^{-0.11(6-1)} \\ &= 35 \cdot e^{-0.11(5)} \\ &= 35 \cdot e^{-0.55} \\ &\approx 35 \cdot 0.5769498 \\ &\approx 20.193243 \end{aligned}$$

Interpretation: We conclude that if the first-ranked person participated 35 times, the sixth-ranked person participated approximately 20 times.

Notice that this function is a decay function, which seems reasonable in this situation. We would expect that as p increases, $N(p)$ will decrease.

EXAMPLE SET B

When a particular amount of money P, called the principal, is invested at the interest rate r and is compounded k times a year, the amount A accumulated after t years is

$$A = P\left(1 + \frac{r}{k}\right)^{kt}$$

Determine the amount of money accumulated after 15 years if \$2000 is invested in an account that pays 8% interest compounded quarterly.

Solution:
Since the interest rate is 8%, we substitute 0.08 for r. Since interest is compounded quarterly, meaning four times a year, we substitute 4 for k. Finally, since the investment is made for 15 years, we substitute 15 for t.

$$\begin{aligned} A &= 2000\left(1 + \frac{0.08}{4}\right)^{4\cdot 15} \\ &= 2000(1 + 0.02)^{60} \\ &= 2000(1.02)^{60} \\ &\approx 2000(3.2810308) \\ &\approx 6562.0616 \end{aligned}$$

Interpretation: We conclude that after 15 years, \$2000 will grow to approximately \$6562.06 if it is invested at 8% interest compounded quarterly.

EXAMPLE SET C

Summarize the behavior of $f(x) = e^{-x^2}$.

Solution:
The domain of $f(x)$ is all real numbers. Note also that this function is always positive; it is never negative or 0.

$$f'(x) = -2xe^{-x^2} \quad \text{and} \quad f''(x) = -2e^{-x^2}(1 - 2x^2)$$

We search for critical and hypercritical values by checking the two conditions.

1. $f'(x) = 0$ when $-2xe^{-x^2} = 0$. Since the exponential function is itself never 0, $f'(x)$ is 0 only when $-2x = 0$. $-2x = 0$ when $x = 0$.

$f''(x) = 0$ when $-2e^{-x^2}(1-2x^2) = 0$. This can only happen when $1 - 2x^2 = 0$. Solve this equation for x.

$$1 - 2x^2 = 0$$

$$1 = 2x^2$$

$$\frac{1}{2} = x^2$$

$$x = \frac{1}{\sqrt{2}}, \frac{-1}{\sqrt{2}}$$

2. Since $f'(x)$ and $f''(x)$ have no denominators, they are never undefined. Thus, $x = 0$ is a critical value and, therefore, a potential relative extreme point, and $x = 1/\sqrt{2}$ and $-1/\sqrt{2}$ are hypercritical values and, therefore, potential points of inflection. Figure 5.12 shows the sign chart for $f(x)$.

f'	(–)(–)(+) = +		(–)(–)(+) = +		(–)(+)(+) = –		(–)(+)(+) = –
f''	(–)(+)(–) = +	$\frac{-1}{\sqrt{2}}$	(–)(+)(+) = –	0	(–)(+)(+) = –	$\frac{1}{\sqrt{2}}$	(–)(+)(–) = +

Figure 5.12

Now, using the sign chart, we construct the summary table for this function (See Table 5.2). Using the summary table, we construct a sketch of $f(x)$ and illustrate it in Figure 5.13.

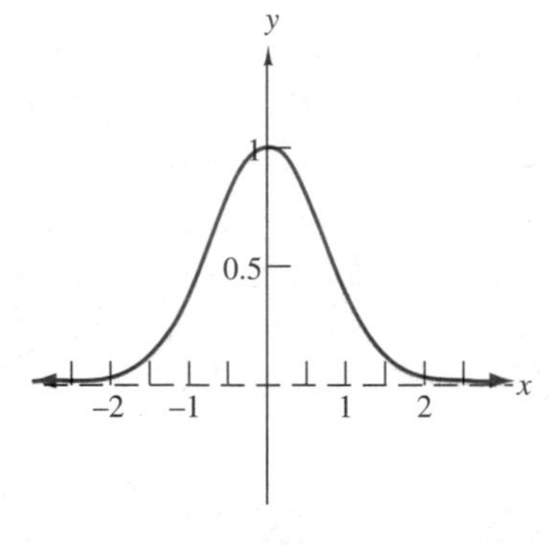

Figure 5.13

Table 5.3

Int/Val	$f'(x)$	$f''(x)$	$f(x)$	**Behavior of** $f(x)$
$\left(-\infty, \frac{-1}{\sqrt{2}}\right)$	+	+		incr at an incr rate
$\frac{-1}{\sqrt{2}}$		0	$\left(\frac{-1}{\sqrt{2}}, .61\right)$	infl at $\approx \left(\frac{-1}{\sqrt{2}}, .61\right)$
$\left(\frac{-1}{\sqrt{2}}, 0\right)$	+	–		incr at a decr rate
0	0		(0, 1)	relative max at (0, 1)
$\left(0, \frac{1}{\sqrt{2}}\right)$	–	–		decr at an incr rate
$\frac{1}{\sqrt{2}}$		0	$\left(\frac{1}{\sqrt{2}}, .61\right)$	infl at $\approx \left(\frac{1}{\sqrt{2}}, .61\right)$
$\left(\frac{1}{\sqrt{2}}, +\infty\right)$	–	+		decr at a decr rate

Summary of behavior: As x increases from negative infinity to $\frac{-1}{\sqrt{2}}$, $f(x)$ is positive and increases at an increasing rate. When $x = \frac{-1}{\sqrt{2}}$, $f(x) \approx 0.607$ and experiences a point of inflection (which alerts to a coming change in concavity). As x increases to 0, $f(x)$ increases, but now does so at a decreasing rate (the rate of increase is slowing). At $x = 0$, $f(x) = 1$ and attains a relative maximum. As x increases beyond 0 to $\frac{1}{\sqrt{2}}$, $f(x)$ decreases at an increasing rate (the function is decreasing and is doing so more quickly as x gets bigger). At $x = \frac{1}{\sqrt{2}}$, $f(x) \approx 0.607$ and experiences a point of inflection (which alerts to a coming change in concavity). Then, as x increases without bound, $f(x)$ decreases, but at a decreasing rate.

Logistic Growth Function

The last example we will examine involves a variation of the exponential growth function called the **logistic growth function**. The logistic growth function is used to describe *restricted* growth; that is, growth that is restricted to some upper bound by factors imposed by some particular conditions. It is used, for example, by biologists to predict or describe the growth of a population in which environmental factors can restrict that growth, and by sociologists to describe the spread of a rumor. It has the form

$$f(t) = \frac{M}{1 + Ae^{-Mkt}}$$

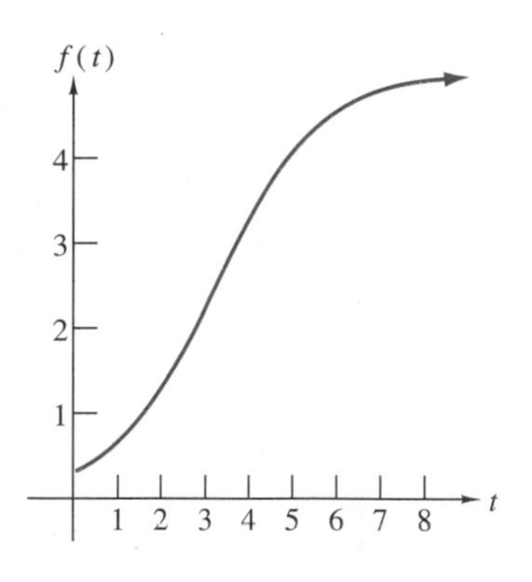

Figure 5.14

The numbers M and A are positive constants, k is the growth constant, and t represents the amount of time the process has been going on. M is the maximum level the population can reach. The graph of the logistic function $f(t) = \dfrac{5}{1 + 15e^{-0.85t}}$ is illustrated in Figure 5.14. Notice that for small values of t, the logistic function closely resembles the exponential growth function. However, as the values of t get larger, the curve levels off to the maximum value prescribed in the numerator. This leveling-off effect reflects restrictions on the population.

EXAMPLE SET D

Let us model the number of automobile sales in a developing country as

$$N(t) = \frac{156{,}000}{1 + 2e^{-1.5(t-1987)}}, \quad t \geq 1987$$

where N is the number of automobiles sold in one year and t is the number of years since 1987.

Find:

1. How many cars will be sold in 1987, the first year of observation?
2. How many cars will be sold in 1995?
3. If the model holds well into the future, at what level will the sale of cars stabilize?

Solution:

1. Since t is the number of years since 1987, the value of t at the beginning of the observation period is 1987. Then

$$N(t) = \frac{156{,}000}{1 + 2e^{-1.5(t-1987)}}$$

$$\begin{aligned} N(1987) &= \frac{156{,}000}{1 + 2e^{-1.5(1987-1987)}} \\ &= \frac{156{,}000}{1 + 2e^{-1.5(0)}} \\ &= \frac{156{,}000}{1 + 2e^{0}} \\ &= \frac{156{,}000}{1 + 2} \\ &= 52{,}000 \end{aligned}$$

Thus, in 1987, about 52,000 cars can be expected to be sold.

2. In 1995,

$$\begin{aligned} N(1995) &= \frac{156{,}000}{1 + 2e^{-1.5(1995-1987)}} \\ &= \frac{156{,}000}{1 + 2e^{-1.5(8)}} \\ &\approx 155{,}998 \end{aligned}$$

Thus, in 1995, about 155,998 cars can be expected to be sold.

3. If the trend continues, t will increase without bound, which alerts us to taking the limit of the function as t approaches infinity. We need to compute

$$\begin{aligned} \lim_{x\to\infty} \frac{156{,}000}{1 + 2e^{-1.5(t-1987)}} &= \lim_{x\to\infty} \frac{156{,}000}{1 + 0} \\ &= 156{,}000 \end{aligned}$$

Thus, if the car sales trend continues, car sales can be expected to stabilize at approximately 156,000 cars per year.

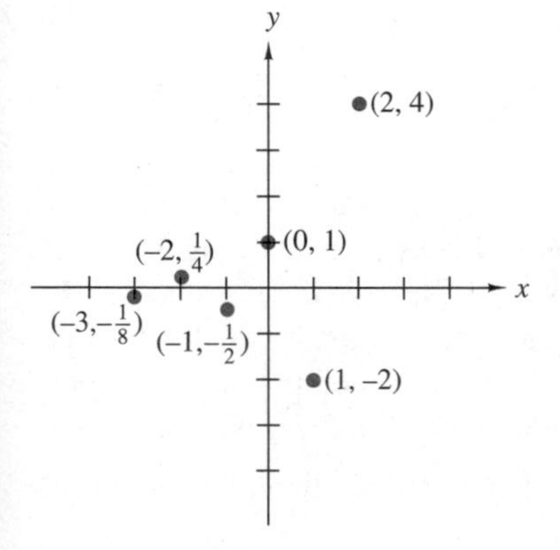

Figure 5.15

We end this section by seeing, graphically, why the base b is restricted to being positive but not equal to 1. The fact that the base b is restricted so that it is greater than 0 but not equal to 1 is responsible for the basic shape of the exponential curve. If b was allowed to be negative or 1, exponential functions would not always exhibit the recognizable shape they now possess.

As examples, the functions $f(x) = (-2)^x$ and $f(x) = 1^x$ are graphed in Figure 5.15 and Figure 5.16, respectively. Notice that the points on the graph of

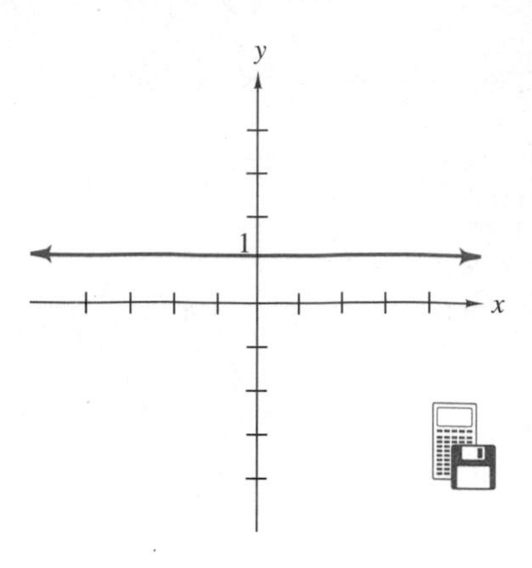

Figure 5.16

$f(x) = (-2)^x$ have not been joined with a smooth curve. Since the expression $(-2)^x$ is not defined for certain exponents (such as $1/2$, $1/4$, $1/6, \dots$, which represent even roots), the graph has infinitely many pieces. The function $f(x) = 1^x$ is actually the constant function $f(x) = 1$, since $1^x = 1$ for all values of x. The graphs of constant functions, as we know, are horizontal lines. Notice that neither graph exhibits the growth nor decay structure apparent when $b > 0$ and $b \neq 1$.

The Exponential Function and Technology

Using Your Calculator You can use your graphing calculator to construct the graph of an exponential function. The following entries illustrate how to construct the graph of the logistic growth function

$$f(t) = \frac{21{,}000}{1 + 165e^{-0.85t}}$$

Y1 = 21000/(1 + 165e^(−0.85X))

Set $0 \leq X \leq 26$ to simulate a 26-week observation period. From your knowledge of the logistic function, you should be able to set the minimum and maximum values of y.

Calculator Exercises

1. Construct the graph of the function

$$f(x) = \left(1 + \frac{1}{x}\right)^x, \qquad x \geq 0$$

Describe the behavior of the function. Include in your description what happens to $f(x)$ as $x \to \infty$.

2. On the same coordinate system, construct the graphs of the functions $f(x) = e^x$, $f(x) = 10e^x$, $f(x) = 20e^x$, $f(x) = -10e^x$, $f(x) = -20e^x$, $f(x) = -30e^x$, and $f(x) = 30e^x$. Based on your graphs, make a statement about the way in which the coefficient of e^x affects the graph of the function $f(x) = e^x$.

3. The function

$$P(t) = Q + (100 - Q)e^{-kt}$$

is called the *Ebbinghaus model of learning* and relates the percentage, P, of knowledge about a subject that is retained t weeks after learning 100% of all there is to know about the subject. The variable Q is the percentage that is never forgotten. Suppose for a particular subject, 20% of the knowledge about it will never be forgotten and that $k = -0.10$. Compute and interpret $P(1)$, $P(10)$, $P(20)$, $P(50)$, $P(100)$, $P(200)$, and $\lim_{t\to\infty} P(t)$. Construct the graph of $P(t)$ over a 100-week period and use it to describe the behavior of the function.

4. A person starts a rumor and t hours later N people have heard it. Suppose N and t are related by the function

$$N(t) = \frac{600}{1 + 400e^{-0.8t}}$$

Construct the graph of this function and use it to describe its behavior over a 24-hour period. Your description should include a statement about the maximum number of people who can hear the rumor. Does this number appear in the function? Is so, where? Try changing this number and see what happens to your graph.

5. In 1626, Native Americans sold Manhattan Island to Peter Minuit, a representative of the West India Company, for \$24 worth of merchandise. If you assume that inflation is exponential at an annual rate of 6%, the worth, W, in dollars, of Manhattan Island t years from 1626 is given by the function

$$W(t) = 24e^{0.06t}$$

Use this equation to determine the worth of Manhattan Island in the year 2000. Change the function to reflect annual inflation rates of first 8% and then 10% and use these new functions to again determine the worth of Manhattan Island in the year 2000.

EXERCISE SET 5.1

A UNDERSTANDING THE CONCEPTS

For Exercises 1–11, identify each function as algebraic or exponential.

1. $f(x) = 5x^3 - 2x + 11$
2. $f(x) = (x+4)^3(x-5)^2$
3. $f(x) = 5^x + 12$
4. $f(x) = x^2(x+1)^8$
5. $f(x) = \left(\frac{x+3}{x-1}\right)$
6. $f(x) = 12^{x-1} + x$
7. $f(x) = 8^{3x}$
8. $f(x) = 4x^e$
9. $f(x) = e^{-2x+1}$
10. $f(x) = (5+e)^{3x}$
11. $f(x) = 2x^{-3} - 2ex^2$
12. A power function has the form (variable)$^{\text{(constant)}}$. Specify the form of the exponential function.
13. Exponential functions are given names according to whether the base is greater than 1 or between 0 and 1. Specify these names.
14. Give a common approximation value for the number e.

B SKILL ACQUISITION

For Exercises 15–25, use a calculator to compute the value of each function. Identify each exponential function as a growth or decay function. Round to three decimal places. (If necessary, approximate e with 2.718.)

15. Find $f(3)$ for $f(x) = 10^x$.
16. Find $f(2)$ for $f(x) = 10^{-2x}$.
17. Find $f(6)$ for $f(x) = (0.99)^{40x}$.
18. Find $f(2)$ for $f(x) = (1.06)^{50x}$.
19. Find $f(5)$ for $f(x) = (1.35)^{3x+4}$.
20. Find $f(55)$ for $f(x) = (1.01)^{10x+120}$.
21. Find $f(3)$ for $f(x) = \left(\frac{1}{2}\right)^{x+1}$.
22. Find $f(4)$ for $f(x) = e^x$.
23. Find $f(2)$ for $f(x) = e^{-x}$.
24. Find $f(6)$ for $f(x) = 2000e^{0.04x}$.
25. Find $f(105)$ for $f(x) = 3560e^{0.06x}$.

For Exercises 26–30, use a calculator to compute the value of each function. Round to three decimal places. (If necessary, approximate e with 2.718.)

26. Find $f(25)$ for $f(x) = 1 - e^{0.11x}$.
27. Find $f(0.10)$ for $f(x) = 1 - e^{-0.3x}$.
28. Find $f(5)$ for $f(x) = 15{,}000 + 20{,}000\left(\frac{3}{8}\right)^{0.75x}$.
29. Find $f(3)$ for $f(x) = 100 - 60\left(\frac{1}{4}\right)^{0.95x}$.
30. Find $f(20)$ for $f(t) = 5{,}500{,}000 \cdot 3^{t/45}$.
31. Use the figure below and, as best you can, match each function with its graph (A, B, C, D).

a. $f(x) = e^{2x/3}$ b. $f(x) = e^x$
c. $f(x) = e^{4x/5}$ d. $f(x) = e^{2x}$

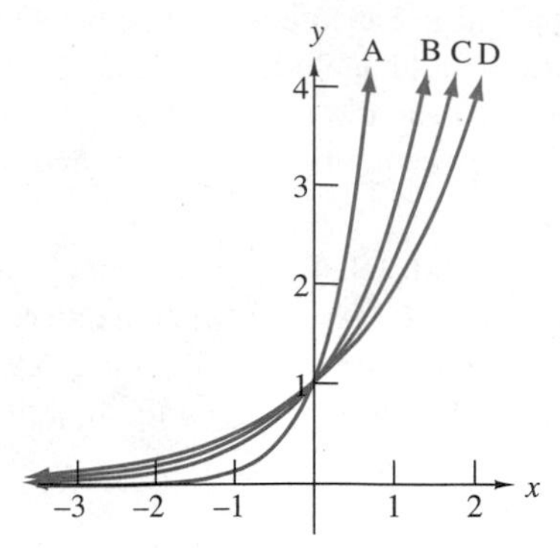

32. Use the figure below and, as best you can, match each function with its graph (A, B, C, D).

a. $f(x) = e^{-x}$ b. $f(x) = e^{-3x/5}$
c. $f(x) = e^{-1x/3}$ d. $f(x) = e^{-2.4x}$

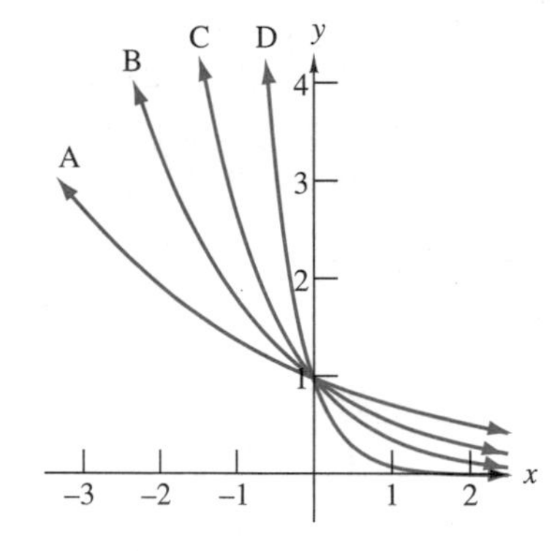

C APPLYING THE CONCEPTS

33. ***Psychology: Forgetting Curves*** In his book *Memory, A Contribution to Experimental Psychology*, published in 1885, psychologist Herman Ebbinghaus presented a *forgetting curve* that describes the relationship between the percentage P of words remembered from a list or story and the time t (in weeks) since memorization took place. The curve is modeled by the exponential decay function

$$P(t) = (100 - a)e^{-kt} + a$$

The numbers a and k are positive constants, and $0 < a < 100$. Suppose that for a particular person under particular conditions,

$$P(t) = (100 - 10)e^{-0.75t} + 10$$

Find and interpret

a. $P(0)$ b. $P(1)$ c. $P(2)$
d. $P(7)$ e. $P(20)$

34. ***Life Science: Bacterial Growth*** The number N of bacteria present in a culture at any time t (in hours) is given by the exponential decay function

$$N(t) = 15{,}000 \cdot 3^{-0.05t}$$

Find and interpret

a. $N(0)$ b. $N(1)$
c. $N(5)$ d. $N(24)$

For Exercises 35 and 36, use the following information. When a particular amount of money P, called the principal, is invested at the interest rate r and is compounded k times a year, the amount A of money accumulated after t years is $A(t) = P\left(1 + \frac{r}{k}\right)^{kt}$.

35. ***Business: Investment*** Find the amount of money accumulated in 5 years if \$12,000 is invested at 8% interest and is compounded

a. annually b. quarterly
c. monthly d. daily

36. ***Business: Investment*** Find the amount of money accumulated in 5 years if \$500 is invested at 11% interest and is compounded

a. annually b. quarterly
c. monthly d. daily

For Exercises 37 and 38, use the following information. We know that when a particular amount of money P, called the principal, is invested at the interest rate r and is compounded k times a year, the amount A of money accumulated after t years is $A(t) = P\left(1 + \frac{r}{k}\right)^{kt}$. If the number k of compoundings increases without bound, that is, if compounding takes place continuously, the above formula reduces to $A(t) = Pe^{rt}$.

37. ***Business: Investment*** Find the amount of money accumulated if \$2000 is invested for 3 years at 7% interest and is compounded (a) quarterly, (b) monthly, (c) daily (assume 365 days in a year), and (d) continuously. Compare the values you find in (a) – (d). Are they close? (They should be.) Since there is such a small difference in the amount of money a bank pays out in interest between compounding quarterly and compounding continuously, one might think that by advertising continuous compounding, a bank would have customers flocking to invest their money, and could, therefore, make quite a bit of money from this advertising ploy. But, although bankers are aware of the continuous compounding formula (from their college calculus course), they rarely, if ever, offer continuous compounding. Can you think of a reason why not?

38. ***Business: Investment*** Find the amount of money accumulated if \$1600 is invested for 5 years at 9% interest and is compounded (a) quarterly, (b) monthly, (c) daily (assume 365 days in a year), and (d) continuously. Compare the values you get in (a)–(d). Are they close? (They should be.)

39. *Business: Market Share* If we model the percentage of the VCR market held by the VHS standard as

$$M(t) = \frac{100}{1 + e^{-0.8(t-1980)}}$$

where M is the percentage of the market in year t, what percentage of the market was held by VHS in 1985?

40. *Business: Customer Base* A restaurant attracts new customers by word-of-mouth. The customer base can be modeled as

$$C(t) = \frac{240}{1 + e^{-0.2t}}$$

where C is the customer base and t is measured in days since the opening of the restaurant. At what level will the customer base stabilize?

41. *Business: Demand for a Product* Demand for a product depends on the price as

$$D(p) = 500 + \frac{1000}{1 + e^{-0.05p}}$$

where D is the quantity demanded and p is the price in dollars. How many units of the product will be sold when the price is $50?

42. *Business: Number of Telephones* The number of telephones per capita in a country depends on the median income of the people living in the country as

$$T(i) = \frac{0.9}{1 + 20e^{-i/400}}$$

where T is the number of telephones per capita and i is the median income.

a. What will be the number of telephones per capita if the median income is $400?

b. What will be the number of telephones per capita if the median income is $4000?

43. *Marketing: Sales* The marketing department of a toy company has determined that t weeks after an advertising campaign ends, the number N of the advertised toys it sells each month decreases according to the exponential decay function $N(t) = 2700e^{-0.12t}$. Find how many of the advertised toys the company will sell (a) immediately at the close of the advertising campaign, (b) one week after the campaign, (c) five weeks after the campaign, and (d) 12 weeks after the end of the campaign.

44. *Medicine: Drug Concentration* When a drug is intravenously administered to a patient on a continuous basis, the amount A (in milligrams) of the drug in the patient's bloodstream t minutes after injection of the drug is given by $A(t) = (1/k)(a - Ce^{-kt})$. In this case, the growth constant k is called the *absorption constant*. The number a is the amount of the drug added to the bloodstream each minute, and the number C is a constant that depends on the values of a and k. For the case when $a = 2$ mg, $C = 35$, and $k = 0.20$, find the amount of the drug in the patient's bloodstream (a) 15 minutes after injection, and (b) 60 minutes after injection.

45. *Medicine: Drug Concentration* When a drug is intravenously administered to a patient on a continuous basis, the amount A (in milligrams) of the drug in the patient's bloodstream t minutes after injection of the drug is given by $A(t) = (1/k)(a - Ce^{-kt})$. In this case, the growth constant k is called the *absorption constant*. The number a is the amount of the drug added to the bloodstream each minute, and the number C is a constant that depends on the values of a and k. For the case when $a = 7$ mg, $C = 40$, and $k = 0.12$, find the amount of the drug in the patient's bloodstream (a) 15 minutes after injection, and (b) 60 minutes after injection.

46. *Medicine: Spread of Disease* Suppose that in a large city, 21,000 people are susceptible to a particular flu virus. When that flu epidemic does break out, the city's health agency determines that t weeks after the beginning of the outbreak the number of cases of flu is given by the function

$$N(t) = \frac{21{,}000}{1 + 165e^{-0.85t}}$$

a. How many people had the flu virus when the health agency began recordings?

b. How many people had the flu virus 5 weeks after the outbreak?

c. If the epidemic continues, how many people can be expected to contract the flu?

47. *Medicine: Contraction of Disease* The health agency of a city has determined that the number N (in thousands) of people contracting a particular disease t weeks after the outbreak of the disease is described by the logistic function $N(t) = \dfrac{45}{1 + 38e^{-0.095t}}$. How many people have contracted the disease (a) when the health agency makes its first recording, (b) 3 weeks after the outbreak, (c) 10 weeks after the outbreak, (d) 26 weeks after the outbreak, and (e) in the long run?

48. *Business: Advertising* When information is diffused through the mass media to a population of size A, the number N of people hearing the news by time t is given by the function $N(t) = A(1 - e^{-kt})$. A clothing company, through frequent daily announcements on television, is advertising, for a three-month period, a new line of clothes. If the total television audience for that three-month period is estimated to be 350,000 people, how many people can be expected to hear about the new clothing line (a) 10 days after the advertisements begin and (b) at the end of the three-month period? Assume $k = 0.059$.

D DESCRIBE YOUR THOUGHTS

49. Describe how the exponential function differs from the power function.

50. Describe how you can distinguish the graph of an exponential growth function from the graph of an exponential decay function.

51. Describe some important properties of the exponential function. (A simple list is not satisfactory.)

52. Use your graphing calculator to graph both $f(x) = 3^x$ and $f(x) = x^3$ on the interval $[0, \infty]$. Compare their behavior.

E REVIEW

53. (2.2) Find, if it exists, $\lim_{x \to 4} \frac{x^2 - x - 12}{x^2 + x - 20}$.

54. (2.2) Find, if it exists, $\lim_{x \to 2} \frac{x^2 - 5x + 6}{x^2 - 4x + 4}$.

55. (2.3) Find, if it exists, $\lim_{x \to \infty} \frac{4x^2 + 8x + 11}{3x^3 - 5x + 1}$.

56. (2.4) Discuss the continuity of

$$f(x) = \begin{cases} 3x - 1, & \text{if } x < 1 \\ 2, & \text{if } x = 1 \\ 5x^2 - 3, & \text{if } x > 1 \end{cases} \quad \text{at } x = 1.$$

57. (2.5) Find the average rate of change of $f(x) = 2x^2 - 5$ as x changes from 3 to 7.

58. (3.1) Use the definition of the derivative to find $f'(x)$ for $f(x) = 3x^2 - 5x - 8$.

59. (3.1) ***Business: Cost of Production*** Suppose that $C(x)$, in dollars, represents the cost of producing x sets of golf gloves (in thousands). Interpret $C'(45) = 0.03$.

60. (3.2) Find the derivative of $f(x) = \frac{3x - 4}{5x + 1}$.

61. (3.4) Find and interpret $f'(0)$ for $f(x) = (3x + 4)^4$.

62. (3.5) Find y' for $3x^2 + 5xy - 3y^2 = 7$.

63. (4.2) Sketch $f(x) = \frac{3x^2}{x^2 - 3}$.

64. (4.3) Find the value of t that minimizes $C(t) = 150t + \frac{4500}{t}$.

5.2 The Natural Logarithm Function

Introduction

Logarithms Are Exponents

The Reversal Effect of the Exponential and Natural Logarithm Functions on Each Other

Properties of the Natural Logarithm

The Graph of the Natural Logarithm Function

The Natural Logarithm Function and Technology

Introduction

In this section we introduce the natural logarithm function and examine its close relationship to the exponential function. A review of general logarithms is provided in Appendix 1.4, where you will find the general definition and properties of all logarithms. This section includes the information provided in Appendix 1.4, but specific to the base e, and will also extend your understanding of the logarithmic function.

Logarithms Are Exponents

In the exponential function, $y = A_0e^{kx}$, that we studied in Section 5.1, we were given a value for the independent variable x and computed the corresponding value of the dependent variable y. For example, when working with wounds to the skin, medical researchers have found that the function $A = A_0e^{-0.11t}$ relates the area A (in square centimeters) of unhealed skin to the number t of days since the skin received the wound. A_0 represents the initial area of the wound. Suppose that someone's skin receives a wound of 2 square centimeters in area. The number of square centimeters of unhealed skin remaining after 8 days is found by substituting 8 for t and 2 for A_0 in the equation $A = A_0e^{-0.11t}$.

$$\begin{aligned} A &= A_0e^{-0.11t} \\ &= 2e^{-0.11(8)} \\ &= 2e^{-0.88} \\ &\approx 2(0.414783) \\ &\approx 0.83 \end{aligned}$$

We conclude that after 8 days there remains approximately 0.83 square centimeters of unhealed skin.

An equally important question is posed by turning the previous question around and asking, "For a skin wound of 2 square centimeters, how many days will have to pass before only about 0.83 square centimeters of unhealed skin remain?" With the question posed this way, we need to find t in the equation $0.83 = 2e^{-0.11t}$.

The variable whose value we wish to determine is in the exponent. However, none of the standard algebraic operations can be used to isolate it and therefore solve for it. But mathematicians have developed a technique for finding the unknown value of an exponent and we can best understand it by considering the exponential function $a = e^t$.

Our problem is: if in the equation $a = e^t$, the value of a is known, what value of t, when used as an exponent on the base e, will produce a? The exponent t used in this way is given a special name, **the natural logarithm**.

The Natural Logarithm (First Form)

In the equation $a = e^t$, the number t when used as an exponent on the base e that produces the number a, is called the **natural logarithm** of a.

The words *natural logarithm* are abbreviated with the symbol ln. Translating from words to symbols, we get

$$t \text{ is the natural logarithm of } a$$

Even more symbolically,

$$t = \log_e x = \ln a$$

Thus, the definition of the natural logarithm can be restated as follows.

Definition of the Natural Logarithm (Final Form)

$$t = \ln a \quad \text{if and only if} \quad a = e^t$$

Argument of the Natural Logarithm

The number a is called the **argument of the natural logarithm**.

We make four important notes:

Note

1. Since t is a logarithm and an exponent, *logarithms are exponents*.
2. $t = \ln a$ if and only if $a = e^t$.
3. We know that e^t is always a positive number. Thus, in $a = e^t$, the number a is *always* positive. Since $a = e^t$ if and only if $t = \ln a$, the argument of the natural logarithm *must always* be a positive number.
4. The natural logarithm is a *function* since for each value of a, $t = \ln a$ produces exactly one value for t.

Illuminator Set A illustrates two important and frequently occurring natural logarithm values.

ILLUMINATOR SET A

1. Find $\ln e$.

   For the moment, we will assign $\ln e$ the value x. That is, we will let

   $$x = \ln e$$

   Now, we will use the notes (1) and (2) above. Note (2) describes how to convert a logarithmic statement to an exponential statement.

   $$(\text{argument}) = e^{(\text{exponent})}$$

   In our case,

   $$e = e^{(\text{exponent})}$$

   Note (1) states that logarithms are exponents. Thus, $x = \ln e$ indicates that x is the logarithm, so x is the exponent. Thus, $x = \ln e$ means that $e = e^x$, and this statement is true (by matching exponents on e) only if $x = 1$. Thus,

   $$\ln e = 1$$

2. Find ln 1.
For the moment, we will assign ln 1 the value x. That is, we will let

$$x = \ln 1$$

Next, we will convert to exponential form (using notes (1) and (2) above).
$x = \ln 1$ means that $1 = e^x$, and this statement is true only when $x = 0$ (since $e^0 = 1$). Thus,

$$\ln 1 = 0$$

Comment: These two examples are two of the basic properties of the natural logarithm.

The Reversal Effect of the Exponential and Natural Logarithm Functions on Each Other

An important and useful feature of the exponential and natural logarithmic functions is the effect they have on each other. This effect is illustrated in Figure 5.17 and Figure 5.18. Figure 5.17 indicates that $e^{\ln x} = x$ and Figure 5.18 indicates that $\ln e^x = x$. Composition of these functions acts to reverse the effect of one on the other. For example, if $x = 8$, then

1. Apply the natural logarithm function to 8, getting ln 8.
2. Now apply the exponential function to ln 8, getting $e^{\ln 8}$.
3. The result is 8, that is, $e^{\ln 8} = 8$.

Also,

1. Apply the exponential function to 8, getting e^8.
2. Now apply the natural logarithm function to e^8, getting $\ln e^8$.
3. The result is 8; that is, $\ln e^8 = 8$.

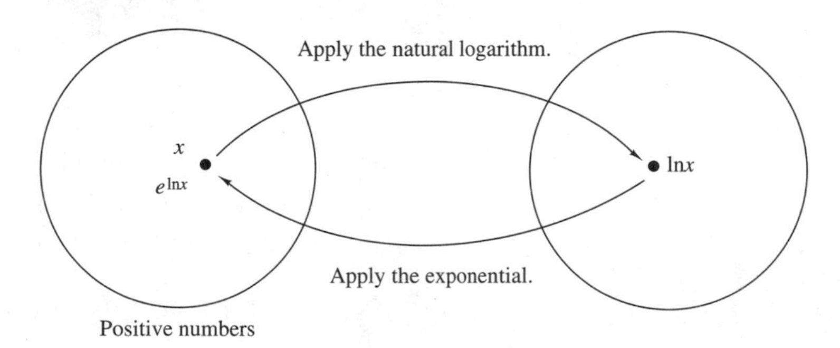

Figure 5.17

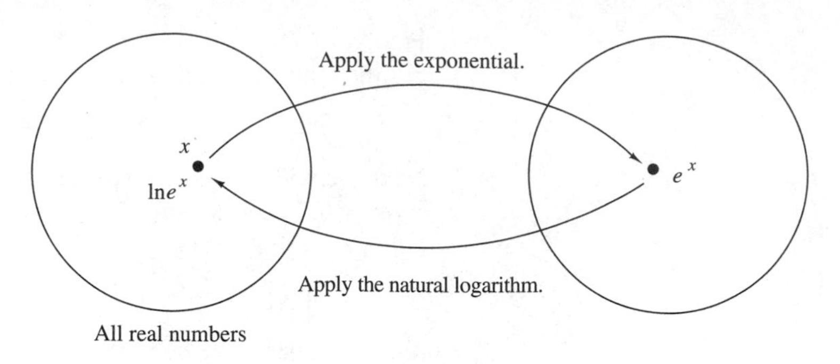

Figure 5.18

If you try these calculations on your calculator, you may start with 8, but you may get back to a number such as 7.9982753. The discrepancy can be explained by the fact that some calculators round when working with the natural logarithm and exponential functions.

To be sure that $e^{\ln x} = x$, note that $\ln x$ equals some number. For the moment, we will call it y; that is, we will let $y = \ln x$. This means, by note (2) on page 274, that $x = e^y$. Then

$$e^{\ln x} = e^y = x$$

A similar argument is used to show that $\ln e^x = x$, and you will be asked to show exactly this in the exercise set.

Properties of the Natural Logarithm

The function $y = \ln x$ possesses several properties that are very useful when solving equations that involve natural logarithms. We list six of them below.

Properties of the Natural Logarithm

1. $\ln e = 1$ (from Illuminator Set A)
2. $\ln 1 = 0$ (from Illuminator Set A)
3. $\ln e^x = x$ (from Section 5.1)
4. $\ln(xy) = \ln x + \ln y$ (Multiplication becomes addition.)
5. $\ln\left(\dfrac{x}{y}\right) = \ln x - \ln y$ (Division becomes subtraction.)
6. $\ln x^r = r \cdot \ln x$ (Exponentiation becomes multiplication.)

EXAMPLE SET A

Use the properties of the natural logarithm to expand each logarithmic expression.

1. $\ln(3x)$

$$\begin{aligned} \ln(3x) &= \ln 3 + \ln x && \text{by property 4} \\ &\approx 1.0986 + \ln x && \text{using a calculator} \end{aligned}$$

2. $\ln\left(\dfrac{8x}{x+4}\right)$

$$\begin{aligned} \ln\left(\frac{8x}{x+4}\right) &= \ln(8x) - \ln(x+4) && \text{by property 5} \\ &= \ln 8 + \ln x - \ln(x+4) && \text{by property 4} \\ &\approx 2.0794 + \ln x - \ln(x+4) && \text{using a calculator} \end{aligned}$$

3. $\ln x^6 = 6 \ln x$ by property 6

4. $12{,}000 \ln(xy^5)$

$$\begin{aligned} 12{,}000 \ln(xy^5) &= 12{,}000 \left[\ln x + \ln y^5\right] && \text{by property 4} \\ &= 12{,}000 \left[\ln x + 5 \ln y\right] && \text{by property 6} \end{aligned}$$

Notice that $\ln(xy^5) \neq 5 \ln(xy)$ since $5 \ln(xy) = \ln\left[(xy)^5\right]$. The exponent 5 in the example is not associated with the *entire* argument of the natural logarithm, and therefore property 6 does not apply.

EXAMPLE SET B

Use the properties of the natural logarithm to write each logarithmic expression as an expression with a single logarithm.

1. $\ln x + 6 \ln y$

$$\begin{aligned} \ln x + 6 \ln y &= \ln x + \ln y^6 && \text{by property 6} \\ &= \ln(xy^6) && \text{by property 4} \end{aligned}$$

2. $\ln a - \ln b - \ln c$

$$\begin{aligned} \ln a - \ln b - \ln c &= (\ln a - \ln b) - \ln c \\ &= \ln \frac{a}{b} - \ln c && \text{by property 5} \\ &= \ln \frac{a}{bc} && \text{by property 5} \end{aligned}$$

EXAMPLE SET C

To four decimal places, approximate the solution to $\ln(x+3)+\ln(x-6)=8$.

Solution:
Equations involving logarithms are commonly solved by combining all logarithms into one single logarithm (using the properties of the natural logarithm), then taking the exponential of each side to reverse the effect of the logarithm.

$$\ln(x+3)+\ln(x-6)=8$$

$$\ln\left[(x+3)(x-6)\right]=8 \qquad \text{by property 4}$$

$$e^{\ln[(x+3)(x-6)]}=e^8 \qquad \text{taking exponentials}$$

$$(x+3)(x-6)=e^8 \qquad \text{by the reversal effect}$$

$$x^2-3x-18\approx 2980.9580$$

$$x^2-3x-2998.9580\approx 0$$

Now, using the quadratic formula, $x=\dfrac{-b\pm\sqrt{b^2-4ac}}{2a}$, with $a=1$, $b=-3$, and $c=-2998.9580$, we get, with the aid of a calculator,

$$x\approx 56.2833 \text{ or } -53.2833$$

We notice, however, that upon substitution into the original equation, -53.2833 produces a negative value for the argument of a natural logarithm. Since arguments of natural logarithms *must always* be positive numbers (note (3) on page 274), we must discard -53.2833. Substitution of 56.2833 into the original equation is acceptable and, in fact, produces 8.00000. The sixth decimal place is a nonzero value and varies with calculators; we credit the discrepancy to calculator rounding.

We conclude that $x\approx 56.2833$ to four decimal places.

EXAMPLE SET D

To four decimal places, approximate the solution to the equation $e^{2x+5}=12$.

Solution:
Exponential equations are commonly solved by taking the natural logarithm of each side. This serves to reverse the effect of the exponential and allows us to isolate the variable and thus solve for it.

$$e^{2x+5}=12$$

$$\ln e^{2x+5}=\ln 12$$

$$2x+5=\ln 12 \qquad \text{by property 3}$$

$$2x=\ln 12-5$$

$$x = \frac{\ln 12 - 5}{2}$$

$$x \approx -1.2575$$

to four decimal places.

EXAMPLE SET E

The radioactive isotope uranium-235, U^{235}, decays according to the formula $A = A_0e^{-0.06t}$, where A_0 is the amount of U^{235} initially present, and A is the amount present after t million years. How many years will have to pass before only 20% of a particular quantity of U^{235} remains? ($0.20A_0$ represents 20% of A_0.)

Solution:
We need to find the value of t when the amount of U^{235} present is 20% of the original amount A_0. That is, we need to find the value of t when $A = 0.20A_0$.

$$A = A_0e^{-0.06t}$$

$$0.20A_0 = A_0e^{-0.06t} \qquad \text{Divide each side by } A_0.$$

$$\frac{0.20A_0}{A_0} = e^{-0.06t}$$

$$0.20 = e^{-0.06t} \qquad \text{Take the natural log of each side.}$$

$$\ln(0.20) = \ln e^{-0.06t}$$

$$\ln(0.20) = -0.06t \qquad \text{Divide each side by } -0.06.$$

$$\frac{\ln(0.20)}{-0.06} = t$$

$$t \approx 26.8$$

We conclude that after approximately 26.8 million years, a quantity of U^{235} will have decayed so that only 20% remains.

EXAMPLE SET F

A piece of equipment depreciates according to the function $A = A_0e^{-0.32t}$. Sandeb Publishing Company has purchased a piece of capital equipment for \$635,000. It will replace that piece of equipment when its value is 10% of its purchase price. How many years will the company own the equipment?

Solution:
Since the initial value of the equipment is \$635,000, we substitute 635,000 for A_0 in the exponential function $A = A_0e^{-0.32t}$. We can also substitute 63,500

for A since that value represents 10% of the original value. These substitutions produce the function

$$63{,}500 = 635{,}000e^{-0.32t}$$

We will solve this equation for t using the method discussed in Example Set E.

$$63{,}500 = 635{,}000e^{-0.32t}$$

$$\frac{63{,}500}{635{,}000} = e^{-0.32t}$$

$$\ln(0.1) = \ln e^{-0.32t}$$

$$-2.302585093 \approx -0.32t$$

$$\frac{-2.302585093}{-0.32} \approx t$$

$$t \approx 7.2$$

Thus, the company will own the machine for approximately 7.2 years.

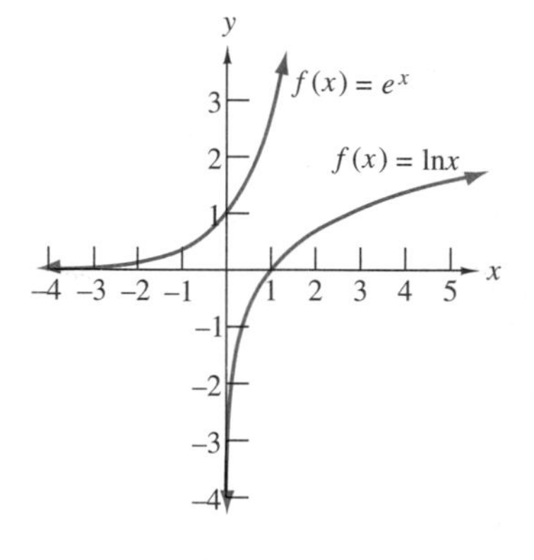

Figure 5.19

The Graph of the Natural Logarithm Function

We have just examined the reversal effect that the exponential and natural logarithm functions have on each other. To generalize this idea, imagine that the points (a_1, b_1), $(a_2, b_2), \ldots, (a_n, b_n)$ are points on the exponential curve $f(x) = e^x$. Reversing the order of the components of each ordered pair results in the set of points (b_1, a_1), $(b_2, a_2), \ldots, (b_n, a_n)$. The graph of these points produces the natural logarithm curve. The exponential function $f(x) = e^x$ and the natural logarithm function $f(x) = \ln x$ are illustrated in Figure 5.19.

Mathematically, the exponential and natural logarithm functions are said to be **inverses** of each other.

Notice the following important features of $f(x) = \ln x$.

1. As x increases, $f(x) = e^x$ increases and does so at an increasing rate.
2. As x increases, $f(x) = \ln x$ increases but does so at a decreasing rate.
3. For $0 < x < 1$, $\ln x < 0$; that is, for input values of x strictly between 0 and 1, the output values $\ln x$ are negative.
4. For $x > 1$, $\ln x > 0$; that is, for input values of x strictly greater than 1, the output values $\ln x$ are positive.
5. $\ln(1) = 0$
6. The domain of the natural logarithm function is the set of all positive numbers. That is, only positive numbers are allowed as input values. The graph shows that $\ln x$ is not defined for negative numbers or zero.

The Natural Logarithm Function and Technology

Using Your Calculator You can use your graphing calculator to help you visualize the inverse relationship between the exponential function $f(x) = e^x$ and the natural logarithm function $f(x) = \ln x$. Begin by setting the mode to parametric (Param). To enter $f(x) = e^x$, enter T for X_{1T} and e^T for Y_{1T} to specify the coordinates of the points to be plotted. Graph this function in the window $-7 \le x \le 25$, $-7 \le y \le 25$. The inverse of a function is obtained by interchanging the components in the ordered pairs. Do this by entering e^T for X_{2T} and T for Y_{2T}. In the Window, set Tmin$= 0$, Tmax$= 6.28$, and Tstep$= 1$. Use the Trace cursor to trace along the curve $f(x) = e^x$. Observe the coordinates of the points. Change curves and observe how the coordinates interchange.

The graphs of functions that are inverses of each other should be symmetric about the line $y = x$ (a 45° line passing through the origin). If your curves do not display this symmetry (because of the nonsquare viewing window), press Zoom Square to square the viewing window.

Calculator Exercises

1. Psychologists and educators have long been interested in *forgetting*. Working together on a particular course, a psychologist and an educator administer an exam to a group of students. At weekly intervals, they administer the exam again to the same group of people. From their data, they produce the *forgetting* function

$$P(t) = 82 - 16\ln(0.8t + 0.6)$$

where $P(t)$ represents the average exam score, in percent, t weeks after the initial exam. Construct the graph of this function and use it to describe its behavior over a 52-week period. Use this function to estimate the average score 10 weeks after the exam was first administered.

2. The population size of a city and average speed at which people walk in that city are related by the function

$$S(p) = 0.39\ln(p) + 0.05$$

where $S(p)$ represents the walking speed in feet per second, and p represents the approximate population of the city. Construct the graph of this function and use it to describe its behavior for cities with populations of up to 5,000,000 people. Use this function to estimate the average walking speed in Madison, Wisconsin, for which the 1995 population was about 195,200.

3. The demand function for a particular product is given by

$$p(q) = \frac{375}{\ln(1.2q + 0.89)}$$

where p represents the price in dollars when q units of the product are in the market. Construct the graph of this function in the window $0 \le q \le 100$, $0 \le p \le 500$ and use it to describe its behavior. Experiment with the constants in the function to determine how to change it so its graph has essentially the same shape, but decreases less rapidly.

EXERCISE SET 5.2

A UNDERSTANDING THE CONCEPTS

In each of Exercises 1–8, an exponential function is given and a question is posed that makes direct use of that exponential relationship. Pose a new question in such a way that a logarithmic relationship must be used. You do not need to make any computations, and answers may vary. (Refer to paragraph two at the beginning of this section.)

1. ***Physics: Atmospheric Pressure*** The function $P(h) = 1.47e^{-0.21h}$ relates the atmospheric pressure P (in pounds per square inch) to the altitude h (in miles above sea level). Find the atmospheric pressure 3.5 miles above sea level.
2. ***Business: Investment*** If \$4000 is invested at 9% interest compounded continuously, the function $A(t) = 4000e^{0.09t}$ relates the amount A (in dollars) accumulated after t years. Find how much money is accumulated in such an account after 6.8 years.
3. ***Economics: City Population*** The function $P(t) = 36e^{0.015t}$ relates the population P (in millions) of a growing city to the time t (in years from now). Find the population of the city 6 years from now.
4. ***Business: Value of an Investment*** The function $PV(t) = 85{,}000e^{t^{2/3}}$ relates the present value (in dollars) of an investment of \$85,000 to the time t (in years from now) the money has been invested. Find the present value of an \$85,000 investment 23 years from now.
5. ***Business: Task Completion*** The function $N(t) = 125 - 125e^{-0.08t}$ relates the number N of tasks a person can complete in an 8-hour working day to the number of days t since the person began performing the tasks. How many tasks can a person who has been working at the job for 21 days be expected to complete in an 8-hour working day?
6. ***Psychology: Memory*** Psychologists believe that the function $f(x) = A(1 - e^{-kt})$ relates the number of symbols f that a person can memorize in a particular time period t (in minutes). How many symbols can a person be expected to memorize in 15 minutes?
7. ***Medicine: Epidemics*** Health agencies are able to predict the percentage $P(t)$ of the population having a particular disease t days after the outbreak of an epidemic using the function $P = Ce^{rt}$. What percentage of the population has the disease 15 days after the outbreak of the epidemic?
8. ***Economics: Price Stability*** Economists interested in price stability have established that the function $p(t) = (p_0 - p_e)e^{k(a-A)t} + p_e$ relates the current price of a commodity to the time t in months since it was initially priced at p_0 dollars. p_e represents the equilibrium price of the commodity, and the numbers a and A are constants that depend on the supply and demand of the item. What is the expected price of the item 14 months after its initial pricing?

For Exercises 9–11, answer the posed question.

9. ***Economics: Market Value of Property*** The market value $MV(t)$ of a piece of city property is related to the time t (in years from now) by the function $MV(t) = 240{,}000e^{0.2t^{2/5}}$. What is the expected market value of the piece of property 15 years from now?
10. ***Physics: Newton's Law of Cooling*** Newton's Law of Cooling relates the temperature T of an object taken from an environment at a temperature of 300°F to an environment of 70°F to the time t (in minutes) that the object has been in the cooler environment by the function $T = 70 + 230e^{-0.19018t}$. What is the temperature of such an object 20 minutes after it is placed into the 70°F environment?
11. ***Marketing: Advertising Campaigns*** The marketing department of a company that produces animal-shaped fruit snacks has determined that t weeks after an advertising campaign ends, the number N of boxes of such snacks it sells each month decreases according to the function $N(t) = 14{,}000e^{-0.23t}$. How many boxes of animal-shaped fruit snacks can the company expect to sell 7 weeks after the close of the advertising campaign?

B SKILL ACQUISITION

For Exercises 12–17, convert each exponential function to the corresponding natural logarithm function.

12. $m = e^{3n}$
13. $a = e^{6x}$
14. $y = e^{3x-5}$
15. $x = e^{-3k+8}$
16. $4 = e^{3x+1}$
17. $8.262 = e^{4x-10}$

For Exercises 18–23, convert each natural logarithm function to the corresponding exponential function.

18. $6 = \ln(4x)$
19. $k = \ln(15)$
20. $x + 3 = \ln(y + 1)$
21. $2y - 4 = \ln(3x + 11)$
22. $-7.18 = \ln(3x + 4)$
23. $-0.002 = \ln(x - 1)$

For Exercises 24–29, solve each exponential equation. If the solution is not an integer, round it to four decimal places.

24. $e^{3x+7} = 12$

25. $e^{-5x+2} = 70$

26. $31e^{8x-6} = 178$

27. $-62e^{3x-10} = -1085$

28. $e^{-0.03t} = 0.4724$

29. $185e^{-1.6t} = 1.5233$

For Exercises 30–36, solve each natural logarithm equation. If the solution is not an integer, round it to four decimal places.

30. $\ln(x-4) + \ln(x+1) = 1.8$

31. $\ln x + \ln(x+6) = -2.4$

32. $\ln(x-3) - \ln(x-4) = 0.6931$

33. $\ln(2x+4) - \ln(x+3) = 0$

34. $\ln(5x+8) - \ln(3x+7) = 0.1335$

35. $\ln(x-12) + \ln(x+15) - \ln(x+4) = 0.5108$

36. $\ln(2x+1) + \ln(x-3) - \ln(2x+5) = 1.5782$

C APPLYING THE CONCEPTS

37. *Physics: Atmospheric Pressure* The function $P(h) = 14.7e^{-0.21h}$ relates the atmospheric pressure P (in pounds per square inch) to the altitude h (in miles above sea level). Find the altitude above sea level at which the atmospheric pressure measures 6.5 pounds per square inch.

38. *Business: Investment* If \$18,500 is invested at 9% interest compounded continuously, the function $A = 18{,}500e^{0.09t}$ relates the amount A (in dollars) accumulated after t years. Approximately how long will it take to accumulate \$27,051.20?

39. *Economics: City Population* The function $P(t) = 36e^{0.015t}$ relates the population P (in thousands) of a growing city to the time t (in years from the present). Approximately how many years from now will the population of this city be 119,175? (Remember, P is in thousands.)

40. *Marketing: Advertising* The marketing department of a company that produces and sells inexpensive but durable dinnerware has determined that t weeks after an advertising campaign ends, the number N of dinnerware sets it sells throughout the country each month decreases according to the function $N = 14{,}800e^{-0.23t}$.

a. How many sets of dinnerware does the company expect to sell immediately at the close of an advertising campaign?

b. Approximately how many weeks after the end of an advertising campaign will the company be selling 4,687 dinnerware sets?

41. *Manufacturing: Assembly Time* Industrial psychologists have determined that the function $N(t) = 125 - 125e^{-0.08t}$ relates the number N of assembly tasks a person can complete each 8-hour working day to the number of days t since the person began doing the tasks. Approximately how many days after starting the particular task will the worker be completing 114 tasks in an 8-hour working day?

42. *Government: Growth of Bureaucracy* The function $N(t) = 14e^{0.019t}$ describes the growth of a county's bureaucracy. N is the size (in thousands) of the bureaucracy, and t is the time (in years) from 1980. Approximately how many years from 1980 will the size of this county's bureaucracy be 20,500? (Remember, P is measured in thousands.)

43. *Medicine: Drug Concentration* When a drug is intravenously administered to a patient on a continuous basis, the amount A (in milligrams) of the drug in the patient's bloodstream t minutes after injection of the drug is given by $A(t) = \frac{1}{0.52}(50 - 50e^{-0.12t})$. Approximately how much time will have to pass after injection of the drug so that 55 milligrams of the drug will be in the patient's bloodstream?

44. *Ecology: Light Intensity in Water* For bodies of water that are relatively clear, light intensity is reduced according to the function $I = I_0e^{-kd}$, where I is the intensity of the light d feet below the surface of the water. One of the clearest saltwater bodies of water is the Sargasso Sea off the West Indies ($k = 0.00942$). At what depth is the light intensity reduced to 60% of that at the surface?

45. *Science: Radioactive Decay* The radioactive isotope Strontium-90 decays according to the formula $A = A_0e^{-0.0248t}$, where A_0 is the amount of Strontium-90 initially present, and A is the amount present after t years. Approximately how many years will have to pass before only 50% of a particular quantity of Strontium-90 remains?

46. *Medicine: Effect of Valium* When in the bloodstream, the drug Valium decays according to the function $A = A_0e^{-0.0578t}$. Approximately how many minutes after taking a prescribed dose of 10 milligrams of Valium will there be less than 2 milligrams in the bloodstream?

47. *Economics: Doubling Time* The amount A of dollars accumulated through an investment of A_0 dollars at $r\%$ interest compounded continuously can be determined from the function $A = A_0e^{rt}$. The *doubling time* of an investment is the time required for an investment to double in value. Find the doubling time for an investment made at (a) 6% interest, (b) 8% interest, and (c) 10% interest.

48. *Medicine: Healing of Wounded Skin* When human skin receives a wound of surface area A, the exponential function $A = A_0e^{-0.11t}$ relates the amount A (in square centimeters) of unhealed skin to the number of days t since the wound was received. A_0 is the initial surface area of the wound. Suppose a person suffers a skin wound measuring 1.24 cm^2. How many days must pass before only 0.5 cm^2 of unhealed skin remains? (This is similar to the question with which we began this section.)

49. *Economics: Mutual Funds* A mutual fund expects annual returns to be modeled by the function $A = A_0e^{0.08t}$. How much would an investor have to invest now to have \$30,000 available in 10 years?

50. *Business: Market Share* A company believes its sales are modeled by the function $A = A_0e^{0.22t}$. The total value of the market is 100 million dollars. The company currently has 2% of the market. How long before the company can expect to have 20% of the market?

51. *Economics: Credit* The function $A = A_0e^{0.195t}$ models the amount an individual owes a credit company t years after some initial debt. If a person currently owes \$2300 and makes no payments, how much time will pass before that person owes \$10,000?

D DESCRIBE YOUR THOUGHTS

52. Describe several important properties of the natural logarithm function.

53. By observing their graphs, compare and contrast the exponential and natural logarithm functions.

54. Describe the strategy you would use to solve problems such as Exercises 24 and 30.

E REVIEW

55. **(2.1)** Determine from the graph below the $\lim_{x\to 5} f(x)$, if it exists. If the limit does not exist, so state.

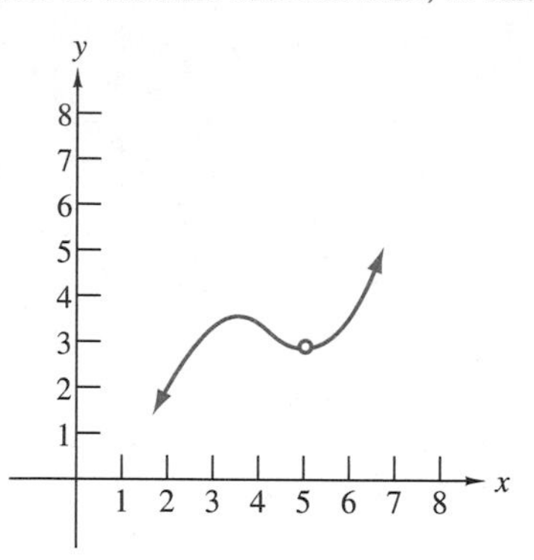

56. **(2.1)** Sketch the graph of a function having the following properties: $f(-1) = 0$, $f(1) = 2$, $f(2) = 0$, and $f(4)$ is not defined, but $\lim_{x\to 4} f(x) = 5$.

57. **(2.2)** Find, if it exists, $\lim_{x\to 5} \dfrac{x^2 + x - 30}{x^2 - 9x + 20}$.

58. **(2.3)** Find, if it exists, $\lim_{x\to\infty} \dfrac{x^3 + 2x - 8}{x^2 + 5x - 4}$.

59. **(3.1)** Use the definition of the derivative to find and interpret $f'(6)$ for $f(x) = x^2 - 2x + 5$.

60. **(3.1)** *Meteorology: Wind Velocity* The function $V(h)$ relates the velocity V of the wind (in centimeters per second) to the height h (in centimeters) above the ground. Interpret $V'(14) = 0.06$.

61. **(3.2)** Find $f'(x)$ for $f(x) = \dfrac{8x - 4}{2x + 6}$.

62. **(3.4)** Find $f'(x)$ for $f(x) = (3x + 1)^4(2x - 7)^3$.

63. **(3.5)** Find y' for $x^3 + 2x^2y^2 + 3y = 8$.

64. **(4.1)** Classify a curve that is increasing at a decreasing rate as concave upward or concave downward.

65. **(4.2)** Sketch the graph of a function that decreases at a decreasing rate when x is less than 3, then decreases but at an increasing rate when x is between 3 and 10, and then decreases at a decreasing rate when x is greater than 10.

66. **(5.1)** Find $f(16)$ for $f(x) = 250e^{-0.05x}$.

5.3 Differentiating the Natural Logarithm Function

Introduction
Differentiation of the Natural Logarithm Function
Differentiation and Technology

Introduction

In many fields of study we are often interested in the rate at which the output variable y changes as the input variable x changes. Of course, when we think *change*, we think *differentiation* (since differentiation is the language of change). In each of the functions introduced in the preceding sections—$y = e^x$ and $y = \ln x$—the output variable y is related to the input variable x by a nonalgebraic function, the exponential and natural logarithm, respectively. Since our current collection of rules of differentiation apply only to algebraic functions, we will need new rules to find $\frac{d}{dx}[e^x]$ and $\frac{d}{dx}[\ln x]$ and, more generally, $\frac{d}{dx}[e^{g(x)}]$ and $\frac{d}{dx}[\ln[g(x)]]$. Since this is an *applied calculus* text, we will go directly to the formulas and examples of their application. In this section we will consider the differentiation of the natural logarithm function.

Differentiation of the Natural Logarithm Function

The Derivative of the Natural Logarithm Function

$$\frac{d}{dx}[\ln x] = \frac{1}{x}$$

A rigorous proof of this rule is beyond the scope of this book; therefore, it will not be presented.

EXAMPLE SET A

Find $f'(x)$ for $f(x) = 7x^3 \ln x$.

Looking at $f(x)$ globally, we see a product. Thus, we will differentiate this function using the product rule.

$$f'(x) = \underbrace{21x^2}_{(\textit{first})'} \cdot \underbrace{\ln x}_{(\textit{second})} + \underbrace{7x^3}_{(\textit{first})} \cdot \underbrace{\frac{1}{x}}_{(\textit{second})'}$$

$$= 21x^2 \ln x + 7x^2$$

$$= 7x^2(3 \ln x + 1)$$

To differentiate $y = \ln[u]$, where $u = g(x)$, such as $y = \ln(3x^2 + 2x - 5)$, we use the chain rule

$$\frac{dy}{dx} = \frac{dy}{du} \cdot \frac{du}{dx}$$

$$\frac{dy}{dx} = \frac{1}{u} \cdot \frac{du}{dx}$$

$$\frac{d}{dx}[\ln[g(x)]] = \frac{1}{g(x)} \cdot g'(x)$$

We have developed the differentiation rule for the generalized natural logarithm function.

The Derivative of the Generalized Natural Logarithm Function

$$\frac{d}{dx}[\ln[g(x)]] = \frac{1}{g(x)} \cdot g'(x)$$

This differentiation rule states that the derivative of the generalized natural logarithm function $f(x) = \ln[g(x)]$ is

(1 over the argument) *times* (the derivative of the argument)

That is,

$$\frac{1}{\text{argument}} \quad \textit{times} \quad (\text{argument})'$$

EXAMPLE SET B

1. Find $f'(x)$ for $f(x) = \ln(5x^2 + 2x - 7)$.

Solution:

$$f'(x) = \overbrace{\frac{1}{5x^2 + 2x - 7}}^{\text{1 over the argument}} \cdot \overbrace{\frac{d}{dx}[5x^2 + 2x - 7]}^{(\text{argument})'}$$

$$= \frac{1}{5x^2 + 2x - 7} \cdot (10x + 2)$$

$$= \frac{10x + 2}{5x^2 + 2x - 7}$$

2. Find $f'(x)$ for $f(x) = \ln(x^2 - 4)^3$.
We will illustrate two ways of differentiating this function, the first using the logarithm property 6, $\ln x^r = r \cdot \ln x$, and the second working from its global form of a power.

CAUTION

First, we will use logarithm property 6 to rewrite $f(x) = \ln(x^2 - 4)^3$ as $f(x) = 3\ln(x^2 - 4)$. (Recall that $\ln(A + B) \neq \ln A + \ln B$; do not mistakenly write $\ln(x^2 - 4)$ as $\ln x^2 - \ln 4$.) Then

$$f'(x) = 3\frac{d}{dx}[\ln(x^2 - 4)]$$

$$= 3 \cdot \frac{1}{(x^2 - 4)} \cdot 2x$$

$$= \frac{6x}{x^2 - 4}$$

Now we will differentiate this function using an approach that does not use the property of logarithms. Globally, this function is a natural logarithm function, and locally, the argument of the logarithm is a power function. Thus, we will be sure to differentiate the argument using the generalized power rule.

$$f'(x) = \frac{1}{(x^2 - 4)^3} \cdot \frac{d}{dx}[(x^2 - 4)^3]$$

$$= \frac{1}{(x^2 - 4)^3} \cdot 3 \cdot (x^2 - 4)^2 \cdot 2x$$

$$= \frac{6x(x^2 - 4)^2}{(x^2 - 4)^3}$$

$$= \frac{6x}{x^2 - 4}$$

Notice that the two methods produce the same results. In this case, the use of the logarithm properties may have saved us from some algebraic manipulations.

3. Find $f'(x)$ for $f(x) = [\ln(3x)]^4$.

Solution:

CAUTION

Looking at this function globally, we see a power. More locally, it is a natural logarithm function. (Recall that $[\ln a]^n \neq \ln a^n$). We will begin differentiating this function using the general power rule.

$$f'(x) = 4 \cdot [\ln(3x)]^3 \cdot \frac{d}{dx}[\ln(3x)]$$

$$= 4 \cdot [\ln(3x)]^3 \cdot \frac{1}{3x} \cdot \frac{d}{dx}[3x]$$

$$= 4 \cdot [\ln(3x)]^3 \cdot \frac{1}{3x} \cdot 3$$

$$= \frac{4[\ln(3x)]^3}{x}$$

4. Find $f'(x)$ for $f(x) = \dfrac{6}{\ln(8x^2)}$.

Solution:
Looking at this function globally, we see a quotient. However, since the numerator is a constant, we will write the function in the more convenient form $f(x) = 6[\ln(8x^2)]^{-1}$. Now we will use the general power rule.

$$f'(x) = (-1)(6)[\ln(8x^2)]^{-2} \cdot \frac{d}{dx}[\ln(8x^2)]$$

$$= -6[\ln(8x^2)]^{-2} \cdot \frac{1}{8x^2} \cdot \frac{d}{dx}[8x^2]$$

$$= -6[\ln(8x^2)]^{-2} \cdot \frac{1}{8x^2} \cdot 16x$$

$$= \frac{-12}{x[\ln(8x^2)]^2}$$

5. Find y' for $3x^4y^2 + \ln(xy^2) = 6$.

Solution:
We can use the logarithm properties $\ln(xy) = \ln x + \ln y$ and $\ln x^r = r \cdot \ln x$ to simplify the term $\ln(xy^2)$ to $\ln x + 2\ln y$. This helps us avoid the product rule for this part of the expression. Thus, the original equation becomes

$$3x^4y^2 + \ln x + 2\ln y = 6$$

Since this function defines y implicitly as a function of x, we will use implicit differentiation.

$$12x^3 \cdot y^2 + 3x^4 \cdot 2yy' + \frac{1}{x} + 2 \cdot \frac{1}{y} \cdot y' = 0$$

$$12x^3y^2 + 6x^4yy' + \frac{1}{x} + \frac{2}{y}y' = 0$$

Eliminate the fractions by multiplying each side by the lowest common denominator, xy.

$$12x^4y^3 + 6x^5y^2y' + y + 2xy' = 0$$

Now, isolate y'.

$$6x^5y^2y' + 2xy' = -12x^4y^2 - y$$

$$y'(6x^5y^2 + 2x) = -y(12x^4y + 1)$$

$$y' = \frac{-y(12x^4y+1)}{6x^5y^2+2x}$$

EXAMPLE SET C

Doubling Time

When P dollars are invested for t years at interest rate r compounded n times annually, the amount of money accumulated in the account is given by the formula $A = P\left(1+\frac{r}{n}\right)^{nt}$. The **doubling time** of an investment is the time required for the investment to double in value. The doubling time for an investment made at the interest rate r, compounded quarterly, is given by $t = \frac{\ln 2}{4\ln\left(1+\frac{1}{4}r\right)}$.

1. Find $\frac{dt}{dr}$; that is, find the rate at which the doubling time changes as the interest rate changes, and
2. Find and interpret the rate at which the doubling time changes as the interest rate changes when the interest rate is 8%.

Solution:

1. Looking at this function globally, we see a quotient. However, since the numerator, ln 2, is a constant, we will write the function in the more convenient form

$$t = \frac{\ln 2}{4}\left[\ln\left(1+\frac{1}{4}r\right)\right]^{-1}$$

Now, we will differentiate each side with respect to r.

$$\begin{aligned}
\frac{dt}{dr} &= -\frac{\ln 2}{4}\left[\ln\left(1+\frac{1}{4}r\right)\right]^{-2}\cdot\frac{1}{1+\frac{1}{4}r}\cdot\frac{1}{4}\\
&= \frac{-\ln 2}{4\left[\ln\left(1+\frac{1}{4}r\right)\right]^2}\cdot\frac{1}{4+r}\\
&= \frac{-\ln 2}{4(4+r)\left[\ln\left(1+\frac{1}{4}r\right)\right]^2}
\end{aligned}$$

2. To find the rate of change when the interest rate is 8%, substitute 0.08 for r into $\frac{dt}{dr}$ and compute.

$$\begin{aligned}
\left.\frac{dt}{dr}\right|_{r=0.08} &= \frac{-\ln 2}{4(4+0.08)\left[\ln(1+\frac{1}{4}(0.08))\right]^2}\\
&\approx -108.31
\end{aligned}$$

or

$$\approx \frac{-108.31 \leftarrow \text{approximate change in doubling time}}{1 \leftarrow 100\% \leftarrow \text{change in interest rate}}$$

$$\approx \frac{-1.0831 \leftarrow \text{approximate change in doubling time}}{1 \leftarrow 1\% \leftarrow \text{change in interest rate}}$$

Interpretation: If the interest rate increases by 1%, from 8% to 9%, the doubling time of an investment will decrease by just slightly more than 1 year.

Differentiation and Technology

Using Your Calculator Your graphing calculator can approximate the derivative of a function that involves an exponential or a natural logarithm function. The following entries illustrate a derivative approximation for the function of Example Set C, $t(r) = \dfrac{\ln 2}{4\ln(1 + \frac{1}{4}r)}$ at $r = 0.08$.

```
Y1 = ln 2/(4ln (1 + (1/4)X))
Y5 = nDeriv(Y1,X,X,.001)
```

In the computation window, enter $Y5(.08)$ and compute, or store 0.08 to X, then access $Y5$ and compute. Your result should be about -108.32.

You might find it interesting to graph the function $Y1$ on $0 \le X \le 0.15$ and $0 \le Y \le 50$, and the function $Y5$ on $0 \le X \le 0.15$ and $-1000 \le Y \le 0$. Is this the curve you would expect for $Y1$? Describe its behavior. What is the approximate doubling time for an investment at 6% (0.06)?

Calculator Exercises

1. Construct the graph of the function $f(x) = (\ln x)^4$. Explain why, although this function involves $\ln x$, it does not exhibit the same type of graph as $f(x) = \ln x$. For example, $f(x) = \ln x$ is negative for $0 < x < 1$, but $f(x) = (\ln x)^4$ is not.
2. On the same coordinate system, construct the graphs of $f(x) = \ln x$, $f(x) = \ln x \cdot e^{-0.1x}$, and $f(x) = \ln x \cdot e^{-0.3x}$. Make a statement about the effect on $f(x) = \ln x$ when it is multiplied by the natural exponential decay function.
3. A politician running for county office claims that since 1980, the number of people working for the county government has been growing outrageously fast. The incumbent says the claim is inaccurate and produces research that shows the number, N, of people working in a county's government t years from 1980 is approximated by the function

$$N(t) = 2.7\ln(82x^{3/5} + 12)$$

where N is in thousands of people. Graph this function and use it to describe its behavior. Who would you say is correct in his claim, the incumbent or the challenger? Explain your decision. Your explanation might include some examples demonstrating the rate of growth.

4. In researching a country's industries, a financial analyst concluded that the total value of all goods produced by a particular industry in the country is approximated by the function

$$V(t) = 925\left[\frac{\ln(0.075t + 0.085)}{0.75t + 0.85}\right]$$

where V is in millions of dollars and t is the number of years since 1945. Construct the graph of this function and use it to describe the function's behavior until the year 2000. Your description should include the approximate maximum value of the industry's production and the approximate year in which it occurred.

EXERCISE SET 5.3

A UNDERSTANDING THE CONCEPTS

In Exercises 1–10, a function is given along with a claimed derivative of the function. State if the indicated derivative is *correct* or *incorrect*.

1. $f(x) = \ln(x^3)$
   $f'(x) = \dfrac{1}{x^3}$
2. $f(x) = \ln(4x^2 + 7x)$
   $f'(x) = \dfrac{1}{4x^2 + 7x}$
3. $f(x) = \ln(3x^2 + 5x - 4)$
   $f'(x) = \dfrac{1}{6x + 5}$
4. $f(x) = \ln(9x - 1)$
   $f'(x) = \dfrac{1}{9x - 1} \cdot \dfrac{1}{9}$
5. $f(x) = 3\ln\left(x^6\right)$
   $f'(x) = 18\ln(x^5)$
6. $f(x) = 4x^2 \ln 6$
   $f'(x) = 8x \ln 6$
7. $f(x) = 10x \ln 8$
   $f'(x) = 10$
8. $f(x) = 5x\ln(x^3)$
   $f'(x) = \dfrac{1}{5x \cdot x^3} = \dfrac{1}{5x^4}$
9. $f(x) = 12\ln 2$
   $f'(x) = 0$
10. $f(x) = \ln\left(e^{5x}\right)$
    $f'(x) = e^{5x} \cdot \dfrac{1}{e^5 x} = \dfrac{e^{5x}}{5x}$

B SKILL ACQUISITION

For Exercises 11–27, find the derivative of each function.

11. $f(x) = \ln(2x - 7)$
12. $f(x) = \ln(5x^2 + 3x)$
13. $f(x) = 8\ln(4x^2)$
14. $f(x) = e^2 \ln(5x^3)$
15. $f(x) = x^2 \ln(3x)$
16. $f(x) = 4x \ln x$
17. $f(x) = \ln(2x + 8)^5$
18. $f(x) = \ln(x^2 + 3x)^2$
19. $f(w) = \left[\ln(w + 4)\right]^3$
20. $f(r) = 5\left[\ln(2r - 1)\right]^6$
21. $f(t) = \ln\sqrt{3t + 5}$
22. $f(c) = \ln\sqrt[3]{7c + 7}$
23. $f(u) = \ln(\ln u)$
24. $f(v) = \ln\left[(\ln v)^2\right]$
25. $f(t) = t^2\left[\ln(t^2)\right]^2$
26. $f(s) = \dfrac{\ln 6}{5\ln(2s + 1)}$
27. $f(t) = 2.75\ln(40t^{3/4} + 15)$

For Exercises 28–31, use implicit differentiation to find y'.

28. $3x^3 + \ln(3xy^2) = 2$
29. $4x + \ln(x^2y^4) = 10$
30. $2xy^2 + \ln(xy) = 1$
31. $3x^3y^4 - \ln(5x^3y) = 5$

For Exercises 32–36, find and interpret each derivative.

32. Find $f'(3)$ for $f(x) = \ln(x^2 + 2x)$.
33. Find $f'(0)$ for $f(x) = \ln(5x + 1)^2$.
34. Find $f'(1)$ for $f(x) = 3x^2 - 4\ln(x^2)$.
35. Find $f'(6)$ for $f(x) = 3\ln(4x) - 2\ln(3x)$.

36. Find $f'(1)$ for $f(x) = \dfrac{\ln(4x)}{\ln(3x)}$.

C APPLYING THE CONCEPTS

37. ***Economics: Doubling Time*** The doubling time t (in years) for an investment of P dollars made at interest rate r compounded semiannually is given by

$$t = \frac{\ln 2}{2\ln(1 + \frac{1}{2}r)}$$

Find and interpret the rate at which the doubling time is changing as the interest rate is changing when the interest rate is 10%.

38. ***Manufacturing: Cost of Production*** The daily cost C (in dollars) for a manufacturer to produce x electronic video components is given by

$$C(x) = 1000 \ln \sqrt[3]{3x^2 - 700}$$

If the manufacturer is currently producing 60 components each day, by how much can the cost be expected to change if production is increased by one component each day to 61 components each day?

39. ***Science: Radioactive Decay*** At the start of an observation, there are 300 grams of a radioactive element present. The number of years t that must pass before N grams remain is given by the natural logarithm equation

$$t = 7549.5 - 1639 \ln N$$

Find and determine $\dfrac{dt}{dN}$ when $N = 120$.

40. ***Biology: Blood Flow*** If the velocity of blood in an artery is too high, the flow of blood will become turbulent. The Reynolds number R is a measure of blood flow and is determined by the radius r of the aorta and positive constants a and b that relate to the density and viscosity of the blood. The Reynolds number related to blood flow is

$$R = a \ln r - br$$

Determine the rate at which the Reynolds number changes as the radius of the aorta changes.

41. ***Business: Textbook Publishing*** The senior editor for the life sciences at a textbook publishing company has determined that the equation

$$N(x) = 1.26 + 1.2 \ln(1.1 + x^3)$$

relates the expected number N (in thousands) of copies of an introductory biology textbook sold in a 1-year period to the number x (in thousands) of complimentary copies sent to professors throughout the country. Find and interpret the rate at which N changes as x changes when $x = 6000$. (Remember that x is in thousands.)

D DESCRIBE YOUR THOUGHTS

42. Describe the strategy you would use in differentiating the function $f(x) = \ln(5x + 3)$.

43. Describe the strategy you would use in differentiating the function $f(x) = \left[\ln(6x^2)\right]^5$.

44. Explain how the derivative formula for $f(x) = \ln x$ is a specific case of the generalized differentiation for $f(x) = \ln\left[g(x)\right]$.

E REVIEW

45. **(2.1)** Consider the graph below.

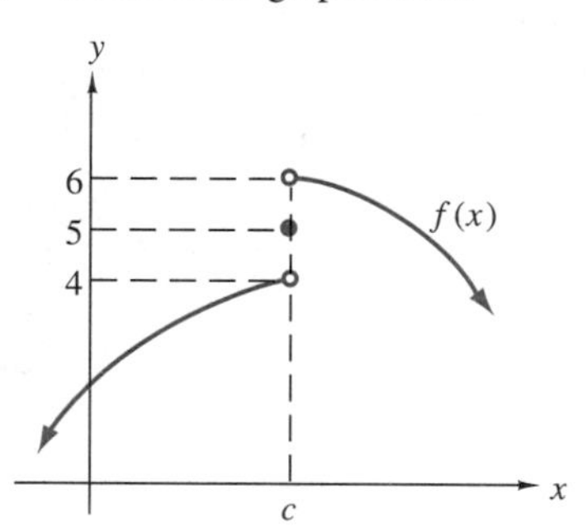

Determine the specified value of c if $\lim_{x\to c} f(x)$ exists. If it does exist, specify its value.

46. **(2.2)** Find, if it exists, $\lim_{x\to 3} 8$.

47. **(2.2)** Find, if it exists, $\lim_{x\to 2} \dfrac{x^2 - 5x + 6}{x^2 - x - 2}$.

48. **(2.3)** Find, if it exists, $\lim_{x\to\infty} \dfrac{x^2 + 3x + 1}{x + 4}$.

49. **(2.5)** Find the average rate of change of the function $f(x) = 2x^2 - x + 4$ from $x = 1$ to $x = 7$.

50. **(3.1)** Use the definition of the derivative to find $f'(x)$ for $f(x) = 5x^2 + 2x - 8$.

51. **(3.2)** Find $f'(x)$ for $f(x) = \dfrac{x+3}{x-4}$.

52. **(3.4)** Find $f'(x)$ for $f(x) = (5x^2 + 3x)^4$.

53. **(3.5)** Use implicit differentiation to find y' for $3x^2y + 2xy^2 = 4$.

54. **(4.1)** Classify a curve that is increasing at a decreasing rate as concave upward or concave downward.

55. **(4.3)** ***Economics: Demand for a Product*** Demand for an electric can opener is related to its selling price p by the equation

$$n = 3200 - 80p$$

where n is the number of can openers that can be sold per month at a price p. Find the selling price that will maximize the revenue received.

56. **(5.1)** Find $P(15)$ for $P(x) = 80 + 150e^{-0.26x}$.

57. **(5.2)** ***Business: Investment*** If \$20,000 is invested at 8% interest compounded continuously, the function $A = 20{,}000e^{0.08t}$ approximates the amount accumulated after t years. Approximately how long will it take to accumulate \$35,011?

5.4 Differentiating the Natural Exponential Function

Introduction

Differentiating the Natural Exponential Function

Differentiating the Generalized Natural Exponential Function

Differentiation and Technology

Introduction

In this section we differentiate the natural exponential function $f(x) = e^x$ and the generalized natural exponential function $f(x) = e^{g(x)}$. We derive the differentiation formula for $f(x) = e^x$ and examine the special, and often occurring, case for $f(x) = e^{ax}$.

Differentiating the Natural Exponential Function

The differentiation rule for the exponential function $f(x) = e^x$ is quickly derived by noting that

$$\ln e^x = x$$

To get $\frac{d}{dx}\left[e^x\right]$, we differentiate each side of $\ln e^x = x$ with respect to x.

$$\frac{d}{dx}\left[\ln e^x\right] = \frac{d}{dx}[x]$$

$$\frac{1}{e^x} \cdot \frac{d}{dx}\left[e^x\right] = 1$$

We then solve for $\frac{d}{dx}\left[e^x\right]$ by multiplying each side by e^x.

$$\frac{d}{dx}\left[e^x\right] = 1 \cdot e^x = e^x$$

This means that e^x is its own derivative!

The Derivative of $f(x) = e^x$

$$\frac{d}{dx}\left[e^x\right] = e^x$$

EXAMPLE SET A

Find $f'(x)$ for $f(x) = 8x^3e^x$.

Solution:
Looking at this function globally, we see a product. We will differentiate this function using the product rule.

$$\begin{aligned} f(x) &= 8x^3e^x \\ f'(x) &= \underbrace{24x^2}_{(\textit{first})'} \cdot \underbrace{e^x}_{(\textit{second})} + \underbrace{8x^3}_{(\textit{first})} \cdot \underbrace{e^x}_{(\textit{second})'} \\ &= 24x^2e^x + 8x^3e^x \qquad \text{Factor out } 8x^2e^x. \\ &= 8x^2e^x(3+x) \end{aligned}$$

Differentiating the Generalized Natural Exponential Function

The differentiation rule for $y = e^{g(x)}$ is derived using the chain rule with $u = g(x)$.

$$\begin{aligned} \frac{dy}{dx} &= \frac{dy}{du} \cdot \frac{du}{dx} \\ &= \frac{d}{dx}\left[e^u\right] \cdot \frac{du}{dx} \\ &= e^u \cdot \frac{du}{dx} \\ &= e^{g(x)} \cdot g'(x) \end{aligned}$$

The Derivative of $f(x) = e^{g(x)}$

$$\frac{d}{dx}\left[e^{g(x)}\right] = e^{g(x)} \cdot g'(x)$$

This differentiation rule states that the derivative of $f(x) = e^{g(x)}$ is

(the function itself) *times* (the derivative of the exponent)

EXAMPLE SET B

1. Find $f'(x)$ for $f(x) = e^{3x^2+5x}$.

Solution:
Globally, this function is an exponential function. So,

$$f'(x) = e^{3x^2+5x} \cdot \frac{d}{dx}[3x^2 + 5x]$$

$$= \underbrace{e^{3x^2+5x}}_{(\textit{function itself})} \cdot \underbrace{(6x+5)}_{(\textit{exponent})'}$$

$$= (6x+5)e^{3x^2+5x}$$

2. Find $f'(3)$ for $f(x) = \ln(2x + e^{-0.05x})$.

Solution:
Globally, this function is a natural logarithm function and we will need the differentiation rule

$$\frac{1}{(\text{argument})} \quad \textit{times} \quad (\text{argument})'$$

$$f'(x) = \frac{1}{2x + e^{-0.05x}} \cdot \frac{d}{dx}[2x + e^{-0.05x}]$$

Now we differentiate the exponential.

$$f'(x) = \frac{1}{2x + e^{-0.05x}} \cdot [2 + e^{-0.05x} \cdot (-0.05)]$$

$$= \frac{2 - 0.05e^{-0.05x}}{2x + e^{-0.05x}}$$

Then, substituting 3 for x,

$$f'(3) = \frac{2 - 0.05e^{-0.05 \cdot 3}}{2 \cdot 3 + e^{-0.05 \cdot 3}}$$

$$\approx 0.2852$$

Other differentiation rules can be used in conjunction with the differentiation rule for exponential functions, as shown in the next example set.

EXAMPLE SET C

1. Find $f''(x)$ for $f(x) = e^{2x^2+5}$.

Solution:
Looking at this function globally, we see an exponential function. So,

$$f'(x) = e^{2x^2+5} \cdot \frac{d}{dx}[2x^2 + 5]$$

$$= e^{2x^2+5} \cdot 4x$$

$$= 4xe^{2x^2+5}$$

We have found the first derivative, so we differentiate $f'(x) = 4xe^{2x^2+5}$ to obtain $f''(x)$. Looking at $f'(x) = 4xe^{2x^2+5x}$ globally, we see a product. We will use the product rule to obtain $f''(x)$.

$$f''(x) = \underbrace{4}_{(first)'} \cdot \underbrace{e^{2x^2+5}}_{(second)} + \underbrace{4x}_{(first)} \cdot \underbrace{e^{2x^2+5} \cdot 4x}_{(second)'}$$

$$f''(x) = 4e^{2x^2+5} + 16x^2e^{2x^2+5} \qquad \text{Factor out } 4e^{2x^2+5}.$$

$$f''(x) = 4e^{2x^2+5}(1 + 4x^2)$$

2. Find $f'(x)$ for $f(x) = 6e^{-7x} \ln 4x$.

Solution:
Looking at this function globally, we see a product. So,

$$f'(x) = \underbrace{-42e^{-7x}}_{(first)'} \cdot \underbrace{\ln 4x}_{(second)} + \underbrace{6e^{-7x}}_{(first)} \cdot \underbrace{\frac{1}{4x} \cdot 4}_{(second)'}$$

$$f'(x) = -42e^{-7x} \ln 4x + \frac{6e^{-7x}}{x}$$

$$f'(x) = 6e^{-7x}(-7 \ln 4x + \frac{1}{x})$$

3. Find and interpret $f'(0)$ for $f(x) = \dfrac{e^x - e^{-x}}{e^x + e^{-x}}$.

Solution:
Looking at this function globally, we see a quotient. So we will begin differentiating $f(x)$ using the quotient rule.

$$f'(x) = \frac{(e^x + e^{-x}) \cdot \frac{d}{dx}[e^x - e^{-x}] - (e^x - e^{-x}) \cdot \frac{d}{dx}[e^x + e^{-x}]}{(e^x + e^{-x})^2}$$

$$= \frac{(e^x + e^{-x})(e^x - e^{-x}(-1)) - (e^x - e^{-x})(e^x + e^{-x}(-1))}{(e^x + e^{-x})^2}$$

$$= \frac{(e^x + e^{-x})(e^x + e^{-x}) - (e^x - e^{-x})(e^x - e^{-x})}{(e^x + e^{-x})^2}$$

$$= \frac{e^{2x} + e^0 + e^0 + e^{-2x} - (e^{2x} - e^0 - e^0 + e^{-2x})}{(e^x + e^{-x})^2}$$

$$= \frac{e^{2x} + 1 + 1 + e^{-2x} - (e^{2x} - 1 - 1 + e^{-2x})}{(e^x + e^{-x})^2}$$

$$= \frac{e^{2x} + 2 + e^{-2x} - e^{2x} + 2 - e^{-2x}}{(e^x + e^{-x})^2}$$

$$= \frac{4}{(e^x + e^{-x})^2}$$

Now we substitute 0 for x and compute to find $f'(0)$.

$$f'(0) = \frac{4}{(e^0 + e^{-0})^2}$$

$$f'(0) = \frac{4}{(1+1)^2}$$

$$f'(0) = \frac{4}{4}$$

$$f'(0) = 1$$

Interpretation: If x increases 1 unit in value, from 0 to 1, $f(x) = \dfrac{e^x - e^{-x}}{e^x + e^{-x}}$ will increase in value by approximately 1 unit. (Remember, we need to say *approximately* because the derivative involves a variable.)

Functions of the form $f(x) = e^{ax}$, where a is a real number, occur frequently and it is convenient to simply write the derivative without writing the individual steps leading up to it. We illustrate those steps by differentiating $f(x) = e^{ax}$.

$$f(x) = e^{ax}$$

$$f'(x) = \underbrace{e^{ax}}_{(\textit{function itself})} \cdot \underbrace{a}_{(\textit{exponent})'}$$

$$= ae^{ax}$$

We have developed the following general rule.

The Derivative of $f(x) = e^{ax}$

$$\frac{d}{dx}\left[e^{ax}\right] = ae^{ax}$$

This differentiation rule states that to differentiate e^{ax}, simply multiply e^{ax} by a, the coefficient of the exponent.

EXAMPLE SET D

1. $\frac{d}{dx}\left[e^{4x}\right] = 4e^{4x}$
2. $\frac{d}{dx}\left[e^{-6x}\right] = -6e^{-6x}$
3. $\frac{d}{dx}\left[5e^{-0.2x}\right] = (-0.2)5e^{-0.2x} = -1e^{-0.2x} = -e^{-0.2x}$

EXAMPLE SET E

The resale value R (in dollars) of a particular type of laser cutting tool t years after its initial purchase is approximated by

$$R(t) = 450{,}000e^{-0.27t}$$

Find and interpret the rate at which the resale value is changing (1) 1 year after the initial purchase, and (2) 5 years after the initial purchase.

We need to find and interpret $R'(1)$, and $R'(5)$. We will begin by finding the general form of the derivative, $R'(t)$.

$$R(t) = 450{,}000e^{-0.27t}$$

$$R'(t) = (-0.27)(450{,}000)e^{-0.27t}$$

$$= -121{,}500e^{-0.27t}$$

Thus, $R'(t) = -121{,}500e^{-0.27t}$.

1.

$$R'(1) = -121{,}500e^{-0.27(1)}$$

$$\approx -92{,}750.61$$

Interpretation: If the number of years since the initial purchase of the laser cutting tool increases by 1, from 1 to 2, the resale value of the tool can be expected to decrease by approximately \$92,750.61.

2.

$$R'(5) = -121{,}500e^{-0.27(5)}$$

$$\approx -31{,}497.69$$

Interpretation: If the number of years since the initial purchase of the laser cutting tool increases by 1, from 5 to 6, the resale value of the tool can be expected to decrease by approximately \$31,497.69. That is, at the end of year 6, the resale value of the tool can be expected to be about \$31,497.69 less than the resale value at the end of year 5.

Differentiation and Technology

Using Your Calculator You can use your graphing calculator to approximate the derivative of a function involving a natural exponential function. The following entries illustrate an approximation for the function of Example Set E, $R(t) = 450{,}000e^{-0.27t}$.

```
Y1 = 450000e^(−0.27X)
Y5 = nDeriv(Y1,X,X,0.001)
```

In the computation window, enter $Y5(5)$ to evaluate $R'(t)$ at $t = 5$. Your result should be about -31497.69, which agrees with the result of the example.

Calculator Exercises

1. A manufacturer believes that the function

$$N(t) = 3250 + 1425e^{-0.15t}$$

models the number of worker hours needed to build the nth unit of a product. Construct the graph of this function and use it to describe the function's behavior. Suppose in writing the function, the $+$ sign between 3250 and 1425 was mistakenly written as a $-$ sign. Describe the effect of this mistake on estimations of necessary worker hours for the building of 1000 or more units. Explain your reasoning.

2. The number N of people in a community that hear a rumor t days after it is started is approximated by the function

$$N(t) = \frac{42{,}750}{1 + e^{-0.65t}}$$

Construct the graph of $N(t)$ and use it to describe the behavior of the function. Experiment with the exponent of the number e to determine how to change the curve so it rises more slowly. Describe, as if you were writing to a friend, what effect the exponent of e has on the behavior of the function.

3. The total cost C (in dollars) to a manufacturer to produce x units of a product is given by

$$C(x) = 4{,}000 + 235xe^{x/575}$$

Construct the graph of this function and use it to describe the function's behavior. Find and interpret both $C(240)$ and $C'(240)$. Find the value of $C''(240)$ and explain what information this value gives you about the behavior of the total cost function.

EXERCISE SET 5.4

A UNDERSTANDING THE CONCEPTS

In Exercises 1–10, a function is given along with a claimed derivative of the function. State if the indicated derivative is *correct* or *incorrect*.

1. $f(x) = e^{4x^2+5}$
   $f'(x) = 8x$
2. $f(x) = e^{3x^5+x-1}$
   $f'(x) = 15x^4 + 1$
3. $f(x) = e^{x^2-6}$
   $f'(x) = e^{x^2-6}(x^2 - 6)$
4. $f(x) = e^{-0.03x}$
   $f'(x) = e^{-0.03x}(-0.03)$
5. $f(x) = 8e^{2x}$
   $f'(x) = 8e^{2x} \cdot 2 = 16e^{2x}$
6. $f(x) = e^{x^2+4}\ln(4x)$
   $f'(x) = e^{x^2+4} \cdot 2x \cdot \frac{1}{4x} = \cdot\frac{e^{x^2+4}}{2}$
7. $f(x) = 2xe^x$
   $f'(x) = 2e^x + 2xe^x$
8. $f(x) = \ln(e^{6x})$
   $f'(x) = \frac{1}{e^{6x}} \cdot e^{6x} \cdot 6 = 6$
9. $f(x) = e^{\ln 8}$
   $f'(x) = \frac{1}{e^{\ln 8}} \cdot \frac{1}{8} = \frac{1}{8e^{\ln x}}$
10. $f(x) = \ln x + e^x$
    $f'(x) = \frac{1}{x} + \frac{1}{e^x}$

B SKILL ACQUISITION

For Exercises 11–30, find the derivative of each function.

11. $f(x) = e^{3x+6}$
12. $f(x) = e^{-4x+5}$
13. $f(x) = e^{6x}$
14. $f(x) = e^{8x}$
15. $f(x) = e^{5x^2+4}$
16. $f(x) = e^{2x^3-x-7}$
17. $f(x) = 9e^{3x+1}$
18. $f(x) = 4x^2 + 5e^{x^2+1}$
19. $f(x) = 5x^3 + 6x - 2e^{-3x}$
20. $f(x) = 6x^3e^{2x+1}$
21. $f(x) = 5x^2e^{x^2+x}$
22. $f(x) = e^{\sqrt{5x+2}}$
23. $f(x) = 3e^{\sqrt[4]{x^2-3x}}$
24. $f(x) = e^{3x}\ln(3x)$
25. $f(x) = 2e^{-x}\ln(x^2 + 2x)$
26. $f(x) = e^{e^{2x}}$
27. $f(x) = 1200 + 50e^{-0.05x}$
28. $f(x) = 3x - 2e^{-0.04x}$
29. $f(x) = \frac{e^{3x}}{e^x + e^{2x}}$
30. $f(x) = \frac{1 + e^{-x}}{1 - e^{-x}}$

For Exercises 31–34, use implicit differentiation to find y'.

31. $2x^3 + e^{3xy} = 120$
32. $7x^2y + e^{4xy^2} = 1$
33. $e^{xy} + \ln(xy) = 0$
34. $e^{y^2} - y^{e^2} = x^2$

For Exercises 35–36, find $f''(x)$.

35. $f(x) = e^{x^2+4}$
36. $f(x) - e^{-x} + e^x$

For Exercises 37–40, find and interpret each derivative.

37. Find $f'(1)$ for $f(x) = 4e^{x^2}$.
38. Find $f'(3)$ for $f(x) = -0.06e^{2x+5}$.
39. Find $f'(40)$ for $f(x) = 25{,}000 + 25{,}000e^{-0.065x}$.
40. Find $f'(1)$ for $f(x) = e^x \ln x$.

C APPLYING THE CONCEPTS

41. *Business: Advertising* The management of a large chain of frozen yogurt stores in New England believes that t days after the end of an advertising campaign, the volume V of sales is approximated by

$$V(t) = 65{,}000 + 65{,}000e^{-0.46t}$$

At what rate is the sales volume changing (a) one day after the end of the advertising campaign, and (b) six days after the end of the advertising campaign?

42. ***Manufacturing: Consumption*** A producer of tetraethylortosilicate (TEOS), a coating for silicon computer wafers, believes, from an extensive study, that the national consumption C (in appropriate units) of TEOS over the next t years is approximated by

$$C(t) = 14.5e^{-1.2t} + 0.14t^2, \quad 0 \le t \le 8$$

The study was prepared when $t = 0$. Find the rate at which the national consumption of TEOS is changing with respect to time (a) 2 years after the preparation of the study, and (b) 5 years after the preparation of the study.

43. ***Medicine: Flu Epidemic*** The total number N of people in a Washington state community contracting a flu virus by the tth day after the outbreak of an epidemic is approximated by

$$N(t) = \frac{14{,}500}{1 + 65e^{-0.4t}}$$

How many more people can be expected to contract the flu virus between days 4 and 5 of the epidemic?

44. ***Psychology: Technical Proofreading*** A psychologist believes the formula

$$N(t) = 85 - 65e^{-0.35t}$$

relates the number N of pages of a technical textbook an average person can proofread and the number of consecutive days t that person has been reading such books. By how many pages can an average person be expected to increase his or her proofreading ability from (a) day 4 to day 5, and (b) from day 10 to day 11?

45. ***Science: Radioactive Decay*** A 120-gram block of radioactive carbon-14, C^{14}, decays according to the function

$$A(t) = 120e^{-0.00012t}$$

where t is measured in years. By how many grams can the block of C^{14} be expected to decrease from year 10 to year 11?

46. ***Medicine: Drug Concentration*** When a particular drug is intravenously administered to a patient on a continuous basis, the amount A (in milligrams) of the drug in the patient's bloodstream t hours after injection is approximated by

$$A(t) = 10 - 175e^{-0.2t}$$

At what rate is the amount of the drug in the patient's bloodstream changing after 10 minutes?

D DESCRIBE YOUR THOUGHTS

47. Describe how the differentiation process of the functions $f(x) = \ln\left[g(x)\right]$ and $f(x) = e^{g(x)}$ are similar.

48. Describe the strategy you would use in differentiating the function $f(x) = e^{5x^2+4}$. Do not perform the differentiation; just describe how you would do it.

49. Explain how the differentiation formula for $f(x) = e^x$ is a specific case of the differentiation formula for $f(x) = e^{g(x)}$.

E REVIEW

50. **(2.1)** Determine from the graph below if $\lim_{x\to 9} f(x)$ exists. If it does exist, specify its value.

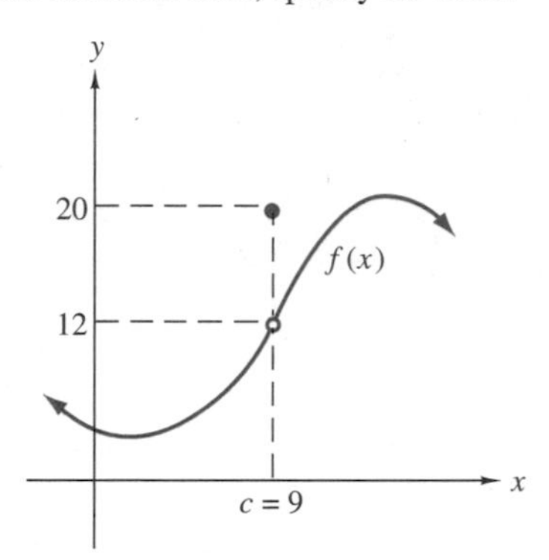

51. **(2.2)** Find, if it exists, $\lim_{x\to -2} \frac{4}{x-2}$.

52. **(2.2)** Find, if it exists, $\lim_{x\to 7} \frac{x^2 - 8x + 7}{x^2 - 5x - 14}$.

53. **(2.3)** Find, if it exists, $\lim_{x\to \infty} \frac{x+4}{x^2 - 4}$.

54. **(2.5)** Find the average rate of change of the function $f(x) = 5x - 6$ from $x = 2$ to $x = 10$.

55. **(3.1)** Use the definition of the derivative to find $f'(x)$ for $f(x) = 3x^2 + 6$.

56. **(3.2)** Suppose that $C(x)$ represents the cost of producing x items. Interpret $C'(250) = 1.48$.

57. **(3.4)** Find $f''(x)$ for $f(x) = 4x^2 - 8x + 11$.

58. **(3.5)** Find $f'(x)$ for $f(x) = (3x^2 + 4)^2(2x - 5)^5$.

59. **(4.3)** ***Business: Revenue*** Suppose that a tool company has determined that its total revenue R for its tools is given by

$$R = -x^3 + 79x^2 + 5000x$$

where R is measured in dollars and x is the number of units in thousands produced. What production level will yield a maximum revenue?

60. **(5.3)** Find $f'(x)$ for $f(x) = \ln(x^2 + 6)$.

Summary

This chapter presented the natural exponential function and the natural logarithm function with their derivatives and applications.

The Exponential Function

Section 5.1 presented the exponential function by first distinguishing the exponential function from the polynomial power function. The function $f(x) = b^x$, where $b > 0$, $b \neq 1$ and x is a variable, is the *exponential function*. It is different in form from the polynomial power function. The power function $f(x) = x^b$, has the form $(\text{variable})^{(\text{constant})}$. The exponential function $f(x) = b^x$, has the form $(\text{constant})^{(\text{variable})}$. This is an important distinction as the derivatives are very different. For example, if $f(x) = x^2$, then $f'(x) = 2 \cdot x^{2-1}$, whereas if $f(x) = 2^x$, then $f'(x) \neq x \cdot 2^{x-1}$.

This section continued with a discussion of the behavior of the exponential function. In particular, for $f(x) = b^x$, when $b > 1$, as x increases, $f(x)$ increases and is called a *growth function*. When $0 < b < 1$, as x increases, $f(x)$ decreases and is called a *decay function*.

The section also presented the most commonly used base in exponential function, the constant e. This irrational number is defined by

$$e = \lim_{x \to \infty} \left(1 + \frac{1}{x}\right)^x$$

Graphs of the more general exponential function $A_0 e^{kx}$ were also examined in this section, and its eight important properties were presented.

The Natural Logarithm Function

Section 5.2 examined the natural logarithm function, $f(x) = \ln x$, $x > 0$. The section began by noting that logarithms are exponents, and then presented the first form of the natural logarithm: In the equation $a = e^t$, the number t, when used as an exponent on the base e that produces the number a, is called the *natural logarithm* of a. Then the final form, the definition of the natural logarithm, was presented: $t = \ln a$ if and only if $a = e^t$.

This section then tied together the actions of the exponential and logarithmic functions by examining their reversal effect on each other. When they have the same base, the exponential function undoes the effect of the logarithmic function, and vice versa. That is, $e^{\ln x} = x$ and $\ln e^x = x$. This reversal property is very important in solving equations involving exponentials or logarithms.

The section continued by presenting six properties of the natural logarithm. These properties are useful when working with functions involving natural logarithms. They are:

1. $\ln e = 1$ (from Example Set A)

2. $\ln 1 = 0$ (from Example Set A)

3. $\ln e^x = x$ (from Section 5.1)

4. $\ln(xy) = \ln x + \ln y$ (Multiplication becomes addition.)

5. $\ln\left(\frac{x}{y}\right) = \ln x - \ln y$ (Division becomes subtraction.)

6. $\ln x^r = r \cdot \ln x$ (Exponentiation becomes multiplication.)

The section ended by examining the graph of the natural logarithm function and pointing out six important features of the graph.

1. As x increases, $f(x) = e^x$ increases and does so at an increasing rate.
2. As x increases, $f(x) = \ln x$ increases but does so at a decreasing rate.
3. For $0 < x < 1$, $\ln x < 0$; that is, for input values of x strictly between 0 and 1, the output values $\ln x$ are negative.
4. For $x > 1$, $\ln x > 0$; that is, for input values of x strictly greater than 1, the output values $\ln x$ are positive.
5. $\ln(1) = 0$
6. The domain of the natural logarithm function is the set of all positive numbers. That is, only positive numbers are allowed as input values. The graph shows that $\ln x$ is not defined for negative numbers or 0.

Differentiating the Natural Logarithm Function

Section 5.3 focused on the derivative of the natural logarithm function. For the natural logarithm function, $f(x) = \ln x$

$$\frac{d}{dx}[\ln x] = \frac{1}{x}$$

and for the generalized natural logarithm function, $f(x) = \ln[g(x)]$,

$$\frac{d}{dx}[\ln[g(x)]] = \frac{1}{g(x)} \cdot g'(x)$$

This differentiation rule states that the derivative of the generalized natural logarithm function $f(x) = \ln[g(x)]$ is

(1 over the argument) *times* (the derivative of the argument)

That is,

$$\frac{1}{\text{the argument}} \quad \textit{times} \quad (\text{argument})'$$

A common mistake when differentiating $f(x) = \ln[g(x)]$ is to neglect to differentiate the argument, $g(x)$.

This section ended with a derivation of the differentiation formula for the natural logarithm function.

Differentiating the Natural Exponential Function

Section 5.4 presented the derivative of the exponential function with base e. For the exponential function $f(x) = e^x$,

$$\frac{d}{dx}\left[e^x\right] = e^x$$

This is the only function that is its own derivative.

Then, for the generalized exponential function, $f(X) = e^{g(x)}$,

$$\frac{d}{dx}\left[e^{g(x)}\right] = e^{g(x)} \cdot g'(x)$$

This differentiation rule states that the derivative of $f(x) = e^{g(x)}$ is

(the function itself) *times* (the derivative of the exponent).

As with the differentiation of the natural logarithm function, a common mistake when differentiating $f(x) = \ln[g(x)]$ is to neglect to differentiate the argument, $g(x)$.

The section ended by discussing the differentiation of the commonly occurring function $f(x) = e^{ax}$.

$$\frac{d}{dx}[e^{ax}] = ae^{ax}$$

This differentiation rule states that to differentiate e^{ax}, simply multiply e^{ax} by a, the coefficient of the exponent. For example, $\frac{d}{dx}\left[e^{-6ax}\right] = -6e^{ax}$.

Supplementary Exercises

For Exercises 1–6, identify each function as algebraic or exponential.

1. $f(x) = x^3 + 7$
2. $f(x) = 5x^2 + x - 4$
3. $f(x) = 3e^{4x+1}$
4. $f(x) = e^{-0.12x}$
5. $f(x) = x^{3e} - 1$
6. $f(x) = e^{-2e+1}$

For Exercises 7–16, use a calculator to compute the value of each function. Identify each function as a growth or decay function. Round to three decimal places.

7. Find $f(2)$ for $f(x) = 10^{0.8x+1}$.
8. Find $f(12)$ for $f(x) = 10^{-0.32x+4}$.
9. Find $f(80)$ for $f(x) = (1.08)^{0.25x+45}$.
10. Find $f(400)$ for $f(x) = (1.065)^{0.11x}$.
11. Find $f(6)$ for $f(x) = e^{-2x+15}$.
12. Find $f(4)$ for $f(x) = e^{-0.41x^2}$.
13. Find $f(16)$ for $f(x) = 14{,}000e^{0.08x}$.
14. Find $f(95)$ for $f(x) = 2800e^{0.075x}$.
15. Find $f(18)$ for $f(x) = 1 - e^{0.22x}$.
16. Find $f(7)$ for $f(x) = 70{,}000 - 24{,}000e^{0.15x}$.

For Exercises 17–19, find the requested values.

17. ***Ergonomics: Aircraft Cabin Pressure*** The function $P(h) = 14.7e^{-0.21h}$ relates the atmospheric pressure (in pounds per square inch) to the altitude h (in miles) above sea level. Find the atmospheric pressure (a) at sea level, (b) 1 mile above sea level, and (c) 6.5 miles above sea level.
18. ***Economics: City Population*** The function $P(t) = 16e^{0.12t}$ relates the population, P in thousands, of a growing city to the time t in years from now. Find the population of the city 4 years from now.
19. ***Economics: Market Value*** The market value $MV(t)$ (in dollars) of a piece of high-tech industrial equipment t years from now is given by the function $MV(t) = 162{,}000e^{-0.21t}$. Find the market value of this piece of equipment 3.5 years from now.

For Exercises 20–25, convert each exponential function to its corresponding natural logarithm function.

20. $a = e^{2b}$
21. $x = e^{-4y}$
22. $x = e^{-2y+5}$
23. $k = e^{3h-1}$
24. $9 = e^{x+6}$
25. $150 = e^{-0.5x+16}$

For Exercises 26–30, convert each natural logarithm function to its corresponding exponential function.

26. $5 = \ln(3x)$

27. $-2 = \ln(x + 6)$

28. $4.08 = \ln(x^2 + x)$

29. $3y + 7 = \ln(2x - 7)$

30. $5x + 3 = \ln(y - 4)$

For Exercises 31–36, solve each exponential equation. If the solution is not an integer, round it to four decimal places.

31. $e^{2x+1} = 19$

32. $e^{4x-3} = 8$

33. $4e^{8-x} = 0.52$

34. $e^{-0.08t} = 0.3263$

35. $e^{-0.55t} = 1$

36. $e^{-0.27t+45} = 105,336$

For Exercises 37–42, solve each natural logarithm equation. If the solution is not an integer, round it to four decimal places.

37. $\ln(x + 4) + \ln(x - 1) = 2.6391$

38. $\ln(x + 6) + \ln(x + 5) = 3.40218$

39. $\ln(2x - 1) + \ln(3x + 1) = 4.5109$

40. $\ln(x + 2) - \ln(x - 8) = 2.3977$

41. $\ln(3x - 5) - \ln(x + 4) = 0.2151$

42. $\ln(8x - 5) + \ln(3x - 3) - \ln(x + 10) = -0.0910$

For Exercises 43–45, find the requested times.

43. ***Manufacturing: Assembly Tasks*** Industrial psychologists have determined that the function $N(t) = 40 - 40e^{-0.07t}$ relates the number N of assembly tasks an average person can complete each 8-hour working day to the number of days t the person has been performing the task. Approximately how many days after starting a particular task will the worker be completing 25 tasks in an 8-hour working day?

44. ***Medicine: Drug Concentration*** When a drug is intravenously administered to a patient on a continuous basis, the amount A (in milligrams) of the drug in the patient's bloodstream t minutes after injection is given by the function $A(t) = \frac{1}{0.15}(100 - 30e^{0.15t})$. Approximately how much time will have to pass so that 175 milligrams of the drug will be in the patient's bloodstream?

45. ***Economics: Tripling Time*** The amount A of dollars accumulated through an investment of A_0 dollars at $r\%$ interest compounded continuously for t years can be determined from the function $A(t) = A_0e^{rt}$. The *tripling time* of an investment is the time required for an investment to triple in value. Find the tripling time for an investment made at 8.5% interest.

For Exercises 46–68, find the derivative of each function.

46. $f(x) = e^{8x}$

47. $f(x) = e^{-3x}$

48. $f(x) = e^{-2x+5}$

49. $f(x) = x^2e^{4x}$

50. $f(x) = \ln(6x)$

51. $f(x) = \ln(4x^2)$

52. $f(x) = \ln(x + 7)$

53. $f(x) = \ln e^{3x}$

54. $f(x) = e^{\ln(3x)}$

55. $f(x) = x^3 \ln x$

56. $f(x) = \ln(x^2 + 2x)$

57. $f(x) = e^{x^2+2x}$

58. $f(x) = e^x \ln x$

59. $f(x) = \left[e^{4x+2}\right]^3$

60. $f(x) = \left[\ln(4x - 5)\right]^3$

61. $f(x) = e^{e^{3x}}$

62. $f(x) = \ln\left[\ln(8x)\right]$

63. $f(x) = \ln\sqrt{5x + 1}$

64. $f(x) = e^{\sqrt{3x+6}}$

65. $f(x) = 121e^{-0.31x}$

66. $f(x) = \dfrac{e^{4x} - e^{-4x}}{e^{4x} + e^{-4x}}$

67. $f(x) = 3\ln(4x + 4) + 10e^{4x+4}$

68. $f(x) = 2e^{-3x}\ln(2x + 7)$

For Exercises 69–72, use implicit differentiation to find y'.

69. $4x^2 + e^{x^2y^2} = 1$

70. $y^3 + x\ln y = 30$

71. $e^{xy} - e^y = e^x$

72. $\ln x - \ln y = e^x \ln y$

For Exercises 73–74, find $f''(x)$.

73. $f(x) = e^{3x}$

74. $f(x) = x^2 \ln x + e^{-2x}$

For Exercises 75–77, find and interpret:

75. $f'(1)$ for $f(x) = 3e^{-x^2}$

76. $f'(10)$ for $f(x) = \ln(3x + 5)$

77. $f'(65)$ for $f(x) = 37{,}500 + 37{,}500e^{-0.085x}$

CHAPTER 6

Integration: The Language of Accumulation

6.1 Antidifferentiation and the Indefinite Integral

Introduction

Differentiation, the language of change, provides us with the capability of determining the rate at which a quantity is changing when we know how large the quantity is at some particular point in time. For example, if $C(x)$ is the cost function associated with the production of x units of a product, $C'(725) = -2.20$ indicates that if the number of units produced is increased by 1, from 725 to 726, the cost per unit will decrease by approximately \$2.20. In applied situations, it is common to observe not the direct relationship between the quantities, but rather the relationship of change. The function observed is the derivative function.

Now, given the derivative function, suppose we would like to find the function that describes the direct relationship between the quantities. The function we have is the derivative of the function we want. To get the function we want, we need to reverse the process of differentiation. That is the process we investigate in this section.

The Antiderivative and the Indefinite Integral

Antidifferentiation, Integration

The process of reversing a differentiation is called **antidifferentiation** or **integration**. The function obtained from this process is called the **antiderivative** or **integral** of the derivative function.

The Antiderivative or Integral

Suppose that $F(x)$ and $f(x)$ are two functions such that $f(x)$ is the derivative of $F(x)$; that is, $F'(x) = f(x)$. Then $F(x)$ is an **antiderivative** or **integral** of $f(x)$.

For example (and you should verify these), if

1. $f(x) = 4x - 5$, then $F(x)$ could be defined by $2x^2 - 5x$.
2. $f(x) = 4x - 5$, then $F(x)$ could be defined by $2x^2 - 5x + 3$.
3. $f(x) = 4x - 5$, then $F(x)$ could be defined by $2x^2 - 5x - 1$.
4. $f(x) = 4x - 5$, then $F(x)$ could be defined by $2x^2 - 5x - 4$.
5. $f(x) = 4x - 5$, then $F(x)$ could be defined by $2x^2 - 5x +$ some constant.

The first four functions are graphed in Figure 6.1.

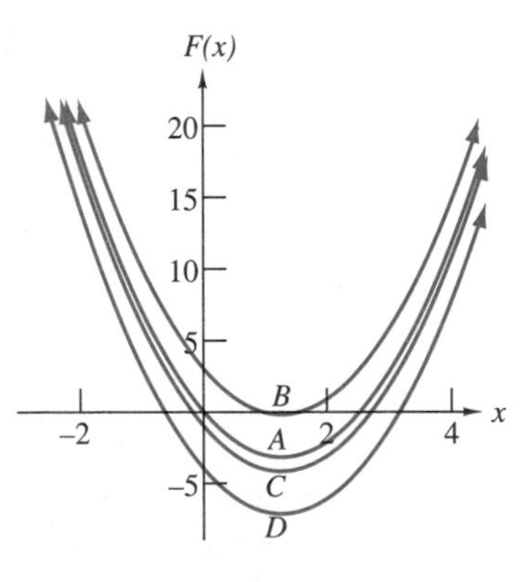

Figure 6.1

Notice that each of the antiderivative functions $F(x)$ above produces the same derivative function $f(x) = 4x - 5$. This means that when the derivative process is reversed, a unique function $F(x)$ is not produced. Rather a *family* of antiderivative functions consisting of infinitely many members is produced. Each member function of the family, however, differs only by a constant.

Suppose that both $F(x)$ and $G(x)$ are antiderivatives of $f(x)$; that is, suppose that $F'(x) = f(x)$ and $G'(x) = f(x)$. We wish to show that the difference of $F(x)$ and $G(x)$ is a constant. Since the difference of two functions is again a function, let

$$H(x) = F(x) - G(x)$$

Then

$$\begin{aligned} H'(x) &= F'(x) - G'(x) \\ &= f(x) - f(x) \qquad \text{by our supposition} \\ &= 0 \end{aligned}$$

But only the constant function has 0 as its derivative. Thus, if c represents some constant,

$$H(x) = c$$

or, since $H(x) = F(x) - G(x)$,

$$F(x) - G(x) = c$$

which means that

$$F(x) = G(x) + c$$

and we have shown that two antiderivatives of the same function differ only by a constant and, therefore, that a derivative function has infinitely many antiderivatives.

The Indefinite Integral

Just as $\frac{d}{dx}f(x)$ or $f'(x)$ symbolizes the derivative of $f(x)$ with respect to x, $\int f(x)\,dx$ symbolizes the antiderivative, or integral of $f(x)$, with respect to x. The symbol $\int dx$ indicates the family of antiderivative functions, and the function we wish to find the antiderivative of must be placed between the $\int$ and the dx.

Indefinite Integral

Suppose that $F(x)$ is any one of the infinitely many members of the family of antiderivatives of $f(x)$. Then,

$$\int f(x)\,dx = F(x) + c$$

where c is an arbitrary constant. The function $F(x) + c$ is called the **antiderivative** or **indefinite integral** of $f(x)$. The number c is called the **constant of integration**, the function $f(x)$ is called the **integrand**, and dx is called the **differential**.

In the notation $\int f(x)\,dx$, the differential dx plays the same role it does in the notation $\frac{d}{dx}$. In $\frac{d}{dx}$, dx indicates that the derivative is to be taken with respect to the independent variable x. In $\int f(x)\,dx$, dx indicates that the antiderivative is to be taken with respect to the independent variable x; likewise, $\int f(y)\,dy$, dy indicates that the antiderivative is to be taken with respect to the independent variable y.

Integration Formulas

The following integration formulas are constructed by reversing the differentiation formulas we have already studied. (You should notice that each one can be proved simply by taking the derivative of the indefinite integral and showing that the integrand results.)

To integrate a constant function, we affix the independent variable to the constant and add an arbitrary constant.

The Integral of a Constant

If k is any constant, then

$$\int k\,dx = kx + c$$

ILLUMINATOR SET A

1. $\int 4\,dx = 4x + c$

Notice that $\frac{d}{dx}(4x + c) = 4$ and that $\int 4\,dx \neq 4x$. Without an arbitrary constant c, infinitely many members have been left out of the family of solutions.

2. $\int du = \int 1\,du = u + c$

Notice that $\frac{d}{du}(u + c) = 1$. $\int 1\,du$ is the same as $\int du$.

The Integral of a Power Function

If r is any real number other than -1, then

$$\int x^r\,dx = \frac{x^{r+1}}{r+1} + c$$

To integrate a power function, we increase the exponent by 1, divide the resulting power by the new exponent, and add an arbitrary constant.

EXAMPLE SET A

1. $\int x^3\,dx$

Solution:

$$\int x^3\,dx = \frac{x^{3+1}}{3+1} + c$$

$$= \frac{x^4}{4} + c$$

Notice that $\frac{d}{dx}\left[\frac{x^4}{4} + c\right] = \frac{4x^3}{4} + 0 = x^3$, the integrand.

2. $\int x^{2/3}\,dx$

Solution:

$$\int x^{2/3}\,dx = \frac{x^{2/3+1}}{2/3+1} + c$$

$$= \frac{x^{5/3}}{5/3} + c$$

$$= \frac{3x^{5/3}}{5} + c$$

3. $\int \frac{1}{x^4}\,dx$

Solution:
This integral does not match any of our integration formulas. We can, however, algebraically manipulate the integrand $\frac{1}{x^4}$ so that it does. Since $\frac{1}{x^4} = x^{-4}$, we will express $\int \frac{1}{x^4} dx$ as $\int x^{-4}\,dx$. Now,

$$\int \frac{1}{x^4}\,dx = \int x^{-4}\,dx$$

$$= \frac{x^{-4+1}}{-4+1} + c$$

$$= \frac{x^{-3}}{-3} + c$$

Since we began with positive exponents, we will express the result using positive exponents. Since $\frac{x^{-3}}{-3} = \frac{1}{-3x^3} = \frac{-1}{3x^3}$,

$$\int \frac{1}{x^4} dx = \frac{-1}{3x^3} + c$$

The Integral of a Constant Times a Function

If k is any constant, then

$$\int kf(x)\,dx = k\int f(x)\,dx$$

To integrate a constant times a function, we integrate the function, then multiply that result by the constant, then add an arbitrary constant. This rule is analogous to the differentiation rule

$$\frac{d}{dx}[k\cdot f(x)] = k\cdot\frac{d}{dx}[f(x)].$$

EXAMPLE SET B

$$\int 6x^2\,dx$$

Solution:

$$\int 6x^2\,dx = 6\int x^2\,dx$$

$$= 6\cdot\frac{x^3}{3} + c$$

$$= 2x^3 + c$$

The Integral of a Sum or Difference

For any two functions $f(x)$ and $g(x)$,

$$\int [f(x) \pm g(x)]\,dx = \int f(x)\,dx \pm \int g(x)\,dx$$

To integrate a sum or difference of two functions, we integrate each function individually, then add or subtract those results. That is, we integrate term by term. This rule is analogous to the derivative rule

$$\frac{d}{dx}[f(x) \pm g(x)] = \frac{d}{dx}f(x) \pm \frac{d}{dx}g(x).$$

EXAMPLE SET C

1. $\int (4x^2 + 8x - 7)\,dx$

Solution:

$$\begin{aligned}\int (4x^2 + 8x - 7)\,dx &= \int 4x^2\,dx + \int 8x\,dx - \int 7\,dx \\ &= 4\int x^2\,dx + 8\int x\,dx - \int 7\,dx \\ &= 4 \cdot \frac{x^3}{3} + c_1 + 8 \cdot \frac{x^2}{2} + c_2 - (7x + c_3) \\ &= \frac{4x^3}{3} + 4x^2 - 7x + c_1 + c_2 - c_3\end{aligned}$$

Each of the three individual integrals produces a constant of integration. We have denoted these by c_1 (produced from the first integral), c_2 (produced from the second integral), and c_3 (produced from the third integral). Since the sum of three (or any number of) constants is again a constant, we will denote $c_1 + c_2 - c_3$ by c. That is, $c_1 + c_2 - c_3 = c$. Thus,

$$\int (4x^2 + 8x - 7)\,dx = \frac{4x^3}{3} + 4x^2 - 7x + c$$

Since the sum of a collection of constants is always another constant, it is convenient to omit the constants that are produced in the intermediate integrations and to simply add an arbitrary constant on at the end of the process.

2. $\int \left[5x^{2/3} - 4\right] dx$

Solution:

$$\begin{aligned}\int \left[5x^{2/3} - 4\right] dx &= 5 \cdot \frac{x^{2/3+1}}{2/3+1} - 4x + c \\ &= 5 \cdot \frac{x^{5/3}}{5/3} - 4x + c \\ &= 5 \cdot \frac{3x^{5/3}}{5} - 4x + c \\ &= 3x^{5/3} - 4x + c\end{aligned}$$

The Integral of the Exponential Function

$$\int e^x\,dx = e^x + c$$

This should be no surprise since the derivative of the exponential function is the exponential function itself.

EXAMPLE SET D

$\int (3e^x - 5x)\, dx$

Solution:

$$\int (3e^x - 5x)\, dx = \int 3e^x\, dx - \int 5x\, dx$$

$$= 3e^x - \frac{5x^2}{2} + c$$

The Integral of the Reciprocal Function

$$\int \frac{1}{x}\, dx = \ln|x| + c$$

Important Note

This formula completes the form $\int x^r\, dx$. If $r = -1$, then

$$\int x^r\, dx = \int x^{-1}\, dx = \int \frac{1}{x}\, dx = \ln|x| + c.$$

The absolute value bars are necessary because when

$$x > 0,\ \frac{d}{dx}[\ln x] = \frac{1}{x}$$

and when

$$x < 0,\ \frac{d}{dx}[\ln(-x)] = \frac{1}{-x} \cdot \frac{d}{dx}[-x]$$

$$= \frac{1}{-x} \cdot (-1)$$

$$= \frac{1}{x}$$

Thus, both $f(x) = \ln x$ and $f(x) = \ln(-x)$ have the same derivative. But, because logarithms are defined only for positive numbers, we need to use the absolute value bars to ensure that resulting integral is defined.

EXAMPLE SET E

1. $\int \frac{7}{x}\,dx$

Solution:

$$\int \frac{7}{x}\,dx = 7\int \frac{1}{x}\,dx$$
$$= 7\ln|x| + c$$

2. Suppose it is known that $x > 0$. Then $\int (6e^x + \frac{4}{x} - 1)\,dx = ?$

Solution:

$$\int (6e^x + \frac{4}{x} - 1)\,dx = 6e^x + 4\ln|x| - x + c$$
$$= 6e^x + 4\ln x - x + c$$

We omit the absolute value bars here since $x > 0$.

3. Suppose it is known that $x > 0$. Then $\int \frac{x^4 + 3x^3 + 5}{x}\,dx = ?$

Solution:

$$\int \frac{x^4 + 3x^3 + 5}{x}\,dx = \int \left(\frac{x^4}{x} + \frac{3x^2}{x} + \frac{5}{x}\right)dx$$
$$= \int \left(x^3 + 3x + \frac{5}{x}\right)dx$$
$$= \frac{x^4}{4} + \frac{3x^2}{2} + 5\ln|x| + c$$
$$= \frac{x^4}{4} + \frac{3x^2}{2} + 5\ln x + c$$

Once again, we omit the absolute value bars here since $x > 0$.

Indefinite Integrals with Initial Conditions

Initial Condition

As we noted at the beginning of the section, when we are given the rate at which a quantity changes and we wish to know how large the quantity is at any time, we need to integrate (antidifferentiate) the derivative function. However, the integral produces an arbitrary constant c. To eliminate c and determine precisely the required individual member of the family of antiderivatives that describes the relationship between the input and output values, we must also know some **initial condition**. An initial condition is some specific information about an input value and its corresponding output value. Example Set F illustrates.

EXAMPLE SET F

Find $f(x)$ if $f'(x) = 3x^3 - 4x + 2$ and $f(1) = 4$.

Solution:

$$f(x) = \int (3x^2 - 4x + 2)\, dx$$

$$= 3 \cdot \frac{x^3}{3} - 4 \cdot \frac{x^2}{2} + 2x + c$$

$$= x^3 - 2x^2 + 2x + c$$

Now, $f(x) = x^3 - 2x^2 + 2x + c$, and we are given the initial condition that $f(1) = 4$. Substituting these values into $f(x) = x^3 - 2x^2 + 2x + c$ produces

$$f(1) = (1)^3 - 2(1)^2 + 2(1) + c$$

$$4 = 1 - 2 + 2 + c$$

$$4 = 1 + c$$

$$c = 3$$

Thus, $f(x) = x^3 - 2x^2 + 2x + 3$.

EXAMPLE SET G

At the beginning of this section, we discussed the marginal cost function $C'(x) = -0.04x + 26$, where x represents the level of production, and $C'(x)$, the marginal cost. If at the production level of 150 units, the cost is known to be \$6555, (1) find the cost function, (2) find the cost of producing 530 items, and (3) find and interpret both $C(530)$ and $C'(530)$.

Solution:

1. Since we are given the marginal cost function (which is the derivative of the cost function), we need to integrate it to find the cost function.

$$C(x) = \int (-0.04x + 26)dx$$

$$= \frac{-0.04x^2}{2} + 26x + c$$

$$= -0.02x^2 + 26x + c$$

Thus, the family of cost functions is $C(x) = -0.02x^2 + 26x + c$. To find the member of the family that describes our particular situation (cost = \$6555 when the level of production is 150 units, or $C(150) = 6555$), we need to find the value of c. We substitute 6555 for $C(x)$ and 150 for x in $C(x) =$

$-0.02x^2 + 26x + c$ and solve for c.

$$6555 = -0.02(150)^2 + 26(150) + c$$

$$c = 3105$$

Therefore, the cost function is $C(x) = -0.02x^2 + 26x + 3105$.

2. The cost of producing 530 units is found by substituting 530 for x in $C(x) = -0.02x^2 + 26x + 3105$.

$$C(530) = -0.02(530)^2 + 26(530) + 3105$$

$$= 11{,}267$$

3. From part 2, $C(530) = 11{,}267$. Substituting 530 for x in $C'(x) = -0.04x + 26$ gives $C'(530) = 4.8$. $C(530) = 11{,}267$ means that the cost of producing all 530 units is \$11,267, whereas $C'(530) = 4.8$ means that the total cost of producing 531 units will be approximately \$4.80 more than the total for producing 530 units.

Integration and Technology

Using Derive You can use Derive to find the symbolic antiderivative of a function as well as to evaluate that antiderivative at a particular value. The following entries show how Derive can be used to solve the problem of Example Set G.

Declare
Function
e1
–0.04x+26
Author
e1(530)
Simplify
Declare
Function
e2
int(e1(x),x)
Manage
Substitute
530
Simplify

Notice: Derive does not affix the arbitrary constant c to the antiderivative. (Did you notice and wonder about this?) You must affix it yourself. This set of Derive directions names the function defined by $-0.04x + 26$ **e1**. The notation **e1(530)** evaluates the function at $x = 530$. The commands **int(e1(x),x)** and **e2** construct the symbolic form of the antiderivative of **e1** and then name this form

e2. The **M**anage **S**ubstitute and **530** commands then evaluate the antiderivative at $x = 530$.

Derive Exercises

1. Use Derive to find the antiderivative of the function $f(x) = 16x^3 - 18x^2 - 10x + 12$. Evaluate both $f(x)$ and the antiderivative of $f(x)$ at $x = 2$.
2. Use Derive to find the particular antiderivative, $F(x)$, of the function $f(x) = \frac{3}{x} + 6x^2$ for which $F(1) = 5$.
3. Use Derive to find the antiderivative of the function $f(x) = (x-2)^3 + (3x+3)^4 + x(4x - 2/5)^2 + (3x^2 - 4)(x-2)$, and then specify the coefficient of the x^3 term.
4. The marginal cost function for a manufacturer to produce x units of product is given by the expression

$$\frac{265}{1.2x^{1.2}}$$

If the fixed costs are \$73,000, find the cost of producing 18,500 units.

EXERCISE SET 6.1

A UNDERSTANDING THE CONCEPTS

1. ***Business: Manufacturing*** Because of the seasonal nature of the business, the number N (in thousands) of men's pants a manufacturer makes and sells is variable and depends on the time t of year (from the number of days since January 1). The function relating N and t is $N(t)$. Interpret both $N'(130) = 8.6$ and its integral $N(130) = 98$.
2. ***Business: Manufacturing*** A manufacturer of modeling clay has determined that if the clay is left uncovered in the open air, it will harden. The clay's hardness H (in the manufacturer's own hardness units) is related to the number of minutes the clay has been in the open air by the function $H(t)$. Interpret both $H'(360) = 2.1$ and its integral $H(360) = 47$.
3. ***Psychology: Attitude*** After studying the habits of a well-known painter, a psychologist believes that the amount A (in the psychologist's own amount units) of work the painter is able to produce in a week's period is related to the painter's average daily attitude coefficient C (in the psychologist's own attitude units) by the function $A(C)$. Interpret both $A'(35) = -0.68$ and its integral $A(35) = 4.6$.
4. ***Life Science: Physiological Stress*** A physiologist's tests on an Olympic platform diver demonstrate a relationship between the amount of stress A (in stress units) placed on the tendons in the diver's legs and the diver's ability to obtain a score S on a scale with 0.00 as low and 10.00 as high. The relationship is given by the function $S(A)$. Interpret both $S'(7.2) = -1.85$ and its integral $S(7.2) = 4.55$.
5. ***Business: Manufacturing*** A manufacturer has determined that the number N of units of her product that fail is related to the speed s (in units per minute) at which the units are produced by the function $N(s)$. Interpret both $N'(170) = 0.008$ and its integral $N(170) = 2.9$.
6. ***Business: Utilization of a Product*** The amount of time T a leisure product is utilized is related to the number of hours h of leisure time each week the user has available by the function $T(h)$. Interpret both $T'(16) = 0.3$ and its integral $T(16) = 6.2$.
7. ***Political Science: Opinion Polls*** A political scientist believes that a political candidate's favorable rating R (in percentage points) in weekly polls is related to the number n of negative advertisements sponsored each week and made public by the candidate's opponent by the function $R(n)$. Interpret both $R'(60) = -0.25$ and its integral $R(60) = 38$.
8. ***Economics: Individual Wealth*** The wealth W (in thousands of dollars) of an individual is related to the time t (in months) since the beginning of a series of investments by the function $W(t)$. Interpret both $W'(30) = -12.25$ and its integral $W(30) = 865$.
9. ***City Management: Traffic*** The amount A (in numbers of cars per 100 yards of roadway) of traffic on a boulevard near a city's business park is related to the time t (in minutes) from 5:00 AM by the function $A(t)$. Interpret both $A'(540) = -.92$ and its integral $A(540) = 7$.
10. ***Economics: Income Distribution*** To measure the inequality of income distribution in a country, economists often use the Gini coefficient of inequality. The Gini coefficient of inequality, in turn, uses the Lorenz function,

a function that relates the percent of the country's income I to the percent P of people in the country. We can express the function as $P(I)$. Interpret both $P'(35) = 0.38$ and its integral $P(35) = 16$.

For Exercises 11–14, verify, by differentiation, that the function $F(x)$ is an antiderivative of the function $f(x)$.

11. $F(x) = \frac{x^5}{5} + \frac{5x^3}{3} + \frac{7x^2}{2} - x + 6$, where $f(x) = x^4 + 5x^2 + 7x - 1$
12. $F(x) = \frac{3x^5}{5} - \frac{5x^4}{4} + 2x^2 - 8$, where $f(x) = 3x^4 - 5x^3 + 4x$
13. $F(x) = \frac{1}{3}e^{3x} - e^{-x} + x^2 + 4$, where $f(x) = e^{3x} + e^{-x} + 2x$
14. $F(x) = (x^2+4)^3(x^2-5)^4 + 8$, where $f(x) = 2x(x^2+4)^2(x^2-5)^3(7x^2+1)$

For Exercises 15–24, determine if the given integral has been evaluated correctly. If it has, write "correctly evaluated." If it has not, write "not correctly evaluated," and then write the correct evaluation.

15. $\int e^{6x}\,dx = \frac{e^{6x+1}}{6x+1} + c$
16. $\int x^{5/2}\,dx = \frac{2}{7}x^{7/2} + c$
17. $\int \sqrt[3]{x}\,dx = \frac{3}{4}x^{4/3}$
18. $\int (16x^3 + 9x^2)\,dx = 48x^2 + 18x + c$
19. $\int (3x+1)^4\,dx = \frac{1}{15}(3x+1)^5 + c$
20. $\int x(x^2+4)^6\,dx = \frac{1}{14}(x^2+4)^7$
21. $\int (e^{3x} - e^{-3x})\,dx = 3e^{3x} + 3e^{-3x} + c$
22. $\int (e^{2x} + \frac{3}{x} + 3)\,dx = \frac{1}{2}e^{2x} + \ln x^3 + 3c$
23. $\int e^{-5x}\ln 4\,dx = \frac{-\ln 4}{5}e^{-5x} + c$
24. $\int \frac{e}{x}\,dx = e\ln|x| + c$

B SKILL ACQUISITION

For Exercises 25–42, evaluate each indefinite integral.

25. $\int 6x^5\,dx$
26. $\int x^{2/3}\,dx$
27. $\int \frac{1}{x^4}\,dx$
28. $\int \frac{-2}{x^{2/3}}\,dx$
29. $\int \frac{8}{\sqrt[5]{x^2}}\,dx$
30. $\int \frac{55}{x}\,dx$
31. $\int 6e^x\,dx$
32. $\int (x^4 + 5x^3 + 2)\,dx$
33. $\int \left(2x^3 + \frac{3}{x^2} + 1\right)\,dx$
34. $\int \left(\frac{4}{x^4} - \frac{3}{x^2}\right)\,dx$
35. $\int (x^{-3} + x^{-2} - x^{-1})\,dx$
36. $\int \left(3e^x + \frac{3}{x} + x^3 + 3\right)\,dx$
37. $\int (0.03 + 0.12x^{-1/2})\,dx$
38. $\int (3x^2 - 6x - 7)\,dx$, where $f(x) = 10$ when $x = 2$
39. $\int (5x^4 - 4x^3 + 8x)\,dx$, where $f(x) = 9$ when $x = 1$
40. $\int \frac{1}{x^{2/3}}\,dx$, where $f(x) = 1$ when $x = 8$
41. $\int \left(3e^x + \frac{4}{x}\right)\,dx$, where $f(x) = 5e$ when $x = 1$
42. $\int (x^2 + 2e^x + 4\sqrt[3]{x})\,dx$, where $f(x) = 1$ when $x = 0$

C APPLYING THE CONCEPTS

43. ***Business: Marginal Cost*** A company's marginal cost C of producing x units of a product is given by the function $C'(x) = -0.104x + 82.12$. At the production level of 600 units, the cost is known to be \$28,752.00,
    a. Find the cost function.
    b. Find the cost of producing 950 items.
    c. Find and interpret both $C'(950)$ and its integral $C(950)$.

44. ***Economics: Population of a City*** The growth rate of a city t years after 1990 is given by the function $P'(t) =$

$240\sqrt[5]{t}+170$. In 1990, the population of the city is 8400.

a. Find the population function.
b. Find the expected population of the city in 1996.
c. Find and interpret both $P'(6)$ and its integral $P(6)$.

45. *Ecology: Environmental Pollution* A city's environmental commission estimates that the amount A (in parts per million) of carbon monoxide in the summer daytime air will, without new protective measures, increase at the rate of $A'(t) = 0.0027t^2 + 0.04t + 0.08$ parts per million t years from now. Currently, the average daytime level of carbon monoxide in the city's air is 3 parts per million. If no air quality protective measures are taken,

a. At what rate will the level of carbon monoxide be changing 5 years from now?
b. What is the expected daytime level of carbon monoxide in this city 5 years from now?

46. *Medicine: Tumor Growth* A medical study has shown that during the first 20 days of radiation therapy, a particular type of malignant tumor decreases according to the function $M'(t) = -0.48t^2 + 1.0t$, where M represents the mass (in grams) of the tumor and t is the time (in days) since the beginning of the radiation treatment. If a tumor of this particular type has a mass of 200 grams just prior to the start of radiation treatment,

a. At what rate will the tumor be decreasing in 10 days?
b. in 15 days?
c. What will be the mass of the tumor in 10 days?
d. in 15 days?

47. *Business: Sales* A manufacturer has determined that the growth rate of sales S (in units) of a newly developed product should be approximated by $S'(t) = \dfrac{2000}{\sqrt[3]{t}}$, where t is the number of years from now. Assuming there are no sales at the introduction of the product to the market,

a. How many units of the product will have been sold 5 years from now?
b. At what rate will the number of sales be changing afater 5 years?
c. Using the result from part (b), estimate how many units of the product will have been sold 6 years from now.
d. Measuring as you did in part (a) and without using the results of part (b), how many units of the product will have been sold 6 years from now?

48. *Biology: Human Surface Area* As a person's mass m (in kilograms) changes, his or her surface area, A (in square meters), changes according to the function $A'(m) = \dfrac{0.073}{\sqrt[3]{m}}$. (A reasonable initial condition here is that a person with no mass has no surface area.)

a. Find the surface area of a person with a mass of 64 kg (64 kg $\approx$ 142 pounds).
b. Find the rate at which the surface area of a 64-kg person is changing.
c. Interpret $A'(64)$.

49. *Medicine: Drug Concentration* For a particular glucose tolerance test for hypoglycemia, physicians believe that the amount A (in milligrams) of glucose remaining in the blood t hours after the ingestion of the glucose changes according to the function $A'(t) = \dfrac{2.4}{\sqrt{t}}$. If for this particular test, 7.9 milligrams of glucose are in the bloodstream 1 hour after the glucose is ingested,

a. At what rate is the glucose in the bloodstream changing after 4 hours?
b. What is the total amount of glucose in the bloodstream after 4 hours?
c. Interpret $A'(4)$.

50. *Business: Inventory Costs* The inventory cost C (in dollars) of stocking x boxes of packages of men's socks is changing at the rate $C'(x) = \dfrac{-203{,}000}{x^2} + 2$. Find and interpret both $C'(350)$ and its integral $C(350)$.Assume it costs \$5000 to store 100 boxes.

51. *Marketing: Advertising and Sales* The management of a chain of frozen yogurt stores believes that t days after the end of an advertising campaign, the rate at which the volume V (in dollars) of sales is changing is approximated by $V'(t) = -29{,}900e^{-0.46t}$. On the day the advertising campaign ends ($t = 0$), the sales volume is \$130,000. Find and interpret both $V'(5)$ and its integral $V(5)$.

52. *Ecology: National Consumption of Toxic Chemicals* A government prepared study shows that the rate at which the national consumption of a particular toxic chemical is changing is given by the function $C'(t) = -17.4e^{-1.2t} + 0.28t$, where C represents the national consumption of the chemical (in consumption units), and t represents the number of years after the study. At the time the study was released, the national consumption of the chemical was 14.5 consumption units. Find and interpret both $C'(2)$ and its integral $C(2)$.

D DESCRIBE YOUR THOUGHTS

53. Explain why $\dfrac{x^4}{4}$ may not be *the* antiderivative of $\int x^3\,dx$.

54. Describe how an initial condition affects an antiderivative.

55. Explain what is special about $\dfrac{d}{dx}\left[e^x\right]$ and $\int e^x\,dx$.

E REVIEW

56. (2.2) Find, if it exists, $\lim_{x\to 7} \dfrac{x^2 - 3x - 28}{x^2 - 6x - 7}$.

57. (2.4) Construct the graph of a function that is discontinuous at $x = 4$ because continuity condition 2 is the first of the three continuity conditions to fail.

58. (3.1) Use the definition of the derivative to find $f'(x)$ for $f(x) = 5x^2 - 8x + 4$.

59. (3.1) Suppose that $P(x)$ (in dollars) represents the profit made on the sale of x units of a product. Interpret $P'(835) = 120.50$.

60. (3.2) Find $f'(x)$ for $f(x) = \dfrac{3x^2 + 1}{4x - 2}$.

61. (3.4) Find $f'(x)$ for $f(x) = (7x + 8)^3(2x - 5)^4$.

62. (4.2) Use the summary table below to construct the graph of the corresponding function.

Int /Val	$f'(x)$	$f''(x)$	$f(x)$	**Behavior of** $f(x)$
$(-\infty, -3)$	$+$	$+$	incr/cu	incr at an incr rate
-3	und	und	$(-3, 6)$	rel max at $(-3, 6)$
$(-3, 0)$	$-$	$+$	decr/cu	decr at a decr rate
0	0		$(0, 3)$	rel min at $(0, 3)$
$(0, 2)$	$+$	$+$	incr/cu	incr at an incr rate
2		0	$(2, 4)$	pt of infl at $(2, 4)$
$(2, 5)$	$+$	$-$	incr/cd	incr at a decr rate
5	0		$(5, 8)$	rel max at $(5, 8)$
$(5, +\infty)$	$-$	$-$	decr/cd	decr at an incr rate

63. (5.1) Suppose x is a variable and b is a constant. Specify the forms of both a power function and an exponential function.

64. (5.3) Find $f'(x)$ for $f(x) = \ln(4x^2 + 1)^3$.

65. (5.4) Find the derivative of the function $f(x) = \dfrac{2 + e^{-2x}}{2 - e^{-2x}}$.

6.2 Integration by Substitution

Introduction

In this section we introduce a process that produces an antiderivative of some functions that do not match any of the basic forms discussed in Section 6.1. For example, the integral $\int(5x^4 + 2x^2 + 7)\,dx$ can be evaluated directly because its form matches one of the basic integral forms presented in Section 6.1. However, the integral $\int \frac{12x+4}{6x^2+4x+1}\,dx$ does not match any of the basic integral forms and therefore cannot by evaluated directly. It can, however, be evaluated indirectly by making a substitution that transforms the integral to a known basic form.

Integration by Substitution

Integration by Substitution

The method of integration called **integration by substitution** involves substituting a variable (any variable other than the original can be used; we often use u) for some part of the integrand and then expressing the rest of the integrand in terms of that variable. This strategy is outlined in Figure 6.2. The flow diagram

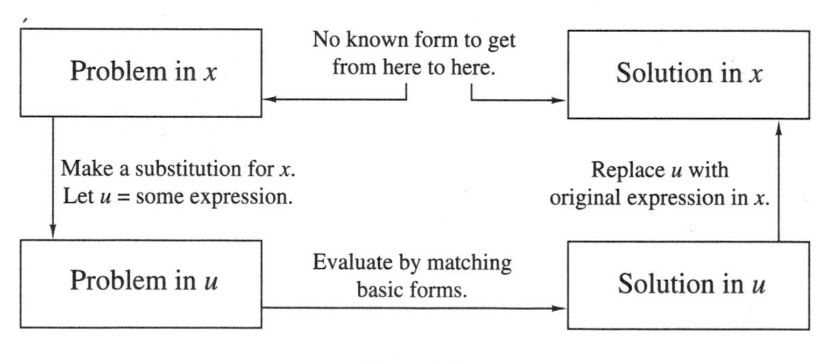

Figure 6.2

shows that we want to get from a problem in x to a solution in x. But because we have no form we can match the integral to, we cannot take this route directly. Instead, we must take another route: We make a substitution that transforms the original integrand to a new integrand that does match one of the basic forms. A beneficial substitution often involves a choice of u for which $\frac{du}{dx}$ is a constant multiple of one of the factors in the integrand. We evaluate that integral, and then resubstitute to get to our destination, a solution in x. We will illustrate the technique in Illuminator Set A.

ILLUMINATOR SET A

Evaluate $\displaystyle\int \frac{12x+4}{6x^2+4x+1}\,dx$.

Solution:

We imagine ourselves in the box at the top left of the flow diagram of Figure 6.2. We have a problem in x with no known basic integral form that will allow us direct passage to a solution in x. We need to take an indirect route via a substitution. In this case, we will let the variable u represent the expression $6x^2+4x+1$ (we will explain why after Illuminator Set B).

Important Note

When making a substitution into an integral involving only one variable, it is very important to make a *complete substitution* so that there is only one variable present at all times. This requirement is the reason why not all integrals can be evaluated by substitution.

With $u = 6x^2+4x+1$, we must still replace $(12x+4)\,dx$ to make the substitution complete. We account for this remaining part of the integrand, $(12x+4)\,dx$, with the differential of u. Since

$$u = 6x^2+4x+1$$

$$\frac{du}{dx} = 12x+4$$

so that the differential of u is

$$du = (12x+4)\,dx$$

Thus, the remaining part of the integrand $(12x+4)\,dx$ is precisely du when $u = 6x^2+4x+1$. Now we can write a complete substitution:

$$\int \frac{12x+4}{6x^2+4x+1}\,dx = \int \frac{du}{u}$$

We are now located in the box at the lower left of the flow diagram; we have the problem in u. Furthermore, we can match this form to a basic integral form to get a solution in u.

$$\int \frac{du}{u} = \ln|u| + c$$

With this solution, we arrive in the box at the lower right of the flow diagram and can use our original substitution to resubstitute to get a solution in x. Since $u = 6x^2+4x+1$, we can replace u in $\ln|u|+c$ with $6x^2+4x+1$, and write

$$\int \frac{12x+4}{6x^2+4x+1}\,dx = \ln|6x^2+4x+1| + c$$

Sometimes, after applying the differential, some adjustments are necessary to get the precise expression that remains in the integrand. Illuminator Set B illustrates this process.

ILLUMINATOR SET B

Evaluate $\int \frac{x}{3x^2+5}\,dx$.

Solution:

We let $u = 3x^2 + 5$. The remaining part of the integrand is $x\,dx$. If integration by substitution is going to work, we should be able to get $x\,dx$ from the differential of u. We let

$$u = 3x^2 + 5$$

so that

$$\frac{du}{dx} = 6x$$

giving us the differential

$$du = 6x\,dx$$

But we want $x\,dx$, not $6x\,dx$. We need to make an adjustment, so we eliminate the 6 by dividing both sides of $du = 6x\,dx$ by 6.

$$du = 6x\,dx$$

$$\frac{du}{6} = \frac{6x\,dx}{6}$$

so that

$$x\,dx = \frac{1}{6}\,du$$

Thus,

$$\begin{aligned}\int \frac{x}{3x^2+5}\,dx &= \int \frac{\frac{1}{6}\,du}{u}\\ &= \frac{1}{6}\int \frac{du}{u}\\ &= \frac{1}{6}\ln|u| + c\\ &= \frac{1}{6}\ln|3x^2+5| + c \quad \text{(since } 3x^2+5 \text{ is always positive)}\\ &= \frac{1}{6}\ln(3x^2+5) + c\end{aligned}$$

You should be able to prove that this result is correct by differentiating $\frac{1}{6}\ln(3x^2+5) + c$ to get $\frac{x}{3x^2+5}$.

A Strategy for Substitution

How do we know which part of an integrand to let u represent? Here are some suggestions that *often* work, but *not always*.

Suggestions for Substitutions

If the integrand involves

1. a quotient, let $u =$ (the denominator).
2. an exponential expression, let $u =$ (the exponent).
3. a power or a radical, let $u =$ (the base or the radicand).

EXAMPLE SET A

Evaluate $\int 3e^{3x+7}\,dx$.

Solution:
First, we will rewrite the integral as $3\int e^{3x+7}\,dx$.

Part of the integrand is an exponential so we will try letting u represent the exponent $3x + 7$. The remaining part of the integrand that involves the variable is dx. If integration by substitution is going to work, we should be able to get dx from the differential of u. We let

$$u = 3x + 7$$

so that

$$\frac{du}{dx} = 3$$

giving us the differential

$$du = 3\,dx$$

Therefore,

$$3\,dx = du$$

Thus,

$$3\int e^{3x+7}\,dx = \int e^{3x+7} \cdot 3\,dx$$

$$= \int e^u\,du \quad \text{(which matches a basic integral form)}$$

$$= e^u + c \qquad \text{replace } u \text{ with } 3x + 7$$

$$= e^{3x+7} + c$$

EXAMPLE SET B

Evaluate $\int 4xe^{5x^2-2}\,dx$.

Solution:
First, we will rewrite the integral as $4\int xe^{5x^2-2}\,dx$. The integrand is an exponential so we will try letting u represent the exponent $5x^2 - 2$. The remaining part of the integrand is $x\,dx$. If integration by substitution is going to work, we should be able to get $x\,dx$ from the differential of u. We let

$$u = 5x^2 - 2$$

so that

$$\frac{du}{dx} = 10x$$

giving us the differential

$$du = 10x\,dx$$

Therefore,

$$x\,dx = \frac{1}{10}\,du$$

Thus,

$$4\int xe^{5x^2-2}\,dx = 4\int e^u \cdot \frac{1}{10}\,du$$

$$= \frac{2}{5}\int e^u\,du \quad \text{(which matches a basic integral form)}$$

$$= \frac{2}{5}e^u + c \qquad \text{replace } u \text{ with } 5x^2 - 2$$

$$= \frac{2}{5}e^{5x^2-2} + c$$

EXAMPLE SET C

Evaluate $\int 10(9x+1)(9x^2+2x+7)^8\,dx$.

Solution:
First, we will rewrite the integral as $10\int(9x+1)(9x^2+2x+7)^8\,dx$. Part of the integrand involves a power so we will try letting u represent the base $9x^2+2x+7$. The remaining part is $(9x+1)\,dx$. If integration by substitution is going to work, we should be able to get $(9x+1)\,dx$ from the differential of u. We let

$$u = 9x^2 + 2x + 7$$

so that

$$\frac{du}{dx} = 18x + 2$$

giving us the differential

$$du = (18x + 2)\,dx$$

which when factored gives

$$du = 2(9x + 1)\,dx$$

We then solve for $(9x + 1)\,dx$.

$$(9x + 1)\,dx = \frac{1}{2}\,du$$

Thus,

$$\begin{aligned} 10\int(9x^2+2x+7)^8(9x+1)\,dx &= 10\int u^8 \cdot \frac{1}{2}\,du \\ &= 5\int u^8\,du \\ &= 5 \cdot \frac{u^9}{9} + c \\ &= \frac{5(9x^2+2x+7)^9}{9} + c \end{aligned}$$

EXAMPLE SET D

Evaluate $\displaystyle\int \frac{x^3}{\sqrt[3]{8x^4-5}}\,dx$.

Solution:
Part of the integrand involves a radical so we will try letting u represent the radicand $8x^4 - 5$. The remaining part is $x^3\,dx$. If integration by substitution is going to work, we should be able to get $x^3\,dx$ from the differential of u. We let

$$u = 8x^4 - 5$$

so that

$$\frac{du}{dx} = 32x^3$$

giving us the differential

$$du = 32x^3\,dx$$

Therefore,

$$x^3\,dx = \frac{1}{32}\,du$$

Thus,

$$\int \frac{x^3}{\sqrt[3]{8x^4 - 5}}\,dx = \int \frac{1}{\sqrt[3]{u}} \cdot \frac{1}{32}\,du$$

$$= \frac{1}{32} \int \frac{1}{\sqrt[3]{u}}\,du$$

$$= \frac{1}{32} \cdot \int \frac{1}{u^{1/3}}\,du$$

$$= \frac{1}{32} \cdot \int u^{-1/3}\,du$$

$$= \frac{1}{32} \cdot \frac{u^{-1/3+1}}{-1/3+1} + c$$

$$= \frac{1}{32} \cdot \frac{u^{2/3}}{2/3} + c$$

$$= \frac{1}{32} \cdot \frac{3u^{2/3}}{2} + c$$

$$= \frac{3}{64}u^{2/3} + c$$

$$= \frac{3}{64}(8x^4 - 5)^{2/3} + c$$

EXAMPLE SET E

Evaluate $\int x\sqrt{x+3}\,dx$.

Solution:
Part of the integrand is a radical so we will try letting u represent the radicand $x+3$. The remaining part is $x\,dx$. If integration by substitution is going to work, we should be able to get $x\,dx$ from the differential of u. We let

$$u = x + 3$$

so that

$$\frac{du}{dx} = 1$$

giving us the differential

$$du = 1\,dx$$

or

$$du = dx$$

But this only gives us dx. We still need to get the remaining x in terms of u. We need an adjustment that is different from the adjustments we made in the previous example sets. By our original substitution, $u = x + 3$, so that by subtracting 3 from each side we get $x = u - 3$, and then we have x in terms of u. To summarize, we now have:

$$u = x + 3$$

$$x = u - 3$$

$$dx = du$$

Thus,

$$\int x\sqrt{x+3}\,dx = \int (u-3)\sqrt{u}\,du$$

$$= \int (u-3)u^{1/2}\,du \qquad \text{multiply}$$

$$= \int (u^{3/2} - 3u^{1/2})\,du$$

$$= \int u^{3/2}du - 3\int u^{1/2}\,du$$

$$= \frac{u^{5/2}}{5/2} - 3\frac{u^{3/2}}{3/2} + c$$

$$= \frac{2u^{5/2}}{5} - 2u^{3/2} + c \qquad \text{replace } u \text{ with } x+3$$

$$= \frac{2(x+3)^{5/2}}{5} - 2(x+3)^{3/2} + c$$

Integrating the Exponential Function

As you may recall we noted that the function $f(x) = e^{ax}$ occurs often in applications and that its derivative, which follows directly from the chain rule, can be obtained quickly by multiplying the expression e^{ax} by a, the coefficient of x. That is, if $f(x) = e^{ax}$, then $f'(x) = ae^{ax}$.

Similarly, since the integral of $f(x) = e^{ax}$ occurs often, it is convenient to have a special rule specifically for it so that we can avoid making a substitution every time we come across it. Therefore, to integrate $f(x) = e^{ax}$, we simply divide the expression e^{ax} by a, the coefficient of x. That is,

The Integral of $f(x) = e^{ax}$

If $f(x) = e^{ax}$, then

$$\int e^{ax}\,dx = \frac{e^{ax}}{a} + c$$

You will be asked to verify this result by substitution in Exercise 37.

EXAMPLE SET F

1. $\int e^{6x}\,dx$

Solution:

$$\int e^{6x}\,dx = \frac{e^{6x}}{6} + c$$

2. $\int e^{\frac{2}{3}x}\,dx$

Solution:

$$\int e^{\frac{2}{3}x}\,dx = \frac{e^{\frac{2}{3}x}}{2/3} + c$$

$$= \frac{3e^{\frac{2}{3}x}}{2} + c$$

Integration by Substitution and Technology

Using Derive You can use Derive to generate the formulas for

$$\frac{d}{dx}\left[e^{ax}\right] \quad \text{and} \quad \int e^{ax}\,dx$$

The following entries show how this is done.

Declare
Function
e1
exp(ax)
Declare
Function
e2

```
dif(e1(x),x)
Simplify
Declare
Function
e3
int(e1(x),x)
Simplify
```

This set of Derive commands names the function defining the expression e^{ax} as **e1**, its derivative as **e2**, and its integral as **e3**.

Derive Exercises

1. Find $\int 3x^2(x^3-7)^5\,dx$ using both pencil and paper and Derive. Compare the results.
2. Use Derive to find $\int (\ln x)^4 x^{-2}\,dx$. Before using Derive, state what you think the proper substitution should be.
3. Use Derive to find $\int \frac{\sqrt{4+5\ln x}}{x}\,dx$. Before using Derive, state what you think the proper substitution should be.
4. Use Derive to find $\int \frac{(x+2)^3}{(x+4)^5}\,dx$. Before using Derive, state what you think the proper substitution should be.
5. Use Derive to find $\int \ln x\,dx$. Before using Derive, state what you think the proper substitution should be.
6. Twice a year, a county treats its lakes with a chemical that reduces the bacteria present in the lakes. According to the chemical manufacturer, the rate, in bacteria per liter of water, at which the concentration of bacteria in a lake changes is a function of the number of days, t, after the chemical is introduced, and is approximated by the expression

$$\frac{-1377t^{1.8}}{1.2+t^{2.8}}$$

If the initial concentration of bacteria in a lake was 5200 per liter of water, what will be the concentration after 10 days?

EXERCISE SET 6.2

A UNDERSTANDING THE CONCEPTS and B SKILL ACQUISITION

For Exercises 1–36, evaluate each integral.

1. $\int (3x-4)^4\,dx$
2. $\int \sqrt[3]{5x+1}\,dx$
3. $\int \frac{6}{7x+5}\,dx$
4. $\int \frac{10}{2x-9}\,dx$
5. $\int 2x\sqrt{x^2+4}\,dx$
6. $\int (9x^2+2)(3x^3+2x)^{3/2}\,dx$
7. $\int \frac{2}{3x^{1/3}(x^{2/3}+4)^3}\,dx$
8. $\int \sqrt{x}\sqrt{x^{3/2}+2}\,dx$
9. $\int \frac{e^{4x}}{e^{4x}+2}\,dx$
10. $\int e^{2x}\sqrt{e^{2x}+1}\,dx$

11. $\int x^2 e^{5x^3-2}\,dx$

12. $\int \frac{e^{\sqrt{x}}}{\sqrt{x}}\,dx$

13. $\int \frac{12x^2+5}{4x^3+5x-6}\,dx$

14. $\int \frac{10x^3+4x}{10x^4+8x^2-1}\,dx$

15. $\int \frac{(7+\ln x)^{3/2}}{x}\,dx$

16. $\int \frac{1+\ln x}{x\ln x}\,dx$

17. $\int (2x\ln x)^4(\ln x+1)\,dx$

18. $\int \frac{x}{x+3}\,dx$

19. $\int \frac{5x}{5x-4}\,dx$

20. $\int \frac{x}{\sqrt{x+4}}\,dx$

21. $\int 4x(2x+5)^5\,dx$

22. $\int \frac{6(\ln x^2)^4}{x}\,dx$

23. $\int \frac{x\ln(x^2+4)}{\ln(x^2+4)}\,dx$

24. $\int (x-7)(x+7)^4\,dx$

25. $\int \frac{x-10}{(x+2)^3}\,dx$

26. $\int 6xe^{-x^2}\,dx$

27. $\int e^{-0.02x}\,dx$

28. $\int e^{-0.12x}\,dx$

29. $\int 45e^{0.15x}\,dx$

30. $\int \frac{e^{1/x}}{x^2}\,dx$

31. $\int \left(e^{3x}+4e^{-3x}\right)dx$

32. $\int \left(x^3e^{x^4}-x^2\right)dx$

33. $\int \left(e^{x-2}+\frac{\ln x}{x}\right)dx$

34. $\int \frac{18x}{9x+5}\,dx$

35. $\int x^5\sqrt[3]{x^3+1}\,dx$

36. $\int \frac{e^{1/x^2}}{x^3}\,dx$

37. Show by substitution that $\int e^{ax}\,dx = \frac{e^{ax}}{a}+c$.

C APPLYING THE CONCEPTS

38. *Manufacturing: Assembly Efficiency* A worker having no experience with a particular process can assemble 60 units of a product. After training and t weeks of experience, a worker can assemble the units at the rate of

$$N'(t) = 450e^{-3t}$$

units per week.

a. Specify the function that describes the worker's weekly output.
b. Find the number of units a worker having 5 weeks of experience can assemble.
c. What is the maximum number any worker can be expected to assemble per week? (Hint: When you verbalize part (c) to yourself, it should spark the thought of a *limiting value*. You can answer this question by taking the proper limit.)

39. *Medicine: Antibody Production* A medical researcher believes that the human immune system produces antibodies to a particular vaccine at the rate of

$$N'(t) = \frac{3000}{10t+15}$$

thousand per day.

a. Specify the function that describes the antibody production in terms of t.
b. Find the number of antibodies that are produced 5 days after vaccination.

40. *Psychology: Proofreading Efficiency* The number of pages N of technical matter a proofreader can read each day depends on the number of consecutive days t the proofreader has been reading such material. A psychologist believes that the number of pages changes according to the function

$$N'(t) = 24.5e^{-0.35t}$$

After reading such material for 10 consecutive days, a particular proofreader can read 50 pages of material.

a. Specify the function that describes the number of pages this proofreader can read in terms of t.
b. Find the number of pages of technical material that can be proofread by this reader after having read for 5 consecutive days.
c. What is the upper limit, if one exists, to the number of pages that can be proofread by this proofreader each day?

41. ***Business: Resale Value*** The resale value R (in dollars) of a laser optical device t months from its time of purchase is decreasing at the rate of

$$R'(t) = -1580e^{-0.004t}$$

dollars each month. Ten months after the time of purchase, the device has a resale value of $370,000.

a. Specify the function that describes the resale value of the device in terms of t.
b. Find the resale value of the device 24 months from its time of purchase.
c. What is the resale value of this device in the long run?

42. ***Ecology: Air Pollution*** The level of pollution P (in parts per million) over a city changes as the temperature T (in degrees Celsius) changes according to the function

$$P'(t) = \frac{0.024T}{\sqrt{0.04T + 180}}$$

When the temperature is 0°, the level of pollution is 4 parts per million.

a. Specify the function that describes the level of pollution in terms of the temperature.
b. Find the level of pollution when the temperature is 25°.
c. What is the upper limit, if one exists, to the level of air pollution?

43. ***Marketing: Advertising and Sales*** A company has determined that the number N of daily sales of one of its products t days after an advertising campaign decreases according to

$$N'(t) = -6875e^{-0.32t}$$

If on the day that the advertising campaign ends ($t = 0$), the number of daily sales is 55,000,

a. Specify the function that describes the number of daily sales in terms of the number of days after the advertising campaign.
b. Find the number of daily sales 10 days after the end of the campaign.
c. What is the expected number of daily sales long after the advertising campaign is over?

44. ***Manufacturing: Worker Production Level*** Data kept by the human resources department of a manufacturing company indicates that t hours after beginning work, the production P (in units completed) of an average worker changes at the rate

$$P'(t) = \frac{0.2(t+2)}{\sqrt{t^2 + 4t}}$$

If 2 hours after beginning work, an average worker can complete 15 units,

a. Specify the function that describes the number of units an average worker can complete in terms of the number of hours after beginning work.
b. Find the number of units an average worker can complete 6 hours after beginning work.
c. Approximate the number of units an average worker can complete from hour 3 to hour 4.

45. ***Economics: Population of a City*** A city's population P is expected to grow at the rate

$$P'(t) = \frac{126e^{18t}}{1 + e^{18t}}$$

where t is the number of months from the present. Presently, the population is 28,000.

a. Specify the function that describes the population of the city in terms of the number of months from the present.
b. Find the population of the city 10 months from now.
c. Approximate the increase in population from month 5 to month 6.

D DESCRIBE YOUR THOUGHTS

46. Describe the strategy you would use to evaluate $\int \frac{2x}{x^2+4}\,dx$.

47. Explain why the substitution method can be used for any function that matches a basic integral form.

48. Explain why the substitution method is an unlikely method for the evaluation of $\int \frac{e^x}{x^3 - 5}\,dx$.

E REVIEW

49. **(2.3)** Find, if it exists, $\lim_{x\to\infty} \frac{x^2 - 4x + 1}{3x^2 - 11x + 6}$.

50. **(2.5)** Find the average rate of change of $f(x) = x^2 - 4x$ with respect to x on the interval $[2, 5]$.

51. (3.2) *Manufacturing: Revenue* A manufacturer of ceramic vases has determined that her weekly revenue and cost functions for the manufacture and sale of x vases are $R(x) = 93x - 0.09x^2$ dollars and $C(x) = 1280 + 32x - 0.04x^2$ dollars, respectively. Determine the marginal profit realized on the 500th vase.

52. (3.4) Find $\frac{dy}{dx}$ if $y = 2u^2 + 4$ and in turn $u = 5x^2 + 1$.

53. (3.5) Find y' for $7x^2y^2 + y^3 = 2$.

54. (4.1) Should a curve that is increasing at a decreasing rate be classified as concave upward or concave downward?

55. (4.2) *Economics: Energy Use* The graph illustrated below displays the relationship between the amount of energy E (in energy units) used in a city and the time t in hours from midnight. Describe the behavior of this function.

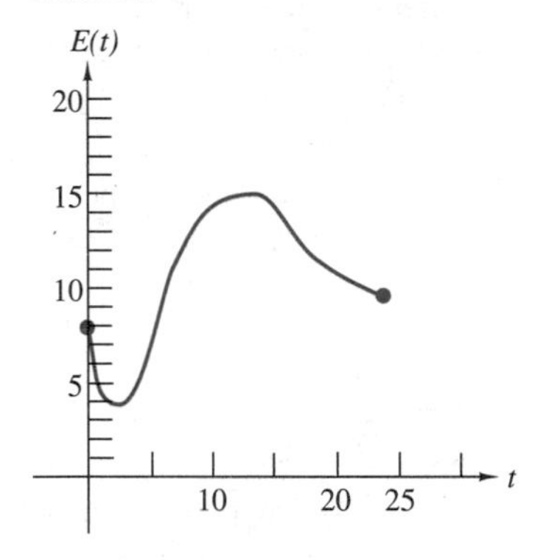

56. (4.3) *Mathematics: Volume* A block of ice is melting so that each edge is decreasing at the rate of 1.5 inches per hour. Find how fast the volume of the ice is decreasing at the instant when each edge is 6 inches long. (Assume the block to be a perfect cube.)

57. (5.4) Find $f'(x)$ for $f(x) = 6x - 4e^{-0.05x}$.

58. (6.1) *Business: Insurance Analysis* An insurance company's analysis of a city's hospital data indicates a relationship between the number of deaths per week due to serious damage to a person's major organ (such as the heart, liver, or brain), and the time t in hours after the damage occurs. If the relationship is expressed by the function $N(t)$, interpret both $N'(4) = -2$ and its integral $N(4) = 8$.

6.3 Integration by Parts

Introduction

Integrating Products of Functions Using Integration by Parts

Integration by Parts and Technology

Introduction

We have seen that sums and differences of functions can be differentiated term by term, but their products and quotients cannot be. That is,

$$\frac{d}{dx}[f(x) \pm g(x)] = \frac{d}{dx}f(x) \pm \frac{d}{dx}g(x), \quad \text{but}$$

$$\frac{d}{dx}[f(x) \cdot g(x)] \neq \frac{d}{dx}f(x) \cdot \frac{d}{dx}g(x), \quad \text{and}$$

$$\frac{d}{dx}\left[\frac{f(x)}{g(x)}\right] \neq \frac{\frac{d}{dx}f(x)}{\frac{d}{dx}g(x)}$$

The situation is similar for integration. That is,

$$\int [f(x) \pm g(x)]\,dx = \int f(x)\,dx \pm \int g(x)\,dx, \quad \text{but}$$

$$\int [f(x) \cdot g(x)]\,dx \neq \int f(x)\,dx \cdot \int g(x)\,dx, \quad \text{and}$$

$$\int \frac{f(x)}{g(x)}\,dx \neq \frac{\int f(x)\,dx}{\int g(x)\,dx}$$

In this section, we will investigate a method for handling integrals of products of functions.

Integrating Products of Functions Using Integration by Parts

The product for derivatives states that for differentiable functions $u = u(x)$ and $v = v(x)$,

$$\frac{d}{dx}[u \cdot v] = u \cdot \frac{dv}{dx} + v \cdot \frac{du}{dx}$$

Integrating both sides of this equation with respect to x produces

$$\int \frac{d}{dx}[u \cdot v]\,dx = \int \left[u \cdot \frac{dv}{dx} + v \cdot \frac{du}{dx}\right] dx$$

$$u \cdot v = \int \left[u \cdot \frac{dv}{dx}\right] dx + \int \left[v \cdot \frac{du}{dx}\right] dx$$

$$u \cdot v = \int u\,dv + \int v\,du$$

Solving this equation for $\int u\,dv$ produces a formula called **integration by parts**.

Integration by Parts

If $u = u(x)$ and $v = v(x)$ are differentiable functions, then

$$\int u\,dv = uv - \int v\,du$$

This formula can be used for integrating products of two functions, one representing u and the other dv. If the selections of u and dv are made carefully, the integral on the right side, $\int v\,du$, will be easier to integrate than the one on the left side, $\int u\,dv$. We will illustrate this by the following examples.

EXAMPLE SET A

Evaluate $\displaystyle\int x^4 \ln x\,dx$.

Solution:
Since this involves a product, we will try to solve it using integration by parts. We need to identify one part as u and the other part as dv; u should be something that is easy to differentiate and dv should be something that is easy to integrate. Since $\ln x$ is easy to differentiate but not to integrate, we will try the following substitutions.

Let $u = \ln x$ and $dv = x^4 dx$, then

$$du = \frac{1}{x}dx \text{ and } v = \int x^4\,dx = \frac{x^5}{5}$$

Now, substituting into the formula $\displaystyle\int u\,dv = uv - \int v\,du$, we get

$$\int x^4 \ln x\,dx = \int \underbrace{\ln x}_{u} \cdot \underbrace{x^4\,dx}_{dv} = \underbrace{\ln x}_{u} \cdot \underbrace{\frac{x^5}{5}}_{v} - \int \underbrace{\frac{x^5}{5}}_{v} \cdot \underbrace{\frac{1}{x}\,dx}_{du}$$

$$= \frac{1}{5}x^5 \ln x - \frac{1}{5}\int x^4\,dx$$

$$= \frac{1}{5}x^5 \ln x - \frac{1}{5} \cdot \frac{x^5}{5} + C$$

$$= \frac{1}{25}x^5(5 \ln x - 1) + C$$

Thus, $\int x^4 \ln x \, dx = \frac{1}{25}x^5(5 \ln x - 1) + C.$

EXAMPLE SET B

Evaluate $\int xe^{4x} \, dx.$

Solution:

Since both parts are easy to both integrate and differentiate, we can make our substitution either way. If we let $u = x$ and $dv = e^{4x} \, dx$, then

$$du = dx \text{ and } v = \int e^{4x} \, dx = \frac{1}{4}e^{4x}$$

Now, substituting into the formula $\int u \, dv = uv - \int v \, du$, we get

$$\int xe^{4x} \, dx = x \cdot \frac{1}{4}e^{4x} - \int \frac{1}{4}e^{4x} \, dx$$

$$= \frac{1}{4}xe^{4x} - \frac{1}{4}\int e^{4x} \, dx$$

$$= \frac{1}{4}xe^{4x} - \frac{1}{4} \cdot \frac{1}{4}e^{4x} + C$$

$$= \frac{1}{16}e^{4x}(4x - 1) + C$$

Thus, $\int xe^{4x} \, dx = \frac{1}{16}e^{4x}(4x - 1) + C.$

EXAMPLE SET C

Evaluate $\int \ln x \, dx.$

Solution:

This integral can also be viewed as a product and, therefore, solved by integration by parts. Since $\ln x$ is what we are trying to integrate, we will let it be the part that we differentiate. If we let $u = \ln x$ and $dv = dx$, then

$$du = \frac{1}{x}dx \text{ and } v = \int dx = x$$

Now, substituting into the formula $\int u\,dv = uv - \int v\,du$, we get

$$\begin{aligned}\int \ln x\,dx &= \ln x \cdot x - \int x \cdot \frac{1}{x}\,dx \\ &= x\ln x - \int 1\,dx \\ &= x\ln x - x + C\end{aligned}$$

Thus, $\int \ln x\,dx = x\ln x - x + C.$

EXAMPLE SET D

A manufacturer estimates that the relationship between the number of months t from now and the number N (in thousands) of units of a product she can produce per month is given by

$$N(t) = 14te^{-0.08t}$$

Find the function that estimates the total production by the manufacturer if the total production initially is 0.

Solution:

The total production is the accumulated production, which we can find by integration.

$$\text{Total Production} = \int N(t)\,dt = \int 14te^{-0.08t}\,dt = 14\int te^{-0.08t}\,dt$$

Integrating by parts, if we let $u = t$ and $dv = e^{-0.08t}\,dt$, then

$$du = dt \text{ and } v = \frac{e^{-0.08t}}{-0.08} = -12.5e^{-0.08t}$$

Now, substituting into the formula $\int u\,dv = uv - \int v\,du$, we get

$$\begin{aligned}\int 14te^{-0.08t}\,dt &= 14\left[t \cdot (-12.5)e^{-0.08t} - \int \left(-12.5e^{-0.08t}\right)\,dt\right] \\ &= 14\left(-12.5te^{-0.08t} + 12.5\int e^{-0.08t}\,dt\right) \\ &= 14\left(-12.5te^{-0.08t} + 12.5\frac{e^{-0.08t}}{-0.08} + C\right) \\ &= 14\left(-12.5te^{-0.08t} - 156.25e^{-0.08t} + C\right) \\ &= -175e^{-0.08t}\,(t + 12.5) + C\end{aligned}$$

Initially, at $t = 0$, the total production is 0. Therefore,

$$\begin{aligned}\text{Total Production} = 0 &= -175e^{-0.08\cdot 0}(0 + 12.5) + C \\ 0 &= -175e^{0}(12.5) + C \\ 0 &= -175 \cdot 1(12.5) + C \\ 0 &= -2187.5 + C \\ 2187.5 &= C\end{aligned}$$

Consequently, Total Production $= -175e^{-0.08t}(t + 12.5) + 2187.5$ is the function we were asked to find.

Integration by Parts and Technology

Using Derive We can use Derive to integrate by parts. To do so you must load certain utilities (just as you needed to do to have Derive perform implicit differentiation). The following Derive instructions demonstrate the evaluation of $\int xe^{4x}\,dx$.

Transfer
Load
Utility
misc
Author
int_parts(x,exp(4x),x)
Simplify

The commands **T**ransfer, **L**oad, **U**tility, and **m**isc load Derive's miscellaneous commands, of which integration by parts is one. You might also try evaluating this integral using the commands presented in Section 6.1.

Derive Exercises

1. Use Derive's integration by parts utility to evaluate $\int x^3 e^x\,dx$. Try to check the result using pencil and paper.

2. Use Derive's integration by parts utility to evaluate $\int \frac{\ln x}{x}\,dx$.

3. Use Derive's integration by parts utility to evaluate $\int x \ln 6x\,dx$.

4. Use Derive's integration by parts utility to evaluate $\int (\ln x)^4\,dx$.

5. Use Derive to construct a formula for $\int x^n e^{ax}\,dx$, where n is a positive integer and a is any real number.

6. A well is expected to produce oil at the rate of $14{,}150te^{-1.85t}$ barrels per year t years from now. Find the total amount of oil this well is expected to produce over the next 5 years.

EXERCISE SET 6.3

A UNDERSTANDING THE CONCEPTS and B SKILL ACQUISITION

Evaluate each integral. (Integration by parts may not be necessary.)

1. $\int xe^x \, dx$
2. $\int xe^{7x} \, dx$
3. $\int xe^{-8x} \, dx$
4. $\int x^2e^{-x} \, dx$
5. $\int x^3e^{3x} \, dx$
6. $\int \ln(4x) \, dx$
7. $\int \ln(x^3) \, dx$
8. $\int 5x^3 \ln x \, dx$
9. $\int (\ln x)^3 \, dx$
10. $\int \frac{\ln x}{x^2} \, dx$
11. $\int \frac{1}{x} \ln x \, dx$
12. $\int x\sqrt{x-1} \, dx$
13. $\int x(\ln x)^3 \, dx$
14. $\int xe^{x^2} \, dx$
15. $\int \ln(7x-4) \, dx$
16. $\int \frac{x}{e^x} \, dx$
17. $\int x(1-x)^{3/2} \, dx$
18. $\int x(3x+4)^2 \, dx$
19. $\int e^{x+3}(2x+1) \, dx$
20. $\int x\ln(x^2) \, dx$
21. $\int (2x+1)^2 \ln(2x+1) \, dx$

C APPLYING THE CONCEPTS

22. ***Business: Rate of Sales*** A manufacturer has determined that t weeks after the end of an advertising campaign, the rate of change of sales (in 100,000 units per week) is given by the function

$$f(t) = te^{-0.2t}$$

At the end of the advertising campaign, sales were 105,000 units per week. Find the function that describes the total number of sales after t weeks after an advertising campaign.

23. ***Business: Marginal Cost*** The marginal cost for a manufacturer for the production of x units of a commodity is given by the function

$$C(x) = x\ln(x+1)$$

Find the function that describes the total change in the cost of production in terms of x if the total change was 0 initially.

24. ***Manufacturing: Demand for a Product*** A manufacturer believes that the demand for one of its products over the next 5 years is given by the function

$$N(t) = 420\left(15 + te^{-0.1t}\right)$$

where t is the number of years from now. Find the function that describes the total number of the product that is demanded after t years when initially the number in demand is 6300.

25. ***Business: Revenue*** The daily revenue R, in dollars, for a company for one of its products is given by the function

$$R(t) = 12{,}500 + 120.5t^2e^{-t/2}, \quad 0 \le t \le 365$$

where t is the number of days from now. Find the function that describes the total revenue realized by the company t days from now if the total revenue was \$12,500 initially.

26. ***Medicine: Drug Assimilation*** The amount A (in milligrams) of a drug assimilated into an adult male's bloodstream t minutes after ingesting the drug is given by the function

$$A(t) = 4te^{-0.8t}$$

Find the function that describes the total amount of the drug assimilated into the male's bloodstream t minutes after ingestion. Of course, initially, nothing was assimilated.

27. ***Economics: Energy Usage*** A city has determined that the number N of kilowatt hours used by a family each day is given by the function

$$N(t) = 8te^{-0.8t}$$

where t is the time in hours from 6:00 AM. Find the function that describes the total usage by the family in terms of t.

D DESCRIBE YOUR THOUGHTS

28. Describe the strategy you would use to evaluate $\int xe^{3x}\,dx$.

29. Describe what you are looking for in an integral that alerts you to possibly applying the method of integration by parts.

E REVIEW

30. **(2.2)** Find, if it exists, $\lim_{x\to 6} \frac{x^2 - 3x + 8}{x^2 - 5x - 6}$.

31. **(2.3)** Find, if it exists, $\lim_{x\to\infty} \frac{2x^2 + 3x + 1}{x^3 + 1}$.

32. **(2.4)** The function $f(x) = \begin{cases} x^2 - 4x + 5, & x \le 2 \\ 3x + 1, & x \ge 2 \end{cases}$

is discontinuous at $x = 2$. Which of the three continuity conditions is the first to be violated?

33. **(2.6)** Find the average rate of change of the function

$$f(x) = x^2 - 5x + 2$$

as x changes from $x = 2$ to $x = 5$.

34. **(3.1)** Use the definition of the derivative to find $f'(x)$ for $f(x) = \frac{x}{x+2}$.

35. **(3.4)** Find $f'(x)$ for $f(x) = (4 - 3x)^6(2x^3 - 3)^3$.

36. **(3.5)** Find $\frac{dy}{dx}$ for $xy^2 - 3y + 2x + 4 = x^3$.

37. **(4.3)** ***Business: Revenue*** A company has determined that the revenue it realizes from the sale of x units of a product is given by the function

$$R(x) = -2x^3 + 96x^2 + 1736x$$

where x is measured in thousands of units. Find the number of units that produce the maximum revenue. What is this maximum revenue?

38. **(5.4)** Find $f'(x)$ for $f(x) = e^{5x}\ln 5x$.

39. **(6.2)** Evaluate $\int \frac{4x}{x^2 + 3}\,dx$.

6.4 The Definite Integral

Introduction

Definite Integration

As we have seen so well, differentiation is the language of change, and as we experienced in Sections 6.1 and 6.2, indefinite integration is the language of antidifferentiation. In this section, we will see that another form of integration, called **definite integration**, is the language of accumulation.

Developing the Definite Integral

Suppose that since 1985, the sales of the Dewtex Publishing Company have been growing according to the function

$$S(t) = 48e^{1.2t}$$

where $S(t)$ is the *rate* (think derivative) of sales (in dollars) t years from 1985. What is the total amount of sales of Dewtex for the years 1988 through 1992? That is, how much money has Dewtex accumulated from sales from 1988 to 1992?

This is an accumulation problem since the total amount of sales from 1988 to 1992 is just the amount of dollars that have accumulated from sales over this interval of time. We will let $T(t)$ represent the total accumulated sales. Then the accumulated sales from 1988 to 1992 is found by evaluating $T(7) - T(3)$. (1992 is 7 years from 1985 and 1988 is 3 years from 1985 so that $T(7)$ represents the accumulated sales from 1985 to 1992, $T(3)$ represents the accumulated sales from 1985 to 1988, and $T(7) - T(3)$ represents the required difference.)

But since $S(t)$ is the *rate of change* of $T(t)$, $T(t)$ is an integral of $S(t)$ and

$$\begin{aligned} T(t) &= \int S(t)\,dt \\ &= \int 48e^{1.2t}\,dt \\ &= \frac{48e^{1.2t}}{1.2} + c \\ &= 40e^{1.2t} + c \end{aligned}$$

Thus,

$$T(t) = 40e^{1.2t} + c$$

and

$$\begin{aligned} T(7) &= 40e^{1.2(7)} \\ &\approx 177{,}882.67 + c \end{aligned}$$

and

$$\begin{aligned} T(3) &= 40e^{1.2(3)} \\ &\approx 1463.93 + c \end{aligned}$$

so that

$$\begin{aligned} T(7) - T(3) &= 177{,}882.67 + c - (1463.93 + c) \\ &= 177{,}882.67 + c - 1463.93 - c \\ &= 176{,}418.74 \end{aligned}$$

Thus, between 1988 and 1992, Dewtex has accumulated \$176,418.74 from sales.

Notice that in each integration, the constant of integration c appeared, but cancels out in the end. When using integration to compute an accumulation, we omit the constant of integration since the final evaluation is independent of the choice of c (it may be any value) and c always cancels out.

Let's analyze what we have done.

1. We wished to find the total amount, $F(x)$, accumulated by a function $f(x)$ over some interval $[a, b]$. (In our example, $f(x)$ was $S(t)$, $a = 3$, and $b = 7$.)
2. Although the function $F(x)$ was not originally known, its derivative $f(x)$ was, and we were able to determine the family to which the function $F(x)$ belonged by integrating the given derivative function.
3. The total accumulation was the difference between $F(b)$ and $F(a)$; that is, $F(b) - F(a)$.

It is common to use $F(x)\Big|_a^b$ to indicate the evaluation $F(b) - F(a)$, and the constant of integration is omitted when writing $F(x)$.

The Fundamental Theorem of Calculus

We now present the Fundamental Theorem of Calculus, a theorem that summarizes this evaluation process.

The Fundamental Theorem of Calculus

If $f(x)$ is a continuous function on an interval $[a, b]$, where $a < b$, and if $F(x)$ is an antiderivative of $f(x)$, then the **definite integral** of $f(x)$ from a to b is

$$\int_a^b f(x)\,dx = F(x)\Big|_a^b = F(b) - F(a)$$

and represents the total amount accumulated by $F(x)$ as x increases from a to b. The number a is called the **lower limit of integration** and the number b the **upper limit of integration**.

The Difference Between Indefinite and Definite Integrals

It is worthwhile noting the difference between the indefinite integral $\int f(x)\,dx$ and the definite integral $\int_a^b f(x)\,dx$.

1. The indefinite integral $\int f(x)\,dx$ *is a function* of x and represents a family of antiderivatives of (f). For example,

$$\int x^3\,dx = \frac{x^4}{4} + c, \qquad \text{which is a function of } x.$$

2. The definite integral $\int_a^b f(x)\,dx$ *is an accumulator* and registers *a real number*, a constant that records the total amount accumulated by a particular member of a family of antiderivatives. For example,

$$\begin{aligned}\int_1^3 x^3\,dx &= \frac{x^4}{4}\Big|_1^3 \\ &= \frac{3^4}{4} - \frac{1^4}{4} \\ &= \frac{81}{4} - \frac{1}{4} \\ &= \frac{80}{4} \\ &= 20, \quad \text{which is a real number}\end{aligned}$$

Properties of the Definite Integral

The definite integral satisfies the following properties.

1. $\int_a^b kf(x)\,dx = k\int_a^b f(x)\,dx$, where k is a constant.

2. $\int_a^a f(x)\,dx = 0$ No accumulation without change.

3. $\int_a^b f(x)\,dx = -\int_b^a f(x)\,dx$

4. $\int_a^b [f(x) \pm g(x)]\,dx = \int_a^b f(x)\,dx \pm \int_a^b g(x)\,dx$

5. $\int_a^b f(x)\,dx = \int_a^c f(x)\,dx + \int_c^b f(x)\,dx$, where $a \le c \le b$.

EXAMPLE SET A

Evaluate $\int_3^5 (x^2 + 3x - 2)\,dx$.

Solution:

$$\begin{aligned}
\int_3^5 (x^2 + 3x - 2)\,dx &= \int_3^5 x^2\,dx + \int_3^5 3x\,dx - \int_3^5 2\,dx \\
&= \left(\frac{x^3}{3} + \frac{3x^2}{2} - 2x\right)\Big|_3^5 \\
&= \left(\frac{5^3}{3} + \frac{3(5)^2}{2} - 2(5)\right) - \left(\frac{3^3}{3} + \frac{3(3)^2}{2} - 2(3)\right) \\
&= \left(\frac{125}{3} + \frac{75}{2} - 10\right) - \left(\frac{27}{3} + \frac{27}{2} - 6\right) \\
&= \frac{158}{3}
\end{aligned}$$

Interpretation: As x increases from 3 to 5, the function $F(x) = \frac{x^3}{3} + \frac{3x^2}{2} - 2x$ accumulates $\frac{158}{3}$ units.

The Definite Integral and Substitution

Sometimes we must use the substitution method to evaluate a definite integral. Example Sets B and C show two different ways. In Example Set B, we show the most common and simplest method, replacing the limits of the original integral with new ones that are appropriate for the new variable from the substitution. The new definite integral can be evaluated without returning to the original variable.

EXAMPLE SET B

Evaluate $\int_4^{10} \frac{1}{x-3}\,dx$.

Solution:
We let $u = x - 3$. Then $du = dx$.

Now we change the limits of integration. (These are the new steps!) Since $u = x - 3$, when $x = 4$, $u = 4 - 3 = 1$, and when $x = 10$, $u = 10 - 3 = 7$. Since all the variables are in terms of u,

$$\int_4^{10} \frac{1}{x-3}\,dx = \int_1^7 \frac{1}{u}\,du$$

$$= \ln|u|\Big|_1^7$$

$$= \ln|7| - \ln|1|$$

$$= \ln 7 - 0 \quad (\text{since } \ln 1 = 0)$$

$$= \ln 7$$

In Example Set C, we will show how this same integral can be evaluated by a process that keeps the limits of integration in terms of x, transforms to u, and then resubstitutes back to x and evaluates. (Note the greater number of steps.)

EXAMPLE SET C

Evaluate $\displaystyle\int_4^{10} \frac{1}{x-3}\,dx$.

Solution:
Since the integrand does not match one of the basic integral forms, we will try the substitution method. Since the integrand is a quotient, we will try letting u represent the denominator, $x - 3$.

We let $u = x - 3$. The remaining part of the integrand is $1 \cdot dx = dx$. If integration by substitution is going to work, we should be able to get dx from the differential of u. We let

$$u = x - 3$$

so that

$$\frac{du}{dx} = 1$$

giving us the differential

$$du = dx$$

so that

$$dx = du$$

which is precisely the remaining part of the integrand.

We need to be careful about what we write at this point. The limits of integration are in terms of x and the integrand is in terms of u. To keep this fact clearly established, we write

$$\int_{x=4}^{x=10} \frac{1}{u}\,du$$

This integrand now matches one of the basic integral forms.

$$\begin{aligned}
\int_{x=4}^{x=10} \frac{1}{u}\,du &= \ln|u|\Big|_{x=4}^{x=10} \quad \text{Now we resubstitute and evaluate} \\
&= \ln|x-3|\Big|_{x=4}^{x=10} \\
&= \ln|10-3| - \ln|4-3| \\
&= \ln 7 - \ln 1 \\
&= \ln 7 - 0 \quad (\text{since } \ln 1 = 0) \\
&= \ln 7
\end{aligned}$$

Thus, $\displaystyle\int_4^{10} \frac{1}{x-3}\,dx = \ln 7 \approx 1.9459.$

Please note that the method used in Example Set B is preferred over that used in this example.

Definite Integration and Technology

Using Your Calculator You can use your graphing calculator in several different ways to approximate a definite integral. The following entries show one way of approximating $\displaystyle\int_4^{10} \frac{1}{x-3}\,dx$.

```
Y1 = 1/(X−3)
Y10 = fnInt(Y1,X,A,B)
```

The syntax for the fnInt command is

```
fnInt(function, variable of integration, lower limit, upper limit)
```

In the computation window, enter

```
4 → A
10 → B
Y10
```

and compute. You should get approximately 1.9459, which agrees with the result of Example Set B.

Entering Y10 = fnInt(Y1,X,A,B) rather than Y10 = fnInt(Y1,X,4,10) keeps your Y10 more general. You can use it to integrate any function stored under Y1 using any limit you choose.

Calculator Exercises

1. A manufacturer believes that the function
$$N(t) = 355 - 145e^{-0.03t}$$
models the rate at which an average worker can assemble N units per month of a product after t weeks of experience. Find and interpret both $N(5)$ and $\int_0^5 N(t)\,dt$.

2. The income from a chain of convenience stores can flow into a company at a yearly rate given by the function
$$T(t) = 14{,}000e^{0.0345t}$$
dollars where t represents the number of years since the company's inception. Find and interpret both $T(5)$ and $\int_0^5 T(t)\,dt$.

3. Economists use the Gini index, G, to measure how a country's money is distributed among its population. As the value of g approaches 0, the money becomes more evenly distributed to the population. The Gini index is given by the function
$$G(x) = 1 - \int_0^1 f(x)\,dx$$
Suppose that for country A, $f(x) = x^{1.28}$, and that for country B, $f(x) = x^{3.28}$. Find the Gini index for each country and specify the country for which the money supply is more evenly distributed.

4. The rate at which the number of pounds N of a pollutant is entering a lake each day is given by the function
$$N(t) = 6.5t^{1.15}$$
Find and interpret both $N(10)$ and $\displaystyle\int_0^{10} N(t)\,dt$.

Using Derive You can use Derive to evaluate the definite integral. Derive will produce exact or approximate results. The following set of instructions demonstrates the evaluation of $\int_2^5 14xe^{-0.08x}\,dx$. The approX command simplifies expressions using decimal approximations of irrational numbers.

Author
int(14xexp(-0.08x),x,2,5)
Simplify
appro**X**

Derive Exercises

1. Use Derive to produce the exact value of both $\int_1^3 210x^{2/5}\,dx$ and $\int_1^3 310x^{2/5}\,dx$. Now use pencil and paper to evaluate the first integral and Derive to produce the approximate value of the second integral.

2. Use Derive to produce the exact evaluation of $\displaystyle\int_0^1 \frac{5}{x-5\,dx}$. If the result involves irrational numbers, use Derive to obtain a two-decimal approximation.

3. Use Derive to produce the exact evaluation of $\int_0^5 \sqrt{x^2 - 15}$. If the result involves irrational numbers, use Derive to obtain a two-decimal approximation.

4. Use Derive to produce the exact evaluation of $\int_{e^2}^{e^3} \frac{5}{\sqrt{x}}$. If the result involves irrational numbers, use Derive to obtain a two-decimal approximation.

5. The total pressure drop from the outer wall of a hurricane to the eye of the hurricane is given by the integral

$$\int_a^c \frac{TR}{p}\, dp$$

where a is the pressure at the outer wall, c is the pressure at the eye, T is the temperature in degrees Celcius, R is a constant, and p is the atmospheric pressure in millibars. Use Derive to evaluate this integral. (This function was presented in the July–August, 1988 edition of *Scientific American* in an article entitled, "Toward a General Theory of Hurricanes," by Kerry Emanuel.)

EXERCISE SET 6.4

A UNDERSTANDING THE CONCEPTS

In each of Exercises 1–4, an error occurs in the evaluation or interpretation of the definite integral. Specify the step number at which the error occurs, then fix it.

1. Evaluate $\int_2^6 (3x - 4)\, dx$.

Step 1. $\int_2^6 (3x - 4)\, dx = \left(\frac{3x^2}{2} - 4x\right)\Big|_2^6$

Step 2. $= \left(\frac{3(2)^2}{2} - 4(2)\right) - \left(\frac{3(6)^2}{2} - 4(6)\right)$

Step 3. $= (6 - 8) - (54 - 24)$

Step 4. $= -2 - 30$

Step 5. $= -32$

2. Evaluate $\int_1^4 (6x - 5)\, dx$.

Step 1. $\int_1^4 (6x - 5)\, dx = (3x^2 - 5x)\Big|_1^4$

Step 2. $= (3 \cdot 4^2) - (5 \cdot 1)$

Step 3. $= (3 \cdot 16) - 5$

Step 4. $= 48 - 5$

Step 5. $= 43$

3. Evaluate $\int_2^3 3e^{3x}\, dx$.

Step 1. $\int_2^3 3e^{3x}\, dx = 3 \cdot 3e^{3x}\Big|_2^3$

Step 2. $= 9e^{3x}\Big|_2^3$

Step 3. $= (9e^{3(3)}) - (9e^{3(2)})$

Step 4. $= 9e^9 - 9e^6$

4. Evaluate $\frac{d}{dx}\left[\int_1^2 x^3\, dx\right]$.

Step 1. $\frac{d}{dx}\left[\int_1^2 x^3 dx\right] = x^3\Big|_1^2$

Step 2. $= 2^3 - 1^3$

Step 3. $= 7$

For Exercises 5–18, interpret each definite integral.

5. ***Business: Investment*** If $A(t)$ represents the rate at which the amount of money in an investment is changing from month a to month b, interpret $\int_a^b A(t)\,dt$.

6. ***Business: Advertising*** If $S(t)$ represents the rate at which a company spends money on advertising from month a to month b, interpret $\int_a^b S(t)\,dt$.

7. ***Economics: Population*** If $P(t)$ represents the rate at which the population of a city is changing between years a and b, interpret $\int_a^b P(t)\,dt$.

8. ***Biology: Blood Pressure*** If $P(t)$ represents the rate of change of blood pressure in the aorta between milliseconds a and b, interpret $\int_a^b P(t)\,dt$.

9. ***Social Science: Spread of News*** If $S(t)$ represents the rate at which a population hears news between days a and b, interpret $\int_a^b S(t)\,dt$.

10. ***Ecology: Toxic Leaks*** If $A(t)$ represents the rate at which a toxic chemical is leaked into a city's water supply from hour a to hour b, interpret $\int_a^b A(t)\,dt$.

11. ***Business: Cost of Production*** If $C(x)$ represents the rate at which the cost of producing x items changes as the number of items produced increases from a to b, interpret $\int_a^b C(x)\,dx$.

12. ***Economics: Personal Wealth*** If $W(t)$ represents the rate at which a person's wealth changes as she ages from a years to b years, interpret $\int_a^b W(t)\,dt$.

13. ***Life Science: Evaporation of a Solution*** If $E(T)$ represents the rate at which a solution evaporates between temperatures a and b, interpret $\int_a^b E(T)\,dT$.

14. ***Economics: Income Streams*** If $I(t)$ represents the rate at which an income stream is increasing from year a to year b, interpret $\int_a^b I(t)\,dt$.

15. ***Business: Service Rate Increases*** If $U(f)$ represents the rate at which a utility company raises the price of its service to the community relative to fuel prices a and b, $a < b$, interpret $\int_a^b U(f)\,df$.

16. ***Business: Depreciation*** If $D(t)$ represents the rate at which the depreciation of a commodity changes over the period from month a to month b, interpret $\int_a^b D(t)\,dt$.

17. ***Business: Air Travel*** If $L(t)$ represents how the decrease in the length of time t a company spends on air travel is changing between years a and b, interpret $\int_a^b L(t)\,dt$.

18. ***Mathematics: Probability*** If $P(t)$ represents the change in the increase in the probability that a particular event occurs between times a and b, interpret $\int_a^b P(t)\,dt$.

B SKILL ACQUISITION

For Exercises 19–35, evaluate each definite integral.

19. $\displaystyle\int_1^3 4x^3\,dx$

20. $\displaystyle\int_0^4 e^{5x}\,dx$

21. $\displaystyle\int_1^e \frac{3}{x}\,dx$

22. $\displaystyle\int_2^9 8\,dx$

23. $\displaystyle\int_0^{15} e^{-0.05}\,dx$

24. $\displaystyle\int_1^3 \left(x^2 + \frac{2}{x} - 5e^{5x}\right)dx$

25. $\displaystyle\int_4^5 \frac{2}{x-3}\,dx$

26. $\displaystyle\int_0^2 x(x^2+1)\,dx$

27. $\displaystyle\int_0^1 (6x+1)e^{3x^2+x}\,dx$

28. $\displaystyle\int_1^2 (5x-3)^4\,dx$

29. $\displaystyle\int_2^3 \frac{x}{\sqrt{x^2-1}}\,dx$

30. $\displaystyle\int_3^8 \frac{4x+5}{2x^2+5x-5}\,dx$

31. $\displaystyle\int_2^5 e^{2x}(e^{2x}-4)\,dx$

32. $\displaystyle\int_1^4 \frac{3+\ln x}{x}\,dx$

33. $\displaystyle\int_{\ln 3}^{\ln 5} 3e^x\,dx$

34. $\displaystyle\int_2^4 \frac{4}{x(\ln x)^2}\,dx$

35. $\displaystyle\int_2^3 \frac{1}{x\ln(3x)}\,dx$

C APPLYING THE CONCEPTS

36. ***Medicine: Spread of Disease*** A city's health department reports that a disease is increasing at the rate

$$N(t) = \frac{900t}{0.45t^2 + 600}$$

What is the total number of people who have contracted this disease during the first 30 days after its detection?

37. ***Business: Investment in Technology*** A company's investment in new technology is expected to save it money at the rate of $S(t) = 350(1 - e^{-0.07t})$ thousands of dollars each year after the technology is purchased. What is the accumulated savings this company would realize for the period of 1 to 3 years after purchase?

38. ***Biology: Population Size*** A biologist believes that the rate at which a colony of bacteria changes when a particular toxin is introduced into it is approximated by the function

$$P(t) = \frac{-800}{\sqrt{14 - 0.4t}}$$

What is the expected total decrease in the population between minutes 12.5 and 25?

39. ***Business: Advertising and Sales*** A company believes that with an aggressive advertising campaign, its number of sales could grow at the rate of $f(t) = 3 + 0.6e^{0.05t}$ units per week t weeks from the beginning of the campaign. If the company's belief is true, how many sales can they expect in the first 28 days after the beginning of the campaign?

40. ***Psychology: Memory Recall*** The probability that a randomly selected person can recall x percent of some given material changes according to the function $R(x) = 4.5x\sqrt{2 - x}$. What is the probability that a randomly selected person can recall between 80 and 100 percent of some given material?

41. ***Business: Resale Value*** The rate at which the resale value of a particular type of machine changes is approximated by the function

$$R(T) = -1280e^{0.4t}$$

What is the total amount of depreciation of this machine over the first 4 years of its life?

42. ***Business: Value of Real Estate*** In a recession, the value of real estate increases slowly or even decreases. Suppose that in a recessionary period, the rate at which the value, in thousands of dollars, of a piece of land changes is given by the function

$$V(t) = \frac{8t^3}{\sqrt{0.4t^4 + 3000}}$$

where t is the number of years from now. How much value has this piece of land accumulated 2 years from now?

43. ***Business: Production of Natural Gas*** A company is producing natural gas from a field at the rate of $R(t) = \dfrac{4.5t}{1.5t^2 + 2}$ million cubic feet per year t years from now. How much natural gas will this company produce over the next 4 years?

44. ***Education: Language Acquisition*** Educational researchers conducted a study of college students who took French in high school and visited France for at least 3 weeks, but have not actively spoken French since. The researchers believe that the students forget vocabulary at the rate of $100x\%$ per year, where $x = \dfrac{0.30}{2t + 0.5}$ and t is the time (in years) since the language was last actively spoken. If the researchers are correct, what percent of the student's original vocabulary can he or she be expected to forget over the next 5 years?

45. ***Business: Cost of Machine Maintenance*** As machines get older, the cost of maintaining them tends to increase. Suppose that for a particular machine, the rate at which the maintenance cost is increasing is approximated by the function $M(t) = (1.5t + 4)\sqrt{1.5t^2 + 8x}$, where t is the number of years from the time of purchase. What is the total cost of maintenance for the period 4 years to 9 years after purchase?

D DESCRIBE YOUR THOUGHTS

46. When a definite integral is evaluated over a closed interval, the constant of integration does not appear in the final result. Explain why.

47. Describe the directions given in the Fundamental Theorem of Calculus for evaluating a definite integral.

48. Explain why $\int_a^a f(x)\,dx = 0$.

E REVIEW

49. **(2.3)** ***Life Science: Population Size and Pollution*** The number of fish in a pond is related to the number p of PCBs (in parts per million) by the function

$$N(p) = \frac{635}{2 + 0.95p}$$

If the level of PCBs were allowed to increase without control, what would happen to the number of fish in the pond in the long run?

50. **(2.3)** Which is the first of the three continuity conditions to fail for the discontinuous function illustrated in the figure below?

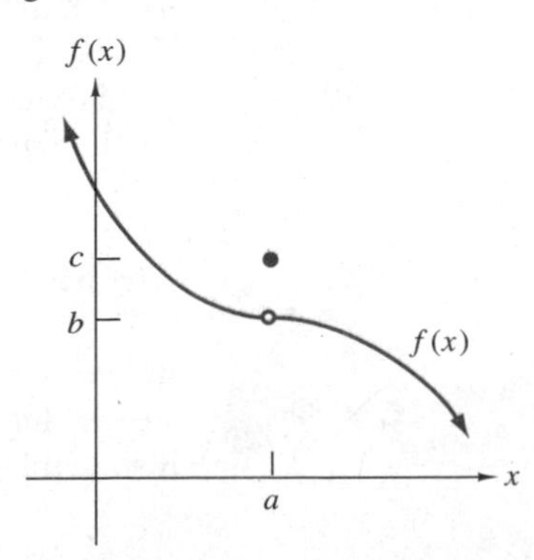

51. **(2.5)** Find the instantaneous rate of change of the function $f(x) = 3x^2 - 2x + 6$ with respect to x at $x = 3$.

52. **(3.3)** ***Business: Manufacturing Cost*** Suppose that $C(x)$ represents the cost (in dollars) to a manufacturer for producing x units of a product. Interpret the information provided by $C'(250) = 0.18$ and $C''(250) < 0$.

53. **(3.4)** Find $f'(x)$ for $f(x) = \left(\frac{3x-1}{2x+2}\right)^5$.

54. **(4.2)** Construct a summary table for the graph pictured in the figure below.

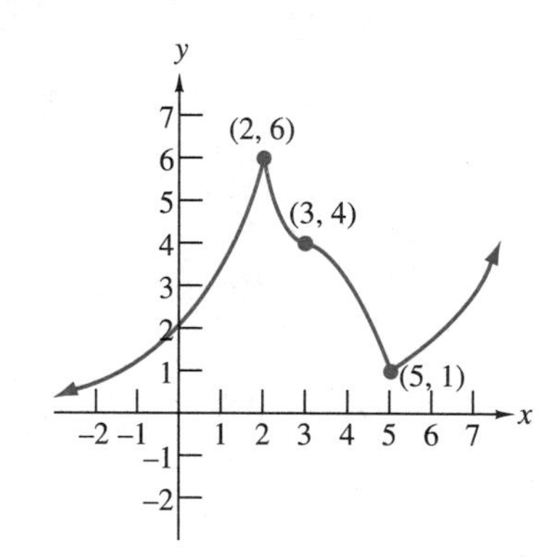

55. **(5.3)** ***Business: Advertising and Sales*** The senior editor for business at a textbook publishing company has determined that the function

$$x = 6.15 - 305 \ln(25 - N)$$

relates the expected number N (in thousands) of copies of an introductory marketing textbook sold in a 1-year period to the number x of complimentary copies sent to professors throughout the country. Find and interpret the rate at which x changes as N increases when $N = 5000$. (Remember that N is in thousands.)

56. **(5.4)** Find and interpret $f'(1)$ for $f(x) = -.02e^{3x-1}$.

57. **(6.2)** Evaluate $\int \frac{5x^4+2}{2x^5+4x-4}\,dx$.

58. **(6.3)** Evaluate $\int x^4 e^{4x}\,dx$.

6.5 The Definite Integral and Area

Introduction

In this section we investigate the relationship between area and accumulation. We tie together the definite integral and the area under a curve.

Area

Area

Before we start computing areas, we need to have some understanding of what area is. Intuitively, the **area** of a region is a *measure* of the size of the region; it is something that can be defined. For example,

1. The area of a rectangle having width x and height h is the product of the width and height. See the leftmost illustration of Figure 6.3.
2. The area of a triangle having base b and height h is one-half the product of the base and the height. See the middle illustration of Figure 6.3.
3. The area of a circle having radius r is the product of π and the square of the radius. See the rightmost illustration of Figure 6.3.

In each case, we use nonnegative real numbers to indicate the size of a planar (flat surface) region.

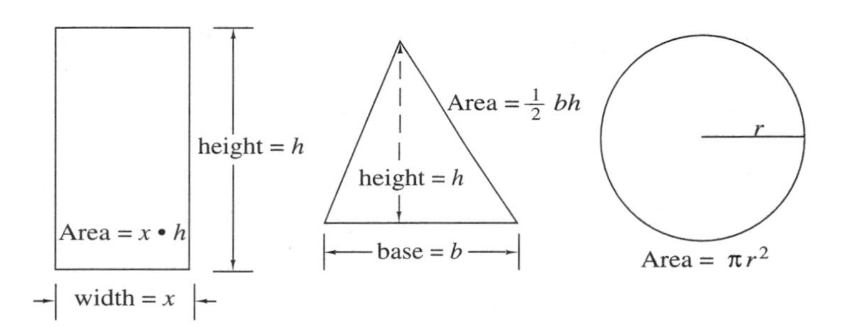

Figure 6.3

Approximating Area

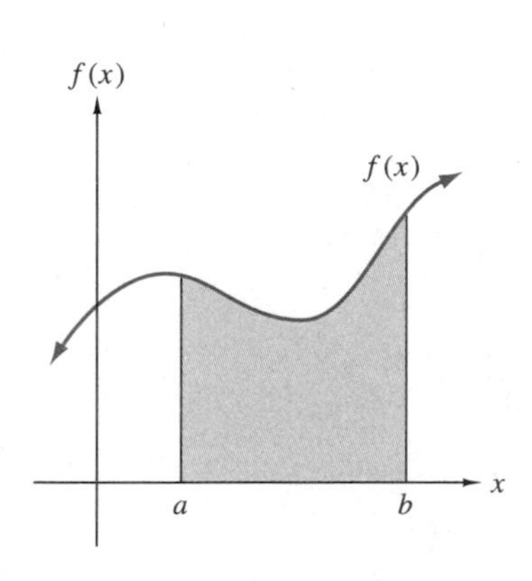

Figure 6.4

As you will see, many applied problems can be solved by relating their characteristics to area. To do so requires that we define the area of more general planar regions. For our purposes, we will begin by defining the area of planar regions that lie in a rectangular coordinate system, above the input axis, below some specified nonnegative function, and between two vertical lines. Figure 6.4 shows such a region. We do something that is common in mathematics: We define the area in terms of something with which we are already familiar. In particular, we define the area of this planar region in terms of rectangles.

We can use one rectangle, having as its width the entire interval $b - a$, and as its height $f(a)$, the output (height) at the left-hand endpoint of the interval $[a, b]$, as a crude approximation to the number we wish to assign as the area of this region. The left side of Figure 6.5 shows this approximation. Using one rectangle, the approximation of the area is $f(a)(b - a)$. This is certainly a poor approximation, and we can do better.

Using more rectangles will give us a better approximation. If we subdivide the interval $[a, b]$ into, say, four subintervals of equal length (for convenience), $\dfrac{b-a}{4}$, and on each subinterval construct a rectangle with height taken as the functional value at the left-hand endpoint of the subinterval, we will get the approximation illustrated in the right side of Figure 6.5.

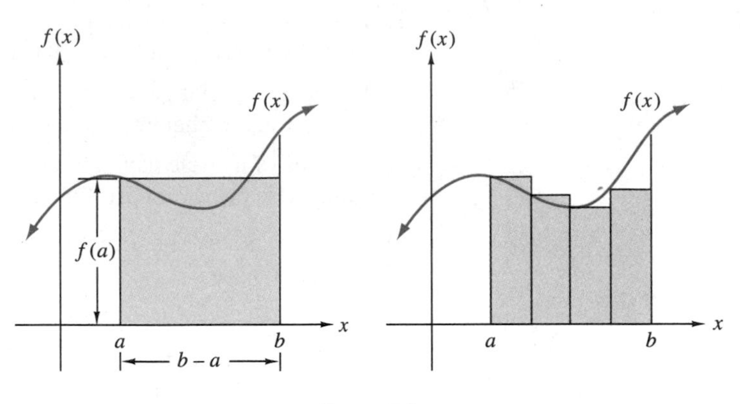

Figure 6.5

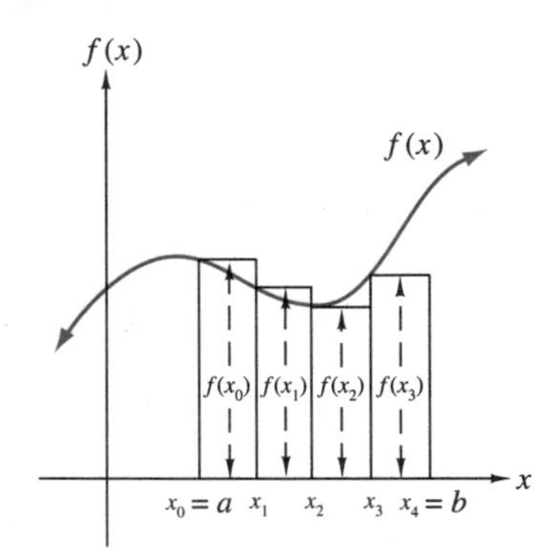

Figure 6.6

So that we can compute the area of these four rectangles, we introduce the following notation (which is illustrated in Figure 6.6).

- Let x_0 represent the left-most endpoint of the 1st subinterval. That is, $x_0 = a$.
- Let x_2 represent the left-most endpoint of the 3rd subinterval.
- Let x_3 represent the left-most endpoint of the 4th subinterval.
- Let x_4 represent the right-most endpoint of the 4th subinterval. That is, $x_0 = b$.

With this notation,

- $f(x_0)$ represents the height of the 1st rectangle.
- $f(x_1)$ represents the height of the 2nd rectangle.
- $f(x_2)$ represents the height of the 3rd rectangle.
- $f(x_3)$ represents the height of the 4th rectangle.

We will also let Δx represent the width of each rectangle. Then, $\Delta x = \dfrac{b-a}{4}$. Then, letting S_4 represent the sum of the areas of all four rectangles, we get

$$\begin{aligned} S_4 &= (\text{area of rectangle 1}) + (\text{area of rectangle 2}) \\ &\quad + (\text{area of rectangle 3}) + (\text{area of rectangle 4}) \\ &= f(x_0)\Delta x + f(x_1)\Delta x + f(x_2)\Delta x + f(x_3)\Delta x \end{aligned}$$

and, factoring Δx from each term

$$= \left[f(x_0) + f(x_1) + f(x_2) + f(x_3)\right]\Delta x$$

Thus, a four-rectangle approximation to the region is given by

$$S_4 = \left[f(x_0) + f(x_1) + f(x_2) + f(x_3)\right]\Delta x$$

We will now look at an actual example of this situation.

EXAMPLE SET A

Use four rectangles to approximate the area of the planar region illustrated in Figure 6.7.

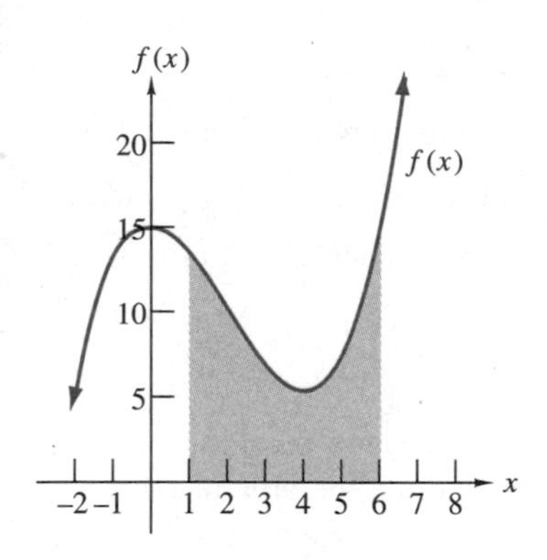

Figure 6.7

Solution:
We have $a = 1$ and $b = 6$, so that the width of each rectangle is $\frac{6-1}{4} = \frac{5}{4} =$ 1.25 units. Then,

$$x_0 = 1$$
$$x_1 = 1 + 1.25 = 2.25$$
$$x_2 = 2.25 + 1.25 = 3.50$$
$$x_3 = 3.50 + 1.25 = 4.75$$
$$x_4 = 4.75 + 1.25 = 6$$

(x_4 is only used to check that the entire interval is accounted for.) See Figure 6.8. We will use the formula

$$S_4 = \left[f(x_0) + f(x_1) + f(x_2) + f(x_3)\right]\Delta x$$

to approximate the area. The computations are shown in the following table. Thus, $S_4 = (35.15)(1.25) = 43.9375$.

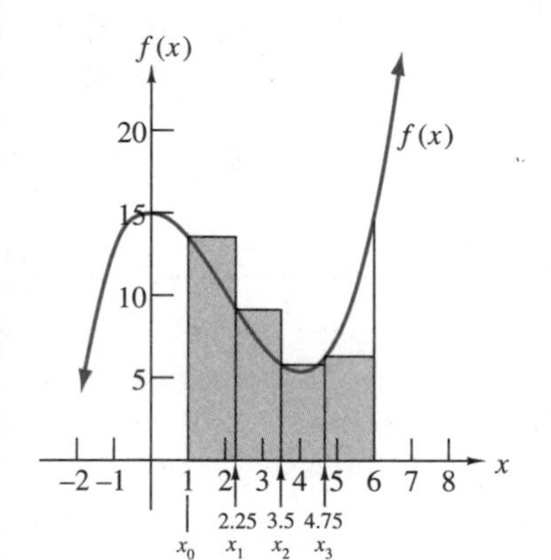

Figure 6.8

x	$f(x) = 0.3x^3 - 1.8x^2 + 15$
1	13.5
2.25	9.3
3.50	5.81
4.75	6.54
Sum	35.15

Then, using four rectangles, we would approximate the area of this planar region as 43.9375 square units.

Using only one rectangle would give

$$\begin{aligned} S_1 &= f(x_0)(b-a) \\ &= f(1)(6-1) \\ &= (13.5)(5) \\ &= 67.5 \end{aligned}$$

The actual area is 43.125 square units (you will be able to compute this for yourself very soon). Comparing the two results, it is apparent that using four rectangles provides a better approximation to the actual area. What if we were to use 40 or 4000 rectangles?

Defining the Area of a Planar Region

Suppose we subdivide the interval $[a, b]$ into n subintervals of equal length, $\Delta x = \dfrac{b-a}{n}$, and construct rectangles using the functional value at c_i in each subinterval, $[x_i, x_{(i+1)}]$, as the height of the rectangle. Then, the approximating area is

$$S_n = [f(c_0) + f(c_1) + \cdots + f(c_{n-1})]\Delta x$$

Riemann Sum

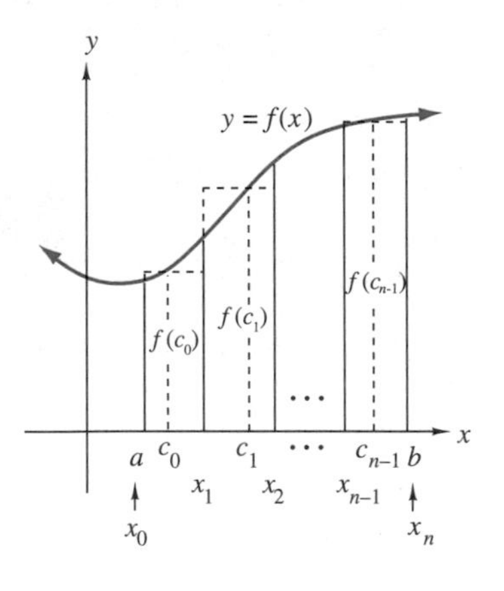

Figure 6.9

This sum is called a **Riemann sum**. The larger the value of n, the better the number S_n approximates what we would consider to be the area of the planar region. Figure 6.9 illustrates this. This leads us to define the area of the planar region below the nonnegative function $f(x)$, about the x–axis, and between the vertical lines $x = a$ and $x = b$, to be

$$\lim_{n\to\infty} S_n = \lim_{n\to\infty} [f(c_0) + f(c_1) + \cdots + f(c_{n-1})]\Delta x$$

provided this limit exists. For the functions we consider in this text, this limit will always exist, but Figure 6.10 shows a planar region for which it does not. The line $x = b$ is a vertical asymptote to the function $f(x)$. The planar region bounded by the function $f(x)$, the x axis, and the lines $x = a$ and $x = b$ is said to have undefined area.

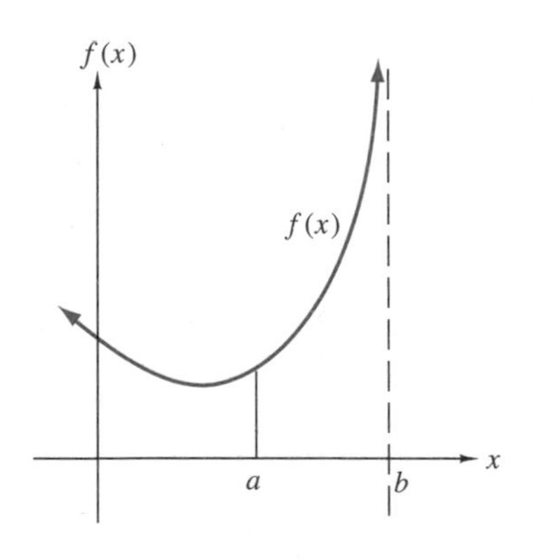

Figure 6.10

Relating Area and the Definite Integral

Let's analyze what we have done so far. We have defined the area under a curve that lies above the input axis and between two vertical lines using the *accumulated* areas of rectangles. The word *accumulated* makes us think of the definite integral, and, in fact, we draw the following relationship between area and the definite integral.

Area and the Definite Integral

If $f(x)$ is a continuous, nonnegative function on the interval $[a, b]$, the area A (in square units) bounded by the function $f(x)$, the x axis, and the vertical lines $x = a$ and $x = b$ is precisely equal to the value of the definite integral of $f(x)$, from $x = a$ to $x = b$. That is,

$$A = \int_a^b f(x)\,dx$$

Notice the relationship between the limit as $n \to \infty$ of the Riemann sum,

$$\lim_{n\to\infty} S_n = \lim_{n\to\infty} [f(c_0) + f(c_1) + \cdots + f(c_{n-1})]\Delta x$$

and the definite integral

$$\int_a^b f(x)\,dx$$

The expression $f(c_0)+f(c_1)+\cdots+f(c_{n-1})$ in the Riemann sum indicates that we are to sum—that is, accumulate—the heights of n rectangles and then multiply that value by the width, Δx, of the rectangles. In the definite integral, the integral sign $\int_a^b f(x)\,dx$ indicates the accumulation, from a to b, of the products of the heights, $f(x)$, and widths, dx. The $\int$ sign is actually a stretched-out S to indicate an infinite sum (or more precisely, a limiting value of a finite sum, in the limit as the number of terms in the sum approaches infinity and the maximum width of any subinterval approaches 0) and the dx represents an infinitesimally small Δx. Thus,

$$\lim_{n\to\infty} S_n = \int_a^b f(x)\,dx$$

The definite integral $\int_a^b f(x)\,dx$ may be computed using the Fundamental Theorem of Calculus, because it may be proven, using the definition of the derivative involving limits, that $\frac{d}{dx}A(x)=f(x)$, where $A(x)$ is the area bounded by the function $f(t)$, the t axis, and the lines $t=a$ and $t=x$, which may be written as $A(x)=\int_a^x f(t)\,dt$. That is,

$$\int_a^b f(x)\,dx = A(x)\Big|_a^b = A(b)-A(a), \quad \text{where} \quad \frac{d(A(x))}{dx}=f(x)$$

EXAMPLE SET B

Find the area bounded by the function $f(x)=0.3x^3-1.8x^2+15$, the x axis, and the lines $x=1$ and $x=6$. (You may remember this function from Example Set A.)

Solution:
The function is displayed in Figure 6.7. Notice that over the interval $[1,6]$, $f(x)$ lies completely above the x axis so that it is always nonnegative here. Since it is nonnegative on $[1,6]$, the required area can be found by evaluating the related definite integral.

$$\begin{aligned} A &= \int_1^6 (0.3x^3-1.8x^2+15)\,dx \\ &= \left(\frac{0.3x^4}{4}-\frac{1.8x^3}{3}+15x\right)\Big|_1^6 \\ &= \left(\frac{0.3\cdot 6^4}{4}-\frac{1.8\cdot 6^3}{3}+15\cdot 6\right)-\left(\frac{0.3\cdot 1^4}{4}-\frac{1.8\cdot 1^3}{3}+15\cdot 1\right) \\ &= (97.2-129.6+90)-(0.075-0.6+15) \\ &= 57.6-14.475 \end{aligned}$$

$= 43.125$

Interpretation: The area bounded by $f(x) = 0.3x^3 - 1.8x^2 + 15$, the x axis, and the lines $x = 1$ and $x = 6$ is exactly 43.125 square units.

If the function $f(x)$ is negative somewhere in $[a, b]$, the definite integral will not record area. The definite integral acts as an accumulator for area *only* when the function is nonnegative throughout the interval. Recall that $f(x)$ represents the heights of the rectangles and if these values are negative, a negative value for the area will result, which would not make sense. So, if we want the area of a region where $f(x)$ is negative, we must take the opposite of $f(x)$; that is, $A = \int_a^b (-f(x))\, dx$.

The evaluation of a definite integral does not necessarily have to involve the computation of an area and, therefore, can result in values less than or equal to zero. But when the definite integral is being used specifically to find an area, we must pay close attention to the graph of the function and account carefully for any areas below the input axis. The next two examples will illustrate this.

EXAMPLE SET C

Find the value of $\displaystyle\int_2^5 (x^2 - 5x + 4)\, dx.$

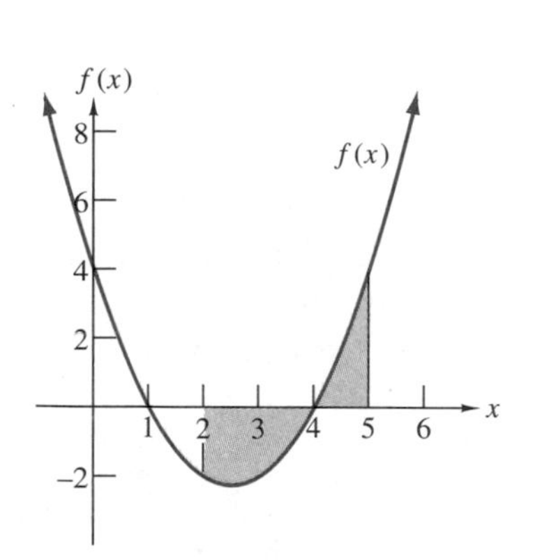

Figure 6.11

Solution:
Notice that we are not asked to find any area. So we will evaluate the integral without looking at its graph.

$$\int_2^5 (x^2 - 5x + 4)\, dx = \left(\frac{x^3}{3} - \frac{5x^2}{2} + 4x\right)\Big|_2^5$$

$$= \left(\frac{5^3}{3} - \frac{5 \cdot 5^2}{2} + 4 \cdot 5\right) - \left(\frac{2^3}{3} - \frac{5 \cdot 2^2}{2} + 4 \cdot 2\right)$$

$$= \left(-\frac{5}{6}\right) - \left(\frac{2}{3}\right)$$

$$= -\frac{3}{2}$$

Interpretation: Now surely the area bounded by this curve from $x = 2$ to $x = 5$ is not $-3/2$ square units. Let's look at the graph of $f(x) = x^2 - 5x + 4$ in Figure 6.12. We see that this definite integral does not accumulate area because the function is not always nonnegative on $[2, 5]$. The function is negative from $x = 2$ to almost $x = 4$. Recall that the relationship between area and the definite integral exists only when the function is nonnegative over the entire given interval (that is, when the function lies on or above the x axis throughout the interval). Consequently, the value of the definite integral is $-3/2$, but it does not represent the area of the associated region. We will look at that area in the next example.

EXAMPLE SET D

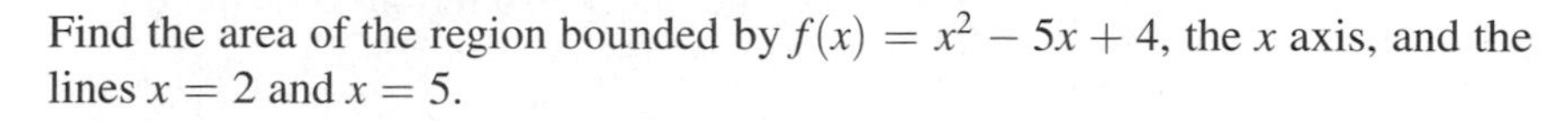

Find the area of the region bounded by $f(x) = x^2 - 5x + 4$, the x axis, and the lines $x = 2$ and $x = 5$.

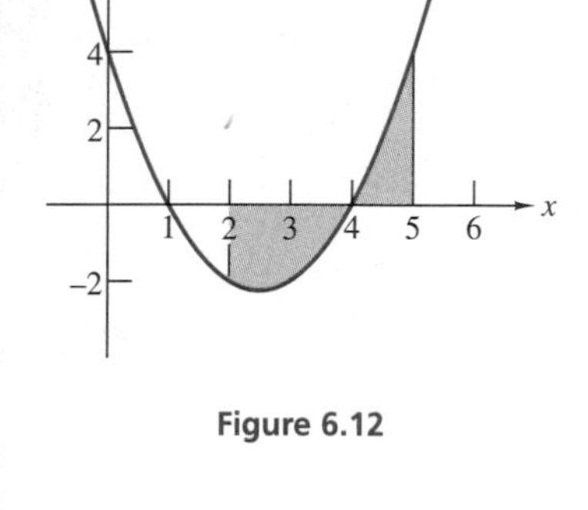

Figure 6.12

Solution:
Figure 6.12 from the last example shows the graph of $f(x)$. To find the area, we will need to write separate integrals for the region below the x axis, where $f(x) \le 0$, and the region above the x axis, where $f(x) \ge 0$.

$$\text{For } f(x) \le 0, A = \int_2^4 [-f(x)]\,dx. \quad \text{For } f(x) \ge 0, A = \int_4^5 f(x)\,dx.$$

The accumulated area A is:

$$\begin{aligned}
A &= \int_2^4 [-f(x)]\,dx + \int_4^5 f(x)\,dx \\
&= \int_2^4 [-(x^2 - 5x + 4)]\,dx + \int_4^5 (x^2 - 5x + 4)\,dx \\
&= \left(-\frac{x^3}{3} + \frac{5x^2}{2} - 4x\right)\Big|_2^4 + \left(\frac{x^3}{3} - \frac{5x^2}{2} + 4x\right)\Big|_4^5 \\
&= \left[\left(-\frac{4^3}{3} + \frac{5\cdot 4^2}{2} - 4\cdot 4\right) - \left(-\frac{2^3}{3} + \frac{5\cdot 2^2}{2} - 4\cdot 2\right)\right] \\
&\quad + \left[\left(\frac{5^3}{3} - \frac{5\cdot 5^2}{2} + 4\cdot 5\right) - \left(\frac{4^3}{3} - \frac{5\cdot 4^2}{2} + 4\cdot 4\right)\right] \\
&= \left[\left(\frac{8}{3}\right) - \left(-\frac{2}{3}\right)\right] + \left[\left(-\frac{5}{6}\right) - \left(-\frac{8}{3}\right)\right] \\
&= \left(\frac{10}{3}\right) + \left(\frac{11}{6}\right) \\
&= \frac{31}{6}
\end{aligned}$$

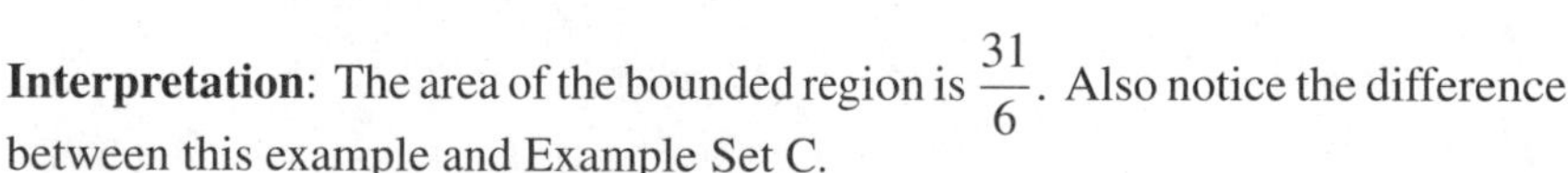

Interpretation: The area of the bounded region is $\frac{31}{6}$. Also notice the difference between this example and Example Set C.

The area under a curve can be used to represent the probability that a quantity will assume a value between a and b, as shown in the next example.

EXAMPLE SET E

The probability function $L(t)$ is the function with the property that the probability, given by $L(t)\Delta t$, that the useful life of a product is between t years and $t + \Delta t$

years as long as Δt is small enough. Through statistical analysis, a manufacturer has determined that for one of its products,

$$L(t) = 0.20e^{-0.20t}$$

Probability Function

The function $L(t)$ is an example of a **probability function** and is displayed in Figure 6.13. Approximate, to two decimal places, the probability that a randomly selected unit will last between 5 and 6 years.

Solution:
Since the function $L(t)$ lies above the t axis on $[5, 6]$, we can relate the definite integral $\int_5^6 L(t)\,dt$ to area and, therefore, to probability.

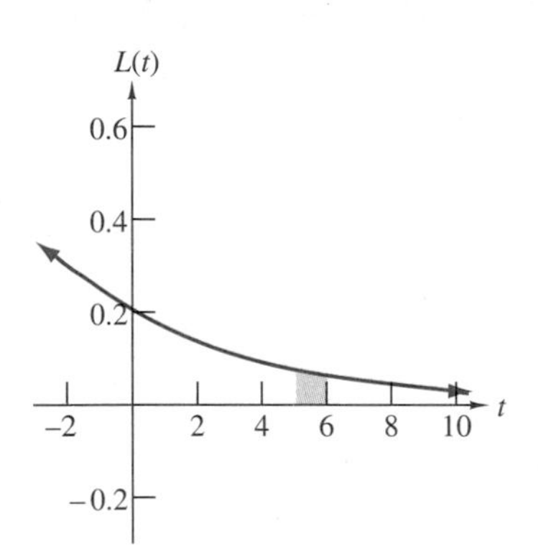

Figure 6.13

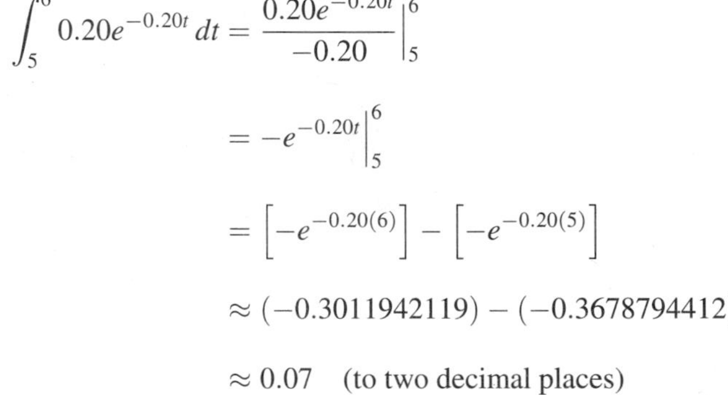

$$\int_5^6 0.20e^{-0.20t}\,dt = \left.\frac{0.20e^{-0.20t}}{-0.20}\right|_5^6$$

$$= \left.-e^{-0.20t}\right|_5^6$$

$$= \left[-e^{-0.20(6)}\right] - \left[-e^{-0.20(5)}\right]$$

$$\approx (-0.3011942119) - (-0.3678794412)$$

$$\approx 0.07 \quad \text{(to two decimal places)}$$

Interpretation: If a unit is randomly chosen from this manufacturer's production lot, there is about a 7% chance that it will have a useful life of between 5 and 6 years. In other words, about 7% of the units produced by this manufacturer will have useful lives of 5 to 6 years.

Area and Technology

Using Your Calculator You can use your graphing calculator to approximate the area under a curve over a closed interval. The following entries show the evaluation of $\int_1^6 (0.3x^3 - 1.8x^2 + 15)\,dx$.

Y1 = 0.3x^3 − 1.8x^2 + 15

You may first want to see the graph of this function on $[1, 6]$. Set the viewing window to $0 \le x \le 10$ and $0 \le y \le 25$ to build a frame around the graph, then graph the function. Now, select the Zoom Integer feature and press the Enter key to activate it. This selection causes both x and y to change in 1-unit increments when tracing the curve. It also rescales both the x and y axes from 1 to 10. It will also cause the viewing to be redefined so that your graph may appear distorted.

Now access the $\int f(x)\,dx$ feature. Move the cursor to $x = 1$ and fix this lower limit of integration by pressing Enter. Move the cursor to $x = 6$ and fix it by pressing Enter. As you press Enter, the region under the curve will become-highlighted and its area will be specified in the calculator graphing window. Are you getting 43.125? (Depending on how you originally set the viewing window, the highlighting of the region under the curve may or may not be visible. For example, if you originally set $0 \leq y \leq 100$ rather than $0 \leq y \leq 25$, you would not see the region highlighted, but you would still see the evaluation of the integral.)

Calculator Exercises

1. The number, N, of tons being dumped into a river t years from 1985 is approximated by the function

$$N(t) = \frac{615}{(0.65x + 3.6)^{2/3}}, \quad 0 \leq t \leq 20$$

where $t = 0$ corresponds to 1985. Construct the graph of this function on $0 \leq N(t) \leq 300$. Evaluate and interpret

$$\int_5^{10} N(t)\, dt$$

Highlight this region on your graph.

2. A manufacturer uses the probability density function

$$P(t) = 0.37e^{-0.26t}$$

to measure the probability, P, that her product lasts between $t = a$ and $t = b$ years. Given that probability is always between and including 0 and 1, construct the graph of this function. Find the probability that the product lasts, first, less than 3 years and then between 5 and 10 years. Shade the region that corresponds to the probability that the product lasts between 5 and 10 years.

3. A time-release vitamin produced by a drug manufacturer keeps A milligrams of the vitamin in the bloodstream t hours after it is first ingested. The relationship between A and T is given by the function

$$A(t) = 26.8t^{2.48} - 206.5t^{1.48} + 437.3t^{0.48}$$

where $0 \leq t \leq 3.5$. Construct the graph of this function. Shade the region that corresponds to the total amount of the vitamin in the bloodstream through the first hour after ingestion. What is this amount? What is the total amount of vitamin in the bloodstream throughout the life of the vitamin?

4. Choose any function, $f(x)$, and any lower and upper limits of integration, a and b. Compute both $\int_a^b f(x)\,dx$ and $\int_b^a f(x)\,dx$. Do this several times for other choices of $f(x)$ and a and b. Use your results to make a conjecture about the relationship between $\int_a^b f(x)\,dx$ and $\int_b^a f(x)\,dx$.

EXERCISE SET 6.5

A UNDERSTANDING THE CONCEPTS

1. Suppose $f(x)$ is the function pictured below. Use two definite integrals to express the area from $x = a$ to $x = b$ using a, b, and c.

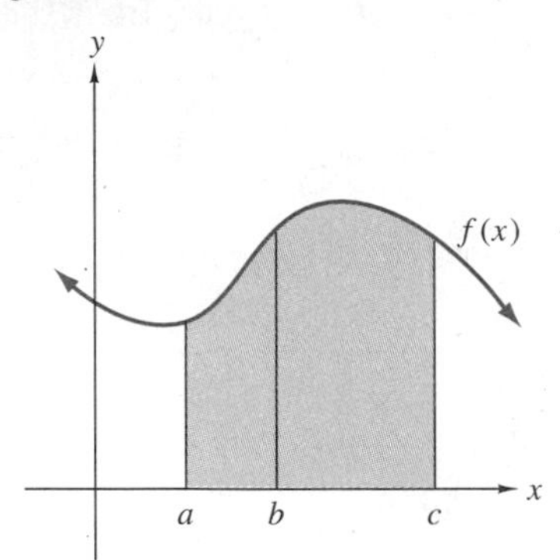

2. Suppose $f(x)$ is the function pictured below. Use two definite integrals to describe the area of the shaded region. (More than one answer is possible).

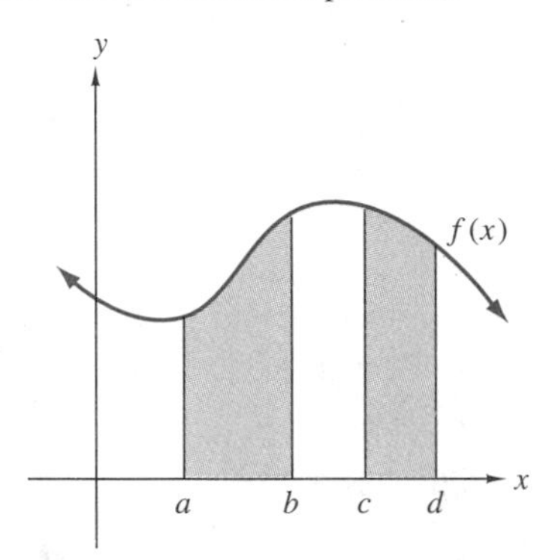

3. Suppose $f(x)$ is the function pictured below. Use two definite integrals to describe the area of the shaded region using a, b, and c as limits of integration.

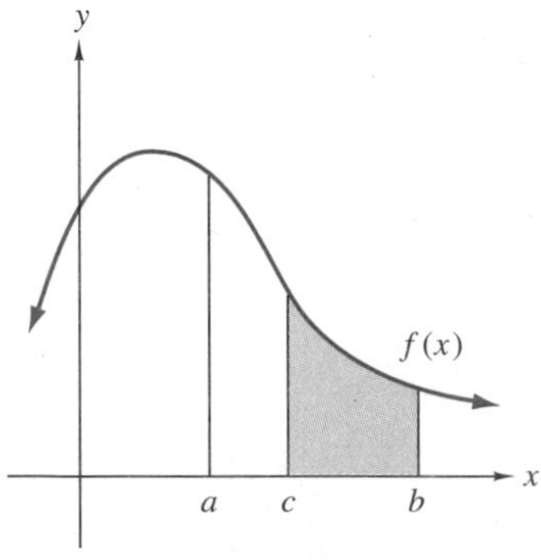

4. The function $f(x)$ pictured below is not always nonnegative over $[a, b]$. It is nonnegative over (a, c), but negative over (c, b). Use two definite integrals to describe the area bounded by the curve $f(x)$ and the vertical lines $x = a$ and $x = b$. (Think about the height of a representative rectangle between c and b. As it is drawn, is it positive or negative?)

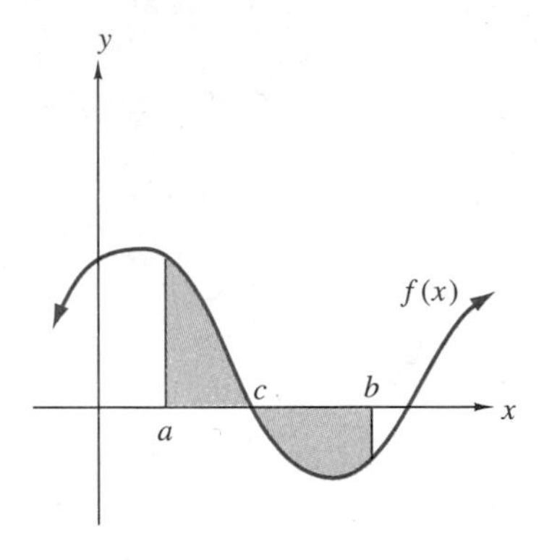

B SKILL ACQUISITION

For Exercises 5–20, determine the area bounded by the given function, the x axis, and the given vertical lines.

5. $f(x) = -6x^2 + 26x, \quad x = 0$ and $x = 3$
6. $f(x) = \dfrac{1}{x^3}, \quad x = 1$ and $x = 2$
7. $f(x) = 3x^2 - 3x + 5, \quad x = 0$ and $x = 2$
8. $f(x) = e^x, \quad x = 0$ and $x = 1$
9. $f(x) = \dfrac{1}{x}, \quad x = 1$ and $x = 4$
10. $f(x) = x^2 - 10, \quad x = 4$ and $x = 6$
11. $f(x) = 2500e^{0.2x}, \quad x = 0$ and $x = 4$
12. $f(x) = 300e^{-0.03x}, \quad x = 5$ and $x = 10$
13. $f(x) = 2\sqrt{x}, \quad x = 1$ and $x = 4$
14. $f(x) = e^x(e^x + 2)^3, \quad x = 0$ and $x = 2$
15. $f(x) = \dfrac{x}{(x^2 + 8)}, \quad x = 3$ and $x = 5$
16. $f(x) = e^{-x+2}, \quad x = 1$ and $x = 5$
17. $f(x) = \dfrac{800}{\sqrt[3]{(x + 3)^2}}, \quad x = 10$ and $x = 25$
18. $f(x) = 420e^{0.01x} - 20 + 0.2x, \quad x = 0$ and $x = 10$
19. $f(x) = \dfrac{x^2}{2} - \dfrac{3x}{3}, \quad x = 5$ and $x = 20$
20. $f(x) = x^{2/3}, \quad x = 0$ and $x = 5$

For Exercises 21–23, some or all of the given function lies below the x axis, making it negative. To find the area of the indicated region, you will have to use the process of Example Set D.

21. The function $f(x) = x^2 - 2x - 3$ is negative on $[0, 3]$. Find the area of the region bounded by $f(x)$, the x axis, and the lines $x = 0$ and $x = 3$.

22. The function $f(x) = x^2 - 8x + 12$ is positive on $[0, 2]$ and negative on $[2, 6]$. Find the area of the region bounded by $f(x)$, the x axis, and the lines $x = 0$ and $x = 6$.

23. The function $f(x) = x^2 - 14x + 24$ is positive on $[0, 2]$ and negative on $[2, 12]$. Find the area of the region bounded by $f(x)$, the x axis, and the lines $x = 0$ and $x = 12$.

C APPLYING THE CONCEPTS

24. ***Economics: Social Utility of Consumption*** To sell x units of a product each month, a company must price each unit according to the demand function

$$D(x) = \frac{350}{1 + 0.04x}$$

dollars. The figure below shows the graph of this demand function. Economists call the shaded region bounded by the function, the x axis, and the vertical lines $x = 0$ and $x = 400$ (or some value a) the **social utility of consumption**, and it is the accumulated value the public places on its consumption of these 400 (or, in general, these a) units per unit of time. Since the social utility of consumption is an accumulated value, it is defined by an integral. In general, the social utility of a units of a product to the public is defined as

$$\text{Social utility of consumption} = \int_0^a D(x)\,dx$$

Determine the social utility of consumption associated with the sale of 400 units of this product.

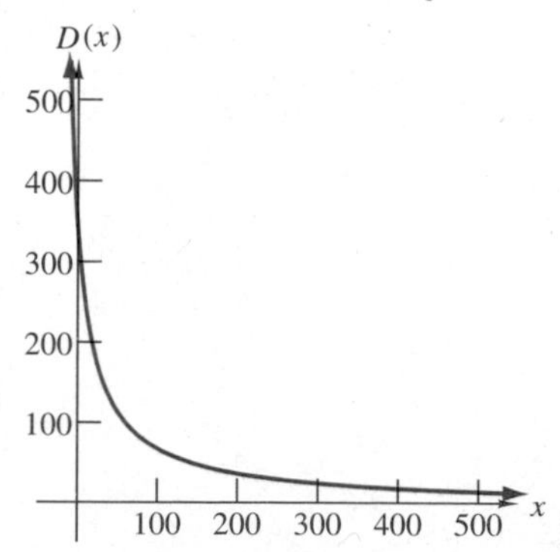

25. ***Economics: Revenue Stream*** A person's investment is expected to produce a continuous stream of revenue R of

$$R(t) = 2500\sqrt{t + 10}$$

dollars over the next 8 years. The accumulated value of this **revenue stream** from time $t = a$ to $t = b$ is given by the integral

$$\int_a^b R(t)\,dt$$

The graph of $R(t)$ appears below. Find the total revenue generated by this revenue stream over the 8-year period.

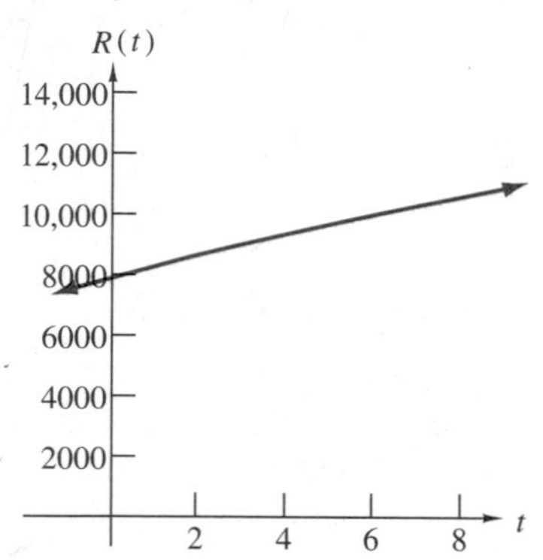

26. ***Business: Revenue Stream*** A person's investment is expected to produce a continuous stream of revenue R of

$$R(t) = 400e^{0.08t}$$

dollars over the next 8 years. The graph of $R(t)$ appears below. Find the total revenue generated by this revenue stream 5 to 8 years into the investment. (Hint: See Exercise 25.)

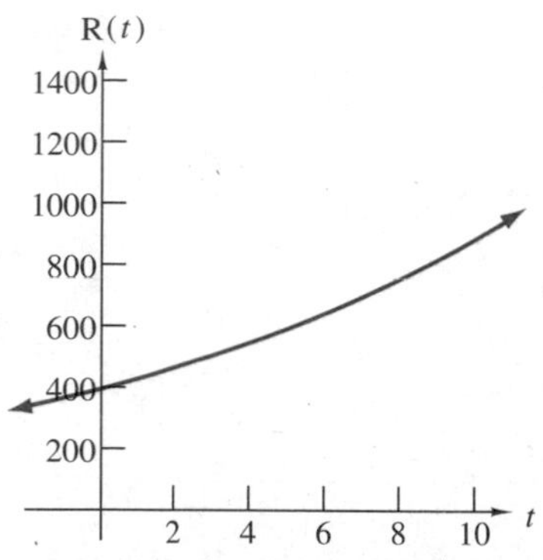

27. ***Education: Accumulation of Chalk Dust*** Students in an applied calculus class notice that as their instructor erases the chalk from a blackboard, chalk dusk falls from the board and settles into the tray at the bottom of the board. Knowing that the definite integral represents accumulation, they are interested in knowing the accumulated amount of dust at the end of one erasing. The instructor, after taking a sample and using some curve-fitting techniques, determines that at each point x feet from the left end of the chalk tray,

$$C(x) = \frac{140}{(x + 80)^2}$$

grams of dust fall to the tray. If the tray is 20 feet long, and the board is written on in such a way that the function approximates the way the chalk falls, how many grams of chalk dust will have accumulated in the tray after one board erasure? The graph of $C(x)$ appears below.

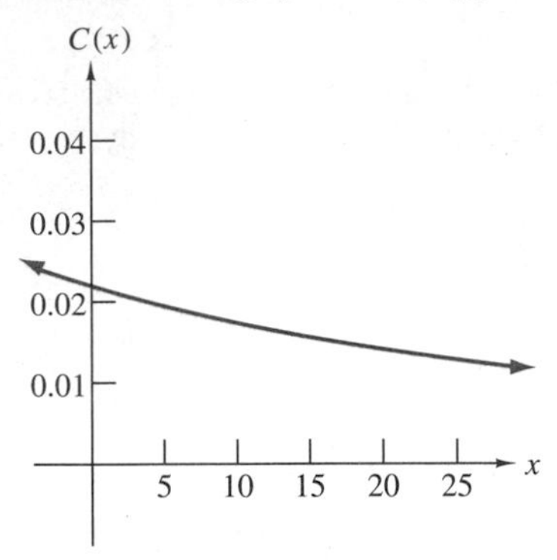

28. ***Business: Capital Accumulation of an Investment*** The **capital accumulation** of an investment over a time period $t = a$ to $t = b$, is the integral, over that time period, of the rate of investment $\frac{dI}{dt}$, if the cash flow into the investment is continuous. Find the capital accumulation of an investment over a 3-year time period if

$$\frac{dI}{dt} = 2650(4x + 2)^{1/3}$$

The graph of $\frac{dI}{dt}$ appears below.

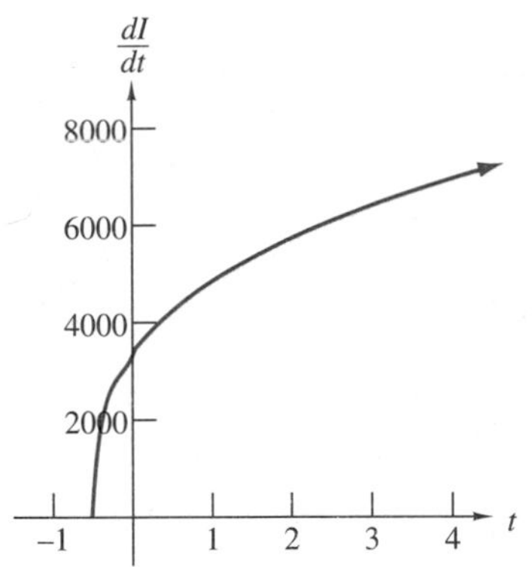

29. ***Social Science: Death Rates*** In a particular country, the function

$$D(t) = -0.027t^2 + 1.4t + 63$$

approximates the death rate (for each 1000 people who are t years of age). Find the total number of deaths per 1000 people of those who are between 5 and 12 years old. The graph of $D(t)$ appears below.

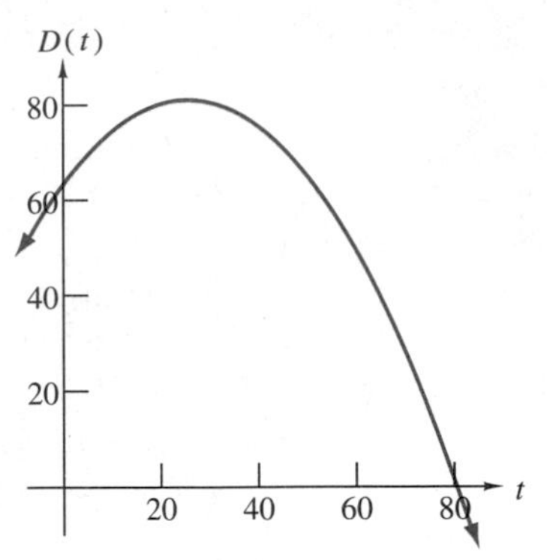

30. ***Manufacturing: Worker Production*** An average worker can assemble $N(t) = -2.7x^2 + 26t$ units of a product each hour. Determine the number of units the average worker will assemble in an 8-hour working day. The graph of $N(t)$ appears below.

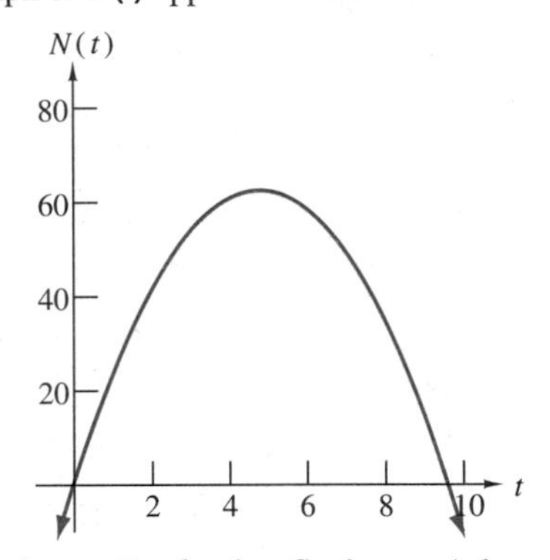

31. ***Business: Production Savings*** A large company has installed new robotic equipment in its assembly areas. This new equipment produces a savings that accumulates at the rate of

$$S(t) = 36e^{-0.18t}$$

thousand dollars per month. Determine the savings over the first 1-year period. The graph of $S(t)$ is displayed below.

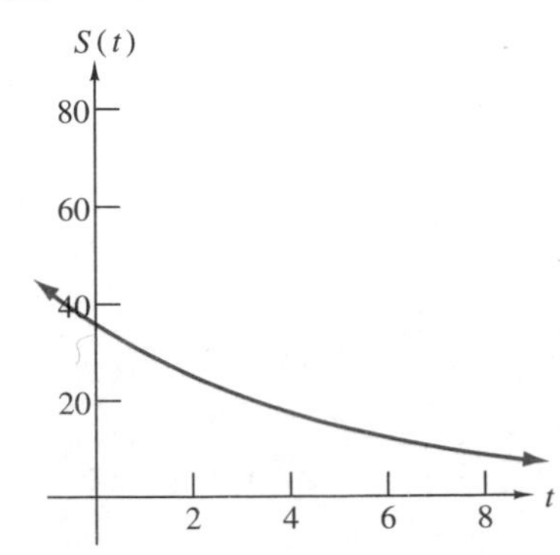

D DESCRIBE YOUR THOUGHTS

32. Describe the information that is being conveyed in the Riemann sum formula

$$S_n = \left[f(x_1) + f(x_2) + \cdots + f(x_{n-1})\right] \Delta x$$

Your description should include not only what the formula does, but what it represents geometrically.

33. Describe how the definite integral and the area under a curve and over an interval are related.

34. Explain why $\int_1^6 (x-3)^3\, dx$ does not represent the area under the curve $f(x) = (x-3)^3$ over the interval $[1, 6]$.

E REVIEW

35. **(2.2)** Find, if it exists, $\lim_{x \to 4} \frac{x^2 - 9x + 18}{x^2 - 3x - 4}$.

36. **(2.3)** The population P (in thousands) of a county t years from the beginning of 1990 is given by the function

$$P(t) = \frac{37t^2 + 85t + 85}{t^2 + 6t + 65}$$

What is the expected population of the county in the long term?

37. **(3.4)** Find and interpret $f'(x)$ for $f(x) = (2x+5)^3(x-1)^3$ at $x = 1$.

38. **(4.3)** ***Manufacturing: Production Costs*** A rectangular storage box with no top and a square base is to have a volume of 4000 cubic inches. What are the dimensions (length of base and height) that will minimize the surface area of the box?

39. **(5.3)** Find $f'(x)$ for $f(x) = \ln(3x + 5)$.

40. **(6.1)** Evaluate the indefinite integral $\int \frac{20x}{5x^2 + 2}\, dx$.

41. **(6.3)** Evaluate the definite integral $\int_1^e x^3 \ln x\, dx$.

42. **(6.4)** Evaluate the definite integral $\int_1^4 \frac{\ln x}{x}\, dx$.

43. **(6.4)** Evaluate the definite integral $\int_1^3 \frac{x}{x+3}\, dx$.

6.6 Improper Integrals

Introduction

Improper Integrals

Improper Integrals and Technology

Introduction

There are many situations in which it is necessary to accumulate quantities over infinitely long intervals. For example, we may wish to determine how much of a pollutant will seep into the ground near a dump site if it is allowed to discharge into the ground indefinitely, or how many barrels of oil a well will produce if it operates indefinitely, or what total profit a company could realize from the sale of an unlimited number of units of some product. Definite integrals involved in such accumulations are called **improper integrals** and differ from the integrals we have worked with so far in that at least one of the limits of integration is infinite.

Improper Integrals

Improper Integrals

b	$1 - \frac{1}{b}$
2	0.5
3	0.666...
5	0.8
10	0.9
100	0.99
1000	0.999
10,000	0.9999
100,000	0.99999

We can motivate a definition for improper integrals by evaluating $\int_1^b \frac{1}{x^2}\,dx$ as b gets bigger and bigger. Figure 6.14 shows, from top-to-bottom and left-to-right, the evaluation, and the accompanying table shows values of $\int_1^b f(x)\,dx$ for increasingly larger values of b.

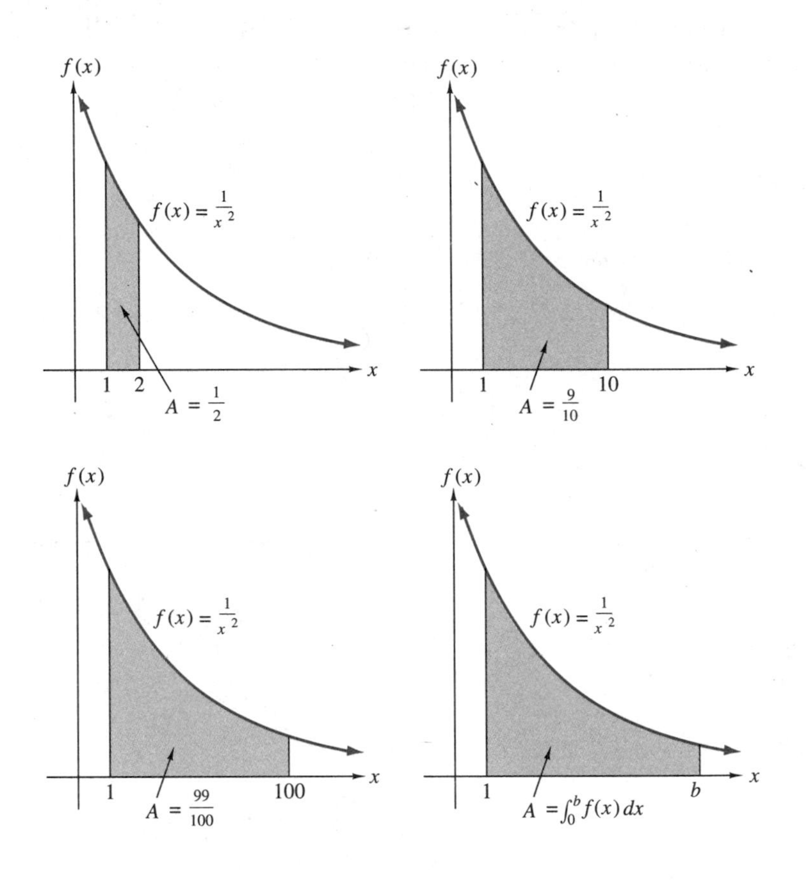

Figure 6.14

Apparently, as b gets bigger and bigger—that is, as b approaches infinity—$\int_1^b \frac{1}{x^2}\,dx$ approaches 1. We can symbolize this as

$$\lim_{b\to\infty} \int_1^b \frac{1}{x^2}\,dx = 1$$

and make the following definitions.

Improper Integrals

1. $\displaystyle\int_a^{\infty} f(x)\,dx = \lim_{b\to\infty}\int_a^b f(x)\,dx$
2. $\displaystyle\int_{-\infty}^{b} f(x)\,dx = \lim_{a\to-\infty}\int_a^b f(x)\,dx$
3. $\displaystyle\int_{-\infty}^{\infty} f(x)\,dx = \lim_{a\to-\infty}\int_a^c f(x)\,dx + \lim_{b\to\infty}\int_c^b f(x)\,dx,$

where $a \le c \le b$

If these limits exist, the improper integral is said to **converge** to the computed value. Otherwise, it **diverges**.

Our first example showed that $\int_1^{\infty} \frac{1}{x^2}\,dx$ converges to 1; that is, though the region goes on forever, its area is finite and never exceeds 1.

However, the integral $\int_1^{\infty} \frac{1}{x}\,dx$ does not converge, since

$$\begin{aligned}\int_1^{\infty} \frac{1}{x}\,dx &= \lim_{b\to\infty}\int_1^b \frac{1}{x}\,dx \\ &= \lim_{b\to\infty} \ln|x|\Big|_1^b \\ &= \lim_{b\to\infty} (\ln|b| - \ln|1|) \\ &= \lim_{b\to\infty} (\ln b - 0) \\ &= \lim_{b\to\infty} \ln b\end{aligned}$$

b	$\int_1^b \frac{1}{x}\,dx \approx \ln b$
2	0.6931
3	1.0986
5	1.6094
10	2.3026
100	4.6052
1000	6.9078
10,000	9.2103
100,000	11.5129

which does not exist; remember, $\ln x$ is an increasing function, as noted in the adjoining table. It is interesting to note that such an apparently small change in the integrand, in this case from $\frac{1}{x^2}$ to $\frac{1}{x}$, can produce such a drastic effect.

$$\int_1^{\infty} \frac{1}{x^2}\,dx \text{ converges, whereas } \int_1^{\infty} \frac{1}{x}\,dx \text{ diverges.}$$

EXAMPLE SET A

Evaluate $\displaystyle\int_5^{\infty} \frac{1}{x^{3/2}}\,dx$.

Solution:

$$\int_5^{\infty} \frac{1}{x^{3/2}}\,dx = \lim_{b\to\infty} \int_5^b \frac{1}{x^{3/2}}\,dx$$

$$= \lim_{b\to\infty} \int_5^b x^{-3/2}\,dx$$

$$= \lim_{b\to\infty} \left(-2x^{-1/2}\right)\Big|_5^b$$

$$= \lim_{b\to\infty} -2\left(b^{-1/2} - 5^{-1/2}\right)$$

$$= -2\left(0 - \frac{1}{5^{1/2}}\right)$$

$$= \frac{2}{\sqrt{5}}$$

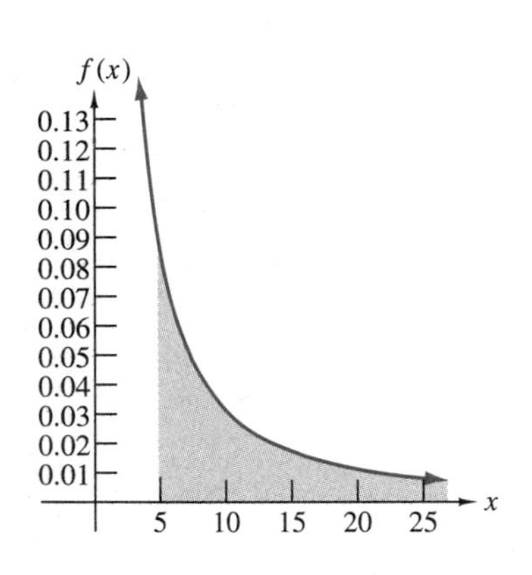

Figure 6.15

Interpretation: The improper integral $\int_5^{\infty} \frac{1}{x^{3/2}}\,dx$ converges to $\frac{2}{\sqrt{5}}$. Figure 6.15 shows the graph of $f(x) = \frac{1}{x^{3/2}}$. Since $f(x)$ lies above the x axis when $x \geq 5$, the integral can be associated with the area under the curve. Therefore, the area of the region below $f(x)$, above the x axis, and to the right of $x = 5$ converges to $\frac{2}{\sqrt{5}} \approx 0.894$ as x increases without bound.

EXAMPLE SET B

Evaluate $\int_{-\infty}^{1} \frac{1}{(x-2)^3}\,dx$.

Solution:

We let $u = x - 2$ and $du = dx$. The new limits of integration are:

when $x = 1$, then $u = 1 - 2 = -1$ and

when $x = -\infty$, then $u = -\infty$.

Therefore,

$$\int_{-\infty}^{1} \frac{1}{(x-2)^3}\,dx = \int_{-\infty}^{-1} \frac{1}{u^3}\,du$$

$$= \int_{-\infty}^{-1} u^{-3}\,du$$

$$= \lim_{b\to-\infty} \int_b^{-1} u^{-3}\,du$$

$$= \lim_{b\to-\infty} \frac{u^{-2}}{-2}\Big|_b^{-1}$$

$$= \lim_{b \to -\infty} \left[\frac{(-1)^{-2}}{-2} - \left(\frac{b^{-2}}{-2} \right) \right]$$

$$= \lim_{b \to -\infty} \left(-\frac{1}{2} + \frac{1}{2b^2} \right)$$

$$= -\frac{1}{2} + 0$$

$$= -\frac{1}{2}$$

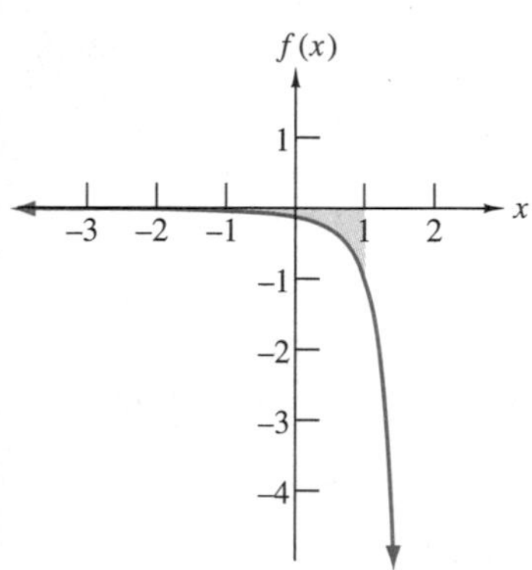

Figure 6.16

Interpretation: The improper integral $\int_{-\infty}^{1} \frac{1}{(x-2)^3}\,dx$ converges to $\frac{1}{2}$. Figure 6.16 shows the graph of $f(x) = \frac{1}{(x-2)^3}$. Since $f(x)$ lies below the x axis when $x \leq 1$, the integral can be associated with the opposite of the area of the region above the curve, below the x axis, and to the left of $x = 1$. Therefore, the opposite of the area of this region converges to $-\frac{1}{2} = -0.5$ as x decreases without bound.

EXAMPLE SET C

Evaluate $\int_{-\infty}^{\infty} x\,dx$.

Solution:

$$\int_{-\infty}^{\infty} x\,dx = \lim_{a \to -\infty} \int_a^0 x\,dx + \lim_{b \to \infty} \int_0^b x\,dx$$

$$= \lim_{a \to -\infty} \frac{x^2}{2}\bigg|_a^0 + \lim_{b \to \infty} \frac{x^2}{2}\bigg|_0^b$$

$$= \lim_{a \to -\infty} \left(\frac{0^2}{2} - \frac{a^2}{2} \right) + \lim_{b \to \infty} \left(\frac{b^2}{2} - \frac{0^2}{2} \right)$$

$$= \lim_{a \to -\infty} \left(-\frac{a^2}{2} \right) + \lim_{b \to \infty} \left(\frac{b^2}{2} \right)$$

$$= -\infty + \infty$$

$$= \text{Does not exist or diverges}$$

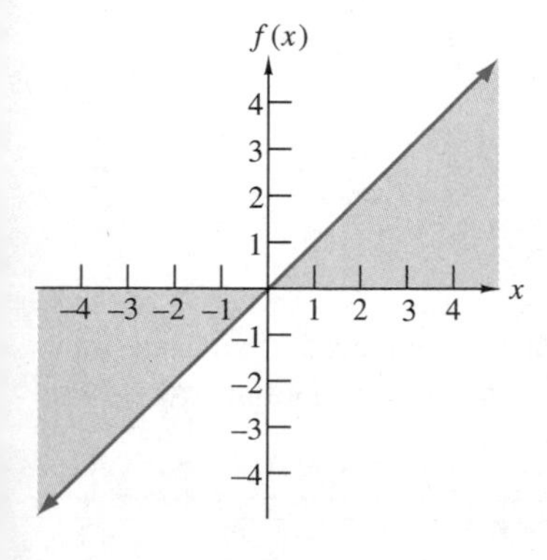

Figure 6.17

Interpretation: The improper integral $\int_{-\infty}^{\infty} x\,dx$ diverges. The graph of $f(x) = x$ is shown in Figure 6.17. Notice that in the interval from 0 to ∞ the curve is above the x axis, and therefore, the integral is associated with the area under the curve, which obviously continues to get larger as x increases without bound. In the interval from $-\infty$ to 0, the curve is always below the x axis, and therefore, the integral is associated with the opposite of the area between the curve and the

x axis that decreases as x decreases without bound. Consequently, the integral does not converge on any particular number; it diverges.

EXAMPLE SET D

Suppose a pollutant is seeping into the ground near a dump site at the rate of $f(t) = \dfrac{5400}{(t+1.5)^3}$ liters per month, where t denotes the time from now in months. If the seepage of the pollutant continues indefinitely into the future, what is the total amount of pollutant that will seep into the ground?

Solution:

Since we wish to find the total amount of pollutant, we will use the definite integral as an accumulator. Furthermore, since we wish to accumulate over an interval with an unspecified upper bound, we will need to use an improper integral.

$$
\begin{aligned}
\text{Total amount of pollutant} &= \int_0^\infty \frac{5400}{(t+1.5)^3}\,dt \\
&= \lim_{b\to\infty} \int_0^b \frac{5400}{(t+1.5)^3}\,dt \\
&= \lim_{b\to\infty} \int_0^b 5400(t+1.5)^{-3}\,dt \\
&= \lim_{b\to\infty} \left. \frac{5400(t+1.5)^{-2}}{-2} \right|_0^b \\
&= \lim_{b\to\infty} \left. \frac{-2700}{(t+1.5)^2} \right|_0^b \\
&= -2700 \lim_{b\to\infty} \left(\frac{1}{(b+1.5)^2} - \frac{1}{(0+1.5)^2} \right) \\
&= -2700 \left(0 - \frac{1}{2.25} \right) \\
&= -2700 \left(-\frac{1}{2.25} \right) \\
&= 1200
\end{aligned}
$$

Interpretation: If the seepage of the pollutant goes unchecked and continues indefinitely into the future, 1200 liters will collect in the ground near the dump site.

Improper Integrals and Technology

Using Your Calculator You can use your calculator to approximate an improper integral by choosing a large value for the upper limit of integration. For example, to approximate

$$\int_5^\infty \frac{1}{x^{3/2}}\,dx$$

you might enter

```
Y1 = 1/X^(3/2)
Y10 = fnInt(Y1,X,A,B)
5 → A
1000 → B
Y10
```

This produces 0.8311. The exact result is $\frac{2}{\sqrt{5}}$, which is approximately 0.8944. You might obtain more accuracy by replacing 1000 with 5000. Try it.

Calculator Exercises

1. An oil well is expected to produce oil at the rate of $6765e^{-0.13t}$ barrels per year for the next t years. Find the total amount of oil the well is expected to produce in its first 10 years of operation. If the well is operated long into the future, find the total number of barrels of oil the well can be expected to produce. Construct the graph of this rate function and shade the region that corresponds to the total amount of oil produced by the well over its first 10 years of operation.

2. A company expects income to flow continuously into it at the rate of $36e^{-0.22t}$ dollars per year for the next t years. Find the total amount of income to the company in its first 5 years of operation, in its first 10 years of operation, and in its first 20 years of operation. If the company operates long into the future, find the total amount of money that can be expected to flow into it. Construct the graph of this rate function and shade the region that corresponds to the total income flow over all the years of the company's operation.

3. Concerned with clean-up costs, the management of a manufacturing facility determines that the rate at which the plant dumps industrial wastes into a lake is approximated by $3.7e^{-0.82t}$ pounds per year, where t is the number of years from now. Find the total amount of waste dumped into the lake over the company's first 5 years of operation. If the company operates indefinitely, find the total amount of waste it can expect to dump into the lake.

4. The integral $\int_1^\infty \frac{1}{x}\,dx$ diverges, as illustrated in Example Set A. Use your calculator to estimate $\int_1^\infty \frac{1}{x}\,dx$ and explain the discrepancy between the computed value and the theoretical value.

EXERCISE SET 6.6

A UNDERSTANDING THE CONCEPTS

1. ***Business: Income Flow*** The function $f(t)$ relates the rate of flow f (in dollars) of income from a rental property to the time t in years from now. Interpret

$$\int_0^\infty f(t)\,dt$$

2. ***Ecology: Pollution*** The function $P(t)$ relates the rate P (in parts per million) of a toxic pollutant that is seeping into the ground near a city's water supply to the time t in years from now. Interpret

$$\int_5^\infty P(t)\,dt$$

3. ***Ecology: Energy Conservation*** The function $E(t)$ relates the rate at which energy E (in kilowatt hours) is being conserved in a community to the time t in years from now. Interpret

$$\int_{1.5}^\infty E(t)\,dt$$

4. ***Political Science: Immigration*** The function $I(t)$ relates the rate I (in thousands of people) at which people from Eastern European countries are immigrating to Western European countries to the time t in years from 1970. Interpret

$$\int_{10}^\infty I(t)\,dt$$

5. ***Business: Quality Control*** The function $Q(x)$ relates the rate Q (in units) at which a company improves the quality of its product to the number x (in units) of the product it produces. Interpret

$$\int_0^\infty Q(x)\,dx$$

6. ***Business: Advertising*** The function $P(t)$ relates the rate P (in thousands of units) at which consumers purchase a product to the number x of people who have been introduced to the product through television advertisements. Interpret

$$\int_0^\infty P(x)\,dx$$

B SKILL ACQUISITION

For Exercises 7–27, evaluate each improper integral if it is convergent. If it is not convergent, write *divergent*.

7. $\int_1^\infty \frac{1}{x^4}\,dx$

8. $\int_3^\infty \frac{5}{x^3}\,dx$

9. $\int_e^\infty \frac{3}{x}\,dx$

10. $\int_0^\infty \frac{1}{x+3}\,dx$

11. $\int_{-1}^\infty \frac{3}{3x+4}\,dx$

12. $\int_0^\infty e^{-3x}\,dx$

13. $\int_0^\infty e^{-4x}\,dx$

14. $\int_{-\infty}^0 e^{-x}\,dx$

15. $\int_{-\infty}^3 e^{2x}\,dx$

16. $\int_5^\infty e^{2x}\,dx$

17. $\int_0^\infty \frac{1}{(x+1)^3}\,dx$

18. $\int_{10}^\infty \frac{8}{(x-8)^2}\,dx$

19. $\int_{-\infty}^\infty xe^{-x^2}\,dx$

20. $\int_{-\infty}^\infty x^{2/3}\,dx$

only 2 problems

21. $\int_0^\infty xe^{-x^2}\,dx$

22. $\int_{100}^\infty \frac{2x+3}{x^2+3x}\,dx$

23. $\int_{50}^\infty \frac{6x^2}{(x^3+6)^{3/2}}\,dx$

24. $\int_{200}^\infty \frac{1}{x\ln x}\,dx$

25. $\int_e^\infty \frac{1}{x(\ln x)^2}\,dx$

26. $\int_{\sqrt{e}}^{\infty} \frac{\ln x}{x}\,dx$

27. $\int_{0}^{\infty} ae^{-ax}\,dx, \quad a > 0$

C APPLYING THE CONCEPTS

28. ***Ecology: Pollution*** The rate A (in tons per year) at which a pollutant is being dumped near a city is given by the function

$$A(t) = 1200e^{-0.04t}$$

where t represents the time in years from now. If the dumping continues unchecked, what will be the total amount of pollutant dumped near the city?

29. ***Business: Income Flow*** The rate I (in dollars per year) at which income flows into an account in which interest is compounded continuously at 8% from a rental property is given by the function

$$I(t) = 200{,}000e^{-0.08t}$$

where t is the time in years from now. If the income is not disrupted and continues indefinitely, how much money will accumulate in this account?

30. ***Political Science: Waiting Time*** Telephone calls to a U.S. Congresswoman's local office between 9:00 A.M. and 12:00 noon are received approximately at the rate of 15 per hour. The probability that there is a time period of 5 minutes or more between calls is given by

$$\int_{1/12}^{\infty} 15e^{-15t}\,dt$$

Determine this probability. (Hint: 5 minutes = $\frac{5}{60} = \frac{1}{12}$ hours.)

31. ***Ecology: Pollution*** In a particular area of the country, manufacturing plants are located alongside a large river that flows into the ocean. As these plants operate, they release chemical pollutants into the river. The river carries the pollutants downstream and eventually into the ocean. If the rate A (in tons per year) at which the chemical pollutants are deposited into the ocean is given by

$$A(t) = \frac{1200}{(0.8 + 0.025t)^2}$$

where t is the number of years from now, and the situation continues unabated, how many tons of the chemical pollutant will be deposited into the ocean?

32. ***Business: Production*** A company claims it can produce a particular product at the rate of

$$N(t) = 400.4e^{-0.26t}$$

units per year, where t is the time in years from now. Assuming the company's claim is correct, how many units of the product will the company eventually produce?

33. ***Business: Marginal Profit*** A retailer believes that the marginal profit P (in dollars) he makes on the sale of x units of a product is given by the function

$$P(x) = 147.24e^{-0.04x}$$

If this retailer can sell what is essentially infinitely many units of the product, what will be his total profit on this product?

D DESCRIBE YOUR THOUGHTS

34. Describe the strategy you would use to evaluate $\int_{1}^{\infty} \frac{1}{(x+3)^2}\,dx$.

35. Explain how you know that $\int_{1}^{\infty} \frac{1}{x+4}\,dx$ diverges.

36. Describe the difference between an improper integral and a definite integral.

E REVIEW

37. **(2.2)** Find, if it exists, $\lim_{x\to 5} \frac{x^2 - 2x - 15}{x^2 - 7x + 10}$.

38. **(2.3)** ***Economics: Population Size of a County*** The population P (in thousands) of a county t years from the beginning of 1990 is given by the function

$$P(t) = \frac{21.6t^2 + 72.4t + 45}{t^2 + 1.6t + 52}$$

What is the expected population of the county in the long run?

39. **(3.4)** Find and interpret $f'(x)$ for $f(x) = (5x-2)^3(x+2)^2$ at $x = 0$.

40. **(4.2)** Sketch a curve over the interval $[1, 7]$ so that it is decreasing at a decreasing rate over $[1, 4]$, has a corner at $(4, 2)$, and has a relative maximum at $(5, 6)$.

41. **(4.3)** ***Manufacturing: Production*** A rectangular storage box with no top and a square base is to have a volume of 4000 cubic inches. What are the dimensions (length of base and height) that will minimize the surface area of the box?

42. **(5.3)** Find $f'(x)$ for $f(x) = e^{-2x}\ln(5x)$.

43. **(6.1)** Evaluate the indefinite integral $\int \frac{5x}{4x^2+1}\,dx$.

44. **(6.3)** Evaluate the definite integral $\int \sqrt{x}\ln x\,dx$.

45. **(6.4)** Evaluate the definite integral $\int_{1}^{3} \frac{\ln x}{2x}\,dx$.

46. **(6.5)** ***Political Science: Voter Registration*** A political party believes that it will be able to register people to vote who support the party's candidate at the rate of

$$R(t) = 15e^{0.15t}$$

individuals each day. If the party's belief is correct, how many supporters can it expect to register in the next 60 days?

Summary

At the beginning of the text, we noted that there are two types of calculus, differential calculus and integral calculus. Differential calculus is the language of change. Integral calculus is the language of accumulation. This chapter introduced integral calculus.

Antidifferentiation and the Indefinite Integral

Section 6.1 introduced the antiderivative and the integral of a function. Suppose that $F(x)$ and $f(x)$ are two functions such that $f(x)$ is the derivative of $F(x)$; that is, $F'(x) = f(x)$. Then $F(x)$ is an *antiderivative* or *integral* of $f(x)$. For example, if $f(x) = 4x - 5$, then $F(x)$ could be defined by $2x^2 - 5x$, since $F'(x) = f(x) = 4x - 5$. It is important to remember that a function has infinitely many antiderivatives, with each pair differing by a constant. This is made clearer in the definition of the indefinite integral. Suppose that $F(x)$ is any one of the infinitely many members of the family of antiderivatives of $f(x)$. Then,

$$\int f(x)\,dx = F(x) + c,$$

where c is an arbitrary constant. The function $F(x) + c$ is called the *antiderivative* or *indefinite integral* of $f(x)$. The number c is called the *constant of integration*, the function $f(x)$ is called the *integrand*, and dx is called the *differential*.

This section continued by presenting the basic integration formulas.

1. If k is any constant, then $\int k\,dx = kx + c$. To integrate a constant function, affix the independent variable to the constant and add an arbitrary constant.
2. If r is any real number other than -1, then $\int x^r\,dx = \frac{x^{r+1}}{r+1} + c$. To integrate a power function, increase the exponent by 1, divide the resulting power by the new exponent, and add an arbitrary constant.
3. If k is any constant, then $\int kf(x)\,dx = k\int f(x)\,dx$. To integrate a constant times a function, integrate the function, then multiply that result by the constant, and add an arbitrary constant.
4. For any two functions $f(x)$ and $g(x)$,

$$\int [f(x) \pm g(x)]\,dx = \int f(x)\,dx \pm \int g(x)\,dx$$

To integrate a sum or difference of two functions, integrate each function individually, then add or subtract those results.

5. For $f(x) = e^x$, $\int e^x\,dx = e^x + c$. The antiderivative, or integral, of the exponential function $f(x) = e^x$ is the function itself.
6. For $f(x) = \frac{1}{x}$, $\int \frac{1}{x}\,dx = \ln|x| + c$.

This section also discussed how the value of the constant of integration can be determined if initial conditions are presented. To determine the value of the constant of integration, which amounts to finding the particular curve in the family of curves, substitute the input and output values into the antiderivative and solve for the constant.

Integration by Substitution

Section 6.2 demonstrated how, in some cases, you can find the antiderivative of a function that is not a basic form. If the function has particular characteristics, you may be able to make a substitution that transforms it to a basic form. Some suggestions for substitutions follow.

If the integrand involves

1. a quotient, let $u =$ (the denominator).
2. an exponential expression, let $u =$ (the exponent).
3. a power or a radical, let $u =$ (the base or the radicand).

This section ended by showing how the derivative and integral of $f(x) = e^{ax}$ are related.

$$\frac{d}{dx}\left[e^{ax}\right] = ae^{ax} \quad \text{and} \quad \int e^{ax}\,dx = \frac{e^{ax}}{a} + c$$

Integration by Parts

Section 6.3 introduced a special technique of integration called *integration by parts*. If an integral is of the form $\int u\,dv$, then integration by parts states that

$$\int u\,dv = u \cdot v - \int v\,du$$

It is important to remember to set the portion of the integrand that is easy to integrate to dv and the portion that is easy to differentiate to u.

The Definite Integral

Section 6.4 introduced the definite integral of a function and presented the Fundamental Theorem of Calculus. If $f(x)$ is a continuous function on an interval $[a, b]$, where $a < b$, and if $F(x)$ is an antiderivative of $f(x)$, then the *definite integral* of $f(x)$ from a to b is

$$\int_a^b f(x)\,dx = F(x)\Big|_a^b = F(b) - F(a)$$

and represents the total amount accumulated by $F(x)$ as x increases from a to b.

You should keep in mind the difference between the indefinite and the definite integral.

1. The indefinite integral $\int f(x)\,dx$ *is a function* of x and represents a family of antiderivatives of (f).
2. The definite integral $\int_a^b f(x)\,dx$ *is an accumulator* and registers *a real number*, a constant that records the total amount accumulated by a particular member of a family of antiderivatives.

The Definite Integral and Area

Section 6.5 demonstrated how the definite integral is related to area. It began by defining the area of a planar region in terms of the Riemann sum,

$$S_n = \left[f(x_0) + f(x_1) + \cdots + f(x_{n-1})\right]\Delta x$$

The area of a planar region was then defined in terms of the Riemann sum. Specifically, we defined the area of the planar region below the nonnegative function $f(x)$, about the x axis, and between the vertical lines $x = a$ and $x = b$ to be

$$\lim_{n\to\infty} S_n = \lim_{n\to\infty}\left[f(x_0) + f(x_1) + \cdots + f(x_{n-1})\right]\Delta x$$

A relationship was then made between the area of a planar region and the definite integral. If $f(x)$ is a continuous, nonnegative function on the interval $[a, b]$, the area A (in square units) bounded by the function $f(x)$, the x axis, and the vertical lines $x = a$ and $x = b$ is precisely equal to the value of the definite integral of $f(x)$, from $x = a$ to $x = b$. That is,

$$A = \int_a^b f(x)\,dx$$

Improper Integrals

Section 6.6 focused on improper integrals; that is, integrals that have one or more limits that are infinite. The integrals

$$\int_a^\infty f(x)\,dx = \lim_{b\to\infty}\int_a^b f(x)\,dx$$

$$\int_{-\infty}^b f(x)\,dx = \lim_{a\to-\infty}\int_a^b f(x)\,dx$$

$$\int_{-\infty}^\infty f(x)\,dx = \lim_{a\to-\infty}\int_a^c f(x)\,dx + \lim_{b\to\infty}\int_c^b f(x)\,dx, \quad \text{where } a \le c \le b$$

are examples of improper integrals. As the forms show, these integrals are evaluated by replacing the infinite limit with a finite value, computing the antiderivative, then evaluating the limit of the antiderivative as the finite value approaches infinity.

Supplementary Exercises

1. ***Business: Quality Control*** The probability P that a retailer will accept a lot of canned goods depends on the net weight w (in ounces) of a sample of ten cans. The function relating probability and weight is $P(w)$. Interpret both $P'(14.5) = 0.18$ and its integral $P(14.5) = 0.63$.
2. ***Business: Quality Control*** Quality control engineers are concerned about the efficiency of a pump. They believe the efficiency E (in percent) of the pump is related to the number of gallons x of fluid being pumped by the function $E(x)$. Interpret both $E'(30) = -3.2$ and its integral $E(30) = 64$.
3. ***Life Science: Evaporation of Water*** The rate of evaporation E (in gallons per hour) of water in a lake is related to the speed s (in miles per hour) of the wind over the lake by the function $E(s)$. Interpret, for a particular lake, both $E'(6.2) = 70$ and its integral $E(6.2) = 1275$.
4. Verify by differentiation that
$$F(x) = -5e^{-2x} + 6\ln x + \frac{2}{x^2} + c$$
is an antiderivative of the function
$$f(x) = 10e^{-2x} + \frac{6}{x} - \frac{4}{x^3}.$$
5. Verify by differentiation that $F(x) = 2\ln x + c$ is an antiderivative of the function $f(x) = \frac{2}{x}$.
6. Evaluate $\int \frac{16}{x}\,dx$.
7. Evaluate $\int (3x^2 - 5x + 2)\,dx$.
8. Evaluate $\int 4\sqrt[5]{x^3}\,dx$.
9. Evaluate $\int 40e^{-0.02x}\,dx$.
10. ***Economics: Population Growth of a City*** The growth rate of a city t years after 1990 is given by the function $P(t) = 180\sqrt[4]{t} + 125$. In 1990, the population of the city is 4200. Find the expected population of this city in 1995.
11. ***Business: Advertising and Sales Volume*** The management of a chain of retail stores believes that t days after the end of an advertising campaign, the rate at which the volume V (in dollars) of sales is changing is approximated by the function $V'(t) = -12{,}480e^{-0.24t}$. On the day the advertising campaign ends, the sales volume is \$104,000. Find and interpret both $V'(5)$ and $V(5)$.
12. Evaluate $\int x(2x^2 - 5)^{2/3}\,dx$.
13. Evaluate $\int \frac{x}{x^2+1}\,dx$.
14. Evaluate $\int 2\sqrt{x^2 + 4x - 1}(x+2)\,dx$.
15. Evaluate $\int (5x^4 + 2)e^{x^5+2x}\,dx$.
16. Evaluate $\int \frac{e^x}{2e^x + 3}\,dx$.
17. Evaluate $\int \frac{5\ln x}{x}\,dx$.
18. Evaluate $\int \frac{x}{x-1}\,dx$.
19. Evaluate $\int \frac{3x}{x-4}\,dx$.
20. Evaluate $\int 42e^{-0.13}\,dx$.
21. Evaluate $\int x^4 e^{x^5}\,dx$.
22. ***Ecology: Pollution*** If $A(t)$ represents the rate at which a toxic chemical is leaked into an adjacent stream from day a to day b, interpret $\int_a^b A(t)\,dt$.
23. ***Psychology: Spread of News*** If $H(t)$ represents the rate at which a population hears news between days a and b, interpret $\int_a^b H(t)\,dt$.
24. ***Education: Intellectual Discussion*** If $D(t)$ represents the increase in the amount of time a college student spends on philosophical discussions with her peers between a and b months after beginning college, interpret $\int_a^b D(t)\,dt$.
25. Evaluate $\int_1^3 (3x-1)^3\,dx$.
26. Evaluate $\int_0^1 e^{3x}\,dx$.
27. Evaluate $\int_1^3 \frac{2x}{x^2-3}\,dx$.
28. Evaluate $\int_2^4 e^{0.5x}\,dx$.
29. Evaluate $\int_0^1 x\sqrt{1-x^2}\,dx$.
30. Evaluate $\int_0^3 x(x-3)^2\,dx$.
31. Evaluate $\int_0^4 xe^{x^2}\,dx$.

32. Evaluate $\int_{e^2}^{e^4} \frac{15}{x}\, dx$.

33. Evaluate $\int_{\ln 2}^{\ln 5} 2e^x\, dx$.

34. Evaluate $\int_{2}^{5} \frac{3}{x \ln 4x}\, dx$.

35. Evaluate $\int_{e}^{e^2} \frac{6}{x \ln x^3}\, dx$.

36. Evaluate $\int_{1}^{3} 24x^{2/5}\, dx$.

37. Evaluate $\int_{1}^{3} f(x)\, dx$, where

$$f(x) = \begin{cases} 6x^2 - 5, & \text{if } 0 \leq x \leq 5 \\ x - 2, & \text{if } x > 5 \end{cases}$$

38. Evaluate $\int_{2}^{3} f(x)\, dx$, where

$$f(x) = \begin{cases} 10e^{-0.2x}, & \text{if } 0 \leq x \leq 1 \\ 4x + 1, & \text{if } 1 < x < 5 \\ 5 \ln x^2 + 3x - 4, & \text{if } x \geq 5 \end{cases}$$

39. ***Business: Investment in Technology*** A company's investment in new technology is expected to save it money at the rate of $S(t) = 220(1 - e^{-0.04t})$ thousands of dollars each year after the technology is purchased. What is the accumulated savings this company would realize for the first 3 years after purchase?

40. ***Business: Marginal Revenue*** A company's marginal revenue function (in thousands of dollars) is $R'(x) = 0.006x^2 + 0.01x + 0.003$, where x and $R(x)$ are in thousands of dollars. What is this company's accumulated revenue if sales increase from 35,000 to 50,000?

For Exercises 41–45, evaluate each integral.

41. $\int 3x^3y^2\, dx$

42. $\int 5xy^5\, dx$

43. $\int 5xe^{-y}\, dx$

44. $\int 10 \ln xe^{-0.5y}\, dy$

45. $\int \frac{28xy}{y^2 + 5}\, dy$

CHAPTER 7

Applications of Integration

7.1 Area of Regions in the Plane

Introduction

In this section we will consider areas of planar regions that are not restricted to areas under a curve as we saw in Section 6.5, but are areas of planar regions between two curves as in Figure 7.2 on the next page. As we noted in Section 6.5, many applied problems can be solved by relating their characteristics to area. In this section we will investigate how the area between two curves relates to problems such as revenue flow, advertising, birth rates of insects, and the savings realized from the purchase of new equipment. In subsequent sections, we will relate area to problems involving consumer's and producer's surplus, annuities and money streams, and probability.

The Area Bounded by Two Curves

We know that when $f(x)$ is nonnegative (not below the x axis) on $[a, b]$, the definite integral $\int_a^b f(x)\,dx$ can be used to determine the area of a planar region that is bounded by the curve $f(x)$, the x axis, and the two vertical lines $x = a$ and $x = b$. Figure 7.1 shows such a case. In Section 6.5, we developed this

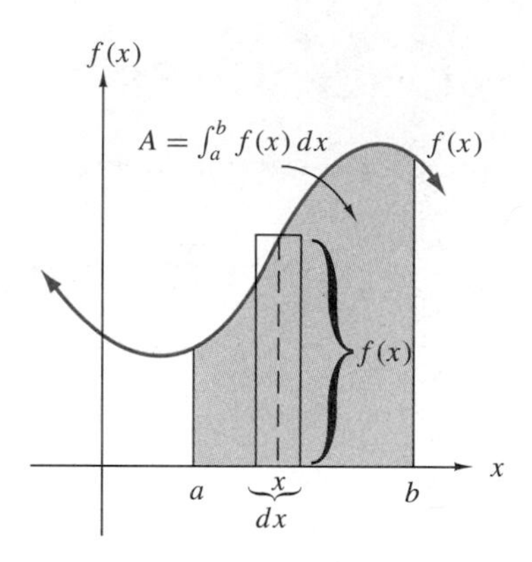

Figure 7.1

relationship using rectangles. Figure 7.1 shows such a planar region with a representative rectangle. The height of the rectangle is $f(x)$ and the width is dx.

We can extend this technique of using the definite integral to determining the area of a planar region bounded by a curve $f(x)$ and the x axis to determining the area of a planar region bounded by a curve $f(x)$ and another curve $g(x)$. Figure 7.2 shows such a region. Look back for a moment at Figure 7.1 and imagine $f(x)$ as the upper curve and the x axis as the lower curve. The x axis is a horizontal line with constant output 0. If we let $g(x)$ represent an output value, the x axis can be described by $g(x) = 0$. Then the height of the representative rectangle in Figure 7.1 can be expressed as

$$f(x) \quad \text{or} \quad f(x) - 0 \quad \text{or} \quad f(x) - g(x)$$

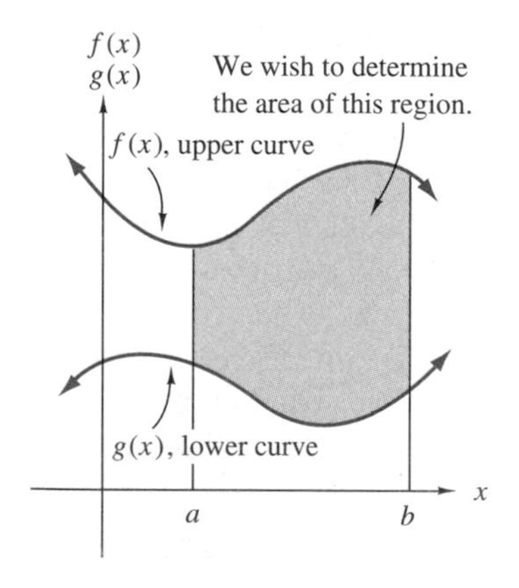

Figure 7.2

The area of the planar region is then expressed as

$$A = \int_a^b \Big[f(x) - g(x)\Big]\,dx$$

The height of the representative strip is then the $\big($height of the top of the rectangle, which is $f(x)\big) - \big($ height of the bottom of the rectangle, which is $g(x)\big)$.

But now, instead of restricting $g(x)$ to be the x axis, if we let $g(x)$ represent a curve that lies below the curve $f(x)$, the area of the more general planar region (Figure 7.2) between the two curves can also be expressed as

$$A = \int_a^b \Big[f(x) - g(x)\Big]\,dx$$

where $f(x)$ is the upper curve and $g(x)$ is the lower curve. Figure 7.3 illustrates this idea using a representative rectangle.

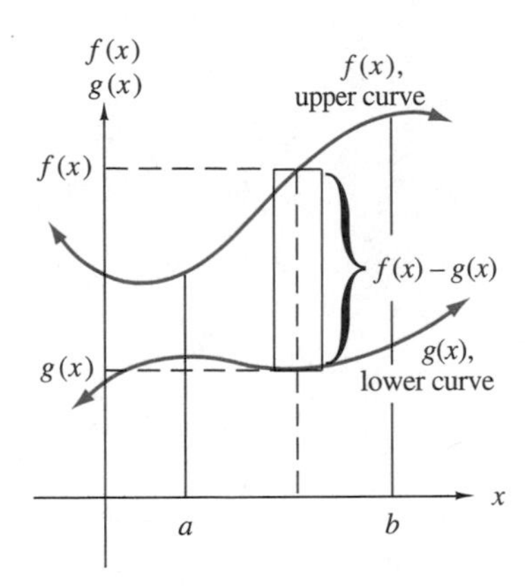

Figure 7.3

Area Bounded by Two Curves

If $f(x)$ and $g(x)$ are two continuous functions for which $f(x) \geq g(x)$ on $[a, b]$, then the area bounded by the two functions and the vertical lines $x = a$ and $x = b$ on $[a, b]$ is

$$A = \int_a^b \left[f(x) - g(x)\right] dx$$

EXAMPLE SET A

Approximate, to two decimal places, the area bounded by the functions $f(x) = 0.05e^{0.05x} + 6$, $g(x) = x^2 - 6x + 11$, and the vertical lines $x = 2$ and $x = 5$.

Solution:

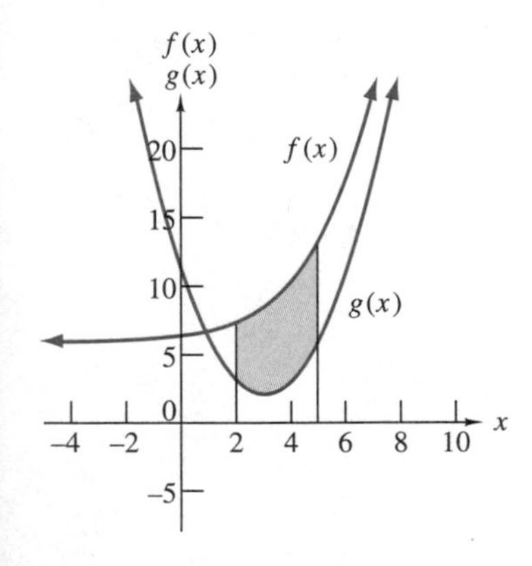

Figure 7.4

Figure 7.4 shows the graphs of both $f(x)$ and $g(x)$. We notice that on the interval $[2, 5]$, $f(x) \geq g(x)$; that is, $f(x)$ lies above $g(x)$, so that $f(x)$ will be the upper curve and $g(x)$ the lower curve.

$$A = \int_2^5 [f(x) - g(x)]\, dx$$

$$= \int_2^5 [(0.05e^{0.05x} + 6) - (x^2 - 6x + 11)]\, dx$$

$$= \int_2^5 \left[0.05e^{0.05x} + 6 - x^2 + 6x - 11\right] dx$$

$$= \left(\frac{0.05e^{0.05x}}{0.05} + 6x - \frac{x^3}{3} + \frac{6x^2}{2} - 11x\right)\Big|_2^5$$

$$= \left(e^{0.05x} + 6x - \frac{x^3}{3} + 3x^2 - 11x \right) \Big|_2^5$$

$$= \left(e^{0.05x} - \frac{x^3}{3} + 3x^2 - 5x \right) \Big|_2^5$$

$$= \left(e^{0.05(5)} - \frac{5^3}{3} + 3(5)^2 - 5(5) \right) - \left(e^{0.05(2)} - \frac{2^3}{3} + 3(2)^2 - 5(2) \right)$$

$$\approx (9.61736) - (0.43850)$$

$$\approx 9.17886$$

$$\approx 9.18$$

Interpretation: The area, to two decimal places, of the planar region bounded by the functions $f(x) = 0.5e^{0.05x}$, $g(x) = x^2 - 6x + 11$, and the vertical lines $x = 2$ and $x = 5$, is 9.18 square units.

EXAMPLE SET B

Find the area, approximate to three decimal places, bounded by the functions $f(x) = x^2 + 3x - 18$, $g(x) = \frac{1}{x}$, and the vertical lines $x = -5$ and $x = -2$.

Solution:
Figure 7.5 shows the graphs of this situation. We notice that on the interval $[-5, -2]$, $g(x) > f(x)$; that is, $g(x)$ lies above $f(x)$.

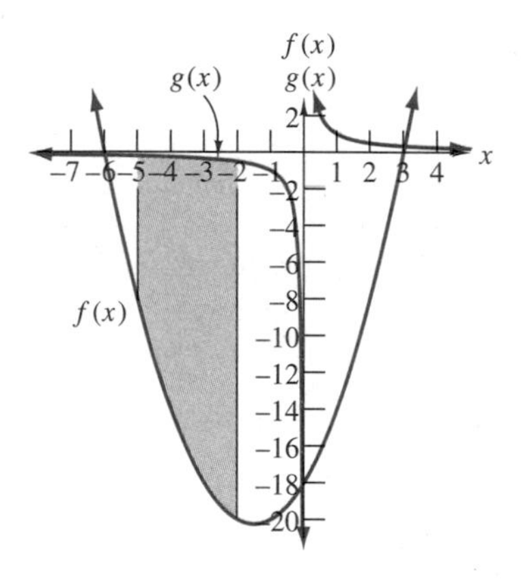

Figure 7.5

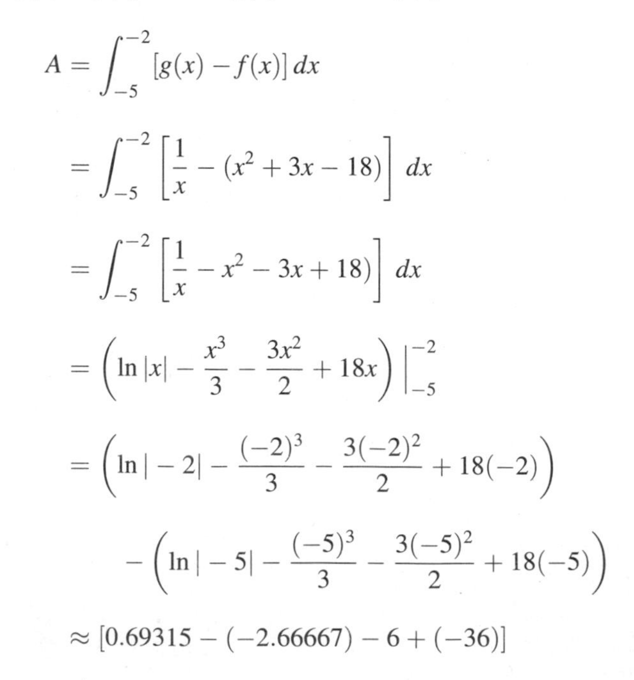

$$A = \int_{-5}^{-2} [g(x) - f(x)]\, dx$$

$$= \int_{-5}^{-2} \left[\frac{1}{x} - (x^2 + 3x - 18) \right] dx$$

$$= \int_{-5}^{-2} \left[\frac{1}{x} - x^2 - 3x + 18) \right] dx$$

$$= \left(\ln|x| - \frac{x^3}{3} - \frac{3x^2}{2} + 18x \right) \Big|_{-5}^{-2}$$

$$= \left(\ln|-2| - \frac{(-2)^3}{3} - \frac{3(-2)^2}{2} + 18(-2) \right)$$

$$- \left(\ln|-5| - \frac{(-5)^3}{3} - \frac{3(-5)^2}{2} + 18(-5) \right)$$

$$\approx [0.69315 - (-2.66667) - 6 + (-36)]$$

$$-[1.60944 - (-41.66667) - 37.5 + (-90)]$$

$$\approx (-38.64018) - (-84.22387)$$

$$\approx 45.584$$

Interpretation: The area, to three decimal places, of the planar region described is 45.584 square units.

In the next example, we will see a situation where two curves bound a region without any vertical lines. They intersect in such a way that they naturally enclose a region. With this situation, we will need to find the limits of integration by finding the x values of the points of intersection between the two curves.

EXAMPLE SET C

Find the area of the planar region bounded by the functions $f(x) = 6 - x - x^2$ and $g(x) = x - 2$.

Solution:

Figure 7.6 shows the graphs of this situation. We notice that there is a region that is bounded strictly by these two curves with $f(x) \geq g(x)$; that is, $f(x)$ lies above $g(x)$. We must next find the interval that defines the enclosed region. We do this by finding the x values of the points of intersection between $f(x)$ and $g(x)$. In other words, when does $f(x) = g(x)$?

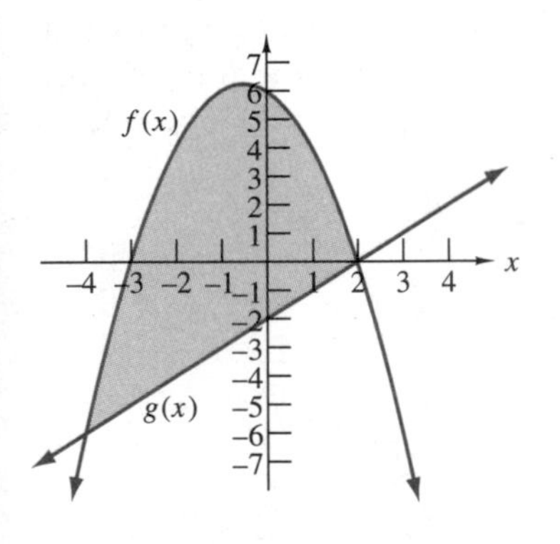

Figure 7.6

$$f(x) = g(x)$$

$$6 - x - x^2 = x - 2 \quad \text{Make the right side of the equation 0.}$$

$$x^2 + 2x - 8 = 0 \quad \text{Solve by factoring.}$$

$$(x + 4)(x - 2) = 0$$

$$x = -4 \text{ or } 2$$

Thus, the interval we are interested in is $[-4, 2]$.

$$A = \int_{-4}^{2} [f(x) - g(x)]\, dx$$

$$= \int_{-4}^{2} [(6 - x - x^2) - (x - 2)]\, dx$$

$$= \int_{-4}^{2} (6 - x - x^2 - x + 2)\, dx$$

$$= \int_{-4}^{2} (8 - 2x - x^2)\, dx$$

$$= \left(8x - \frac{2x^2}{2} - \frac{x^3}{3}\right)\Big|_{-4}^{2}$$

$$= \left(8 \cdot 2 - 2^2 - \frac{2^3}{3}\right) - \left(8 \cdot -4 - (-4)^2 - \frac{(-4)^3}{3}\right)$$

$$= \left(16 - 4 - \frac{8}{3}\right) - \left(-32 - 16 + \frac{64}{3}\right)$$

$$= \frac{28}{3} + \frac{80}{3}$$

$$= \frac{108}{3} = 36$$

Interpretation: The area of the planar region described is 36 square units.

Area and Technology

Using Your Calculator You can use your graphing calculator to approximate the area bounded by two curves. The following entries show how to approximate the area bounded by the curves $f(x) = 0.5e^{0.05x} + 6$, $g(x) = x^2 - 6x + 11$, and the lines $x = 2$ and $x = 5$.

```
Y1 = .5e^(.05X) + 6
Y2 = X^2 − 6X + 11
Y3 = Y1 − Y2
Y0 = fnInt(Y3,X,A,B)
```

Now, in the computation window, enter

```
2 → A
5 → B
Y0
```

The result is approximately 9.18.

Calculator Exercises

1. On the same coordinate system, construct the graphs of the functions $f(x) = 200e^{-0.35(x+2.5)}$ and $g(x) = 200e^{-0.45(x+2.5)}$. Shade the region that corresponds to the area between the two curves over the interval $[0, 5]$. Determine this area. Compute the area between these curves from $x = 0$ to $x = 25$; from $x = 0$ to $x = 100$; from $x = 0$ to $x = 1000$. Use the information you just generated to make a conjecture about the total area between these two curves.
2. A casino finds that t hours after midnight, income, I, in dollars, from a slot machine flows in continuously at the rate of

$$I(t) = 15e^{-0.1t}$$

The cost of operating the machine is constant at \$23.00 per hour. On the same coordinate system, construct the graphs of these curves. About how many hours after midnight does the machine make a profit? What is the total amount of money brought in by such a machine from noon to midnight?

3. A county found that in 1990 the number, N_1, of accidents along its roadways was given by the function

$$N_1(t) = 0.026x^2 + 0.48x + 12, \quad 0 \le t \le 36$$

After a new safety program in which the county added more traffic lights and stop signs, and improved roads went into effect, the number, N_2, of accidents was approximated by the function

$$N_2(t) = -0.041x^2 + 2.5x + 7, \quad t > 36$$

On the same coordinate system, construct the graphs of these curves. Shade the region between the curves from $t = 36$ to $t = 48$. Find and interpret this value.

EXERCISE SET 7.1

A UNDERSTANDING THE CONCEPTS

In Exercises 1–6, set up an integral that represents the area of the planar region shaded in each figure.

1.

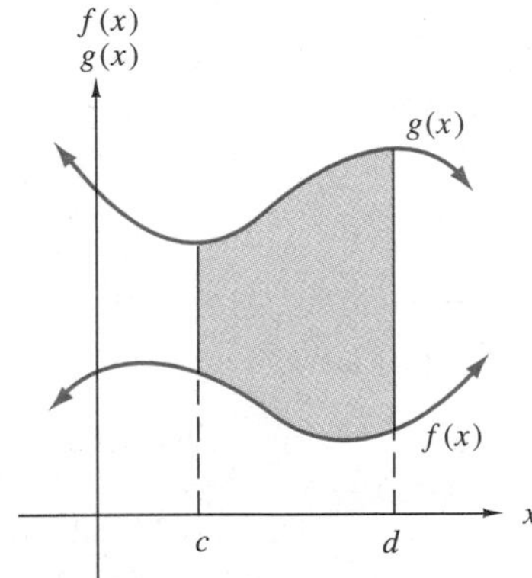

2.

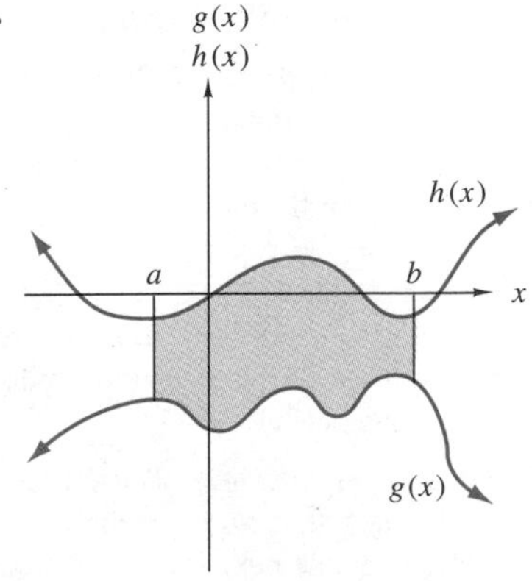

3.

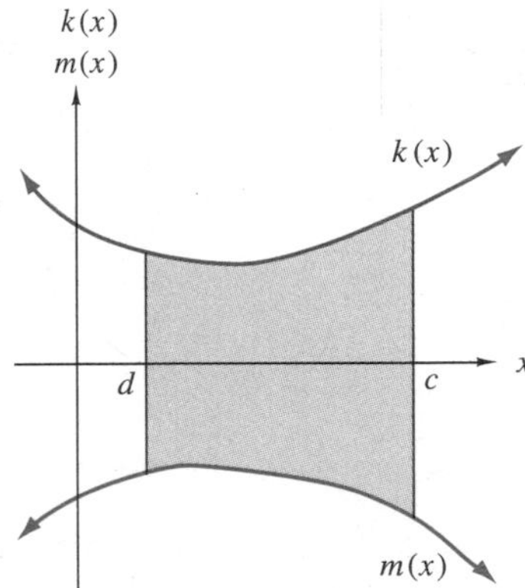

4.

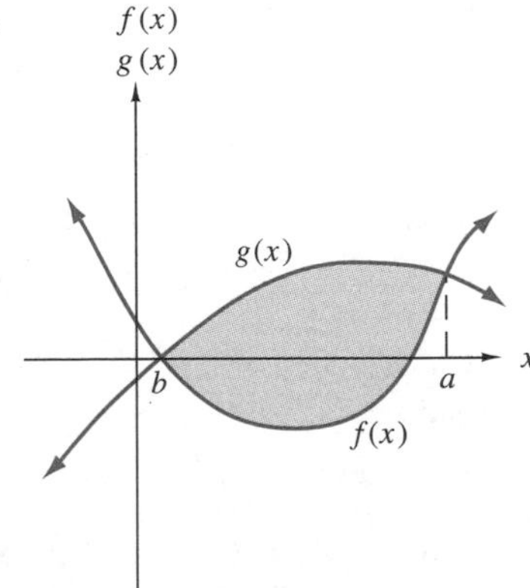

5.

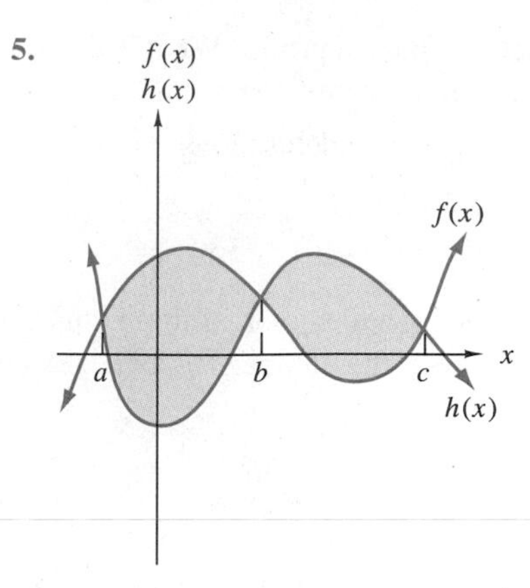

6.

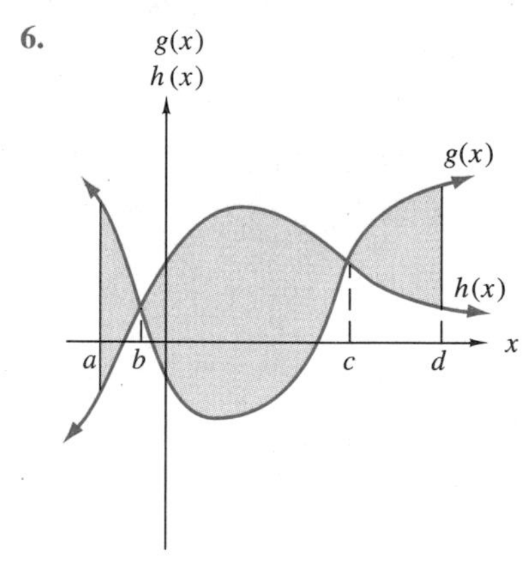

B SKILL ACQUISITION

For Exercises 7–20, find, or approximate to two decimal places, the described area.

7. The area bounded by the functions $f(x) = x + 3$, $g(x) = x^2 + 1$, and the lines $x = 0$ and $x = 2$.

8. The area bounded by the functions $f(x) = x + 5$, $g(x) = x^2 - 2x + 2$, and the lines $x = 0$ and $x = 3$.

9. The area bounded by the functions $f(x) = x + 4$, $g(x) = e^{0.5x}$, and the lines $x = 1$ and $x = 3$.

10. The area bounded by the functions $f(x) = -3x^2 + 6x + 4$, $g(x) = -x + 3$, and the lines $x = 0$ and $x = 2$.

11. The area bounded by the functions $f(x) = -x^2 + 8x - 10$, $g(x) = e^{-0.2x}$, and the lines $x = 3$ and $x = 5$.

12. The area bounded by the functions $f(x) = 9 - x^2$, $g(x) = x^2 + 1$, and the lines $x = 0$ and $x = 2$.

13. The area bounded by the functions $f(x) = x^2 - 5x + 11$, $g(x) = 0.7e^{0.7x}$, and the lines $x = 0$ and $x = 2$.

14. The area bounded by the functions $f(x) = e^{2x}$, $g(x) = e^x$, and the lines $x = 0$ and $x = 2$.

15. The area bounded by the functions $f(x) = x^4$, $g(x) = x^3$, and the lines $x = 0$ and $x = 1$.

16. The area bounded by the functions $f(x) = e^x$, $g(x) = e^{-x}$, and the lines $x = 0$ and $x = 1$.

17. The area completely enclosed by $f(x) = x + 2$ and $g(x) = x^2 - 4$.

18. The area completely enclosed by $f(x) = 2x - 3$ and $g(x) = -x^2 + 4x$.

19. The area completely enclosed by $f(x) = 2 - x^2$ and $g(x) = x$.

20. The area completely enclosed by $f(x) = x$ and $g(x) = \sqrt[3]{x}$.

C APPLYING THE CONCEPTS

21. *Business: Revenue Flow* Over the next 10 years, a company projects its continuous flow of revenue to be $R(t) = 54e^{0.06t}$ and its costs $C(t) = 0.6x^2 + 20$, where t is in years from now and both $R(t)$ and $C(t)$ are in millions of dollars. Approximate the profit this company can expect over the next 10 years.

22. *Business: Savings Realized from New Equipment* A company can save money by buying new equipment, but at the same time will have to spend money to maintain it. The savings realized by the new equipment are

$$S(t) = 3.6t + 8$$

and the cost of maintaining the equipment is

$$C(t) = 4.8t$$

where t is the number of years since the purchase of the equipment and $S(t)$ and $C(t)$ are in thousands of dollars. Find the area between the curve $S(t)$ and $C(t)$ and interpret the result. (Hint: To find the area you need to establish the limits of integration. Determine where the curves intersect by solving the equation $S(t) = C(t)$.)

23. *Business/Life Science: Birth Rate of Insects* The research staff at a chemical company has determined that, over a period of a few days and in a particular southwestern region, the birth rate of an insect is approximated by $B_1(t) = 16e^{0.04x} - 5$. After an application of its new pesticide, the birth rate is approximated by $B_2(t) = 16e^{0.02x} - 5$. By how many insects will an application of the pesticide have reduced the population over a 3-day period?

24. *Business: Advertising and Sales* The owner of a hardware store estimates that with extensive radio advertising, store sales over the next few months could be increasing at the rate of $500e^{0.4t}$ dollars per month t months from now, instead of at the current rate of $(500 + e^{0.9t})$ dollars per month t months from now. Find the additional amount of money the owner would expect

to get in sales over the next 12 months by implementing the radio advertising.

25. ***Economics: Population*** The population of Honduras in 1993 was growing at a rate of $5.24e^{0.028t}$ million people per year, t years from 1993. With efforts to control their growth they estimate that they can decrease the growth to a rate of $4.9e^{0.028t}$ million people a year, t years from 1993. Determine the estimated decrease in people by the year 2001 if the controls are put into place in 1993.

D DESCRIBE YOUR THOUGHTS

26. The formula $\int_a^b [f(x) - g(x)]\,dx$ is used to find the area between two curves $f(x)$ and $g(x)$. Describe how you would determine which curve to represent with $f(x)$ and which to represent with $g(x)$.

27. Explain why the formula $\int_a^b [f(x) - g(x)]\,dx$ cannot be used to find the area between the curve $f(x) = x - 2$ and $g(x) = 0.5x^2 + 7$, from $x = 0$ to $x = 2$.

28. Explain how the area bounded by a curve, the x axis, and the vertical lines $x = a$ and $x = b$ can be thought of as a specific case for a region bounded by two curves and the lines $x = a$ and $x = b$.

E REVIEW

29. (2.2) Find, if it exists, $\lim_{x \to 3} \dfrac{x^2 - x - 6}{x^2 - 5x + 6}$.

30. (3.2) Find and interpret $f'(x)$ for $f(x) = (3x + 4)^5$ at $x = 1$.

31. (3.5) Find and interpret y' for $y^2 - 3y = 2x - 7$ at $(3, \frac{7}{2})$.

32. (4.1) Should a curve that is decreasing at an increasing rate be classified as concave upward or concave downward?

33. (5.3) Use implicit differentiation to find y' for $8\ln(xy^2) + 2y - x = 1$.

34. (6.1) Evaluate the indefinite integral $\int xe^{5x^2}\,dx$.

35. (6.3) Evaluate the indefinite integral $\int xe^{4x}\,dx$

36. (6.4) Evaluate the definite integral $\int_2^7 \dfrac{5e^{\ln x}}{x}\,dx$.

37. (6.5) ***Economics: Income Stream*** A person's investment is expected to produce a continuous income stream of revenue of

$$R(t) = 240e^{0.05t}$$

dollars over the next 25 years. Find the total revenue generated by this revenue stream over the first 10 years of the investment.

38. (6.6) Evaluate the improper integral $\int_4^{\infty} \dfrac{5}{3x^{2/3}}\,dx$.

7.2 Consumer's and Producer's Surplus

Introduction
Consumer's Surplus
Producer's Surplus
Surplus and Technology

Introduction

Both the supply and demand for an item depend on the price of the item. As the price p increases, the number of units x supplied (the supply) tends to increase, and the number of units x sought after for purchase (the demand) tends to decrease. In this sense, price drives both supply and demand so that p is the input variable and x is the output variable, and x is a function of p. Notice, however, that as the supply x of an item increases, the price p of the item tends to decrease. Also, as the demand x of an item increases, the price p tends to increase. In this sense, supply and demand drive price so that x is the input variable and p is the output variable, and p is a function of x.

Let S represent supply and D represent demand,

$$p = S(x) \quad \text{(price is a function of supply)}$$

$$p = D(x) \quad \text{(price is a function of demand)}$$

Supply Function
Demand Function
Equilibrium Point

The function $p = S(x)$ is called a **supply function** and represents the price per unit at which producers are willing to produce and supply x units of a product. The supply function is of interest to producers.

The function $p = D(x)$ is called a **demand function** and represents the price per unit consumers are willing to pay when x units of the item are made available in the market. The demand function is of interest to consumers.

The point (x_e, p_e) at which supply equals demand is called the **equilibrium point** and represents the price p_e consumers are willing to pay when x_e units of the item are produced. Figure 7.7 shows the equilibrium point for a pair of supply and demand functions. The equilibrium point (x_e, p_e) is a point of satisfaction for both the consumer and supplier. As the graph in Figure 7.7 indicates, when the number of items supplied to the market is less than the equilibrium number, both the consumer and supplier benefit. The consumer benefits since, although he was willing to buy at a price higher than the equilibrium price, he was able to buy at a lower price and thus record a savings, or surplus. This surplus is called the **consumer's surplus**, and we will denote it by *CS*. The supplier benefits since, although he was willing to sell at a price lower than the equilibrium price, he was able to sell at a higher price, and thus record a profit, or surplus. This surplus is called the **producer's surplus**, and we will denote it by *PS*.

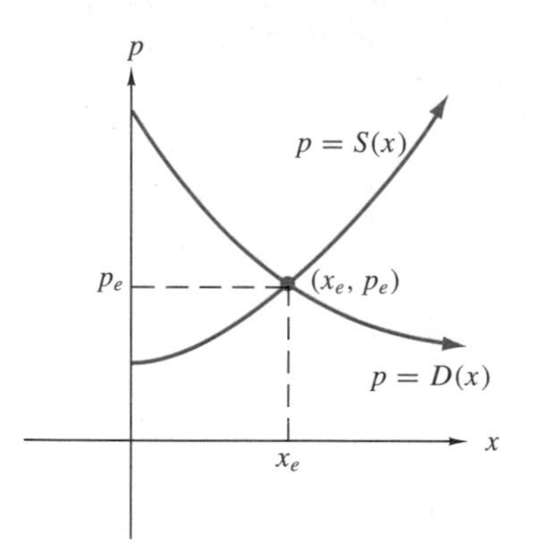

Figure 7.7

We can measure these surpluses using the definite integral as an accumulator of surplus. First for the consumer, for a production of x units up to and including the equilibrium value, the total cost, *TC*, to the consumer is the area of the rectangular region shown in the leftmost illustration of Figure 7.8. This illustration

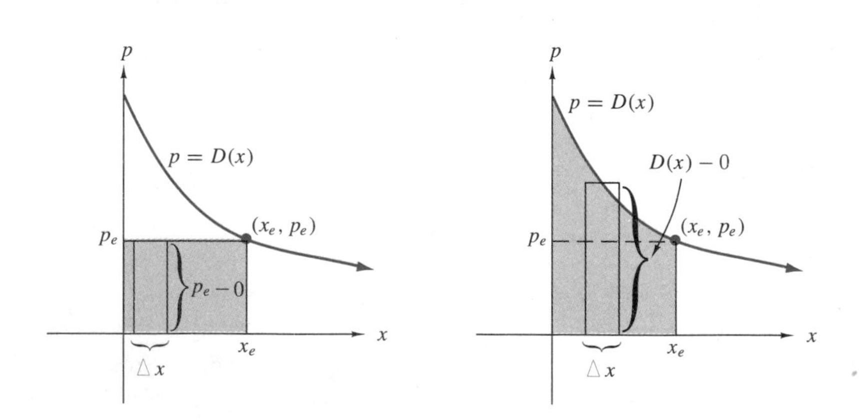

Figure 7.8

shows a representative rectangle of height p_e, which is actually $p_e - 0$, and width

Δx. The area of this rectangle is $[p_e - 0]\Delta x$. The sum of n such rectangles is a Riemann sum and the infinite limit is the definite integral

$$TC = \int_0^{x_e} [p_e - 0]\, dx$$

Similarly, the definite integral

$$TW = \int_0^{x_e} [D(x) - 0]\, dx$$

measures the total amount consumers are willing to spend on up to x_e units. The rightmost illustration of Figure 7.8 shows the area that corresponds to the consumer's total willingness to buy, and also shows a representative rectangle of height $D(x) - 0$ and width Δx.

Consumer's Surplus

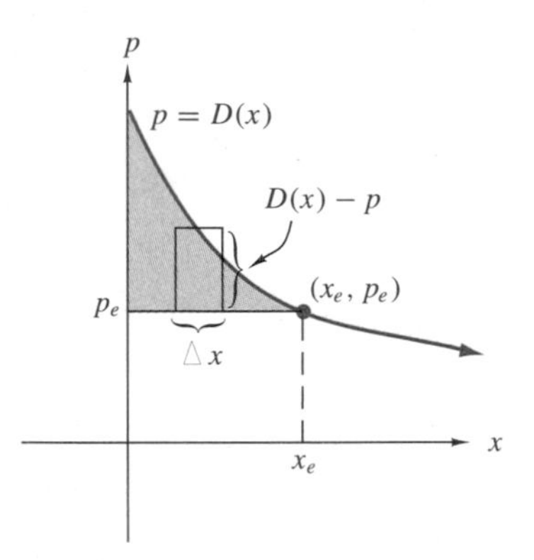

Figure 7.9

The difference between the total amount a consumer is willing to spend and the total amount actually expended is called the **consumer's surplus** and measures the amount of surplus money that results when a purchase is made at a price that is lower than the maximum the consumer was willing to pay. Figure 7.9 shows the area that corresponds to the consumer's surplus. The figure also shows a representative rectangle of height $D(x) - p_e$ and width Δx. The definite integral associated with this area, that is, the measure of consumer's surplus, is given as follows.

Consumer's Surplus

The **consumer's surplus** is defined as the definite integral

$$CS = \int_0^{x_e} \left[D(x) - p_e\right] dx$$

and measures the amount of surplus money that results when a purchase is made at a price that is lower than the maximum the consumer was willing to pay.

Notice that in computing this integral, the demand function $D(x)$ must be known as well as the equilibrium point (x_e, p_e).

EXAMPLE SET A

The demand for a particular item is given by the function $D(x) = 1450 - 3x^2$. Find the consumer's surplus if the market price (equilibrium price) of a unit is \$250.

Solution:
We will use the definite integral as an accumulator of surplus. Here $D(x) = 1450 - 3x^2$, and $p_e = 250$. But p_e is only one coordinate of the equilibrium point. We need the other coordinate, x_e. When $p = 250$,

$$250 = 1450 - 3x^2$$

$$3x^2 = 1200$$

$$x^2 = 400$$

$$x = -20,\ 20$$

Since $x > 0$, we will choose $x = 20$. Thus, the equilibrium point is $(20, 250)$, so that when $p_e = 250$ and $x_e = 20$, we have

$$= \int_0^{x_e} \left[D(x) - p_e\right] dx$$

$$= \int_0^{20} \left[1450 - 3x^2 - 250\right] dx$$

$$= \int_0^{20} \left[1200 - 3x^2\right] dx$$

$$= \left[1200x - x^3\right]\Big|_0^{20}$$

$$= \left[1200(20) - 20^3\right] - \left[1200(0) - 0^3\right]$$

$$= \left[24{,}000 - 8{,}000\right] - [0]$$

$$= 16{,}000$$

Interpretation: When the cost to the consumer of a unit of a particular product is \$250 and the demand for the product is given by $D(x) = 1450 - 3x^2$, the consumer's surplus is \$16,000. Put another way, when $x_e = 20$ units of a product are demanded and purchased and the equilibrium price is $p_e = \$250$, the total savings to the consumer is \$16,000.

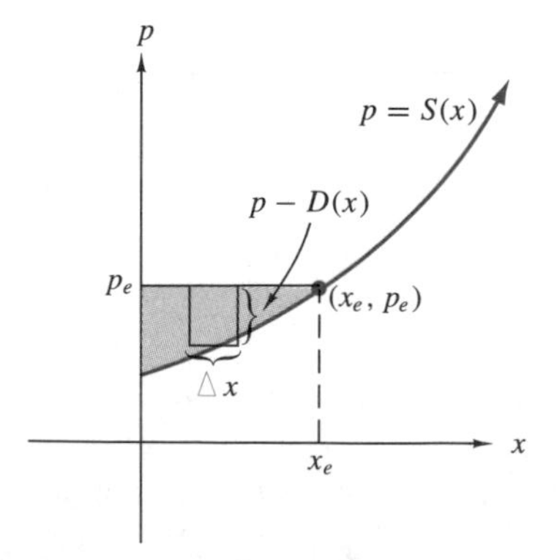

Figure 7.10

Producer's Surplus

Similar to the consumer's surplus, the **producer's surplus** is a measure of the additional money contributed by consumers on the sale of x_e units of a product. The producer's surplus is the difference between the total amount consumers actually spend and the total amount producers are willing to accept for x_e units. Figure 7.10 displays the region that corresponds to the producer's surplus. The figure also displays a representative rectangle with height $p_e - S(x)$ and width Δx.

The definite integral associated with this area, that is, the measure of producer's surplus, is given as follows.

Producer's Surplus

The **producer's surplus** is defined as the definite integral

$$PS = \int_0^{x_e} \left[p_e - S(x)\right] dx$$

and measures the additional money contributed by consumers on the sale of x_e units of a product.

Notice that in computing this integral, the supply function $S(x)$ must be known as well as the equilibrium point (x_e, p_e).

EXAMPLE SET B

Find both the consumer's and producer's surplus if for a product $D(x) = 25 - 0.005x^2$ and $S(x) = 0.004x^2$.

Solution:
Since both surpluses require that we know the equilibrium point, we will start by finding their coordinates, x_e and p_e. The equilibrium point is located where supply equals demand; that is, where $S(x) = D(x)$.

$$S(x) = D(x)$$

$$0.004x^2 = 25 - 0.005x^2$$

$$0.009x^2 = 25$$

$$x^2 = \frac{25}{0.009}$$

$$x \approx -52.70,\ 52.70$$

Since x must be positive, we will choose $x \approx 52.70$. We may substitute 52.70 for x into either $S(x)$ or $D(x)$ (since $p = S(x)$ and $p = D(x)$). We will use $S(x)$ here.

$$p = 0.004x^2$$

$$p \approx 0.004(52.70)^2$$

$$p \approx 11.11$$

Therefore, $x_e \approx 52.70$ and $p \approx 11.11$.

We will begin with the consumer's surplus.

$$
\begin{aligned}
CS &= \int_0^{x_e} [D(x) - p_e]\, dx \\
&\approx \int_0^{52.70} \left[25 - 0.005x^2 - 11.11\right] dx \\
&\approx \int_0^{52.70} \left[13.89 - 0.005x^2\right] dx \\
&\approx \left[13.89x - 0.005\frac{x^3}{3}\right] \Big|_0^{52.70} \\
&\approx [488.06] - [0] \\
&\approx 488.06
\end{aligned}
$$

Interpretation: When the price to the consumer for a unit of a particular product is about \$11.11, and the demand for the product is given by $D(x) = 25 - 0.005x^2$, the consumer's surplus is approximately \$488.06.

We will now find the producer's surplus.

$$
\begin{aligned}
PS &= \int_0^{x_e} [p_e - S(x)]\, dx \\
&\approx \int_0^{52.70} \left[11.11 - 0.004x^2\right] dx \\
&\approx \left[11.11x - 0.004\frac{x^3}{3}\right] \Big|_0^{52.70} \\
&\approx [390.35] - [0] \\
&\approx 390.35
\end{aligned}
$$

Interpretation: When the price to the consumer for a unit of a particular product is about \$11.11, and the supply function for the product is given by $S(x) = 0.004x^2$, the producer's surplus is approximately \$390.35.

Surplus and Technology

Using Your Calculator You can use your graphing calculator to find the consumer's surplus. The following entries show how the solution to the problem of Example Set A is found.

Since $D(x) = 1450 - 3x^2$ and $p_e = 250$, the function you want for Y1 is $250 = 1450 - 3x^2$, which simplifies to $D(x) = 1200 - 3x^2$.

```
Y1 = 1200 − 3x^2
```

```
Y7 = solve(Y1,X,G)
```

where G is the guess for the zero of Y1.

Graph Y1 on $0 \le x \le 25$ and $-500 \le y \le 1500$. The curve intersects the x axis at approximately 20. Use $x = 20$ as x_e, the guess for the zero of Y1.

```
20 → G
Y7
```

The value 22 is a good estimate.

```
Y0 = fnInt(Y1,X,A, B)
0 → A
20 → B
Y0
```

The result of 16,000 agrees with the example.

Calculator Exercises

1. The demand for a particular item is given by the function $D(x) = -0.97x^2 - 8.4x + 1162$. Find the consumer's surplus if the market price (equilibrium price) of a unit is \$824. Construct a graph of this function and shade the region that corresponds to the consumer's surplus.
2. The supply for a particular item is given by the function $S(x) = 32.5 + 94.2(x+1.26)^{2.1}$. Find the producer's surplus if the equilibrium point (x_e, p_e) is $(17.5, 44{,}479)$.
3. The demand for a particular item is given by the function $D(x) = \ln(1.28x + 8.8)$ and the supply by the function $S(x) = \ln\left(\dfrac{721}{1.28x + 3.97}\right)$. Find both the consumer's and producer's surplus. On the same coordinate system, construct graphs of these functions and use different shading schemes to indicate the regions that correspond to the consumer's surplus and the producer's surplus.
4. The demand for a particular item is given by the function $D(x) = e^{4.26-0.124x}$ and the supply by the function $S(x) = e^{1.88+0.214x}$. Find both the consumer's and producer's surplus. On the same coordinate system, construct graphs of these functions and use different shading schemes to indicate the regions that correspond to the consumer's surplus and the producer's surplus.

EXERCISE SET 7.2

A UNDERSTANDING THE CONCEPTS

1. The demand for a particular product is given by the function $D(x)$. The equilibrium point $(x_e, p_e) = (14, 85)$. Interpret

$$\int_0^{14} \left[D(x) - 85\right] dx = 8{,}500$$

2. The demand for a particular product is given by the function $D(x)$. The equilibrium point $(x_e, p_e) = (46, 240)$. Interpret

$$\int_0^{46} \left[D(x) - 240\right] dx = 581{,}250$$

3. The supply for a particular product is given by the function $S(x)$. The equilibrium point $(x_e, p_e) = (35, 12)$. Interpret

$$\int_0^{35} \left[12 - S(x)\right] dx = 1645.15$$

4. The supply for a particular product is given by the function $S(x)$. The equilibrium point $(x_e, p_e) = (925, 1226)$. Interpret

$$\int_0^{925} \left[1226 - S(x)\right] dx = 45.85$$

5. Complete the figure below by filling in the appropriate English phrase at the top of the drawing, and the appropriate integral at the bottom of the drawing.

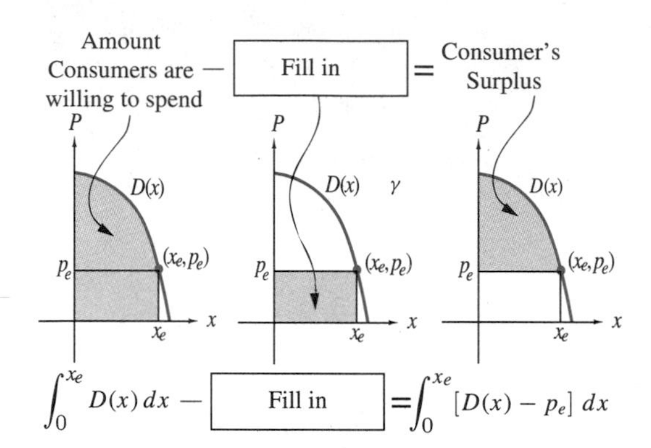

B SKILL ACQUISITION

6. The demand for a particular item is given by the demand function $D(x) = 750 - 2x^2$. Find the consumer's surplus if the equilibrium point (x_e, p_e) is $(15, 300)$.

7. The demand for a particular item is given by the demand function $D(x) = 300 - x^2$. Find the consumer's surplus if the equilibrium point (x_e, p_e) is $(10, 200)$.

8. The supply for a particular item is given by the supply function $S(x) = x^2 + 5x + 20$. Find the producer's surplus if the equilibrium point (x_e, p_e) is $(18, 434)$.

9. The supply for a particular item is given by the supply function $S(x) = \sqrt{x + 15}$. Find the producer's surplus if the equilibrium point (x_e, p_e) is $(35, 7.07)$.

10. Find both the consumer's and producer's surplus if for a product $D(x) = 50 - 2x$, $S(x) = 12 + x$, and the equilibrium point is $\left(12\frac{2}{3}, 24\frac{2}{3}\right)$.

11. Find both the consumer's and producer's surplus if for a product $D(x) = 500 - 5x$, $S(x) = 10x + 200$, and the equilibrium point is $(20, 400)$.

C APPLYING THE CONCEPTS

12. *Business: Consumer's Surplus* The demand for a certain product is given by the function $D(x) = 2250 - 3.2x^2$. Find the consumer's surplus if the market price (equilibrium price) of a unit is \$306.

13. *Business: Consumer's Surplus* The demand for a product is $D(x) = 26 - 6e^{0.08x}$. Find the consumer's surplus if the market price (equilibrium price) of a unit is \$15. (Hint: You will need to use logarithms to solve for the equilibrium point.)

14. *Business: Consumer's Surplus* The demand for a product is $D(x) = \dfrac{200}{x+2}$. Find the consumer's surplus if the market price (equilibrium price) of a unit is \$8.

15. *Business: Producer's Surplus* The supply of a product is $S(x) = 02.6e^{0.02x}$. Find the producer's surplus if the market price (equilibrium price) of a unit is \$135. (Hint: You will need to use logarithms to solve for the equilibrium point.)

16. *Business: Producer's Surplus* The supply of a certain product is given by the function $S(x) = 0.012x^2$. Find the producer's surplus if the market price (equilibrium price) of a unit is \$500.

17. *Business: Producer's Surplus* The supply of a certain product is given by the function $S(x) = 10e^{x/3}$. Find the producer's surplus when 15 units of the product are sold.

18. *Business: Producer's Surplus* The supply of a certain product is given by the function $S(x) = 8e^{x/50}$. Find the producer's surplus if the market price (equilibrium price) of a unit is \$18. (Hint: Use logarithms to solve for the equilibrium point.)

19. *Business: Consumer's and Producer's Surplus* Find both the consumer's and producer's surplus if $D(x) = 60 - 0.04x^2$ and $S(x) = 0.06x^2$, respectively.

20. *Business: Consumer's and Producer's Surplus* Find both the consumer's and producer's surplus if $D(x) = \dfrac{5510}{x+25}$ and $S(x) = 0.03x^2 + 15$, respectively,

21. *Business: Consumer's and Producer's Surplus* Find both the consumer's and producer's surplus if $D(x) = \dfrac{10{,}000}{\sqrt{x+10}}$ and $S(x) = 100\sqrt{10 + 0.05x}$, respectively.

22. *Business: Consumer's and Producer's Surplus* Find both the consumer's and producer's surplus if $D(x) = \dfrac{6000}{\sqrt{x+60}}$ and $S(x) = 20\sqrt{60 + 3x}$, respectively.

D DESCRIBE YOUR THOUGHTS

23. Compare and contrast consumer's and producer's surplus.

24. Describe what the equilibrium point is and its importance to both consumers and producers.

25. Describe how the definite integral relates to consumer's surplus and producer's surplus.

E REVIEW

26. (2.3) Find, if it exists, $\lim_{x\to\infty} \dfrac{3x^2 - 3x + 40}{7x^2 + x - 500}$.

27. (2.4) Discuss, by writing an analysis, the continuity of the function

$$f(x) = \frac{x^2 - x - 5}{x - 5} \text{ at } x = 5$$

28. (3.4) Find the derivative of the function

$$A(s) = s^4(3s + 7)^{1/3}$$

29. (4.2) Find the derivative of the function

$$g(x) = \ln x^2 - 3x$$

30. (4.4) Find the derivative of the function

$$f(x) = \frac{3 - e^{-x}}{3 + e^{-x}}$$

31. (5.3) Find $f'(x)$ for $f(x) = \ln \dfrac{6x - 4}{2x + 7}$.

32. (6.2) Evaluate the indefinite integral

$$\int \left(\frac{3}{5x} + e^{-7x}\right) dx$$

33. (6.3) Evaluate the definite integral

$$\int_0^{\ln 2} xe^{2x}\, dx$$

34. (6.4) Evaluate the definite integral

$$\int_1^4 \frac{3x}{6 - 5x}\, dx$$

35. (7.1) Find the area under the curve $f(x) = 2x - 5$ and between $x = 7$ and $x = 12$.

7.3 Annuities and Money Streams

Introduction

Accumulated Value of a Non-Interest-Earning Stream

The Accumulated Value of a One-Time, Lump-Sum, Interest-Earning Investment

The Accumulated Value of an Interest-Earning Stream: Annuities

The Accumulated Value of an Interest-Earning Continuous Stream

The Present and Future Values of a Perpetual Income Stream

Income Systems and Technology

Introduction

Typically, a company will reinvest its income. The reinvestment may be into an interest-earning account or a non-interest-earning account. Also, the reinvestment may be made in one lump sum, or in equal sums over specified time intervals, or continuously over some time interval. The continuous investment into an interest-earning account is the most interesting investment, and that is our main focus in this section.

Accumulated Value of a Non-Interest-Earning Stream

In most cases, the company's income does not come in all at once at the end of the year, but rather flows in somewhat continuously throughout the year. For example, money flows into fast-food restaurants, retail stores, airline travel operations, toll roads and bridges, hotels and rental properties, and savings accounts and investments, somewhat continuously and in differing amounts throughout the year. Regardless of how the money flows in, if it is reinvested into a non-interest-earning account, the accumulated amount at the end of some specified time period can be determined by simply adding the individual amounts.

Continuous Stream of Money
Rate of Flow Function

If there are a great many individual amounts, such as the individual transactions at a fast-food restaurant throughout a one-year period, it is convenient to think of the income flow as a **continuous stream of money**. In some cases, it is possible to approximate the rate of flow of the stream by a function $f(t)$. The function $f(t)$ is called the **rate of flow function** and the definite integral can be used as an accumulator to determine the accumulated value of the stream over any specified time period. It is worth noting that the definite integral only approximates the accumulated value since, although the income flow is *actually* discrete (in individual parts), it is being described using a continuous function.

Accumulated Value of a Non-Interest-Earning Stream

If over some interval $0 \leq t \leq T$, money flows continuously into a pool at a rate approximated by the function $f(t)$ dollars per time period, and if the money is reinvested in a non-interest-earning account, then the total amount of money A accumulated at the end of T years is

$$A = \int_0^T f(t)\,dt$$

dollars.

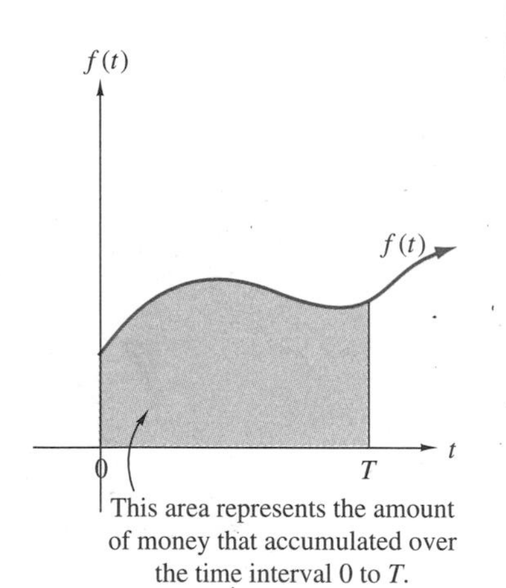

Figure 7.11

The total accumulated amount of money corresponds to the area bounded by the curve $f(t)$, the t axis, and the vertical lines $t = 0$ and $t = T$. Figure 7.11 illustrates this idea.

EXAMPLE SET A

Revenue flows continuously into a retail operation at the constant rate of $f(t) = 128{,}000$ dollars per year. As the money is received, it is placed into a non-interest-earning account. Determine how much money has accumulated in the operation's account at the end of a 3-year time period.

Solution:

The total amount of money accumulated is

$$A = \int_0^T f(t)\,dt$$

$$= \int_0^3 128{,}000\,dt$$

$$= 128{,}000t\Big|_0^3$$

$$= 128{,}000(3) - 128{,}000(0)$$

$$= 384{,}000$$

Thus, $\int_0^3 128{,}000\,dt = 384{,}000$.

Interpretation: If revenue flows into the account of a retail operation at the constant rate of $f(t) = 128{,}000$ dollars per year, and if it is deposited into a non-interest-earning account, the amount of money that will have accumulated at the end of a 3-year period will be \$384,000. Notice that this is the same computation as 128,000 per year for 3 years: $3 \cdot 128{,}000 = 384{,}000$. In other words, this situation was so basic ($f(t) =$ a constant) that the integration was unnecessary.

The Accumulated Value of a One-Time, Lump-Sum, Interest-Earning Investment

We will consider now the case where money is reinvested at some annual interest rate r that is compounded continuously. When P dollars is invested at an annual interest rate r that is compounded continuously, the total amount A of money that will have accumulated at the end of t years is given by

$$A = Pe^{rt}$$

Future Value of the Investment

The amount A is called the **future value of the investment.**

EXAMPLE SET B

If a one-time, lump-sum investment of $P = 128{,}000$ is made into an account paying 7% interest compounded continuously, the amount A of money that will have accumulated at the end of 3 years is

$$A = Pe^{rt}$$

$$= 128{,}000e^{0.07(3)}$$

$$\approx 157{,}910.79$$

Interpretation (1): If a one-time, lump-sum investment of $P = 128{,}000$ is made into an account paying 7% interest compounded continuously, the amount A of money that will have accumulated in the account at the end of 3 years is approximately \$157,910.79.

Interpretation (2): The future value of \$128,000 placed into an account paying 7% interest compounded continuously for 3 years is approximately \$157,910.79. The \$128,000 is the amount of money that would have to be invested now to get \$157,910.79 three years from now. Therefore, we say the future value of \$128,000 is \$157,910.79.

The Accumulated Value of an Interest-Earning Stream: Annuities

We will now consider the case where money is reinvested periodically at some rate of interest r that is compounded continuously. As we have just noted, when P dollars is invested at an annual interest rate r that is compounded continuously into an interest-earning account, the total amount A of money that will have accumulated at the end of t years is given by

$$A = Pe^{rt}$$

where the amount A is called the future value of the investment.

Annuity
Payment Period

If the same amount of money P (P for *principal*) is regularly deposited into an interest-earning account at the end of equal periods of time, the sequence of payments is called an **annuity**. The **amount of the annuity**, or the **future value of the annuity**, is the final amount in the account at the end of all the time periods. It is the total amount of the payments plus the total amount of interest earned from all the payments. The time between payments is the **payment period**.

ILLUMINATOR SET A

Suppose that at the end of each month, for 6 months, \$100 is put into an account paying 6% annual interest compounded continuously. Except for the last \$100, which earns no interest at all, each \$100 earns interest over a different period of time. Using the continuous compounding formula $A = Pe^{rt}$, we can find the total amount of money in the account at the end of 6 months. The interest rate r is the annual rate and, since we are working in months, the corresponding monthly rate is $\frac{0.06}{12} = 0.005$.

1. The first \$100 earns interest for 5 months, so at the end of 6 months it is worth
$$A_1 = Pe^{rt} = 100e^{0.005(5)} \approx 100(1.0253) \approx \$102.53$$

2. The second \$100 earns interest for 4 months, so at the end of 6 months it is worth
$$A_2 = Pe^{rt} = 100e^{0.005(4)} \approx 100(1.0202) \approx \$102.02$$

3. The third \$100 earns interest for 3 months, so at the end of 6 months it is worth
$$A_3 = Pe^{rt} = 100e^{0.005(3)} \approx 100(1.0151) \approx \$101.51$$

4. The fourth \$100 earns interest for 2 months, so at the end of 6 months it is worth
$$A_4 = Pe^{rt} = 100e^{0.005(2)} \approx 100(1.0101) \approx \$101.01$$

5. The fifth \$100 earns interest for 1 month, so at the end of 6 months it is worth
$$A_5 = Pe^{rt} = 100e^{0.005(1)} \approx 100(1.0050) \approx \$100.50$$

6. The last \$100 earns interest for 0 months, so at the end of 6 months it is worth
$$A_6 = Pe^{rt} = 100e^{0.005(0)} \approx 100(1) \approx \$100.00$$

Then the total amount of money that has accumulated at the end of the 6-month period is just the sum of the individual totals, A_1, A_2, A_3, A_4, A_5, and A_6. Thus,

$$\begin{aligned} A &= A_1 + A_2 + A_3 + A_4 + A_5 + A_6 \\ &= 102.53 + 102.02 + 101.51 + 101.01 + 100.50 + 100 \\ &= 607.57 \end{aligned}$$

Interpretation: The future value of \$100 invested at the end of each month, for 6 months, in an account paying 6% annual interest compounded continuously is \$607.57. That is, the \$600 investment is actually now worth \$607.57. (If the \$100 had been deposited into a non-interest-earning account each month, it would be worth \$600 at the end of 6 months.) Figure 7.12 illustrates the process.

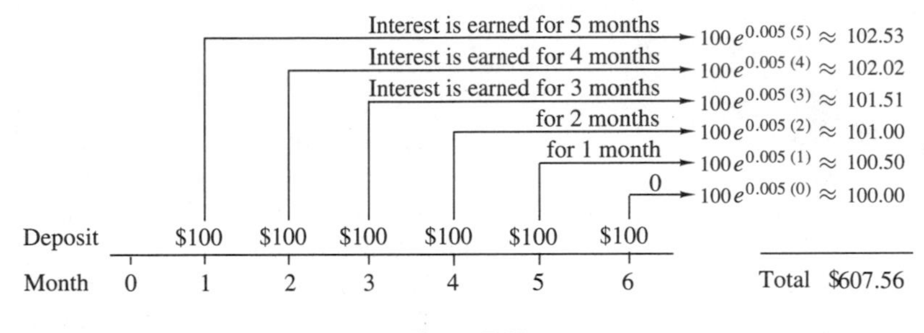

Figure 7.12

Notice that the amount of the annuity is a sum. In fact, it is a Riemann sum with $\Delta t = 1$ month. This means that the amount of the annuity in Illuminator Set A can be approximated by the definite integral $\int_0^6 100e^{0.005t}\,dt$.

Future Value of an Annuity

The amount or **future value** A of an annuity at the end of n payment periods is approximated by

$$A = \int_0^n Pe^{rt}\,dt$$

where P is the number of dollars invested at the end of each payment period, and r is the interest rate per time period. The integral measures the value of the $\$P \cdot n$ investment in the future.

EXAMPLE SET C

Approximate the future value of the annuity in Illuminator Set A over (1) a 6-month period, and (2) a 20-year period.

Solution:

1. Over a 6-month period, the future value of the annuity is

$$\begin{aligned} A &= \int_0^6 100e^{0.005t}\,dt \\ &= 20{,}000e^{0.005t}\Big|_0^6 \\ &= 20{,}000(e^{0.005(6)} - e^{0.005(0)}) \\ &\approx 20{,}000(1.03045 - 1) \\ &\approx 609 \end{aligned}$$

Interpretation: The future value of \$100 invested at the end of each month, for 6 months, in an account paying 6% annual interest compounded continuously is approximately \$609. That is, the \$600 investment is actually worth approximately \$609.

Comparing the actual sum to the approximated sum we see we are in error by \$609 − \$607.56 = \$1.44. In this case the error is somewhat large compared to the amount of interest that the investment generated, but over longer periods of time, like 5, 10, or 20 years, the error becomes negligible.

2. Over a 20-year period, the future value of the annuity is approximated by the integral $\int_0^{240} e^{0.06t}\,dt$. Notice that the upper limit of integration is 240. This comes from the fact that there are $20 \cdot 12 = 240$ payment periods.

$$\begin{aligned} A &= \int_0^{240} 100e^{0.005t}\,dt \\ &= 20{,}000e^{0.005t}\Big|_0^{240} \\ &= 20{,}000(e^{0.005(240)} - e^{0.005(0)}) \\ &\approx 20{,}000(3.320117 - 1) \\ &\approx 46{,}402.34 \end{aligned}$$

Interpretation: The future value of \$100 invested at the end of each month, for 20 years (240 months), in an account paying 6% annual interest compounded continuously is approximately \$46,402.34. That is, the \$24,000 investment is actually worth approximately \$46,402.34.

To obtain the future value of a one-time, lump-sum, interest-earning investment, we used the formula $A = Pe^{rt}$, where A denoted the future value, P the principal (amount invested), r the annual interest rate, and t the time period. If the formula is solved for P, we get

$$P = Ae^{-rt}$$

Present Value

(Can you show this?) The value P is called the **present value** of A, and it represents the amount, P, that must be invested now ($t = 0$) so that the amount A will be accumulated in the future. This situation would be of interest to us, for example, if we wanted to know how much to invest now so that we could have a particular amount of money after a specific time period to invest in a child's education.

EXAMPLE SET D

How much money must be invested now at 8% interest compounded continuously, so that \$15,000 will be available in 10 years?

Solution:
We wish to determine the present value of \$15,000.

$$\begin{aligned} P &= Ae^{-rt} \\ &= 15{,}000e^{-0.08\cdot 10} \\ &\approx 6{,}739.93 \end{aligned}$$

Thus, the present value of \$15,000 is \$6,739.93.

Interpretation: If \$6,739.93 is invested now in an account that pays 8% interest compounded continuously, it will be worth \$15,000 in 10 years.

The Accumulated Value of an Interest-Earning Continuous Stream

We can extend the concept of the future value of an annuity and the present value of an investment to a continuous income stream.

The future value measures the total accumulated amount of income that will be realized at the end of the time period $[0, T]$. The present value measures the amount of money that would have to be deposited now to produce the future value.

The future and present values of a continuous income stream are useful in the analysis of investment opportunities. In Example Set E, we will interpret the present and future value of an income stream in terms of the opportunity for a company to buy a machine that will generate income.

Present and Future Value of an Income Stream

If $f(t)$ is the rate of flow of money into an account that pays an annual interest rate r compounded continuously during some time interval $0 \le t \le T$, then

1. The **future value** of the income stream is

$$\text{Future value} = e^{rT} \int_0^T f(t)e^{-rt}\,dt$$

2. The **present value** of the income stream is

$$\text{Present value} = \int_0^T f(t)e^{-rt}\,dt$$

EXAMPLE SET E

A company can buy a machine that is expected to increase the company's net income by \$10,000 each year for the 5-year life of the machine. The company also estimates that for the next 5 years, money will be worth 7% compounded continuously. Find and interpret (1) the total money flow over the 5-year life of the machine, (2) the future value of the income stream, and (3) the present value of the income stream. (4) Use these values to determine the conditions under which the company should and should not buy the machine.

Solution:
In this situation, the rate of flow of income is given by $f(t) = 10{,}000$.

1. In 5 years, and without any interest, the \$10,000 income stream will produce

$$\int_0^5 10{,}000\,dt = 10{,}000t\Big|_0^5 = \$50{,}000.$$

   Interpretation: Without being invested into an interest-earning account, the \$10,000 five-year income stream will produce \$50,000.

2. The future value of the income stream will tell us how much money, including direct income and interest, i machine will generate.

$$\begin{aligned}\text{Future value} &= e^{rT} \int_0^T f(t)e^{-rt}\,dt \\ &= e^{0.07(5)} \int_0^5 10{,}000e^{-0.07t}\,dt \\ &= e^{0.35} \cdot \frac{10{,}000}{-0.07} e^{-0.07t}\Big|_0^5 \\ &= e^{0.35} \cdot \frac{10{,}000}{-0.07}\left(e^{-0.07(5)} - 1\right)\end{aligned}$$

$$\approx 59{,}866.79$$

Interpretation: If the $10,000 continuous stream is invested immediately as it is received into an account paying 7% interest compounded continuously for 5 years, it will be worth $59,866.79 at the end of 5 years. (This is similar to an annuity, but rather than investing a specific amount at the end of each time period, the investment is made continuously.)

3. The present value of the continuous income stream that the machine will produce is the value of the stream *right now* to the company.

$$\begin{aligned}
\text{Present value} &= \int_0^T f(t)e^{-rt}\,dt \\
&= \int_0^5 10{,}000e^{-0.07t}\,dt \\
&= \frac{10{,}000}{-0.07}\cdot e^{-0.07t}\Big|_0^5 \\
&= \frac{10{,}000}{-0.07}\cdot\left(e^{-0.07(5)}-1\right) \\
&\approx 42,187.42
\end{aligned}$$

Interpretation: The present value of this income stream is $42,187.42. This means that $42,187.42 would have to be invested today to produce $59,866.79 at the end of 5 years. That is, at the end of 5 years, the machine will bring in $50,000 as direct income, but if the income is invested, the machine will bring in $59,866.79. The present value of $42,187.42 is the value of the income stream to the company *right now* in the following sense: If the company were to make a one-time, lump-sum investment of $42,187.42 now into an account paying 7% interest compounded continuously for 5 years, the total amount of money it would have would be $59,866.79. (Verify this using the method of Example Set B.)

4. Should the company buy the machine? It depends on the price, of course. (a) If the machine sells for exactly $42,187.42, the present value of the income stream, the company may realize a net income, including interest, of $59,866.79, from the income stream it generates. But, this stream is not guaranteed. The company can guarantee the $59,866.79 income by placing the $42,187.42 into a 7% interest-earning account for 5 years. (b) If the machine sells for more than $42,187.42, the company would do better to invest the income in a 7% interest-earning account. Otherwise, it has to invest more than $42,187.42 to generate the same $59,866.79 income. For example, if the company could buy the machine for $45,000, it could generate $59,866.79 from that $45,000. But it can generate the $59,866.79 from only $42,187.42. Consequently, a $45,000 investment for the machine is not worth it. (c) If the machine sells for less than $42,187.42, the company should buy it because it can generate the $59,866.79 income using less than $42,187.42. For example, if the company could buy the machine for $25,000, it could generate

\$59,866.79 from that \$25,000. Thus, the \$25,000 presently has the same earning capacity to the company and therefore the same value as \$42,187.42 invested in the bank.

The Present and Future Values of a Perpetual Income Stream

In some situations, it is reasonable to consider the income as flowing forever. In such cases, we can approximate the present value of the flow using the improper integral.

EXAMPLE SET F

Rental income from a piece of property, upon which there is an indefinite lease, flows at the rate of \$70,000 per year. The income is invested immediately into an account paying 6.5% annual interest compounded continuously. Find and interpret the present value of the flow.

Solution:
The rate of flow function is $f(t) = 70{,}000$, and since the lease is indefinite, we can, for all practical purposes, consider the time period infinite. Then,

$$
\begin{aligned}
\text{Present value} &= \int_0^T f(t)e^{-rt}\,dt \\
&= \int_0^\infty 70{,}000e^{-0.065t}\,dt \\
&= \lim_{b\to\infty} \int_0^b 70{,}000e^{-0.065t}\,dt \\
&= \frac{70{,}000}{-0.065} \cdot \lim_{b\to\infty} e^{-0.065t}\Big|_0^b \\
&= \frac{70{,}000}{-0.065} \cdot \lim_{b\to\infty} \left(e^{-0.065b} - e^{-0.065(0)}\right) \\
&= \frac{70{,}000}{-0.065} \cdot \lim_{b\to\infty} \left(\frac{1}{e^{0.065b}} - 1\right) \\
&= \frac{70{,}000}{-0.065} \cdot (0 - 1) \\
&= 1{,}076{,}923.077
\end{aligned}
$$

Interpretation: To match the total income realized from the lease of this property, approximately \$1,076,923.08 would have to be invested now into an account that pays annual interest at 6.5% compounded continuously forever.

Income Systems and Technology

Using Your Calculator You can use your graphing calculator to determine the future value of an annuity. The following entries show how the future value problem of Example Set E is solved.

```
Y1 = 10000e^(−0.07X)
Y0 = fnInt(Y1,X,A,B)
0 → A
5 → B
e^(0.07*5)*Y0
```

The result, 59,866.79, agrees with the example.

Calculator Exercises

1. Income flows continuously into a manufacturing company at the rate of \$124,000 per year. Find the total money flow over the next 5 years. Find the future value of the money stream if the income is invested at 5.5% compounded continuously. Find the present value of the money stream if the income is invested at 5.5% compounded continuously.
2. Income flows continuously into a company at the rate of $36{,}450e^{-0.052t}$ dollars per year, where t is the number of years from now. Find the total money flow over the next 5 years. Find the future value of the money stream if the income is invested at 8.5% compounded continuously. Find the present value of the money stream if the income is invested at 8.5% compounded continuously.
3. Money flows continuously into a mutual fund at the rate of \$1000 per month. Find the total money flow over the next 36 months. If the interest rate is 11%, find the future value of the annuity over (a) the first 12-month period, and (b) the first 20-year period.
4. Rental income from a piece of property, upon which there is an indefinite lease, flows at the rate of \$62,575 per year. The income is invested immediately into an account paying 8.5% annual interest compounded continuously. Find and interpret the present value of the flow.
5. A company expects its profits over the next 6 years to be $f(t) = 22 + 8.6e^{0.03t}$ thousand dollars per year. If they immediately reinvest their profits at 7.4% interest, compounded continuously, what is the present value of the company?

EXERCISE SET 7.3

A UNDERSTANDING THE CONCEPTS

1. A company considers buying a machine that will increase its annual net income by an estimated \$70,000 per year for the next 6 years. Over the next 6 years, company economists believe money will be worth 5%. The future and present values of this stream over the next 6 years are \$489,802.33 and \$362,854.49, respectively. If the cost of the machine is \$375,000, what should the company do and why?
2. A company considers buying a machine that will increase its annual net income by an estimated \$106,500 per year for the next 8 years. Over the next 8 years, company economists believe money will be worth 7.25%. The future and present values of this stream over the next 8 years are \$1,154,663.35 and \$646,494.12, respectively. If the cost of the machine is \$550,750, what should the company do and why?
3. A company considers buying a machine that will increase its annual net income by an estimated \$35,000 per year for the next 5 years. Over the next 5 years, company economists believe money will be worth 6%. The future and present values of this stream over the next 5 years are \$192,199.36 and \$151,189.37, respectively. What should be the company's strategy?

4. A company considers buying a machine that will increase its annual net income by an estimated \$84,000 per year for the next 12 years. Over the next 12 years, company economists believe money will be worth 5.5%. The future and present values of this stream over the next 12 years are \$919,479.75 and \$737,899.78, respectively. What should be the company's strategy?

5. For 10 months, a person invests \$300 at the end of each month into an account that pays an annual interest rate of 7% compounded continuously. The future and present values are, respectively, \$3089.23 and \$2914.18. Interpret these values.

6. For 24 months, a person invests \$500 at the end of each month into an account that pays an annual interest rate of 5.5% compounded continuously. The future and present values are \$12,684.88 and \$11,363.55, respectively. Interpret these values.

7. For 12 years, a person invests \$100 at the end of each month into an account that pays an annual interest rate of 6% compounded continuously. The future and present values are \$29,192.06 and \$11,868.61, respectively. Interpret these values.

8. For 10 years, a person invests \$1500 at the end of each month into an account that pays an annual interest rate of 8% compounded continuously. The future and present values are \$275,746.71 and \$123,900.98, respectively. Interpret these values.

B SKILL ACQUISITION and C APPLYING THE CONCEPTS

9. For 36 months, a person invests \$200 at the end of each month into an account that pays an annual interest rate of 10% compounded continuously. Find and interpret both the future and present values of this investment.

10. For 60 months, a person invests \$150 at the end of each month into an account that pays an annual interest rate of 4% compounded continuously. Find and interpret both the future and present values of this investment.

11. For 12 years, a person invests \$400 at the end of each month into an account that pays an annual interest rate of 8% compounded continuously. Find and interpret both the future and present values of this investment.

12. For 30 months, a person invests \$2500 at the end of each month into an account that pays an annual interest rate of 9% compounded continuously. Find and interpret both the future and present values of this investment.

13. A company considers buying a machine that it believes will increase its net annual income by \$60,000 per year for each of the next 8 years. Company economists believe that money will be worth 6.6%. If the cost of the machine is \$351,275, what should the company do and why?

14. A company considers buying a machine that it believes will increase its net annual income by \$125,000 per year for each of the next 4 years. Company economists believe that money will be worth 10%. If the cost of the machine is \$400,000, what should the company do and why?

15. A company considers buying a machine that it believes will increase its net annual income by \$360,000 per year for each of the next 6 years. Company economists believe that money will be worth 8%. What should be the company's strategy?

16. A company considers buying a machine that it believes will increase its net annual income by \$1,200,000 per year for each of the next 2 years. Company economists believe that money will be worth 6.5%. What should be the company's strategy?

17. Rental income from a piece of property, upon which there is an indefinite lease, flows at the rate of \$16,000 per year. The income is invested immediately upon receipt into an account that pays 7% annual interest compounded continuously. Find and interpret the present value of this money flow.

18. Rental income from a piece of property, upon which there is an indefinite lease, flows at the rate of \$450,000 per year. The income is invested immediately upon receipt into an account that pays 9% annual interest compounded continuously. Find and interpret the present value of this money flow.

D DESCRIBE YOUR THOUGHTS

19. We have not discussed the future value of continuous money flow over an indefinite time period. Can you argue why we have taken this position?

20. Describe why the definite integral approximates the accumulated value of an income stream over some specified time period.

21. Explain what an annuity is.

22. Explain what the future value of an annuity is.

E REVIEW

23. **(2.2)** Find, if it exists, $\lim_{x\to 5} \dfrac{x^2 - 10x + 24}{x^2 - 5x + 4}$.

24. **(2.4)** Discuss, by writing an analysis, the continuity of the function

$$f(x) = \begin{cases} \dfrac{x^2 - 2x - 3}{x - 3}, & \text{if } x \neq 3 \\ 2, & \text{if } x = 3 \end{cases}$$

at $x = 3$.

25. **(2.6)** Find the average rate of change of the function $f(x) = -2x^2 - 5x + 6$ as x changes from $x = 4$ to $x = 9$.

26. **(3.3)** In the late 1980s and early 1990s the United States experienced a slowdown in the economy. In fact, during a several-year period, the slowdown increased. If $E(t)$ is a function that approximates the economy at time t, describe this situation in terms of $E'(t)$ and $E''(t)$.
27. **(3.4)** Find $S'(t)$ for the function $S(t) = t^3(6t - 30)^4$.
28. **(4.2)** Sketch the graph of a function that increases at an increasing rate when x is less than 5, then increases at a decreasing rate when x is between 5 and 8, and then decreases at an increasing rate when x is greater than 8. The function also approaches a horizontal asymptote as x approaches negative infinity.
29. **(5.1)** Suppose x is a variable and b is a constant. Specify the forms of a power function and an exponential function.
30. **(5.3)** Find $f'(x)$ for $f(x) = \ln(x^2 + 7)$.
31. **(6.6)** ***Ecology: Pollution*** The rate $f(t)$ (in tons per year) at which a pollutant is being released into the atmosphere is given by $f(t) = 1400e^{-0.02t}$, where t represents the time in years from now. If no controls are placed on the system, find the total amount of pollutant that will be released into the atmosphere in the future.
32. **(7.2)** ***Business: Consumer's Surplus*** Find and interpret the consumer's surplus if for a product $D(x) = 2000 - 3x^2$ and the market (equilibrium) price is $280.

7.4 Differential Equations

Introduction

Terminology

Separation of Variables

General Solutions

Initial Values and Particular Solutions

Differential Equations and Technology

Introduction

Many phenomena in business, life science, and the social sciences are governed by certain principles. So that we can predict the behavior of these phenomena, we wish to discover and then mathematically model these underlying principles. Discovery is often through observation and data collection. We can then construct mathematical models (functions or equations) by applying statistical or advanced mathematical methods to the collected data, or by assuming the quantities involved, and/or their rates of change, are related by some proportion. (A quantity y is said to be proportional to a quantity x if $y = kx$, where k is a constant called the constant of proportionality. The quantity y is jointly proportional to x and z if $y = kxz$, and inversely proportional to x if $y = k/x$.)

Terminology

We often find that observations or assumptions of proportionality involve changes of functions so that the direct modeling function is unknown and must be found. Modeling functions or equations that involve one dependent variable and one or more of its derivatives (or differentials) with respect to an independent variable are called **differential equations**. A function is a solution to the differential

Differential Equation

equation if it (1) relates the variables involved in the phenomena, and (2) satisfies the differential equation.

The following equations are examples of differential equations.

$$\frac{dy}{dx} = -\frac{2y}{3x}$$

$$(1 + x^2)y' = x$$

$$ty^3\,dt + e^{t^2}\,dy = 0$$

Ordinary Differential Equation

In each of these equations, the dependent variable is y, and the independent variable is either x or t. Also, each equation is an **ordinary differential equation** because there is only *one* independent variable. We will restrict our attention to ordinary differential equations.

Separation of Variables

Separable Variables

Since solutions to differential equations are generally difficult to find, we will examine only a special class of differential equations: the **separable variables** class. The solutions to these equations can be found relatively easily by separating the dependent and independent variables to opposite sides of the equal sign and integrating.

EXAMPLE SET A

Solve $(1 + x^2)\dfrac{dy}{dx} = x$ and verify the solution.

Solution:
Dividing each side by $1 + x^2$ and multiplying each side by dx will separate the variables so that the dependent variable y appears on the left side of the equal sign and the independent variable x appears on the right side.

$$dy = \frac{x}{1 + x^2}\,dx$$

Now, we integrate each side.

$$\int dy = \int \frac{x}{1 + x^2}\,dx$$

$$y + c_1 = \int \frac{x}{1 + x^2}\,dx$$

To evaluate the integral $\displaystyle\int \frac{x}{1 + x^2}\,dx$, we let $u = 1 + x^2$, so that $du = 2x\,dx$ and $x\,dx = \dfrac{1}{2}\,du$. Then

$$y + c_1 = \int \frac{\frac{1}{2}\,du}{u}$$

$$y + c_1 = \frac{1}{2}\ln u + c_2$$

$$y + c_1 = \ln u^{1/2} + c_2$$

$$y + c_1 = \ln(1 + x^2)^{1/2} + c_2$$

$$y = \ln(1 + x^2)^{1/2} + \underbrace{c_2 - c_1}_{c}$$

$$y = \ln(1 + x^2)^{1/2} + c$$

To verify that this is indeed a solution, we will differentiate it.

$$\frac{dy}{dx} = \frac{1}{(1 + x^2)^{1/2}} \cdot \frac{1}{2}(1 + x^2)^{-1/2} \cdot 2x$$

$$= \frac{x}{1 + x^2}$$

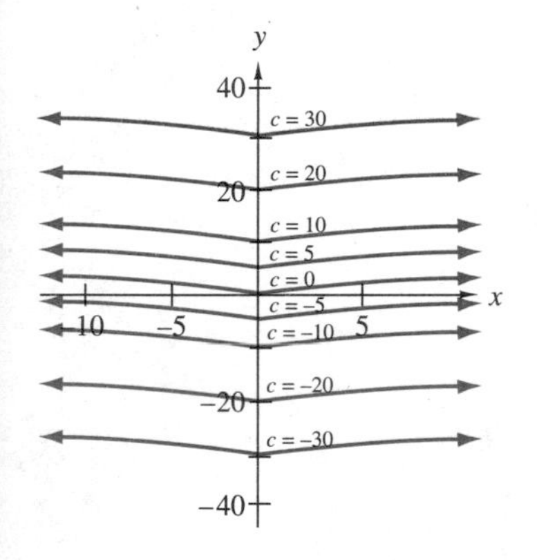

Figure 7.13

Multiplying each side by $1+x^2$ produces the original equation $(1+x^2)\frac{dy}{dx} = x$, and we conclude that $y = \ln(1 + x^2)^{1/2} + c$ is a solution to the differential equation.

A few examples of differential equations that are not separable are:

$$\frac{dy}{dx} = \frac{3x^2 - y^2}{2xy} \quad \text{and} \quad \frac{dy}{dx} = 2^{xy}$$

You would find that it is not possible to separate the variables in these equations. Our work in this course is only concerned with the separable variable class of differential equations.

General Solutions

General Solutions

In Example Set A, the solution $y = \ln(1 + x^2)^{1/2} + c$ involves an arbitrary constant c, which, we know from our examination of the indefinite integral, results in a *family of curves*. Solutions of differential equations that involve one or more arbitrary constants are called **general solutions** and represent a family of solutions. Figure 7.13 shows some of the members of the family $y = \ln(1 + x^2)^{1/2} + c$; we have shown the graphs for which $c = 0, 5, -5, 10, -10, 20, -20, 30$, and -30.

Initial Values and Particular Solutions

More often than not in applied situations, we will know some output value that corresponds to some input value; that is, we will know some initial condition. When this is the case, these values can be substituted into the general solution to obtain a value for the arbitrary constant c. Then, from all of the members of the family of solutions, we can specify the one that suits our particular situation. Solutions that arise from substituting initial conditions into general solutions are called **particular solutions** and the problems that produce them are called **initial-value problems**.

Particular Solutions
Initial-Value Problems

EXAMPLE SET B

Solve the differential equation $y' = 3x^2y^2$ subject to the condition that $y = -0.1$ when $x = 2$.

Solution:

We will begin by writing y' as $\frac{dy}{dx}$, then we will separate the variables and integrate.

$$y' = 3x^2y^2$$

$$\frac{dy}{dx} = 3x^2y^2$$

$$\frac{dy}{y^2} = 3x^2\,dx$$

$$\int y^{-2}\,dy = \int 3x^2\,dx$$

$$\frac{y^{-1}}{-1} = x^3 + c$$

$$\frac{-1}{y} = x^3 + c$$

Now we substitute 2 for x and -0.1 for y to find the value of c, and hence, the particular solution.

$$\frac{-1}{-0.1} = 2^3 + c$$

$$10 = 8 + c$$

$$2 = c$$

Thus,

$$\frac{-1}{y} = x^3 + 2 \quad \text{so that} \quad y = \frac{-1}{x^3 + 2}$$

Therefore, $y = \frac{-1}{x^3 + 2}$ is the particular solution.

EXAMPLE SET C

Solve the differential equation $y' = \dfrac{5x}{y^3}$ subject to the condition that $y = 3$ when $x = -4$.

Solution:
We will begin by writing y' as $\dfrac{dy}{dx}$, then we will separate the variables and integrate.

$$y' = \frac{5x}{y^3}$$

$$\frac{dy}{dx} = \frac{5x}{y^3}$$

$$y^3\,dy = 5x\,dx$$

$$\int y^3\,dy = \int 5x\,dx$$

$$\frac{y^4}{4} = \frac{5x^2}{2} + c$$

$$y^4 = 10x^2 + c$$

Now we substitute -4 for x and 3 for y to find c.

$$3^4 = 10(-4)^2 + c$$

$$81 = 10 \cdot 16 + c$$

$$81 = 160 + c$$

$$-79 = c$$

Thus,

$$y^4 = 10x^2 - 79$$

$$y = \sqrt[4]{10x^2 - 79}$$

Therefore, the particular solution is $y = \sqrt[4]{10x^2 - 79}$.

EXAMPLE SET D

Solve the differential equation $y' = x^2 y$ subject to the condition that $y = e^5$ when $x = 0$.

Solution:

We will begin by writing y' as $\frac{dy}{dx}$, then we will separate the variables and integrate.

$$y' = x^2 y$$

$$\frac{dy}{dx} = x^2 y$$

$$\frac{dy}{y} = x^2\,dx$$

$$\int \frac{dy}{y} = \int x^2\,dx$$

$$\ln y = \frac{x^3}{3} + c$$

Now we substitute 0 for x and e^5 for y to find c.

$$\ln e^5 = \frac{0^3}{3} + c$$

$$5 = 0 + c$$

$$5 = c$$

Thus,

$$\ln y = \frac{x^3}{3} + 5$$

$$y = e^{x^3/3+5}$$

Therefore, the particular solution is $y = e^{x^3/3+5}$ or $y = e^5 e^{x^3/3}$.

EXAMPLE SET E

The growth of a new restaurant depends on the number of its regular customers as

$$p'(t) = 0.01\sqrt{p(t)}$$

where $p(t)$ represents the number of regular customers at any time t and $p'(t)$ represents the rate at which the number of regular customers is changing at any time t. Solve this differential equation subject to the condition that $p = 100$ when $t = 0$.

Solution:

We will begin by writing $p'(t)$ as $\frac{dp}{dt}$ and $p(t)$ as p, then we will separate the variables and integrate.

$$p'(t) = 0.01\sqrt{p}$$

$$\frac{dp}{dt} = 0.01p^{1/2}$$

$$100\frac{dp}{dt} = p^{1/2}$$

$$100\frac{dp}{p^{1/2}} = dt$$

$$100\int p^{-1/2}\,dp = \int dt$$

$$100 \cdot \frac{p^{1/2}}{1/2} = t + c$$

$$200p^{1/2} = t + c$$

Now we substitute 0 for t and 100 for p to find c.

$$200 \cdot 100^{1/2} = 0 + c$$

$$c = 2000$$

Thus,

$$200p^{1/2} = t + 2000$$

$$p = \left[\frac{t + 2000}{200}\right]^2$$

Therefore, the particular solution is $p(t) = \left[\frac{t + 2000}{200}\right]^2$.

Differential Equations and Technology

Using Derive You can use Derive to solve differential equations. The following entries show how to solve the differential equation of Example Set D. As you did with implicit differentiation (Section 3.5) and integration by parts (Section 6.3), you will have to load a utility program, named *ordinary differential equations 1* (**ode1**), to make the necessary computations. Remember to press Enter at the end of each line.

Transfer
Load
Utility
ode1
Author
separable(x^2,y,x,y,0,#e^5)
Simplify
SoLve
y

The result, $t = e^{x^3/3+5}$, agrees with the example. The **separable** and SoLve commands instruct Derive to solve the differential equation $y' = x^2y$ for y, under the conditions that when $x = 0$, $y = e^5$.

Derive Exercises

1. Use Derive to solve the differential equation $y' = -\frac{y}{x^2}$ subject to the condition that when $x = 1$, $y = e$. Use pencil and paper to verify Derive's result.
2. Use Derive to solve the differential equation $2x^2y' = y^3$ subject to the condition that when $x = 1$, $y = 0$. Use pencil and paper to verify Derive's result.
3. You can modify the Derive instructions to produce the general solution to the equation $(x^2y + x^2)y' = x^3$. Do so by omitting the parameters *x*, *y*, *number*, and *number* following the differential equation.
4. A wholesale chain believes the differential equation
$$\frac{dv}{dp} = -\frac{3}{7}\left(\frac{v}{p+7}\right)$$
relates the sales volume, *v*, in thousands of dollars, to the price, *p*, in dollars, of one of its products. Use Derive to solve this differential equation if the sales volume is \$9500 when the price is \$12.

EXERCISE SET 7.4

A UNDERSTANDING THE CONCEPTS

For Exercises 1–8, verify that each function is a solution to the corresponding differential equation.

1. $y = 2x^2 + c, \quad \frac{dy}{dx} = 4x$
2. $y = 2x^4 + x^2 + c, \quad \frac{dy}{dx} = 8x^3 + 2x$
3. $y = ce^{x^2}, \quad \frac{dy}{dx} = 2xy$
4. $y^2 = x^2 + c, \quad \frac{dy}{dx} = \frac{x}{y}$
5. $y = ce^x, \quad \frac{dy}{dx} = y$
6. $y = ce^{kx}, \quad \frac{dy}{dx} = ky$
7. $\ln x + \frac{1}{y} = c, \quad \frac{dy}{dx} = \frac{y^2}{x}$
8. $y = e^{-4/5\cdot x} + 10, \quad 5y' + 4y = 0$

For Exercises 9–16, state whether the differential equation is in the separable variable class or not.

9. $\frac{dy}{dx} = 7x$
10. $\frac{dy}{dx} = 4xy^3$
11. $\frac{dy}{dx} = 3x^2y - 5$
12. $\frac{dz}{dw} = \frac{-2w^4z}{z+w}$
13. $\frac{dr}{dt} = \frac{3t}{r}$
14. $\frac{dA}{dr} = 2\pi r$
15. $\frac{dy}{dx} = e^{5xy}$

16. $\frac{dy}{dx} = y(x^2 - 3x + 5)(3y - 2)^2$

B SKILL ACQUISITION

For Exercises 17–28, find the general solution to each differential equation.

17. $y' = x^4$
18. $y' = 7$
19. $\frac{dy}{dx} = 0$
20. $\frac{dy}{dx} = 25 - 4x$
21. $y\frac{dy}{dx} = 2x$
22. $\frac{4}{3}\frac{dy}{dx} = e^{3x-3}$
23. $y' = xe^{x^2}$
24. $xy' = 1$
25. $(1+x)\frac{dy}{dx} = x$
26. $(1+x)\frac{dy}{dx} + (1+y) = 0$
27. $x\frac{dy}{dx} = y \ln y$
28. $y' = \frac{xy}{x-1}$

For Exercises 29–41, find the particular solution to each differential equation.

29. $y' = \frac{x^2}{y^2}$, and $y = 4$ when $x = 0$
30. $y' = \frac{x^2}{y^3}$, and $y = 1$ when $x = 1$
31. $y' = \frac{1}{xy}$, and $y = 4$ when $x = 1$
32. $y' = \frac{x}{x^2+1}$, and $y = 2$ when $x = 0$

to find C

33. $y' = xe^x - x$, and $y = -1$ when $x = 0$
34. $(x^2 - 3)y' = 4x$, and $y = 5$ when $x = 2$
35. $2xyy' = 1 + y^2$, and $y = 3$ when $x = 5$
36. $\frac{dR}{dS} = \frac{k}{S}$, and $R = 0$ when $S = S_0$. (This equation is called the **Weber-Fechner law** and describes the relationship between a stimulus S and a response R.)
37. $\frac{dP}{dt} = kt$, and $P = P_0$ when $t = 0$. (This equation models unlimited growth and decay.)
38. $\frac{dP}{dt} = k(L - P)$, where L is a constant, and $P = P_0$ when $t = 0$. (This equation models limited growth and decay.)

C APPLYING THE CONCEPTS

39. ***Education: Rate of Learning*** In a community college's court reporting program, students are believed to progress at the rate of

$$\frac{dQ}{dt} = k(160 - Q)$$

where Q represents the number of words per minute a student can type t weeks after the start of the program. If the average student cannot type any words initially and can type 35 words per minute 4 weeks into the program, how many words per minute can the average student be expected to type after 18 weeks into the program?

40. ***Economics: Investment*** Two years ago, \$6000.00 was placed into an investment in which the amount of money present, P, grows according to

$$\frac{dP}{dt} = kt$$

Today, the investment is worth \$7328.00. How much will the investment be worth 5 years from now?

41. ***Social Science: White-Collar Crime*** A county's statistics indicate that the number of white-collar crimes is growing at the constant rate of 4% per year. If 135 white-collar crimes were reported in 1991, how many can be expected to be reported in 1996 if this trend continues?

42. ***Manufacturing: Newton's Law of Cooling*** Newton's Law of Cooling states that

$$\frac{dT}{dt} = k(T - S)$$

where T represents the temperature of an object t minutes after it is removed from a heater (or refrigerator), and S is the temperature of the surrounding medium. If an object is 225°F when it is removed from an oven and placed into a room with a constant temperature of 70°F, and has cooled to 200°F 20 minutes later, what will be its temperature 60 minutes from the time it is removed from the oven?

43. ***Ergonomics: Aircraft Cabin Pressure*** As one leaves the surface of the earth and rises into the atmosphere, the pressure exerted by the atmosphere diminishes according to

$$\frac{dP}{dh} = kP$$

where P represents pressure (in pounds per square inch), and h represents the height (in feet) above sea level. At 18,000 feet above sea level, the pressure is half of what it is at sea level; that is, at 18,000 feet above sea level, $P = \frac{1}{2}P_0$, where P_0 is the pressure at sea level. Find the pressure, as a percentage of P_0, at 10,000 feet.

44. ***Medicine: Healing Time for Wounds*** Medical researchers have found that the rate at which the number

of square centimeters A of unhealed skin changes t days after receiving a wound is given by

$$\frac{dA}{dt} = kA$$

If a wound is initially 2 square centimeters, and only 1.3 square centimeters after 3 days, what will be the size of the wound 7 days after it is first received?

45. ***Ecology: Pollution*** The rate of change of the concentration C (in parts per million, ppm) of pollutants that enter a particular river each day is related to the number x of homes near the river by the differential equation

$$\frac{dC}{dx} = kx$$

When there were no homes near the river the number of pollutants entering the river each day is believed to have been 4 ppm, and when there were 50 homes near the river, the number of pollutants entering the river each day was 10 ppm. A contractor is contemplating a new housing site with 20 homes. With these additional homes, how many pollutants will be entering the river each day?

46. ***Business: Marginal Cost*** A company's marginal cost for producing x units of a product is given by

$$\frac{dC}{dx} = \frac{k}{x^{1/3}}$$

The company's fixed costs are \$2300, and it costs \$8500 to make 25 units.

a. Find a function that gives the cost of producing x units of the product.
b. Find how much it costs to produce 100 units.

47. ***Political Science: Voting*** In a particular county, the percentage P of registered voters who do not vote in presidential elections is changing at the rate of

$$\frac{dP}{dt} = k\sqrt[3]{t}$$

where t is the time in years, with $t = 0$ being 1982. If in 1982 60% of all registered voters voted, but in 1986 only 53% voted, what percentage can be expected to vote in 1996?

48. ***Biology: Cell Nutrients*** A cell receives nutrients through its surface. The function $w'(t) = k\left[w(t)\right]^{2/3}$ expresses the relationship between $w(t)$ (the weight of the cell at time t) and $w'(t)$ (the rate at which the weight is changing), where k is a positive constant that depends on the type of cell. Solve this differential equation subject to the condition that the initial weight is 2 units.

49. ***Economics: Reinvestment*** A farmer reinvests his profits in buying neighboring land. The growth rate of the land he owns is given by $a'(t) = 0.2a(t)^{3/4}$, where a is the area, in acres, of the land he owns. Solve this differential equation for a, subject to the condition that $a = 20$ when $t = 0$.

50. ***Business: Loss of Employees*** A business is losing employees due to falling morale. The loss rate of employees is related to the current number of employees by the function $n'(t) = 0.067n(t) - 3.685$, where e is the current number of employees, and t is the number of months from now.

a. Solve this equation for $n(t)$, subject to the condition that $n = 95$ when $t = 0$.
b. When will the company have lost all its employees?

51. ***Business: Aerospace Employment*** Employment in the aerospace industry in a state varies as $e'(t) = 125{,}000 - 0.05e(t)$, where e is the number of employees t years after 1985.

a. Solve this equation for $e(t)$, subject to the condition that $e = 300{,}000$ in 1985.
b. At what rate will aerospace employment be changing in 1997?

52. ***Economics: Population of a City*** The population of a city grows according to $P'(t) = 0.003\sqrt[3]{P(t)}$, where P represents the population since 1940.

a. Solve this equation for $P(t)$, subject to the condition that $P = 100{,}000$ in 1940.
b. What was the population of this city in 1957?

D DESCRIBE YOUR THOUGHTS

53. Explain what a differential equation is.

54. When integrating both sides of an equation, two constants of integration result, one on each side. Explain why the general solution to a differential equation involves only one constant of integration.

55. Describe the strategy you would use to solve the equations in Exercises 17 or 18.

E REVIEW

56. **(2.2)** Find, if it exists, $\lim\limits_{x \to 2} \dfrac{x^2 - 4x - 32}{x^2 - 6x - 16}$.

57. **(2.4)** Construct the graph of a function that is discontinuous at $x = 3$ because continuity condition 3 is the first condition to fail.

58. **(3.1)** Use the definition of the derivative to find $f'(x)$ for $f(x) = -3x^2 - 5x + 1$.

59. **(3.1)** ***Business: Revenue from Production*** Suppose that $R(x)$ (in dollars) represents the revenue realized from the production and sale of x units of a product. Interpret $R'(1500) = 225$.

60. **(3.2)** Find $f'(x)$ for $f(x) = \dfrac{x^2 - 1}{x^2 + 1}$.

61. **(3.4)** Find $f'(x)$ for $f(x) = (x^2 + 1)^4(x^2 - 1)^2$.

62. **(4.1)** On an interval $[a, b]$, the function $f(x)$ is concave upward and $f(a) = f(b)$. Relate both $f'(x)$ and $f''(x)$ to 0.

63. **(5.3)** Find $f'(x)$ for $f(x) = 2xe^{2x}$.

64. (5.4) Find $f'(x)$ for $f(x) = e^{-3x+1}$.

65. (6.4) Evaluate $\int_0^1 3x^2 e^{x^3}\, dx$.

7.5 Applications of Differential Equations

Introduction

Models of Unlimited Growth and Decay

Models of Limited Growth and Decay

Models of Logistic Growth

Introduction

As we noted at the beginning of Section 7.4, in many phenomena one quantity changes proportionally to another. The examples of this section illustrate how mathematical models that can be used to predict the behavior of these phenomena can be constructed using the idea of proportionality. We will develop three models: the unlimited growth model, the limited growth model, and the logistic growth model.

Models of Unlimited Growth and Decay

Law of Natural Growth

Many populations (not just biological ones) grow in such a way that the larger the population is, the faster it grows. The **law of natural growth** states that *the rate at which a population changes over time is proportional to the current size of the population.* We can symbolize this law and construct a mathematical model that represents it by letting P represent the size of the population at time t. Since the law assumes no limit to the size of the population, we refer to it as the **unlimited growth model**.

Unlimited Growth Model

If P represents the size of a population, t represents time, and k is a constant, then

$$\frac{dP}{dt} = kP$$

We can solve for the unknown function P by separating the variables.

$$\frac{dP}{dt} = kP$$

$$\frac{dP}{P} = k\,dt$$

$$\int \frac{dP}{P} = \int k\,dt$$

$$\ln P = kt + c_1$$

To solve for P, we need to eliminate the ln. So we take exponentials of each side.

$$e^{\ln P} = e^{kt+c_1}$$

Then, using the properties $e^{\ln a} = a$ and $x^{m+n} = x^n \cdot x^m$, we get

$$P = e^{kt} \cdot e^{c_1}$$

Since e^{c_1} represents a constant raised to a constant, it is itself a constant that we will call C. This gives us

$$P = Ce^{kt}$$

Thus, the solution to the differential equation $\frac{dP}{dt} = kP$ is the exponential function $P = Ce^{kt}$.

Growth Constant

The constant k is called the **growth constant**. If $k > 0$, the model represents growth, and if $k < 0$, the model represents decay (negative growth). Figure 7.14 shows the graphs of each model. (These should look familiar to you; you studied them in Chapter 5.)

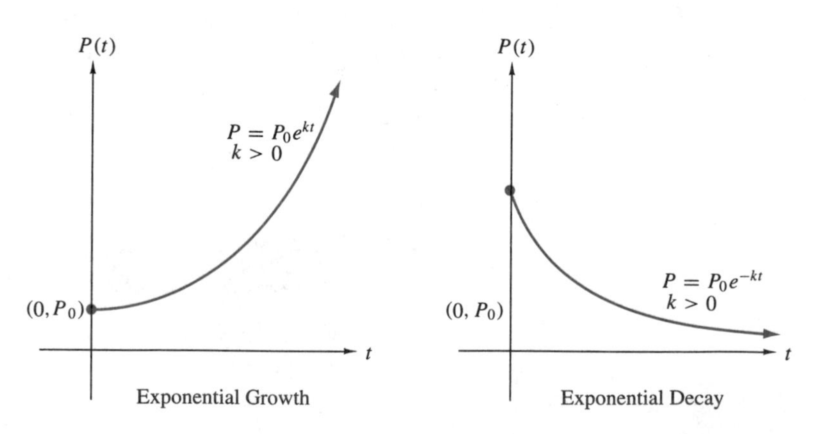

Figure 7.14

If, at some initial observation time t, the population size P is P_0—that is, if $P = P_0$ when $t = 0$—then $P = P_0e^{kt}$. This fact is illustrated in Example Set A and you will be asked to prove it in general in the exercises. Now the size of the population at $t = 0$ can be immediately substituted for the arbitrary constant.

EXAMPLE SET A

The value of real estate in a city is increasing, without limitations, at a rate proportional to its current value. A piece of property in the city appraised at \$240,000 4 months ago appraises at \$242,000 today. At what value will the property be appraised 12 months from now?

Solution:

The unlimited growth model applies to this situation. Let P represent the value of the property at time t, where $t = 0$ is the time at which the property was appraised at \$240,000. Then the solution to the natural growth differential equation $\frac{dP}{dt} = kP$ is $P = 240{,}000e^{kt}$.

We can solve for k, the growth constant, using the fact that when $t = 4$ (which corresponds to today—4 months from the initial appraisal), $P = 242{,}000$. Then the solution to this particular situation (the particular solution) can be specified.

$$242{,}000 = 240{,}000e^{k(4)}$$

$$\frac{242{,}000}{240{,}000} = e^{4k}$$

$$\ln\frac{242{,}000}{240{,}000} = \ln e^{4k}$$

$$\ln\frac{242{,}000}{240{,}000} = 4k$$

$$k = \frac{\ln\frac{242{,}000}{240{,}000}}{4}$$

$$k \approx 0.0021$$

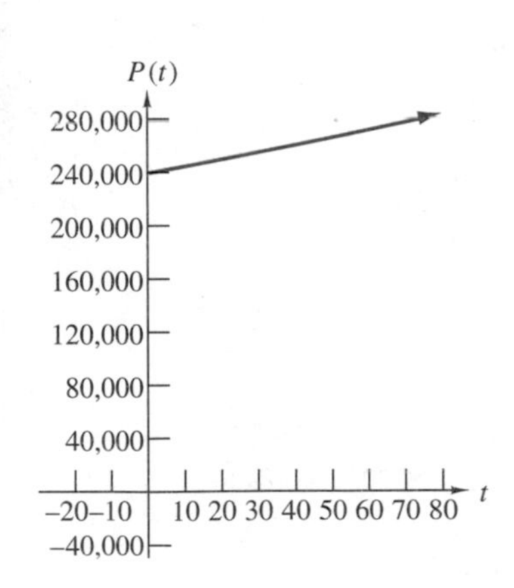

Figure 7.15

Thus, the particular solution is $P = 240{,}000e^{0.0021t}$. Figure 7.15 shows $P(t)$.

Now, using this mathematical model we can predict the value of this piece of real estate at any time t. Specifically, 12 months from now is $t = 4 + 12 = 16$ months from when $t = 0$, so that $t = 16$. Then

$$P = 240{,}000e^{0.0021(16)}$$

$$P \approx 248{,}201$$

Interpretation: Twelve months from now, this piece of property will appraise at approximately \$248,201. The growth constant k in this case is 0.0021. This means that the value of this property is increasing (growing) at the rate of 0.21% per month.

Models of Limited Growth and Decay

Carrying Capacity

Many populations grow in such a way that the rate of increase is large when the population is small and then decreases (because of limitations on growth) as the population increases. Let P represent the size of the population at time t and L be a fixed upper (or lower) limit to the size of the population. When the number L represents an upper limit, it is called the **carrying capacity** of the population and represents the maximum size of the population.

Limited Growth Model

When the rate at which a population changes with respect to time is proportional to the difference between the carrying capacity and the current size of the population, the mathematical model for growth (or decay) is

$$\frac{dP}{dt} = k(L - P)$$

If P_0 is the size of the population at some initial observation time $t = 0$, then the particular solution is

$$P = L + (P_0 - L)e^{-kt}$$

We can derive this particular solution by separating the variables.

$$\frac{dP}{dt} = k(L - P)$$

$$\frac{dP}{L - P} = k\,dt$$

$$\int \frac{dP}{L - P} = \int k\,dt$$

$$-\ln(L - P) + c_1 = kt + c_2$$

$$\ln(L - P) + c_1 = -kt - c_2$$

$$\ln(L - P) = -kt \underbrace{-c_2 - c_1}_{c_3}$$

$$\ln(L - P) = -kt + c_3$$

$$e^{\ln(L-P)} = e^{-kt+c_3}$$

$$L - P = e^{-kt} \cdot \underbrace{e^{c_3}}_{C}$$

$$L - P = Ce^{-kt}$$

$$-P = -L + Ce^{-kt}$$

$$P = L - Ce^{-kt}$$

Now, from the initial condition that $P = P_0$ when $t = 0$,

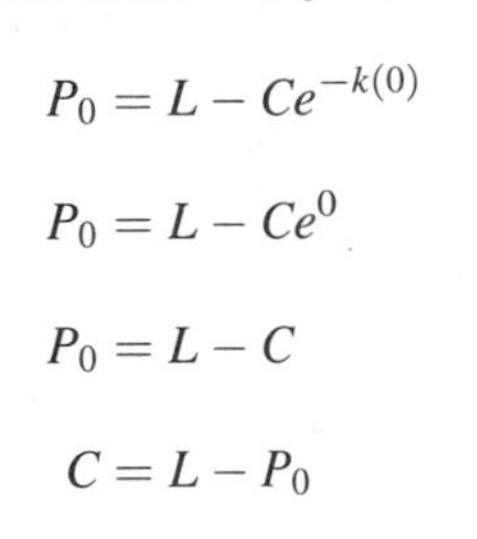

$$P_0 = L - Ce^{-k(0)}$$

$$P_0 = L - Ce^0$$

$$P_0 = L - C$$

$$C = L - P_0$$

and thus,

$$P = L - (L - P_0)e^{-kt}$$

$$P = L + (P_0 - L)e^{-kt}$$

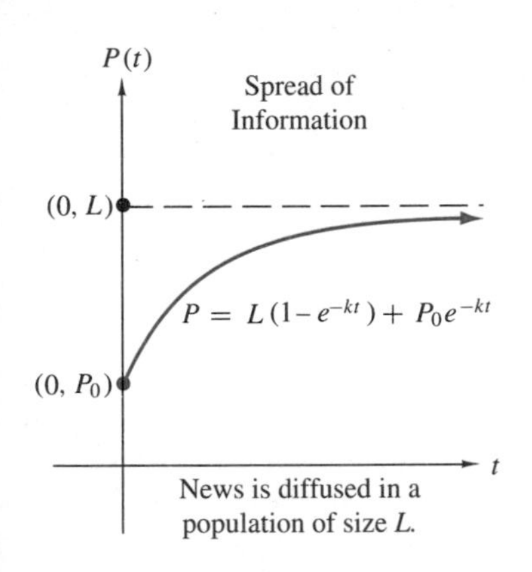

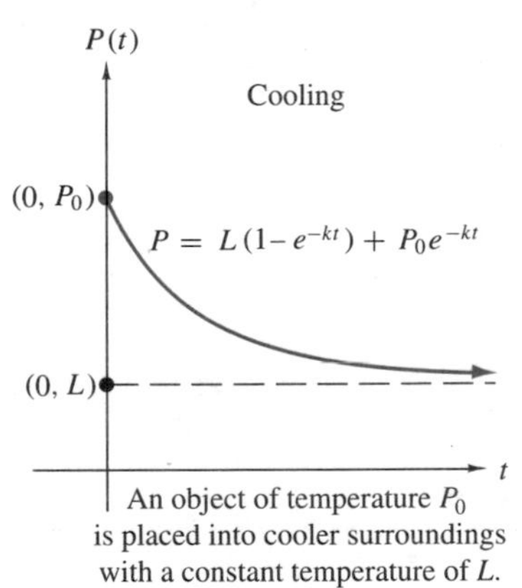

Figure 7.16

Figure 7.16 shows the graph of this particular solution for the cases when L is a fixed upper limit and when L is a fixed lower limit, along with phenomena these cases might model.

EXAMPLE SET B

The temperature of a machine when it is first shut down after operating is 220°C. The surrounding air temperature is 30°C. After 20 minutes, the temperature of the machine is 160°C. Find a function that gives the temperature of the machine at any time t and then find the temperature of the machine 30 minutes after it is shut down.

Solution:

The limited growth (decay) model applies in this situation because there is a lower limit to the temperature of the machine. Let P represent the temperature of the machine at time t and let $t = 0$ be the time when the machine is shut down after operating. At $t = 0$, the initial temperature of the machine is $P_0 = 220°C$, and the limiting temperature is 30°C, the temperature of the surrounding air. Then

$$P = 30 + (220 - 30)e^{-kt}$$

$$= 30 + 190e^{-kt}$$

We can use the initial condition that $P = 160$ when $t = 20$ to find the value of the growth constant (decay constant, in this case) k.

$$160 = 30 + 190e^{-k(20)}$$

$$130 = 190e^{-20k}$$

$$\frac{130}{190} = e^{-20k}$$

$$\ln \frac{13}{19} = \ln e^{-20k}$$

$$\ln \frac{13}{19} = -20k$$

$$\frac{\ln \frac{13}{19}}{-20} = k$$

$$0.019 \approx k$$

Then $P = 30 + 190e^{-0.019t}$ models this particular situation and gives the temperature of the machine at any time t. Figure 7.17 shows $P(t)$. Now, using this mathematical model we can predict the temperature of the machine any time t. Specifically, 30 minutes from the time the machine is shut down means that $t = 30$. Then

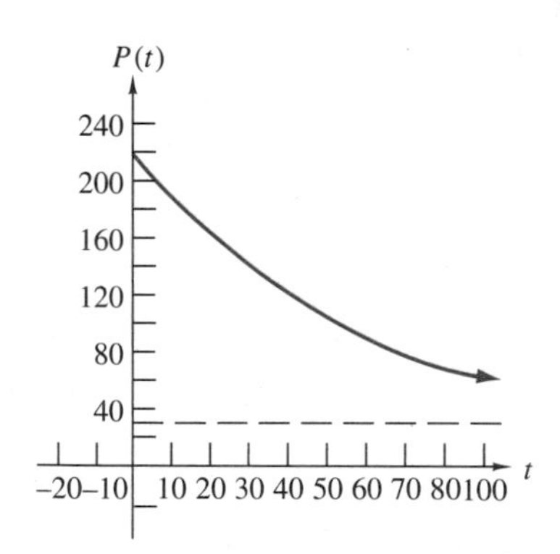

Figure 7.17

$$P = 30 + 190e^{-0.019(30)}$$

$$P \approx 137.45$$

Interpretation: Thirty minutes after the machine is shut down, its temperature will be approximately 137.45°C. The growth constant k in this case is 0.019. This means that the temperature of the machine is decreasing at the rate of 1.9% each minute.

The differential equation $\frac{dP}{dt} = k(L - P)$ also models phenomena such as the diffusion of information into a population and the learning of new tasks. Both these phenomena grow quickly at first, then more slowly. Information spreads quickly at first and then more slowly as more people hear it. In a similar way, people learn or become better at a new task quickly at first and then more slowly as time goes by.

EXAMPLE SET C

A government agency has introduced a new software application on 4000 of its personal computers. Initially, 80 employees knew how to use the application very well, but after two weeks, 2800 employees knew how to use it very well. The number of people who learn how to use the application increases at a rate that is proportional to the difference between the total number of employees at the agency and the number of employees who have learned to use it at any time t (in weeks). Find a function that gives the number of employees who have learned to use the software at any time t, then find the number of employees who have learned to use it 6 weeks after its introduction into the agency.

Solution:
The differential equation $\frac{dP}{dt} = k(L - P)$ with solution $P = L + (P_0 - L)e^{-kt}$ is the appropriate model for this application. In this case, $L = 4000$, and $P_0 = 80$. Thus,

$$P = 4000 + (80 - 4000)e^{-kt}$$

$$= 4000 - 3920e^{-kt}$$

We can solve for k using the initial condition that $P = 2800$ when $t = 2$. (You should be able to solve for k using Example Sets A and B as guides.)

$$2800 = 4000 - 3920e^{-k(2)}$$

$$k \approx 0.592$$

Then, $P = 4000 - 3920e^{-0.592t}$ models this particular situation and gives the number of employees who have learned the software application at any time t. Figure 7.18 shows the graph of $P(t)$.

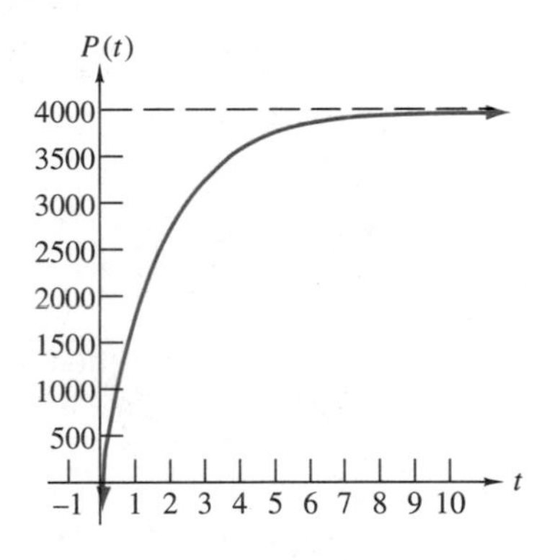

Figure 7.18

Specifically, in 6 weeks, $t = 6$, so that

$$P = 4000 - 3920e^{-0.592(6)}$$

$$P \approx 4000 - 3920(0.0287)$$

$$P \approx 4000 - 112.3756$$

$$P \approx 3888$$

Interpretation: Six weeks from the time the software application is introduced into the agency, approximately 3888 of the 4000 employees have learned to use it. The growth constant k in this case is 0.592. This means that the number of employees who know how to use the software application is increasing at the rate of 59.2% each week.

Models of Logistic Growth

Logistic Growth

When bounds are placed on population growth, the **logistic growth** function may best model the phenomena. Phenomena that exhibit logistic growth grow exponentially at first and then more slowly, as in the limited growth model. Figure 7.19 shows the graph of the logistic growth curve.

As in limited growth, the number L is called the *carrying capacity* of the population.

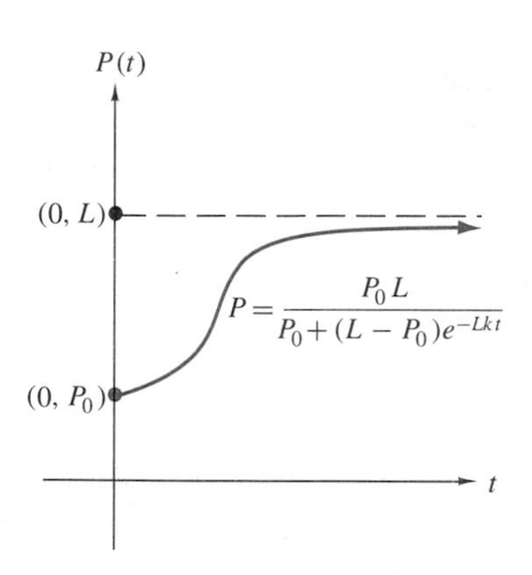

Figure 7.19

Logistic Growth Model

When the rate at which a population changes is proportional to the current size of the population and the difference between the carrying capacity and the current size of the population, the mathematical model is

$$\frac{dP}{dt} = kP(L - P)$$

If P_0 is the size of the population at time $t = 0$, then the particular solution is

$$P = \frac{P_0 L}{P_0 + (L - P_0)e^{-Lkt}}$$

The derivation of this model is beyond the scope of this course, so we will instead illustrate its use with an example.

EXAMPLE SET D

Twenty-four months ago, the membership of a business marketer's professional organization was 3800. Today it is 4500. The number of members increases at a rate that is proportional to the current size of the membership and the difference between the total number of potential members and the current membership. If there are 30,000 potential members, find a function that gives the number of members 4 months from now.

Solution:

The differential equation $\frac{dP}{dt} = kP(L - P)$ with the particular solution $P = \frac{P_0 L}{P_0 + (L - P_0)e^{-Lkt}}$ is the appropriate model for this situation. In this case, $L = 30{,}000$, $P_0 = 3800$, and, since t is measured from 24 months ago, $t = 24$ is the present. Thus,

$$P = \frac{3800(30{,}000)}{3800 + (30{,}000 - 3800)e^{-30{,}000kt}}$$

$$= \frac{114{,}000{,}000}{3800 + 26,200e^{-30{,}000kt}}$$

We can solve for k using the initial condition that $P = 4500$ when $t = 24$.

$$4500 = \frac{114{,}000{,}000}{3800 + 26{,}200e^{-30{,}000k(24)}}$$

$$4500 = \frac{114{,}000{,}000}{3800 + 26{,}200e^{-720{,}000k}}$$

$$4500(3800 + 26{,}200e^{-720{,}000k}) = 114{,}000{,}000$$

$$17{,}100{,}000 + 117{,}900{,}000e^{-720{,}000k} = 114{,}000{,}000$$

$$e^{-720{,}000k} = \frac{114{,}000{,}000 - 17{,}100{,}000}{117{,}900{,}000}$$

$$e^{-720{,}000k} \approx 0.8219$$

$$-720{,}000k \approx \ln(0.8219)$$

$$k \approx \frac{\ln(0.8219)}{-720,000}$$

$$k \approx 0.000\,000\,272$$

Then $P \approx \dfrac{114{,}000{,}000}{3800 + 26{,}200e^{-30{,}000(0.000\,000\,272)t}}$, so that,

$$P \approx \frac{114{,}000{,}000}{3800 + 26{,}200e^{-0.00816t}}$$

$$\approx \frac{1{,}140{,}000}{38 + 262e^{-0.00816t}}$$

models this particular situation and gives the number of members at any time t.

Four months from now, $t = 24 + 4 = 28$, so that

$$P \approx \frac{1{,}140{,}000}{38 + 262e^{-0.00816(28)}}$$

$$P \approx 4625$$

Interpretation: Four months from now, the professional organization can expect to have approximately 4625 members.

EXERCISE SET 7.5

A UNDERSTANDING THE CONCEPTS

For Exercises 1–8, choose the model (unlimited growth, limited growth, or logistic growth) that is appropriate for the situation and explain your answer.

1. ***Economics: Investment*** Five thousand dollars is deposited into an account that pays 7.5% interest per year,

compounded quarterly, for t years. We are interested in knowing how much money has accumulated in the account at any time t.

2. *Economics: Population Growth of a City* The population of a particular city is increasing at a rate proportional to itself. We are interested in knowing the size of the population at any time t.

3. *Manufacturing: Newton's Law of Cooling* An object is heated to 140°C and then placed into a water bath that is 6°C. We are interested in knowing the temperature of the object at any time t.

4. *Biology: Growth of a Mold* A mold is growing at a rate that is proportional to itself. We are interested in knowing the size of the mold at any time t.

5. *Physical Science: Radioactive Decay* A radioactive substance is decaying at a rate proportional to the current amount present. We are interested in knowing how much of the substance remains at any time t.

6. *Medicine: Spread of a Virus* Five hundred fifty students, all of whom are susceptible to an infectious virus, live in a college dormitory. The virus grows at a rate that is jointly proportional to the number of infected students and the number of uninfected students. We are interested in knowing how many students have been infected at any time t.

7. *Chemistry: Chemical Conversion* A substance A is being converted, through a chemical reaction, to another substance B. The rate of the conversion of substance A is proportional to that amount of substance A. We are interested in knowing how much of substance A has been converted at any time t.

8. *Business: Learning a Technical Procedure* A company believes that the rate at which the company's employees are learning a new, complicated technical procedure is increasing at a rate that is proportional to the total number of employees who need to learn the procedure and the number of employees who have already learned it. We are interested in knowing how many employees have learned the new procedure at any time t.

B SKILL ACQUISITION

9. In the derivation of the solution of the unlimited growth model, we stated that $P = Ce^{kt}$ became $P = P_0e^{kt}$ if $P = P_0$ when $t = 0$. Show that this is true.

10. In the derivation of the solution to the limited growth model, the integral

$$\int \frac{dP}{L-P}$$

occurred. Show that this integral is equivalent to

$$-\ln(L-P) + c.$$

11. In Example Set C, the differential equation $\frac{dP}{dt} = k(L - P)$ with solution $P = L + (P_0 - L)e^{-kt}$ occurred. The conditions $L = 4000$ and $P_0 = 80$ resulted in the equation $P = 4000 - 3920e^{-kt}$. Then using the initial conditions that $P = 2800$ when $t = 2$, the value of k was determined to be approximately 0.592. Show that this is true.

C APPLYING THE CONCEPTS

12. *Manufacturing: Newton's Law of Cooling* An object that is heated by a manufacturing process to 90°C is placed into the surrounding air which is 25°C. Twenty minutes after being removed from the heater, the temperature of the object is 75°C.
    a. Find a function that gives the temperature of the object at any time t after removal from the heater.
    b. Find the temperature of the object 60 minutes after it is removed from the heater.

13. *Manufacturing: Newton's Law of Cooling* A chemical company is experimenting with a chemical that it hopes will keep liquids that have been refrigerated, then placed into a warmer environment, from warming too rapidly. A liquid that has been treated with the company's chemical is refrigerated to 40°F and then placed into the surrounding air, which is 68°F. Fifteen minutes later, the liquid has warmed to 45°F.
    a. Find a function that gives the temperature of the liquid at any time t after removal from the refrigerator.
    b. Find the temperature of the liquid 30 minutes after it is removed from the refrigerator and placed into the surrounding air.

14. *Medicine: Spread of Disease* A college dormitory houses 700 students, all of whom are susceptible to a particular infectious virus. On a particular day, the infectious virus had been contracted by 25 students. Three days later, 125 students had contracted the virus. The number of infected students grows at a rate that is proportional to the current number of students who have contracted the virus and the difference between the total number of students in the dorm and the number who have contracted the virus.
    a. Find a function that gives the number of students who have contracted the virus at any time t.
    b. Find the number of students who have contracted the virus 5 days after its initial outbreak.

15. *Business: Worker Production* A worker's production increases at a rate that is proportional to the difference between 650 units per month and the number of units the worker currently produces per month. A worker with no experience can produce 45 units per month and after 2 week's experience, can produce 160 units per month.
    a. Find a function that gives the number of units an experienced worker can produce at any time t.

b. Find the number of units a worker with 1-month experience can produce.

16. ***Physical Science: Radioactive Decay*** A radioactive substance decays at a rate that is proportional to the current amount present. There are initially 1500 grams of the substance present, but 1350 grams present after 1 hour.

a. Find a function that gives the number of grams present at any time t.
b. Find the number of grams present after 10 hours.

17. ***Social Science: Spread of a Rumor*** In group of 2500 people, one person starts a rumor. Two hours later, 40 people have heard the rumor. The rumor spreads at a rate that is proportional to the number of people who have heard it and the number of people who have not yet heard it.

a. Find a function that gives the number of people who have heard the rumor at any time t.
b. Find the number of people who have heard the rumor 6 hours after it was started.

18. ***Political Science: Political Polls*** A polling group has determined that the percentage of voters registered as members of a new political party is increasing at a rate that is proportional to the percentage of registered voters who are not members of the new party. In 1990, 0.2% of all registered voters were members of the new party and 0.6% were members in 1992.

a. Find a function that gives the percentage of all registered voters who are members of the new party at any time t.
b. Find the percentage of all registered voters that the new party can expect to have registered in the year 2000.

19. ***Archaeology: Carbon-14 Dating*** The half-life of carbon-14, a radioactive isotope of carbon, is 5770 years. This means that if P_0 is the current amount of carbon-14, then the amount present in 5770 years will be $\frac{1}{2}P_0$. Suppose an archaeologist finds a human bone that has only one-fifth the amount of carbon-14 it originally contained. If it is known that carbon-14 decays at a rate that is proportional to the current amount present, how old is the bone? (Hint: You need to first use the information about the half-life of carbon-14 to establish the value of k, then the fact that one-fifth the original amount P_0 can be expressed as $\frac{1}{5}P_0$.)

20. ***Business: Investment of Capital*** In their investment of capital, a company uses the strategy that whenever the capital C decreases below a baseline value of ten million dollars, it is invested at a rate proportional to the difference between ten million and C. Suppose that the initial capital is two million dollars and that 6 months later it is five million dollars. What is the expected capital 12 months later if this investment strategy is adhered to?

21. ***Social Science: Spread of News*** The news media of a city announced an important story at noon, and 4 hours later 60% of the city's people had heard it. How long will it take for 99% of the city's people to hear the story?

D DESCRIBE YOUR THOUGHTS

22. Describe the strategy you would use to solve Exercises 20 or 21.

23. Compare and contrast the unlimited growth and logistic growth models.

24. As if you were explaining it in a note to a friend, explain what the carrying capacity of a system is and how to recognize its value in an equation.

E REVIEW

25. **(2.2)** Find, if it exists, $\lim_{x\to 7} \frac{x^2 - 5x + 6}{x^2 - 10x + 16}$.

26. **(2.4)** Construct the graph of a function that is discontinuous at $x = 3$ because continuity condition 3 is the first to fail.

27. **(3.1)** Use the definition of the derivative to find $f'(x)$ for $f(x) = 4x^2 - x + 10$.

28. **(3.1)** Suppose that $C(x)$ (in dollars) represents the cost of producing x units of a product. Interpret $C'(125) = -1.40$.

29. **(3.2)** Find $f'(x)$ for $f(x) = \frac{2x + 4}{6x - 7}$.

30. **(3.4)** Find $f'(x)$ for $f(x) = (9x + 3)^4(2x + 7)^4$.

31. **(4.2)** Use the summary table, Table 7.1 located at the end of the exercise set on the next page, to construct the graph of the corresponding function.

32. **(5.3)** Find $f'(x)$ for $f(x) = \ln(5x^2 + 3)^3$.

33. **(6.4)** Evaluate $\int_e^{e^5} \frac{5}{x(\ln x)^2}\,dx$.

34. **(7.4)** Solve the differential equation

$$\frac{dy}{dx} = \frac{4y}{x - 1}$$

Table 7.1

Int / Val	$f'(x)$	$f''(x)$	$f(x)$	Behavior of $f(x)$
$(-\infty, 2)$	+	+	incr/cu	incr at a decr rate
2	und	und	$(2, 8)$	rel max at $(2, 8)$
$(2, 5)$	−	+	decr/cu	decr at a decr rate
5	0		$(5, 4)$	rel min at $(5, 4)$
$(5, 8)$	+	+	incr/cu	incr at an incr rate
8	und	und	$(8, 8)$	rel max at $(8, 8)$
$(8, +\infty)$	−	−	decr/cd	decr at an incr rate

7.6 Probability

Introduction
Fundamental Concepts of Probability
Continuous and Discrete Random Variables
Probability Density Function
Probability Distributions
The Uniform Probability Distribution
The Exponential Probability Distribution
The Normal Probability Distribution
Probability and Technology

Introduction

Probability

We have seen that differentiation measures the amount of change in a physical system and that integration measures the amount of accumulation. We will now see that integration can be used to measure the *chance* that a physical system will assume a particular state when the system is allowed to operate in some prescribed manner that allows for some degree of variability (that is, due to chance). The measure of chance is called **probability** and its value reflects the percentage of a large number of trials that can be expected to result in a particular state of a system.

Fundamental Concepts of Probability

If an automatic coffee dispensing machine is adjusted to drop a paper cup that will hold 7 ounces of coffee and fill it with 6 ounces of coffee, we expect each cup to be filled with a little more or a little less than 6 ounces, with variation due to the randomness of the dispensing device.

We might wish to find the probability that a randomly selected cup is filled with between 5.7 and 6.1 ounces of coffee. That is, we might wish to measure the chance that the system assumes the state $5.7 \le X \le 6.1$, where X represents the number of ounces of coffee dispensed to the cup. If we determined through a great number of trials that $5.7 \le X \le 6.1$ occurred 72% of the time, we would write $P(5.7 \le X \le 6.1) = 0.72$ (read, "the probability of X being between 5.7 and 6.1, inclusive, is 0.72"). This means that we expect that out of every 100 cups, the machine would fill about 72 of them with between 5.7 and 6.1 ounces of coffee. Thus, probability is viewed as long-term relative frequency.

Continuous and Discrete Random Variables

Continuous Random Variable

Discrete Random Variable

The variable X in the coffee example is a **continuous random variable** because it can assume, in a random way, any value in the entire interval of values (0 to 7 ounces, in this case). If, on the other hand, X represented the number of games in a best-of-seven series until a winner was determined, X could only assume the specific values 4, 5, 6, and 7. This variable X is still a random variable because (if the games are not "fixed") it could assume any of these values in a random way, but now it is a **discrete random variable**. Continuous random variables result from measures upon a system and can take any value in an entire interval. Discrete random variables result from a count made upon a system and can take on only particular values. In this text, we will restrict our attention to continuous random variables.

Probability Density Function

Experiments with a system may generate data which, in turn, may sometimes be used to generate a function that approximately describes the system.

Suppose that $p(X)$ is a data-generated function that describes the probability that a system is in some particular state, that state being some interval of real numbers. A possible graph of $p(X)$ is pictured in Figure 7.20. In the figure, the horizontal axis records the values of the random variable X, and the vertical axis records the probability of those values. Since $p(X)$ is the probability that the system is in state X (or that some event X occurs), the graph shows that as X increases, the probability increases first at an increasing rate, then at a decreasing rate, then reaches a maximum value, then decreases at an increasing rate, and finally decreases at a decreasing rate and approaches zero. When working with continuous random variables, we concern ourselves only with intervals of values and not individual values. Thus, intervals form states of the system.

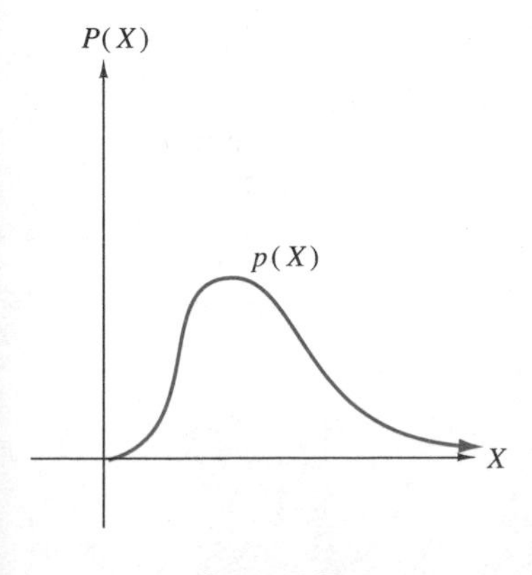

Figure 7.20

Notice the difference between the meanings of $P(X)$ and $p(X)$. $P(X)$ denotes the probability of the event X, whereas $p(X)$ denotes a probability function.

The probability that a randomly chosen value of X lies in some particular interval $[a, b]$ is the area of the region that lies below the curve $p(X)$ and above the interval $[a, b]$. Figure 7.21 shows this relationship for the function $p(X)$. The notation $P(a \leq X \leq b)$ represents the probability that X takes on a value between a and b, inclusive. But this area, we know, is the value of a definite integral over this interval. Thus,

$$P(a \leq X \leq b) = \int_a^b p(X)\,dX$$

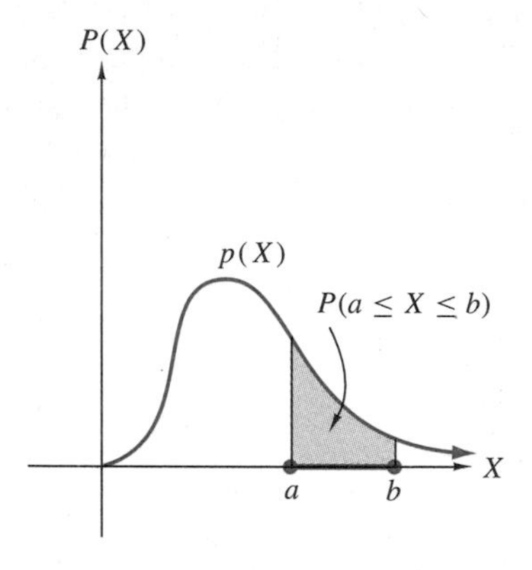

Figure 7.21

$p(X)$ describes a **probability density function** if it records the probability that a system is in some particular state and if there is some closed interval $[a, b]$ for which $\int_a^b p(X)\,dX = 1$ as summarized by:

Probability Density Function

A function $p(X)$ such that $\int_a^b p(X)\,dX$ gives the probability that a system is in a state X with $a \leq X \leq b$ is a **probability density function** (commonly referred to as a **probability function**) if

1. $p(X) \geq 0$ for each X, and
2. $\displaystyle\int_{-\infty}^{\infty} p(X)\,dX = 1$

The first condition states that a probability is always between 0 and 1; it is never negative nor greater than 1. The second condition states that something (any possible outcome of the experiment) must happen.

If there is an interval $[a, b]$ within the interval $[c, d]$ for which $p(X) = 0$ for each X inside $[c, d]$ but outside $[a, b]$, the interval $[a, b]$ is called the domain of the probability function. For all the probability functions that follow, we will assume that $p(X) = 0$ for each X outside the associated interval.

EXAMPLE SET A

Show that $p(X) = \dfrac{3X^2}{124}$ for X in $[1, 5]$ is a probability function. (Keep in mind that we are assuming that $p(X) = 0$ for each X outside the interval $[1, 5]$.)

Solution:
To demonstrate that $p(X)$ is a probability function, we must show that both conditions for a probability density function hold.

1. We will first demonstrate that $p(X) \geq 0$ for each X in $[1, 5]$.

   Since X^2 is always positive in $[1, 5]$, $\dfrac{3X^2}{124} \geq 0$ in $[1, 5]$.
   Thus, for each X in $[1, 5]$, $p(X) \geq 0$ and condition 1 is met.
2. We will now demonstrate that $\int_1^5 p(X)\,dX = 1$.

$$\int_1^5 p(X)\,dX = \int_1^5 \frac{3X^2}{124}\,dX$$

$$= \frac{3}{124} \int_1^5 X^2 \, dX$$

$$= \frac{3}{124} \cdot \frac{X^3}{3} \bigg|_1^5$$

$$= \frac{X^3}{124} \bigg|_1^5$$

$$= \frac{125}{124} - \frac{1}{124}$$

$$= \frac{124}{124}$$

$$= 1$$

Since both conditions for a probability density function are satisfied, $p(X) = \frac{3X^2}{124}$ X in $[1, 5]$ is a probability density function.

EXAMPLE SET B

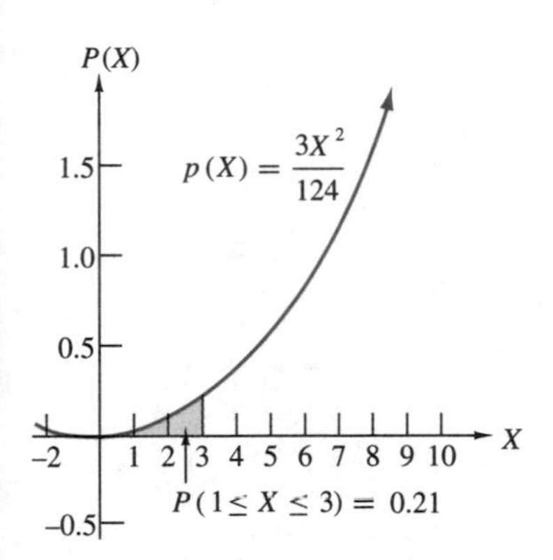

Figure 7.22

For the probability density function $p(X) = \frac{3X^2}{124}$, find the probability that a randomly selected value of X lies in the interval $[1, 3]$. (Notice that this interval lies within the domain, $[1, 5]$, of the probability density function from the last example.)

Solution:
The probability that a randomly selected value of X lies in $[1, 3]$ is denoted by $P(1 \leq X \leq 3)$, and the area under the curve and directly above the interval is a measure of this probability. Figure 7.22 shows this.

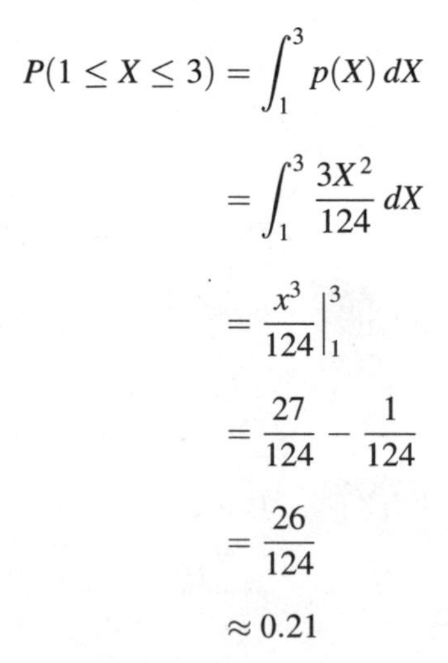

$$P(1 \leq X \leq 3) = \int_1^3 p(X) \, dX$$

$$= \int_1^3 \frac{3X^2}{124} \, dX$$

$$= \frac{x^3}{124} \bigg|_1^3$$

$$= \frac{27}{124} - \frac{1}{124}$$

$$= \frac{26}{124}$$

$$\approx 0.21$$

Interpretation: If the experiment described by the function $p(X) = \frac{3X^2}{124}$ were performed a great many times, we would expect that in about 13 out of every 62 trials, X would assume a value in the interval $[1, 3]$. Put another way, if the experiment were performed a great many times, X would, approximately 21% of those times, take on a value between 1 and 3.

Probability Distributions

Certain probability density functions have been found that approximate, fairly nicely, many physical phenomena. In fact, many seemingly different phenomena are described by the same type of probability model. Figure 7.23 shows some of the commonly used probability models. The probability functions on the left

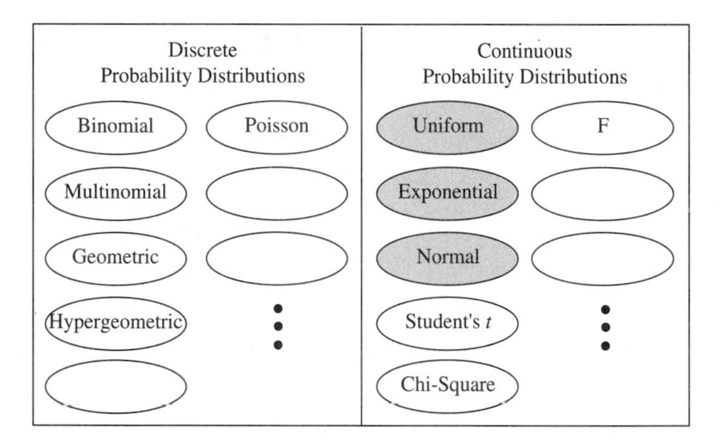

Figure 7.23

side of the diagram are functions of discrete random variables and are studied in some detail in introductory statistics courses. The probability functions on the right side of the diagram are functions of continuous random variables. In this section, we will examine the ones that are highlighted. The others are studied in both introductory and advanced statistics courses. Since the probability density function determines how the probability is distributed over various intervals, these models are called **probability distributions**.

Probability Distribution

The Uniform Probability Distribution

Uniformly Distributed

Loosely, a random variable is **uniformly distributed** if the probability that it assumes any one value in an interval $[a, b]$ is equal to the probability that it assumes any other value in $[a, b]$. That is, if X, Y, Z, and W are all in $[a, b]$, then $P(X) = P(Y) = P(Z) = P(W)$. The leftmost figure in Figure 7.24 illustrates this idea graphically. We can be more precise by saying that a random variable is uniformly distributed if the probability that it lies in any subinterval, of a particular length, of $[a, b]$, is equal to the probability that it lies in any other subinterval of the same length of $[a, b]$. The rightmost figure in Figure 7.24 illustrates this idea. In the diagram, the subintervals $[c, d]$ and $[e, f]$ have the same length, $d - c = f - e$. These illustrations help us to understand the following definition of the uniform probability distribution.

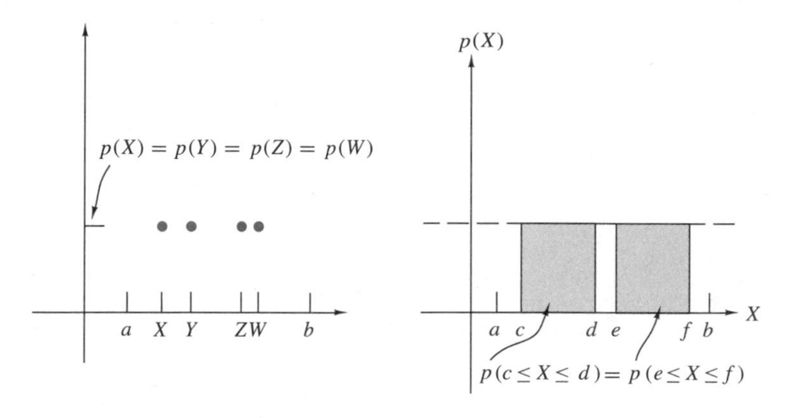

Figure 7.24

Uniform Probability Distribution

A random variable X is uniformly distributed on $[a, b]$ if its probability density function, $p(X)$, is defined by

$$p(X) = \frac{1}{b - a}$$

for every X in $[a, b]$.

EXAMPLE SET C

A state's weather service provides a continuously running 120-second recorded telephone message on weather conditions throughout the state. If a person calls and gets connected to the message, what is the probability that he or she will hear at most 20 seconds of the message before it repeats?

Solution:

Let X represent the amount of time the message is heard. Since a phone call can be connected any time within the 120-second message, all connect times are equally likely so that X is uniformly distributed on $[0, 120]$, with the probability density function $P(x) = \frac{1}{(120 - 0)} = \frac{1}{120}$.

A caller will hear at most 20 seconds of the message if he or she is connected any time after the message has played 100 seconds. Thus, we want to compute the probability that X lies in the interval $[100, 120]$.

$$\begin{aligned} P(100 \le X \le 120) &= \int_{100}^{120} \frac{1}{120}\,dX \\ &= \frac{1}{120}X\Big|_{100}^{120} \\ &= \frac{1}{120}(120) - \frac{1}{120}(100) \\ &\approx 1 - 0.833 \\ &\approx 0.167 \quad \left(\frac{167}{1000}\right) \end{aligned}$$

Thus, $P(100 \le X \le 120) \approx 0.167$.

Interpretation: If a person calls and is connected to the weather service message, there is about a 0.167 chance that he or she will hear at most 20 seconds of the message before it repeats. That is, out of every 1000 calls, we would expect about 167 people to hear at most 20 seconds of the message before it repeats.

The Exponential Probability Distribution

Figure 7.25 shows the graph of an exponential probability distribution. The

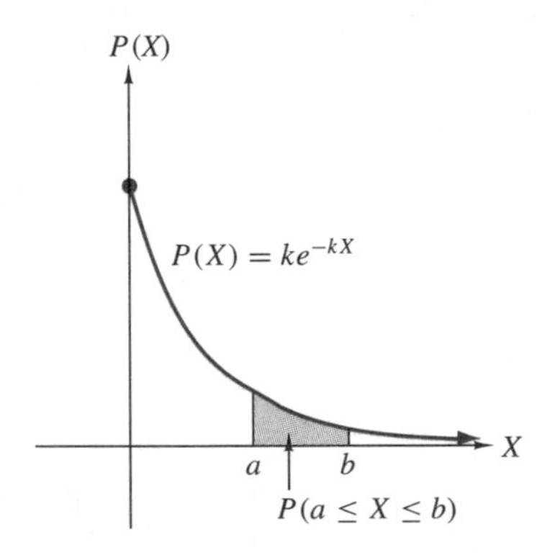

Figure 7.25

exponential probability density function is often an appropriate model for situations in which the values of the random variable are more likely to be small than large.

Notice from the graph that the exponential probability distribution is defined for only nonnegative values of X (that is, for $X \geq 0$) and that the curve (the density function) exhibits the same decreasing behavior as the exponential decay function. In fact, the exponential probability distribution is an exponential decay function and is defined in the following way.

Exponential Probability Distribution

A random variable X is exponentially distributed over the interval $[0, +\infty)$ if its probability density function is

$$P(X) = ke^{-kX}$$

where k is a constant determined by the situation.

Waiting Time Model

Exponentially distributed random variables are characterized by a **waiting time model** and include such models as the reliability or lifespan of a product (waiting time for failure), the duration of a signal (the waiting time for a signal to start or stop), the intervals of time between successive parts appearing on an assembly line (the waiting time between parts), and the amount of time it takes to learn a task (the waiting time until a task is successfully completed). We will investigate such a situation in Example Set D.

EXAMPLE SET D

The research and development department of a lightbulb manufacturer has determined that its new 100-watt low-energy bulb has a life span (in weeks) that is exponentially distributed with probability density function $p(X) = 0.01e^{-0.01X}$. Find the probability that a randomly selected bulb will last (1) at most 75 weeks under continuous use, and (2) between 120 and 150 weeks under continuous use.

Solution:

1. Since X is exponentially distributed,

$$\begin{aligned} P(0 \leq X \leq 75) &= \int_0^{75} 0.01e^{-0.01X}\,dX \\ &= -e^{-0.01X}\Big|_0^{75} \\ &= -e^{-0.01(75)} - (-e^{-0.01(0)}) \\ &\approx -0.4724 + 1 \\ &\approx 0.5276 \quad \left(\frac{5276}{10{,}000}\right) \end{aligned}$$

Thus, $P(0 \leq X \leq 75) = 0.5276$.

Interpretation: The company can expect that approximately 52.76% of its new 100-watt low-energy bulbs will last less than 75 weeks under continuous burning. Put another way, the company can expect that about 5276 out of every 10,000 of its new 100-watt low-energy bulbs will last less than 75 weeks under continuous burning.

2. Since X is exponentially distributed,

$$\begin{aligned} P(120 \le X \le 150) &= \int_{120}^{150} 0.01e^{-0.01X}\,dX \\ &= -e^{-0.01X}\Big|_{120}^{150} \\ &= -e^{-0.01(150)} - (-e^{-0.01(120)}) \\ &\approx -0.2231 + 0.3012 \\ &\approx 0.0781 \quad \left(\frac{781}{10{,}000}\right) \end{aligned}$$

Thus, $P(120 \le X \le 150) = 0.0781$.

Interpretation: The company can expect that approximately 7.81% of its new 100-watt low-energy bulbs will last between 120 and 150 weeks under continuous burning. Put another way, the company can expect that about 781 out of every 10,000 of its new 100-watt low-energy bulbs will last between 120 and 150 weeks under continuous burning.

The Normal Probability Distribution

The normal probability distribution is one of the most used distributions in statistics and arose from investigations by Pierre Laplace (1749–1827), Karl Freidrich Gauss (1777–1855), and Abraham deMoivre (1667–1745) into the nature of errors in experiments. The normal probability distribution models many types of phenomena such as physiological measurements, quality control measurements, IQ scores, and various psychological measurements.

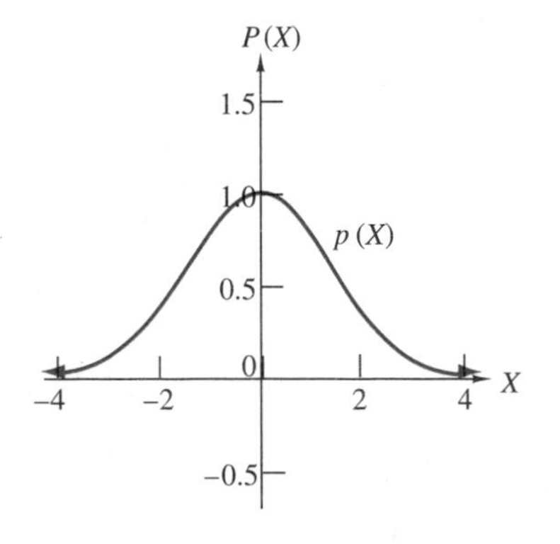

Figure 7.26

The normal probability density function graphs as a bell-shaped curve like the one pictured in Figure 7.26. It is characterized by two numbers, the mean μ, which indicates the center of the distribution, and the standard deviation σ, which indicates how the data is scattered about the mean. Notice that the curve is symmetric about its mean.

Normal Probability Distribution

A random variable X, with mean μ and standard deviation σ, is normally distributed over the interval $(-\infty, +\infty)$ if its probability density function is

$$p(X) = \frac{1}{\sigma\sqrt{2\pi}} e^{\frac{-(X-\mu)^2}{2\sigma^2}}$$

The probability that a randomly selected value of X lies in the interval $[a, b]$ is given by

$$P(a \leq X \leq b) = \frac{1}{\sigma\sqrt{2\pi}} \int_a^b e^{\frac{-(X-\mu)^2}{2\sigma^2}}\, dX$$

The evaluation of the integral is so difficult that we use computer-generated tables or calculator or computer technology to approximate the probability. Table 7.2, located at the end of this section after the exercise set, is a computer-generated table of normal probabilities. Since so many normal curves exist (one for each each pair of numbers μ and σ) and we want only one table, we *standardize* all normal curves to one with $\mu = 0$ and $\sigma = 1$. This curve is called the **standard normal curve** and the formula

Standard Normal Curve

$$Z = \frac{X - \mu}{\sigma}$$

is used to convert values of the random variable X to values of a new random variable Z. Figure 7.27 shows both a data-generated normal curve, its corresponding standardized normal curve, and the relationship between the corresponding probabilities. The figure shows that the probabilities determined by X and Z are equal.

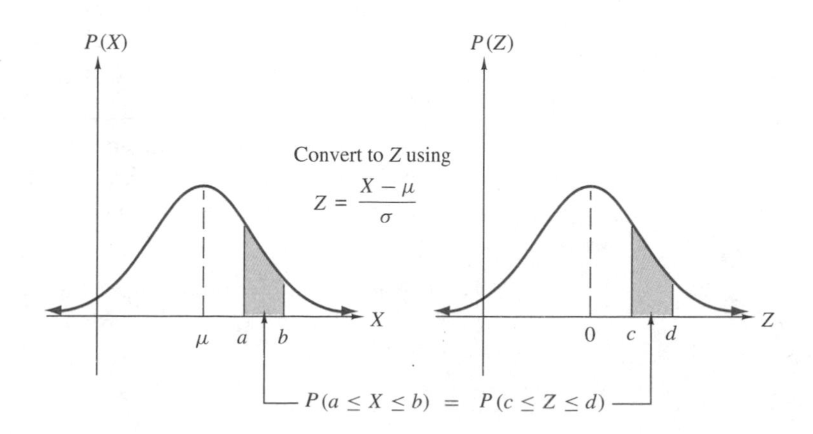

Figure 7.27

EXAMPLE SET E

A machine is adjusted to dispense 6 ounces of coffee into a 7-ounce cup. The amount dispensed is normally distributed with mean $\mu = 6$ ounces and standard deviation $\sigma = 0.5$ ounces. Find the probability that a randomly selected cup is filled with between 5.5 and 5.8 ounces of coffee.

Solution:
Let X represent the amount of coffee dispensed into a cup. We wish to find $P(5.5 \leq X \leq 5.8)$. First, we convert both X values, 5.5 and 5.8, to Z values using $Z = \dfrac{X - \mu}{\sigma}$.

$$P(5.5 \leq X \leq 5.8) = P\left(\frac{5.5 - 6.0}{0.5} \leq Z \leq \frac{5.8 - 6.0}{0.5}\right)$$

$$= P(-1.00 \leq Z \leq -0.40)$$

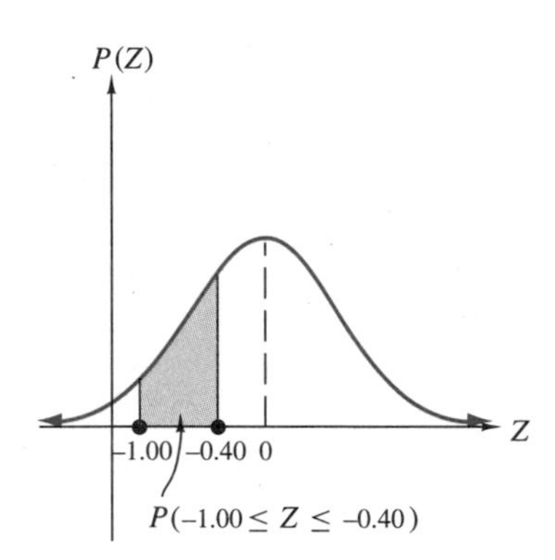

Figure 7.28

Using Table 7.2, we find this area using the fact that the curve is symmetric about 0. The area to the right of 0 is 0.5, and to the left of 0 is 0.5 (so that they combine to produce 1). We subtract the area for the interval 0 to -0.44 from the area for the interval 0 to -1.11. (See Figure 7.28.)

$$\text{(Area from 0 to } -1.00) - \text{(Area from 0 to } -0.40) = 0.3413 - 0.1554$$

$$= 0.1859 \quad \left(\frac{1859}{10{,}000}\right)$$

Thus, $P(5.5 \leq X \leq 5.8) = 0.1859$.

Interpretation: With its present adjustment, the machine dispenses between 5.5 and 5.8 ounces of coffee into 7-ounce cups about 18.59% of the time. That is, out of every 10,000 cups of coffee dispensed, the machine dispenses between 5.5 and 5.8 ounces into about 1,859 of them.

EXAMPLE SET F

A one-hour photo processing lab develops rolls of 36 prints with a mean time of 18.20 seconds and a standard deviation of 2.28 seconds. Find the probability that a randomly selected 36-print roll of film takes more than 25 seconds to process.

Solution:
Let X represent the number of seconds needed to process a 36-print roll of film. We wish to find $P(X > 25)$. We convert the X value, 25, to a Z value using $Z = \dfrac{X - \mu}{\sigma}$.

$$P(X > 25) = P\left(Z > \frac{25 - 18.20}{2.28}\right)$$

$$= P(Z > 2.98)$$

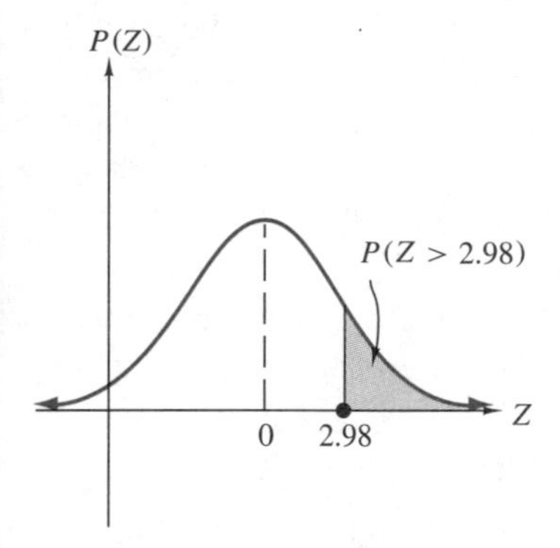

Figure 7.29

Using Table 7.2, we find this area by subtracting the area for the interval 0 to 2.98 from 0.5000, the area of the right half of the distribution. (See Figure 7.29.)

$$0.5000 - (\text{Area from 0 to 2.98}) = 0.5000 - 0.4986$$

$$= 0.0014 \quad \left(\frac{14}{10{,}000}\right)$$

Thus, $P(X > 25) = 0.0014$.

Interpretation: Approximately 0.14% of the time, it will take more than 25 seconds to process a 36-print roll of film. That is, out of every 10,000 rolls of film that this lab processes, about 14 of them will take more than 25 seconds to process.

Probability and Technology

Using Your Calculator The graphing calculator can be very effective and efficient in approximating probabilities. The following entries demonstrate the solution to the problem of Example Set E. The solution technique of the example set involved converting the X scores to Z scores, then using a table of probabilities. The conversion to Z scores is not necessary when using the calculator.

Begin by entering the normal curve probability density function into Y1. Use the letter U to represent μ and the letter S to represent σ.

Y1 = (1/(S $\sqrt{2\pi}$)*e^(−(X−U)^2/(2S^2))

Be sure to use the $*$ sign as you enter the expression for Y1. Now enter the integral into Y9.

Y9 = fnInt(Y1,X,A,B)

In the computation window, enter

6 → U
0.5 → S
5.5 → A
5.8 → B

Now access and compute Y9. The calculator returns approximately 0.1859, which agrees with the example.

Calculator Exercises

1. The probability, P, that a person has to wait t minutes to be seated at a restaurant is

$$P(t) = \frac{16.535}{16(t+1)^2}, \quad 0 \leq t \leq 60$$

Find the probability that any person entering the restaurant will have to wait (a) between 10 and 15 minutes to be seated, (b) at most 5 minutes to be seated, and (c) between 30 and 60 minutes to be seated.

2. As part of a research effort, a medical researcher introduces cholesterol into blood samples in such a way that the cholesterol levels are uniformly distributed over the interval $[50, 500]$. What is the probability that a randomly selected sample has a cholesterol level bewteen 175 and 250? Construct the graph of this probability function and shade the region that corresponds to the probability you just computed.
3. A manufacturer claims that the probability that one of its machine components will last t months after it is put into use is given by the exponential probability function

$$P(t) = 0.024e^{-0.024t}$$

What is the probability that a randomly selected component lasts for at most 3 years? What is the probability that a randomly selected component lasts for at least 10 years? Construct the graph of this function and shade the region that corresponds to the probability of the component lasting at least 10 years.
4. The seeds of plants in a certain area are dispersed by the wind in such a way that the probability, P, that a seed travels x feet from the plant is given by the function

$$p(X) = 0.013e^{-0.013X}$$

Find the probability that a seed is not blown more than 10 feet from the plant. Find the probability that a seed is blown from 25 to 50 feet from the plant. Find the probability that a seed is blown more than 1000 feet from the plant.
5. A company sells a product that it claims has an expected life span, X, that is normally distributed with a mean of 12 years and a standard deviation of 3.3 years. Find the probability that a randomly selected unit lasts (a) less than 5 years, (b) lasts between 8 and 12 years, and (c) lasts more than 20 years.
6. The value of σ in the formula for the normal probability distribution affects the dispersion of probabilities for events. To see this, compute $P(-1 \leq X \leq 1)$ for $\sigma = 0.5$, $\sigma = 1$, $\sigma = 2$, and $\sigma = 3$. In each case and using the same coordinate system, construct the graph of the corresponding normal probability function. Use the window $-5 \leq X \leq 5$ and $0 \leq p(X) \leq 0.5$. Make a conjecture about how increasing values of σ affect probabilities.

EXERCISE SET 7.6

A UNDERSTANDING THE CONCEPTS

For Exercises 1–8, interpret each probability statement in terms of the given situation.

1. ***Life Science: Exposure to Radiation*** The probability that a randomly selected air traveler is exposed to between 5.00 and 5.50 mrem of radiation while flying across the continental United States is 0.0952.
2. ***Business: Effectiveness of Insecticide*** The probability that a newly developed insecticide will kill at least 85 of 100 insects is 0.8661.
3. ***Life Science: Rainfall*** The probability that a particular county will receive between 2.4 and 4.4 inches of rain during the month of February is 0.5526.
4. ***Medicine: Blood Clotting*** The probability that, after receiving a wound in which blood flows, a person's blood clots within 35 seconds is 0.7218.
5. ***Business: Waiting Time*** The probability that, in a certain city, a telephone call will last longer than 16 minutes is 0.1512.
6. ***Social Science: Dinner at Home*** The probability that, of a group of working adults who had eaten dinner at home the previous night, the dinner had been prepared at home is 0.7335.
7. ***Psychology: Reaction Time*** The probability that a person's reaction time to a particular stimulus is more than 0.2 seconds is 0.0059.

8. ***Business: Germination of Seedlings*** The probability of germination for a particular type of seedling is 0.8989.

9. Explain why the function $p(X) = \frac{3}{100}X^2$ over the interval $[2, 5]$ is *not* a probability function.

10. Explain why the function $p(X) = \frac{3}{e}e^{2X-1}$ over the interval $[0, 1]$ is *not* a probability function.

B SKILL ACQUISITION

For Exercises 11–16, determine if each function over the given interval is a probability function.

11. $p(X) = \frac{X-2}{16}$ on $[1, 5]$

12. $p(X) = \frac{5X^{2/3}}{96}$ on $[0, 4]$

13. $p(X) = 4X^2$ on $[-2, 2]$

14. $p(X) = \frac{X^2}{21}$ on $[-1, 1]$

15. $p(X) = \frac{25}{9}X^4$ on $[1, \sqrt[5]{2}]$

16. $p(X) = \frac{1}{5}X$ on $[-2, 2]$

For Exercises 17–22, find a value for k that will make $p(X)$ a probability function on the given interval.

17. $p(X) = kX^{1/3}$ on $[1, 8]$

18. $p(X) = kX^{2/3}$ on $[0, 8]$

19. $p(X) = kX$ on $[1, 5]$

20. $p(X) = kX$ on $[-3, 0]$

21. $p(X) = \frac{k}{\sqrt{X}}$ on $[4, 9]$

22. $p(X) = \frac{k}{\sqrt[5]{X^3}}$ on $[1, 32]$

For Exercises 23–30, the given function is a probability function over an appropriate interval. Find the indicated probability.

23. $p(X) = \frac{Xe^{-X^2/2}}{2}$. Find $P(0 \le X \le 1)$.

24. $p(X) = e^{-X}$. Find $P(0 \le X \le 1)$.

25. $p(X) = \frac{1}{9X}$. Find $P(1 \le X \le 10)$.

26. $p(X) = \frac{1}{X^6}$. Find $P(1 \le X \le 32)$.

27. $p(X) = 0.04e^{-0.04X}$. Find $P(5 \le X \le 10)$.

28. $p(X) = 0.02e^{-0.02X}$. Find $P(0 \le X \le 20)$.

29. $p(X) = \frac{3}{40}(X^2 - 2X)$. Find $P(2 \le X \le 4)$.

30. $p(X) = \frac{1}{10}(1 - e^{-3X})$. Find $P(0 \le X \le 2)$.

C APPLYING THE CONCEPTS

31. ***Business: Life Span of Equipment*** The probability density function for the operating life span of a mechanical device is

$$p(X) = \frac{6}{1127}(X^2 + 3X)$$

for X, the operating life span of the device, in $[0, 7]$. Find and interpret the probability that a randomly selected device lasts (a) between 0 and 3 years, and (b) less than 1 year.

32. ***Business: Life Span of Equipment*** The probability density function for the life span of an electronic computer component is

$$p(X) = 0.002e^{-0.002X}$$

for $X > 0$, the number of hours of operating life. Find and interpret the probability that a randomly selected component operates for (a) between 100 and 500 hours, and (b) less than 200 hours.

33. ***Transportation: Distance Between Consecutive Cars*** A city's transportation department believes that the distance X (in feet) between consecutive cars on the city's main freeway during the morning commute has the probability density function

$$p(X) = 0.03e^{-0.03X}$$

where $X > 0$. Find the probability that for a randomly selected time during the morning commute the distance between consecutive cars is (a) between 40 and 50 feet, and (b) less than 100 feet.

34. ***Business: Waiting Time for Service*** A retail store claims that no customer will have to wait more than 4 minutes before being served. If X represents the waiting time, the probability density function is

$$p(X) = \frac{3}{128}(16 - X^2)$$

where X is in $[0, 4]$. Find the probability that a randomly selected customer waits (a) between 1 and 2 minutes for service, and (b) more than 3 minutes for service.

35. ***Business: Waiting Time for Service*** The time (in minutes) that a customer waits in line at a retail store is exponentially distributed with the probability function

$$p(X) = 0.14e^{-0.14X}$$

What is the probability that a randomly selected customer has to wait in line (a) between 5 and 6 minutes, and (b) no more than 1 minute?

36. ***Business: Weight of Bags of Fertilizer*** A chemical company produces a nonorganic lawn fertilizer for home

use. It is sold in bags which vary in weight from 20 pounds to 50 pounds. If the weight of the bags is uniformly distributed, what is the probability that a randomly selected bag contains (a) between 40 and 50 pounds of fertilizer, and (b) less than 25 pounds of fertilizer?

37. ***Business: Construction of Hamburger Patties*** A fast-food franchise operation uses automatic processors to construct its hamburger patties. The processors create patties that range in weight from 3.1 ounces to 4.5 ounces, and does so uniformly.

a. What is the probability that a randomly selected patty weighs between 4 and 4.3 ounces?

b. What percentage of the patties weighs more than a quarter of a pound?

38. ***Business: Weight of Cereal in Boxes*** The weights of cereal in boxes are uniformly distributed over 10 ounces to 18 ounces.

a. What is the probability that a randomly selected box contains between 10 and 12 ounces of cereal?

b. What percentage of the boxes contains more than 16 ounces of cereal?

39. ***Transportation: Waiting Time*** A shuttle bus leaves from point A to point B every 15 minutes and does so uniformly. If a person arrives at the shuttle stop and finds no bus there, what is the probability that he or she will have to wait (a) no more than 2 minutes, and (b) at least 10 minutes?

40. ***Transportation: Waiting Time*** A particular traffic light stays red for 50 seconds and does so uniformly. If a person arrives at the light and finds it red, what is the probability that he or she will have to wait (a) less than 10 seconds for it to turn green, and (b) more than 45 seconds for it to turn green?

41. ***Business: Life Span of a Product*** A manufacturer of video disc players believes that the operating life (in years) of her product is exponentially distributed with the probability density function

$$p(X) = 0.07e^{-0.07X}$$

What is the probability that a randomly selected video disc player will operate (a) between 5 and 8 years, (b) less than 2 years, and (c) more than 10 years? (Hint: For part (c), $P(X \geq 10) = 1 - P(0 \leq X \leq 10)$.)

42. ***Business: Quality Control*** One type of O-ring made for the space shuttle must be 5 centimeters in diameter so as to fit properly. The O-ring can vary up to 0.25 centimeters without producing dangerous leaks. The manufacturer of this O-ring produces them with a mean of 5 centimeters and standard deviation of 0.14 centimeters. Assuming the O-ring's diameters are normally distributed, find the probability that a randomly selected O-ring (a) has a diameter of between 5.0 and 5.15 centimeters, (b) has a diameter of between 4.8 and 5.2 centimeters, (c) has a diameter of greater than 5.4 centimeters, and (d) fits properly.

43. ***Education: Graduate Management Exam*** In a certain year, the scores of students taking the Graduate Management Test averaged 826 with a standard deviation of 175. Assuming scores are normally distributed, what is the probability that a randomly selected test has a score of between 900 and 1000?

44. ***Economics: Consensus Forecasting*** One method of arriving at an economic forecast is to collect forecasts from a large number of economic analysts and average them. This method is called the **consensus** method of forecasting. Suppose that the individual 1993 January forecasts of the prime interest rate for all the analysts are approximately normally distributed with a mean of 4.5% and a standard deviation of 0.9%. What is the probability that a randomly selected analyst's forecast was (a) between 6% and 7%, and (b) less than 5%?

45. ***Life Science: Breathing*** The number of breaths per minute taken by an adult when he or she is at rest depends on the age of the person and varies a great deal from person to person. Assume that the number of breaths is approximately normally distributed with a mean of 18 breaths per minute with a standard deviation of 6 breaths per minute. What is the probability that a randomly selected adult at rest will take (a) between 12 and 15 breaths per minute, and (b) more than 23 breaths per minute?

46. ***Life Science: Waiting Time*** According to a medical researcher, the time between successive reports of a particular disease is exponentially distributed with the probability density function

$$p(X) = 0.008e^{-0.008X}$$

What is the probability that the time between successive reports of the disease is (a) between 30 and 40 days, and (b) greater than 50 days?

47. ***Business: Life Span of a Product*** The operating life of an electronic component is normally distributed with a mean of 350 hours and standard deviation of 45 hours. What is the probability that a randomly selected component has an operating life of (a) between 315 and 320 hours, (b) less than 300 hours, and (c) more than 475 hours?

D DESCRIBE YOUR THOUGHTS

48. Explain, as if you were explaining to someone you were tutoring, why it is not sufficient to have only the condition that $\int_a^b p(X)\,dX = 1$ for $p(X)$ to be a probability density function.

49. Explain, as if you were explaining to someone you were tutoring, how you distinguish between the uniform, exponential, and normal probability distributions.

50. Describe the strategy you would use to solve Exercise 41 or 42.

E REVIEW

51. (2.2) Find, if it exists, $\lim_{x \to 8} \dfrac{x^2 - 12x + 32}{x^2 - 9x + 8}$.

52. (2.3) *Economics: Housing Starts* The number of housing starts x (in thousands) each month in a country is related to the time t (in months) from the end of a recession, by the function

$$x(t) = \frac{4t^2 + 250t + 525}{3t^2 + 60t + 340}$$

What is the expected number of housing starts in the long term?

53. (2.4) Construct the graph of a function that is discontinuous at $x = 5$ because continuity condition 1 is the first condition to fail.

54. (3.1) Use the definition of the derivative to find $f'(x)$ for $f(x) = 5x^2 + 5x + 5$.

55. (3.1) *Ecology: Pollution* Suppose that $N(x)$ represents the number of pollutants, in parts per million (ppm), entering a lake each day when there are x homes within a radius of 2 miles from the lake. Interpret $N'(800) = 0.06$.

56. (3.2) Find $f'(x)$ for $f(x) = \dfrac{2x^2 + 3x - 1}{x - 4}$.

57. (3.3) Find $f''(x)$ for $f(x) = \dfrac{x + 4}{2x - 3}$.

58. (3.4) Find $f'(x)$ for $f(x) = (x - 6)^4(5x^2 - 1)^2$.

59. (3.5) *Business: Container Construction* The manufacturer of cardboard packing boxes has determined that its monthly cost C (in thousands of dollars) of producing such boxes is related to the price p (in dollars) it has to pay for a unit of cardboard by the function

$$C(p) = 25{,}000 + 12{,}000\left(\frac{175p + 15}{50p + 70}\right)^{3/4}$$

If the current price of a unit of cardboard is \$0.35 but is increasing at the rate of \$0.03 per month, at what rate is the cost of producing the boxes changing?

60. (5.3) Find $f'(x)$ for $f(x) = 5e^{-5x}\ln(5x)$.

Table 7.2 Areas of a Standard Normal Probability Distribution

The entries in the table on the next page represent the area under the standard normal probability curve from 0 to a selected value of z, where z is computed to two decimal places using the conversion formula

$$z = \frac{x - \mu}{\sigma}$$

Figure 7.30, below, illustrates the area designated in the table on the next page.

To use the table, locate the units and tenths digits of the z value in the left column and the hundredths digit along the top row. The four–digit decimal number in that row and column is the approximate probability that a randomly chosen value of Z lies between 0 and z. For example, if $z = 2.37$, then $P(0 \le Z \le 2.37) = .4911$.

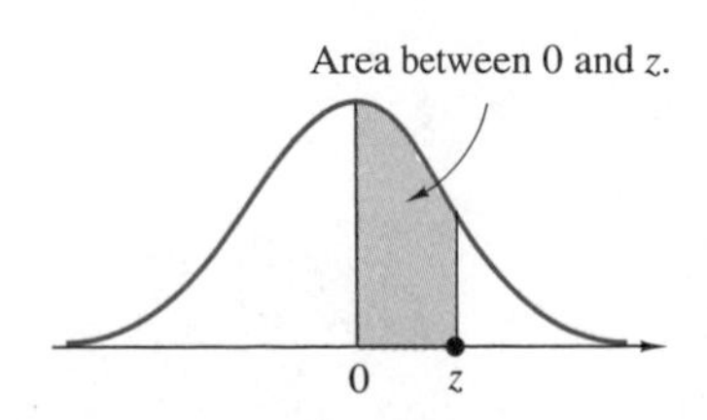

Figure 7.30

Table 7.2 Areas of a Standard Normal Probability Distribution

z	.00	.01	.02	.03	.04	.05	.06	.07	.08	.09
.0	.0000	.0040	.0080	.0120	.0160	.0199	.0239	.0279	.0319	.0359
.1	.0398	.0438	.0478	.0517	.0557	.0596	.0636	.0675	.0714	.0753
.2	.0793	.0832	.0871	.0910	.0948	.0987	.1026	.1064	.1103	.1141
.3	.1179	.1217	.1255	.1293	.1331	.1368	.1406	.1443	.1480	.1517
.4	.1554	.1591	.1628	.1664	.1700	.1736	.1772	.1808	.1844	.1879
.5	.1915	.1950	.1985	.2019	.2054	.2088	.2123	.2157	.2190	.2224
.6	.2257	.2291	.2324	.2357	.2389	.2422	.2454	.2486	.2517	.2549
.7	.2580	.2611	.2642	.2673	.2704	.2734	.2764	.2794	.2823	.2852
.8	.2881	.2910	.2939	.2967	.2995	.3023	.3051	.3078	.3106	.3133
.9	.3159	.3186	.3212	.3238	.3264	.3289	.3315	.3340	.3365	.3389
1.0	.3413	.3438	.3461	.3485	.3508	.3531	.3554	.3577	.3599	.3621
1.1	.3643	.3665	.3686	.3708	.3729	.3749	.3770	.3790	.3810	.3830
1.2	.3849	.3869	.3888	.3907	.3925	.3944	.3962	.3980	.3997	.4015
1.3	.4032	.4049	.4066	.4082	.4099	.4115	.4131	.4147	.4162	.4177
1.4	.4192	.4207	.4222	.4236	.4251	.4265	.4279	.4292	.4306	.4319
1.5	.4332	.4345	.4357	.4370	.4382	.4394	.4406	.4418	.4429	.4441
1.6	.4452	.4463	.4474	.4484	.4495	.4505	.4515	.4525	.4535	.4545
1.7	.4554	.4564	.4573	.4582	.4591	.4599	.4608	.4616	.4625	.4633
1.8	.4641	.4649	.4656	.4664	.4671	.4678	.4686	.4693	.4699	.4706
1.9	.4713	.4719	.4726	.4732	.4738	.4744	.4750	.4756	.4761	.4767
2.0	.4772	.4778	.4783	.4788	.4793	.4798	.4803	.4808	.4812	.4817
2.1	.4821	.4826	.4830	.4834	.4838	.4842	.4846	.4850	.4854	.4857
2.2	.4861	.4864	.4868	.4871	.4875	.4878	.4881	.4884	.4887	.4890
2.3	.4893	.4896	.4898	.4901	.4904	.4906	.4909	.4911	.4913	.4916
2.4	.4918	.4920	.4922	.4925	.4927	.4929	.4931	.4932	.4934	.4936
2.5	.4938	.4940	.4941	.4943	.4945	.4946	.4948	.4949	.4951	.4952
2.6	.4953	.4955	.4956	.4957	.4959	.4960	.4961	.4962	.4963	.4964
2.7	.4965	.4966	.4967	.4968	.4969	.4970	.4971	.4972	.4973	.4974
2.8	.4974	.4975	.4976	.4977	.4977	.4978	.4979	.4979	.4980	.4981
2.9	.4981	.4982	.4982	.4983	.4984	.4984	.4985	.4985	.4986	.4986
3.0	.4987	.4987	.4987	.4988	.4988	.4989	.4989	.4989	.4990	.4990

For $z > 3.10$, approximate the area with 0.4999.

Summary

Chapter 7 considered a variety of applications of integration.

Areas of Regions in the Plane

Section 7.1 discussed how the definite integral can be used to find the area bounded by two curves. If $f(x)$ and $g(x)$ are two continuous functions for which $f(x) \geq g(x)$ on $[a, b]$, then the area bounded by the two functions and the vertical lines $x = a$ and $x = b$ on $[a, b]$ is

$$A = \int_a^b [f(x) - g(x)] \, dx$$

When preparing to make this computation, we choose $f(x)$ to be the higher curve and $g(x)$ to be the lower curve.

Consumer's and Producer's Surplus

Section 7.2 introduced, in some detail, the concepts of consumer's and producer's surplus. The section began with the *supply function* $p = S(x)$, which represents the price per unit at which producers are willing to produce and supply x units of a product. The supply function is of interest to producers.

This was followed by a discussion of the *demand function* $p = D(x)$, which represents the price per unit consumers are willing to pay when x units of the item are made available in the market. The demand function is of interest to consumers.

The point (x_e, p_e) at which supply equals demand is called the *equilibrium point* and represents the price p_e consumers are willing to pay when x_e units of the item are produced.

Next, *consumer's surplus* was defined as the definite integral

$$CS = \int_0^{x_e} [D(x) - p_e] \, dx$$

which measures the amount of surplus money that results when a purchase is made at a price that is lower than the maximum the consumer was willing to pay.

Then, *producer's surplus* was defined as the definite integral

$$PS = \int_0^{x_e} [p_e - S(x)] \, dx$$

which measures the additional money contributed by consumers on the sale of x_e units of a product.

Annuities and Money Streams

Section 7.3 was devoted to a discussion of annuities and money streams. It began by defining *accumulated value of a non-interest-earning stream* in the following way. If over some interval $0 \leq t \leq T$, money flows continuously into a pool at a rate approximated by the function $f(t)$ dollars per time period, and if the money is reinvested in a non-interest-earning account, then the total amount of money A accumulated at the end of T years is

$$A = \int_0^T f(t)\,dt$$

dollars.

Next, *future value of the investment* was defined as:

$$A = Pe^{rt}$$

where P dollars is invested at an annual interest rate r that is compounded continuously into an interest-earning account, and A is the total amount of money that will have accumulated at the end of t years.

Annuities were then described as the sequence of payments where an amount of money is regularly deposited into an interest-earning account at the end of equal periods of time. The *amount of the annuity*, or the *future value of the annuity*, is the final amount in the account at the end of all the time periods. It is the total amount of the payments plus the total amount of interest earned from all the payments. The time between payments is the *payment period.*

The future value A of an annuity at the end of n payment periods is approximated by

$$A = \int_0^n Pe^{rt}\,dt$$

where P is the number of dollars invested at the end of each payment period, and r is the interest rate per time period. It is the value of the \$ $P \cdot n$ investment in the future.

The *present value* of an annuity,

$$P = Ae^{-rt}$$

represents the amount, P, that must be invested now ($t = 0$) so that the amount A will be accumulated in the future.

The *present and future values of an income stream* were defined as:

If $f(t)$ is the rate of flow of money into an account that pays an annual interest rate r compounded continuously during some time interval $0 \le t \le T$, then

1. The *future value* of the income stream is

$$\text{Future value} = e^{rT} \int_0^T f(t)e^{-rt}\,dt$$

2. The *present value* of the income stream is

$$\text{Present value} = \int_0^T f(t)e^{-rt}\,dt$$

The future value measures the total accumulated amount of income that will be realized at the end of the time period $[0, T]$. The present value measures the amount of money that would have to be deposited now to produce the future value.

Finally, this section concluded with *the present and future values of a perpetual income stream.* In some situations, it is reasonable to consider the income as flowing forever. In such cases, we can approximate the present value of the flow using the improper integral:

$$\text{Present value} = \int_0^{\infty} f(t)e^{-rt}\,dt$$

Differential Equations

Section 7.4 introduced *differential equations* and considered only a special class called the *separable variables* class. The solutions to these equations can be found relatively easily by separating the dependent and independent variables to opposite sides of the equal sign and integrating.

Solutions of differential equations that involve one or more arbitrary constants are called *general solutions* and represent a family of solutions. Solutions that arise from substituting initial conditions into general solutions are called *particular solutions* and the problems that produce them are called *initial-value problems*.

Applications of Differential Equations

Section 7.5 considered applications of differential equations by developing three models: the unlimited growth model, the limited growth model, and the logistic growth model.

The *unlimited growth model* states that if P represents the size of a population, t represents time, and k is a constant, then

$$\frac{dP}{dt} = kP$$

It states that the rate at which a population changes over time is proportional to the current size of the population. The solution to the differential equation $\frac{dP}{dt} = kP$ is the exponential function $P = Ce^{kt}$, where the constant k is the *growth constant*. If $k > 0$, the model represents growth, and if $k < 0$, the model represents decay (negative growth).

The *limited growth model* states that when the rate at which a population changes with respect to time is proportional to the difference between the *carrying capacity*, L, and the current size of the population, P, the mathematical model for growth (or decay) is

$$\frac{dP}{dt} = k(L - P)$$

If P_0 is the size of the population at some initial observation time $t = 0$, then the particular solution is

$$P = L + (P_0 - L)e^{-kt}$$

The *logistic growth model* states that when the rate at which a population changes is proportional to the current size of the population and the difference between the carrying capacity and the current size of the population, the mathematical model is

$$\frac{dP}{dt} = kP(L - P)$$

If P_0 is the size of the population at time $t = 0$, then the particular solution is

$$P = \frac{P_0 L}{P_0 + (L - P_0)e^{-Lkt}}$$

Probability

Section 7.6 presented concepts from the theory of probability. The section began by distinguishing between discrete and continuous random variables. Discrete random variables take on only particular values and result from a count made upon a system, while continuous random variables can assume any value in an entire interval of values and result from measures made upon the system.

The section then presented the probability density function. A function $P(X)$ is a probability density function if it records the probability that a system is in some particular state and if there is some closed interval $[a, b]$ for which $\int_{-\infty}^{\infty} p(X)\,dX = 1$.

Section 7.6 continued by examining three probability distributions, the *uniform* probability distribution, the *exponential* probability distribution, and the *normal* probability distribution.

1. The *uniform* probability distribution is characterized as follows: A random variable X is uniformly distributed on $[a, b]$ if its probability density function, $p(X)$, is defined by

$$p(X) = \frac{1}{b - a} \qquad \text{for every } X \text{ in } [a, b]$$

2. The *exponential* probability distribution is characterized as follows: A random variable X is exponentially distributed over the interval $[0, +\infty)$ if its probability density function is

$$P(X) = ke^{-kX}$$

where k is a constant determined by the situation.

3. The *normal* probability distribution is characterized as follows: A random variable X, with mean μ and standard deviation σ, is normally distributed over the interval $(-\infty, +\infty)$ if its probability density function is

$$p(X) = \frac{1}{\sigma\sqrt{2\pi}} e^{\frac{-(X-\mu)^2}{2\sigma^2}}$$

Supplementary Exercises

For Exercises 1–7, find the area of the region bounded by the given function, the x axis, and the given vertical lines.

1. $f(x) = x^2 - 4x + 3$, $x = 0$ to $x = 3$
2. $f(x) = \frac{3}{x - 6}$, $x = 7$ to $x = 9$
3. $f(x) = 50 - x^4$, $x = 0$ to $x = 2$
4. $f(x) = 4 - e^{-x}$, $x = 0$ to $x = 5$
5. $f(x) = 3e^{-x/2}$, $x = 0$ to $x = 2$

6. $f(x) = \dfrac{x}{x^2+1}$, $x = 0$ to $x = 2$

7. $f(x) = x^{1/3}$, $x = 0$ to $x = 8$

8. Find the area bounded by the curves $f(x) = \frac{1}{2}x + 3$ and $g(x) = x^2$ from $x = 0$ to $x = 2$.

9. Find the area bounded by the curves $f(x) = -x^2 + 4x$ and $g(x) = -\frac{1}{4}x + 6$ from $x = 2$ to $x = 4$.

10. Find the area bounded by the curves $f(x) = 10\sqrt{x-2}$ and $g(x) = x^2 - 5x + 14$ from $x = 3$ to $x = 6$.

11. Find the area bounded by the curves $f(x) = 1.6e^{-0.32x}$ and $g(x) = x^2 - 7x + 15$ from $x = 3$ to $x = 6$.

12. ***Business: Consumer's Surplus*** The demand function for a product is $D(x) = 35e^{-0.05x}$. Find the consumer's surplus if the market price of a unit is \$15.

13. ***Business: Consumer's Surplus*** The demand function for a product is $D(x) = \dfrac{450}{(0.2x+1)^2}$. Find the consumer's surplus if the market price of a unit is \$65.

14. ***Business: Producer's Surplus*** The supply function for a product is $S(x) = 55\sqrt{x+4}$. Find the producer's surplus if the market price of a unit is \$200.

15. ***Business: Producer's Surplus*** The supply function for a product is $S(x) = 25e^{0.125x}$. Find the producer's surplus if the market price of a unit is \$315.

16. Verify, by differentiation, that $y = (x^2+1)^3 + C$ is a solution to the differential equation $\dfrac{dy}{dx} = 6x(x^2+1)^2$.

17. Verify, by differentiation, that $y = \dfrac{x-5}{x+6} + C$ is a solution to the differential equation $\dfrac{dy}{dx} = \dfrac{11}{(x+6)^2}$.

18. Verify, by differentiation, that $y^2 - 2y = x^2 + 2x + C$ is a solution to the differential equation $\dfrac{dy}{dx} = \dfrac{x+1}{y-1}$.

19. Verify, by differentiation, that $y = C(x+4) - 4$ is a solution to the differential equation $\dfrac{dy}{dx} = \dfrac{y+4}{x+4}$.

20. Verify, by differentiation, that $y = \dfrac{-\ln(C - e^{3x})}{3}$ is a solution to the differential equation $\dfrac{dy}{dx} = \dfrac{e^{3x}}{e^{-3y}}$.

For Exercises 21–27, find the general solution to each differential equation.

21. $\dfrac{dy}{dx} = 2x^5$

22. $\dfrac{dy}{dx} = 15 - 0.03e^{0.03x}$

23. $\dfrac{dy}{dx} = xe^{-y}$

24. $y' = 2xe^y$

25. $y' = xy(\ln x)(\ln y)$

26. $y' = 4ye^{2x}$

27. $\dfrac{dy}{dx} = \dfrac{xy^2}{x^2+4}$

For Exercises 28–34, find the particular solution to each differential equation.

28. $(x+1)y' = 2y$, and $y = 1$ when $x = 0$

29. $2xy' = 3y$, and $y = 4$ when $x = 1$

30. $y' = y^2(1+e^x)$, and $y = 1$ when $x = 0$

31. $y' = ye^{-x} + e^{-x}$, and $y = 0$ when $x = 0$

32. $(x^4 - 1)y' = 4x^3y$, and $y = 90$ when $x = 2$

33. $y' = (x+5)^2e^y$, and $y = 0$ when $x = 0$

34. $y' = xe^{-x}e^y$, and $y = 0$ when $x = 0$

35. ***Business: Investment*** Two years ago, \$20,000 was placed into an investment in which the amount of money present P grows according to

$$\frac{dP}{dt} = kt$$

Today, the investment is \$27,680. How much will the investment be worth 5 years from now?

36. ***Social Science: Voting*** In a particular county, the percentage P of voters who do not vote in presidential elections is changing at the rate of

$$\frac{dP}{dt} = k\sqrt[3]{t^2}$$

where t is the time in years with $t = 0$ being 1972. If in 1972, 50% of all registered voters voted, but in 1980 only 46% voted, what percentage can be expected to vote in 1996?

37. ***Economics: Value of Real Estate*** The value of real estate in a city is increasing, without limitation, at a rate proportional to its current value. A piece of property in the city appraised at \$460,000 2 years ago is appraised at \$490,000 today. At what value will the property be appraised 1 year from now?

38. ***Manufacturing: Newton's Law of Cooling*** A chemical company is experimenting with a chemical it hopes will keep liquids that have been refrigerated, then placed into a warmer environment, from warming too rapidly. A liquid that has been treated with the company's chemical is refrigerated to 35°F and then placed into the surrounding air, which is 72°F. Twenty minutes later, the liquid has warmed to 38°F.

a. Find a function that gives the temperature of the liquid at any time t after removal from the refrigerator.
b. Find the temperature of the liquid 30 minutes after its removal from the refrigerator.

39. ***Political Science: Political Polls*** A polling group has determined that the percentage of voters registered as members of a new political party is increasing at a rate that is proportional to the percentage of registered voters who are not members of the new party. They found that 0.4% of all registered voters were members of the new party in 1984 and 0.6% were members in 1988.
 a. Find a function that gives the percentage of all registered voters who are members of the new party at any time t.
 b. Find the percentage of all registered voters that the new party can expect to have registered by the year 2000.

40. ***Social Science: Spread of a Rumor*** In a group of 5000 people, 1 person starts a rumor. One hour later 25 people have heard the rumor. The rumor spreads at a rate that is proportional to the number of people who have heard it and the number of people who have not yet heard it.
 a. Find a function that gives the number of people who have heard the rumor at any time t.
 b. Find the number of people who have heard the rumor 24 hours after it was started.

41. ***Business: Probability and Advertising*** The probability that a randomly selected male in the age group 15–25 will buy a particular style of advertised clothes is 0.011. Interpret this value.

42. ***Business: Life Span of a Product*** The probability that an all-weather paint lasts more than 5 years when applied to the outside of a randomly selected home is 0.3925. Interpret this value.

43. ***Business: Life Span of a Product*** The probability that an interstate highway lane reflector lasts 3 years or less is 0.085. Interpret this value.

44. Determine if the function $P(X) = \frac{1}{2}(X^3 - X)$ is a probability function.

45. Determine if the function

$$P(X) = \frac{1}{\ln 67 - \ln 28} \cdot \frac{X}{X^2+3}$$

is a probability function.

46. Find a value of k that will make $P(X) = k(2X - 6)$ a probability function over $[3, 6]$.

47. Find a value of k that will make $P(X) = k(X + \sqrt{X})$ a probability function over $[0, 9]$.

48. Find a value of k that will make $P(X) = \frac{k\sqrt{2}}{X^2}$ a probability function over $[5, 10]$.

For Exercises 49–53, the given function is a probability function over an appropriate interval. Find the indicated probability.

49. $P(X) = 2e^{-2X}$. Find $P(0 \le X \le 3)$.

50. $P(X) = 0.06e^{-0.06X}$. Find $P(20 \le X \le 28)$.

51. $P(X) = \frac{1}{70}(X - 3)$. Find $P(3 \le X \le 10)$.

52. $P(X) = \frac{2}{117}(X + \sqrt{X}$. Find $P(0 \le X \le 1)$.

53. $P(X) = \frac{12}{135}$. Find $P(40 \le X \le 50)$.

For Exercises 54–55, use the standard normal probability table (Table 7.1) to find each probability. Assume the random variable X is normally distributed.

54. $P(12 \le X \le 15)$, where $\mu = 14$ and $\sigma = 1.3$

55. $P(130 \le X \le 160)$, where $\mu = 140$ and $\sigma = 6$

56. ***Business: Life Span of a Product*** A company that makes yellow paint that a city uses on its streets for lane dividers, claims that the deterioration time of the paint is normally distributed with a mean life of 225,000 vehicle crossings with a standard deviation of 23,000 vehicle crossings. Find the probability that a randomly selected line will withstand between 170,000 and 190,000 vehicle crossings before it deteriorates and must be restored. Based on this line (this sample of size 1), do you think the company's claim is valid or not?

57. ***Business: Quality Control*** A manufacturer that produces aircraft engine bearings claims that the diameters of the bearings are normally distributed with a mean of 0.805 cm and standard deviation of 0.0035 cm. Find the probability that a randomly selected bearing has a diameter in the range 0.795 cm to 0.815 cm. Based on this single bearing (this sample of size one), do you think the manufacturer's claim is valid or not?

58. ***Business: Life Span of a Product*** The probability density function for the operating life of an electronic video camera component is

$$P(X) = \frac{-6}{1552}(x^2 - 25x)$$

for X, the operating life in years, in $[1, 5]$. Find and interpret the probability that a randomly selected device lasts (a) between 2 and 3 years, (b) less than 3 years, and (c) more than 4 years.

59. ***Business: Life Span of a Product*** The probability density function of a low-amp boat battery is

$$P(X) = \frac{4}{789}\left(15 - \sqrt[3]{X}\right)$$

for X in $[0, 15]$ where X represents hours of use. Find and interpret the probability that a randomly selected battery lasts (a) between 10 and 15 hours, (b) less than 5 hours, and (c) more than 12 hours.

60. ***Business: Waiting Time for Service*** The probability density function for the length of time (in minutes) a customer must wait to be connected to a service representative of the Dewtex Business Machine Company is

$$P(X) = 0.36e^{-0.36X}$$

What is the probability that a randomly selected customer is kept waiting (a) for at most 4 minutes, (b) between 1 and 2 minutes, and (c) at most one-half a minute?

61. ***Transportation: Waiting Time at a Signal*** A traffic light stays red for 35 seconds and does so uniformly. If a person arrives at the light and finds it red, what is the probability that he or she will have to wait (a) less than 15 seconds for it to turn green, (b) between 5 and 10 seconds for it to turn green, and (c) more than 25 seconds for it to turn green?

CHAPTER 8

Calculus of Functions of Several Variables

8.1 Functions of Several Variables

Introduction
Notation
Evaluation of Functions of Several Variables
Graphs of Functions of Several Variables
Interpretations
Partial Derivatives and the Total Differential
Cobb–Douglas Production Functions
Functions of Several Variables and Technology

Introduction

The functions we have worked with up to this point have all been functions of one variable. For example, the sales function $N(x) = 2400x - 3x^2$ is a function of the one variable x. If x represents the amount of money a company spends on television advertisements, then $N(x)$ indicates that the number N of sales depends *only* on the amount of money spent on television advertisements. In familiar terminology, $N(x)$ indicates that the dependent variable N depends only on the one independent variable x.

Suppose, however, that the company not only advertises on television, but also on radio and in the newspapers. If x represents the amount of money spent on television advertisements, y the amount of money spent on radio advertisements, and z the amount of money spent on newspaper advertisements, then $N(x, y, z)$ indicates that N depends on all three variables x, y, and z. In familiar terminology, $N(x, y, z)$ indicates that the dependent variable N depends on the three independent variables x, y, and z.

Notation

When working with functions of one variable, we saw (in Chapter 1) that a function such as $y = 3x + 4$ was better described using the f notation. Rather than writing $y = 3x + 4$, we replaced y with $f(x)$ and wrote $f(x) = 3x + 4$ and kept in mind that y and $f(x)$ both represented the same rule; that is, that $y = f(x)$. Sometimes it is convenient to use y, other times, $f(x)$. Similarly, a function of two variables x and y might be represented by $f(x, y)$ or, maybe, just z. If both notations can be used, then we keep in mind that z and $f(x, y)$ both represent the same rule; that is, that $z = f(x, y)$. Likewise, if w and $f(x, y, z)$ represent the same rule, then $w = f(x, y, z)$. These alternate notations are often useful in labeling axes when graphing.

Evaluation of Functions of Several Variables

Suppose that the sales function $N(x, y, z)$ is defined by the expression $2300x + 1400y + 900z - 3x^2 - y^2 - z^2$. Since a function is a rule that prescribes how an output value is obtained from an input value — or from several input values — we can evaluate a function of several variables by substituting the input values for their corresponding variables and computing.

EXAMPLE SET A

For the sales function $N(x, y, z) = 2300x + 1400y + 900z - 3x^2 - y^2 - z^2$, find the number of sales N if, in thousands of dollars, $x = 80$, $y = 16$, and $z = 20$.

We substitute 80 for x, 16 for y, and 20 for z.

$$N(80, 16, 20) = 2300(80) + 1400(16) + 900(20) - 3(80)^2 - (16)^2 - (20)^2$$

$$= 204{,}544$$

Interpretation: We conclude that if \$80,000 is spent on television ads, \$16,000 on radio ads, and \$20,000 on newspaper ads, the company will make 204,544 sales of the advertised item.

Graphs of Functions of Several Variables

We know that a function of one variable requires two coordinate axes, a horizontal one from which the input values are obtained, and a vertical one from which the output values are read. The two axes are constructed perpendicular to each other and form a flat surface (a plane), and the graph is some type of curve. For example, the graph of $f(x) = x^2 - 4x + 1$ appears in Figure 8.1 in its $f(x)$ and y form, respectively.

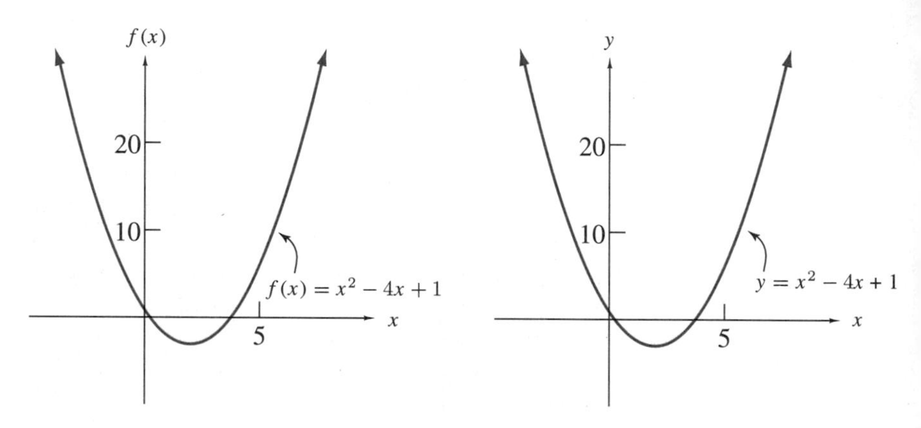

Figure 8.1

To graph a function of two variables requires three coordinate axes, two from which the values of the independent variables are obtained and one from which the values of the dependent variable are read. The three axes are constructed mutually perpendicular to each other and form what we call **space**. Figure 8.2 shows this 3-dimensional coordinate system in both its $f(x, y)$ and z forms, respectively. Try to imagine these 2-dimensional drawings as 3-dimensional.

Space

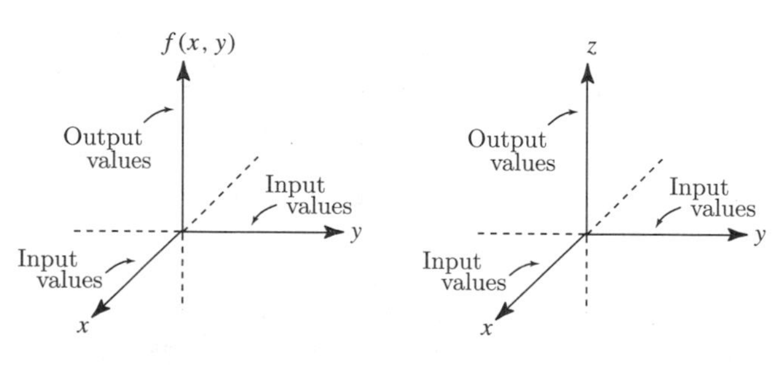

Figure 8.2

Graphs in space are no longer just curves, but various types of surfaces. Since they can be hard to draw with pencil and paper, we usually produce them using computers. Graphs of functions of two variables can be very interesting to look at but at the same time difficult to interpret. Figure 8.3 illustrates some graphs of functions of two variables.

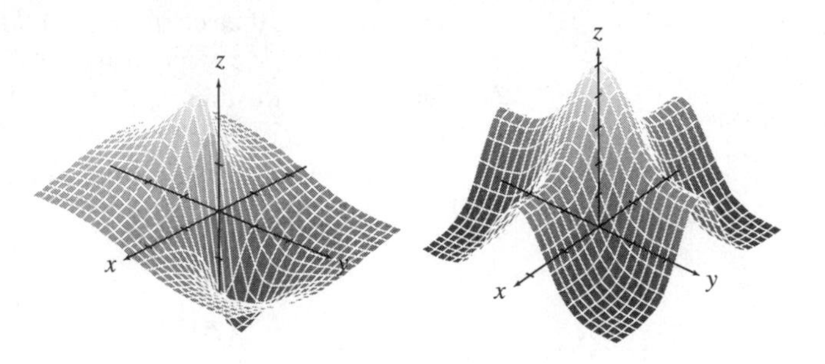

Figure 8.3

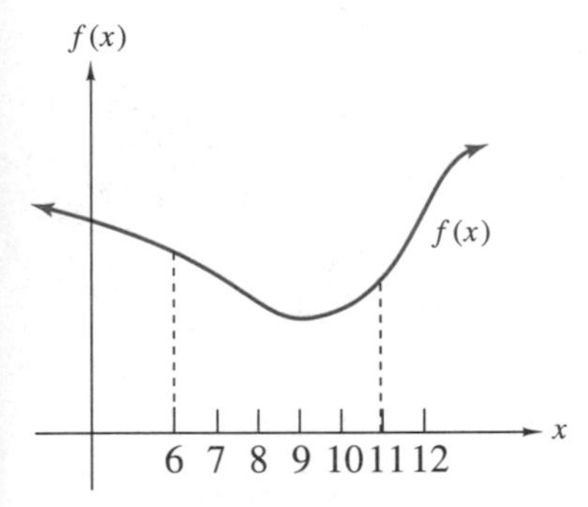

Figure 8.4

Interpretations

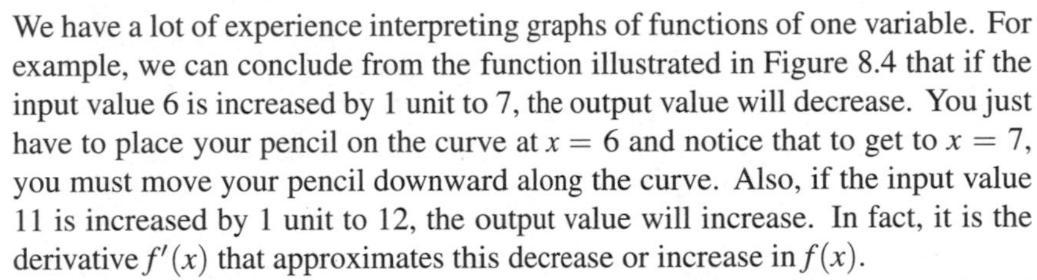

We have a lot of experience interpreting graphs of functions of one variable. For example, we can conclude from the function illustrated in Figure 8.4 that if the input value 6 is increased by 1 unit to 7, the output value will decrease. You just have to place your pencil on the curve at $x = 6$ and notice that to get to $x = 7$, you must move your pencil downward along the curve. Also, if the input value 11 is increased by 1 unit to 12, the output value will increase. In fact, it is the derivative $f'(x)$ that approximates this decrease or increase in $f(x)$.

To interpret graphs of functions of two variables we will have to look and think harder. For example, consider the graph of the function $f(x, y)$ illustrated in Figure 8.5. The dot on the surface represents the output value $z = 3$ corre-

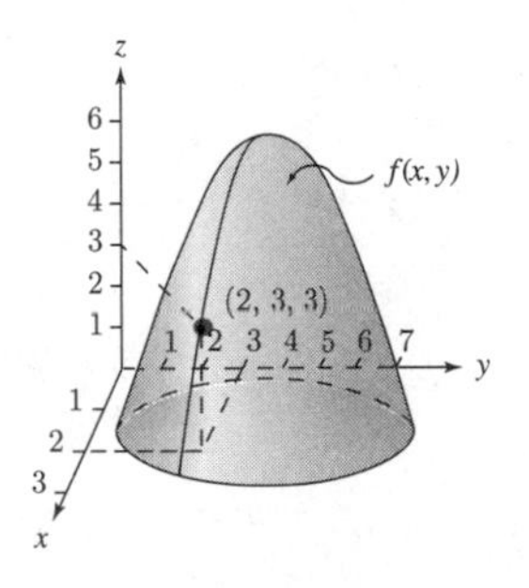

Figure 8.5

sponding to the input values $x = 2$ and $y = 3$. To see this, place your pencil at the origin and then move it 2 units along the x axis. Now move your pencil 3 units in the y direction (parallel to the y axis). This point in the xy plane represents the input values. Now move your pencil vertically 3 units, stopping when you get to the surface (at the dot). This vertical distance is the output value associated with the two input values. The **ordered triple** associated with this point is $(2, 3, 3)$.

Ordered Triple

Now, located at this point on the surface, we can change our position in one of three ways. We can move in the x direction only, thus changing only the x coordinate and keeping the y coordinate constant. Or, we can move in the y

direction only, thus changing only the y coordinate and keeping the x coordinate constant. Or, we can move diagonally, thus changing both the x and y coordinates. The leftmost illustration in Figure 8.6 shows a change of 5 units in the x direction only, and the rightmost illustration a change of 3 units in the y direction only. Figure 8.7 illustrates changes in both the x and y directions, a 5-unit change in

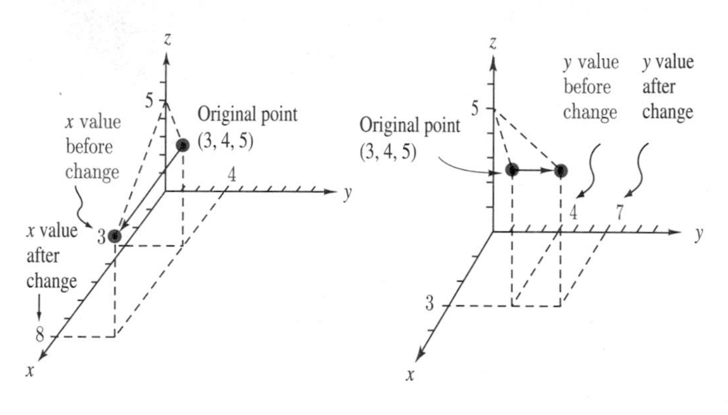

Figure 8.6

the x direction and a 3-unit change in the y direction.

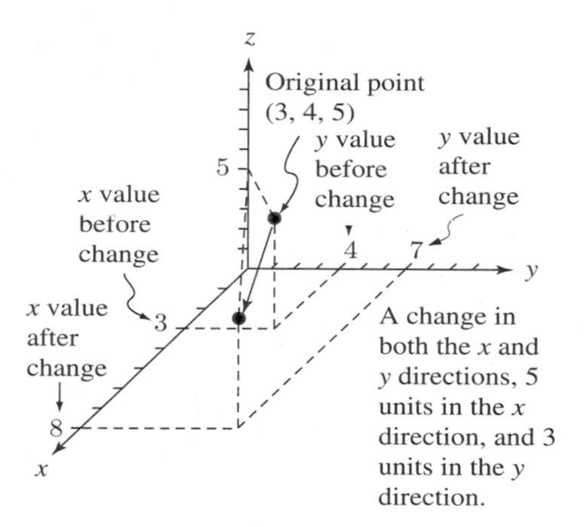

Figure 8.7

You can interpret the effect of one of these input changes on the output by physically making the change with your pencil. As you try to physically work your way through the next few examples, you may find it difficult to actually see where you are located on the surface; fortunately, we seldom actually have to do this. Just try your best.

1. *Changing x and keeping y constant* Refer to Figure 8.8 and place your pencil on the surface at the dot. To change only the x value by 1 unit, move your pencil parallel to the x axis. Move 1 unit in the positive x direction by moving parallel to the x axis and the same distance as the distance between

two consecutive tick marks on the x axis. To get back onto the surface, you must move your pencil downward $1\frac{1}{4}$ units.

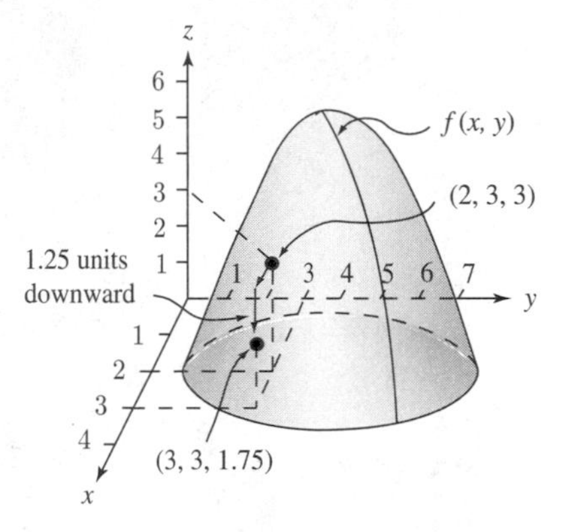

Figure 8.8

Interpretation: If, when $x = 2$ and $y = 3$, so that $z = 3$, x is increased by 1 unit and y is kept constant, then the function decreases by $1\frac{1}{4}$ units.

2. *Changing y and keeping x constant* Refer to Figure 8.9 and place your pencil on the surface at the dot. To change only the y value by 1 unit, move your pencil parallel to the y axis. Move 1 unit in the positive y direction by moving parallel to the y axis and the same distance as the distance between two consecutive tick marks on the y axis. To get back onto the surface, you must move your pencil downward $1\frac{3}{4}$ units.

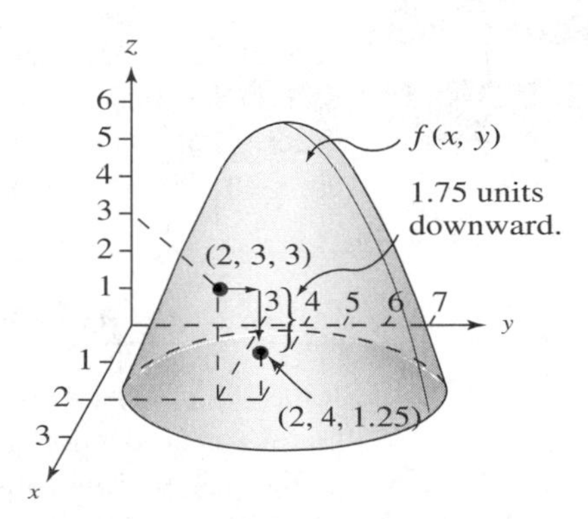

Figure 8.9

Interpretation: If, when $x = 2$ and $y = 3$, so that $z = 3$, y is increased by 1 unit and x is kept constant, then the function decreases by $1\frac{3}{4}$ units.

3. *Changing both the x and y coordinates* Refer to Figure 8.10 and place your pencil on the surface at the dot. To change both the input values, change one of them first, and then before moving to the surface, change the other. Then after both changes are made, move vertically to the surface. Place your pencil

on the surface at the dot. To change the x value by 1 unit and the y value by 1 unit, first change x according to (1) above and then change y according to (2) above. To get back onto the surface you must move your pencil downward 2 units.

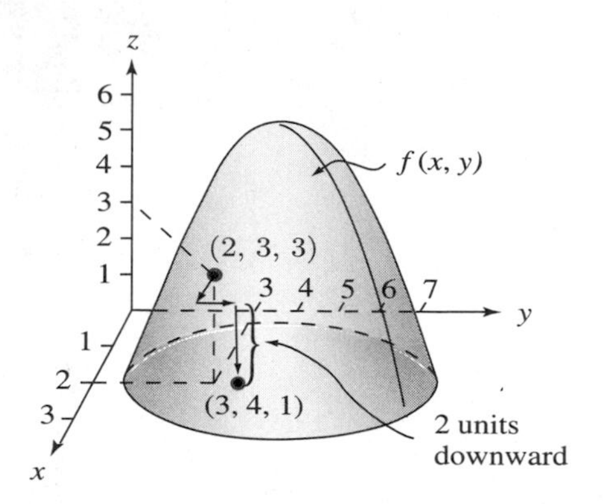

Figure 8.10

Interpretation: If, when $x = 2$ and $y = 3$, so that $z = 3$, x is increased by 1 unit and y is increased by 1 unit, then the function decreases by 2 units.

Partial Derivatives and the Total Differential

Partial Derivative

Total Differential

The derivative of a function of several variables approximates the change in the output value for a change in one or all of the input values. If only one of the input values is changed (as in (1) and (2) above), the derivative is called a **partial derivative** of the function. We will see how to compute partial derivatives in Section 8.2. If all the input values are changed (as in (3) above), we get the **total differential**. (There is no total derivative.) We will see how to compute the total differential in Section 8.5.

Cobb–Douglas Production Functions

In the 1920s, Charles W. Cobb and Paul H. Douglas constructed a power function useful for production studies. The function relates the number of units of a product a company can produce to the number of units of labor and capital it uses to produce this product. The function is of the form

$$f(x, y) = Cx^a y^b$$

where x represents the number of units of labor used to produce the product and y the number of units of capital used. $f(x, y)$ represents the number of units of the product produced. The numbers C, a, and b are constants.

Cobb–Douglas Production Functions Returns to Scale

The impact of the work done by Cobb and Douglas was so great that these power functions are now referred to as **Cobb-Douglas production functions**. These power functions answer questions of **returns to scale**; that is, questions that ask how a proportionate increase or decrease in all the input values will affect the output value, the total production. If the proportional increase in all the input values is equal to the proportional increase in the output, then the *returns to scale are constant.* For example, if returns to scale are constant, a doubling of the number of units of both labor and capital results in a doubling of production. If the proportional increase in output is greater than the proportional increase in all the inputs, then the *returns to scale are increasing.* Finally, if the proportional increase in output is less than the proportional increase in all the inputs, then the *returns to scale are decreasing.* Cobb–Douglas production functions having the form

$$f(x, y) = Cx^a y^{1-a}$$

where the exponents on the input variables add to 1, always exhibit a constant return to scale.

EXAMPLE SET B

The manufacturing process of a company is described by the Cobb–Douglas production function $f(x, y) = 4x^{3/5}y^{2/5}$.

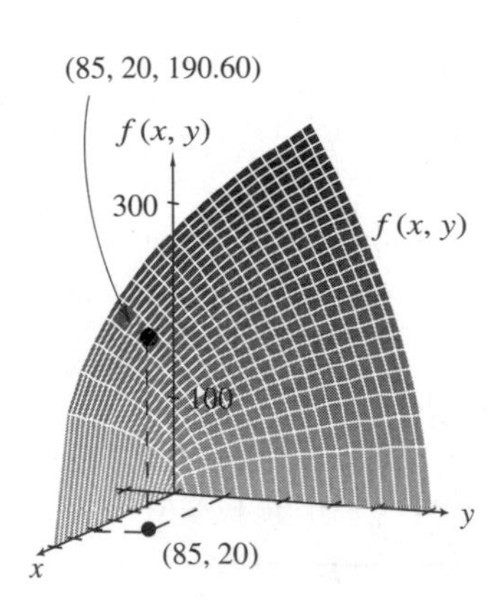

Figure 8.11

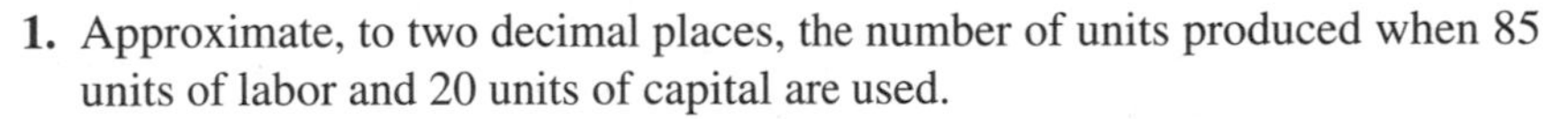

1. Approximate, to two decimal places, the number of units produced when 85 units of labor and 20 units of capital are used.
2. Find the change in the level of production if the number of units of labor is decreased by 1 from 85 units to 84 units, and the number of units of capital is increased by 5 from 20 to 25.
3. Show that this function exhibits a constant return to scale.

Solution:

1. Substituting 85 for x and 20 for y gives

$$f(85, 20) = 4(85)^{3/5}(20)^{2/5}$$
$$\approx 190.60$$

Figure 8.11 shows the graph of this Cobb–Douglas production function.

Interpretation: When 85 units of labor and 20 units of capital are used in a manufacturing process, approximately 190.6 units will be produced.

2. Substituting 84 for x and 25 for y gives

$$f(84, 25) = 4(84)^{3/5}(25)^{2/5}$$
$$\approx 206.92$$

Then, the change in the production level is

$$f(84, 25) - f(85, 20) \approx 206.92 - 190.60$$
$$\approx 16.32$$

Interpretation: If the number of units of labor is decreased by 1 from 85 to 84, and the number of units of capital is increased by 5 from 20 to 25, the level of production will increase by approximately 16.32 units.

3. To show that the return to scale is constant, we need to show that if both the input values x and y are multiplied by some constant, say k, then the output value $f(x, y)$ is multiplied by the same constant. We begin by noting that $f(x, y) = 4x^{3/5}y^{2/5}$. We now multiply each input value by k.

$$\begin{aligned} f(kx, ky) &= 4(kx)^{3/5}(ky)^{2/5} \\ &= 4k^{3/5}x^{3/5}k^{2/5}y^{2/5} \\ &= 4k^{3/5}k^{2/5}x^{3/5}y^{2/5} \\ &= 4k^{3/5+2/5}x^{3/5}y^{2/5} \\ &= 4k^{5/5}x^{3/5}y^{2/5} \\ &= 4kx^{3/5}y^{2/5} \\ &= k \cdot 4x^{3/5}y^{2/5} \\ &= k \cdot f(x, y) \end{aligned}$$

Thus, if both the input values x and y are multiplied by some constant, say, k, then the output value $f(x, y)$ is multiplied by the same constant, and we conclude that this Cobb–Douglas function exhibits a constant return to scale.

Finally, we make a general comment about functions of more than two variables. These functions are very difficult to interpret graphically (try to imagine a fourth axis perpendicular to the other three, etc.). Fortunately for our work and your work, most functions that you will need to deal with are of only one or two variables.

Functions of Several Variables and Technology

Using Your Calculator Your graphing calculator can evaluate functions of several variables. The following example illustrates the evaluation of $f(x, y, z) = 2x^3 - 5y^2 + 3z$ at $x = 2$, $y = 1$, and $z = -5$.

```
Y1 = 2x^3 - 5y^2 + 3z
```

In the computation window, enter

```
2 → X
1 → Y
−5 → Z
Y1
```

The result is -4. That is, $f(2, 1, -5) = -4$.

Using Derive You can also use Derive to evaluate functions of several variables. The following example illustrates the evaluation of $f(x, y, z) = 2x^3 - 5y^2 + 3z$ at $x = 2$, $y = 1$, and $z = -5$.

Author
f(x,y,z) := 2x^3 – 5y^2 + 3z
Author
f(2,1,–5)
Simplify

The first **A**uthor command creates the function $2x^3 - 5y^2 + 3z$, and the second **A**uthor command creates the function value $f(x, y, z)$ at $(2, 1, -5)$.

Derive can also plot the graph of a function of several variables. The following example shows the graph of $f(x, y) = x^2 + y^2$ on $-4 \leq x \leq 5$ and $-5 \leq y \leq 5$.

Author
x^2 + y^2
Options
Display
Graphics
Plot
Overlay
Plot

The **O**verlay command directs Derive to *overlay* a coordinate system onto the computer screen.

Calculator/Derive Exercises

1. A stockbroker charges commission, C, based on the number, x, of shares purchased or sold and the price, y, of each share according to the function

$$C(x, y) = 30 + 0.90x + 0.003xy$$

What is the stockbroker's commission on the purchase of 1500 shares of a stock that sells for \$48.50 per share?

2. A manufacturer subjects its product to three inspections. The cost, C, of repairing a unit depends on the number of defective parts found in each inspection and is given by the function

$$C(x, y, z) = 0.12x^{1.6} + 1.1x^{0.45}y^{0.68}z^{0.33} + 1.45x + 0.25y^{0.25} + 3.4z^{1.6}$$

Find the cost of repairing a unit if 15 defective parts are found in the first inspection, 7 in the second, and 3 in the third.

3. A retail department store bases its projections of customer motivation, M, to purchase on the function

$$M(x, y) = -2.4x^2 - 3.1y^2 + 205x + 217y + 1.6xy - 8{,}955$$

where x represents the Fahrenheit temperature of the store and y the decibel level of music in the store. Find a customer motivation index if the store temperature is 70°F and the music level is 45 decibels.

4. Use Derive to plot the graph of the function $f(x, y) = xy\frac{x^2 - y^2}{x^2 + y^2}$. Begin using the intervals $-10 \leq x \leq 10$ and $-10 \leq y \leq 10$. If these intervals do not produce a satisfactory graph, change them.
5. Use Derive to plot the graph of the function $f(x, y) = \ln(x^2 + 2y^2)$. Begin using the intervals $0 \leq x \leq 10$ and $0 \leq y \leq 10$. If these intervals do not produce a satisfactory graph, change them.
6. Use Derive to plot the graph of the function $f(x, y) = ey - (x^2 + y^2)$. Begin using the intervals $-5 \leq x \leq 10$ and $-5 \leq y \leq 10$. If these intervals do not produce a satisfactory graph, change them.

EXERCISE SET 8.1

A UNDERSTANDING THE CONCEPTS

For Exercises 1–4, answers may vary.

1. *Medicine: AIDS* The number N of AIDS cases in a city each year is known to be a function of the number x of homosexual men living in the city. But it is also a function of other variables. Specify two, name them y and z, and then use several variable function notation to indicate that the number of AIDS cases in a city depends on the identified variables.
2. *Economics: Education* The number N of students attending a community college is a function of the number x of unemployed people in the community. But it is also a function of other variables. Specify another one, name it y, and then use several variable function notation to indicate that the number of students attending a community college depends on the identified two variables.
3. *Animal Science: Anatomy* The surface area A of a mammal is a function of the weight w of the mammal. But it is also a function of other variables. Specify another one, name it y, and then use several variable function notations to indicate that the surface area of a mammal depends on the identified variables.
4. *Manufacturing: Cost* The cost C of producing a precision timing instrument is a function of the cost u of unskilled labor and the cost i of an integrated circuit chip. But it is also a function of other variables. Specify another one, name it with a letter, and then use several variable function notations to indicate that the cost of producing a precision timing instrument depends on the identified variables.

For each of Exercises 5–10, construct a 3-dimensional coordinate system and plot the given point.

5. $(2, 3, 4)$
6. $(4, 2, 2)$
7. $(5, 5, 1)$
8. $(-2, 3, 4)$
9. $(-1, -3, -5)$
10. $(0, 0, 4)$

B SKILL ACQUISITION

For Exercises 11–22, find the exact or approximate (to two decimal places) value of each function.

11. Find $f(1, 4)$ if $f(x, y) = 2x^3 - y^2$.
12. Find $f(3, 2)$ if $f(x, y) = 50x + 25y - 2x^2 - 3y^2$.
13. Find $f(200, 60)$ if $f(x, y) = 4x^{1/3}y^{2/3}$.
14. Find $f(465, 106)$ if $f(x, y) = 25x^{2/5}y^{3/5}$.
15. Find $P(20, 27)$ if $P(s, t) = \sqrt{3s^3 - 2t^2}$.
16. Find $N(15, 7)$ if $N(x, y) = \ln(2x + y) + \frac{y}{x}$.
17. Find $f(5, 8)$ if $f(x, y) = e^{0.01(3x-5y)}\ln(5y - 3x)$.
18. Find $f(2, 1, 3)$ if $f(x, y, z) = e^x e^y \ln(z)$.
19. Find $P(12, 2, 20)$ if $P(x, y, z) = e^{\sqrt{x+3z}} - e^{\sqrt{y+\ln(y)}}$.
20. Find $f(10, 4, 35, 0)$ if $f(x, y, z, w) = \sqrt[3]{\ln(x) + 200e^{-y} + 8z - w}$.
21. Find $f(0, 12)$ if $f(x, y) = y^{e^x + \ln(y)} - 5750$.
22. Find $P(2000, 10)$ if $P(A, t) = Ae^{0.08t}$.

For Exercises 23–24, your answer will be in terms of h.

23. Find $\frac{f(x + h, y) - f(x, y)}{h}$ if $f(x, y) = x^2 - y^2$.
24. Find $\frac{f(x, y + h) - f(x, y)}{h}$ if $f(x, y) = 3x + 5y - 8x^2 - y^2$.

For Exercises 25–26, find the requested limit.

25. Find $\lim_{h\to 0} \frac{f(x, y + h) - f(x, y)}{h}$ if $f(x, y) = y^2 + 4xy$.

26. Find $\lim_{h\to 0} \dfrac{f(x+h,y)-f(x,y)}{h}$ if $f(x,y) = 5x^2 + 2y^2 - 6x + 20$.

C APPLYING THE CONCEPTS

27. ***Economics: Cobb–Douglas Production*** For a particular product, the Cobb–Douglas production function is

$$f(x,y) = 600x^{0.4}y^{0.6}$$

a. Find the number of items produced if 120 units of labor, x, and 50 units of capital, y, are used.
b. Find the change in the level of production if the number of units of labor is increased by 1 unit from 120 to 121 units.
c. Find the change in the level of production if the number of units of capital is decreased by 5 from 50 to 45 units.
d. This function exhibits a constant return to scale. Show that, if both the costs of labor and capital are tripled, the level of production will also be tripled.

28. ***Medicine: Drug Effect*** The effect E of a drug on a human patient is a function of both the amount (in milligrams) of the drug administered and the amount of time t (in hours) that has passed since the drug was administered. For a particular drug, the function that relates these quantities is

$$E(x,t) = 18.5x^{1.20}e^{-0.04t}$$

a. Find the effect that 200 mg of the drug has on a patient 2 hours after it has been administered.
b. Find the change in the effect of the drug on the patient if the dosage had been 190 mg rather than 200 mg.
c. Find the change in the effect of the drug 3 hours after 200 mg has been administered.

29. ***Psychology: Depression*** A psychology researcher believes that the function

$$D(x,y,z) = 10x^{2/3} + 8y^{4/3} - 7z^{3/2} - \ln(x^2 + y^2 - z^2)$$

relates the number D of units (on the researcher's scale) of depression a college student feels on a particular day to the number x of hours over 6 of sleep the student had the previous night, the number y of semester units beyond 15 the student is enrolled in, and the number z of good quality conversations with friends or family the during previous 2 days.

a. Find the number of depression units for a student who the previous night had 13 hours of sleep, who is carrying 21 units, and who had no good quality conversations for the past 2 days with friends or family.
b. What would be the change in the number of depression units for this student if, in the next 2 days, he has 5 good quality conversations with friends or family, but his amount of sleep and number of units he is enrolled in remain constant?

30. ***Business: Waiting Time*** The function

$$W(a,s) = \frac{1}{2(s-a)}, \quad a < s$$

relates the average waiting time W, in hours, in a line to the average service rate s, expressed in the units of the number of customers each hour, and the average customer arrival rate a, expressed in the number of customers per hour.

a. What is the average waiting time in a line if the average service rate is 20 customers per hour and the average arrival rate is 14 customers per hour?
b. What would be the change in the average waiting time if the average service rate were to increase by 3 customers per hour, from 20 to 23, and the average arrival rate were to increase by 5 customers per hour, from 14 to 19 customers per hour?

D DESCRIBE YOUR THOUGHTS

31. Suppose x represents the number of salespeople employed by a company and y the number of units of a particular commodity in stock. If P represents the profit made by this company, explain the difference in the meaning of $P(x)$ and $P(x,y)$.

32. Suppose x represents the daily number of units of fat intake and y the number of daily units of fiber intake. If C represents the cholesterol level of a person, explain the difference in the meaning of $C(x)$ and $C(x,y)$.

33. Suppose x represents the number of housing construction companies in a state, y the number of people in the state, z the rate of inflation in the state, and w the amount of confidence the public has in the economy. If H represents the number of housing sales in a year, explain the difference in the meaning of $H(w)$ and $H(x,y,z,w)$.

34. Suppose x represents the number of units of a toxic chemical introduced into a lake, y the length of the fishing season, z the number of fishermen fishing the lake each season, and w, the amount of rainfall at the lake each year. If N represents the number of trout in the lake at the end of the year, explain the difference in the meaning of $N(x,y)$ and $N(x,y,z,w)$.

E REVIEW

35. **(2.2)** Find, if it exists, $\lim_{x\to 8} \dfrac{x^2 - 9x + 8}{x^2 - 10x + 16}$.

36. **(3.2)** Find $f'(x)$ for $f(x) = \dfrac{3x^2 - 3}{x^2 + 5}$.

37. **(4.1)** Classify, as concave upward or concave downward, a curve that is decreasing at an increasing rate.

38. **(4.1)** Use the graph illustrated below to find the value of x that minimizes the function
$C(x) = \frac{1250}{x} + 3.5x^2.$

C(x)

C(x)

x

5 6

39. **(4.2)** Sketch the graph of a function that increases at an increasing rate when x is less than 3, then decreases at a decreasing rate when x is between 3 and 6, and then increases at an increasing rate when x is greater than 6. At $x = 3$, the function is defined but has an undefined first derivative. The function also approaches a horizontal asymptote as x approaches negative infinity.

40. **(5.3)** Find $f'(x)$ for $f(x) = \ln(4x^2 + 3)$.

41. **(5.4)** Find $f'(x)$ for $f(x) = e^{8x+4}$.

42. **(6.1)** Evaluate $\int \frac{22}{x}\,dx$.

43. **(6.2)** Evaluate $\int \frac{18x}{9x+5}\,dx$.

44. **(6.3)** Evaluate $\int 12xe^{-7x}\,dx$.

8.2 Partial Derivatives

Introduction
The Partial Derivative
Notation for Partial Derivatives
The Geometry of Partial Derivatives
Higher-Order Partial Derivatives
Second-Order Partial Derivatives as Rates of Change
Partial Derivatives and Technology

Introduction

As we know, differentiation is the language of change. A natural question is, "What effect on the output of a function of more than one variable does an increase (or decrease) in *one* of the input variables have?" The answer is given by the *partial derivative* of the function. In this section we examine the notation of a partial derivative and the process of finding one.

The Partial Derivative

We will investigate partial differentiation by finding the partial derivatives of the function $f(x, y) = 7x^2 + 3y^2 + 5xy + 12x - 8y$.

Suppose we wish to know how the function changes as x changes and y is held constant at some particular value, say, k. Substituting k for y in the function produces the function of the one variable x

$$f(x,k) = 7x^2 + 3k^2 + 5xk + 12x - 8k$$

Now, keeping in mind that the derivative of a constant is zero,

$$\frac{d}{dx}f(x,k) = 14x + 0 + 5k + 12 + 0$$

$$= 14x + 5k + 12$$

Now replace k with y and obtain

$$\frac{d}{dx}f(x,y) = 14x + 5y + 12$$

Of course, we could have saved ourselves some energy by not replacing y with k at all, and just *visualizing* y as a constant. We will do this in all the following examples. Since it may be hard to visualize y as a constant (since all through algebra it was a variable), you may want to replace it with k until you feel more comfortable with the process. After all, k looks more like a constant than does y.

Notation for Partial Derivatives

The derivative notations $\frac{d}{dx}$ and $\frac{d}{dy}$ are usually reserved for functions of one variable. For functions of more than one variable, we use

$$\frac{\partial f}{\partial x} \quad \text{or} \quad f_x \quad \text{or} \quad f_x(x,y)$$

to indicate the partial derivative of f with respect to x, and

$$\frac{\partial f}{\partial y} \quad \text{or} \quad f_y \quad \text{or} \quad f_y(x,y)$$

to indicate the partial derivative of f with respect to y.

$\frac{\partial f}{\partial x}$ represents the change in the value of the function with respect to x when the value of x is changed and the value of y is held constant; that is, it represents the rate of change of the function *in the x direction*. $\frac{\partial f}{\partial y}$ represents the change in the value of the function with respect to y when the value of y is changed and the value of x is held constant; that is, it represents the rate of change of the function *in the y direction.*

EXAMPLE SET A

The following examples will illustrate the meaning of $\frac{\partial}{\partial x}f(x,y) = 14x + 5y + 12$ for several different values of x and y.

1.
$$\frac{\partial}{\partial x} f(2, 3) = 14(2) + 5(3) + 12 = 55 \quad \left(\frac{55}{1}\right)$$

Interpretation: If, when $x = 2$ and $y = 3$, x is increased by 1 unit, from 2 to 3, then the value of the function will increase by approximately 55 units.

2.
$$\frac{\partial}{\partial x} f(8, 3) = 14(8) + 5(3) + 12 = 139 \quad \left(\frac{139}{1}\right)$$

Interpretation: If, when $x = 8$ and $y = 3$, x is increased by 1 unit, from 8 to 9, then the value of the function will increase by approximately 139 units.

For the above derivative, we held y constant and differentiated with respect to x. As noted, we can also hold x constant. Using the function from the beginning of this section, $f(x, y) = 7x^2 + 3y^2 + 5xy - 8y$, and differentiating with respect to y, we have

$$\frac{\partial}{\partial y} f(x, y) = 0 + 6y + 5x - 8$$
$$= 6y + 5x - 8$$

(You may find it helpful to verify this by replacing x with k, differentiating with respect to the variable y, and then replacing k with x.)

The Geometry of Partial Derivatives

Figure 8.12 illustrates the geometrical meaning of $\frac{\partial f}{\partial x}$ and $\frac{\partial f}{\partial y}$.

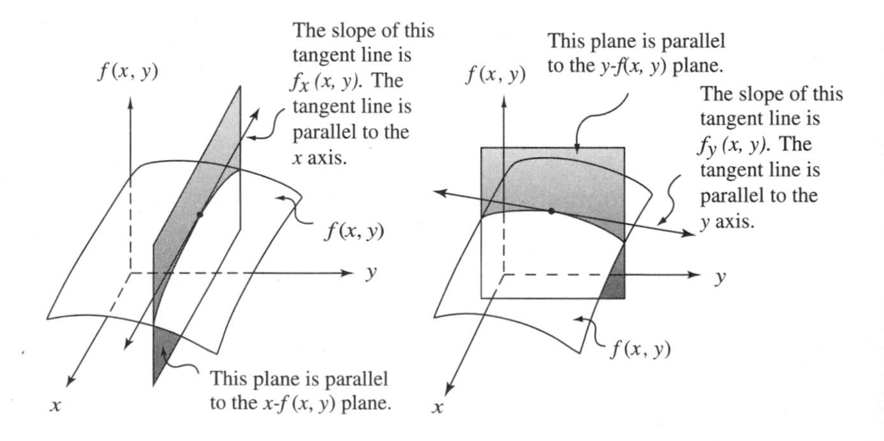

Figure 8.12

In the leftmost illustration of Figure 8.12, the slope of the tangent line to the curve $f(x, y)$ — that lies in a plane that is parallel to the x-$f(x, y)$ plane and is also parallel to the x axis — is given by $f_x(x, y)$ or $\frac{\partial f}{\partial x}$. Similarly, in the rightmost illustration, the slope of the tangent line to the curve $f(x, y)$ — that lies in a plane that is parallel to the y-$f(x, y)$ plane and is also parallel to the y axis — is given by $f_y(x, y)$ or $\frac{\partial f}{\partial y}$.

As you study each of the following examples, you may wish to follow along with pencil and paper. Until you feel confident that you can treat a letter as a constant, you may wish to include the step of substituting k for the value to be held constant.

EXAMPLE SET B

Find $\frac{\partial f}{\partial x}$ and $\frac{\partial f}{\partial y}$ for $f(x, y) = 8x^3 + 5y^2 + 6x - 2y$ and evaluate each at $(2, 7)$.

Solution:

1. To find $\frac{\partial f}{\partial x}$, we keep y constant and differentiate with respect to x.

$$\begin{aligned}\frac{\partial f}{\partial x} &= \frac{\partial}{\partial x}[8x^3] + \frac{\partial}{\partial x}[5y^2] + \frac{\partial}{\partial x}[6x] - \frac{\partial}{\partial x}[2y] \\ &= 24x + 0 + 6 - 0 \\ &= 24x + 6\end{aligned}$$

Thus, $\frac{\partial f}{\partial x} = 24x + 6$, and this rate of change depends only on x. Then

$$\begin{aligned}\left.\frac{\partial f}{\partial x}\right|_{(2,7)} &= 24(2) + 6 \\ &= 54\end{aligned}$$

Interpretation: If, when $x = 2$ and $y = 7$, the value of x is increased by 1 unit, from 2 to 3, the value of the function will increase by approximately 54 units. (Remember, the change in the function value is only approximate since the derivative involves a variable.)

2. To find $\frac{\partial f}{\partial y}$, we keep x constant and differentiate with respect to y.

$$\begin{aligned}\frac{\partial f}{\partial y} &= \frac{\partial}{\partial y}[8x^3] + \frac{\partial}{\partial y}[5y^2] + \frac{\partial}{\partial y}[6x] - \frac{\partial}{\partial y}[2y] \\ &= 0 + 10y + 0 - 2 \\ &= 10y - 2\end{aligned}$$

Thus, $\frac{\partial f}{\partial y} = 10y - 2$, and this rate of change depends only on y. Then

$$\left.\frac{\partial f}{\partial y}\right|_{(2,7)} = 10(7) - 2$$

$$= 68$$

Interpretation: If, when $x = 2$ and $y = 7$, the value of y is increased by 1 unit, from 7 to 8, the value of the function will increase by approximately 68 units.

EXAMPLE SET C

Find and interpret $f_x(2,3)$ and $f_y(2,3)$ for $f(x,y) = x^4 + 8x^2y^3 + 5y^4$.

Solution:

1. To find f_x, we keep y constant and differentiate with respect to x.

$$f_x = \frac{\partial}{\partial x}\left[x^4\right] + \frac{\partial}{\partial x}\left[8x^2y^3\right] + \frac{\partial}{\partial x}\left[5y^4\right]$$

$$= 4x^3 + 8y^3 \cdot 2x + 0$$

$$= 4x^3 + 16xy^3$$

Thus, $f_x(2,3) = 4x^3 + 16xy^3$, and this rate of change depends on the values of both x and y.

$$f_x(2,3) = 4(2)^3 + 16(2)(3)^3$$

$$= 896$$

Interpretation: When the input values are $x = 2$ and $y = 3$, if x is increased by 1 unit from 2 to 3, the output value will increase by approximately 896 units.

2. To find f_y, we keep x constant and differentiate with respect to y.

$$f_y = \frac{\partial}{\partial y}\left[x^4\right] + \frac{\partial}{\partial y}\left[8x^2y^3\right] + \frac{\partial}{\partial y}\left[5y^4\right]$$

$$= 0 + 8x^2 \cdot 3y^2 + 20y^3$$

$$= 24x^2y^2 + 20y^3$$

Thus, $f_y(2,3) = 24x^2y^2 + 20y^3$, and this rate of change depends on the values of both x and y.

$$f_y(2,3) = 24(2)^2(3)^2 + 20(3)^3$$

$$= 1404$$

Interpretation: When the input values are $x = 2$ and $y = 3$, if y is increased by 1 unit from 3 to 4, the output value will increase by approximately 1404 units.

EXAMPLE SET D

Find f_x and f_y for $f(x, y) = (5x^3 - 8y^2)^4$.

Solution:

1. To find f_x, we will keep y constant and differentiate with respect to x. Viewing this function globally, we see a power, so we will begin by using the general power rule.

$$f_x = \underbrace{4 \cdot (5x^3 - 8y^2)^3}_{(outside)'} \cdot \underbrace{(15x^2 - 0)}_{(inside)'}$$

$$= 4(5x^3 - 8y^2)^3(15x^2)$$

$$= 60x^2(5x^3 - 8y^2)^3$$

Thus, $f_x = 60x^2(5x^3 - 8y^2)^3$, and this derivative depends on both x and y.

2. To find f_y, we will keep x constant and differentiate with respect to y. Viewing this function globally, we see a power, so we will begin by using the general power rule.

$$f_y = \underbrace{4 \cdot (5x^3 - 8y^2)^3}_{(outside)'} \cdot \underbrace{(0 - 16y)}_{(inside)'}$$

$$= 4(5x^3 - 8y^2)^3(-16y)$$

$$= -64y(5x^3 - 8y^2)^3$$

Thus, $f_x = -64y(5x^3 - 8y^2)^3$, and this derivative depends on both x and y.

EXAMPLE SET E

Find f_x, f_y, and f_z for $f(x, y, z) = e^{x+3z} + 5\ln(xyz)$.

Solution:

1. To find f_x, we will treat both y and z as constants and differentiate with respect to x.

$$f_x = \frac{\partial}{\partial x}\left[e^{x+3z}\right] + \frac{\partial}{\partial x}[5\ln(xyz)]$$

$$= e^{x+3z} \cdot \frac{\partial}{\partial x}[x + 3z] + 5 \cdot \frac{1}{xyz} \cdot \frac{\partial}{\partial x}[xyz]$$

$$= e^{x+3z} \cdot (1) + 5\frac{1}{xyz} \cdot (yz)$$

$$= e^{x+3z} + \frac{5}{x}$$

Thus, $f_x = e^{x+3z} + \frac{5}{x}$, and its value depends on the values of both x and z.

2. To find f_y, we will treat both x and z as constants and differentiate with respect to y.

$$f_y = \frac{\partial}{\partial y}\left[e^{x+3z}\right] + \frac{\partial}{\partial y}[5\ln(xyz)]$$

$$= e^{x+3z} \cdot \frac{\partial}{\partial y}[x+3z] + 5 \cdot \frac{1}{xyz} \cdot \frac{\partial}{\partial y}[xyz]$$

$$= e^{x+3z} \cdot (0) + 5\frac{1}{xyz} \cdot (xz)$$

$$= \frac{5}{y}$$

Thus, $f_y = \frac{5}{y}$, and its value depends only on the value of y.

3. To find f_z, we will treat both x and y as constants and differentiate with respect to z.

$$f_z = \frac{\partial}{\partial z}\left[e^{x+3z}\right] + \frac{\partial}{\partial z}[5\ln(xyz)]$$

$$= e^{x+3z} \cdot \frac{\partial}{\partial z}[x+3z] + 5 \cdot \frac{1}{xyz} \cdot \frac{\partial}{\partial z}[xyz]$$

$$= e^{x+3z} \cdot (3) + 5\frac{1}{xyz} \cdot (xy)$$

$$= 3e^{x+3z} + \frac{5}{z}$$

Thus, $f_z = 3e^{x+3z} + \frac{5}{z}$, and its value depends on the values of both x and z.

EXAMPLE SET F

Find all the points (x, y) for which both f_x and f_y equal zero, where

$$f(x,y) = x^2 + y^2 - xy + y - 8$$

Solution:
We begin by finding each partial derivative.

$$f_x = 2x - y \text{ and } f_y = 2y - x + 1$$

To find the points where $f_x = 0$ and $f_y = 0$, we need to solve the system

$$\begin{cases} f_x = 0 & \ldots(1) \\ f_y = 0 & \ldots(2) \end{cases} \quad \text{that is,} \quad \begin{cases} 2x - y = 0 & \ldots(1) \\ 2y - x + 1 = 0 & \ldots(2) \end{cases}$$

We will solve this system using the method of elimination by substitution. Solving equation (1) for y, we get $y = 2x$. Then, substituting $2x$ for y in equation (2), we get

$$2(2x) - x + 1 = 0$$

$$x = -\frac{1}{3}$$

Then, since $y = 2x$ and $x = -\frac{1}{3}$, we get $y = 2(-\frac{1}{3}) = -\frac{2}{3}$. Thus, for the function $f(x, y) = x^2 + y^2 - xy + y - 8$, both partial derivatives f_x and f_y equal zero at the point $\left(-\frac{1}{3}, -\frac{2}{3}\right)$.

Higher-Order Partial Derivatives

Just as it was useful to differentiate the derivative $f'(x)$ of a function of one variable $f(x)$, it is useful to differentiate the partial derivatives of a function of two variables $f(x, y)$. Since a function of two variables has *two* partial derivatives, and each partial derivative may, in turn, be a function of two variables, each partial derivative can be differentiated. Thus, a function $f(x, y)$ of two variables has four partial derivatives.

Notation:

$$\frac{\partial}{\partial x}\left(\frac{\partial f}{\partial x}\right) = \frac{\partial^2 f}{\partial x^2} = f_{xx}$$

$$\frac{\partial}{\partial x}\left(\frac{\partial f}{\partial y}\right) = \frac{\partial^2 f}{\partial x \partial y} = f_{yx}$$

$$\frac{\partial}{\partial y}\left(\frac{\partial f}{\partial y}\right) = \frac{\partial^2 f}{\partial y^2} = f_{yy}$$

$$\frac{\partial}{\partial y}\left(\frac{\partial f}{\partial x}\right) = \frac{\partial^2 f}{\partial y \partial x} = f_{xy}$$

Second-Order and Mixed Partial Derivatives

The partial derivatives f_{xx} and f_{yy} are called **second-order partial derivatives** of f with respect to x and y, respectively. The partial derivatives f_{yx} and f_{xy} are called **mixed partial derivatives**.

It is important to note the order of differentiation implied by the notations $\frac{\partial^2 f}{\partial y \partial x}$, f_{xy} and $\frac{\partial^2 f}{\partial x \partial y}$, f_{yx}.

1. $\underbrace{f_{xy}}_{left-to-right} = \underbrace{\frac{\partial^2 f}{\partial y \partial x}}_{right-to-left}$ and indicates that f is to be differentiated first with respect to x, then with respect to y, and

2. $\underbrace{f_{yx}}_{left-to-right} = \underbrace{\frac{\partial^2 f}{\partial x \partial y}}_{right-to-left}$ and indicates that f is to be differentiated first with respect to y, then with respect to x.

EXAMPLE SET G

Find the second partial derivatives of $f(x, y) = x^3 + 2x^2y - 5xy^2 + 3y^3$.

Solution:
We begin by finding the first partial derivatives.

$$f_x = \frac{\partial}{\partial x}(x^3 + 2x^2y - 5xy^2 + 3y^3) = 3x^2 + 4xy - 5y^2, \quad \text{and}$$

$$f_y = \frac{\partial}{\partial y}(x^3 + 2x^2y - 5xy^2 + 3y^3) = 2x^2 - 10xy + 9y^2, \quad \text{so that}$$

$$f_x = 3x^2 + 4xy - 5y^2 \quad \text{and} \quad f_y = 2x^2 - 10xy + 9y^2$$

Then, taking the derivatives of these functions with respect to x and y,

$$f_{xx} = \frac{\partial}{\partial x}(f_x) = \frac{\partial}{\partial x}\left(3x^2 + 4xy - 5y^2\right) = 6x + 4y, \quad \text{and}$$

$$f_{yy} = \frac{\partial}{\partial y}(f_y) = \frac{\partial}{\partial x}\left(2x^2 - 10xy + 9y^2\right) = -10x + 18y, \quad \text{and}$$

$$f_{xy} = \frac{\partial}{\partial y}(f_x) = \frac{\partial}{\partial y}\left(3x^2 + 4xy - 5y^2\right) = 4x - 10y, \quad \text{and}$$

$$f_{yx} = \frac{\partial}{\partial x}(f_y) = \frac{\partial}{\partial x}\left(2x^2 - 10xy + 9y^2\right) = 4x - 10y$$

Thus, $f_{xx} = 6x + 4y$, $f_{yy} = -10x + 18y$, $f_{xy} = 4x - 10y$, $f_{yx} = 4x - 10y$.

Notice that f_{xy} and f_{yx} are equal. This is no coincidence; it turns out that all functions of two variables for which the mixed partial derivatives are continuous on an open set (such as the interior of a circle) have equal mixed partial derivatives at any point in the open set. All the functions we will examine in this text meet such conditions and will have equal mixed partial derivatives.

Second-Order Partial Derivatives as Rates of Change

As the second derivative $f''(x)$ measures the concavity of the curve $f(x)$, f_{xx} and f_{yy} measure the concavity in the x direction and the y direction, respectively, of the surface $f(x, y)$. In fact, the signs of f_x and f_{xx}, and f_y and f_{yy}, give us information about the behavior of the function. (This is analogous to our discussion in Section 3.3. You may wish to examine that discussion again.)

Two partial derivatives of the same kind, f_x, f_{xx}, and f_y, f_{yy}, with the *same sign* indicate that the function f changes at an *increasing rate.*

$f_x > 0,\ f_{xx} > 0$ $\rightarrow f$ is increasing at an increasing rate.
$f_y > 0,\ f_{yy} > 0$ $\rightarrow f$ is increasing at an increasing rate.
$f_x < 0,\ f_{xx} < 0$ $\rightarrow f$ is decreasing at an increasing rate.
$f_y < 0,\ f_{yy} < 0$ $\rightarrow f$ is decreasing at an increasing rate.

Two partial derivatives of the same kind, f_x, f_{xx}, and f_y, f_{yy}, with *opposite signs* indicate that the function f changes at a *decreasing rate.*

$f_x > 0,\ f_{xx} < 0$ $\rightarrow f$ is increasing at a decreasing rate.
$f_y > 0,\ f_{yy} < 0$ $\rightarrow f$ is increasing at a decreasing rate.
$f_x < 0,\ f_{xx} > 0$ $\rightarrow f$ is decreasing at a decreasing rate.
$f_y < 0,\ f_{yy} > 0$ $\rightarrow f$ is decreasing at a decreasing rate.

EXAMPLE SET H

To produce batteries, a manufacturer uses x units of chemical A and y units of chemical B. The amount P of pollution washed into the cleaning water is given by the pollution function $P(x, y) = 0.05x^2 + 0.009xy + 0.03y^2$. If the company is currently using 16 units of chemical A and 10 units of chemical B, find the rate at which the amount of pollution washed into the cleaning water is changing as the number of units of chemical B changes. Determine if this rate is changing at an increasing or a decreasing rate.

Solution:

1. To find the rate at which the amount of pollution is changing as the amount of chemical B changes, we need to find

$$\frac{\partial}{\partial y} P(x, y) = \frac{\partial}{\partial y}(0.05x^2 + 0.009xy + 0.03y^2).$$

$$\frac{\partial}{\partial y} P(x, y) = \frac{\partial}{\partial y}(0.05x^2 + 0.009xy + 0.03y^2)$$

$$= 0.009x + 0.06y \qquad \text{and}$$

$$\frac{\partial}{\partial y} P(16, 10) = 0.009(16) + 0.06(10)$$

$$= 0.744$$

Thus, $\frac{\partial}{\partial y} P(16, 10) = 0.744$.

2. To find if this rate is changing at an increasing or a decreasing rate, we need to find and interpret $\frac{\partial^2}{\partial y^2} P(16, 10)$.

$$\frac{\partial^2}{\partial y^2} P(x, y) = \frac{\partial}{\partial y}(0.009x + 0.06y)$$
$$= 0.06$$

Thus, $\frac{\partial^2}{\partial y^2} P(16, 10) = 0.06$. Since 0.06 is always positive, the rate is changing at an increasing rate.

Interpretation: If, when 16 units of chemical A and 10 units of chemical B are being used to produce batteries, the number of units of chemical B is increased by 1, from 10 units to 11, then the amount of pollutant washed into the cleaning water will increase by approximately 0.744 units. Furthermore, at this point, the amount of pollution is increasing at an increasing rate.

Partial Derivatives and Technology

Using Your Calculator You can use your graphing calculator to approximate partial derivatives. The following entries show the computation of $f(4, 7)$, $f_x(4, 7)$, and $f_y(4, 7)$ for the function $f(x, y) = x^2 + y^2 + 6$.

```
Y1 = x^2 + y^2 + 6
Y8 = nDeriv(Y1,X,X,.001)
Y9 = nDeriv(Y1,Y,Y,.001)
```

Then, in the computation window, enter

```
4 → X
7 → Y
Y1
Y8
Y9
```

The results are, respectively, 71, 8, and 14.

Calculator Exercises

1. For a manufacturer, the Cobb–Douglas production function is

$$f(x, y) = 35x^{0.43}y^{0.57}$$

Find and interpret both $\frac{\partial f}{\partial x}$ and $\frac{\partial f}{\partial y}$ when 15 units of labor are used ($x = 15$) and 21 units of capital are used ($y = 21$). Explain why it is better to increase the amount spent on labor by 1 unit rather than the amount spent on capital by 1 unit. Estimate the change in the level of production if the amount spent on labor increases by 3 units.

2. A manufacturer produces x units of product X and y units of product Y. The cost function is

$$C(x, y) = \sqrt{0.6x^{1.2}y^{0.8} + 14.2}$$

and the revenue function is

$$R(x, y) = 2.6x^{1.1} + 1.3y^{1.2}$$

The profit function is given by $P(x, y) = R(x, y) - C(x, y)$. Find and interpret $P(600, 420)$, $\frac{\partial P}{\partial x}$, and $\frac{\partial P}{\partial y}$ for the function when $x = 600$ and $y = 420$.

3. A county estimates that the percentage P of residents applying for welfare in a particular year depends on both the rate of unemployment x, in percent, in the county and the average number of years of education y of the residents. The percentage is given by the function

$$P(x, y) = e^{0.213x} - e^{0.013y}$$

Find and interpret $P(6, 13)$, $\frac{\partial P}{\partial x}$, and $\frac{\partial P}{\partial y}$ for the function when $x = 6$ and $y = 13$.

4. Find and interpret both $\frac{\partial f}{\partial x}$ and $\frac{\partial f}{\partial y}$ for the function

$$f(x, y) = x^{2/3} + y^{-1/3}$$

when $x = 10$ and $y = 2$.

Using Derive Derive can produce the symbolic form of the derivative of a function and then evaluate it at given values. The following set of commands illustrate the computation of $f(2, 3), f_x(x, y), f_y(x, y), f_x(2, 3)$, and $f_y(2, 3)$ for the function $f(x, y) = 5x^2 - 2y^2 + 5$. We have broken these commands into several sets so we can talk about what Derive is doing. However, you should enter them as one set.

Algebra
Author
f(x,y) := 5x^2 – 2y^3 + 5
Calculus
Differentiate (*Press* Enter *twice*)
Simplify

These first few commands access Derive's algebra package, define the function, access Derive's calculus package, and then differentiate and simplify the function.

Manage
Substitute
2
Simplify

The above set of commands substitutes 2 for x and then simplifies the expression.

Calculus
Differentiate (*Press the number of the expression* $F(x, y) := 5x^2 - 2y^3$ *and press* Enter)
y (*Press* Enter)
Simplify
Manage
Substitute

3
Simplify
Author
Substitute
f(2,3)
Simplify

Derive Exercises

1. For the function $f(x, y) = 2x^4 + 6y^2 + 3x - y + 5$, use Derive to produce $f(4, 2)$, $f_x(x, y)$, $f_y(x, y)$, $f_x(4, 2)$, and $f_y(4, 2)$.
2. For the function $f(x, y) = e^{-xy} + e^{y/x}$, use Derive to produce $f(1, 0)$, $f_x(x, y)$, $f_y(x, y)$, $f_x(1, 0)$, and $f_y(1, 0)$.
3. The commission, C, charged by a stockbroker is related to the number of shares x traded and the share selling price y by the function

$$C(x, y) = 35 + 0.12x + 0.008xy$$

Find an expression that indicates how the broker's commission changes as the number of shares traded changes. Find an expression that indicates how the broker's commission changes as the price of a share changes. If the broker sells 350 shares at $45 per share, how would his commission change if the price per share increased by $1 and the number of shares traded remained the same?

EXERCISE SET 8.2

A UNDERSTANDING THE CONCEPTS

1. Suppose that $P(x, y)$ is a function of the two variables x and y. Interpret

$$\frac{\partial}{\partial x}P(3, 8) = 12 \text{ and } \frac{\partial}{\partial y}P(5, 6) = -9.2$$

2. Suppose that $N(x, y)$ is a function of the two variables x and y. Interpret

$$\frac{\partial}{\partial x}N(100, 600) = 30 \text{ and}$$

$$\frac{\partial}{\partial y}N(80, 420) = 16$$

3. Suppose that $T(x, y, z)$ is a function of the three variables x, y, and z. Interpret

$$\frac{\partial}{\partial x}T(2, 3, 11) = 0.06,$$

$$\frac{\partial}{\partial y}T(3, 4, 10) = 0.008, \text{ and } \frac{\partial}{\partial z}T(5, 5, 8) = 0.4$$

4. Suppose that $S(x, y, z, w, m)$ is a function of the five variables x, y, z, w, and m. Interpret

$$\frac{\partial}{\partial y}S(2, 1, 2, 3, 3) = -65$$

5. ***Business: Profit*** Suppose $P(x, y)$ represents the monthly profit for a company when x thousands of dollars are spent on advertising and y number of salespeople are working. What information about the profit of this company is contained in the inequalities $\frac{\partial}{\partial x}P(15, 6) > 0$ and $\frac{\partial^2}{\partial x^2}P(15, 6) < 0$?
6. ***Economics: World Bank Loans*** A scientific organization believes that the function $N(x, y)$ relates the number of acres of destroyed rain forest to x, the amount of money (in millions of dollars) the World Bank loans to the country in which the forest is located, and y, the population (in millions of people) of the country. What information about the rate of destruction of the country's rain forest is contained in the inequalities $\frac{\partial}{\partial y}N(40, 70) > 0$ and $\frac{\partial^2}{\partial y^2}N(40, 70) > 0$?
7. ***Biology: Blood Flow*** The function $R(l, r)$ relates the resistance to blood flow in a vessel to the length l (in centimeters) and the radius r (in centimeters) of the vessel. What information about blood flow in the vessel is contained in the inequalities $\frac{\partial}{\partial r}R(10, 0.2) < 0$ and $\frac{\partial^2}{\partial r^2}R(10, 0.2) > 0$?
8. ***Psychology: IQ*** A person's IQ (intelligence quotient) is a function of the person's mental age M (in years)

and chronological age C (in years); that is, $IQ(M, C)$. What information about a person's IQ is contained in the inequalities $\frac{\partial}{\partial c} IQ(M, C) < 0$ and $\frac{\partial^2}{\partial c^2} IQ(M, C) > 0$?

9. ***Business: Quality Control Costs*** For a high-tech electronics company, the cost C of quality control is related to the number of inspections x, y, and z made at three inspection points A, B, and C. What information about the cost C of quality control is contained in the inequalities $C_y(x, y, z) > 0$ and $C_{yy}(x, y, z) > 0$?

10. ***Medicine: Illness Recovery Time*** The time T (in days) it takes a person to recover from pneumonia is related to the age t (in years) of the person and the number n of units of penicillin the person receives each day; that is, $T(t, n)$. What information about the time it takes a person to recover from pneumonia is contained in the inequalities $T_t(t, n) < 0$ and $T_{tt}(t, n) < 0$?

11. ***Social Science: Welfare*** The proportion P of the population applying for welfare each month in a particular state is related to the monthly unemployment rate x and the average number of years of education y of the adults in the state; that is, $P(x, y)$. What information about the proportion of the population applying for unemployment is contained in the inequalities $P_x(x, y) > 0$ and $P_{xx}(x, y) > 0$?

12. ***Social Science: City Arson*** Suppose a study shows that the number N of arsons in a city is related to the concentration x of residents of public housing in the city and the amount y (in thousands of dollars) spent by the city on maintenance of the housing. What information about the number of arsons is contained in the inequalities $\frac{\partial}{\partial x} N(x, y) > 0$ and $\frac{\partial^2}{\partial x^2} N(x, y) > 0$?

13. ***Medicine: Radiation Treatment*** In the treatment of cancer, the dose D of radiation is related to the width w of the radiating ray and the depth d of penetration of the ray; that is, $D(w, d)$. What information about the dose of radiation in the treatment of cancer is contained in the inequalities $\frac{\partial}{\partial w} D(w, d) > 0$ and $\frac{\partial^2}{\partial w^2} D(w, d) > 0$?

For Exercises 14 and 15, recall that the Cobb–Douglas production function $f(x, y) = Cx^a y^{1-a}$ relates the number of units of labor x and the number of units of capital y used in a particular production process to the number of units f that are produced from the process.

14. ***Business: Cobb–Douglas Production*** Suppose that for a company that produces kitchen sink faucets, the production function is $f(x, y) = x^{0.3}y^{0.7}$. For this function, $\frac{\partial}{\partial x} f = 0.3x^{-.7}y^{0.7}$ and $\frac{\partial^2}{\partial x^2} f = -0.21x^{-1.7}y^{0.7}$. What information about production is contained in the partial derivatives $\frac{\partial}{\partial x} f(20, 40)$ and $\frac{\partial^2}{\partial x^2} f(20, 40)$? Round your answer to two decimal places.

15. ***Business: Cobb–Douglas Production*** Suppose that for a company that produces tool sets, $f(x, y) = x^{0.25}y^{0.75}$. For this function, $\frac{\partial}{\partial y} f = 0.75x^{0.25}y^{-.25}$ and $\frac{\partial^2}{\partial x^2} f = -0.1875x^{0.25}y^{-1.25}$. What information about production is contained in the partial derivatives $\frac{\partial}{\partial y} f(30, 100)$ and $\frac{\partial^2}{\partial y^2} f(30, 100)$? Round your answer to three decimal places.

16. ***Economics: Demand/Price*** The demand for a compact disc player is given by $D(p_1, p_2)$, where p_1 is the price of the compact disc player and p_2 is the price of compact discs. Interpret $\frac{\partial D}{\partial p_1} < 0$ and $\frac{\partial D}{\partial p_2} < 0$.

17. ***Economics: Demand/Price*** The demand for a particular type of graphing calculator depends on both the price p_1 of the calculator and the price p_2 of its only competitor. The demand is given by $D(p_1, p_2)$. Interpret $\frac{\partial D}{\partial p_1} < 0$ and $\frac{\partial D}{\partial p_2} > 0$.

18. ***Business: Number of Customers*** The number of customers at a large supermarket depends on the average price p_1 of its items, the average number n_1 of checkers, the average price p_2 of its nearest competitor's items, and average number n_2 of its competitor's checkers. The number of customers is given by $N(p_1, n_1, p_2, n_2)$. Construct inequality statements regarding the first-order partial derivatives of $N(p_1, n_1, p_2, n_2)$.

19. ***Psychology: Student Performance*** A psychologist believes that a student's performance, P, on a standardized exam depends on the number n of practice exams similar to the standardized exam the student takes and the student's anxiety level a. The student's performance is given by the function $P(n, a)$. The psychologist believes that performance is enhanced by using practice exams and is harmed by heightened levels of anxiety. Construct inequality statements regarding the first-order partial derivatives of $P(n, a)$.

20. ***Economics: Consumption*** A family's annual consumption of fish depends on the family's total income i, the average price p of fish, the average price m of red meat, and the number n of people in the family. The annual consumption is given by the function $C(i, p, m, n)$. Construct inequality statements regarding the first-order partial derivatives of $C(i, p, m, n)$.

B SKILL ACQUISITION

21. For $f(x, y) = 5x^2 + 6xy + 8y^3$, find

a. $\frac{\partial f}{\partial x}$ b. $\frac{\partial f}{\partial y}$
c. $\frac{\partial}{\partial x} f(2, 1)$ d. $\frac{\partial}{\partial y} f(4, 2)$

22. For $f(x, y) = 6x^2 + 2y^2 - 2xy + 25$, find
a. $\frac{\partial f}{\partial x}$ b. $\frac{\partial f}{\partial y}$
c. $\frac{\partial}{\partial x} f(-1, 2)$ d. $\frac{\partial}{\partial y} f(2, -2)$

23. For $f(x, y) = x^3 - 4y^2 + 3x^3y^2$, find
a. f_x b. f_y
c. $f_x(0, -1)$ d. $f_y(-2, 1)$

24. For $f(x, y) = 10x + 2y - x^2 - y^2 + 4x^2y^4$, find
a. f_x b. f_y
c. $f_x(4, 1)$ d. $f_y(0, 2)$

25. For $f(x, y) = e^{x+y}$, find
a. f_x b. f_y
c. $f_x(1, 1)$ d. $f_y(2, 1)$

26. For $f(x, y) = e^{2x-y}$, find
a. f_x b. f_y
c. $f_x(2, 0)$ d. $f_y(2, 0)$

27. For $f(x, y) = \ln(2 + 4x^2y^2)$, find
a. f_x b. f_y
c. $f_x(0, 0)$ d. $f_y(0, 0)$

28. For $f(x, y) = \ln(3x^4 - 4x^3)$, find
a. f_x b. f_y
c. $f_x(1, 1)$ d. $f_y(-1, -1)$

29. For $f(x, y) = \frac{x^2 + 3y}{5x - 3y^2}$, find
a. f_x b. f_y
c. $f_x(0, 2)$ d. $f_y(0, 1)$

30. For $f(x, y) = \frac{6x^2 - 6y^2}{x^2 + y^2}$, find
a. f_x b. f_y
c. $f_x(1, 1)$ d. $f_y(1, 1)$

31. For $f(x, y) = x^2e^{3y}$, find
a. f_x b. f_y
c. $f_x(0, 0)$ d. $f_y(0, 0)$

32. For $f(x, y) = (y^2 - 3)\ln(x + e^y)$, find
a. f_x b. f_y
c. $f_x(0, 0)$ d. $f_y(0, 0)$

33. For $f(x, y) = 2x^2 + 5y^2 - 3x - 4y + 10$, find
a. f_{xx} b. f_{yy}
c. f_{xy} d. f_{yx}

34. For $f(x, y) = 8x^3 + 5y^3 - 8x^2 - y^2 + 100$, find
a. f_{xx} b. f_{yy}
c. f_{xy} d. f_{yx}

35. For $f(x, y) = 10x^2 + 5y^2 + 4xy - 8$, find
a. f_{xx} b. f_{yy}
c. f_{xy} d. f_{yx}

36. For $f(x, y) = x^2 - 25xy + 2y^2 + 180$, find
a. f_{xx} b. f_{yy}
c. f_{xy} d. f_{yx}

37. For $f(x, y) = 9x^2 - 6xy + y^2 + 1400$, find
a. f_{xx} b. f_{yy}
c. f_{xy} d. f_{yx}

38. For $f(x, y) = -5xe^y$, find
a. f_{xx} b. f_{yy}
c. f_{xy} d. f_{yx}

39. For $f(x, y) = 8ye^{2x}$, find
a. f_{xx} b. f_{yy}
c. f_{xy} d. f_{yx}

40. For $f(x, y) = 3x^2 + 5xz + 2z^3$, find
a. f_{xx} b. f_{yy}
c. f_z d. f_{zy}

41. For $f(x, y) = 6y^3 - 6xy - z^2 + 15$, find
a. f_{xx} b. f_{yy}
c. f_z d. f_{zy}

42. For the function $f(x, y) = 4x^2 + 3y^2 + 2xy - 1$, find the values of x and y so that both $f_x(x, y) = 0$ and $f_y(x, y) = 0$.

43. For the function $f(x, y) = x^2 + 7y^2 + 3x + 4y + 10$, find the values of x and y so that both $f_x(x, y) = 0$ and $f_y(x, y) = 0$.

44. For the function $f(x, y) = 5x^2 + 3y^2 - x^3 - y^3$, find the values of x and y so that both $f_x(x, y) = 0$ and $f_y(x, y) = 0$.

45. For the function $f(x, y) = 500 + 3x + 2y - x^3 - y^3$, find the values of x and y so that both $f_x(x, y) = 0$ and $f_y(x, y) = 0$.

C APPLYING THE CONCEPTS

46. ***Business: Revenue*** The total revenue realized by a company for the sale of x units of product A and y units of product B is $R(x, y) = 350x + 600y - 4x^2 - 3y^2$.

a. Specify the total revenue if 20 units of product A and 15 units of product B are sold.
b. How will the company's revenue change if x increases one unit, from 20 to 21, while y remains constant at 15?
c. How will the company's revenue change if y decreases one unit, from 15 to 14, while x remains constant at 20?

47. ***Business: Number of Sales*** The number of units of a particular product that a company sells each month depends on the number of thousands of dollars spent on advertisements in newspapers x, on television y, and on radio z, and is given by the function $N(x, y, z) = 400x + 550y + 120z - 20x^2 - 20y^2 - 10z^2$.
   a. Specify the total number of units sold if the company spends \$2000 on newspaper ads, \$8000 on TV ads, and \$3000 on radio ads.
   b. How will the number of units sold change if the amount spent on newspaper ads is increased by \$1000, from \$2000 to \$3000?
   c. How will the number of units sold change if the amount spent on radio ads is decreased by \$1000, from \$3000 to \$2000?

48. ***Animal Science: Oxygen Consumption*** The oxygen consumption C of a well insulated, nonsweating animal is approximately related to the animal's internal body temperature, T (in °C), the temperature F (in °C) of the animal's fur, and the animal's weight W (in kilograms (kg)) by the function $C(T, F, W) = 2.5(T - F)W^{-0.67}$.
   a. Find the oxygen consumption of an animal weighing 36 kg, and having an internal temperature of 40°C, and a fur temperature of 22°C.
   b. How will the animal's oxygen consumption change if its fur temperature is increased by 1°C, from 22°C to 23°C, and its internal temperature and weight remain constant at 40°C and 36 kg, respectively?
   c. How will the animal's oxygen consumption change if its fur and internal temperature remain constant at 22°C and 40°C, respectively, but its weight decreases by 1 kg, from 36 kg to 35 kg?

49. ***Psychology: IQ*** A person's IQ (intelligence quotient) is a function of the person's mental age m (in years) and chronological age c (in years) and is defined by the function $IQ(m, c) = \dfrac{100m}{c}$.
   a. Find the IQ of a 22-year-old person who has a mental age of 30 years.
   b. How will a person's IQ change if his chronological age increases by one year, from 22 years to 23 years, but his mental age remains constant at 30 years?

50. ***Business: Cobb–Douglas Production*** Suppose that x units of labor and y units of capital are needed to produce $f(x, y) = 75x^{2/3}y^{1/3}$ units of a particular commodity.
   a. Find the number of units produced if 27 units of labor and 64 units of capital are used.
   b. How will the number of units produced change if the number of units of labor used increases by 1 unit, from 27 units to 28?
   c. How will the number of units produced change if the number of units of capital used decreases by 1 unit, from 64 units to 63?

D DESCRIBE YOUR THOUGHTS

51. Describe how a partial derivative differs from an ordinary derivative.

52. Explain what is meant about a function $f(x, y)$ if $f_x > 0$ and $f_{xx} > 0$.

53. Describe how $f_x(x, y)$ differs geometrically from $f_y(x, y)$.

E REVIEW

54. **(2.2)** Find, if it exists, $\lim\limits_{x \to 3} \dfrac{x^2 - 3x - 4}{x - 3}$.

55. **(2.6)** Find the average rate of change of the function $f(x) = 5x^2 - 2x - 12$ as x changes from $x = 5$ to $x = 9$.

56. **(3.2)** Find $f'(x)$ for $f(x) = \dfrac{3x + 5}{2x - 1}$.

57. **(3.3)** Find $f'(x)$ and $f''(x)$ for the function $f(x) = 6x^3 - 8x^2 + 7x - 4$

58. **(4.1)** Construct a summary table for the function illustrated below.

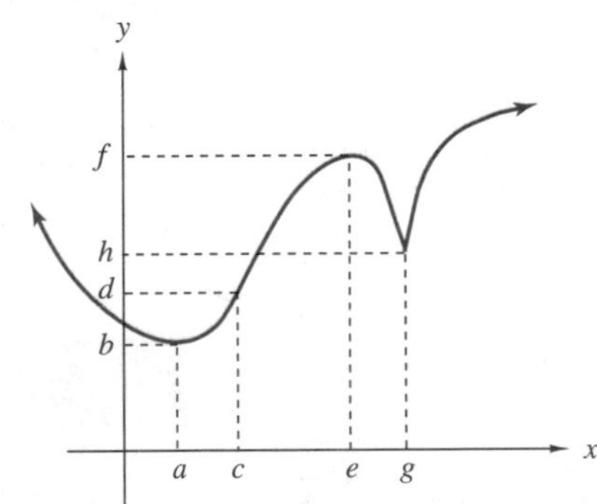

59. **(4.4)** A rectangular box with no top is to be made so that it has a square base and a volume of 108 cubic inches. What should be the dimensions of the box so that it has a minimal surface area, and what is that surface area?

60. **(5.4)** Find $f'(x)$ for $f(x) = e^{3x+1} + 2\ln x^3$.

61. **(7.1)** Find the area completely enclosed by $f(x) = x+2$ and $g(x) = x^2 - 4$.

62. **(7.2)** The demand for a particular item is given by the demand function $D(x) = 200 - x^2$. Find the consumer's surplus if the equilibrium point (x_e, p_e) is $(10, 100)$.

63. **(8.1)** Compute $f(1, 0, 1)$ for $f(x, y, z) = xe^{y/3z}$.

64. **(8.1)** Suppose that $R(x, y) = 90x + 135y - 0.04x^2 - 0.02y^2$ represents the revenue realized by a company on the sale of x units of product A and y units of product B. Find and interpret $R_x(20, 30)$.

8.3 Optimization of Functions of Two Variables

Introduction
Relative Extrema
Location of Relative Extrema/Critical Points
The Second Derivative Test
Optimization and Technology

Introduction

Just as derivatives of one-variable functions are helpful in locating relative extreme points, partial derivatives are helpful in locating relative extreme points of functions of several variables. Since it is functions of two variables that most often occur in business, life science, and social science situations, we will restrict our attention to two-variable functions for which all second-order partial derivatives exist and are continuous.

Relative Extrema

We begin with the definition of relative extrema for a function of two variables.

Relative Extrema

The point (a, b) produces a **relative maximum** of the function $f(x, y)$ if for every point (x, y) *near* $(a, b), f(a, b) \geq f(x, y)$.
The point (a, b) produces a **relative minimum** of the function $f(x, y)$ if for every point (x, y) *near* $(a, b), f(a, b) \leq f(x, y)$.

The leftmost graph in Figure 8.13 illustrates a function with a relative maximum at the point $(a, b, f(a, b))$. Notice that all the points (x, y) near (in the small disk) the point (a, b) produce function values smaller than that produced by the point (a, b). The rightmost picture in Figure 8.13 illustrates a function with a relative minimum at $(a, b, f(a, b))$. Notice that all the points (x, y) near (in the small disk) the point (a, b) produce function values greater than that produced by the point (a, b).

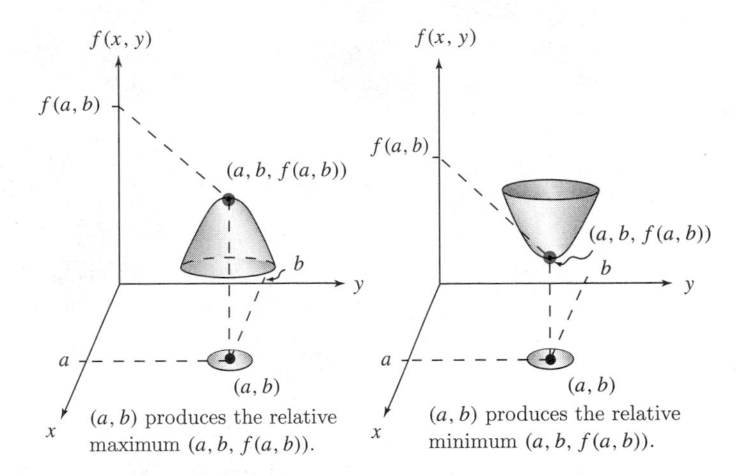

Figure 8.13

Figure 8.14 displays a function of two variables with a relative maximum at $(a, b, f(a, b))$ and in which the tangent lines in both the x and y directions have been sketched in. Notice that each of the tangent lines is horizontal relative to the xy plane. This means, of course, that each has zero slope; that is, that $f_x(a, b) = 0$ and $f_y(a, b) = 0$.

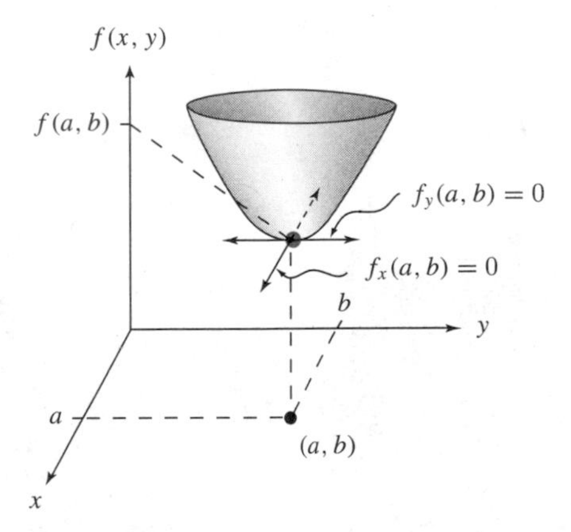

Figure 8.14

This is an important fact that we will use when we attempt to locate relative extreme points and critical points of functions of two variables.

Location of Relative Extrema /Critical Points

The fact that at a relative extreme point, both partial derivatives must simultaneously be zero helps us to locate such points.

Location of Relative Extrema/Critical Points

If $f(x, y)$ is a function with a relative extreme point at $(a, b, f(a, b))$ and both $f_x(a, b)$ and $f_y(a, b)$ exist, then

$$f_x(a, b) = 0 \quad \text{and} \quad f_y(a, b) = 0$$

The points (a, b) for which $f_x(a, b) = 0$ and $f_y(a, b) = 0$ are called **critical points** of the function $f(x, y)$.

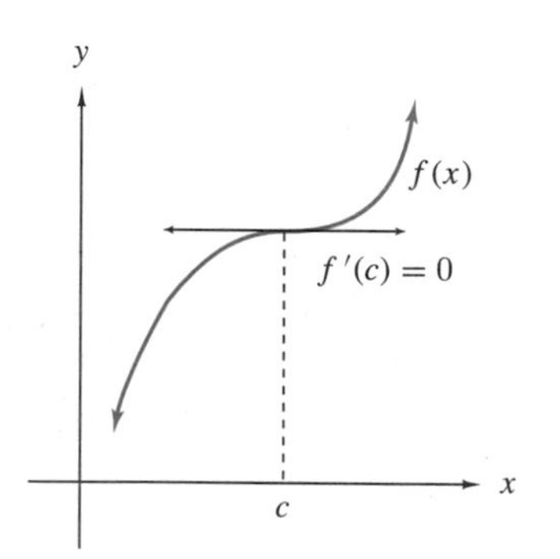

Figure 8.15

The fact that the partial derivatives are both zero at a point (a, b) is not sufficient to guarantee a relative extremum at that point. Recall that for a function of one variable, it was possible that at some point $x = c, f'(c) = 0$, but $x = c$ did not produce a relative extremum. Instead, it produced a point of inflection. Figure 8.15 shows just such a case.

An analogous situation is possible for functions of two variables. Figure 8.16 illustrates a function for which both $f_x(a, b) = 0$ and $f_y(a, b) = 0$, but the point (a, b) does not produce a relative extreme point. The point (a, b) is called a *saddle point.*

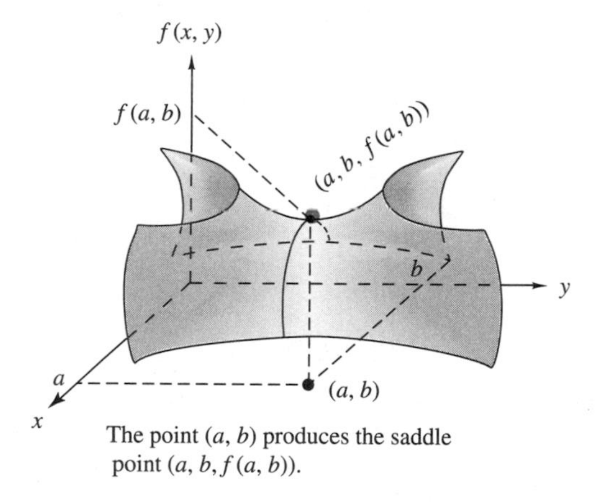

The point (a, b) produces the saddle point $(a, b, f(a, b))$.

Figure 8.16

Saddle Point

A **saddle point** is a point that is a minimum in one direction (the y direction, in this case) and a maximum in the other (the x direction, in this case).

The following Example Sets A–D illustrate how to locate critical points. A graph of each function is presented to better convince you that the method we use actually does produce all the critical points of a function.

EXAMPLE SET A

Find the critical points of the function $f(x, y) = \frac{3}{2}x^2 + y^2 + 6x - 8y + 9$.

Solution:
By the previous discussion, critical points are points (a, b) for which $f_x(a, b) = 0$ and $f_y(a, b) = 0$. We begin by finding f_x and f_y.

$$f_x(x, y) = 3x + 6 \qquad \text{and} \qquad f_y(x, y) = 2y - 8$$

Since these partial derivatives need to be zero simultaneously, we need to solve the system

$$\begin{cases} 3x + 6 &= 0 \\ 2y - 8 &= 0 \end{cases}$$

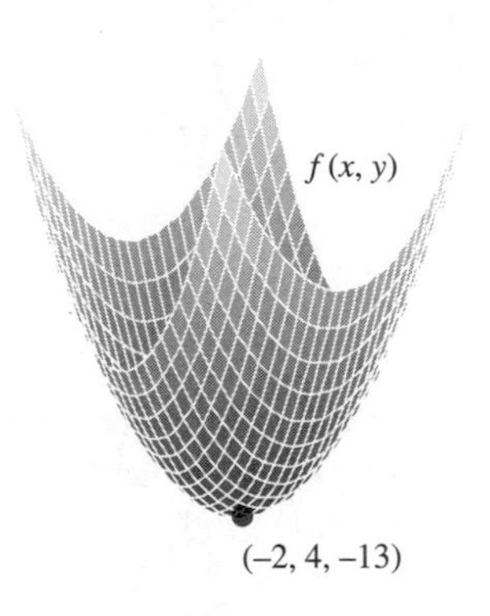

Figure 8.17

Now,

$$\begin{aligned} 3x + 6 &= 0 \quad \text{when} \\ 3x &= -6 \\ x &= -2 \end{aligned}$$

and

$$\begin{aligned} 2y - 8 &= 0 \quad \text{when} \\ 2y &= 8 \\ y &= 4 \end{aligned}$$

This function has only one critical point at $(-2, 4)$.

Interpretation: The critical point $(-2, 4)$ may produce a relative maximum, a relative minimum, or a saddle point. The graph of $f(x, y)$ is displayed in Figure 8.17. It seems to indicate that the critical point $(-2, 4)$ produces a relative minimum.

EXAMPLE SET B

Find the critical points of the function $f(x, y) = \frac{x^2}{2} + \frac{y^2}{2} - 3x + 9y + 5xy + 6$.

Solution:
We begin by finding f_x and f_y.

$$f_x(x, y) = x - 3 + 5y \qquad \text{and} \qquad f_y(x, y) = y + 9 + 5x$$

Since critical points are located where these partial derivatives are simultaneously zero, we need to solve the system

$$\begin{cases} x - 3 + 5y &= 0 \\ y + 9 + 5x &= 0 \end{cases} \longrightarrow \begin{cases} x + 5y &= 3 & \ldots(1) \\ y + 5x &= -9 & \ldots(2) \end{cases}$$

Solving equation (1) for x produces $x = 3 - 5y$. Substituting $3 - 5y$ for x in equation (2) will give us the value of y.

$$y + 5(3 - 5y) = -9$$

$$y + 15 - 25y = -9$$

$$-24y = -24$$

$$y = 1$$

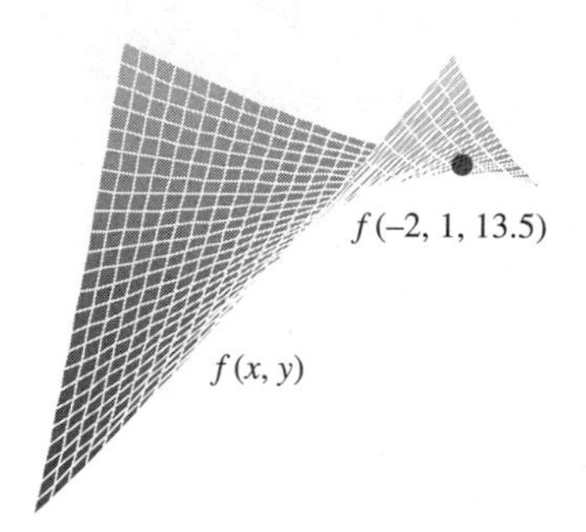

Figure 8.18

Now, $x = 3 - 5y$ and with $y = 1$, $x = 3 - 5(1)$ so that $x = -2$. This function has only one critical point at $(-2, 1)$.

Interpretation: The critical point $(-2, 1)$ may produce a relative maximum, a relative minimum, or a saddle point. The graph of $f(x, y)$ is displayed in Figure 8.18. It seems to indicate that the critical point $(-2, 1)$ produces a saddle point.

EXAMPLE SET C

Find the critical points of the function $f(x, y) = \dfrac{3x^2}{2} + \dfrac{y^4}{2} - y^2 - 3$.

Solution:
We begin by finding f_x and f_y

$$f_x(x, y) = 3x \qquad \text{and} \qquad f_y(x, y) = 2y^3 - 2y$$

Since critical points are located where these partial derivatives are simultaneously zero, we need to solve the system

$$\begin{cases} 3x & = 0 \quad \ldots(1) \\ 2y^3 - 2y & = 0 \quad \ldots(2) \end{cases}$$

Solving equation (1) for x produces $x = 0$. Solving equation (2) for y produces

$$2y^3 - 2y = 0$$

$$2y(y^2 - 1) = 0$$

$$2y(y + 1)(y - 1) = 0$$

$$y = 0, \ -1, \ 1$$

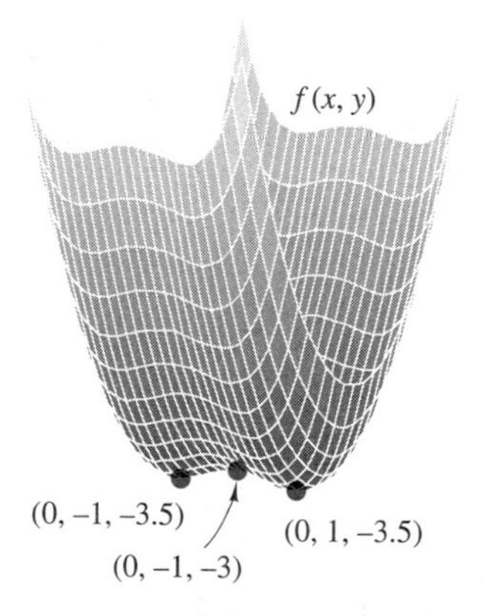

Figure 8.19

Thus, f_x and f_y are zero when $x = 0$ and $y = 0, \ -1, \ 1$, and consequently, this function has three critical points, $(0, 0)$, $(0, -1)$ and $(0, 1)$.

Interpretation: The critical points could produce relative maxima, relative minima, saddle points, or some combination of these. The graph of $f(x, y)$ is displayed in Figure 8.19. It seems to indicate that the critical point $(0, 0)$ produces a saddle point and that $(0, -1)$, and $(0, 1)$ produce relative minima.

The next example illustrates how to optimize a function of two variables when the first partial derivatives are constants.

EXAMPLE SET D

Find thc critical points of the function $f(x, y) = 4x - 9y + 2$.

Solution:
We begin by finding f_x and f_y.

$$f_x(x, y) = 4 \qquad \text{and} \qquad f_y(x, y) = -9$$

Since critical points are located where these partial derivatives are simultaneously zero, we need to solve the system

$$\begin{cases} 4 & = 0 \quad \ldots (1) \\ -9 & = 0 \quad \ldots (2) \end{cases}$$

But this system has no solutions; that is, f_x and f_y are never zero.

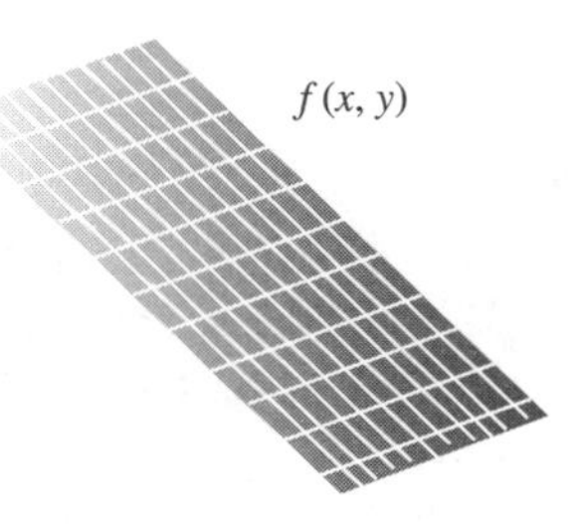

Figure 8.20

Interpretation: This function has no critical points. The graph of $f(x, y)$ is displayed in Figure 8.20. It seems to indicate that the function is a plane (a flat surface) which would lead us to believe that no relative extreme points could exist.

The Second Derivative Test

The Second Derivative Test provides a way of classifying critical points as relative maxima, relative minima, or saddle points.

Second Derivative Test

Suppose that (a, b) is a critical point of the function $f(x, y)$ and that $D(x, y)$ represents $f_{xx}(a, b) \cdot f_{yy}(a, b) - f_{xy}^2(a, b)$. Then,

1. if $D(a, b) > 0$ and $f_{xx}(a, b) > 0$, (a, b) produces a relative minimum.
2. if $D(a, b) > 0$ and $f_{xx}(a, b) < 0$, (a, b) produces a relative maximum.
3. if $D(a, b) < 0$, (a, b) is a saddle point.
4. if $D(a, b) = 0$, the test provides no information about (a, b).

Note that if D, when evaluated at a critical point (a, b), is positive, the critical point will necessarily be a relative extreme point. The critical point will be a saddle point only when D is negative.

The following examples illustrate the use of the Second Derivative Test. The critical points are located using the techniques demonstrated in Example Sets A – D.

EXAMPLE SET E

Locate the relative extrema, if any exist, of the function $f(x, y) = 2x^2 + 3y^2 + 8x - 12y + 3$.

Solution:
The Second Derivative Test indicates that we need to determine the partial derivatives f_x, f_y, f_{xx}, f_{yy}, and f_{xy}.

$$f_x(x, y) = 4x + 8, \quad f_y(x, y) = 6y - 12, \quad f_{xx} = 4, \quad f_{yy} = 6, \quad f_{xy} = 0$$

We will use the first-order partials, f_x and f_y, to generate any critical points. Setting $f_x = 0$ and $f_y = 0$ produces the system

$$\begin{cases} 4x + 8 &= 0 \\ 6y - 12 &= 0 \end{cases}$$

which produces the single critical point $(-2, 2)$.

Now, let $D(x, y) = f_{xx}(a, b) \cdot f_{yy}(a, b) - f_{xy}^2(a, b)$. $D(x, y) = 4 \cdot 6 - 0^2$, so that $D(x, y) = 24$. Since $D(x, y) = 24$, it is *always* positive. In particular, $D(-2, 2) = 24 > 0$, so that the critical point $(-2, 2)$ is necessarily a relative extreme point. To determine which type (max or min), we compute $f_{xx}(-2, 2)$.

$f_{xx} = 4$ for all points in this example; therefore, $f_{xx} > 0$, leading us to conclude that $(-2, 2)$ produces a relative minimum. To locate the point, we must compute the output value for the input values $x = -2$ and $y = 2$. Substituting -2 for x and 2 for y into the original function, we get

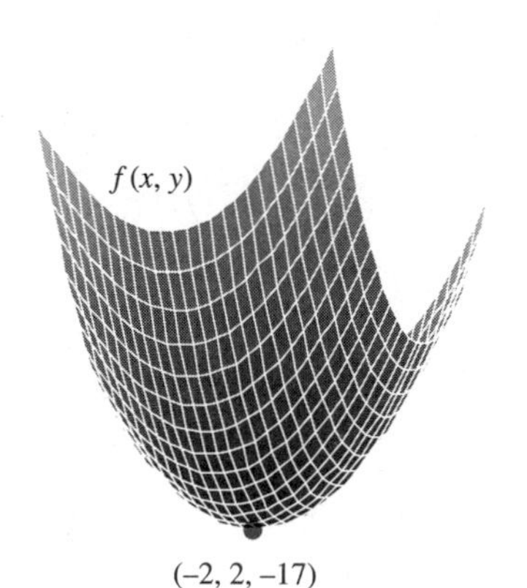

Figure 8.21

$$f(-2, 2) = 2(-2)^2 + 3(2^2) + 8(-2) - 12(2) + 3$$

$$= -17$$

Interpretation: The critical point $(-2, 2)$ produces the relative minimum point $(-2, 2, -17)$. The graph of $f(x, y)$ is displayed in Figure 8.21.

Notice that the partial derivatives f_{xx}, f_{yy}, and f_{xy} are each constant. This makes their evaluation at the critical points very easy (since no substitution is necessary).

EXAMPLE SET F

Locate the relative extrema, if any exist, of $f(x, y) = \dfrac{2x^3}{3} + \dfrac{4y^3}{3} - 8y^2 - 50x + 1$.

Solution:
The Second Derivative Test indicates that we need to determine the partial derivatives f_x, f_y, f_{xx}, f_{yy}, and f_{xy}.

$$f_x(x, y) = 2x^2 - 50, \quad f_y(x, y) = 4y^2 - 16y, \quad f_{xx} = 4x, \quad f_{yy} = 8y - 16, \quad f_{xy} = 0$$

We will use the first-order partials, f_x and f_y, to generate any critical points. Setting $f_x = 0$ and $f_y = 0$ produces the system

$$\begin{cases} 2x^2 - 50 &= 0 \\ 4y^2 - 16y &= 0 \end{cases}$$

which produces the critical points $(-5,0)$, $(-5,4)$, $(5,0)$, and $(5,4)$.
Now, compute $D(x,y)$.

$$D(x,y) = f_{xx}(a,b)\cdot f_{yy}(a,b) - f_{xy}^2(a,b)$$
$$= 4x\cdot(8y-16) - 0^2$$
$$= 32xy - 64x$$

To classify these four critical points as relative maxima, minima, or saddle points, we will employ the Second Derivative Test. To do so, we need to compute $D(x,y)$ and $f_{xx}(x,y)$ at each of these points. We summarize our computations in the following tables.

Table 8.1

CP	$f'(x)$	$f_{xx}(x)$	**Conclusion**
$(-5,0)$	$32(-5)(0)-64(-5)>0$	$4(-5)<0$	Produces rel max
$(-5,4)$	$32(-5)(4)-64(-5)<0$		Produces saddle pt
$(5,0)$	$32(5)(0)-64(5)<0$		Produces saddle pt
$(5,4)$	$32(5)(4)-64(5)>0$	$4(5)>0$	Produces rel min

Placing these input values into the function produces the output values and the relative extrema and saddle points.

Table 8.2

$f(a,b)$	**Conclusion**
$f(-5,0)\approx 167.67$	Rel max at approx $(-5,0,167.67)$
$f(-5,4)=125$	Saddle pt at $(-5,4,125)$
$f(5,0)\approx -165.67$	Saddle pt at approx $(5,0,-165.67)$
$f(5,4)\approx -208.33$	Rel min at approx $(5,4,-208.33)$

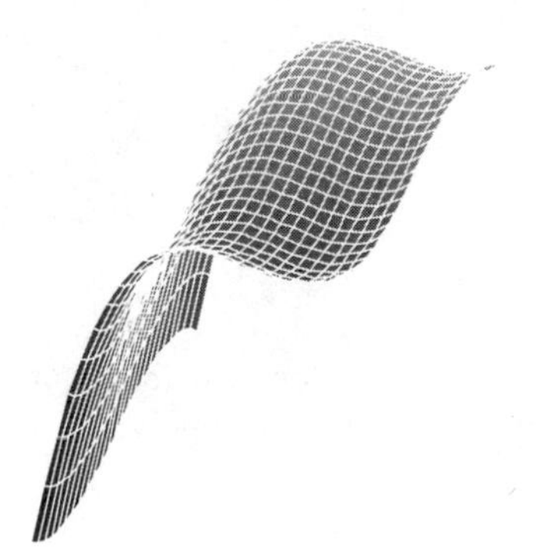

Figure 8.22

The graph of $f(x,y)$ is displayed in Figure 8.22.

EXAMPLE SET G

A manufacturer markets a product in two states, A and B, and, because of the different economies of the states, must price the product differently in each. (Such pricing is called *price discrimination* or *differential pricing*.) The manufacturer wishes to sell x units of the product in state A and y units of the product in state B. To do so, the manufacturer must set the price in state A at $86 - \frac{x}{18}$ dollars, and in state B at $122 - \frac{y}{28}$ dollars. The cost of producing all $x + y$ items is $45{,}000 + 4(x + y)$ dollars. How many items should be produced for states A and B, respectively, to maximize the manufacturer's profit, and what is that maximum profit?

Solution:
Since we wish to maximize profit, we need to construct a profit function.

$$\text{Profit} = (\text{revenue}) - (\text{cost})$$

$$= [(\text{revenue from state A}) + (\text{revenue from state B})] - (\text{cost})$$

Since (revenue) = (price) · (quantity), we have

$$P(x,y) = \left[\left(86 - \frac{x}{18}\right)x + \left(122 - \frac{y}{28}\right)y\right] - [45{,}000 + 4(x+y)]$$

$$P(x,y) = 86x - \frac{x^2}{18} + 122y - \frac{y^2}{28} - 45{,}000 - 4x - 4y$$

$$P(x,y) = 82x - \frac{x^2}{18} + 118y - \frac{y^2}{28} - 45{,}000$$

To use the Second Derivative Test to locate the extreme points of this function we need P_x, P_y, P_{xx}, P_{yy}, and P_{xy}.

$$P_x(x,y) = 82 - \frac{x}{9}, \qquad P_y(x,y) = 118 - \frac{y}{14}, \qquad P_{xx}(x,y) = -\frac{1}{9},$$

$$P_{yy}(x,y) = -\frac{1}{14}, \qquad P_{xy}(x,y) = 0$$

We use the first partial derivatives P_x and P_y to find any critical points. Setting $P_x = 0$ and $P_y = 0$ produces the following equivalent systems

$$\begin{cases} 82 - \frac{x}{9} = 0 \\ 118 - \frac{y}{14} = 0 \end{cases} \longrightarrow \begin{cases} \frac{x}{9} = 82 \\ \frac{y}{14} = 118 \end{cases} \longrightarrow \begin{cases} x = 738 \\ y = 1652 \end{cases}$$

which produce the single critical point $(738, 1652)$.

We now compute $D(x,y) = P_{xx}(x,y) \cdot P_{yy}(x,y) - P_{xy}^2(x,y)$ to determine its sign (+ or −) at the critical point $(738, 1652)$. $D(738, 1652) = (-\frac{1}{9}) \cdot (-\frac{1}{14}) - 0^2 > 0$, so that the critical point $(738, 1652)$ necessarily produces an extreme point.

To classify which type (maximum or minimum), we compute $P_{xx}(738, 1652)$. $P_{xx}(738, 1652) = -\frac{1}{9} < 0$, leading us to conclude that $(738, 1652)$ produces a relative maximum. Then, substituting 738 for x and 1652 for y into the original function, we get

$$P(738, 1652) = 82(738) - \frac{738^2}{18} + 118(1652) - \frac{1652^2}{28} - 45{,}000$$

$$= 82{,}726$$

Interpretation: The manufacturer should produce and sell 738 units of the product in state A and 1652 units of the product in state B to obtain the maximum profit of $82,726.

Optimization and Technology

Using Derive Derive can optimize functions of two variables. The following set of commands demonstrates the procedure using the function $f(x, y) = e^{xy}$. To make these computations, you should put all the following commands together. We have broken them up only to make comments about what each subset of these commands does. Remember to press Enter at the end of each line.

Declare
Function
f
exp(xy)
These commands declare the function $f(x, y) = e^{xy}$.
Declare
Function
dx
dif(f(x,y),x)
Simplify

These commands declare and compute $f_x(x, y)$.

Declare
Function
dy
dif(f(x,y),y)
Simplify

These commands declare and compute $f_y(x, y)$.

Declare
Function
dxx
dif(dx(x,y),x)
Simplify

These commands declare and compute $f_{xx}(x, y)$.

Declare
Function

dyy
dif(dy(x,y),y)
Simplify

These commands declare and compute $f_{yy}(x, y)$.

Declare
Function
dxy
dif(dx(x,y),y)
Simplify

These commands declare and compute $f_{xy}(x, y)$.

Now, find the critical points by solving the system

$$\begin{cases} dx & = 0 \\ dy & = 0 \end{cases}$$

The equations of a system are entered into Derive by enclosing them in square brackets and separating them with commas. Although this is the most natural way to try to solve this system, unfortunately, it does not work. Derive has problems solving systems involving exponential equations.

Author
[dx = 0, dy = 0]
so**L**ve

By solving the system algebraically yourself, you can conclude the point $(0, 0)$ is the solution to the system and the only critical point.

Find the value of $D = f_{xx} \cdot f_{yy} - f^2{}_{xy}$ at the point $(0, 0)$.

Declare
Function
D
dxx(x,y)dyy(x,y)–dxy(x,y)^2
Simplify
Manage
Substitute
0
0
Simplify

These commands declare and compute $D = f_{xx} \cdot f_{yy} - f^2{}_{xy}$. The **M**anage and **S**ubstitute commands evaluate D at the point $(0, 0)$. The value of D is shown to be -1, indicating that the critical point $(0, 0)$ produces a saddle point.

Derive Exercises Use Derive to optimize each function.

1. $f(x, y) = x^2 + 3y^3 + 4x - 9y + 11$

2. $f(x, y) = x^3 + 2x^2y + 4x - 8y + 55$

3. $f(x, y) = 2xy + 4\ln x + y^2 - 10$

4. The profit $P(x)$, in dollars, from the sales of x units of product A and y units of product B is given by the function

$$P(x, y) = 22x + 73y - 1.2x^2 - 1.5y^2$$

How many units of each product should be sold to maximize profit?

5. A manufacturer can produce

$$N(x, y) = -1.4x^2 - 1.3y^2 + 15x + 12y$$

thousand units of a product using x units of labor and y units of capital. Find the maximum number of units produced by the manufacturer.

EXERCISE SET 8.3

A UNDERSTANDING THE CONCEPTS

In each of Exercises 1–7, the first-order, second-order, and mixed partial derivatives of a function are given along with the critical point(s) of the function. Use this information to classify, if possible, the critical point(s) as a point(s) that produces a relative maximum, relative minimum, or a saddle point.

1. $f_x(x, y) = 3x + 12$
   $f_y(x, y) = 2y - 8$
   $f_{xx} = 3, \quad f_{yy} = 2, \quad f_{xy} = 0$
   $(-4, 4)$
2. $f_x(x, y) = 5x - 15$
   $f_y(x, y) = 3y + 6$
   $f_{xx} = 5, \quad f_{yy} = 3, \quad f_{xy} = 0$
   $(3, -2)$
3. $f_x(x, y) = 9x - 18$
   $f_y(x, y) = -4y + 20$
   $f_{xx} = 9, \quad f_{yy} = -4, \quad f_{xy} = 0$
   $(2, 5)$
4. $f_x(x, y) = -8x + 16$
   $f_y(x, y) = y + 6$
   $f_{xx} = -8, \quad f_{yy} = 1, \quad f_{xy} = 0$
   $(2, -6)$
5. $f_x(x, y) = 3x + 2y - 19$
   $f_y(x, y) = 4x - 2y - 2$
   $f_{xx} = 3, \quad f_{yy} = 4, \quad f_{xy} = 2$
   $(3, 5)$
6. $f_x(x, y) = -2x + 8y - 2$
   $f_y(x, y) = 4x + 2y + 4$
   $f_{xx} = -2, \quad f_{yy} = 2, \quad f_{xy} = 8$
   $(-1, 0)$
7. $f_x(x, y) = -2x - 4$
   $f_y(x, y) = -2y$
   $f_{xx} = -2, \quad f_{yy} = -2, \quad f_{xy} = 0$
   $(-2, 0)$

B SKILL ACQUISITION

For Exercises 8–20, find and classify, if possible, all the relative extreme points and saddle points.

8. $f(x, y) = \frac{3}{2}x^2 + y^2 + 15x - 8y + 6$
9. $f(x, y) = 3x^2 - y^2 - 12x + 16y + 21$
10. $f(x, y) = \frac{5}{2}y^2 - 2x^2 - 12x - 20y + 7$
11. $f(x, y) = -x^2 - \frac{3}{2}y^2 + 6x + 21y + 8$
12. $f(x, y) = -x^2 - y^2 - xy + x + 6y + 12$
13. $f(x, y) = x^2 + \frac{3}{2}y^2 - 5xy + 11x + 3y - 8$
14. $f(x, y) = x^3 - y^2 - 3x + 4y + 5$
15. $f(x, y) = x^3 - 2y^3 + 6y + 8$
16. $f(x, y) = x^2 - 2y^3 + 6y - 10$
17. $f(x, y) = x^3 - 2xy + 4y + 6$
18. $f(x, y) = x^3 - 3xy + y^3$
19. $f(x, y) = e^{xy}$
20. $f(x, y) = x^{2/3} + y^{2/3}$

C APPLYING THE CONCEPTS

21. ***Marketing: Revenue from Advertising*** The marketing department of a company has determined that if it spends x thousands of dollars on radio advertisements and y thousands of dollars on newspaper advertisements, the company's revenue (in thousands of dollars) will be $R(x, y) = -0.07x^2 - 100y^2 + 4x + 5y + 2xy$. How much money should this company spend on radio advertisements and newspaper advertisements to maximize its revenue?

22. ***Manufacturing: Cost of Production*** A manufacturer markets a product in two states, A and B, and prices it differently in each state. The manufacturer wishes to sell x units of the product in state A and y units in state B. To do so, it must set the price in state A at $678 - \frac{x}{22}$

dollars per unit and in state B at $151 - \frac{y}{16}$ dollars per unit. The cost of producing all $x + y$ units of the product is $\$85,000 + 6(x + y)$. How many units of the product should be produced for states A and B, respectively, to maximize the manufacturer's profit?

23. ***Manufacturing: Cost of Production*** A manufacturer of boxes wishes to construct a rectangular box having a volume of 64 cubic inches. What dimensions of the box will have the minimal surface area?

24. ***Business: Construction Costs*** A building is to be constructed so as to enclose 31,250 cubic feet. The cost of the roof is \$8.00 per square foot, the cost of the sides is \$8.00 per square foot, and the cost of the floor is \$16.00 per square foot. What dimensions of the building will minimize construction costs?

25. ***Business: Postal Costs*** Postal regulations require that the (length) + (girth) of a rectangular package be no more than 84 inches. (The girth is the distance around the middle of the package.) Find the dimensions of the rectangular package of largest volume that meet these postal requirements.

D DESCRIBE YOUR THOUGHTS

26. The definition of a relative maximum of the function $f(x, y)$ states that if the point (a, b) produces a relative maximum of $f(x, y)$, then for every point (x, y) near the point (a, b), $f(a, b) > f(x, y)$. Explain what this means geometrically.

27. Explain how it is possible for both $f_x(x, y)$ and $f_y(x, y)$ to equal zero at a point (a, b), but for the point (a, b) to not produce a relative maximum or a relative minimum.

28. The Second Derivative Test for a function of two variables states that for the critical point (a, b), if $D(a, b) > 0$ and $f_{xx}(a, b) > 0$, the point (a, b) produces a relative minimum. Explain why $f_{xx}(a, b) > 0$ makes this so.

E REVIEW

29. **(2.2)** Find, if it exists, $\lim_{x \to 4} \frac{x^2 + 2x + 1}{x^2 - 3x - 4}$.

30. **(2.4)** Discuss, by writing an analysis, the continuity of the function

$$f(x) = \frac{x^2 - 2x - 3}{x - 3} \quad \text{at } x = 3$$

31. **(3.2)** Find and interpret $f'(x)$ for $f(x) = \frac{4x - 1}{3x + 2}$ at $x = 2$.

32. **(3.5)** Find and interpret y' for $y^2 + y^3 = 12x + 12$ at $(2, 3)$.

33. **(4.3)** ***Medicine: Drug Concentration*** A drug manufacturer claims that t hours after a particular drug is administered to a male weighing between 135 and 160 pounds, the concentration C (in milligrams per liter) is given by the function

$$C(t) = \frac{5t}{2t^2 + 8}$$

Find the time when the concentration is maximum.

34. **(5.4)** Use implicit differentiation to find y' for $8e^{x}y^2 + 2y - x = 1$.

35. **(7.4)** Solve for y: $(x^2 - 1)y' = 4x$, and $y = 5$ when $x = 2$.

36. **(7.4)** Solve for y: $y' = \frac{1}{xy}$, and $y = 4$ when $x = 1$.

37. **(8.1)** Evaluate the function $f(x, y) = 2x^2 - 3y^2 + 5xy + 9x - 12y + 25$ at the point $(2, -3)$.

38. **(8.2)** Find the first, second, and mixed partial derivatives of the function $f(x, y) = 5x^3 - 7y^4 - 6x^2y + 2x - 8y + 105$.

8.4 Constrained Maxima and Minima

Introduction

An important application of differentiation is optimization. Using differentiation, we are able to determine if any relative extrema of a function exist, and if they do, to locate and classify them as relative maxima, relative minima, or saddle points. In Section 8.3 we used the Second Derivative Test to find all the relative extreme points of a function of two variables. This method involved no restriction on the values of the input variables. In this section we study the method of Lagrange multipliers, a method for optimizing a function when there *are* restrictions on the values of the input variables.

Unconstrained and Constrained Relative Extrema

Unconstrained Relative Extrema

Figure 8.23 shows the graph of a function $f(x, y)$ of two variables with a relative maximum at the point (a, b, c_1). This relative maximum is called an **unconstrained relative maximum** because there are no constraints (restrictions) on the values of the input variables x and y.

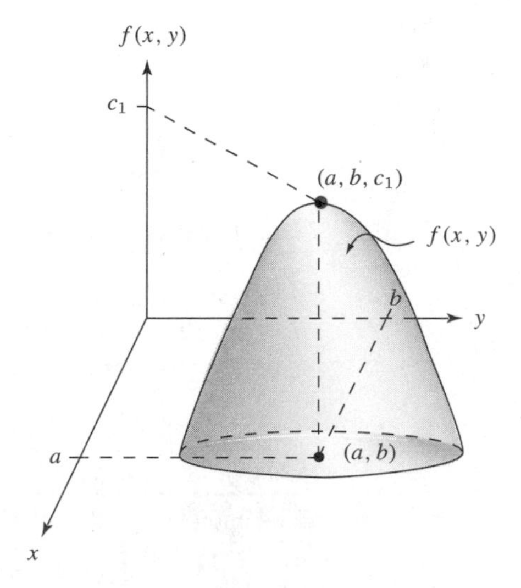

Figure 8.23

However, very often in applied situations, there is a constraint on the input variables that must be considered when optimizing a function $f(x, y)$. A constraint is a restriction that limits the values of the input variables and, therefore, the value of the output variable. Constraints occur when the input variables are related.

There are many different types of constraints, but an example in business might involve constraints due to a certain amount of money budgeted for labor costs and investments. Constraints can, therefore, be described as functions, and Figure 8.24 shows the graph of a function $f(x, y)$ with a constrained relative maximum at the point (a, b, c_2). The relative maximum occurs at (a, b, c_2) rather than at (a, b, c_1) because the input variables are related by the linear function of two variables $g(x, y)$. The output values, the points on the surface $f(x, y)$, are confined to the points that lie directly above the constraint function. Relative extrema such as the one in Figure 8.24 are called **constrained relative extrema**.

Constrained Relative Extrema

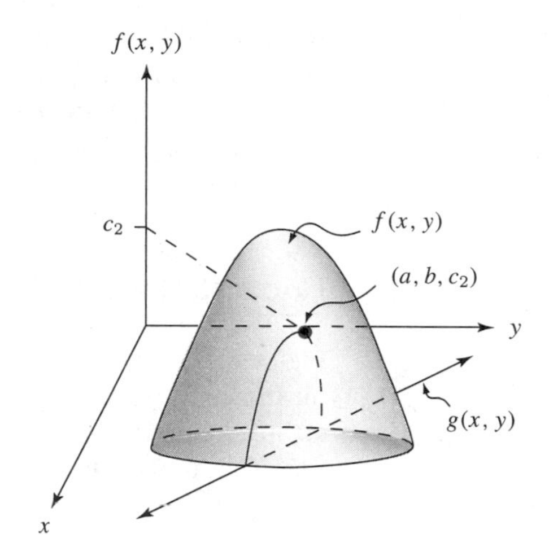

Figure 8.24

The Method of Algebraic Substitution

If the constraint is easily solved for one of the variables, any relative extreme points can be located and classified using substitution and the optimization technique discussed in Section 4.3. Example Set A illustrates this method.

EXAMPLE SET A

Find and classify the relative extreme points of the function

$$f(x, y) = -3x^2 - 2y^2 + 4xy - 13x - 8y + 80$$

subject to the constraint that $x + 2y = 20$.

Solution:
Notice first that the input variables are related by the function $x + 2y = 20$. That is, the value of x plus twice the value of y is always 20. We can solve for x in terms of y. Subtracting $2y$ from each side produces $x = -2y + 20$. We then substitute the expression $-2y + 20$ for x in the original function. Since all the x's will be eliminated and replaced with y's, the original function of two variables will be reduced to a function of only one variable.

$$f(x, y) = -3x^2 - 2y^2 + 4xy - 13x - 8y + 80$$

$$\begin{aligned} f(y) &= -3(-2y + 20)^2 - 2y^2 + 4(-2y + 20)y - 13(-2y + 20) - 8y + 80 \\ &= -22y^2 + 338y - 1380 \end{aligned}$$

We can now use the First Derivative Test method introduced in Section 4.3. We find $f'(y)$ and determine where it is zero or undefined, if at all.

$$f'(y) = -44y + 338$$

1. Where is $f'(y) = 0$?
   $-44y + 338 = 0 \quad \longrightarrow \quad y = \dfrac{169}{22}$
2. Where is $f'(y)$ undefined?
   Since $f'(y)$ is a polynomial function, it is never undefined.
   Thus, the only critical value is $y = \dfrac{169}{22}$ (≈ 7.7).

   To use the First Derivative Test, we choose a y value to the left of $\dfrac{169}{22}$, like 0, and then to the right of $\dfrac{169}{22}$, like 8, and compute $f'(y)$ at each to determine its sign at these values.

   $f'(0) = 338 > 0 \quad \longrightarrow \quad f(y)$ is increasing to the left of $\dfrac{169}{22}$.

   $f'(8) = -14 < 0 \quad \longrightarrow \quad f(y)$ is decreasing to the right of $\dfrac{169}{22}$.

   We conclude that $y = \dfrac{169}{22}$ produces a relative maximum. The corresponding value of x can be found by substituting $\dfrac{169}{22}$ for y in $x = -2y + 20$, which we obtained from the constraint function. This produces $x = \dfrac{51}{11}$ (≈ 4.6). Thus, the critical point is $\left(\dfrac{169}{22}, \dfrac{51}{11}\right)$, and it produces the constrained relative maximum of $\left(\dfrac{169}{22}, \dfrac{51}{11}, \dfrac{-65{,}105}{484}\right)$.

The Method of Lagrange Multipliers

Sometimes it may be difficult (or even impossible) to solve the constraint for one of the variables. Also, even when the constraint can be solved for one of the variables, a complicated function may result from the substitution. In the eighteenth century, the French mathematician Joseph Lagrange (1736–1813) constructed a method for optimizing functions of several variables that are subject to a constraint, $g(x, y) = 0$, that eliminates the need for the substitution. In Lagrange's honor, the method is called the **method of Lagrange multipliers**. The method involves creating a new function, called the **Lagrangian function**, which, for a function of the two variables x and y, is defined as

Method of Lagrange Multipliers

Lagrangian Function

$$F(x, y, \lambda) = f(x, y) + \lambda \cdot g(x, y)$$

Lagrange Multiplier

The Greek letter lambda, λ, is called the **Lagrange multiplier** and serves to eliminate the dependency of one variable on another. The Lagrange multiplier always multiplies the constraint function $g(x, y)$. The Lagrangian function, $F(x, y, \lambda)$, is an unconstrained function of the three independent variables x, y, and λ. Because of the independence of the variables, a form of the Second Derivative Test may be used to locate and classify any critical points as relative extrema.

The Method of Lagrange Multipliers

To locate potential relative extreme points of a function $f(x, y)$ subject to the constraint $g(x, y) = 0$,

1. Set the constraint equal to zero so that it is in the form $g(x, y) = 0$.
2. Construct the Lagrangian function
$$F(x, y, \lambda) = f(x, y) + \lambda g(x, y)$$
3. Find the first-order partial derivatives F_x, F_y, and F_λ.
4. Locate critical points by solving the system of equations
$$\begin{cases} F_x = 0 & \ldots(1) \\ F_y = 0 & \ldots(2) \\ F_\lambda = 0 & \ldots(3) \end{cases}$$
for x, y, and λ. Each ordered pair solution, (x, y), is a critical point. The relative extreme points are among these critical points.
5. Use the Second Derivative Test for the method of Lagrange multipliers to classify each critical point as a relative maximum or a relative minimum. (The Second Derivative Test is presented directly after the following note.)

Note: The system in step (4) is often easily solved using equations (1) and (2) and eliminating λ. This will produce a new equation that involves only x and y. Then, using equation (3) and this new equation, we obtain the values of x and y. These values can then be substituted into either of equations (1) or (2) to obtain the corresponding value of λ. Although it is not necessary to obtain the value of λ to locate and classify relative extreme points, λ has an important and interesting interpretation (which we will discuss after we present an example).

The following test is the Second Derivative Test for Lagrange multipliers.

Second Derivative Test

Suppose that (a, b) is a critical point of the function $f(x, y)$ and that $L(x, y)$ represents $F_{xx}(a, b) \cdot F_{yy}(a, b) - F_{xy}^2(a, b)$. Then,

1. If $L(a, b) > 0$ and $F_{xx}(a, b) > 0$, (a, b) is a constrained relative minimum.
2. If $L(a, b) > 0$ and $F_{xx}(a, b) < 0$, (a, b) is a constrained relative maximum.
3. If $L(a, b) \leq 0$, then the test fails, and it is necessary to investigate values of $f(x, y)$ near (a, b).

The test provides a method of classifying critical points as constrained relative maxima or constrained relative minima. It is nearly identical to the unconstrained Second Derivative Test for relative extrema. The only difference is that no saddle point can exist.

EXAMPLE SET B

Find and classify all the relative extreme points of the function $f(x, y) = x^2 + 10y^2$ subject to the constraint $x - y = 18$.

Solution:
Although this constraint makes the function a good candidate for the method of algebraic substitution, we will use the method of Lagrange multipliers as a point of illustration.

1. We begin by writing the constraint in the form $g(x, y) = 0$.

$$x - y = 18 \quad \longrightarrow \quad \underbrace{x - y - 18}_{g(x,y)} = 0, \text{ so that } g(x, y) = x - y - 18$$

2. Next, we construct the Lagrangian function $F(x, y, \lambda) = f(x, y) + \lambda g(x, y)$.

$$F(x, y, \lambda) = x^2 + 10y^2 + \lambda(x - y - 18)$$

or

$$F(x, y, \lambda) = x^2 + 10y^2 + \lambda x - \lambda y - \lambda 18$$

3. Now we find the first-order partial derivatives F_x, F_y, and F_λ.

$$F_x = 2x + \lambda, \quad F_y = 20y - \lambda, \quad F_\lambda = x - y - 18$$

Notice that F_λ equals $g(x, y)$. The Lagrangian function is constructed so that this will always be true. (Try differentiating $F(x, y, \lambda) = f(x, y) + \lambda g(x, y)$ with respect to λ.)

4. We solve the system

$$\begin{cases} F_x = 0 & \ldots(1) \\ F_y = 0 & \ldots(2) \\ F_\lambda = 0 & \ldots(3) \end{cases} \longrightarrow \begin{cases} 2x + \lambda & = 0 & \ldots(1) \\ 20y - \lambda & = 0 & \ldots(2) \\ x - y - 18 & = 0 & \ldots(3) \end{cases}$$

We will eliminate λ by adding equations (1) and (2). Symbolically, our method will be $(1) + (2)$, and we will label the resulting equation (4).

$$\begin{cases} 2x + \lambda = 0 & \ldots(1) \\ 20y - \lambda = 0 & \ldots(2) \end{cases}$$

$$2x + 20y = 0 \qquad \ldots(4)$$

Now we form a system that involves only x and y using equations (3) and (4).

$$\begin{cases} x - y - 18 = 0 & \ldots(3) \\ 2x + 20y \;\; = 0 & \ldots(4) \end{cases} \longrightarrow \begin{cases} x - \;\; y = 18 & \ldots(3) \\ 2x + 20y = 0 & \ldots(4) \end{cases}$$

Next we eliminate x by adding -2 times equation (3) to equation (4). This will produce the value of y.

$$\begin{cases} -2x + \;\; 2y = -36 & (-2 \text{ times equation } (3)) \\ \;\; 2x + 20y = \;\;\; 0 & \ldots(4) \end{cases}$$

$$22y = -36$$

$$y = \frac{-36}{22}$$

$$y = \frac{-18}{11}$$

Then, if $y = \dfrac{-18}{11}$, substitution into equation (3) produces

$$x - \frac{-18}{11} - 18 = 0$$

$$x = \frac{180}{11}$$

Then, if $x = \dfrac{180}{11}$ and $y = \dfrac{-18}{11}$, substitution into equation (1) produces

$$2 \cdot \frac{180}{11} + \lambda = 0$$

$$\lambda = \frac{-360}{11}$$

Thus, $x = \dfrac{180}{11}$, $y = \dfrac{-18}{11}$, $\lambda = \dfrac{-360}{11}$; and $\left(\dfrac{180}{11}, \dfrac{-18}{11}\right)$ is the only critical point. (We will discuss the meaning of λ in the next subsection.)

5. Finally, we classify the critical point $\left(\dfrac{180}{11}, \dfrac{-18}{11}\right)$ using the Second Derivative Test. To use the test, we need to find F_{xx}, F_{yy}, and F_{xy}.

Since $F_x = 2x + \lambda$, $F_{xx} = 2$.
Since $F_y = 20y - \lambda$, $F_{yy} = 20$.
Since $F_x = 2x + \lambda$, $F_{xy} = 0$.
For this test to succeed we must have $L(x, y) > 0$, so we next find $L(x, y)$ evaluated at $x = \frac{180}{11}$ and $y = \frac{-18}{11}$.

$$\begin{aligned} L\left(\frac{180}{11}, \frac{-18}{11}\right) &= F_{xx}\left(\frac{180}{11}, \frac{-18}{11}\right) \cdot F_{yy}\left(\frac{180}{11}, \frac{-18}{11}\right) - F_{xy}^2\left(\frac{180}{11}, \frac{-18}{11}\right) \\ &= (2) \cdot (20) - 0^2 \\ &= 40 \end{aligned}$$

Thus, $L\left(\frac{180}{11}, \frac{-18}{11}\right) > 0$ so that $\left(\frac{180}{11}, \frac{-18}{11}\right)$ is either a constrained relative maximum or minimum. To determine which, we need to determine the sign of $F_{xx}\left(\frac{180}{11}, \frac{-18}{11}\right)$.

Since $F_{xx} = 2 > 0$, it is always positive, so $\left(\frac{180}{11}, \frac{-18}{11}\right)$ produces a constrained relative minimum.

To find the constrained relative minimum value, we substitute $x = \frac{180}{11}$ and $y = \frac{-18}{11}$ into the original function $f(x, y) = x^2 + 10y^2$ and compute. Doing so produces

$$\begin{aligned} f\left(\frac{180}{11}, \frac{-18}{11}\right) &= \left(\frac{180}{11}\right)^2 + 10\left(\frac{-18}{11}\right)^2 \\ &= \frac{32{,}400}{121} + 10\left(\frac{324}{121}\right) \\ &= \frac{32{,}400}{121} + \frac{3240}{121} \\ &= \frac{35{,}640}{121} \end{aligned}$$

Interpretation: The function $f(x, y) = x^2 + 10y^2$, when subjected to the constraint $x - y - 18 = 0$, has a relative minimum at $\left(\frac{180}{11}, \frac{-18}{11}, \frac{35{,}640}{121}\right)$.

The Significance of the Lagrange Multiplier

In Example Set B, we found the value of the Lagrange multiplier, λ, but did not discuss its significance. The Lagrange multiplier can, however, provide useful information.

The Significance of the Lagrange Multiplier λ

Suppose that M is the constrained maximum or minimum value of the function $f(x, y)$ when it is subjected to the constraint $g(x, y) = c$, where c is some constant. Then,

$$\frac{dM}{dc} = \lambda$$

That is, λ is the rate at which the constrained maximum or minimum value of the function $f(x, y)$ changes with respect to c.
Interpretation: If c is increased by 1 unit, from c to $c + 1$, then M changes by approximately λ units.

ILLUMINATOR SET A

In Example Set B, the function to be optimized was $f(x, y) = x^2 + 10y^2$, subject to the constraint $x - y = 18$. The maximum value of $f(x, y)$ was $\frac{35{,}640}{121}$ and λ was $\frac{-360}{11}$. In this case, $c = 18$, and we conclude that if the constant in the constraint is increased by 1 unit, from 18 to 19, then the maximum value of $f(x, y)$ will decrease by approximately $\frac{360}{11}$ units, from $\frac{35{,}640}{121}$ to $\frac{35{,}280}{121}$.

The next example illustrates how a critical point can be established as a constrained relative maximum or minimum when the Second Derivative Test for Lagrange multipliers fails (that is, whcn $L(x, y) \leq 0$).

EXAMPLE SET C

The production of a manufacturer is given by the Cobb–Douglas production function

$$f(x, y) = 30x^{1/5}y^{4/5}$$

where x represents the number of units of labor (in hours) and y represents the number of units of capital (in dollars) invested. Labor costs \$15 per hour and there are 8 hours in a working day, and 250 working days in a year. The manufacturer has allocated \$4,500,000 this year for labor and capital. How should the money be allocated to labor and capital to maximize productivity this year?

Solution:
The function to be maximized is $f(x, y) = 30x^{1/5}y^{4/5}$. We need to develop the appropriate constraint function. The number of units of labor to be used is x. The annual cost of labor is then (\$15 dollars per hour)(8 hours per day)(250 days)x = \$30,000x. Thus, labor's contribution to the total cost is \$30,000x. Capital's

contribution to the total cost is simply y (since y represents the number of dollars invested). Therefore, the constraint function is

$$30{,}000x + y = 4{,}500{,}000.$$

Hence, we need to maximize $f(x, y) = 30x^{1/5}y^{4/5}$ subject to the constraint $30{,}000x + y = 4{,}500{,}000$.

1. We express the constraint function in the form $g(x, y) = 0$.

$$30{,}000x + y = 4{,}500{,}000 \quad \longrightarrow \quad 30{,}000x + y - 4{,}500{,}000 = 0$$

so that $g(x, y) = 30{,}000x + y - 4{,}500{,}000$.

2. We construct the Lagrangian function $F(x, y, \lambda) = f(x, y) + \lambda g(x, y)$.

$$F(x, y, \lambda) = 30x^{1/5}y^{4/5} + \lambda(30{,}000x + y - 4{,}500{,}000)$$

or

$$F(x, y, \lambda) = 30x^{1/5}y^{4/5} + 30{,}000\lambda x + \lambda y - 4{,}500{,}000\lambda$$

3. We find the first-order partial derivatives F_x, F_y, and F_λ.

$$F_x = 6x^{-4/5}y^{4/5} + 30{,}000\lambda$$

$$F_y = 24x^{1/5}y^{-1/5} + \lambda$$

$$F_\lambda = 30{,}000x + y - 4{,}500{,}000$$

4. We solve the system

$$\begin{cases} F_x = 0 & \ldots(1) \\ F_y = 0 & \ldots(2) \\ F_\lambda = 0 & \ldots(3) \end{cases} \quad \longrightarrow \quad \begin{cases} 6x^{-4/5}y^{4/5} + 30{,}000\lambda = 0 & \ldots(1) \\ 24x^{1/5}y^{-1/5} + \lambda = 0 & \ldots(2) \\ 30{,}000x + y - 4{,}500{,}000 = 0 & \ldots(3) \end{cases}$$

Since this system involves an xy term, we will need to solve it differently than we did the system in Example Set B. We will solve equations (1) and (2) for λ, set those resulting expressions equal, and establish a relation between x and y.

From (1),

$$30{,}000\lambda = \frac{-6y^{4/5}}{x^{4/5}}$$

$$\lambda = \frac{-y^{4/5}}{5000x^{4/5}}$$

From (2),

$$\lambda = \frac{-24x^{1/5}}{y^{1/5}}$$

Therefore, $\dfrac{-y^{4/5}}{5000x^{4/5}} = \dfrac{-24x^{1/5}}{y^{1/5}}$.

We eliminate the denominators by multiplying both sides by $5000x^{4/5}y^{1/5}$.

$$5000x^{4/5}y^{1/5} \cdot \frac{-y^{4/5}}{5000x^{4/5}} = 5000x^{4/5}y^{1/5} \cdot \frac{-24x^{1/5}}{y^{1/5}}$$

$$-y = -120{,}000x$$

$$y = 120{,}000x$$

Substitute $120{,}000x$ for y in equation (3) to obtain a single equation in x.

$$30{,}000x + 120{,}000x - 4{,}500{,}000 = 0$$

$$150{,}000x - 4{,}500{,}000 = 0$$

$$150{,}000x = 4{,}500{,}000$$

$$x = 30$$

so that

$$y = 120{,}000(30) = 3{,}600{,}000$$

The only critical point is $(30, 3{,}600{,}000)$.

5. We suspect that the critical point $(30, 3{,}600{,}000)$ is a constrained relative maximum, but we need to prove it, so we apply the Second Derivative Test. To use the test, we need to find F_{xx}, F_{yy}, and F_{xy}.
Since

$$F_x = 6x^{-4/5}y^{4/5} + 30{,}000\lambda,$$

$$F_{xx} = \frac{-24}{5}x^{-9/5}y^{4/5} \longrightarrow F_{xx} = \frac{-24y^{4/5}}{5x^{9/5}}$$

Since

$$F_y = 24x^{1/5}y^{-1/5} + \lambda$$

$$F_{yy} = \frac{-24}{5}x^{1/5}y^{-6/5} \longrightarrow F_{yy} = \frac{-24x^{1/5}}{5y^{6/5}}$$

Since

$$F_x = 6x^{-4/5}y^{4/5} + 30{,}000\lambda$$

$$F_{xy} = \frac{24}{5}x^{-4/5}y^{-1/5} \longrightarrow F_{xy} = \frac{24}{5x^{4/5}y^{1/5}}$$

Now,

$$L(x, y) = F_{xx}(x, y) \cdot F_{yy}(x, y) - F_{xy}^2(x, y)$$

$$= \frac{-24y^{4/5}}{5x^{9/5}} \cdot \frac{-24x^{1/5}}{5y^{6/5}} - \left[\frac{24}{5x^{4/5}y^{1/5}}\right]^2$$

$$= \frac{24^2}{25x^{8/5}y^{2/5}} - \frac{24^2}{25x^{8/5}y^{2/5}}$$

$$= 0 \longrightarrow \text{The test failed!}$$

The test failed, and no information is provided by the Second Derivative Test. In fact, the Second Derivative Test will always fail when it is applied to Cobb–Douglas production functions (see Exercise 16).

We can still classify the point using the fact that Lagrange's method guarantees that the optimal solution is among the critical points (see the method of Lagrange multipliers, step 4). Since we have only one critical point, it must be optimal. But is it a relative maximum or a relative minimum? To determine which, we will evaluate the original function at the critical point and one other point near the critical point (that also satisfies the constraint). If the function value at the critical point is larger than the function value at a nearby point, we will know we have a relative maximum, and vice versa for a relative minimum.

$$f(30, 3{,}600{,}000) = 30(30)^{1/5}(3{,}600{,}000)^{4/5}$$

$$\approx 10{,}413{,}279.04$$

To obtain another point on the surface $f(x, y)$ that also satisfies the constraint, we will choose $x = 31$ and substitute it into the constraint to find the corresponding value of y.

$$30{,}000(31) + y = 4{,}500{,}000$$

$$930{,}000 + y = 4{,}500{,}000$$

$$y = 4{,}500{,}000 - 930{,}000$$

$$y = 3{,}570{,}000$$

We will evaluate $f(x, y)$ at $(31, 3{,}570{,}000)$.

$$f(31, 3{,}570{,}000) = 30(31)^{1/5}(3{,}570{,}000)^{4/5} \approx 10{,}411{,}856.32$$

Now we have that $f(30, 3{,}600{,}000) > f(31, 3{,}570{,}000)$ and we can confidently claim that there is a constrained relative maximum of $f(x, y)$ at $(30, 3{,}600{,}000)$.

Also, substituting 30 for x and 3,600,000 for y in $\lambda = \dfrac{-y^{4/5}}{5000x^{4/5}}$ produces $\lambda \approx -2.31$.

Interpretation: To maximize production, the manufacturer should allocate 30 hours of labor to each working day of the year, and \$3,600,000 of capital. Also, if the number of dollars allocated for labor and capital is increased by 1, from 4,500,000 to 4,500,001, the number of units produced will decrease by approximately 2.31 units, from approximately 10,413,279.04 to 10,413,276.73.

$|\lambda|$ = Marginal Productivity of Money

When the Lagrange multiplier is used with a Cobb–Douglas production function, its absolute value, $|\lambda|$, is called the **marginal productivity of money**. For the function of Example Set C, when 30 is substituted for x and 3,600,000 for y in $\lambda = \dfrac{-y^{4/5}}{5000x^{4/5}}$, $\lambda \approx -2.31$. Then $|\lambda| = |-2.13| = 2.13$, and approximately, for each additional dollar allocated for production, 2.13 units of a product can be manufactured. For example, for the function of Exercise Set C, if the number of dollars allocated for production is increased by 1, from 3,600,000 to 3,600,001, the number of units produced would increase by approximately 2.31, from 10,413,279.04 to 10,413,281.35. Likewise, if the number of dollars allocated for production were decreased by 10,000, the number of units produced would decrease by approximately $(10,000)(2.13) = 21{,}300$, from 10,413,279.04 to 10,390,979.04.

EXERCISE SET 8.4

A UNDERSTANDING THE CONCEPTS

In each of Exercises 1–5, a value of λ is given along with a critical point that produces a constrained relative maximum or minimum, and the corresponding maximum or minimum value. Interpret the meaning of λ as it applies to the given situation.

1. ***Education: Achievement*** In a large, four-unit general anthropology course given at a college, the function $f(x, y)$ measures student achievement. The number of hours devoted to lecture is represented by x and the number of hours devoted to discussion by y. The course requires 72 hours of instruction. The constraint function is $x+y = 80$, achievement is maximized at 85 at $(54, 18)$, and the corresponding value of λ is 1.3.
2. ***Economics: Pesticide Application*** Scientists believe that the function $N(x, y)$ approximates the number (in thousands) of insects killed when sprayed with x pounds of pesticide A and y pounds of pesticide B. Because too much pesticide is harmful to the plants that live in the vicinity of the insects, no more than 250 pounds of the pesticide can be used in one spraying. The constraint function is then $x + y = 250$. $N(x, y)$ is maximized at $(180, 70)$ and, furthermore, $N(180, 70) = 250$. The corresponding value of λ is 3.2.
3. ***Business: Container Size*** The typical beverage can has a volume of about 21.66 cubic inches. The function $S(h, r)$ describes the surface area of the can in terms of the can's height and radius. A radius of 1.51 inches and a height of 3.02 inches produces the minimum surface area of approximately 42.98 square inches. The constraint function is $\pi r^2 h = 21.66$, and the corresponding value of λ is 1.32.
4. ***Business: Cobb–Douglas Production*** For a certain company, the Cobb–Douglas production function $f(x, y)$ is maximum at $(1500, 3500)$, when the constraint is $3x + 11y = 43{,}000$. The corresponding value of λ is 5.2.
5. ***Business: Cost of Quality Control*** The cost, $C(x, y)$ (in dollars), of quality control at a manufacturing company is a function of the number x of items tested at point A, and the number y of items tested at point B. Company policy specifies that 35 items be tested each day. The constraint function is $x + y = 35$, and the cost function is minimized at \$355 when $x = 12$ and $y = 23$. The corresponding value of λ is 4.09.

B SKILL ACQUISITION

For Exercises 6–15, optimize each function given that it is subjected to the noted constraint.

6. $f(x, y) = x^2 + y^2 - 10$ subject to $x + y = 3$
7. $f(x, y) = 3x^2 + 4y^2 + 1$ subject to $x + 2y = 4$
8. $f(x, y) = x^2 + 5xy + 2y^2$ subject to $2x + 3y - 4 = 0$
9. $f(x, y) = x^2 y$ subject to $2x + y - 6 = 0$
10. $f(x, y) = x^2 + xy + y^2 - 2x - 5y - 4$ subject to $y = x - 1$
11. $f(x, y) = 25 - x^2 - y^2$ subject to $y = -x - 1$
12. $f(x, y) = y^3 + 6x$ subject to $y - 2x = 0$
13. $f(x, y) = 2x^2 - y^2$ subject to $x - y = 2$
14. $f(x, y) = 6x^{1/3}y^{2/3}$ subject to $4x + 3y = 36$

15. $f(x, y) = 21x^{3/7}y^{4/7}$ subject to $9x + 2y - 20 = 0$

16. ***Business: Cobb–Douglas Production*** Show that the Second Derivative Test for Lagrange multipliers always fails for the Cobb–Douglas production function $f(x, y) = Cx^a y^{1-a}$ when it is subjected to the linear constraint $px + qy = k$.

17. ***Business: Cobb–Douglas Production*** It is true, although we will not prove it here, that a constrained relative extreme point of a Cobb–Douglas production function is always a maximum. Show that the constrained relative maximum of the Cobb–Douglas production function $f(x, y) = Cx^a y^{1-a}$, when it is subjected to the linear constraint $px + qy = k$, is $\left(\frac{ak}{p}, \frac{(1-a)k}{q}\right)$.

C APPLYING THE CONCEPTS

18. ***Manufacturing: Cobb–Douglas Production*** The production of a manufacturer is given by the Cobb–Douglas production function

$$f(x, y) = 12x^{2/3}y^{1/3}$$

where x represents the number of units of labor (in hours) and y represents the number of units of capital (in dollars) invested. Labor costs \$8 per hour, there are 8 hours in a working day and 260 working days in a year. The manufacturer has allocated \$1,622,400 this year for labor and capital. How should the money be allocated to labor and capital to maximize productivity this year? Find and interpret the marginal productivity of money.

19. ***Manufacturing: Cobb–Douglas Production*** The production of a manufacturer is given by the Cobb–Douglas production function

$$f(x, y) = 4x^{5/8}y^{3/8}$$

where x represents the number of units of labor (in hours) and y represents the number of units of capital (in dollars) invested. Labor costs \$9.50 per hour, there are 8 hours in a working day and 255 working days in a year. The manufacturer has allocated \$2,232,576 this year for labor and capital. How should the money be allocated to labor and capital to maximize productivity this year? Find and interpret the marginal productivity of money.

20. ***Manufacturing: Cobb–Douglas Production*** The production of a manufacturer is given by the Cobb–Douglas production function

$$f(x, y) = x^{3/5}y^{2/5}$$

where x represents the number of units of labor (in hours) and y represents the number of units of capital (in dollars) invested. Labor costs \$12 per hour, there are 8 hours in a working day and 22 working days in a month. The manufacturer has allocated \$48,032 this month for labor and capital. How should the money be allocated to labor and capital to maximize productivity this month? Find and interpret the marginal productivity of money.

21. ***Economics: Optimum Level of Production*** It is an economic law (that we will not prove) that at the optimum level of production, the ratio of unit costs of labor and capital is equal to the ratio of their marginal productivities. Symbolically, $\frac{\text{unit cost of labor}}{\text{unit cost of capital}} = \frac{f_x}{f_y}$. Illustrate this law using the following situation. Suppose that x units of labor and y units of capital can produce $f(x, y) = 20x^{3/5}y^{2/5}$ units of a product. Suppose that labor costs \$50 per unit and capital costs \$300 per unit. Assume that \$50,000 is available to spend on production.

22. ***Manufacturing: Production Costs*** A manufacturer has two sites, A and B, at which it can produce a product, and because of certain conditions, site A must produce three times as many units as site B. The total cost of producing the units is given by the cost function $C(x, y) = 0.3x^2 - 150x - 610y + 150{,}000$, where x represents the number of units produced at site A and y represents the number of units produced at site B. How many units should be produced at each site to minimize the cost? What is the minimal cost? Find and interpret the value of the Lagrange multiplier.

23. ***Manufacturing: Quality Assurance Costs*** A manufacturer finds that the cost of quality assurance is related to the number of inspections, x and y, it makes each day at two points, A and B, respectively, in its assembly process. The manufacturer must make 5 times as many inspections at point A than at point B. The cost of making these inspections is approximated by the function $C(x, y) = 6x^2 + 3y^2 - 612x$. How many inspections at each location should the manufacturer make to minimize costs?

24. ***Business: Fencing Costs*** Determine the dimensions of the largest rectangular area that can be fenced using 340 feet of fencing material.

25. ***Business: Publishing Costs*** The typical sheet of typing paper is rectangular and measures 8.5 inches by 11 inches. These dimensions produce a sheet of paper having a perimeter (length around) of 39 inches and an area of 93.5 square inches. The 93.5 square inches is not the maximum area possible for a rectangular sheet of paper having a 39-inch perimeter. What dimensions, subject to the constraint that they produce a perimeter of 39 inches, of a rectangular sheet of paper would result in the maximum area?

26. ***Manufacturing: Container Size*** The volume of a right circular cylinder with radius r and height h is given by $\pi r^2 h$, and the circumference of a circle with radius r is $2\pi r$. Use these facts to find the dimensions of a 10-ounce (approximately 5.74π cubic inches) can with a top

and bottom that can be made using the least amount of material.

27. ***Business: Advertising*** A franchise operation in a city has determined that if it buys x radio and y newspaper advertisements each day, the number of people it can reach and possibly influence is approximated by the function $N(x, y) = 30x(4y + 40)^{2/3}$. If the franchise operation has $42,000 budgeted each day for advertisements and each radio advertisement costs $100 per day and each newspaper advertisement costs $50 per day, how many radio and newspaper ads should it buy each day to reach the maximum number of people? What is that maximum number of people? Find and interpret the value of the Lagrange multiplier.

28. ***Business: Publishing Revenue*** A publisher believes it can sell 26,600 copies of its new book. The publisher will first make the book available in hardback and then, some time later, in paperback. If it sells each paperback book for x dollars, and each hardback book for y dollars, the publisher believes it can sell $5{,}150 - 350x + 700y$ paperback copies and $9{,}200 + 1050x - 350y$ hardback copies. Find the price of each type of book that will maximize the publisher's revenue. Find and interpret the value of the Lagrange multiplier. (Hint: Recall that revenue equals price times quantity.)

29. ***Business: Agricultural Pesticide*** To keep pests from destroying a particular crop, agriculturists spray a mix of x pounds of pesticide A and y pounds of pesticide B onto the crop. They have determined that the percentage of pests killed by this application is approximated by the function $f(x, y) = 1 - \frac{1}{4}e^{-x/5} - \frac{1}{4}e^{-y/70}$. If 2000 pounds of pesticide are to be applied, how many pounds of each type of pesticide should be used to maximize the percentage of pests killed by an application of the pesticide? (Hint: You will need to use the natural logarithm to solve for x or y.)

D DESCRIBE YOUR THOUGHTS

30. Describe the difference between unconstrained and constrained relative extrema.
31. Describe the strategy you would use to solve a problem such as Exercise 6 or 7.
32. Explain, as if you were writing a letter to a friend who is also studying calculus, the meaning of the Lagrange multiplier λ.

E REVIEW

33. **(2.2)** Find, if it exists, $\lim_{x \to 4} \dfrac{x^2 - 8x + 16}{x - 4}$.

34. **(3.2)** Find and interpret $f'(3)$ for $f(x) = \dfrac{x - 1}{x + 2}$.

35. **(3.3)** Find $f'(0)$ for $f(x) = \dfrac{5}{(x^2 + 1)^4}$.

36. **(3.3)** Find $f'(4)$ for $f(x) = (x + 1)^3(x - 5)^4$.

37. **(3.4)** Find $f^{(4)}(x)$ for $f(x) = 5x^3 - 3x^2 + 2x - 8$.

38. **(4.3)** ***Business: Revenue*** Suppose a dry food company has determined that its total revenue for its food is given by

$$R = -x^3 + 27x^2 + 900x$$

where R is measured in dollars and x is the number of units in thousands produced. What production level will yield a maximum revenue?

39. **(5.4)** Find $f'(x)$ for $f(x) = 2x\ln(2x)$.

40. **(6.2)** Evaluate $\int e^{-0.14x}\,dx$.

41. **(6.3)** Evaluate $\int 4x^5 e^{12x}\,dx$.

42. **(8.2)** Find the second-order partial derivatives of the function $f(x, y) = 4x^2 + 3y^2 + 6xy + 8x - 4y - 3$.

43. **(8.3)** Find, if any exist, the relative extreme points of the function $f(x, y) = x^3 + y^3 - 9xy + 1$.

8.5 The Total Differential

Introduction

For a function $f(x, y)$, the partial derivative f_x can be interpreted as the approximate change in the function value as the value of x changes and the value of y remains constant. The partial derivative for f_y can be similarly interpreted. In this section we investigate how to approximate the change in the function value as both x and y change.

The Total Differential

Just as there is a tangent line at a particular x value associated with a function $f(x)$ of one variable, there is a tangent plane at a particular point associated with a function $f(x, y)$ of two variables. Imagine being located at a point (x, y) in the xy plane and moving small distances dx and dy in the x and y directions, respectively, to the new point $(x + dx, y + dy)$. To get to the new point, we could first move dx units parallel to the x axis, and from there, dy units parallel to the y axis. The product

$$f_x(x, y) \cdot dx$$

approximates the change in the function as x changes value, and the product

$$f_y(x, y) \cdot dy$$

approximates the change in the function as y changes value.

The total change in the function is then approximated by

$$f_x(x, y)dx + f_y(x, y)dy$$

We denote this sum of individual differentials by df (f being the name of the function) and call it the **total differential** of the function $f(x, y)$.

The Total Differential

The total differential of a function of two variables $f(x, y)$ is

$$df = f_x(x, y)dx + f_y(x, y)dy$$

and it approximates the change in the function as x changes a small amount by dx units and y changes a small amount by dy units.

The exact change in the function is $f(x+dx, y+dy) - f(x, y)$, and thus,

$$df \approx f(x+dx, y+dy) - f(x, y)$$

Extending the Differential

The differential formula can be extended to functions of more than two variables. For example, for a function of three variables, x, y, and z, the differential formula is

$$df = f_x\,dx + f_y\,dy + f_z\,dz$$

EXAMPLE SET A

The productivity (the number of units of a product a company is able to produce) of a company is described by the Cobb-Douglas production function

$$f(x, y) = 42x^{3/7}y^{4/7}$$

where x represents the number of units of labor and y the number of units of capital utilized. Approximate the change in output if the number of units of labor is decreased from 2187 to 2183, and the number of units of capital is increased from 128 to 130.

Solution:
We will begin by finding the partial derivatives, f_x and f_y, and the changes in x and y.

$$f_x(x, y) = \frac{18y^{4/7}}{x^{4/7}} \quad \text{and} \quad f_y(x, y) = \frac{24x^{3/7}}{y^{3/7}}$$

$$dx = 2183 - 2187 = -4 \quad \text{and} \quad dy = 130 - 128 = 2$$

Substituting the partial derivatives into the differential formula $df = f_x\,dx + f_y\,dy$ produces the differential for the function.

$$df(x, y) = \frac{18y^{4/7}}{x^{4/7}}dx - \frac{24x^{3/7}}{y^{3/7}}dy$$

We then evaluate the differential at $(x, y) = (2187, 128)$, $dx = -4$, and $dy = 2$.

$$\begin{aligned} df(2187, 128) &= \frac{18(128)^{4/7}}{(2187)^{4/7}} \cdot (-4) + \frac{24(2187)^{3/7}}{(128)^{3/7}} \cdot (2) \\ &\approx -14.2 + 162 \\ &\approx 147.8 \end{aligned}$$

Interpretation: If the number of units of labor is decreased by 4 units, from 2187 to 2183, and the number of units of capital is increased by 2 units, from 128 to 130, the number of units of output will increase by approximately 147.8. (Remember, the change is approximate because there is a variable in the derivative expression.)

The exact value of the change in the function is given by the functional value at the new point minus the functional value at the original point.

$$f(2183, 130) - f(2187, 128) = 42(2183)^{3/7}(130)^{4/7} - 42(2187)^{3/7}(128)^{4/7}$$
$$\approx 147.1$$

Thus, the difference between the approximate value and the actual value is $147.8 - 147.1 = 0.7$, a relatively small error.

The Total Differential and Technology

Using Your Calculator You can use your graphing calculator to find the total differential of a function of two variables. The following entries illustrate the procedure for the function $f(x, y) = 42x^{3/7}y^{4/7}$.

```
Y1 = 42x^(3/7) y^(4/7)
Y5 = nDeriv(Y1,X,X,.001)
Y7 = nDeriv(Y1,Y,Y,.001)
Y9 = Y5*C + Y7*D
```

The functions Y5 and Y7 are, respectively, f_x and f_y. The entries C and D represent, respectively, the differentials dx and dy. In the computation window, store 2187 to X, 128 to Y, −4 to C, and 2 to D. Then compute Y9. Your result should be about 147.77 . . .

Calculator Exercises

1. Wind chill is often used to measure how cold it is in the wintertime. Suppose a wind chill function is

$$W(v, T) = 30 - 0.045(10.5 + 9\sqrt{v} - 0.085v)(30 - T)$$

The variables W, v, and T measure the wind chill, the wind velocity in miles per hour, and the temperature in degrees Celsius, respectively. Approximate the change in the wind chill if the wind velocity increases from 22 miles per hour to 24.5 miles per hour and the temperature decreases from $-6.5°$ to $-7.5°$C.

2. An industrial psychologist has determined that reaction time R, in reaction units, to x units of a particular drug t hours after it is introduced into the bloodstream is approximated by the function

$$R(x, t) = 3.2x(8.6 - 1.3x)t^{1.2}e^{-1.8t}$$

Find the approximate change in a person's reaction if the number of units of the drug is decreased from 150 units to 145 units and the time since introduction increases from 0.75 hours to 0.88 hours.

3. For a manufacturer, the Cobb–Douglass production function is

$$f(x, y) = 245x^{0.36}y^{0.64}$$

Approximate the change in production if the number of units of labor is decreased from \$650 to \$645 and the number of units of capital is decreased from \$825 to \$820.

4. The number N, in thousands of people reached by an advertiser each week through the use of x radio advertisements and y television advertisements, is approximated by the function

$$N(x, y) = 16.5x(9.3y + 18.2)^{0.45}$$

Approximate the change in the number of people reached by the advertiser if the number of radio ads is decreased by 3 from 20 and the number of TV ads is increased by 4 from 40.

EXERCISE SET 8.5

A UNDERSTANDING THE CONCEPTS

1. Suppose that for a function $f(x, y)$, $dx = 1$ and $dy = 2$ at the point (a, b). Determine, by considering $f_x(a, b)$ and $f_y(a, b)$, if it is possible for $df(a, b) < 0$.
2. Suppose that for a function $f(x, y)$, $f_x(a, b) > 0$ and $dx > 0$ at the point (a, b). What conditions do $f_x(a, b)$ and dy have to satisfy so that $df(a, b) < 0$?
3. Suppose that $f(x, y) = e^{x+y}$, $dx < 0$, and $dy < 0$ at the point (a, b). Is it possible that $df(a, b) > 0$? If so, how?
4. If you think the differential of a function f of four variables x, y, z, and w could exist, specify its form.

B SKILL ACQUISITION

For Exercises 5–16, find the differential of the given function.

5. $f(x, y) = 8x^2 - 3y^2 + 4x + 5y + 6$, if $dx = 1$ and $dy = 2$
6. $f(x, y) = 16x^2 + 4y^2 - 10xy$, if $dx = 3$ and $dy = 1$
7. $f(x, y) = x^2 - 15y^2 + 4e^{xy}$, if x increases from 3 to 4, and y increases from 1 to 2
8. $f(x, y) = 3x + 2y + 6\ln(x^2y)$, if x decreases from 5 to 4, and y increases from 10 to 11
9. $f(x, y) = \sqrt{xy}$, if x increases from 9 to 10, and y decreases from 4 to 3
10. $f(x, y, z) = x^2 + y^2 - z^2$, if x increases from 10 to 11, y increases from 25 to 27, and z decreases from 1 to 0
11. $f(x, y, z) = 5x^2e^{2y-z^2}$, if x decreases from 3 to 1, y increases from 1 to 2, and z decreases from 11 to 8
12. $f(x, y) = 300x^{2/3}y^{1/3}$, if x increases from 27 to 28, and y decreases from 64 to 61
13. $f(x, y) = \dfrac{y}{x}$, if x increases from 150 to 155, and y increases from 20 to 21
14. $f(x, y) = \ln(y - x)$, if x decreases from 5 to 4, and y increases from 60 to 61
15. $f(x, y, z) = y\ln x$, if x increases from 1 to 2, y increases from 0 to 2, and z decreases from 0 to 1
16. $f(x, y, z) = \dfrac{x+z}{y-z}$, if x decreases from 100 to 96, y decreases from 25 to 24, and z increases from 10 to 11

C APPLYING THE CONCEPTS

17. ***Business: Cobb–Douglas Production*** The output of a company is given by the Cobb–Douglas production function $f(x, y) = 90x^{2/3}y^{1/3}$, where x represents the number of thousands of dollars invested in labor and y the number of thousands of dollars invested in capital.
    a. What is the approximate change in the output if the amount of money invested in labor is decreased from \$30,000 to \$27,000 and the amount invested in capital is decreased from \$90,000 to \$85,000?
    b. Compare this approximate change to the actual change.
18. ***Economics: City Transportation*** The number N (in thousands) of riders of a city's public transportation system each day is related to both the number x (in thousands) of people living in the city and the price y (in dollars) of a ticket by the function $N(x, y) = \dfrac{0.1x}{0.5 + \ln y}$.
    a. What is the approximate change in the number of riders if the number of people living in the city increases from 163,000 to 164,500 and the price of a ticket increases from \$0.80 to \$0.90?
    b. Compare this approximate change to the actual change.
19. ***Business: Quality Assurance Costs*** The cost C (in dollars) of quality assurance at a company is related to the number of inspections x made at point A in a manufacturing process, and the number y made at point B by the function $C(x, y) = 6x^2 + 4y^2 - x - 3y$.
    a. What is the approximate change in the cost of quality assurance if the number of inspections at point A is increased from 14 to 16, and the number at point B is decreased from 12 to 10?

b. Compare this approximate change to the actual change.

20. ***Business: Cost of Production*** A company produces two products, A and B. The cost C of producing x units of A and y units of B is $C(x, y) = \frac{x^3}{600} + \frac{y^3}{300} + 25x + 10y + 10xy$.

a. What is the approximate change in the cost of producing these two products if the company increases production of product A from 45 units to 50 units, and decreases the production of product B from 30 units to 28 units?

b. Compare this approximate change to the actual change.

21. ***Business: Production Output*** The output (in thousands of units each week) of a company is given by the production function $f(x, y) = 0.8xe^{0.03y}$, where x represents the number of hours allocated each week to labor and y the number of thousands of dollars allocated each week to capital. What is the approximate change in the output of the company if the number of weekly hours allocated to labor is decreased from 200 to 195, and the amount of money allocated for capital is increased from $40,000 to $40,500?

D DESCRIBE YOUR THOUGHTS

22. Describe how the total differential of a function $f(x, y)$ differs from a partial derivative of $f(x, y)$.

23. Describe the strategy you would use to solve a problem such as Exercise 9 or 10.

E REVIEW

24. **(2.2)** Find, if it exists, $\lim_{x\to 6} \frac{x^2 - 8x + 12}{x^2 + 2x - 48}$.

25. **(2.3)** Find, if it exists, $\lim_{x\to \infty} \frac{5x^4 - 3x^3 + x - 10}{x^2 + 80}$.

26. **(2.4)** The function

$$f(x) = \begin{cases} x^2 - 6x + 13, & x \leq 3 \\ 5, & x = 3 \\ x - 2, & x \geq 3 \end{cases}$$

is discontinuous at $x = 3$. Which of the three continuity conditions is the first to be invalid?

27. **(3.1)** Use the definition of the derivative to find $f'(x)$ for $f(x) = 5x^2 - 2x + 4$.

28. **(3.5)** Use implicit differentiation to find $f'(x)$ for $2x^3 + 3y^2 + 4x^2y + 3x = 2$.

29. **(5.4)** Find $f'(x)$ for $f(x) = x^2e^{2x+7}$.

30. **(6.6)** ***Economics: Income Flow*** The rate I (in dollars per year) at which income from a rental property flows into an account in which interest is compounded continuously at 7% is given by the function

$$I(t) = 20,000e^{-0.07t}$$

where t is the time in years from now. If the income is not disrupted and continues indefinitely, how much money will accumulate in this account?

31. **(7.1)** ***Business: Revenue Flow*** Over the next 10 years, a company projects its continuous flow of revenue to be $R(t) = 69e^{0.08t}$, and its costs $C(t) = 0.4t^2 + 28$, where t is in years from now and both $R(t)$ and $C(t)$ are in millions of dollars. Approximate the profit this company can expect over the next 10 years.

32. **(8.3)** Find the relative extrema of the function $f(x, y) = x^3 + y^3 - 6xy + 4$.

33. **(8.4)** Find the absolute extreme points of the function $f(x, y) = 2x^2y - 50$ when it is subjected to the constraint $x^2 + y^2 = 1$.

8.6 Double Integrals

Introduction
Double Indefinite Integrals
Double Definite Integrals
Double Integrals and Technology

Introduction

When we worked with functions of several variables in Section 8.2, we noted that when we performed a differentiation such as

$$\frac{\partial}{\partial x}\left[x^2+y^2+3x+8y+5xy-2\right]$$

the symbol $\frac{\partial}{\partial x}$ indicated that the variable was x. Any other letters were considered constants.

In the differentiation $\frac{\partial^2}{\partial x \partial y}\left[x^2+y^2+3x+8y+5xy-2\right]$, the symbols $\partial x \partial y$ indicate that two derivatives are to be found. The function is differentiated first with respect to y and then with respect to x.

We will now turn our attention to the integration of functions of more than one variable. We will make our study an introductory one by restricting our attention to functions of only two variables. You will see that the symbols $dx\,dy$ or $dy\,dx$ indicate that two integrals are to be found.

Double Indefinite Integrals

Suppose $f(x, y)$ is a function of the two variables x and y. Then, because of the dx, $\int f(x, y)\,dx$ indicates that $f(x, y)$ is to be integrated with respect to x, and that the y is to be treated as a constant. Since this integral is an indefinite integral, a constant of integration must appear. For example,

$$\int [x^2+y^2+3x+8y+5xy-2\,dx$$

$$=\frac{x^3}{3}+y^2x+\frac{3x^2}{2}+8yx+\frac{5x^2y}{2}-2y+g(y)$$

Function of Integration

The function $g(y)$ represents the constant of integration and is called the **function of integration**. Remember that since y is treated as a constant, the constant may be more than just a real number. It may include terms involving y. The notation $g(y)$ indicates this possibility. We can check the integration by differentiating with respect to x. Showing most of the steps,

$$\frac{\partial}{\partial x}\left[\frac{x^3}{3}+y^2x+\frac{3x^2}{2}+8yx+\frac{5x^2y}{2}-2y+g(y)\right]$$

$$= x^2 + y^2 + 3x + 8y + 5xy - 2 + 0$$

$$= x^2 + y^2 + 3x + 8y + 5xy - 2$$

which is the original function, so we conclude that the integration is correct. The 0 in the differentiation process comes from $\frac{\partial}{\partial x} g(y)$, since the derivative of a constant is 0. Thus, $\int f(x, y)\, dx$ results in another function of two variables that can itself be integrated. Let's consider the meaning of

$$\int \left[\int f(x, y)\, dx \right] dy$$

Double Indefinite Integrals

Integrals such as $\int \left[\int f(x, y)\, dx \right] dy$ are called **double indefinite integrals** and, just as do ordinary integrals, they represent functions.

This notation first directs us to integrate $f(x, y)$ with respect to x, getting some new function of x and y, and then to integrate this new function with respect to y. For example,

$$\int \left[\int f(x, y)\, dx \right] dy = \int \left[\frac{x^3}{3} + y^2 x + \frac{3x^2}{2} + 8yx + \frac{5x^2 y}{2} - 2y + g(y) \right] dy$$

$$= \frac{x^3 y}{3} + \frac{y^3 x}{3} + \frac{3x^2 y}{2} + \frac{8y^2 x}{2} + \frac{5x^2 y^2}{4} - \frac{2y^2}{2} + \int g(y)\, dy + h(x)$$

$$= \frac{x^3 y}{3} + \frac{y^3 x}{3} + \frac{3x^2 y}{2} + \frac{8y^2 x}{2} + \frac{5x^2 y^2}{4} - \frac{2y^2}{2} + G(y) + h(x)$$

In this result, the $\int g(y)\, dy$ represents some function of y, which we decided to call $G(y)$, and the $h(x)$ appears as the function of integration in the same way $g(y)$ appeared in the previous example.

EXAMPLE SET A

1. Evaluate $\int \int xy^2\, dx\, dy$.

Solution:

$$\int \int xy^2\, dx\, dy = \int \left[\int xy^2\, dx \right] dy$$

$$= \int \left[\frac{x^2}{2} \cdot y^2 + g(y) \right] dy$$

$$= \frac{x^2}{2} \cdot \frac{y^3}{3} + \int g(y)\, dy + h(x)$$

$$= \frac{1}{6} x^2 y^3 + G(y) + h(x)$$

2. Evaluate $\int\int (x^2 + y^2 - 3xy)\, dy\, dx$.

Solution:

$$\int\int (x^2 + y^2 - 3xy)\, dy\, dx = \int \left[\int (x^2 + y^2 - 3xy)\, dy\right] dx$$

$$= \int \left[x^2 y + \frac{y^3}{3} - 3x \cdot \frac{y^2}{2} + g(x)\right] dx$$

$$= \frac{x^3}{3} \cdot y + \frac{y^3}{3} \cdot x - \frac{3x^2}{2} \cdot \frac{y^2}{2} + \int g(x)\, dx + h(y)$$

$$= \frac{1}{3}x^3 y + \frac{1}{3}xy^3 - \frac{3}{4}x^2 y^2 + G(x) + h(y)$$

Double Definite Integrals

The process of double integration can be extended to definite integrals. Just as the constant of integration cancels in ordinary definite integrals, the functions of integration cancel in double definite integrals, and there is no need to write them.

You may recall that in the differentiation of a function of two variables, if $f(x, y)$ is continuous on $a \leq x \leq b$ and $c \leq y \leq d$, then

$$\frac{\partial^2 f}{\partial x \partial y} = \frac{\partial^2 f}{\partial y \partial x}$$

That is, the order of differentiation did not matter. An analogous situation exists for the integration of functions of two variables.

Integrating Double Definite Integrals

If $f(x, y)$ is continuous on the region R defined by $a \leq x \leq b$ and $c \leq y \leq d$, then

$$\int_c^d \int_a^b f(x, y)\, dx\, dy = \int_a^b \int_c^d f(x, y)\, dy\, dx$$

This fact allows us to express

$$\int_c^d \left[\int_a^b f(x, y)\, dx\right] dy \quad \text{as} \quad \int_c^d \int_a^b f(x, y)\, dx\, dy$$

and

$$\int_a^b \left[\int_c^d f(x, y)\, dx\right] dy \quad \text{as} \quad \int_a^b \int_c^d f(x, y)\, dy\, dx$$

since the order of integration does not affect the final answer when integrating continuous functions. Also, like their ordinary definite integral counterparts, double definite integrals represent real numbers. Notice also that the integrals are evaluated *inside out*.

Example Set B illustrates the process of double integration with definite integrals.

EXAMPLE SET B

1. Evaluate $\int_0^2 \int_0^1 (24x^3y^2 - 20xy)\, dx\, dy$.

Solution:

The notation $dx\, dy$ indicates that we should first integrate with respect to x and then with respect to y.

$$\int_0^2 \int_0^1 (24x^3y^2 - 20xy)\, dx\, dy$$

$$= \int_0^2 \left[\frac{24x^4y^2}{4} - \frac{20x^2y}{2}\right]\Big|_0^1 dy$$

$$= \int_0^2 \left[6x^4y^2 - 10x^2y\right]\Big|_0^1 dy \quad \text{Substitute 1 and 0 for } x.$$

$$= \int_0^2 \left(\left[6(1)^4y^2 - 10(1)^2y\right] - \left[6(0)^4y^2 - 10(0)^2y\right]\right) dy$$

$$= \int_0^1 \left(\left[6y^2 - 10y\right] - [0 - 0]\right) dy$$

$$= \int_0^2 (6y^2 - 10y)\, dy$$

$$= \left[\frac{6y^3}{3} - \frac{10y^2}{2}\right]\Big|_0^2$$

$$= \left[2y^3 - 5y^2\right]\Big|_0^2 \quad \text{Now substitute 2 and 0 for } y.$$

$$= \left[2(2)^3 - 5(2)^2\right] - \left[2(0)^3 - 5(0)^2\right]$$

$$= [16 - 20] - [0 - 0]$$

$$= [-4] - 0$$

$$= -4$$

Thus, $\int_0^2 \int_0^1 (24x^3y^2 - 20xy)\, dx\, dy = -4$.

Example 2 will be the same integral as Example 1, but the order of integration will be changed.

2. Evaluate $\int_0^2 \int_0^1 (24x^3y^2 - 20xy)\, dx\, dy.$

Solution:

This is the double integral we evaluated in Example 1. Since we know we can change the order of integration, let's do so and then compare the result with the result of Example 1.

$$\int_0^2 \int_0^1 (24x^3y^2 - 20xy)\, dx\, dy$$

$$= \int_0^1 \int_0^2 (24x^3y^2 - 20xy)\, dy\, dx$$

$$= \int_0^1 \left[\frac{24x^3y^3}{3} - \frac{20xy^2}{2}\right]\Big|_0^2 dx$$

$$= \int_0^1 \left[8x^3y^3 - 10xy^2\right]\Big|_0^2 dx \quad \text{Substitute 2 and 0 for } y.$$

$$= \int_0^1 \left(\left[8x^3(2)^3 - 10x(2)^2\right] - \left[8x^3(0)^3 - 10x(0)^2\right]\right) dx$$

$$= \int_0^1 \left(\left[64x^3 - 40x\right] - [0 - 0]\right) dx$$

$$= \int_0^1 \left(64x^3 - 40x\right) dx$$

$$= \left[\frac{64x^4}{4} - \frac{40x^2}{2}\right]\Big|_0^1$$

$$= \left[16x^4 - 20x^2\right]\Big|_0^1 \quad \text{Now substitute 1 and 0 for } x.$$

$$= \left[16(1)^4 - 20(1)^2\right] - \left[16(0)^4 - 20(0)^2\right]$$

$$= [16 - 20] - [0 - 0]$$

$$= -4$$

Thus, $\int_0^2 \int_0^1 (24x^3y^2 - 20xy)\, dx\, dy = -4.$

Comparing Examples 1 and 2, we see, as we expected, that the order of integration did not affect the result and that

$$\int_0^2 \int_0^1 (24x^3y^2 - 20xy)\, dx\, dy = \int_0^1 \int_0^2 (24x^3y^2 - 20xy)\, dy\, dx$$

EXAMPLE SET C

Evaluate $\int_0^2 \int_0^3 xy\, dy\, dx$.

Solution:

$$\begin{aligned}
\int_0^2 \int_0^3 xy\, dy\, dx &= \int_0^2 \left[\int_0^3 xy\, dy\right] dx \\
&= \int_0^2 \left[x \cdot \frac{y^2}{2}\right]\Big|_0^3 dx \\
&= \int_0^2 \frac{x}{2}\left(3^2 - 0^2\right) dx \\
&= \int_0^2 \frac{9x}{2}\, dx \\
&= \left(\frac{9}{2} \cdot \frac{x^2}{2}\right)\Big|_0^2 \\
&= \frac{9}{4}x^2\Big|_0^2 \\
&= \frac{9}{4}\left(2^2 - 0^2\right) \\
&= \frac{9}{4} \cdot 4 \\
&= 9
\end{aligned}$$

In the next two examples, we will look at situations where the limits on one of the integrals involves variable expressions.

EXAMPLE SET D

Evaluate $\int_0^3 \int_{2y}^2 dx\, dy$.

Solution:

$$\int_0^3 \int_{2y}^2 dx\,dy = \int_0^3 \left[\int_{2y}^2 dx\right] dy$$

$$= \int_0^3 x\Big|_{2y}^2 dy$$

$$= \int_0^3 (2 - 2y)\,dy$$

$$= \left(2y - \frac{2y^2}{2}\right)\Big|_0^3$$

$$= (2y - y^2)\Big|_0^3$$

$$= (2 \cdot 3 - 3^2) - (2 \cdot 0 - 0^2)$$

$$= (6 - 9) - (0 - 0)$$

$$= -3$$

EXAMPLE SET E

Evaluate $\int_0^2 \int_{3x^2-6x}^{2x-x^2} 2x\,dy\,dx$.

Solution:

$$\int_0^2 \int_{3x^2-6x}^{2x-x^2} 2x\,dy\,dx = \int_0^2 \left[\int_{3x^2-6x}^{2x-x^2} 2x\,dy\right] dx$$

$$= \int_0^2 2xy\Big|_{3x^2-6x}^{2x-x^2} dx$$

$$= \int_0^2 2x\left[(2x - x^2) - (3x^2 - 6x)\right] dx$$

$$= \int_0^2 2x\left(8x - 4x^2\right) dx$$

$$= \int_0^2 \left(16x^2 - 8x^3\right) dx$$

$$= \left(\frac{16x^3}{3} - \frac{8x^4}{4}\right)\Big|_0^2$$

$$= \left(\frac{16x^3}{3} - 2x^4\right)\Big|_0^2$$

$$= \left[\frac{16(2)^3}{3} - 2(2)^4\right] - \left[\frac{16(0)^3}{3} - 2(0)^4\right]$$

$$= \left(\frac{128}{3} - 32\right) - (0 - 0)$$

$$= \frac{32}{3}$$

The following example will illustrate how to use substitutions with definite double integrals.

EXAMPLE SET F

Evaluate $\int_0^2 \int_0^1 xye^{-(x^2+y^2)}\, dy\, dx$.

Solution:
Let

$$u = -\left(x^2 + y^2\right)$$

and

$$du = -2y\, dy$$

$$-\frac{1}{2}\, du = y\, dy$$

Then (watch how the limits on the inner integral change with the substitution),

$$\int_0^2 \int_0^1 xye^{-(x^2+y^2)}\, dy\, dx = \int_0^2 \int_{-x^2}^{-(x^2+1)} -\frac{1}{2}xe^u\, du\, dx$$

$$= \int_0^2 -\frac{1}{2}xe^u \Big|_{-x^2}^{-(x^2+1)}\, dx$$

$$= -\frac{1}{2}\int_0^2 x\left[e^{-(x^2+1)} - e^{-x^2}\right] dx$$

$$= -\frac{1}{2}\int_0^2 xe^{-x^2}\left(e^{-1} - 1\right) dx$$

$$= -\frac{1}{2}\left(e^{-1} - 1\right)\int_0^2 xe^{-x^2}\, dx$$

Now let

$$w = -x^2$$

$$dw = -2x\,dx$$

$$-\frac{1}{2}\,dw = x\,dx$$

Therefore,

$$= -\frac{1}{2}\left(e^{-1} - 1\right)\int_{0}^{-4} -\frac{1}{2}e^{w}\,dw$$

$$= \frac{1}{4}\left(e^{-1} - 1\right)e^{w}\Big|_{0}^{-4}$$

$$= \frac{1}{4}\left(e^{-1} - 1\right)\left(e^{-4} - e^{0}\right)$$

$$= \frac{1}{4}\left(e^{-1} - 1\right)\left(e^{-4} - 1\right)$$

Double Integrals and Technology

Using Derive Derive can evaluate double definite integrals. The following entries show the evaluation of

$$\int_{0}^{2}\int_{0}^{1}\left(24x^{3}y^{2} - 20xy\right)\,dx\,dy$$

Remember to press Enter at the end of each line.

Author
int(int(24x^3y^2 - 20xy,x,0,1),y,0,2)
Simplify

These commands author the function $f(x, y)$, then the **S**implify command evaluates it.

Derive Exercises Use Derive to evaluate each double integral.

1. $\int_{0}^{3}\int_{0}^{5}(2x - 3y + 5)\,dx\,dy$

2. $\int_{2}^{5}\int_{0}^{3}\left(e^{2x-y}\right)\,dx\,dy$

3. $\int_{0}^{2}\int_{x^2}^{x}\left(x^{4}y^{2}\right)\,dy\,dx$

4. $\int_{0}^{3}\int_{x^2-2x}^{x-3x^2} 7x\,dy\,dx$

5. $\int_{2}^{3}\int_{y^4}^{2y^4}\frac{1}{2x}\,dy\,dx$

6. A *triple iterated integral* is an expression of the form

$$\int_a^b \int_c^d \int_e^f f(x, y, z)\, dx\, dy\, dz$$

Modify the above set of Derive commands to evaluate

$$\int_0^2 \int_1^2 \int_{-1}^2 (3x + 2y - 2z)\, dx\, dy\, dz$$

EXERCISE SET 8.6

A UNDERSTANDING THE CONCEPTS, B SKILL ACQUISITION, and C APPLYING THE CONCEPTS

For Exercises 1–6, evaluate each indefinite integral.

1. $\int 24x^2y^3\, dx$

2. $\int (5xy^3 + 7xy)\, dy$

3. $\int (x^2y^4 - xy^5)\, dy$

4. $\int 4xe^x\, dy$

5. $\int 9x^2e^{3y}\, dx$

6. $\int \frac{10xy}{x^2+2}\, dx$

For Exercises 7–12, evaluate each indefinite double integral.

7. $\int\int 2x\, dx\, dy$

8. $\int\int y^2\, dx\, dy$

9. $\int\int xy\, dy\, dx$

10. $\int\int \left(x^2 - 2x + y\right)\, dy\, dx$

11. $\int\int e^x\, dx\, dy$

12. $\int\int \frac{1}{x}\, dy\, dx$

For Exercises 13–20, evaluate each definite double integral.

13. $\int_0^1 \int_1^2 (1 + x^2)\, dy\, dx$

14. $\int_1^2 \int_0^1 xy\, dy\, dx$

15. $\int_1^3 \int_0^2 dy\, dx$

16. $\int_0^2 \int_0^1 xe^{xy}\, dy\, dx$

17. $\int_1^2 \int_0^1 (y - x)\, dx\, dy$

18. $\int_0^1 \int_0^1 2xye^{-x^2}\, dx\, dy$

19. $\int_1^2 \int_2^3 \frac{2y}{x}\, dy\, dx$

20. $\int_3^4 \int_1^e \frac{x}{3y}\, dy\, dx$

For Exercises 21–35, evaluate each definite double integral.

21. $\int_1^2 \int_0^y (x + 3y)\, dx\, dy$

22. $\int_0^2 \int_0^{3x} xy^2\, dy\, dx$

23. $\int_0^1 \int_0^y e^{y^2}\, dx\, dy$

24. $\int_0^2 \int_y^{y^2} (6x^2 + y^2)\, dx\, dy$

25. $\int_0^1 \int_0^{x/2} e^{2y-x}\, dy\, dx$

26. $\int_0^2 \int_x^{3x} xy\, dy\, dx$

27. $\int_0^3 \int_0^{x+2} \frac{y}{x+2}\, dy\, dx$

28. $\int_0^1 \int_0^x \sqrt{1 - x^2}\, dy\, dx$

29. $\int_0^3 \int_y^{2y} \left(1 + 3x^2 + 3y^2\right) dx\, dy$

30. $\int_0^1 \int_0^{\sqrt{1-x^2}} (2x + y)\, dy\, dx$

31. $\int_1^3 \int_x^{9-x^2} x^2\, dy\, dx$

32. $\int_2^3 \int_0^1 e^{-(x+y)/2}\, dx\, dy$

33. $\int_0^1 \int_{x^2}^{\sqrt{x}} \left(1 + x^2 - y^2\right) dy\, dx$

34. $\int_0^2 \int_{-y}^0 e^{x-y}\, dx\, dy$

35. $\int_0^1 \int_y^{y^2} x^2 y^2\, dx\, dy$

D DESCRIBE YOUR THOUGHTS

36. Explain why, in the evaluation of $\int f(x, y)\, dx$, the constant of integration is denoted as $g(y)$ rather than c as it is for functions of one variable.

37. Describe the fundamental difference between

$$\int \int f(x, y)\, dx\, dy \quad \text{and} \quad \int \int f(x, y)\, dy\, dx$$

38. Explain why a graphing calculator may not be able to approximate a double definite integral.

E REVIEW

39. **(2.2)** Find, if it exists, $\lim_{x \to 3} \dfrac{x^2 + 3x - 18}{x - 3}$.

40. **(2.4)** The function

$$f(x) = \begin{cases} x^2 - 8x + 19, & x \le 4; \\ 6 & x = 4 \\ -x^2 + 8x - 25, & x \ge 2 \end{cases}$$

is discontinuous at $x = 4$. Which of the three continuity conditions is the first to be violated?

41. **(3.4)** Find $f'(x)$ for $f(x) = \dfrac{(3x^2 - 1)^3}{x^2 + 5}$.

42. **(4.1)** Classify, as concave upward or concave downward, a curve that is decreasing at a decreasing rate.

43. **(6.1)** Evaluate the indefinite integral

$$\int 12xe^{3x^2-4}\, dx$$

44. **(6.4)** Evaluate the definite integral

$$\int_e^{e^4} \frac{3}{x(\ln x)^2}\, dx$$

45. **(6.5)** *Business: Income Stream* A person's investment is expected to produce a continuous income stream of revenue of

$$R(t) = 360e^{0.06t}$$

dollars over the next 10 years. Find the total revenue generated by this revenue stream over the first 3 years of the investment.

46. **(8.2)** Suppose that $P(x, y)$ is a function of the two variables x and y. Interpret $\dfrac{\partial}{\partial x} P(4, 10) = 20$.

47. **(8.3)** Find all the relative extreme points of the function $f(x, y) = x^2 + y^2 + 6xy + 16x$.

8.7 The Average Value of a Function

Introduction

Average Value of a Function of One Variable

Average Value of a Function of Two Variables

Average Value and Technology

Introduction

You are surely familiar with the concept of average. In this section we extend the concept of average to functions of one and two variables. We develop the average value geometrically and then examine a problem in which the average value of the function offers useful information.

Average Value of a Function of One Variable

Suppose that the function displayed in Figure 8.25 represents a person's wealth, $f(t)$, at time t (in years from 1980). We could use the definite integral $\int_a^b f(t)\,dt$

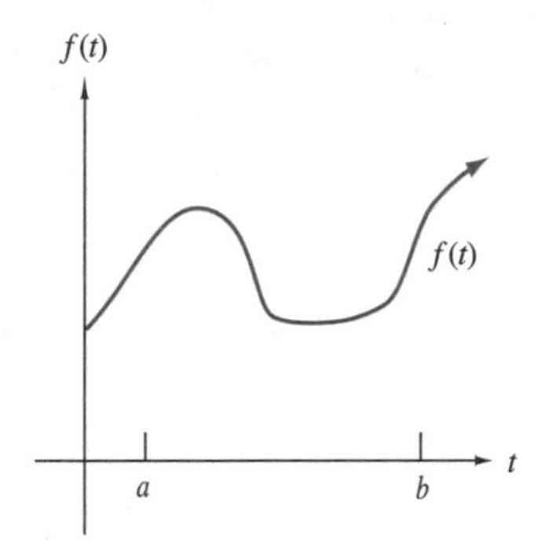

Figure 8.25

as an accumulator to determine the wealth this person accumulated between years a and b.

We could also ask, what is this person's *average* wealth over these $(b - a)$ years?

You probably know that the average of a discrete collection of numbers is the sum (accumulated value) of those numbers divided by the number of numbers. That is, to determine the average value of the numbers $y_1, y_2, y_3, \ldots, y_n$, you would compute

$$\text{Average} = \frac{y_1 + y_2 + y_3 + \cdots + y_n}{n}$$

To determine the average value of a continuous function on some interval $[a, b]$, we make a similar computation. Choose n evenly spaced values of x in $[a, b]$, say, $x_1, x_2, \ldots, x_n$, and compute their corresponding output values $f(x)$, say, $f(x_1), f(x_2), f(x_3), \ldots, f(x_n)$. The average value of these output values is then the sum (accumulated value) of the output values divided by the number of output values. That is,

$$\text{Average} = \frac{f(x_1) + f(x_2) + f(x_3) + \cdots + f(x_n)}{n}$$

This will serve as an approximation to our expectation of the actual average value of the function. Also, as n, the number of selected x values, gets larger and larger, the approximation gets better and better.

In both cases, the average is computed by considering an accumulated value. This makes us think of the definite integral. In fact, we can use the definite integral of a function to produce the definition of the average value of a function in the following way.

Subdivide the interval $[a, b]$ into n subintervals of length $\Delta x = \frac{b-a}{n}$. Choose an x in each subinterval, say, x_1 in the first, x_2 in the second, ... , and x_n in the nth. The x's may be the left endpoints of the subintervals or any point in the interval. The average associated with these x values is

$$\begin{aligned}\text{Average value} &= \frac{f(x_1)+f(x_2)+f(x_3)+\cdots+f(x_n)}{n} \\ &= f(x_1)\cdot\frac{1}{n}+f(x_2)\cdot\frac{1}{n}+f(x_3)\cdot\frac{1}{n}+\cdots+f(x_n)\cdot\frac{1}{n}\end{aligned}$$

where $f(x_1), f(x_2), f(x_3), \ldots,$ and $f(x_n)$ represent the heights of rectangles. We can get the widths, Δx, of the rectangles into this expression using the fact that Δx and n are related by $\Delta x = \frac{b-a}{n}$.

$$\Delta x = \frac{b-a}{n}$$

$$\Delta x = (b-a)\cdot\frac{1}{n}$$

so that

$$\frac{1}{n} = \frac{\Delta x}{b-a}$$

Then, the average value of the function $f(x)$

$$\begin{aligned}&= f(x_1)\cdot\frac{1}{n}+f(x_2)\cdot\frac{1}{n}+f(x_3)\cdot\frac{1}{n}+\cdots+f(x_n)\cdot\frac{1}{n} \\ &= f(x_1)\cdot\frac{\Delta x}{b-a}+f(x_2)\cdot\frac{\Delta x}{b-a}+f(x_3)\cdot\frac{\Delta x}{b-a}+\cdots+f(x_n)\cdot\frac{\Delta x}{b-a} \\ &= f(x_1)\Delta x\cdot\frac{1}{b-a}+f(x_2)\Delta x\cdot\frac{1}{b-a}+f(x_3)\Delta x\cdot\frac{1}{b-a}+\cdots \\ &\qquad + f(x_n)\Delta x\cdot\frac{1}{b-a} \\ &= \frac{1}{b-a}\left[f(x_1)\Delta x+f(x_2)\Delta x+f(x_3)\Delta x+\cdots+f(x_n)\Delta x\right]\end{aligned}$$

The expression $[f(x_1)\Delta x+f(x_2)\Delta x+f(x_3)\Delta x+\cdots+f(x_n)\Delta x]$ is a Riemann sum and approaches $\int_a^b f(x)\,dx$ as $n\to\infty$.

Considering the infinitely many points in the interval $[a, b]$ helps us make the following definition of the average value of a function.

Average Value of a Function of One Variable

If $f(x)$ is a continuous function on $[a, b]$, then the **average value** of $f(x)$ on $[a, b]$ is

$$\text{Average value} = \frac{1}{b-a}\int_a^b f(x)\,dx$$

Geometrically, $f(x)$ represents the height of a rectangle from a point x in the interval $[a, b]$ to the curve. The average value of $f(x)$ is then the average of all such heights and represents the average height of the graph over $[a, b]$.

EXAMPLE SET A

The function $S(x) = 20{,}380e^{-0.25x}$ represents the number of people per week surging to shopping malls to buy a new product x weeks after it is made available to the market. (1) Approximate the total number of people surging to buy the new product between weeks 3 and 7, and (2) approximate the average number of people surging to buy the new product each week for weeks 3 through 7.

Solution:

1. To find the total, we can use the definite integral as an accumulator.

$$\begin{aligned}
\text{Total} &= \int_3^7 S(x)\,dx \\
&= \int_3^7 20{,}380e^{-0.25x}\,dx \\
&= \frac{20{,}380}{-0.25}e^{-0.25x}\Big|_3^7 \\
&= -81{,}520e^{-0.25x}\Big|_3^7 \\
&= \left[-81{,}520e^{-0.25(7)}\right] - \left[-81{,}520e^{-0.25(3)}\right] \\
&\approx 24{,}341
\end{aligned}$$

Interpretation: Between the 3rd and 7th weeks, 24,341 people can be expected to surge into the shopping malls to buy the new product.

2. The approximate average number of people surging into the shopping malls each week between weeks 3 and 7 is

$$\text{Average value} = \frac{1}{7-3}\int_3^7 20{,}380e^{-0.25x}\,dx$$

$$= \frac{1}{4}(24{,}341)$$

$$\approx 6085$$

Interpretation: On the average, between weeks 3 and 7 after the new product is introduced on the market, approximately 6085 people per week can be expected to surge to the malls to buy the new product.

Average Value of a Function of Two Variables

Thus far, we have defined the average value of a function of one variable over an interval $[a, b]$ to be the definite integral

$$\frac{1}{b-a}\int_a^b f(x)\,dx$$

and have interpreted it as the average height of the graph over $[a, b]$. The formula represents the sum of all the heights over $[a, b]$ divided by the length of the interval $[a, b]$.

It is natural to want to extend this idea to functions of two variables. We wish to know the average height of the surface $f(x, y)$ over the rectangular region R. Geometrically, $f(x, y)$ represents the height of a rectangular box from a point (x, y) in a rectangular region R to a surface over R. Extending the definition of average value of functions of one variable to functions of two variables, we present a formula that sums the heights of all such rectangular boxes over R and divides it by the *area* of the region R.

Average Value of a Function of Two Variables

The average value of the function $f(x, y)$ over a rectangular region R is given by

$$\text{Average value} = \frac{1}{(b-a)(d-c)}\int_c^d \int_a^b f(x, y)\,dx\,dy$$

where $a \leq x \leq b$ and $c \leq y \leq d$, and $(b-a)(d-c)$ represents the area of the rectangular region R.

EXAMPLE SET B

The weekly revenue realized by a company is approximated by the revenue function $R(x, y) = 350x + 600y - 4x^2 - 3y^2$ for the sale of x units per week of product A and y units per week of product B. Over the year the company produces between 20 and 30 units of product A each week and between 80 and 110 units of product B each week. Approximate this company's average weekly revenue from products A and B over the year.

Solution:
To determine the average weekly revenue, we need to find the average value of a function of two variables.

$$\text{Average value} = \frac{1}{(b-a)(d-c)} \int_c^d \int_a^b f(x,y)\,dx\,dy$$

$$= \frac{1}{(30-20)(110-80)} \int_{80}^{110} \int_{20}^{30} [350x + 600y - 4x^2 - 3y^2]\,dx\,dy$$

$$= \frac{1}{(10)(30)} \int_{80}^{110} \left[175x^2 + 600xy - \frac{4x^3}{3} - 3xy^2\right]\Big|_{20}^{30} dy$$

$$= \frac{1}{300} \int_{80}^{110} \left[6000y + \frac{186,500}{3} - 30y^2\right] dy$$

$$= \frac{1}{300} \left[3000y^2 + \frac{186,500}{3}y - 10y^3\right]\Big|_{80}^{110}$$

$$= \frac{1}{300} \cdot 10,775,000$$

$$= \frac{107,750}{3}$$

$$\approx 35,916.67$$

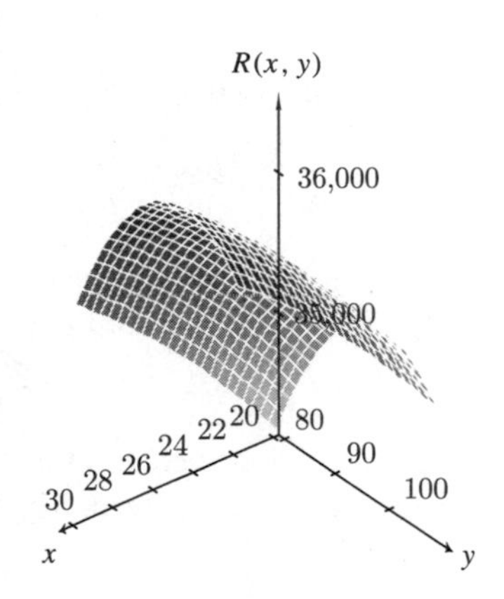

Figure 8.26

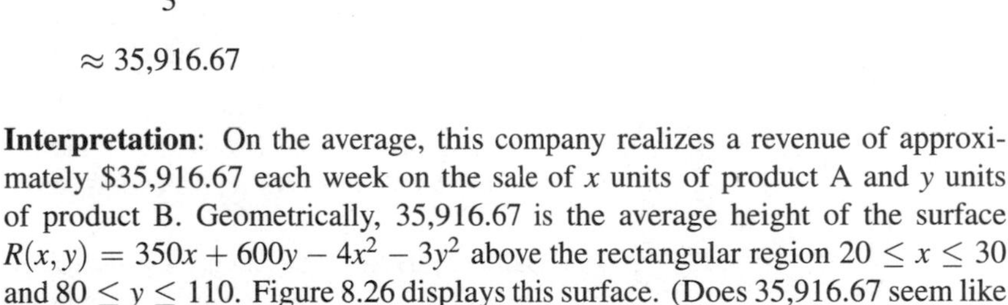

Interpretation: On the average, this company realizes a revenue of approximately \$35,916.67 each week on the sale of x units of product A and y units of product B. Geometrically, 35,916.67 is the average height of the surface $R(x, y) = 350x + 600y - 4x^2 - 3y^2$ above the rectangular region $20 \leq x \leq 30$ and $80 \leq y \leq 110$. Figure 8.26 displays this surface. (Does 35,916.67 seem like an average height to you?)

Average Value and Technology

Using Your Calculator You can use your graphing calculator to solve problems involving the average value of functions of one variable. The following example shows how the problem of Example Set A is solved. The problem involves finding the average value of the function $S(x) = 20,380e^{-0.25x}$ in the interval $[3, 7]$.

```
Y1 = 20380e^(-0.25x)
Y9 = (1/(B-A))*fnInt(Y1,X,A,B)
3 → A
7 → B
Y9
```

The result is 6085.317377, which agrees, approximately, with the result obtained in the example.

Calculator Exercises

1. The population P of a town is related to the number of years t from now by the function

$$N(t) = 2745e^{0.071t}, \quad 0 \leq t \leq 10$$

Construct the graph of this function. Find the population of the town 4 years from now. Find the average population of the town over this 10 year period.

2. The amount A in milliliters of a particular drug in the human body t hours after injection is given by the function

$$A(t) = 2.5e^{-0.07t}$$

Construct the graph of this function. What is the average amount of the drug in the body over the first 12 hours? Over the last 12 hours?

3. The level P of air pollution d miles from a factory in a small town is given by the function

$$P(d) = e^{-0.18d} + 3.7$$

Construct the graph of this curve. At what rate is the level of air pollution changing 5 miles from town? What is the average level of air pollution within 5 miles of the factory?

4. The weekly output Q of units produced by a company t days after the start of a production run is given by the function

$$Q(t) = 245 - 245e^{-0.65t}$$

Construct the graph of this curve. At what rate is the level of production changing after the 7th week? What is the average level of production over the first 10 weeks of a production run?

Using Derive You can use Derive to solve problems involving average value. The following example shows how the solution to Example Set B is solved. The problem involves evaluating

$$\frac{1}{(30-20)(110-80)} \int_{20}^{30} \int_{80}^{110} \left(350x + 600y - 4x^2 - 3y^2\right) dx\, dy$$

Author
1/((30−20)(110−80))int(int(sqrt(350x+600y-4x^2-3y^2),x,20,30),y,80,110)
Simplify

Derive Exercises Find the average value of each function over the specified region.

1. $x + 3y$ over $1 \leq x \leq 4$ and $2 \leq y \leq 5$
2. $x^2 + y^2$ over $0 \leq x \leq 2$ and $0 \leq y \leq 3$
3. xe^y over $0 \leq x \leq 2$ and $0 \leq y \leq \ln x$
4. By using x thousands of worker-hours of labor and y millions of dollars of capital, a manufacturer is able to produce

$$N(x, y) = 2650x^{0.35}y^{0.65}$$

units of a product. If the manufacturer spends between 5 and 10 thousand dollars on worker-hours of labor and between 3 and 5 million dollars on capital, determine the manufacturer's average level of production.

5. The amount A, in parts per million, of pollution produced by the facilities of a company is approximated by the function

$$A(x, y) = \frac{16x}{1.2y + 0.75}$$

where x is the kilowatt capacity of the company, and y is the dollar amount spent during construction of the company on antipollution equipment. If the kilowatt capacity varies between 250 and 1000 hours and the construction costs vary between 1,000,000 and 15,000,000, find the average amount of pollution produced by this company's facilities.

EXERCISE SET 8.7

A UNDERSTANDING THE CONCEPTS

1. *Ecology: Pollution from Manufacturing* The function $P(x)$ describes the level of air pollution emitted from a manufacturing facility. The variable x represents the distance in miles from the facility and P represents the number of pollution units. Interpret

$$\frac{1}{10-3}\int_3^{10} P(x)\,dx$$

2. *Management: Stress in Sales* The human resources department of a manufacturing company believes that the function $A(t)$ relates the amount of stress (in stress units) experienced by pharmaceutical salespeople after visiting a medical agency for a sale of pharmaceutical products and the time t (in days) until they hear that they have or have not made the sale. Interpret

$$\frac{1}{16-9}\int_9^{16} A(t)\,dt$$

3. *Economics: Depreciation of Equipment* The function $D(t)$ represents the depreciation of a piece of equipment t months after it was purchased. Interpret

$$\frac{1}{24-12}\int_{12}^{24} D(t)\,dt$$

4. *Economics: Population of a City* City supervisors believe that the function $P(t)$ approximately describes the population (in thousands of people) of their city t years from now. Interpret

$$\frac{1}{10-5}\int_5^{10} P(t)\,dt$$

5. *Economics: Value of Art* A piece of art is thought to increase in value according to the function $V(t)$, where V is in dollars and t is the time in years since the piece was introduced into the market. Interpret

$$\frac{1}{20-0}\int_0^{20} V(t)\,dt$$

6. *Business: Employee's Earnings* The management of a company believes that, for its sales division, the function $E(x)$ approximates its employee's earnings, where E is in dollars and x is the number of sales made each week. Interpret

$$\frac{1}{35-25}\int_{25}^{35} E(t)\,dt$$

7. *Business: Worker Production* For a particular company, the average worker's production level L (in units of one) depends on the number of days t the worker has been on the job, and is approximated by the function $L(t)$. Interpret

$$\frac{1}{120-60}\int_{60}^{120} L(t)\,dt$$

8. *Biology: Pulse Rate* During heavy physical activity, a person's pulse rate R (in beats per minute) is determined by the amount of time t (in minutes) since the beginning of the activity. Interpret

$$\frac{1}{40-30}\int_{30}^{40} R(t)\,dt$$

B SKILL ACQUISITION

For Exercises 9–16, find the average value of each function on the indicated interval and interpret the result.

9. $f(x) = x^2 + 1$ on $[0, 4]$
10. $f(x) = 8 - x^3$ on $[0, 2]$
11. $f(x) = \sqrt{x}$ on $[0, 16]$
12. $f(x) = \sqrt[3]{x}$ on $[1, 8]$

13. $f(x) = 440e^{-0.04x}$ on $[20, 50]$
14. $f(x) = \frac{10}{x}$ on $[1, 5]$
15. $f(x) = x\sqrt{x^2 - 1}$ on $[1, 5]$
16. $f(x) = \frac{x^2}{(1 - x^3)^{4/3}}$ on $[100, 150]$

C APPLYING THE CONCEPTS

17. ***Economics: Property Value*** In recessionary times, the value of a house in a particular county is approximated by the function $V(t) = 230{,}000e^{-0.02t}$, where t is the number of years since the beginning of the recession. Find and interpret the average value of a house in this county over the first 4 years of a recession.

18. ***Business: Revenue*** A small business estimates that its revenue over the next 5 years will be $R(t) = 120{,}000e^{0.08t}$ dollars. Find and interpret the average revenue for this business over the next 5 years.

19. ***Business: Marginal Cost*** The marginal cost associated with a particular product is $MC(x) = -0.06x + 35$ dollars. Find and interpret the average marginal cost for units 400 to 500 in a production run of 500 units.

20. ***Medicine: Drug Concentration*** The amount A (in milligrams) of a drug in a patient's bloodstream t hours after it is injected into the patient's body is $A(t) = 200e^{-0.4t}$. Find and interpret the average amount of drug in the bloodstream (a) during the first three hours after injection, and (b) over hours eight, nine, and ten after injection.

21. ***Business: Cobb–Douglas Production*** Using x thousand worker-hours of labor and y million dollars of capital, a manufacturer can produce $f(x, y) = 800x^{2/3}y^{1/3}$ units of a product each month. What is the average number of units that can be produced each month if the number of thousands of worker-hours ranges from 8 to 12 each month, and the number of millions of dollars of capital ranges from 1 to 3 each month?

22. ***Business: Cobb–Douglas Production*** Using x thousand worker-hours of labor and y million dollars of capital, a manufacturer can produce $f(x, y) = 3500x^{0.6}y^{0.4}$ units of a product each month. What is the average number of units that can be produced each month if the number of thousands of worker-hours ranges from 4 to 10 each month, and the number of millions of dollars of capital ranges from 50 to 63 each month?

23. ***Business: Average Revenue*** A company sells two products, A and B, having demand functions $x_A = 430 - 4p_A$ and $x_B = 560 - 3p_B$, respectively, where p_A is the price per unit of product A, and p_B is the price per unit of product B. The total revenue realized each month on the sale of these two products is then $R(x_A, x_B) = x_A p_A + x_B p_B$. Approximate the average revenue if the price of product A ranges between \$80 and \$110, and the price of product B ranges between \$40 and \$50.

24. ***Economics: Average Property Value*** From a particular intersection of two streets A and B just outside a city, the property value at any point (x, y) in a rectangular region is

$$V(x, y) = 1000xe^{2x^2 - 0.2y}$$

dollars per square mile. Approximate the average property value if the rectangular region is bounded by a street 2 miles due east of the intersection of streets A and B, and by a street 1 mile due north of the intersection of streets A and B.

25. ***Ecology: Average Amount of Pollution from Production*** To produce safety glass, a manufacturer uses x units of chemical A and y units of chemical B. The amount P of pollution washed into the cleaning water each day during the manufacturing process is given by the function

$$P(x, y) = 0.04x^2 + 0.08xy + 0.006y^2$$

What is the average amount of pollution washed into the cleaning water each day if over a 1-year period the amount of chemical A used varies between 4 and 5 units each day, and the amount of chemical B varies between 16 and 20 units each day?

26. ***Medicine: Recovery Time from Illness*** A medical journal article stated that for a particular illness, the recovery time R (in days) for a person t years old who is given d milligrams each day of medication is approximated by the function

$$R(t, d) = \frac{500t^{1.2}}{d^{1.4}}$$

Approximate the average recovery time for people between 20 and 30 years old who are given between 250 and 400 milligrams per day of the medication.

27. ***Business: Average Revenue*** The annual revenue realized by a company for the sale of x units of product A and y units of product B is $R(x, y) = 120x + 270y - x^2 - y^2$. Over the year the company produces between 50 and 70 units of product A each week and between 200 and 350 units of product B each week. Approximate this company's average weekly revenue from products A and B over the year.

28. ***Business: Advertising and Average Revenue*** The marketing department of a company has determined that if it spends x thousands of dollars per month on radio advertisements and y thousands of dollars per month on newspaper advertisements, it will realize a revenue of

$$R(x, y) = -0.8x^2 - y^2 + 4x + 5y + 2xy$$

thousands of dollars. If over a 1-year period the company spends between \$10,000 and \$15,000 each month on radio advertisements and between \$4,000 and \$9,000 each month on newspaper advertisements, what will be its average monthly revenue over the 1-year period?

29. *Business: Advertising and Influence* A franchise operation in a city believes that if it buys x thousands of dollars of radio advertisements each week and y thousands of dollars of newspaper advertisements each week, it can reach and possibly influence

$$N(x, y) = 20x(4y + 20)$$

people each week. If over a 1-year period, the franchise operation buys between 12 and 20 thousand dollars of radio advertisements each week and between 20 and 24 thousand dollars of newspaper advertisements each week, what is the average number of people this operation can expect to reach and influence each week?

30. *Economics: Welfare* Because of the state of the economy in the country, a state has determined that approximately $P(x, y) = 1.6e^{0.04x} - 1.4e^{0.002y}$ percent of its population will apply for welfare benefits each month. In the function, x represents the unemployment rate during the month, and y is the average number of years of education of people of working age in the state. Approximate the average percentage of the state's population that will apply for welfare benefits each month if unemployment varies between 4% and 7% each month, and the average number of years of education varies between 11 and 14.

31. *Business: Average Weekly Production* The output (in thousands of units each week) of a company is given by the production function

$$f(x, y) = 0.6xe^{0.04y}$$

where x represents the number of hundreds of hours allocated each week to labor, and y is the number of thousands of dollars allocated each week to capital. Find the average weekly output if the number of hours allocated each week to labor ranges from 150 to 200 and the number of dollars allocated each week to capital ranges from 30,000 to 38,000. (Watch the units!)

D DESCRIBE YOUR THOUGHTS

32. Describe, in geometric terms, the meaning of the formula

$$\text{Average value} = \frac{1}{(b-a)(d-c)} \int_c^d \int_a^b f(x, y)\, dx\, dy$$

33. Describe the strategy you would use to solve Exercise 13 or 14.

E REVIEW

34. (2.3) Find, if it exists, $\lim_{x\to\infty} \dfrac{2x^5 + x}{3x^5 + 5x - 8}$.

35. (2.6) Find the average rate of change of the function $f(x) = 4x^2 - 5x + 10$ as x changes from $x = 5$ to $x = 8$.

36. (3.2) Find $f'(x)$ for $f(x) = \dfrac{5x^2 - 3x + 4}{3x + 5}$.

37. (4.1) Classify, as concave upward or concave downward, a curve that is decreasing at a decreasing rate.

38. (6.2) Evaluate the indefinite integral

$$\int \left(6x^2 - 3x + e^{-2x}\right) dx.$$

39. (6.4) Evaluate the definite integral

$$\int_0^5 \frac{x}{2x + 3}\, dx.$$

40. (8.2) Find f_y for $f(x, y) = (3x^4 - 9y^3)^4$.

41. (8.3) Locate all the relative extrema, if any exist, of $f(x, y) = 5x^2 + y^2 - 10x - 6y + 4$.

42. (8.4) *Business: Print Region* Find the dimensions (length and width) of a piece of paper that has a perimeter of 39 inches for its maximum typing area. (A standard 8.5 in. by 11 in. piece of writing paper has a perimeter of 39 inches. Is this maximal?)

43. (8.5) *Business: Cost of Production* A company produces two products, A and B. The cost C of producing x units of product A and y units of product B is approximated by the function $C(x, y) = \dfrac{x^3}{150} + \dfrac{y^3}{180} + 40x + 20y + 20xy$. What is the approximate change in the cost of producing these two products if the company decreases the production of product A from 20 units to 17 units and increases the production of product B from 30 units to 31 units?

8.8 Volumes of Solid Regions

Introduction

We have seen how integration can be used to find areas of planar regions. In this section we will see how double integrals can be used to find volumes of solid regions.

Double Integrals as Volumes over Rectangular Regions

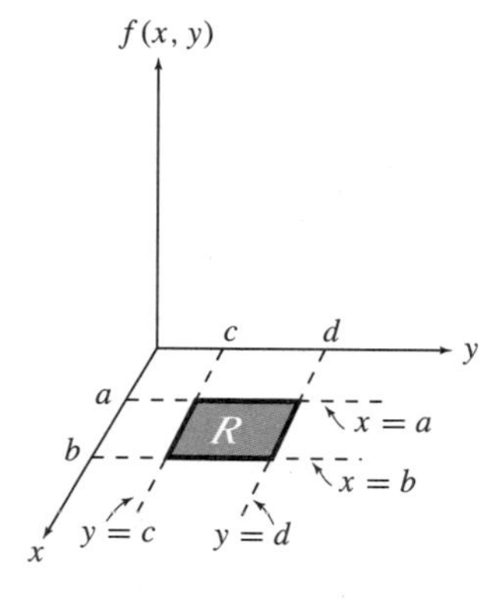

Figure 8.27

In the double definite integral

$$\int_c^d \int_a^b f(x,y)\,dx\,dy$$

the integral with respect to x is over the interval $[a, b]$, and the integral with respect to y is over the interval $[c, d]$. These two intervals, taken together, define a rectangular region R in the xy plane. It is the region bounded by the vertical lines $x = a$ and $x = b$, and the horizontal lines $y = c$ and $y = d$. Figure 8.27 shows such a region, and since $f(x, y)$ represents a surface, the region R is shown in a 3-dimensional coordinate system.

If $f(x, y)$ is nonnegative over the region R, the graph might look something like that pictured in Figure 8.28.

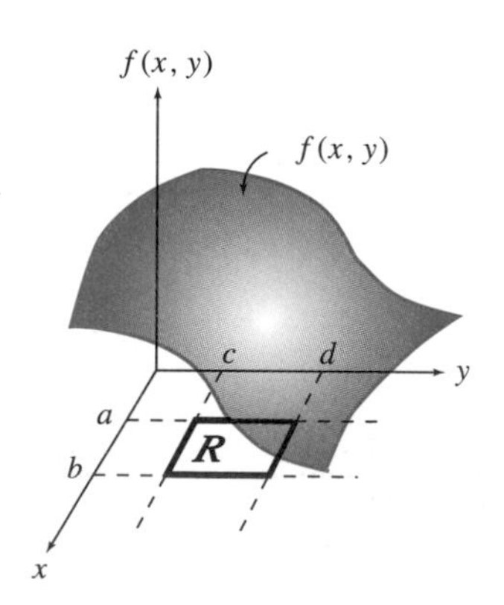

Figure 8.28

In an ordinary definite integral, $\int_a^b f(x)\,dx$, $f(x)$ represents the height of a rectangle and dx the length of the base of the rectangle. Thus, $f(x)\,dx$ represents (height) $\cdot$ (length) $=$ **area** of a rectangle. In the double integral

$$\int_c^d \int_a^b f(x,y)\,dx\,dy$$

$f(x,y)$ represents the height of a rectangular box and dx and dy the length and width, respectively, of the bases of the box. Thus, $f(x,y)\,dx\,dy$ represents (height) $\cdot$ (length) $\cdot$ (width) $=$ **volume** of a rectangular box. Figure 8.29 shows such a box over a region R.

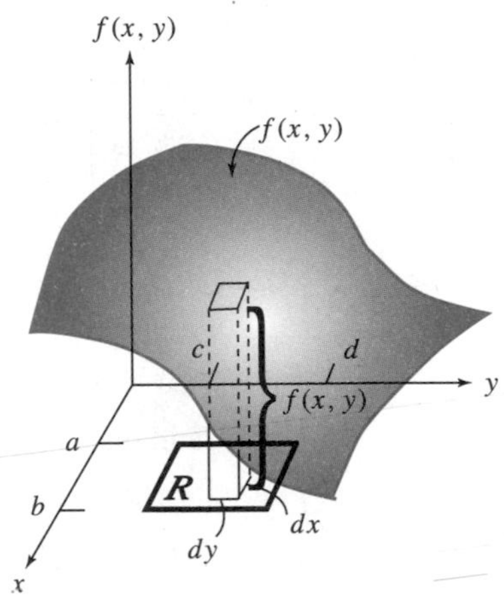

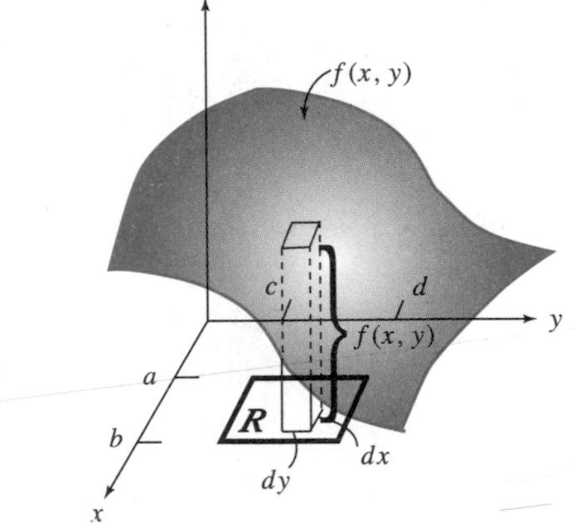

Figure 8.29

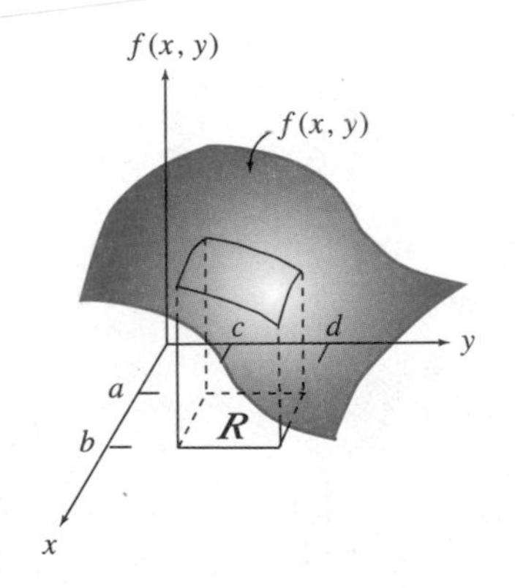

Figure 8.30

The double integrals over $[a,b]$ and $[c,d]$ accumulate all such volumes and hence represent the volume of the 3-dimensional object below the function $f(x,y)$, and over the region R. Figure 8.30 shows such an object.

EXAMPLE SET A

Find the volume of the object between the surface $f(x,y) = x^2 + y^2 + x$ and the xy plane over the region $0 \le x \le 4$ and $1 \le y \le 2$.

Solution:
The volume of this object is given by a double integral.

$$\int_1^2 \int_0^4 \left(x^2 + y^2 + x\right)\,dx\,dy$$

$$= \int_1^2 \left[\frac{x^3}{3} + y^2x + \frac{x^2}{2}\right]\Big|_0^4\,dy \quad \text{Substitute 4 and 0 for } x.$$

$$= \int_1^2 \left(\left[\frac{4^3}{3} + y^2(4) + \frac{4^2}{2}\right] - \left[\frac{0^3}{3} + y^2(0) + \frac{0^2}{2}\right]\right)\,dy$$

$$= \int_1^2 \left(\left[\frac{64}{3} + 4y^2 + 8 \right] - [0] \right) dy$$

$$= \int_1^2 \left[\frac{88}{3} + 4y^2 \right] dy$$

$$= \left[\frac{88}{3} y + \frac{4y^3}{3} \right] \Big|_1^2$$

$$= \left[\frac{88}{3}(2) + \frac{4(2)^3}{3} \right] - \left[\frac{88}{3}(1) + \frac{4(1)^3}{3} \right]$$

$$= \left[\frac{176}{3} + \frac{32}{3} \right] - \left[\frac{88}{3} + \frac{4}{3} \right]$$

$$= \left[\frac{208}{3} \right] - \left[\frac{92}{3} \right]$$

$$= \frac{116}{3}$$

Interpretation: The volume of the object between the surface $f(x, y) = x^2 + y^2 + x$ and the xy plane over the region $0 \le x \le 4$ and $1 \le y \le 2$ is $\frac{116}{3}$ cubic units. Figure 8.31 shows this object.

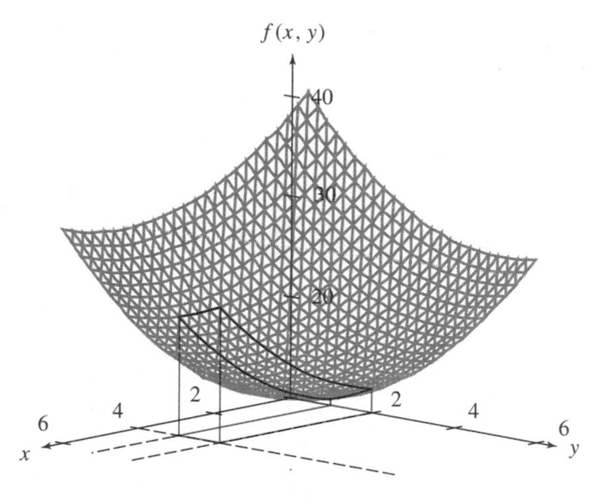

Figure 8.31

Double Integrals as Volumes over Nonrectangular Regions

The region R over which a double integral is evaluated does not have to be rectangular. It could be bounded by two curves, $h_1(x)$ and $h_2(x)$, and two lines,

$x = a$ and $x = b$, as illustrated in Figure 8.32. Alternately, the region R could

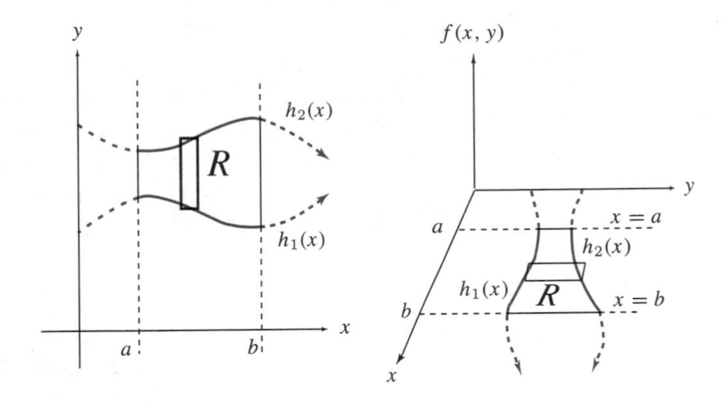

Figure 8.32

be bounded by two curves, $h_1(y)$ and $h_2(y)$, and two lines, $y = c$ and $y = d$, as illustrated in Figure 8.33.

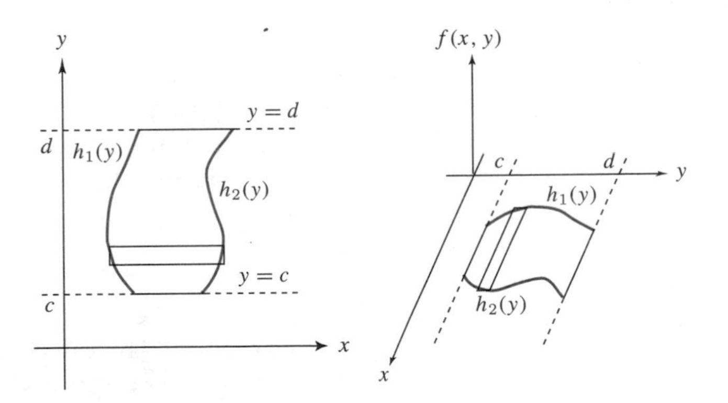

Figure 8.33

Figure 8.32 shows a nonrectangular region R with a representative rectangle in the 2-dimensional xy plane and then the 3-dimensional coordinate system. Notice that the vertical lines are in terms of x and that the curves are functions of x, $h_1(x)$, and $h_2(x)$. We have denoted the upper curve—that is, the curve farthest from the input axis—by h_2, and the lower curve—that is, the curve nearest the input axis—by h_1. With this notation, x is always between a and b, and y is always between $h_1(x)$ and $h_2(x)$ (look at the picture). This means the rectangular region can be described as $a \leq x \leq b$ and $h_1(x) \leq y \leq h_2(x)$. If $f(x, y)$ is nonnegative over this region, then the volume of the object bounded by $f(x, y)$ over R is

$$\int_a^b \int_{h_1(x)}^{h_2(x)} f(x, y)\, dy\, dx$$

Figure 8.33 also shows a nonrectangular region R with a representative rectangle in the 2-dimensional xy plane and then the 3-dimensional coordinate system. Notice that the horizontal lines are in terms of y and that the curves are functions of y, $h_1(y)$, and $h_2(y)$. We have denoted the right-most curve — that is, the curve farthest from the input axis — by h_2, and the left-most curve — that is, the curve nearest the input axis — by h_1. With this notation, x is always between $h_1(y)$ and $h_2(y)$, and y is always between c and d (look at the picture). This means the rectangular region can be described as $h_1(y) \leq x \leq h_2(y)$ and $c \leq y \leq d$. If $f(x, y)$ is nonnegative over this region, then the volume of the object bounded by $f(x, y)$ over R is

$$\int_c^d \int_{h_1(y)}^{h_2(y)} f(x, y)\, dx\, dy$$

Notice in each case, the double integrals are evaluated with the functions as limits on the inside integral; that is, the functions are the limits of the integral that is evaluated first.

EXAMPLE SET B

Find the volume of the object bounded by the surface $\sqrt{\frac{y}{x}}$ over the region defined by $y^2 \leq x \leq y$ and $0 \leq y \leq 1$.

Solution:
The fact that x ranges between y^2 and y, and y between 0 and 1, means the surface is over a nonrectangular region. The variable limits are placed on the inside integral, and since they are in terms of y, and x is between them, the integral has the form

$$\int_0^1 \int_{y^2}^{y} \sqrt{\frac{y}{x}}\, dx\, dy$$

We will begin by eliminating the radical sign and expressing the quotient as a product with negative exponents.

$$\int_0^1 \int_{y^2}^{y} \sqrt{\frac{y}{x}}\, dx\, dy = \int_0^1 \int_{y^2}^{y} x^{-1/2} y^{1/2}\, dx\, dy \quad \text{Integrate with respect to } x.$$

$$= \int_0^1 \left[2x^{1/2}y^{1/2}\right]\Big|_{y^2}^{y} dy$$

$$= \int_0^1 \left(\left[2y^{1/2}y^{1/2}\right] - \left[2yy^{1/2}\right]\right) dy$$

$$= \int_0^1 \left([2y] - \left[2y^{3/2}\right]\right) dy \quad \text{Now integrate with respect to } y.$$

$$= \left[y^2 - \frac{4y^{5/2}}{5}\right]\Big|_0^1$$

$$= \left[1 - \frac{4}{5}\right] - [0 - 0]$$

$$= \frac{1}{5}$$

Interpretation: The volume of the object bounded by the surface $\sqrt{\frac{y}{x}}$ over the region defined by $y^2 \leq x \leq y$ and $0 \leq y \leq 1$ is $\frac{1}{5}$ cubic units.

Volumes of Solids and Technology

Using Derive You can use Derive to find the volume of a solid over a rectangular or nonrectangular region. The following entries show how the volume of the solid of Example Set B is found. Be sure to press Enter at the end of each line.

Author
int(intIsqrt(y/x),x,y^2,y),y,0,1)
Simplify

Derive Exercises Find the volume of the solid bounded above by the function $f(x, y)$ and below by the given region.

1. $\int\int 8xy^2\,dx\,dy \quad x^2 \leq y \leq 1, 0 \leq x \leq 1$
2. $\int\int ye^{x^2}\,dx\,dy \quad 0 \leq y \leq 1, y^2 \leq x \leq 2$
3. $\int\int e^{-2y}\,dx\,dy \quad 0 \leq y \leq \ln x, 0 \leq x \leq e$
4. $\int\int (x^2 + y^2)\,dx\,dy \quad 1 \leq y \leq 5, 1 \leq x \leq 3$

EXERCISE SET 8.8

A UNDERSTANDING THE CONCEPTS

In the text, we presented double definite integrals that represented the volume of an object that was bounded by a surface over a nonrectangular region. That nonrectangular region, however, was itself bounded by two straight lines.

1. Write a double definite integral that represents the volume of an object that is bounded by a surface over a nonrectangular region in which no straight line appears, so that the double integral has the form

$$\int_?^? \int_?^? f(x, y)\,dy\,dx$$

The next figure illustrates such a region. Notice that the figure includes a representative rectangle. The position of that rectangle should help you determine the limits of integration to place on the integrals.

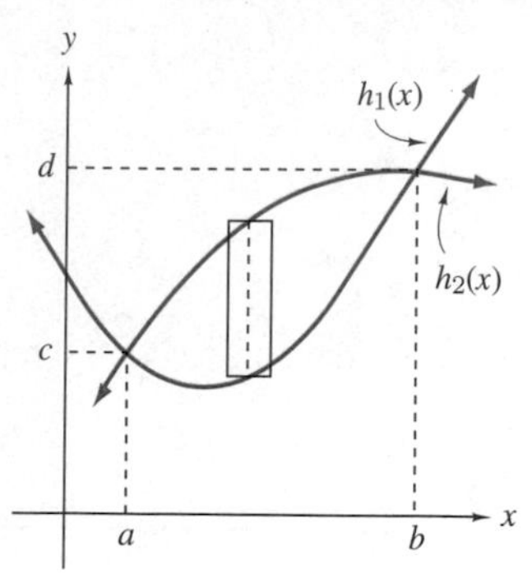

2. Write a double definite integral that represents the volume of an object that is bounded by a surface over a nonrectangular region in which no straight line appears, so that the double integral has the form

$$\int_?^? \int_?^? f(x, y)\, dx\, dy$$

The next figure illustrates such a region. Notice that the figure includes a representative rectangle. The position of that rectangle should help you determine the limits of integration to place on the integrals.

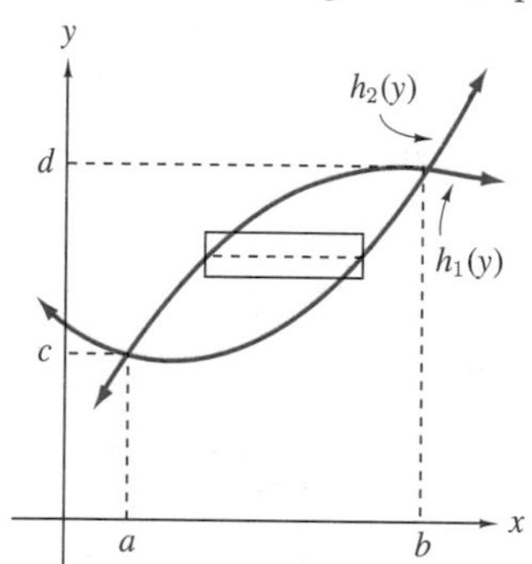

B SKILL ACQUISITION

For Exercises 3–14, find the volume indicated by the following integrals.

3. $\int_0^2 \int_0^1 7xy\, dy\, dx$

4. $\int_1^3 \int_0^2 (x + y)\, dy\, dx$

5. $\int_1^3 \int_0^2 (x - y)\, dy\, dx$

6. $\int_0^2 \int_0^1 xe^{xy}\, dy\, dx$

7. $\int_1^2 \int_0^1 (y - x)\, dx\, dy$

8. $\int_1^2 \int_0^y (x + 3y)\, dx\, dy$

9. $\int_0^2 \int_0^{3x} xy^2\, dy\, dx$

10. $\int_0^2 \int_y^{y^2} (6x^2 + y^2)\, dx\, dy$

11. $\int_0^2 \int_x^{3x} xy\, dy\, dx$

12. $\int_0^3 \int_0^{x+2} \frac{y}{x+2}\, dy\, dx$

13. $\int_0^1 \int_0^x \sqrt{1 - x^2}\, dy\, dx$

14. $\int_0^3 \int_y^{2y} \left(1 + 3x^2 + 3y^2\right)\, dx\, dy$

C APPLYING THE CONCEPTS

For Exercises 15–20, find the volume of the object between the surface $f(x, y)$ and the xy plane over the region defined by R.

15. $f(x, y) = 2x + y$;
R: $1 \le x \le 2$ and $0 \le y \le 1$

16. $f(x, y) = \frac{y}{2}$;
R: $0 \le x \le 4$ and $0 \le y \le 2$

17. $f(x, y) = 6 - 2y$;
R: $0 \le x \le 4$ and $0 \le y \le 2$

18. $f(x, y) = xy^2$;
R: $-1 \le x \le 1$ and $0 \le y \le 1$

19. $f(x, y) = \ln x$;
R: $1 \le x \le e$ and $0 \le y \le 1$

20. $f(x, y) = \frac{y}{x}$;
R: $1 \le x \le e^2$ and $-1 \le y \le 2$

For Exercises 21–28, use double integrals to find the volume of the object bounded by the graphs of the equations.

21. $f(x, y) = xy, f(x, y) = 0, y = 0, y = 4, x = 0, x = 1$

22. $f(x, y) = x^2, f(x, y) = 0, y = 0, y = 4, x = 0, x = 2$

23. $f(x, y) = x + y, y = 0, y = 4, x = 0, x = \sqrt{y}$

24. $f(x, y) = x^2y^2, x = 0, x = 1, y = x^2, y = x^3$

25. $f(x, y) = x, y = 0, y = \sqrt{4 - x^2}, x = 0, x = 2$

26. $f(x, y) = 2y, y = e^{2x}, y = x, x = 0, x = 1$

27. $f(x, y) = ye^x, y = x, y = \sqrt{x}$

28. $f(x, y) = x, y = 1, y = e, x = 0, x = \ln y$

D DESCRIBE YOUR THOUGHTS

29. Describe the strategy you would use to solve Exercises 5 or 6.

30. Describe, in geometric terms, the meaning of the expression $f(x, y)\,dx\,dy$.

E REVIEW

31. (2.2) Find, if it exists, $\lim_{x \to 5} \dfrac{x^2 - 9x + 20}{x^2 - 4x - 4}$.

32. (3.3) *Economics: Wealth Distribution* In the 1980s and 1990s, the distribution of wealth in the United States has become more and more unequal, with the wealthy attaining even more wealth. If $W(t)$ describes the wealth of the upper 2% of the country's wealthy people, describe the situation in terms of the first and second derivatives, and draw a rough sketch of a corresponding curve.

33. (3.4) Find the derivative of the function $S(t) = t^5(4t - 3)^{1/4}$.

34. (6.2) Evaluate $\displaystyle\int \frac{10x^3 + 4x}{10x^4 + 8x^2 - 1}\,dx$.

35. (7.4) Solve the differential equation $\dfrac{dy}{dx} = x^2 y^5$.

36. (8.2) *Business: Cost of Manufacturing* The average cost $\overline{C}$ (in dollars) to a manufacturer who makes x units of product A and y units of product B is given by

$$\overline{C}(x, y) = 5x + 6y + \frac{80}{x} + \frac{216}{y} + 30.$$

How many units of each product should the company produce to minimize its average cost? What is the minimum average cost?

37. (8.4) *Manufacturing: Production* The economist for a manufacturer believes that the function $f(x, y)$ describes the number of units of a product that can be manufactured when x units of labor and y units of capital are used. For a particular month, the company can use 385,000 units of labor and capital (so that the constraint function is $x + y = 385{,}000$). The function $f(x, y)$ is maximized at $(190, 79)$ and the corresponding value of λ, the Lagrange multiplier, is -0.41. Interpret this value of λ.

38. (8.5) *Business: Production* Suppose that a company's production function is $f(x, y)$. If x is increased from 300 to 310 and y is decreased from 850 to 830, interpret the statement

$$df(300, 850) = 235.$$

39. (8.7) Find the average value of $g(x) = 3x^3 - 4x + 2$ over the interval $[-2, 2]$.

40. (8.7) *Business: Average Revenue* The annual revenue realized by a company for the sales of x units of product A and y units of product B is $R(x, y) = 200x + 550y - 5x^2 - 5y^2$. Over the year, the company produces between 25 and 40 units of product A each week and between 85 and 100 units of product B each week. Approximate this company's average weekly revenue from products A and B over the year.

Summary

This chapter presented the important concept of a function of several variables. It discussed the notation, the method of evaluation, the method of partial differentiation, and optimization with and without constraints.

Functions of Several Variables

Section 8.1 was primarily concerned with introducing functions of more than one variable. It began by introducing the method of notation for functions in which there is more than one input variable. For example, $f(x, y, z)$ means that the output f depends on three inputs: x, y, and z.

This section presented several graphs of functions of more than one variable. When there are two input variables and one output variable, we are working in 3-dimensional space. Graphs of functions of two variables are surfaces. When we are working with three input variables and one output variable, we are working in 4-dimensional space, and so on.

This section then discussed partial derivatives and the total differential. The derivative of a function of several variables approximates the change in the output value for a change in one or all of the input values. If only one of the input values is changed, the derivative is called a *partial derivative* of the function. If all the input values are changed, we get the *total differential*.

This section ended with a discussion of the Cobb–Douglas production function. This function relates the number of units of a product a company or country can produce to the number of units of labor and capital it uses to produce this product. The function is of the form $f(x, y) = Cx^a y^b$, where x represents the number of units of labor used to produce the product, and y represents the number of units of capital used. $f(x, y)$ represents the number of units of the product produced. The numbers C, a, and b are constants. These power functions answer questions of *returns to scale*; that is, questions that ask how a proportionate increase or decrease in all the input values will affect the output value, or total production. Cobb–Douglas production functions having the form

$$f(x, y) = Cx^a y^{1-a}$$

where the exponents on the input variables add to 1, always exhibit a constant return to scale.

Partial Derivatives

Section 8.2 presented the notation, geometry, and method of finding partial derivatives of functions. We were reminded that differentiation is the language of change. For functions of more than one variable, we use

$$\frac{\partial f}{\partial x} \quad \text{or} \quad f_x \quad \text{or} \quad f_x(x, y)$$

to indicate the partial derivative of f with respect to x, and

$$\frac{\partial f}{\partial y} \quad \text{or} \quad f_y \quad \text{or} \quad f_y(x, y)$$

to indicate the partial derivative of f with respect to y.

$\frac{\partial f}{\partial x}$ represents the change in the value of the function when the value of x is changed and the value of y is held constant; that is, the rate of change of the function *in the x direction.* $\frac{\partial f}{\partial y}$ represents the change in the value of the function when the value of y is changed and the value of x is held constant; that is, the rate of change of the function *in the y direction.*

This section then discussed higher-order partial derivatives and introduced second partial derivatives and mixed partial derivatives. A function $f(x, y)$ of two variables has four partial derivatives:

$$\frac{\partial}{\partial x}\left(\frac{\partial f}{\partial x}\right) = \frac{\partial^2 f}{\partial x^2} = f_{xx}$$

$$\frac{\partial}{\partial x}\left(\frac{\partial f}{\partial y}\right) = \frac{\partial^2 f}{\partial x \partial y} = f_{yx}$$

$$\frac{\partial}{\partial y}\left(\frac{\partial f}{\partial y}\right) = \frac{\partial^2 f}{\partial y^2} = f_{yy}$$

$$\frac{\partial}{\partial y}\left(\frac{\partial f}{\partial x}\right) = \frac{\partial^2 f}{\partial y \partial x} = f_{xy}$$

The partial derivatives f_{xx} and f_{yy} are called *second partial derivatives* of f with respect to x and y, respectively. The partial derivatives f_{yx} and f_{xy} are called *mixed partial derivatives*. Second partial derivatives can be interpreted in terms of rates of change.

Two partial derivatives of the same kind, f_x, f_{xx}, and f_y, f_{yy}, with the *same sign* indicate that the function f changes at an *increasing rate.*

$f_x > 0,\ f_{xx} > 0$ $\rightarrow f$ is increasing at an increasing rate.
$f_y > 0,\ f_{yy} > 0$ $\rightarrow f$ is increasing at an increasing rate.
$f_x < 0,\ f_{xx} < 0$ $\rightarrow f$ is decreasing at an increasing rate.
$f_y < 0,\ f_{yy} < 0$ $\rightarrow f$ is decreasing at an increasing rate.

Two partial derivatives of the same kind, f_x, f_{xx}, and f_y, f_{yy}, with *opposite signs* indicate that the function f changes at a *decreasing rate.*

$f_x > 0,\ f_{xx} < 0$ $\rightarrow f$ is increasing at a decreasing rate.
$f_y > 0,\ f_{yy} < 0$ $\rightarrow f$ is increasing at a decreasing rate.
$f_x < 0,\ f_{xx} > 0$ $\rightarrow f$ is decreasing at a decreasing rate.
$f_y < 0,\ f_{yy} > 0$ $\rightarrow f$ is decreasing at a decreasing rate.

Optimization of Functions of Two Variables

Section 8.3 centered on how to optimize functions of several variables. It began by defining relative extrema of functions of two variables. This definition is much the same as the definition for relative extrema of functions of one variable presented in Chapter 4. The point (a, b) produces a *relative maximum* of the function $f(x, y)$ if for every point (x, y) *near* (a, b), $f(a, b) \geq f(x, y)$. The point (a, b) produces a *relative minimum* of the function $f(x, y)$ if for every point (x, y) *near* (a, b), $f(a, b) \leq f(x, y)$.

Once we know what relative extreme points are, we need a way of locating them. Again, there is a similarity with functions of one variable. If $f(x, y)$ is a function with a relative extreme point at $\big(a, b,\ f(a, b)\big)$ and both $f_x(a, b)$ and $f_y(a, b)$ exist, then

$$f_x(a, b) = 0 \quad \text{and} \quad f_y(a, b) = 0$$

The points (a, b) for which $f_x(a, b) = 0$ and $f_y(a, b) = 0$ are called *critical points* of the function $f(x, y)$.

Functions of two variables can have points that satisfy the above conditions but do not produce relative extrema. Such points are called *saddle points.*

Also, just as the First Derivative Test tests critical values of functions of one variable to see if they produce relative maxima or relative minima, so the Second Derivative Test tests critical points of functions of two variables to see if they produce relative maxima or relative minima.

Suppose that (a, b) is a critical point of the function $f(x, y)$ and that $D(x, y)$ represents $f_{xx}(a, b) \cdot f_{yy}(a, b) - f_{xy}^2(a, b)$. Then,

1. if $D(a, b) > 0$ and $f_{xx}(a, b) > 0$, (a, b) is a relative minimum.
2. if $D(a, b) > 0$ and $f_{xx}(a, b) < 0$, (a, b) is a relative maximum.
3. if $D(a, b) < 0$, (a, b) is a saddle point.
4. if $D(a, b) = 0$, the test provides no information about (a, b).

The Second Derivative Test is useful for optimizing functions of several variables when there are no constraints on the function. If there are constraints, the method of Lagrange multipliers is useful. This method was presented in Section 8.4.

Constrained Maxima and Minima

Section 8.4 showed how to optimize functions of several variables upon which there are constraints. The method used is that of Lagrange multipliers. This method begins by setting the constraints equal to zero and forming the Lagrangian function $F(x, y, \lambda) = f(x, y) + \lambda g(x, y)$. Then, the first-order partial derivatives, F_x, F_y, and F_λ, are found and are used to locate the critical points by solving the system

$$\begin{cases} F_x = 0 & \ldots(1) \\ F_y = 0 & \ldots(2) \\ F_\lambda = 0 & \ldots(3) \end{cases}$$

for x, y, and λ. Each ordered pair solution, (x, y), is a critical point. The relative extreme points are among these critical points.

The Second Derivative Test can be used for constrained functions to classify the critical points. Suppose that (a, b) is a critical point of the function $f(x, y)$ and that $L(x, y)$ represents $F_{xx}(a, b) \cdot F_{yy}(a, b) - F_{xy}^2(a, b)$. Then,

1. If $L(a, b) > 0$ and $F_{xx}(a, b) > 0$, (a, b) is a constrained relative minimum.
2. If $L(a, b) > 0$ and $F_{xx}(a, b) < 0$, (a, b) is a constrained relative maximum.
3. If $L(a, b) \leq 0$, then the test fails and it is necessary to investigate values of $f(x, y)$ near (a, b).

This section ended with an interpretation of the Lagrange multiplier, λ.

The Total Differential

Section 8.5 discussed the total differential of a function of several variables. Partial derivatives tell approximately how much the output changes if *one* of the input variables is changed and the others are kept constant. The total differential tells approximately how much the output changes if all of the input variables change in value. The total differential of a function of two variables $f(x, y)$ is

$$df = f_x(x, y)dx + f_y(x, y)dy$$

and it approximates the change in the function as x changes a small amount by dx units and y changes a small amount by dy units.

Double Integrals

Section 8.6 began with the *double indefinite integral*. Integrals such as

$$\int \left[\int f(x, y)\, dx \right] dy$$

are called double indefinite integrals and, just as ordinary integrals, they represent functions. This notation first directs us to integrate $f(x, y)$ with respect to x, getting some new function of x and y, and then to integrate this new function with respect to y.

We then considered *integrating double definite integrals*. If $f(x, y)$ is continuous on the region R defined by $a \leq x \leq b$ and $c \leq y \leq d$, then

$$\int_c^d \int_a^b f(x, y)\, dx\, dy = \int_a^b \int_c^d f(x, y)\, dy\, dx$$

This fact allows us to express

$$\int_c^d \left[\int_a^b f(x, y)\, dx \right] dy \quad \text{as} \quad \int_c^d \int_a^b f(x, y)\, dx\, dy$$

and

$$\int_a^b \left[\int_c^d f(x, y)\, dx \right] dy \quad \text{as} \quad \int_a^b \int_c^d f(x, y)\, dy\, dx$$

since the order of integration does not affect the final answer when integrating continuous functions. Also, like their ordinary definite integral counterparts, double definite integrals represent real numbers. Notice also that the integrals are evaluated *inside out*.

The Average Value of a Function

Section 8.7 defined the *average value of a function* of one or two variables. The *average value of a function of one variable* is defined as:

If $f(x)$ is a continuous function on $[a, b]$, then the average value of $f(x)$ on $[a, b]$ is

$$\text{Average value} = \frac{1}{b - a} \int_a^b f(x)\, dx$$

Geometrically, $f(x)$ represents the height of a rectangle from a point x in the interval $[a, b]$ to the curve. The average value of $f(x)$ is then the average of all such heights and represents the average height of the graph over $[a, b]$.

The *average value of a function of two variables* is defined as:

The average value of the function $f(x, y)$ over a rectangular region R is given by

$$\text{Average value} = \frac{1}{(b - a)(d - c)} \int_c^d \int_a^b f(x, y)\, dx\, dy$$

where $a \leq x \leq b$ and $c \leq y \leq d$, and $(b - a)(d - c)$ represents the area of the rectangular region R.

Volumes of Solid Regions

Section 8.8 showed how double integrals can be used to find volumes of solids. Beginning with volumes over rectangular regions, an ordinary definite integral, $\int_a^b f(x)\,dx$, $f(x)$ represents the height of a rectangle and dx the length of the base of the rectangle. Thus, $f(x)\,dx$ represents (height) $\cdot$ (length) $=$ **area** of a rectangle. In the double integral $\int_c^d \int_a^b f(x, y)\,dx\,dy$, $f(x, y)$ represents the height of a rectangular box and dx and dy the length and width, respectively, of the bases of the box. Thus, $f(x, y)\,dx\,dy$ represents (height) $\cdot$ (length) $\cdot$ (width) $=$ **volume** of a rectangular box.

Finally, we considered volumes over nonrectangular regions.

Supplementary Exercises

1. *Business: Advertising Revenue* Suppose x represents the number of dollars spent by a company on newspaper advertisements and y the number of dollars spent on radio advertisements. If R represents the revenue realized by the company from advertising, explain the difference between $R(x)$ and $R(x, y)$.
2. *Economics: Grant Funding* Suppose x represents the number of small businesses applying for government grants and y the number of university business education researchers applying for grants. If A represents the amount of money the government provides the business industry through grants, explain the difference between $A(x)$ and $A(x, y)$.
3. Find the value of $f(x, y) = 3x^2 - 2y^2 + 6xy + 3x + 2y + 5$ at $(1, 3)$.
4. Find $f(0, 2)$ if $f(x, y) = 4xe^{x+y}$.
5. Find $\lim_{h \to 0} \frac{f(x, y + h) - f(x, y)}{h}$ if $f(x, y) = x^2 + 2xy - 60$.
6. *Medicine: Drug Effects* The effect E of a drug on a human patient is a function of both the amount (in milligrams) x of the drug administered and the amount of the time t (in hours) that has passed since the drug was administered. For a particular drug the function relating these quantities is $E(x, t) = 17.2x^{1.16}e^{-0.02t}$.
   a. Find the effect that 170 mg of the drug has on a patient 1 hour after it has been administered.
   b. Find the change in the effect of the drug on the patient if the dosage had been 185 mg rather than 170 mg.
7. Suppose $N(x, y)$ is a function of the two variables x and y. Interpret $\frac{\partial}{\partial x}P(4, 10) = 15$ and $\frac{\partial}{\partial y}P(10, 11) = 4$.
8. Suppose $K(s, t, r)$ is a function of the three variables s, t, and r. Interpret
$$\frac{\partial}{\partial s}K(4, 4, 7) = 5 \text{ and } \frac{\partial}{\partial t}K(10, 12, 8) = -18.$$
9. *Political Science: Elections* The proportion P of the population voting for a particular candidate in an upcoming election is related to the number x (in thousands) of unemployed people in the state and the average number of years y of education of the adults in the state; that is, $P(x, y)$. What information about the proportion of the population voting for the candidate is contained in the inequalities $P_x(x, y) > 0$, $P_y(x, y) > 0$, $P_{xx}(x, y) < 0$, and $P_{yy}(x, y) < 0$?
10. *Economics: Demand/Price* The demand D for a particular product depends on both the price x of the product and the price y of its only competitor. The demand is given by $D(x, y)$. Interpret $D_x(x, y) < 0$ and $D_y(x, y) < 0$.
11. For $f(x, y) = 6x^2 - 8y^2 + 5xy$, find $f_x(2, 1)$ and $f_y(2, 2)$.
12. For $f(x, y) = e^{2x+5y}$, find $f_x(0, 1)$ and $f_y(1, 0)$.
13. For $f(x, y) = x^2 + e^{3x} + e^{3y}$, find f_{xx}, f_{yy}, f_{xy}, and f_{yx}.
14. For the function $f(x, y) = x^2 + 7y^2 + 3x + 4y + 60$, find the values of x and y so that both $f_x(x, y) = 0$ and $f_y(x, y) = 0$.
15. *Economics: Cobb-Douglas Production* Suppose that x units of labor and y units of capital are needed to produce $f(x, y) = 60x^{1/3}y^{2/3}$ units of a particular product.
    a. How will the number of units produced change if the number of units of labor is increased by 1, from 64 to 65?
    b. How will the number of units produced change if the number of units of capital is decreased by 1, from 27 to 26?

In each of Exercises 16 and 17, the first-order, second-order, and mixed partial derivatives of a function are given along with the critical point of the function. Use this information to classify, if possible, the critical point as a relative maximum, a relative minimum, or a saddle point.

16. $f_x(x, y) = 5x - 1$
$f_y(x, y) = y + 7$
$f_{xx}(x, y) = 5, f_{yy}(x, y) = 1, f_{xy}(x, y) = 0$
$(1/5, -7)$

17. $f_x(x, y) = 7x + 4$
$f_y(x, y) = -6y + 5$
$f_{xx}(x, y) = 7, f_{yy}(x, y) = -6, f_{xy}(x, y) = 0$
$(-4/7, 5/6)$

For Exercises 18–23, find and classify, if possible, all the relative extrema and saddle points of the given function.

18. $f(x, y) = 3x^2 - y^2 - 12x + 16y + 5$
19. $f(x, y) = x^2 + 3y^2 + 12x - 6y + 15$
20. $f(x, y) = x^4 - y^4 - 8x^2 + 2y^2 - 10$
21. $f(x, y) = x^3 + y^3 - 9xy + 25$
22. $f(x, y) = e^{(x+2)(y-3)}$
23. $f(x, y) = 4x^{1/3} + 4y^{1/3}$

In each of Exercises 24–26, a value of λ is given along with a critical point that produces a constrained relative maximum or minimum, and the corresponding maximum or minimum value. Interpret the meaning of λ as it applies to the given situation.

24. ***Business: Production*** Economists believe that, for a particular company, the function $f(x, y)$ describes the number of units of a commodity produced when x units of labor and y units of capital are used. For this company, the number of units of labor and capital that can be used is 531,000, so the constraint function is $x+y = 531{,}000$. The function $f(x, y)$ is maximized at $(420, 121)$ and the corresponding value of λ is -0.76.

25. ***Business: Revenue*** A company believes that its annual revenue depends on both the number x of millions of dollars it spends on research, and the number y of millions of dollars it spends on advertising, and is approximated by the function $R(x, y)$. The company has allocated 9.5 million dollars to research and advertising, so the constraint function is $x + y = 9.5$. The revenue function is maximized at $(6.4, 3.1)$, and the corresponding value of λ is -0.43.

26. ***Business: Revenue*** The weekly revenue realized by a company depends on the number x of unfinished wooden items and the number y of finished wooden items it makes and sells each week, and is approximated by the function $R(x, y)$. The company has committed to produce 360 items each week, so the constraint function is $x + y = 360$. The revenue is maximized at $(60, 300)$ and the corresponding value of λ is -4.75.

27. Maximize the function $f(x, y) = -x^2 - y^2 + 2x + 3y + 10$ subject to the constraint $x + 2y = 9$.

28. Minimize the function $f(x, y) = xy$ subject to the constraint $x^2 + y^2 - 8 = 0$.

29. Maximize the function $f(x, y) = 300x^{1/3}y^{2/3}$ subject to the constraint $x + 2y = 300$.

30. Maximize the function $f(x, y) = x^3y^2$ subject to the constraint $6x + 3y - 12 = 0$.

31. ***Economics: Cobb–Douglas Production*** The Cobb–Douglas production function $40x^{1/4}y^{3/4}$ describes the production of a company for which each unit of labor x costs \$75 and each unit of capital y costs \$125. The company has allocated \$60,000 for labor and capital. Find the amount of money that should be allocated to labor and capital to maximize production. What is the maximum level of production? Find and interpret the value of the Lagrange multiplier.

For Exercises 32–34, find the differential of each given function.

32. $f(x, y) = 5x^2 - 4y^2 + 11x - 6y + 4$ as x changes from 3 to 3.2 and y changes from 8 to 7.9.

33. $f(x, y) = e^{3x+2y}$ as x changes from 1 to 0.97 and y changes from 0 to 1.

34. $f(x, y) = \dfrac{2x + 5}{2y + 5}$ as x changes from 55 to 54 and y changes from 83 to 85.

35. ***Business: Revenue*** The revenue R (in dollars) realized by a company from the sale of two of its products, A and B, is given by the function $f(x, y) = 25x + 25y - 10x^2 - 8y^2 + 12xy$, where x is the price of product A and y is the price of product B.
    a. What is the approximate change in the revenue of this company if the price of product A is increased from \$35 to \$36 and the price of product B is decreased from \$15 to \$13?
    b. Compare this approximate change to the actual change.

36. ***Safety: Stopping Distance*** The stopping distance S for a certain vehicle on a dry city highway depends on both the vehicle's speed v (in miles per hour) and weight w (in pounds), and is approximated by the function $S(v, w) = 0.000015wv^2$.
    a. What is the approximate change in this vehicle's stopping distance if the vehicle's speed is decreased from 60 miles per hour to 58 miles per hour, and its weight is decreased from 3240 pounds to 3200 pounds?
    b. Compare this approximate change to the actual change.

For Exercises 37–42, evaluate each double integral.

37. $\int_1^5 \int_1^2 2x^2 y \, dx \, dy$

38. $\int_1^3 \int_1^3 (x^2 - 4y^2) \, dx \, dy$

39. $\int_2^4 \int_0^2 (xy + xy^2) \, dy \, dx$

40. $\int_1^3 \int_1^x xy \, dy \, dx$

41. $\int_0^1 \int_0^y (x + y^2) \, dx \, dy$

42. $\int_1^e \int_0^{\ln x} \frac{y}{x} \, dy \, dx$

43. ***Business: Cost of Production*** The function $C(x)$ describes the cost (in dollars) to a company to produce x units of a product. Interpret

$$\frac{1}{250 - 150} \int_{150}^{250} C(x) \, dx$$

44. ***Economics: Population*** The function $P(t)$ describes the population of third-world countries since 1980. Interpret

$$\frac{1}{20 - 10} \int_{10}^{20} P(t) \, dt$$

45. ***Business: Temperature of a Room*** The temperature T (in degrees Fahrenheit) in a section of a conference room in a business plaza t hours from 7:00 A.M. is given by the function

$$T(t) = -0.3t^2 + 6t + 53.$$

Find the average temperature from 10:00 A.M. to 2:00 P.M.

46. ***Medicine: Drug Concentration*** For a male weighing between 135 and 175 pounds, the amount A (in milligrams) of a drug in his bloodstream t hours after it is administered is given by the function

$$A(t) = 21e^{-0.4t}$$

Find the average amount of the drug in his bloodstream over the first 6 hours after administration.

47. ***Business: Magazine Subscriptions*** The number N of subscribers to a music magazine t years from now is given by the function

$$N(t) = \frac{-26{,}000}{\sqrt{2 + 0.4t}} + 35{,}000$$

Find the average number of subscribers over the next 5 years.

48. ***Economics: Average Concentration of Population*** A concentration of population in a rectangular region of a county is approximated by the function

$$P(x, y) = 425 - x^2 - y^2$$

thousands of people per square mile. The point $(0, 0)$ is taken to be the center of the county. What is the average population per square mile for a rectangular region that extends 5 miles to the north and south of the center and 7 miles to the east and west of the center? (If you draw this rectangular region, you may more readily see what the limits of integration should be.)

49. ***Manufacturing: Cobb–Douglas Production*** Using x thousand worker-hours and y million dollars of capital, a manufacturer can produce $f(x, y) = 660x^{2/5}y^{3/5}$ units of a product. What is the average number of units that can be produced if the number of thousands of worker-hours ranges from 6 to 10 each month, and the number of millions of dollars of capital ranges from 4 to 5 each month?

50. ***Business: Advertising and Revenue*** The marketing department of a company has determined that if the company spends x thousands of dollars on radio advertisements and y thousands of dollars on newspaper advertisements each month, it will realize a revenue R of

$$R(x, y) = -0.6x^2 - 1.2y^2 + 6x + 6y + 4xy$$

thousands of dollars. If over a 1-year period, the company spends between \$15,000 and \$20,000 on radio advertisements each month and between \$10,000 and \$14,000 on newspaper advertisements each month, what is the company's expected average revenue over the 1-year period?

Appendix: Algebra Review

A.1 Properties of Exponents and Radicals

Introduction

The properties that govern operations with exponential and radical expressions are reviewed here. These properties are useful for simplifying an expression after a calculus operation has been performed on a given expression.

Properties of Exponents

The following laws are true for any real numbers a, b, m, and n.

1. **Law for Multiplication:** $a^m \cdot a^n = a^{m+n}$
2. **Law for Powers:** $(a^m)^n = a^{mn}$
3. **Law for Division:** $\dfrac{a^m}{a^n} = a^{m-n} \quad a \neq 0$

4. Law of a Power of a Product $(ab)^n = a^n \cdot b^n$

5. Law of a Power of a Quotient $\left(\frac{a}{b}\right)^n = \frac{a^n}{b^n} \quad b \neq 0$

The following definitions are true for any real number $a \neq 0$, and any positive integer n.

1. Definition of a Zero Exponent: For any real number $a \neq 0$, $a^0 = 1$

2. Definition of a Negative Exponent: $a^{-n} = \frac{1}{a^n}$

The following definitions are true for any real number $a > 0$, and any positive integers n and m.

1. Definition of a Fractional Exponent of the Form $\frac{1}{n}$: $a^{1/n} = \sqrt[n]{a}$

2. Definition of a Fractional Exponent of the Form $\frac{m}{n}$: $a^{m/n} = \sqrt[n]{a^m} = \left(\sqrt[n]{a}\right)^m$

EXAMPLE SET A

1. In simplifying $4x^3 \cdot 2x^5$, we are multiplying two powers with the same base, so we add the exponents.

$$4x^3 \cdot 2x^5 = 8x^{3+5}$$
$$= 8x^8$$

2. In simplifying $\left(3x^4\right)^3$, we are raising a power to a power, so we multiply the exponents.

$$\left(3x^4\right)^3 = 3^3x^{4\cdot3}$$
$$= 27x^{12}$$

3. In simplifying $\frac{20x^5y^3}{8x^3y^3}$, we are dividing powers with the same base, so we subtract the exponents.

$$\frac{20x^5y^3}{8x^3y^3} = \frac{20}{8} \cdot x^{5-3}y^{3-3}$$
$$= \frac{5}{2}x^2y^0$$
$$= \frac{5}{2}x^2 \cdot 1$$
$$= \frac{5}{2}x^2$$

4. In simplifying $\left(x^5y^{-3}\right)^{-4}$, we are raising a power to a power, so we multiply the exponents. Since negative exponents are involved, we eliminate them and express the solution so that only positive exponents appear.

$$\begin{aligned}\left(x^5y^{-3}\right)^{-4} &= x^{(5)(-4)}y^{(-3)(-4)}\\ &= x^{-20}y^{12}\\ &= \frac{1}{x^{20}}\cdot y^{12}\\ &= \frac{y^{12}}{x^{20}}\end{aligned}$$

Properties of Radicals

The following properties of radicals are true for any real numbers $a \geq 0$ and $b \geq 0$.

1. $\sqrt[n]{a^n} = a$

2. $\sqrt[n]{ab} = \sqrt[n]{a}\cdot\sqrt[n]{b}$

3. $\sqrt[n]{\dfrac{a}{b}} = \dfrac{\sqrt[n]{a}}{\sqrt[n]{b}}$ for $b \neq 0$.

EXAMPLE SET B

1. To simplify $\sqrt[3]{64x^6y^9z^3}$, we convert from radical form to exponential form and then simplify the fractional exponents. We will convert back to radical form if fractional exponents remain in the result.

$$\begin{aligned}\sqrt[3]{64x^6y^9z^3} &= 64^{1/3}x^{6/3}y^{9/3}z^{3/3}\\ &= 4x^2y^3z\end{aligned}$$

2. To simplify $\sqrt[4]{a^7}$, we convert from radical form to exponential form and then simplify the fractional exponents. We will convert back to radical if fractional exponents remain in the result.

$$\begin{aligned}\sqrt[4]{a^7} &= \sqrt[4]{a^4a^3}\\ &= a^{4/4}a^{3/4}\\ &= a\sqrt[4]{a^3}\end{aligned}$$

3. To simplify $\sqrt[5]{96x^7y^{11}z^{15}}$, we convert from radical form to exponential form and then simplify the fractional exponents. We will convert back to radical form if fractional exponents remain in the result.

$$\begin{aligned}\sqrt[5]{96x^7y^{11}z^{15}} &= \sqrt[5]{2^5 \cdot x^5y^{10}z^{15} \cdot 3x^2y} \\ &= 2^{5/5}x^{5/5}y^{10/5}z^{15/5} \cdot \sqrt[5]{3x^2y} \\ &= 2xy^2z^3\sqrt[5]{3x^2y}\end{aligned}$$

4. To simplify $\dfrac{\sqrt[3]{-8x^6}}{\sqrt{16x^8}}$, we convert from radical form to exponential form and then simplify the fractional exponents. We will convert back to radical form if fractional exponents remain in the result.

$$\begin{aligned}\frac{\sqrt[3]{-8x^6}}{\sqrt{16x^8}} &= \frac{(-8)^{1/3}x^{6/3}}{(16)^{1/2}x^{8/2}} \\ &= \frac{-2x^2}{4x^4} \\ &= -\frac{2}{4} \cdot x^{2-4} \\ &= -\frac{1}{2}x^{-2} \\ &= -\frac{1}{2x^2}\end{aligned}$$

Sometimes it is necessary to remove all radical expressions from the denominator of a fractional expression. The process is called **rationalizing the denominator** and is illustrated in the following examples.

EXAMPLE SET C

1. To simplify $\dfrac{5}{\sqrt{2x}}$, we rationalize the denominator.

$$\begin{aligned}\frac{5}{\sqrt{2x}} &= \frac{5}{\sqrt{2x}} \cdot \frac{\sqrt{2x}}{\sqrt{2x}} \\ &= \frac{5\sqrt{2x}}{2x}\end{aligned}$$

2. We simplify $\frac{9y}{\sqrt[3]{y}}$ by rationalizing the denominator.

$$\begin{aligned}\frac{9y}{\sqrt[3]{y}} &= \frac{9y}{\sqrt[3]{y}} \cdot \frac{\sqrt[3]{y^2}}{\sqrt[3]{y^2}} \\ &= \frac{9y\sqrt[3]{y^2}}{\sqrt[3]{y^3}} \\ &= \frac{9y\sqrt[3]{y^2}}{y} \\ &= 9\sqrt[3]{y^2}\end{aligned}$$

Exponents and Technology

Using Your Calculator You can use your graphing calculator to evaluate expressions that involve exponents. The following entries show how to evaluate the function $f(x) = 4x^{2/3}$ at $x = 2$ and then at $x = -2$. Begin by entering the function.

Y1 = 4X^(2/3)

Now, in the computation window store 2 to x and access and compute Y1.

2 → X
Y1

The calculator returns 6.349604200.

To compute $f(-2)$, store -2 to x and compute Y1.

−2 → X

When you try to compute Y1, the calculator responds with an error. The error is caused because the calculator uses logarithms to compute expressions involving bases having fractional exponents, and logarithms are not defined for negative numbers. (See Section A.4 for a review of logarithms.) You will have to enter this expression using a different, but equivalent description. By the properties of exponents,

$$x^{2/3} = x^{2\cdot 1/3} = (x^2)^{1/3}$$

The advantage of this description of $x^{2/3}$ is that the base is squared first, making it positive, so that the fractional exponent is applied to a positive number. Computing gives $f(-2) = 6.349604200$.

Calculator Exercises Use your graphing calculator to evaluate and graph each function.

1. Evaluate $f(x) = 3x^{2/5}$ at $x = -5$ and then construct the graph in the window $-10 \le x \le 10$ and $-5 \le y \le 10$.

2. A government study estimates that consumers will buy a quantity $Q(p)$, in pounds, of imported spices when the price per pound is p dollars. If $Q(p) = \frac{4525}{p^{1.74}}$, find the quantity purchased when the price per pound is \$18. Construct the graph of this function in the window $0 \le p \le 50$ and $0 \le y \le 250$. Describe the behavior of this function in terms of the problem situation.

3. In the previous exercise, suppose that the price per pound of an imported spice is a function of the number of weeks, t, from the beginning of the year and is given by $p(t) = 0.03t^{15/7} + 0.09t^{15/14} + 5.2$. Entering $p(t)$ into $Y1$ and $Q(p)$ into $Y2$, create a composite function from which you can estimate the quantity of spice consumers will purchase 20 weeks from the beginning of the year.

4. By turning off the graphing capability of one of the functions in the previous exercise, construct the graph of the composite function $Q[p(t)]$ in the window $0 \le t \le 52$ and $0 \le Q \le 250$. Sketch your graph and describe its behavior in terms of the problem situation.

EXERCISE SET A.1

In Exercises 1–18, simplify the expressions.

1. $64^{3/2}$
2. $27^{-2/3}$
3. $2^4 \cdot 2^3$
4. $3x^4 \cdot 12x^5$
5. $(x^3y^{-7})(x^{-4}y^5)$
6. $\frac{24x^{-4}y^5z^{-8}}{32x^8y^{-5}z^{-7}}$
7. $\frac{x^{-2/5}}{x^{3/5}}$
8. $\frac{-4^{-2}}{-3^{-3}}$
9. $\frac{(3x^2y^{-1})^3(x^3y^3)^5}{27x^5y^{-2}(xy^2)^{-1}}$
10. $5x^0$
11. $(5x)^0$
12. $\sqrt[4]{a}\sqrt[6]{a}$
13. $\sqrt[5]{x^3y^2}\,\sqrt[10]{xy}$
14. $\sqrt[3]{x^2}\sqrt[4]{x^3}$
15. $\sqrt[3]{\frac{8a^3}{64b^6}}$
16. $\frac{\sqrt[8]{y^{20}}}{\sqrt[6]{y^{15}}}$
17. $\left(m^{3/5}\right)^{1/2}$
18. $\left[\left(\frac{y}{x}\right)^{-2}\right]^{-4}$

For Exercises 19–22, rationalize the denominators of each expression.

19. $\frac{3x}{\sqrt[4]{x}}$
20. $\frac{9a}{\sqrt[5]{a^3}}$
21. $\frac{8x}{\sqrt{xy}}$
22. $\frac{-6y}{\sqrt[7]{y^4}}$

A.2 Factoring Polynomial Expressions

Introduction

Factoring Polynomial Expressions

Factoring and Technology

Introduction

Once a calculus operation has been applied to a function, the new resulting function may need to be algebraically simplified. A common technique for simplifying function-defining expressions is factoring.

Factoring Polynomial Expressions

Factoring a polynomial expression is a process that changes the polynomial from a sum or difference of terms to a product of other polynomial expressions. To factor a polynomial expression completely, which means all the factors are **prime** or **irreducible**, follow the following steps.

1. Factor out the greatest common factor.
2. If the polynomial is a binomial, it may fit one of these patterns:
   a. **Difference of Perfect Squares:** $x^2 - y^2 = (x+y)(x-y)$
   b. **Difference of Perfect Cubes:** $x^3 - y^3 = (x-y)(x^2+xy+y^2)$
   c. **Sum of Perfect Cubes:** $x^3 + y^3 = (x+y)(x^2-xy+y^2)$
3. If the polynomial is a trinomial, it may fit one of these patterns:
   a. **Perfect-Square Trinomial:** $x^2 + 2xy + y^2 = (x+y)^2$
   b. **Perfect-Square Trinomial:** $x^2 - 2xy + y^2 = (x-y)^2$
4. If a polynomial is a **general trinomial** it will be of the form

$$Fx^{2n} + Gx^n + H$$

   where F, G, and H are integers and n is a positive integer. If it is not prime, then its factors must be of the form $(ax^n+b)(cx^n+d)$, where $ac = F$, $bd = H$, and $ad + bc = G$. The process for finding the factors is a trial-and-error process and will be demonstrated by example.
5. If the polynomial has more than three terms, then we use group-factoring techniques, which will be illustrated by example.

EXAMPLE SET A

1. To factor $4x^2 + 16x$, recognize it first as a binomial with the common factor $4x$.

$$4x^2 + 16x = 4x(x+4)$$

2. To factor $21a^3b^2 - 15a^2b^2 - 18ab^6$, recognize it is a trinomial with common factor $3ab^2$.

$$21a^3b^2 - 15a^2b^2 - 18ab^6 = 3ab^2(7a^2 - 5a - 6b^4)$$

3. To factor $m^2 - 16$, recognize it as a binomial that is the difference of perfect squares.

$$m^2 - 16 = (m + 4)(m - 4)$$

4. To factor $4x^4 - 9$, recognize it as a binomial that is the difference of perfect squares.

$$4x^4 - 9 = (2x^2 + 3)(2x^2 - 3)$$

5. To factor $y^6 + 27$, recognize it as a binomial that is the sum of perfect cubes.

$$y^6 + 27 = (y^2 + 3)(y^4 - 3y^2 + 9)$$

6. To factor $m^6 - n^6$, recognize it as a binomial that is the difference of perfect squares and also perfect cubes. We begin by using the difference of perfect squares pattern and finish by using the sum and difference of perfect cubes pattern.

$$\begin{aligned} m^6 - n^6 &= (m^3 + n^3)(m^3 - n^3) \\ &= (m + n)(m^2 - mn + n^2)(m - n)(m^2 + mn + n^2) \end{aligned}$$

7. To factor $x^2 - 6x + 9$, recognize it as a perfect-square trinomial.

$$x^2 - 6x + 9 = (x - 3)^2$$

EXAMPLE SET B

1. Factor $x^2 - 7x + 10$, a general trinomial.

$$\begin{aligned} x^2 - 7x + 10 &= (\quad)(\quad) && \text{general trinomial.} \\ &= (x \quad)(x \quad) && \text{factors for } x^2 \end{aligned}$$

We need two numbers whose product is 10, the last term, and whose sum is -7, the coefficient of the middle term. The numbers -2 and -5 meet these conditions.

$$x^2 - 7x + 10 = (x - 2)(x - 5)$$

2. Factor $m^2 - 6m - 16$, a general trinomial.

$$\begin{aligned} m^2 - 6m - 16 &= (\quad)(\quad) \\ &= (m \quad)(m \quad) && \text{factors for } m^2 \end{aligned}$$

3. Factor $x^2 - 4xy - 21y^2$, a general trinomial.

$$x^2 - 4xy - 21y^2 = (\quad)(\quad)$$

$$= (x \quad y)(x \quad y) \quad \text{factors for } x^2 \text{ and factors for } y^2$$

We need two numbers whose product is -21, the coefficient of the last term, and whose sum is -4, the coefficient of the middle term. The numbers 3 and -7 meet these conditions.

$$x^2 - 4xy - 21y^2 = (x + 3y)(x - 7y)$$

4. Factor $12x^2 + 20x + 3$, a general trinomial.

$$12x^2 + 20x + 3 = (\quad)(\quad)$$

The possible factors for $12x^2$ are $12x$ and x, $6x$ and $2x$, and $4x$ and $3x$. The last term, 3, has only the factors 1 and 3. By trial-and-error, we try these different combinations.

$$12x^2 + 20x + 3 = (6x + 1)(2x + 3)$$

5. Factor $6a^2x - 3a^2y + 8x^2 - 4xy$. Recognizing this is neither a binomial nor trinomial, we consider factoring by grouping.

$$6a^2x - 3a^2y + 8x^2 - 4xy = (6a^2x - 3a^2y) + (8x^2 - 4xy)$$

$$= 3a^2(2x - y) + 4x(2x - y)$$

$$= (2x - y)(3a^2 + 4x)$$

Factoring and Technology

Using Your Calculator You can use your graphing calculator to approximate the factors of polynomial functions. To do so, it is helpful to recall the relationship between the zeros of a function and the function's factored form. First, the zeros of a function are the input values that produce 0 as the output value. That is, zeros are values of x such that $f(x) = 0$. When the defining expression of a function is in factored form, each of its factors looks like (x – a zero). Graphically, zeros of a function occur at x intercepts. Thus, zeros can be approximated by graphing the function and noting the x values of the x intercepts, and then placing them into the form (x – a zero), and writing the defining expression as a product of these forms.

For example, the graph of $f(x) = 2x^2 + 3x - 20$ intercepts the x-axis at the x values $5/2$ and -4. (Graph this function yourself to be sure.) Thus, the zeros are $5/2$ and -4, so that the factors are $(x - 5/2)$ and $(x - (-4))$, which is more easily expressed as $(x + 4)$. If we were to use these two factors to represent $f(x)$, we would discover that the first product $x \cdot x$ produces x^2 rather than $2x^2$, the first term of $f(x)$. Thus, we need to multiply these factors by an additional constant factor of 2. Then the function $f(x) = 2x^2 + 3x - 20$ can be expressed in factored easily expressed as $(x+4)$. If we were to use these two factors to represent $f(x)$, we would discover that the first product $x \cdot x$ produces x^2 rather than $2x^2$, the first term of $f(x)$. Thus, we need to multiply these factors by an additional constant factor of 2. Then the function $f(x) = 2x^2 + 3x - 20$ can be expressed in factored form as $f(x) = 2(x - 5/2)(x + 4)$. By distributing the 2 into the first factor we can eliminate the fraction, giving us $f(x) = (2x - 5)(x + 4)$.

Using Derive Derive can factor function-defining expressions symbolically. The following commands demonstrate how this is done for the function $f(x) = 2x^2 + 3x - 20$. Be sure to press Enter at the end of each line.

Declare
Function
f
2x^2 + 3x - 20
Factor
Rational

The **D**eclare and **F**unction commands allow you to name @function and enter its defining expression. The **R**ational command instructs Derive to factor the defined expression using only rational numbers. If it cannot do so, it will tell you.

Derive Exercises Use Derive to factor each function-defining expression.

1. $f(x) = 12x^2 - 5x - 28$
2. $f(x) = x^2 - 3$
3. $f(x) = 18x^3 + 18x^2 - 32x - 24$
4. The expression $3x^2 - 18x + 33$ is developed by a manufacturing economist in determining the marginal cost of a commodity. Use Derive to factor this expression.

EXERCISE SET A.2

Factor each expression completely.

1. $a^3 - a^2b$
2. $-14x^5 - 21x^4y^2 + 35x^6y$
3. $x^2 - 5x - 50$
4. $m^2 - 49$
5. $x^8 - y^8$
6. $n^3 + 64$
7. $x^6 + 1$
8. $y^2 + 14y + 49$
9. $m^2 - 13m + 42$
10. $-x^2 + x + 20$
11. $8a^3 - 24a^2 + 16a$
12. $a^2 - 12a + 36$
13. $x^2 - 15 + 2x$
14. $x^2y^4 + 7xy^4 + 10y^4$
15. $4z^2 + 16z + 15$
16. $3x^2 + 12x + 12$
17. $5t^2 - 25t + 30$
18. $9x^2 - 9xy - 4y^2$
19. $a^6 - 6a^3 - 16$
20. $u^3 - 8v^3$
21. $8x^2 + 16xy - 5x - 10y$
22. $9ar - 9as - 21br + 21bs$
23. $4a^3bc - 14a^2bc^3 + 10abc^2 - 35bc^4$
24. $8x^2z - 32x^2 - 12x^3z + 48z^3$

A.3 Rational Expressions

Introduction
Rational Expressions
Rational Functions and Technology

Introduction

Once a calculus operation has been applied to a function, the new resulting function may need to be algebraically simplified. A common technique for simplifying functions-defining expressions is adding or subtracting rational expressions.

Rational Expressions

Rational expressions are ratios between polynomial expressions, such as

$$\frac{3x+2}{4x-1} \quad \text{and} \quad \frac{3}{2x^2-5x+1}$$

Since they are algebraic fractions, they are operated on using the same rules that are used to operate on arithmetic fractions.

We reducc rational expressions to lowest terms by removing any common factors, other than 1 and -1, between the numerator and denominator. Removing common factors is accomplished by factoring the numerator and denominator and dividing out (canceling) factors that are common. If a common factor contains a variable and could take on the value of zero, then a restriction on all values that would cause this factor to become zero must be stated.

EXAMPLE SET A

1. Simplify $\frac{y^2-9}{y^2-4y+3}$ by reducing it to lowest terms. State all restrictions on the variables.

$$\frac{y^2-9}{y^2-4y+3} = \frac{(y+3)(y-3)}{(y-1)(y-3)}$$

$$= \frac{(y+3)}{(y-1)}, \quad y \neq 3$$

2. Simplify $\frac{4-7y}{14y^2-y-4}$ by reducing it to lowest terms. State all restrictions on the variables.

$$\frac{4-7y}{14y^2-y-4} = \frac{4-7y}{(7y-4)(2y+1)}$$

$$= \frac{-(7y-4)}{(7y-4)(2y+1)}$$

$$= \frac{-1}{(2y+1)}, \quad y \neq \frac{4}{7}$$

We multiply rational expressions by multiplying their numerators and multiplying their denominators and then simplifying the resulting expression as we did above. Division of rational expressions is converted into multiplication by using the reciprocal of the second expression.

EXAMPLE SET B

1. Perform the multiplication $\dfrac{x^2-12x+35}{x^2-5x-24} \cdot \dfrac{x^2-4x-32}{x^2-5x-14}$.

$$\frac{x^2-12x+35}{x^2-5x-24} \cdot \frac{x^2-4x-32}{x^2-5x-14} = \frac{(x^2-12x+35)(x^2-4x-32)}{(x^2-5x-24)(x^2-5x-14)}$$

$$= \frac{(x-5)(x-7)(x-8)(x+4)}{(x+3)(x-8)(x-7)(x+2)}$$

$$= \frac{(x-5)(x+4)}{(x+3)(x+2)} \quad x \neq 7, 8$$

2. Perform the division $\dfrac{y-2}{y^2-y-12} \div \dfrac{y^2-5y+6}{y^2-9}$.

$$\frac{y-2}{y^2-y-12} \div \frac{y^2-5y+6}{y^2-9} = \frac{y-2}{y^2-y-12} \cdot \frac{y^2-9}{y^2-5y+6}$$

$$= \frac{(y-2)(y^2-9)}{(y^2-y-12)(y^2-5y+6)}$$

$$= \frac{(y-2)(y+3)(y-3)}{(y+3)(y-4)(y-2)(y-3)}$$

$$= \frac{1}{y-4} \quad y \neq \pm 3, 2$$

We add or subtract rational expressions by finding a common denominator, just as with arithmetic fractions, and then we add the numerators together, leaving the denominator as the common denominator. Finally, we simplify the resulting expression by reducing it to lowest terms.

EXAMPLE SET C

1. Perform the subtraction $\dfrac{3x-7}{4x+1} - \dfrac{2x-15}{4x+1}$.

$$\begin{aligned}\frac{3x-7}{4x+1} - \frac{2x-15}{4x+1} &= \frac{(3x-7)-(2x-15)}{4x+1}\\ &= \frac{3x-7-2x+15}{4x+1}\\ &= \frac{x+8}{4x+1}\end{aligned}$$

2. Perform the addition $\dfrac{5x}{x-7} + \dfrac{3x}{x-3}$.

$$\begin{aligned}\frac{5x}{x-7} + \frac{3x}{x-3} &= \frac{5x(x-3)}{(x-7)(x-3)} + \frac{3x(x-7)}{(x-7)(x-3)}\\ &= \frac{5x^2-15x}{(x-7)(x-3)} + \frac{3x^2-21x}{(x-7)(x-3)}\\ &= \frac{5x^2-15x+3x^2-21x}{(x-7)(x-3)}\\ &= \frac{8x^2-36x}{(x-7)(x-3)}\end{aligned}$$

which can also be expressed as

$$= \frac{4x(2x-9)}{(x-7)(x-3)}$$

3. Perform the addition $\dfrac{y+2}{y-1} + \dfrac{y+8}{y^2-5y+4}$.

$$\begin{aligned}\frac{y+2}{y-1} + \frac{y+8}{y^2-5y+4} &= \frac{y+2}{y-1} + \frac{y+8}{(y-4)(y-1)}\\ &= \frac{(y+2)(y-4)}{(y-1)(y-4)} + \frac{y+8}{(y-4)(y-1)}\\ &= \frac{(y+2)(y-4)+(y+8)}{(y-1)(y-4)}\\ &= \frac{y^2-2y-8+y+8}{(y-1)(y-4)}\\ &= \frac{y^2-y}{(y-1)(y-4)}\\ &= \frac{y(y-1)}{(y-1)(y-4)}\end{aligned}$$

$$= \frac{y}{y-4} \quad y \neq 1$$

Rational Functions and Technology

Using Derive Derive can perform algebraic operations on rational functions as well as determine the zeros of the numerator and denominator. (This is useful in optimizing functions.) The following commands show the determination of the zeros of the sum of $(f+g)(x)$ for the functions $f(x) = \frac{x+3}{x+4}$ and $g(x) = (f+g)(x)$. When entering these commands, remember to press Enter at the end of each line.

Declare
Function
f
(x+3)/(x+4)
Declare
Function
g
(x+5)/(x-7)
Declare
Function
h
f(x)+g(x)
Simplify

You have added the two functions and simplified the result. Now find the zeros of the numerator and the zeros of the denominator.
Press the right arrow key once to highlight the numerator of h.
So**L**ve (*Remember to Press* L *and then* Enter *twice to select* Exact. *You will get two solutions.*)
Press the up arrow key twice and then the right arrow key until the denominator is highlighted.
So**L**ve (*Press* Enter *twice*)

The exact zeros are displayed.

Derive Exercises Use Derive to determine the zeros of each sum, $(f+g)(x)$, or difference, $(f-g)(x)$.

1. $f(x) = \dfrac{x+2}{x-4}$ and $g(x) = \dfrac{x+6}{x+7}$

2. $f(x) = \dfrac{x-6}{(x+4)(x-5)}$ and $g(x) = \dfrac{x+1}{x-5}$

3. $f(x) = \dfrac{4x+1}{3x^2+5x-2}$ and $g(x) = \dfrac{2x+2}{3x^2-10x+3}$

4. $f(x) = 6x$ and $g(x) = \dfrac{2x+3}{x-2}$

EXERCISE SET A.3

Simplify the following expressions.

1. $\dfrac{-28}{7x-21}$
2. $\dfrac{-12x^5(x-1)^4(x-6)^5}{6x^3(x-1)(x-6)^2}$
3. $\dfrac{y^2+2y-8}{2-y}$
4. $\dfrac{6x^2-x-2}{2x^2-15x-8}$
5. $\dfrac{x-6}{x+4} \cdot \dfrac{x+4}{x-1}$
6. $\dfrac{y-5}{3-y} \cdot \dfrac{y-3}{y-2}$
7. $\dfrac{y+8}{y-9} \div \dfrac{8+y}{9-y}$
8. $\dfrac{x^2+4x+3}{x^2-x-6} \cdot \dfrac{x^2+x-12}{x^2+x+2}$
9. $\dfrac{x^2-x-20}{x^2+11x+24} \div \dfrac{x^2-11x+30}{x^2+x-6}$
10. $\dfrac{4}{x-2} + \dfrac{3}{x-2}$
11. $\dfrac{6x+4}{5x-2} - \dfrac{3x-2}{5x-2}$
12. $\dfrac{y+4}{y-8} + \dfrac{y+2}{y+1}$
13. $\dfrac{n-6}{n+6} - \dfrac{n+6}{n-6}$
14. $\dfrac{2z+3}{z-6} + \dfrac{3z-1}{z^2-4z-12}$
15. $\dfrac{2k-7}{k+2} - \dfrac{k^2+5k-8}{k^2+9k+14}$
16. $\dfrac{m^2-m-6}{6m^2-7m-3} - \dfrac{m^2-m-1}{2m-3}$
17. $\dfrac{15x^2-17x-4}{12x^2+x-6} \div \dfrac{3x^2-7x+4}{4x^2-x-4}$
18. $\dfrac{7m^2+9m+2}{m^2+6m+5} \cdot \dfrac{2m^2+8m+15}{2m^2-11m-21}$
19. $(y+6)^3 \div \dfrac{(y+6)^2}{y-6}$
20. $\dfrac{3k^2+k-1}{k^3-27} - \dfrac{k+2}{k^2+3k+9}$

A.4 Logarithms

Introduction
Logarithms
Logarithms, Exponential Equations, and Technology

Introduction

Many theoretical and applied phenomena are modeled by exponential and logarithmic functions. This section helps you recall logarithms and reminds you of some of their properties. A more detailed examination of natural logarithms is made in Sections 5.2 and 5.3. Logarithms are used for solving exponential equations such as $e^{2x+5} = 12$.

Logarithms

We will begin our recollection of logarithms with the definition of a logarithm.

Logarithm

A **logarithm** is an exponent. Symbolically,

$$y = \log_b x \quad \text{if and only if} \quad b^y = x$$

where $b > 0$, $b \neq 1$, and $x > 0$.

The definition indicates that **a logarithm is the exponent** on a number b that produces the number x.

EXAMPLE SET A

1. $\log_3 81 = 4$ since $3^4 = 81$. The exponent 4 on the number 3 produces the number 81. Therefore, 4 is the logarithm of 81, base 3.
2. $\log_2 \frac{1}{8} = -3$ since $2^{-3} = \frac{1}{8}$. The exponent -3 on the number 2 produces the number $\frac{1}{8}$. Therefore, -3 is the logarithm of $\frac{1}{8}$, base 2.
3. $\log_{10} 1 = 0$ since $10^0 = 1$. The exponent 0 on the number 10 produces the number 1. Therefore, 0 is the logarithm of 1, base 10.

There are some basic properties of logarithms similar to those for exponents (remember, logarithms are exponents). The following properties are true for $b > 0$ and $b \neq 1$, $m > 0$ and $n > 0$, and any real number r.

1. **Logarithm Property 1:** $\log_b 1 = 0$ (since $b^0 = 1$)
2. **Logarithm Property 2:** $\log_b b = 1$ (since $b^1 = b$)
3. **Logarithm Property 3:** $\log_b b^r = r$ (since $b^r = b^r$) (Note: In this text the implied order of performing operations is such that exponentiations are applied before function evaluations; thus, $\log_b b^r = \log_b(b^r)$.)
4. **Logarithm Property 4:** $\log_b(mn) = \log_b m + \log_b n$
5. **Logarithm Property 5:** $\log_b \left(\frac{m}{n}\right) = \log_b m - \log_b n$
6. **Logarithm Property 6:** $\log_b m^r = r \log_b m$

The proofs of these properties can be found in any algebra book. The following examples illustrate these properties.

EXAMPLE SET B

1. $\log_5 1 = 0$ (Property 1)
2. $\log_5 5 = 1$ (Property 2)
3. $\log_5 5^7 = 7$ (Property 3)
4. $\log_4(3x) = \log_4 3 + \log_4 x$ (Property 4)
5. $\log_7 \frac{3}{5} = \log_7 3 - \log_7 5$ (Property 5)

6. $\log_{10} 5^7 = 7\log_{10} 5$ (Property 6)

7. $\log_{14}\sqrt{3} = \log_{14} 3^{1/2} = \frac{1}{2}\log_{14} 3$ (Property 6)

EXAMPLE SET C

1. Expand $\log_6 \frac{10mn}{7abc}$ as much as possible.

$$\begin{aligned}\log_6 \frac{10mn}{7abc} &= \log_6(10mn) - \log_6(7abc)\\ &= \log_6 10 + \log_6 m + \log_6 n - (\log_6 7 + \log_6 a + \log_6 b + \log_6 c)\\ &= \log_6 10 + \log_6 m + \log_6 n - \log_6 7 - \log_6 a - \log_6 b - \log_6 c\end{aligned}$$

2. Expand $\log_2\left(\frac{x^2\sqrt{y}}{\sqrt[5]{w^4}}\right)$ as much as possible.

$$\begin{aligned}\log_2\left(\frac{x^2\sqrt{y}}{\sqrt[5]{w^4}}\right) &= \log_2(x^2\sqrt{y}) - \log_2(\sqrt[5]{w^4})\\ &= \log_2 x^2 + \log_2\sqrt{y} - \log_2 w^{4/5}\\ &= 2\log_2 x + \log_2 y^{1/2} - \frac{4}{5}\log_2 w\\ &= 2\log_2 x + \frac{1}{2}\log_2 y - \frac{4}{5}\log_2 w\end{aligned}$$

EXAMPLE SET D

1. Write $2\log_2 x + 5\log_2 y$ as a single logarithmic expression.

$$2\log_2 x + 5\log_2 y = \log_2 x^2 + \log_2 y^5 = \log_2 (x^2y^5)$$

2. Write $\frac{3}{4}\log_5 x - \frac{2}{7}\log_5 y - \frac{1}{4}\log_5 z$ as a single logarithmic expression.

$$\begin{aligned}\frac{3}{4}\log_5 x - \frac{2}{7}\log_5 y - \frac{1}{4}\log_5 z &= \log_5 x^{3/4} - \log_5 y^{2/7} - \log_5 z^{1/4}\\ &= \log_5\left(\frac{x^{3/4}}{y^{2/7}z^{1/4}}\right)\\ &= \log_5\left(\frac{\sqrt[4]{x^3}}{\sqrt[7]{y^2}\sqrt[4]{z}}\right)\end{aligned}$$

Two logarithms that are most commonly used are base 10 logarithms, referred to as **common logarithms** ($\log_{10} x$ is expressed as $\log x$) and base e logarithms, referred to as **natural logarithms** ($\log_e x$ is expressed as $\ln x$). Of course, these logarithms have the same properties as mentioned above.

Both common logarithms and natural logarithms are used to solve exponential equations. Some are demonstrated in the next example set.

EXAMPLE SET E

1. Solve $8^x = 7$ and state its meaning in exponential terms.

$$8^x = 7$$

$$\log 8^x = \log 7$$

$$x \log 8 = \log 7$$

$$x = \frac{\log 7}{\log 8} \quad \text{(You can approximate the solution with a calculator.)}$$

$$x \approx \frac{0.8451}{0.9031}$$

$$x \approx 0.9358$$

This means that $8^{0.9358} \approx 7$. Try this computation on your calculator to see how close the approximation is.

2. Solve $3^{4x-1} = 4$ and state its meaning in exponential terms.

$$3^{4x-1} = 4$$

$$\ln 3^{4x-1} = \ln 4$$

$$(4x-1)\ln 3 = \ln 4$$

$$4x - 1 = \frac{\ln 4}{\ln 3}$$

$$4x = \frac{\ln 4}{\ln 3} + 1$$

$$x = \frac{1}{4} \cdot \left(\frac{\ln 4}{\ln 3} + 1\right)$$

$$x \approx \frac{1}{4} \cdot \left(\frac{1.3863}{1.0986} + 1\right)$$

$$x \approx 0.5655$$

This means that $3^{4(0.5655)-1} \approx 4$.

3. Solve $e^{7x-5} = 2$ and state its meaning in exponential terms.

$$e^{7x-5} = 2$$

$$\ln e^{7x-5} = \ln 2$$

$$(7x - 5) \ln e = \ln 2$$

$$(7x - 5) \cdot 1 = \ln 2$$

$$7x - 5 = \ln 2$$

$$7x = \ln 2 + 5$$

$$x = \frac{\ln 2 + 5}{7}$$

$$x \approx 0.8133$$

This means that $e^{7(0.8133)} \approx 2$.

Logarithms, Exponential Equations, and Technology

Using Your Calculator You can use your graphing calculator to compute the logarithm of any positive real number. To compute $f(x) = 1.6 \ln(214)$, enter

```
Y1 = ln(X)
214 → X
Y1
```

The calculator returns 3.365976015, or, rounded to four decimal places, 3.3660. Since logarithms are exponents and $\ln(214) = \log_e(214)$, 3.3660 is the number to which e must be raised to obtain 214. That is, $e^{3.3660} \approx 214$.

Calculator Exercises For Exercises 1–2, use your graphing calculator to find each logarithm to four decimal places. Once you have found the logarithm, write the corresponding exponential form.

1. $\ln(52)$
2. $\ln(0.072)$
3. Construct a sketch of the function $f(x) = \ln(x)$ in the window $-1 \leq x \leq 5$, $2 \leq y \leq 2$.
4. Compute $f(x) = \ln(x)$ for five values of x that are strictly between 0 and 1. Are the values positive or negative? Compute $f(x) = \ln(x)$ for five values of x that are strictly greater than 1. Are the values positive or negative? Compute $\ln(1)$. How does the graph you sketched in the previous exercise validate these computations?

Using Derive You can use Derive to solve exponential equations. The following commands demonstrate how the solution to $e^{2x+5} = 12$ is obtained.

Author
exp(2x+5)=12
soLve
Press **a** *to access the approximate command.*

Derive returns -1.2575, which means that $e^{2(-1.2575)+5} \approx 12$.

Derive Exercises For Exercises 1–4, use Derive to solve each exponential equation.

1. $e^{3x-1} = 10$
2. $e^{4x+3} = -6$
3. $5^x = 7$
4. $6^{x+3} = 20$
5. Use Derive to try to solve the equation $y = \dfrac{e^{2x} - e^{-2x}}{2}$ for x.
6. The number N of bacteria in a culture t hours after the start of an observation period is given by the function

$$N(t) = N_0 6^{0.02t}$$

where N_0 is the number of bacteria initially present. If there are 5000 bacteria present at the beginning of an observation period, how long will it take until there are 5771 bacteria present?

EXERCISE SET A.4

In Exercises 1–4, write each expression in logarithmic form.

1. $3^4 = 81$
2. $m^p = n$
3. $64^{1/3} = 4$
4. $10^{2x-3} = 7$

In Exercises 5–8, write each expression in exponential form.

5. $\log_2 4 = 2$
6. $\log_m y = x$
7. $a = \log_c 2n$
8. $\log_{3/7} \dfrac{343}{27} = -3$

In Exercises 9–12, simplify each expression.

9. $\log_3 3^5$
10. $\log_4 4^{3x-1}$
11. $\log_5(\log_2 32)$
12. $\log_2(\log_8 64)$

In Exercises 13–16, expand each logarithmic expression as much as possible.

13. $\log_3 \dfrac{5x}{y}$
14. $\log_m \dfrac{5^2}{2x+11}$
15. $\log_2 \dfrac{x^4 \sqrt[3]{y}}{\sqrt[5]{z^2}}$
16. $\log_b \dfrac{6m^{2/3} \sqrt[4]{n^3}}{\sqrt[6]{y^5} \sqrt{a}}$

In Exercises 17–18, express each logarithmic expression as a single logarithm.

17. $\log_2 x + \log_2 y - \log_2 z$
18. $5 \log_4 x + \frac{1}{2} \log_4 y - 4 \log_4 z - 6 \log_4 w$

In Exercises 19–22, solve each logarithmic equation.

19. $\log_7(x+10) = 2$
20. $\log(2y-5) = 1$
21. $\log_5(3x^2) = -2$
22. $\ln e^{3x-1} = 4$

In Exercises 23–28, solve each exponential equation.

23. $7^x = 9$

24. $15^x = 3$

25. $4^{x-7} = 10$

26. $e^x = 4$

27. $8^{4x-3} = 21$

28. $e^{-x} = 37$

SECTION 1.1

13. $V(h)=5h,\ h>0$

15. $V(r)=\frac{4}{3}\pi r^3,\ r>0$

17. $P(s)=4s,\ s>0$

19. $A(a)=\frac{3}{2}a,\ a>0$

21. $A(b,h)=bh,\ b>0,\ h>0$

23. $C(n,p)=np,\ n\geq 0,\ p\geq o$

25. time in hours, rate in buttons per hour (answers will vary)

29.

Input	Output
3	-5
4	-2
5	3
b	b^2-4b-2
x	x^2-4x-2
x+3	$(x+3)^2$-4(x+3)-2
x+h	$(x+h)^2$-4(x+h)-2

SECTION 1.2

1. No 3. Yes
5. Yes 7. Yes
9. No 11. Yes
13. Yes 15. Yes

17. Domain is $\{x|x\neq 8\}$.

19. Domain is the set of all real numbers.

21. Domain is $\{x|x\geq -2,\ x\neq 2\}$.

23. Domain is $\{x|x>-4\}$.

25. $D(x)=\frac{48.26}{\sqrt{x}}$

Range is approximately $16\leq D\leq 48$

27. $W(t)=0.033t^2-0.3974t+7.3032$

Range is approximately $6.1\leq W\leq 8.2$

29. $T(t)=\frac{580}{t^2+10.3t+33.665}$

Range is approximately $0\leq T\leq 17$

SECTION 1.3

1. 31 3. 25 5. 12

7. f(0) = $\frac{0^3+64}{0+4}=16$
9. f(h) = $4h^2+3h-1$
11. $x^2+2xh+h^2+x+h-5$
13. f(x) = 3x-7
15. -f(x) = -5x-6
17. 4h
19. $2xh+h^2+h$
21. $16xh+8h^2-3h$
23. 6
25. 6x+3h+7
27. (a) Approximately 3 hours to reach maximum level of 89%.
    (b) The drug will reach a level of approximately 50% in about 1.1 hours.
29. (a)&(b)
    Plant approximately 150 trees to maximize yield at about 4480 avocados.
    (c) 3500 avocados are produced by 80 trees or by 218 trees.
    (d) No avocados are produced by 0 trees or by 300 trees.

SECTION 1.4

1. $(f\circ g)(x)=2x^2-2x+5$, Domain: all real numbers
   $(g\circ f)(x)=4x^2+18x+20$, Domain: all real numbers
3. $(f\circ g)(x)=4x^4-18x^2+16$, Domain: all real numbers
   $(g\circ f)(x)=2x^4+4x^3-14x^2-16x+27$, Domain: all real numbers

5. $(f\circ g)(x)=\frac{3}{5x-20}$, Domain: $\{x|x\neq 4\}$

   $(g\circ f)(x)=\frac{3}{5x}-4$, Domain: $\{x|x\neq o\}$

7. $(f\circ g)(x)=\sqrt{x+1}$, Domain: $\{x|x\geq -1\}$

   $(g\circ f)(x)=\sqrt{x}+1$, Domain: $\{x|x\geq 0\}$

9. $(f\circ g)(x)=\frac{25-x^2}{5x+x^2}$, Domain: $\{x|x\neq 0,-5\}$

   $(g\circ f)(x)=\frac{5(x+1)}{x^2-1}$,

   Domain: $\{x|x\neq 1,-1\}$

11. $(f\circ g)(x)=\frac{4x+50}{7x+10}$, Domain: $\{x|x\neq \frac{-10}{7},3\}$

    $(g\circ f)(x)=\frac{28x+41}{-2x-14}$,

Domain: $\{x|x\neq-2,-7\}$

13. $f(x) = \{(2,0),(1,-2),(3,2),(4,4)\}$
    $g(x) = \{(-5,4), (-1,3), (0,2), (4,1)\}$
    Domain of $(f\circ g)(x) = \{-5,-1,0,4\}$
15. $(g\circ h)(x)= -5$
    $(f\circ g\circ h)(0) = 24$
17. $(f\circ g)(x)=x \quad (g\circ f)(x)=x$
19. $(f\circ f)(x)=x$ Domain: $\{x|x\neq1\}$
21. (a) $(C\circ x)(5) = \$37,196.10$
    (b) $(C\circ x)(6) = \$54,131.10$
    (c) If C = \$20,000, t is approximately 3.67 hours.
23. (a) $Q(10) \approx 890$ pounds

    (b) $Q(20) \approx 250$ pounds

SECTION 1.0

1. $V(x)=x^3,\ x>0$
3. $A(r)=\pi r^2, r>0$
5. $P(s)=4s,\ s>0$
7. $A(h)=\frac{5}{2}h,\ h>0$
9. Not a function
11. Function
13. Function
15. All real numbers
17. $\{x|x\neq\frac{-4}{5}\}$
19. $\{x|x\leq\frac{-7}{4}\}$
21. $\{x|x\geq-\frac{1}{2}\}\bigcap\{x|x\neq-\frac{1}{3}\}$
23. $\{x|\frac{1}{2}\leq x<\frac{6}{7}\}$
25. -13
27. $f(-z)=-z^3+11$
29. $H(-1)=9$
31. $P(3+c)=c^2+6c+2$
33. $E(a^3)=a^9-2a^6+4a^3-12$
35. $f(x+h)=3x^2+6xh+3h^2+4x+4h-2$
37. $f(x-h)=5x-5h+34$
39. $-f(x)=-x^2+x-2$
41. $f(x+h)-f(x)=4h$
43. $\frac{x+h+5}{x+h+10}-\frac{x+5}{x+10}$
45. $\frac{f(x+h)-f(x)}{h}=4x+2h-5$
47. $\frac{f(x+h)-f(x)}{h}=4x+2h+5$
49. $(f\circ g)(x)=7x^2-2x+25$, Domain: all real numbers
    $(g\circ f)(x)=7x^2+348x+4325$, Domain: all real numbers
51. $(f\circ g)(x)=2x^4+12x^3-2x^2-60x+46$, Domain: all real numbers
    $(g\circ f)(x)=4x^4-10x^2-1$, Domain: all real numbers
53. $(f\circ g)(x)=\frac{1}{9x-21}$ Domain: $\{x|x\neq\frac{7}{3}\}$

    $(g\circ f)(x)=\frac{1}{x}-7$, Domain: $\{x|x\neq0\}$
55. $(f\circ g)(x)=\sqrt{6x+2}$, Domain: $\{x|x\geq-\frac{1}{3}\}$

    $(g\circ f)(x)=2\sqrt{3x-1}+1$, Domain: $\{x|x\geq\frac{1}{3}\}$
57. $(f\circ g)(x)=\frac{x-2}{-x+8}$, Domain: $\{x|x\neq5,8\}$

    $(g\circ f)(x)=\frac{x-1}{-2x+8}$, Domain: $\{x|x\neq3,4\}$
59. $(g\circ h)(x)=2x+1$
    $(f\circ g\circ h)(7)=78$
61. $(f\circ g)(x)=\frac{12x+39}{5}$

    $(g\circ f)(x)=\frac{12x+17}{5}$
63. $(f\circ f)(x)=x$ Domain: all real numbers
65. $(f\circ f)(x)=\frac{-2\left(\frac{-2x}{3-x}\right)}{3-\left(\frac{-2x}{3-x}\right)}=\frac{4x}{9-x}$

    Domain: $\{x|x\neq3,9\}$,
67. $m=2$: As the input value increases by 1 unit, the output value increases by 2 units.
69. $m=\frac{10}{13}$: As the input value increases by 1 unit, the output value increases by 10/13 units.
71. $m=-\frac{1}{5}$, y-int = -8

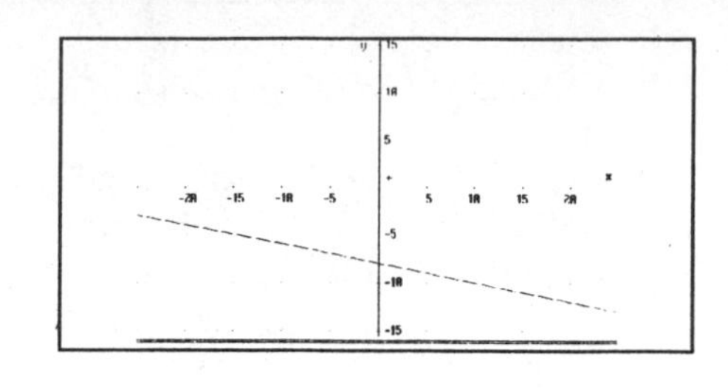

73. $m = \frac{3}{12}$, y-int = 3

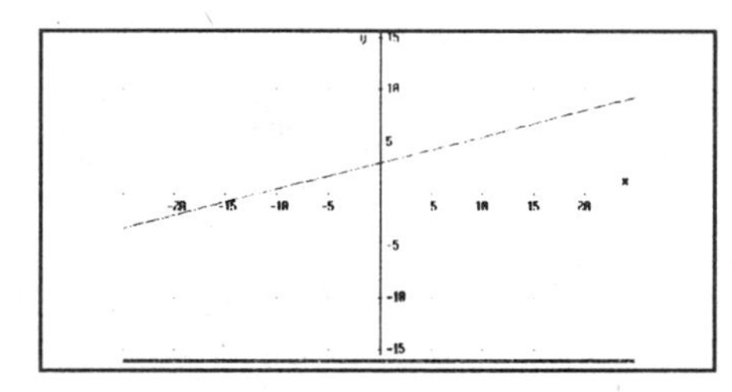

75. m = 3, y-int = -1

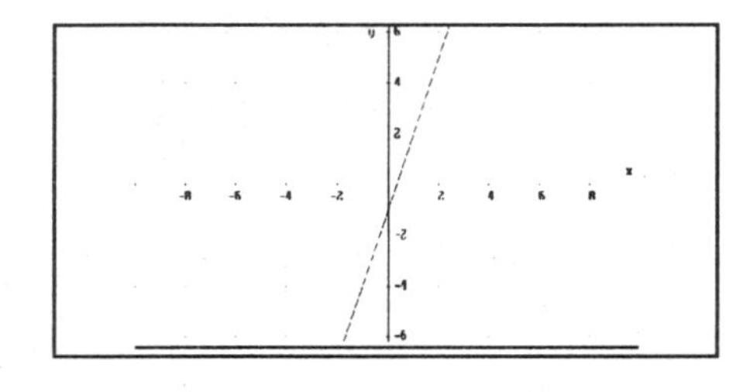

SECTION 2.1

1. y increases on $(-\infty,a)$ y decreases on (a,∞)
   y has a maximum P at x=a

   $\lim_{x\to a} y=P$

3. y decreases on $(-\infty,a)$ and on (b,∞)
   y increases on (a,b)
   y has a minimum P at x=a
   y has a maximum Q at x=b

   $\lim_{x\to a} y=P$ $\qquad \lim_{x\to b} y=Q$

5. y increases on $(-\infty,5)$
   y decreases on $(5,\infty)$
   y has a maximum 12 at x=5

   $\lim_{x\to 5} y=12$

7. y increases on $(-\infty,b)$
   y decreases on (b,∞)
   y has a maximum Q at x=b
   y is undefined at x=a and at x=c

   $\lim_{x\to a} y=P$ $\qquad \lim_{x\to b} y=Q$ $\qquad \lim_{x\to c} y=R$

9. y increases on $(-\infty,3)$ and on $(3,5)$
   y decreases on $(5,\infty)$

   $\lim_{x\to 3^-} y=4$ $\qquad \lim_{x\to 3^+} y=5$

   $\lim_{x\to 5^-} y=7$ $\qquad \lim_{x\to 5^+} y=5$

   Note that y(5)=6

11. $\lim_{x\to 5^-} f(x)=-5$ $\qquad \lim_{x\to 5^+} f(x)=-5$

    $\lim_{x\to 5} f(x)=-5$

13. $\lim_{x\to -8^-} f(x)=6$ $\qquad \lim_{x\to -8^+} f(x)=5$

    $\lim_{x\to -8} f(x)$ does not exist

15. $\lim_{x\to 5} f(x)$ does not exist

17. $\lim_{x\to 5} f(x)=2$

19. $\lim_{x\to 2} f(x)$ does not exist

21. $\lim_{x\to 1^-} f(x)=-3$ $\qquad \lim_{x\to 1^+} f(x)=-3$

    $\lim_{x\to 1} f(x)=-3$

23. $\lim_{x\to -3^-} f(x)=-23$ $\qquad \lim_{x\to -3^+} f(x)=41$

    $\lim_{x\to -3} f(x)$ does not exist

25. $\lim_{x\to 4^-} f(x)=4$ $\qquad \lim_{x\to 4^+} f(x)=2$

$\lim_{x \to 4} f(x)$ does not exist

27. $\lim_{x \to 4} f(x)$ does not exist

29. $\lim_{x \to 2} f(x)$ does not exist

31. $\lim_{x \to -1} f(x) = -4$

41. 2x+h+3

SECTION 2.2

1. a) 11 b) -1
   c) 30 d) 5/6
3. $\lim_{x \to 3} f(x) = 2$
5. a) True b) True
   c) False d) True
   e) False f) True
   g) True h) True
7. 5
9. 2
11. -3
13. -11
15. 1
17. -1/3
19. 0
21. UNDEFINED
23. -1/3
25. -6
27. 1
29. -1/12
31. 2/15
33. 1/2
35. DOES NOT EXIST
37. DOES NOT EXIST
39. 5
41. 12x-4
43. $\dfrac{-6}{(x+4)^2}$
45. a) -7
    b) -7
    c) -7
47. a) 3
    b) 1
    c) DOES NOT EXIST
49. $\lim_{x \to 15} C(x)$ does not exist, because

    $\lim_{x \to 15^-} C(x) \neq \lim_{x \to 15^+} C(x)$
51. a) $\lim_{x \to 10{,}000} T(x)$ does not exist

    b) $\lim_{x \to 25{,}000} T(x) = 3750$, because

    $\lim_{x \to 25{,}000^-} T(x) = \lim_{x \to 25{,}000^+} T(x) = 3750$

    Tax on $25,000 earned is $3750.

    c) $\lim_{x \to 40{,}000} T(x)$ does not exist

    d) $\lim_{x \to 60{,}000} T(x) = 12{,}000;$

    Tax on $60,000 earned is $12,000.
57. $f(x)=x^2-x-2$
59. a) -8
    b) 1
    c) 10
61. $(g \circ f)(x) = \dfrac{2}{\sqrt{x+3}-1}$

    $(g \circ f)(6) = 1$

SECTION 2.3

1. 1
3. a)-1
   b) DOES NOT EXIST
5. a) DOES NOT EXIST
   b) DOES NOT EXIST
   c) 4
7. 2/3
9. 0
11. DOES NOT EXIST
13. 5
15. -12
17. DOES NOT EXIST
19. 0
21. DOES NOT EXIST
23. DOES NOT EXIST
25. 5/2
27. $\dfrac{\sqrt{3}}{5}$
29. 0

31. $\lim_{x \to \infty} \overline{C}(x) = 0.89$

As the company produces more and more tubes of toothpaste, the average cost per tube approaches $.89.

33. $\lim_{p \to \infty} N(p) = 0$

All of the fish would die.
35. The average cost approaches $240 as the number of computers increases without bound.

37. $\lim_{t \to \infty} (t^{\frac{1}{5}} + 5)^4 = \infty$

The amount of pollution the lake increases without bound.
43. $8x+4h-4$
45. 2

SECTION 2.4

1. Discontinuous;

   $\lim_{x \to a} f(x)$ does not exist
3. Discontinuous;

   $f(a)$ does not exist
5. Continuous at $x=a$
7. Continuous at $x=a$
9. Discontinuous;

   $\lim_{x \to a} f(x)$ does not exist
11. Discontinuous;

   $\lim_{x \to a} f(x)$ does not exist
19. Continuous at x=-1.
21. Continuous at x=3.
23. Discontinuous at x=2 because f(2) does not exist.
25. f(x) is discontinuous at x=1 because f(1) does not exist.
27. f(x) is continuous at x=0.
29. f(x) is discontinuous at x=-4 because f(-4)=0 but

   $\lim_{x \to -4} f(x) = -3$
31. f(x) is continuous at x=12.
39. N(t) is discontinuous at t=a. A war, for instance, might cause an instantaneous drop in human population.
41. D(t) is discontinuous at t=a. The jump at t=a might be public response to a new microchip.
43. The discontinuities at t=a, t=b, and t=c might represent the beginning of new semesters when parking fees are raised.
47. $(f \circ g)(x) = 48^2+168x+152$
49. -11
51. 0

SECTION 2.5

1. Average rate of change = 4;
   For values of x between 18 and 21, y increases an average of 4 units for every unit increase in x.
3. Average rate of change = -2;
   For values of x between 25 and 40, y decreases an average of 2 units for every unit increase in x.
5. Average rate of change = -3/4;
   For values of x between 12 and 16, y decreases an average of 3/4 for every unit increase in x.
7. Average rate of change = -.0875;
   For the first 8 hours after the injection, the concentration decreases an average of 0.0875 mg/cc for every hour.
9. For a selling price between 75 cents and 80 cents, the average profit increases 60 cents for every 1 cent increase in price.
11. Instantaneous rate of change of y when x=4 is 8(4)-1, or 31. When x=4, y increases by 31 for every unit increase in x.
13. Instantaneous rate of change of y when x=0 is -2. When x=0 (or any other value), y decreases by 2 for every unit increase in x.
15. Approximate
17. Exact
19. Instantaneous rate of change; Measures the approximate rate of change.
21. Average rate of change.
23. Instantaneous rate of change; Measures the approximate rate of change.
29. 5
31. 4
33. 1
35. At x=3, the instantaneous rate of change is 2(3), or 6.
37. At x=0, the instantaneous rate of change is 10(0)+4, or 4.
39. At x=-2, the instantaneous rate of change is 4.
41. At x=2, the instantaneous rate of change is -5.
43. Average rate of change is 692;
   Between 1984 and 2000, it is anticipated that there will be an average increase of 692,000 jobs per year in marketing and sales.

45. (a) At x=10, the instantaneous rate of change is -2(10)+38, or 18. Then the profit increases by approximately \$18 per week when the selling price increases from \$10 to \$11.
    (b) At x=20, the instantaneous rate of change is -2(20)+38, or -2. Then the profit decreases by approximately \$2 per week when the selling price increases from \$20 to \$21.
47. (a) Since the instantaneous rate of change = 0.08, a constant, then from 1975 to 1976 the median age increases by exactly 0.08 years.
    (b) From 1990 to 1991, the median age also increases by exactly 0.08 years.
53. 22
55. 5/4

SECTION 2.0

1. $\lim_{x\to a} f(x) = P$

3. $\lim_{x\to a^-} f(x) = P$ $\lim_{x\to a^+} f(x) = Q$

5. $\lim_{x\to 13^-} f(x) = 2$ $\lim_{x\to 13^+} f(x) = 2$

7. $\lim_{x\to 2^-} f(x) = 6$

   $\lim_{x\to 2^+} f(x) = 5.\overline{8}$

9. $\lim_{x\to 1} f(x)$ does not exist

15. $\lim_{x\to 1} 6 = 6$

17. 76
19. 7
21. 8
23. 5
25. -8
27. 1/5
29. -3
31. 2
33. 2x
35. 2x+6
37. $\dfrac{-5}{(x-3)^2}$
39. $\dfrac{2}{(x+4)^2}$
41. 7/8
43. 0
45. DOES NOT EXIST
47. 6
49. 1
51. DOES NOT EXIST
53. (a) $\lim_{w\to 8} P(w) = 8$

    (b) $\lim_{w\to 16} P(w)$ DOES NOT EXIST

    (c) $\lim_{w\to 32} P(w) = 50$
55. f(x) is continuous at x=a.
57. f(x) is discontinuous at x=a because f(a) does not exist
63. continuous at x=1.
65. continuous at x=6.
67. continuous at x=2.
73. N(t) is discontinuous at t=a. The police officer is off duty for a few days!
75. Average rate of change = 5;
    For values of x between 35 and 42, y increases an average of 5 units for every unit increase in x.
77. Average rate of change = 46;
    The officers wrote an average of 46 tickets per day between January and August.
79. Approximate
81. Exact
83. Approximate
85. Instantaneous rate of change: Exact
87. Average rate of change.
91. 6
93. 1/7
95. At x=2, the instantaneous rate of change is 2(2)+1=5.
97. At x=7, the instantaneous rate of change is -9.
99. At x=5, the instantaneous rate of change is -8.
101.
    (a) For heart weights between 195 mg and 235 mg, the body weight increases an average of 0.213 mg for every 1 mg increase in heart weight.
    (b) The body weight increases by a constant value of 0.213 mg per mg of heart weight. Then 0.213 mg is the increase at a heart weight of 210 mg.
    (c) 0.213 mg per mg of heart weight.
103.
    (a) The city loses an average of 4 people per year.
    (b) At t=3, the instantaneous rate of change is 2(3)-14=-8. The city is losing population at a rate of 8 people per year when t=3.
    (c) P(4)-P(3) = -7. The city actually lost 7 people.

SECTION 3.1

1. P(60)=40 means that when 60 thousand dollars is spent on advertising, the resulting profit is 40 thousand dollars. $\frac{dp}{dx}\Big|_{x=60} = 2$ means that if the amount spent on advertising is increased from 60 to 61 thousand, profit will increase by approximately $2000.
3. N(150)=1200 means that it requires 150 hours to build 1200 objects. N′(150)=21.25 means that if the number of hours increases from 150 to 151, the number of objects built increases by approximately 21.25.
5. S(44)=87 means that if there are 44 teachers with degrees in their teaching area, the average exam score will be 87.
   S′(44)=1.5 means that if the number of teachers with degrees increases from 44 to 45, the average score on the exam will increase by about 1.5 points per teacher.
7. f(2)=7 means that the cloud has traveled 7 miles in 2 hours since its release. f′(2)=3 means that between t=2 hours and t=3 hours the cloud travels approximately 3 miles.
9. I(6) gives the rate of inflation in June.
11. f(a) does not exist; therefore,
    f′(a) does not exist.
    (b,f(b)) is a cusp; therefore, f′(b) does not exist.
13. There is a vertical tangent line at (a,f(a)); therefore, f′(a) does not exist.
    (b,f(b)) is a cusp; therefore, f′(b) does not exist.
21. 10x-1
23. $-\frac{3}{x^2}$
25. $f'(x) = 16x^3+6x^2+16x-1$
27. $6x^{-\frac{1}{3}}+1$
29. $-25x^{\frac{-7}{2}}-7x^{\frac{-9}{2}}+18x^{-\frac{1}{2}}$
31. $\frac{-15}{x^4}-\frac{1}{x^2}$
33. $\frac{1}{3x^{\frac{2}{3}}}+\frac{3}{4x^{\frac{1}{4}}}-1$
35. 0
37. 44
39. 16
41. -1/4
43. (a) 520 gallons
    (b) Approximately 36 1/4 gallons of wine will flow into the vat between minutes 16 and 17.
45. 1.44 gm increase
47. (a) If advertising is raised from 6 to 7 thousand, revenue will decrease by approximately $243.
    (b) $87,910
49. By increasing the order from 317 to 318 packages, inventory costs decrease by approximately $0.02.
55. -1
57. Continuous at x=3.

SECTION 3.2

1. $12x^3(x^3+1)+3x^4(3x^2)$
3. $(5x^4+2)(x^3-1)+(x^5+2x+6)(3x^2)$
5. $\frac{x}{\sqrt{x^2+6}}\sqrt[3]{2x-9}+\sqrt{x^2+6}\left(\frac{2}{3\sqrt[3]{(2x-9)^2}}\right)$
7. $\frac{-37}{(2x-5)^2}$
9. $\frac{1}{2}x(2x+7)(x^2+3)^{-\frac{3}{4}}+2(x^2+3)^{\frac{1}{4}}$
11. $18x^2+22x+9$
13. $2x+5x^{\frac{2}{3}}-2x^{-\frac{1}{3}}+5$
15. $\frac{-31}{(2x-7)^2}$
17. $\frac{-47}{(5x-8)^2}$
19. $\frac{-x^4-18x^2-2x}{(x^3-1)^2}$
21. $\frac{3x^4-29x^2-2x+28}{(x^2-4)^2}$
23. $\frac{x^6+4x^3+18x^2}{(x^3+1)^2}$
25. $\frac{28x^3+5x^2-4x+55}{(7x+2)^2}$
27. $\frac{-16x-24}{(x^2+3x+4)^2}$

29. $\dfrac{-\dfrac{\sqrt{x}}{2}-\dfrac{2}{\sqrt{x}}}{(x-4)^2}$

31. At x=4, the function is increasing 20 units for each unit increase in x.
33. At t=10, the function is decreasing 0.00027 units for each unit increase in t.
35. One hour after memorizing them, a person remembers about 3 facts, and 3 hours after memorizing, only 1 fact is remembered.
    One hour after memorizing, a person is forgetting about 10.24 facts per hour, and 3 hours after memorizing, the person is forgetting about 0.115 facts per hour.
37. decrease in average daily cost of about $0.02.
39. At a price of $2.50 per tube, there is a decrease in sales of about 4 tubes per penny increase in price.
41. a) $P'(x)=60-0.08x$
    b) As production increases, marginal revenue is decreasing.
    c) As production increases, marginal cost is decreasing.
    d) As production increases, marginal profit is decreasing.
47. 0
49. 20

SECTION 3.3

1. Since the first derivative is positive and the second derivative is positive when 4500 pillows are produced, the cost is increasing at an increasing rate.
3. Since the first derivative and the second derivative are both negative 230 days after the beginning of the year, the flow rate is decreasing at an increasing rate.
5. Since the first derivative is positive and the second derivative is negative at a complexity index of 7, the useful life of the toy is increasing at a decreasing rate.
7. Since the first and second derivatives are both negative at month 15, the number of alcohol related accidents is decreasing at an increasing rate.
9. Since the first and second derivative are both positive when sales are 60 per year, the percent of commission is increasing at an increasing rate.
11. Since the first derivative is positive and the second derivative is negative 3.5 hours after the shift begins, the efficiency is increasing at a decreasing rate.
13. Since the first and second derivatives are both positive when the intensity is 0.8, the amount of photosynthesis occurring is increasing at an increasing rate.
15. Since the first derivative is positive and the second derivative is negative 3 minutes after system began operating, the amount of carbon monoxide ventilated from the lab is increasing at a decreasing rate.
17. $G'(t)<0$ and $G''(t)>0$
19. $A'(t)>0$ and $A''(t)<0$
21. $P'(x)>0$ and $P''(x)<0$
23. $Y'(t)<0$ and $Y''(t)<0$
25. $N'(t)<0$ and $N''(t)>0$
27. $f^{(4)}(x) = 120x$
29. $f^{(4)}(x) = -72+96x^{-5}$
31. $f^{(4)}(x) = \dfrac{-75}{8}x^{-\frac{7}{2}}$
33. $f''(x) = -10x^{-3}$
35. $f''(x) = 18x^{-4}$
37. $f'(x) = -10x+1;\ f'(1) = -9$

    means that at x=1 the function is decreasing 9 units per unit increase in x.

    $f''(x) = -10;\ f''(1) = -10$

    means that at x=1 the function is decreasing at a decreasing rate.
39. $S(40) \approx 297,$

    $S'(40) \approx 6.75, \quad S''(40) \approx -0.035$

    The strength of the reaction is increasing, but at a decreasing rate.
41. $N(12) \approx 7.64,$

    $N'(12) \approx -1.27, \quad N''(12) \approx 0.32$

    The rate at which hair dryers are sold is decreasing, but the rate of decrease is slowing down.
43. The rate at which students input words is increasing and at a faster rate.
47. 22
49. -1/3
51. Revenue is increasing at a decreasing rate.

SECTION 3.4

1. yes 3. yes 5. no 7. no
9. At x=2, y changes 4 units for each unit change in x.
11. At E=204 billion dollars, the population increases 4800 for each billion dollar increase in the economy.
13. global quotient
15. global power
17. 36
19. -20x
21. $300x^3+240x$
23. $750x^5-600x^3+180x$
25. $6(5x+1)(5x^2+2x-6)^2$
27. $\dfrac{6x}{5\left(\sqrt[5]{(x^2+7)^2}\right)}$
29. $2(x^2+3)^2(2x-1)^4(11x^2-3x+15)$
31. $\dfrac{-1}{(x+3)^2}$
33. $\dfrac{-24}{(x-4)^4}$
35. $\dfrac{-24}{(2x+7)^5}$
37. $\dfrac{(x^2+6)^6(13x^2+28x-6)}{(x-2)^2}$
39. $\dfrac{3(-x^2+4x-1)}{(x+2)^4}$
41. $\dfrac{x^3(21x+8)}{(5x+2)^{\frac{4}{5}}}$
43. $\dfrac{2x^6(31x+7)}{(4x+1)^{\frac{1}{4}}}$
45. $\dfrac{90(9x-4)^2}{(3x+2)^4}$
47. A decrease in government spending of approximately $453,000,000.
49. a) $\dfrac{2800}{p^3(420p^{-2})^{\frac{4}{3}}}$

    b) $\dfrac{dc}{dp}\Big|_{p=50} \approx 0.242;$ Demand is increasing at a rate of approximately 0.242 units per dollar increase in price.
51. a) $-4.5C+4.5$

    b) $\dfrac{dN}{dC}\Big|_{C=1.70} = -3.15;$

    A decrease of approximately 3150 pads.
53. $D'(40)\approx-2.25$; Demand will decrease by approximately 2250 tablets.
59. 0
61. a) $\dfrac{f(2)-f(-1)}{2-(-1)} = 9$

    b) At x=0, the instantaneous rate of change is 6.

SECTION 3.5

1. (a) implicit (b) implicit
   (c) implicit (d) explicit
   (e) explicit
3. (Step 1 is in error)
5. (Step 1 is in error)
7. (Step 2 is in error)
9. (Step 1 is in error)
11. $y'=\dfrac{-x^2}{2y^3}$
13. $y' = \dfrac{1-20x^3}{9y^2-4y}$
15. $y' = \dfrac{8x+1}{8(y+6)^7}$
17. $y'=\dfrac{2x-5x(2y^3+5x^2)^2}{3y^2(2y^3+5x^2)^2}$
19. $y' = \dfrac{7-12x-8xy^4}{16x^2y^3}$
21. $y' = \dfrac{6xy^2-10x+2}{-6x^2y-3y^2}$
23. $y'-10(y+1)^{\frac{3}{5}}$
25. $y' = 0$
27. $y'(-2,1)=\dfrac{96}{13}$

29. $y'(-2,-3) = -\frac{81}{16}$

31. Sales will decrease by approximately 1150 collars.
33. $85,050 approximate increase per month.
35. Monthly revenue is increasing by approximately $467.71.
37. Revenue is increasing approximately 0.94 dollars per month.

43. $f'(x) = \frac{-7}{(x-1)^2}$

45. $f'(x) = (3x+3)^2(2x-1)(6x+3)$

SECTION 3.0

1. f(35)=18 means that an $18,000 profit is realized when $35,000 is spent on advertising. f'(35)=2.4 means that at an advertising rate of $35,000, the profit is increasing $2400 per thousand increase in advertising.
3. f(x)=92 means that 4 hours after death the body temperature is 92°F. f'(4)=-1.8 means that at 4 hours, the body temperature is falling 1.8°F per hour.
5. f'(a) does not exist because there is a cusp at x=a.

11. $\frac{dy}{dx}=5$ when x=9 means that y increases 5 units per unit increase in x when x=9.

13. $\frac{dy}{dx}=-2$ when x=4 means that y decreases 2 units per unit increase in x when x=4.

15. implicit 17. implicit
19. implicit 21. explicit

23. $dx=1.2,\ dy=0.7,\ \Delta y=1.1$

25. $dx=0.6,\ dy=-0.2,\ \Delta y=-0.5$

27. $dx=-0.9,\ dy=0.2,\ \Delta y=1.1$

29. global power 31. global power

33. When x=1, f is increasing 90 units per unit increase in x.
35. When x=0, f remains constant for a unit increase in x.

37. $f'(x) = 3x^2+5$

39. $f'(x) = -3$

41. $f'(x) = 1-2x$

43. $f'(x) = 3-8x+3x^2$

45. $f'(x) = -1$

47. $f'(x) = 10x+26$

49. $f'(x) = 5x^4+3x^2$

51. $f'(x) = 3x(5x+4)^2(25x+8)$

53. $f'(x) = 9(x+6)(x+2)$

55. $f'(x) = 5(x-4)(x+1)^2(x-2)$

57. $f'(x) = -\frac{1}{(x-4)^2}$

59. $f'(x) = -\frac{3}{(x+6)^2}$

61. $f'(x) = -\frac{6}{(3x+4)^2}$

63. $f'(x) = \frac{4}{(2x+3)^2}$

65. $f'(x) = \frac{2}{(x+3)^2}$

67. $f'(x) = \frac{-11}{(x-1)^2}$

69. $f'(x) = \frac{-11}{(2x+3)^2}$

71. $f'(x) = \frac{-12(x+2)^2}{(x-4)^3}$

73. $f'(x) = \frac{3}{\sqrt{6x+1}}$

75. $f'(x) = \frac{1}{\sqrt[3]{(3x+1)^2}}$

77. $\frac{dy}{dx}=3$

79. $\frac{dy}{dx}=8x+12$

81. $y' = \frac{-2x}{9y^2}$

83. $y' = \frac{-2x-1}{1-2y}$

85. $y' = \frac{-5y}{2y+5x}$

87. $f^{(4)}(x) = 240x$

89. $f''(x) = 20(x+4)^3$

91. At x=50, f is increasing 7498 units per unit increase in x.

93. At x=15, f decreases $\frac{1}{484}$ units per unit increase in x.

95. At x=3, f increases 6/25 units per unit increase in x.

SECTION 4.1

1. Critical values at x=6, and x=15.
   Increasing on (-∞,6) and (15,∞).
   Decreasing on (6,15).
3. Critical values at x=10, x=23, and x=32.
   Increasing on [0,10) and (23,32).
   Decreasing on (10,23) and (32,38].
5. Critical value at x=9.
   Increasing on (9,∞).
   Decreasing on [0,9).
7. Hypercritical value at x=10.
   Concave up on (10, ∞).
   Concave down on (-∞,10).
9. No hypercritical values.
   Concave down on (0,23) and (23,38).
11. Hypercritical value at x=12.
   Concave up on (9,12).
   Concave down on (12, ∞).
13. Relative maximum at x=6.
   Relative minimum at x=15.
   No absolute maximum or minimum.
15. Relative maximums at x=10 and x=32.
   Relative minimum at x=23.
   Absolute maximum at x=32.
   Absolute minimum at x=23.
17. Relative minimum at x=9.
   No relative maximums.
   No absolute maximum.
   Absolute minimum at x=9.
19. Concave up
21. Concave down
23. Concave down

29.

Test Point	0	2	4	5½	6½	8
$f'(x)$	(+)	(-)	(+)	(-)	(-)	(+)
x	1	3	5	6	7	
$f''(x)$	(-)	(-)	(-)	(-)	(+)	(+)

——————1—3———5———6———7——— x

31.

Test Point	3	5	6½	8	10
$f'(x)$	(-)	(+)	(+)	(-)	(-)
$f''(x)$	(+)	(+)	(-)	(-)	(+)

——————4—6——7———9——————— x

33.

Int/Val	f′(x)	f″(x)	f(x)	Behavior of f(x)
(-∞,5)	-	+	dec, conc up	dec at dec rate
5	0		(5,4)	rel min at (5,4)
(5,∞)	+	+	inc, conc up	inc at inc rate

35.

Int/Val	f′(x)	f″(x)	f(x)	Behavior of f(x)
0			(0,16)	
(0,6)	-	-	dec, conc down	dec at inc rate
6		0	(6,10)	pt of infl (6,10)
(6,11)	-	+	dec, conc up	dec at dec rate
11	0		(11,3)	rel min (11,3)
(11,14)	+	+	inc, conc up	inc at inc rate
14		0	(14,7)	pt of infl (14,7)
(14,∞)	+	-	inc, conc down	inc at dec rate

37. A radius of approximately 2 inches will minimize the cost of the can.
39. A width of approximately 150 feet will maximize the area.
41. The number of people contracting the flu is greatest after approximately 45 days.
43. A width of approximately 7 inches maximizes area.
45. P(t) is maximized when t is approximately 4 weeks.
51. Limit does not exist
53. 3/100
55. f(x) is increasing at a decreasing rate.
57. $\frac{dy}{dx} = \frac{-6x-5y^2}{10xy+3y^2}$

SECTION 4.2

11.

Int/Val	f′(x)	f″(x)	f(x)	Behavior of f(x)
(-∞,3)	-	+	dec, conc up	dec at dec rate
3	0		(3,-2)	rel min at (3,-2)
(3,6)	+	+	inc, conc up	inc at inc rate
6		0	(6,4)	pt of infl at (6,4)
(6,9)	+	-	inc, conc down	inc at dec rate
9	0		(9,7)	rel max (9,7)
(9,∞)	-	-	dec, conc down	dec at inc rate

13.

Int/Val	f′(x)	f″(x)	f(x)	Behavior of f(x)
(-∞,2)	+	-	inc, conc down	inc at dec rate
2	0		(2,8)	rel max at (2,8)
(2,5)	-	-	dec, conc down	dec at inc rate

5	und		(5,2)	rel min at (5,2)
(5,7)	+	-	inc, conc down	inc at dec rate
7		0	(7,5)	pt of infl at (7,5)
(7,∞)	+	+	inc, conc up	inc at inc rate

15.

Int/val	f′(x)	f″(x)	f(x)	Behavior of f(x)
0	und	und	und	VA
(0,3)	-	+	dec, conc up	dec at dec rate
3		0	(3,3)	pt. of infl at (3,3)
(3,6)	-	-	dec, conc down	dec at inc rate
6	und	und	und	VA
(6,9)	-	+	dec, conc up	dec at dec rate
9		0	(9,3)	pt of infl at (9,3)
(9,12)	-	-	dec, conc down	dec at inc rate
12	und	und	und	VA

17.

Int/val	f′(x)	f″(x)	f(x)	Behavior of f(x)
(-∞,-3)	+	-	inc, conc down	inc at dec rate
-3	0		(-3,5)	rel max at (-3,5)
(-3,-1)	-	-	dec, conc down	dec at inc rate
-1		0	(-1,3)	pt of infl (-1,3)
(-1,0)	-	+	dec, conc up	dec at dec rate
0		0	(0,0)	pt of infl at (0,0)
(0,1)	-	-	dec, conc down	dec at inc rate
1		0	(1,-3)	pt of infl at (1,-3)
(1,3)	-	+	dec, conc up	dec at dec rate
3	0		(3,-5)	rel min at (3,-5)
(3,∞)	+	+	inc, conc up	inc at inc rate

19. f(x) decreases on (-∞, 2), increases on (2,∞) and is concave up on (-∞,∞). The point (2,-3) is a relativ minimum.

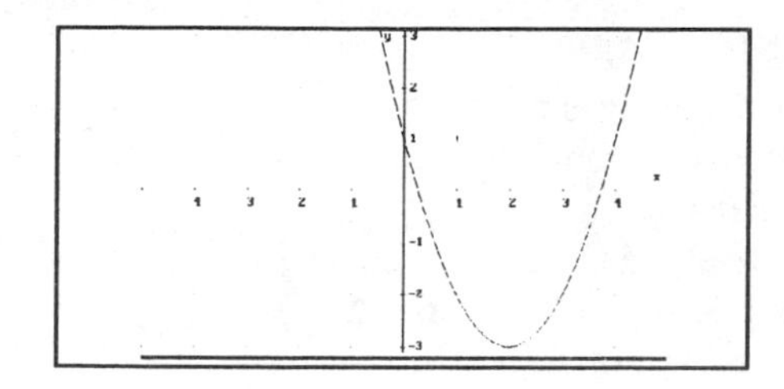

21. f(x) increases on $(-\infty,0)$, and $(2,\infty)$ decreases on $(0,2)$, and is concave down on $(-\infty,1)$ and concave up on $(1,\infty)$. The point $(0,3)$ is a relative maximum, $(2,-1)$ is a relative minimum, and $(1,1)$ is an inflection point.

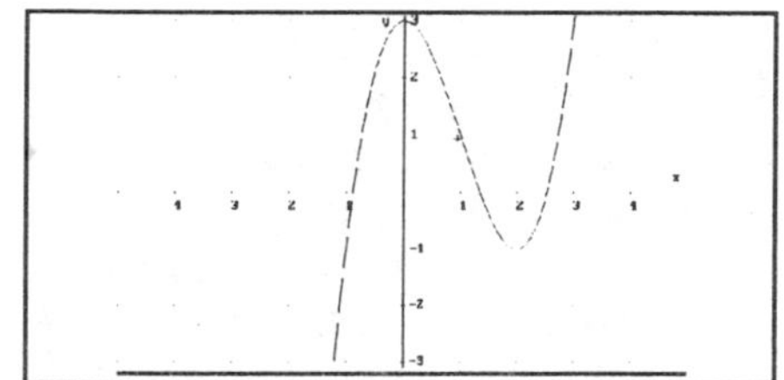

23. f(x) decreases on $(-\infty,1)$, increases on $(1,\infty)$, and is concave up on $(-\infty,0)$ and $(2/3,\infty)$, and concave down on $(0,2/3)$. The point $(1,-1)$ is a relative minimum and $(0,0)$ and $\left(\frac{2}{3},\frac{-16}{27}\right)$ are inflection points.

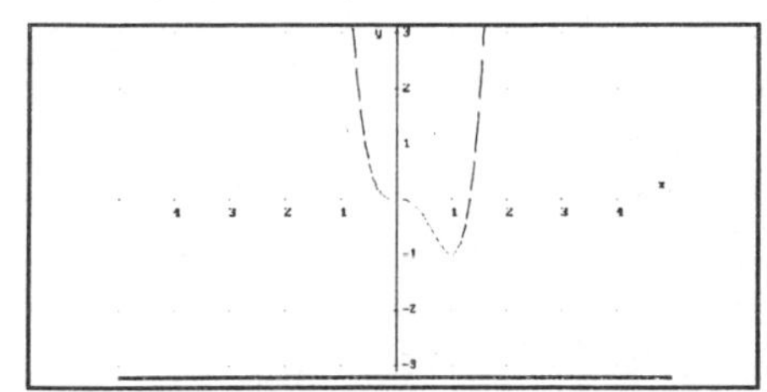

25. f(x) decreases on $(-\infty,0)$, increases on $(0,\infty)$, is concave down on $(-\infty,-1)$ and $(1,\infty)$ and concave up on $(-1,1)$. The point $(0,0)$ is a relative minimum and the points $(-1,\frac{1}{4})$ and $(1,\frac{1}{4})$ are inflection points.

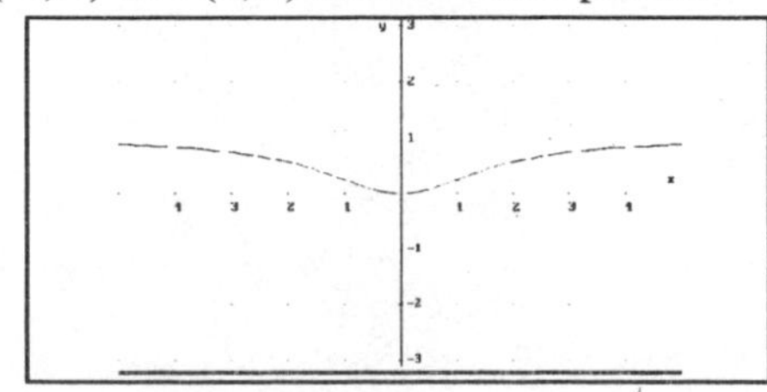

27. f(x) decreases on $(-\infty,0)$ and $(1,\infty)$, increases on $(0,1)$ and is concave down on $(-\infty,\infty)$. The point $(0,0)$ is a relative minimum and $(1,1)$ is a relative maximum.

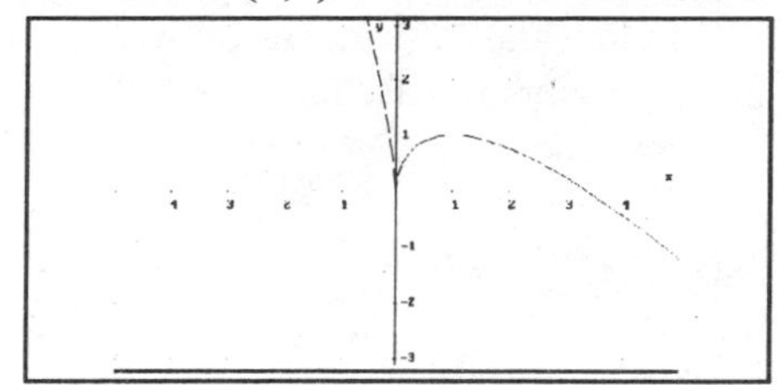

29. f(x) decreases on $(-\infty,\infty)$, is concave down on $(-\infty,0)$ and concave up on $(0,\infty)$. $x=0$ is a vertical asymptote.

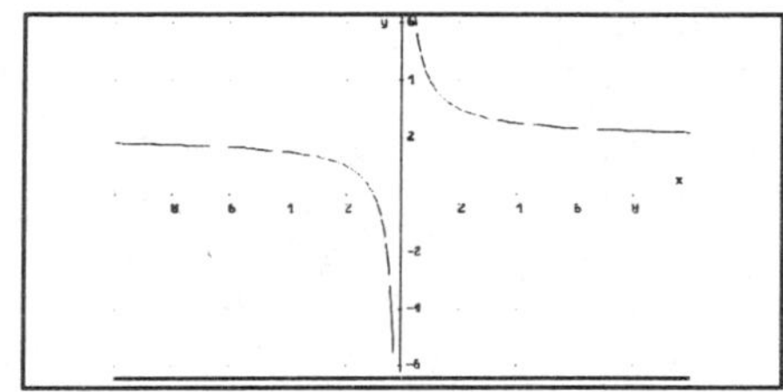

31. C decreases on $(0,1500)$ and increases on $(1500,\infty)$. C is concave up on $(-\infty,\infty)$. A minimum average cost of \$25 occurs when 1500 items are produced.
33. E increases on $(0,\infty)$. E is concave up on $(0,70)$ and concave down on $(70,\infty)$. The enrollment increases towards 37,000 as time goes by, but the rate of increases begins to slow after a 70% tuition increase.

43. $\lim_{x\to\infty} \frac{5}{x^3} = 0$

45. $\frac{f(7)-f(5)}{7-5}=14$

49. $y'=\frac{-4x-18x^3y^2}{9x^4y+2y}$

SECTION 4.3

1. 1975
3. Minimum occurs when L=10. Since the sum of the two legs is 20, the other side is also 10.
5. 2500 magazines
7. Concentration is maximum after 4 minutes.
9. r=5.33 inches minimizes the cost.
11. Revenue is maximized when 50,000 units are produced.
13. The enclosures should be 150 feet by 200 feet.
15. A room rate of \$60 maximizes the revenue.
17. Plant 35 trees per acre to maximize yield.
19. A price of \$16 will maximize revenue.

21. The can should be approximately 2.39 inches in radius and 3.57 inches tall.
23. The room rental should be $90 per night to maximize revenue.

29. $\lim\limits_{x \to \infty} \frac{5}{x} = 0$

31. $f'(5) = \frac{18}{169}$

33. (1,-4) is the point of inflection.

SECTION 4.4

1. C(250)=12,000 means that it costs $12,000 to produce 250 units. C′(250)=-85 means that cost would decrease approximately $85 if production increases from 250 to 251 units.
3. P(55)=16 means that profit is $16,000 when the product is sold for $55. P′(55)=-4 means that profit will decrease approximately $4000 if the price is increased from $55 to $56.
5. C(200)=1500 means cost is $1500 when lot size is 200. C′(200)=300 means cost increases approximately $300 when lot size increases from 200 to 201.
7. Elastic, Revenue decreases.
9. Unit elastic, Revenue remains the same.
11. C(55)=$7732.50
13. An advertising expenditure of $160,000 gives a maximum profit.
15. A production level of 4332 units, maximizes profit at $234,227.80.
17. 1414 orders with 1415 per order OR 1415 orders with 1414 per order minimizes cost.
19. Revenue decreases by $23.04.
    Revenue decreases by $188.80.
21. Revenue increases by $4,593,200.
23. Revenue will decrease by approximately $1120.
25. Revenue will increase by approximately $33,550.
27. R=px[1+0.01a(1-b)]
31. 5
33. The slope of the tangent line from the left and right side of a cusp does not converge.
35. If x increases from 2 to 3, f increases by approximately 5/9.

37. $f'(x) = 4(2x-1)^3(2x+1)(6x+1)$

39. $y' = \frac{6x^2y^2-5y^3-3}{-4x^3y+15xy^2-4}$

SECTION 4.0

1. Critical values at x=2 and x=7
   Decreasing on (-∞,2) and (7,∞)
   Increasing on (2,7)
3. Critical values at x=2 and x=5
   Decreasing on(-∞,2) and (5,∞)
   Increasing on (2,5)
5. No critical values
   Decreasing on [0,∞)
7. Hypercritical value at x=5
   Concave up on (-∞,5)
   Concave down on (5,∞)
9. Hypercritical value at x=7
   Concave down on (-∞,2) and (2,7)
   Concave up on (7,∞)
11. Hypercritical value at x=6
    Concave down on [0,6)
    Concave up on (6,∞)
13. Relative minimum at x=2, relative maximum at x=7, no absolute extrema.
15. Relative and absolute minimum at x=2, relative and absolute maximum at x=5.
17. No relative extrema. Absolute maximum at x=0.
19. Concave down
21. Concave up

23.

Test Value	5		7
$f'(x)$	(-)		(+)
		6	
$f''(x)$	(+)		(-)

25.

Test Value	4		6		12		17		21
$f'(x)$	(-)		(+)		(+)		(+)		(+)
		5		11		16		20	
$f''(x)$	(+)		(-)		(-)		(+)		(-)

27.

Int/val	f′(x)	f″(x)	f(x)	Behavior of f(x)
(-∞,1600)	+	-	inc, conc down	inc at dec rate
1600	0			rel max
(1600,∞)	-	-	dec, conc down	dec at inc rate

29.

Int/val	f′(x)	f″(x)	f(x)	Behavior of f(x)
(-∞,10)	-	-	dec, conc down	dec at inc rate
10		0		pt of infl
(10,20)	-	+	dec, conc up	dec at dec rate
20	0			rel min
(20,35)	+	+	inc, conc up	inc at inc rate
35		0		pt of infl
(35,40)	+	-	inc, conc down	inc at dec rate
40	0			rel max
(40,43)	-	-	dec, conc down	dec at inc rate
43		0		pt of infl
(43,∞)	-	+	dec, conc up	dec at dec rate
y=20				HA
y=32				HA

31. Approximately x=4
33. Poll approximately 900 peopl for a minimum cost of approximately $27,000.
35. Saturation quanitity is approximately 3.
37.

Int/val	f′(x)	f″(x)	f(x)	Behavior of f(x)
(-∞,8)	+	+	inc, conc up	inc at inc rate
8	und		(8,15)	rel max at (8,15)
(8,∞)	-	+	dec, conc up	dec at dec rate

39.

Int/val	f′(x)	f″(x)	f(x)	Behavior of f(x)
$(-\infty,-5)$	+	+	inc, conc up	inc at inc rate
-5		0		pt of infl
(-5,9)	+	-	inc, conc down	inc at dec rate
9	0			rel max
$(9,\infty)$	-	-	dec, conc down	dec at inc rate

47. (0,2) is a relative maximum, (2,-2) is a relative minimum, and (1,0) is a point of inflection.

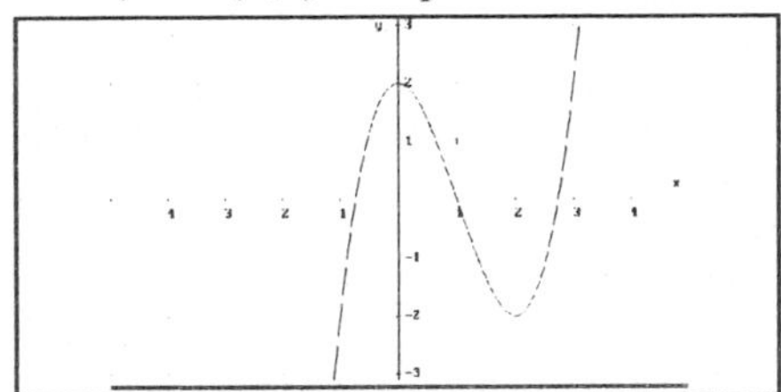

49. (4,400) is a relative maximum, (18,-972) is a relative minimum, and (11,-286) is a point of inflection.

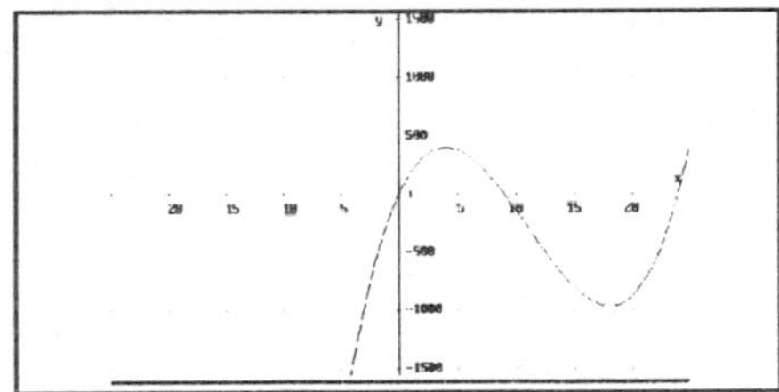

51. There are no critical values or hypercritical values. y=1 is a horizontal asymptote and x=2 is a vertical asymptote.

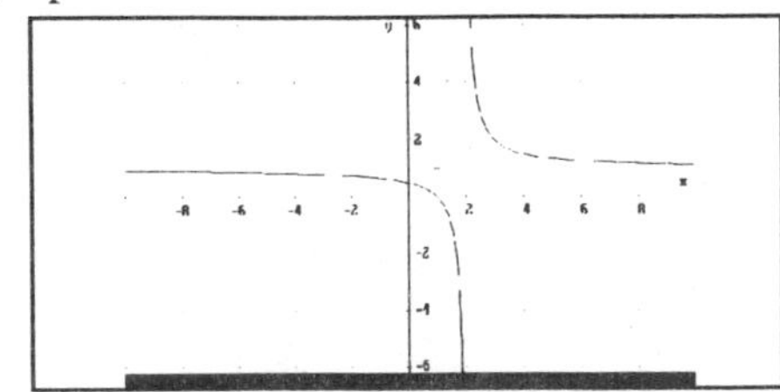

53. (3,0) is point of inflection.

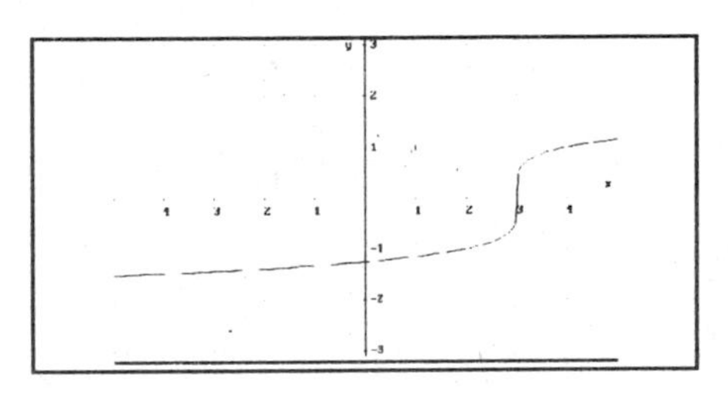

55. (-4/5, 1.03) is a relative maximum, (0,0) is a relative minimum, and (2/5, 1.3) is a point of inflection.

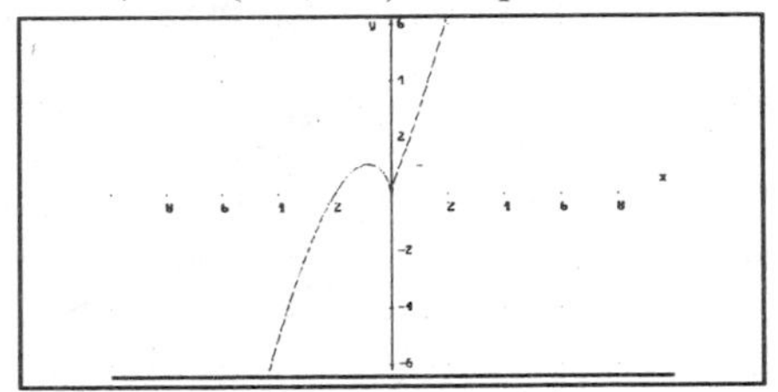

57. The amount of money in the account increases at an increasing rate.
59. From t=0 to t=1 hour, the rate of production increases at an increasing rate. From about t=1 to t=3, the rate of production increases at a decreasing rate to a maximum of slightly over 1 unit per hour. From t=3 to t=4, the rate of production decreases at an increasing rate, after which it decreases at a decreasing rate towards a horizontal asymptote at zero units per hour.
61. The number of products produced decreases at a decreasing rate as assembly time increases.
63. x=3/4k produces a maximum decrease.
65. x should be 8.73 miles to minimize cost.
67. The top and bottom should have a radius of 2 inches, and the height should be 4 inches to minimize material used.
69. x=¼P is a maximum (the rectangle must be a square).

SECTION 5.1

1. Algebraic
3. Exponential
5. Algebraic
7. Exponential
9. Exponential
11. Algebraic
13. $b>1$; exponential growth
    $0<b<1$; exponential decay
15. $f(3)=1000$, growth
17. $f(6)\approx 0.0896$, decay
19. $f(5)\approx 195.391$, growth
21. $f(3)=0.0625$, decay
23. $f(2)\approx 0.135$, decay
25. $f(105)\approx 1,938,676$, growth
27. $f(0.10)\approx 0.0296$
29. $f(3)\approx 98.846$
31. $A\to f(x)=e^{2x}$

    $B\to f(x)=e^{x}$

    $C\to f(x)=e^{\frac{4x}{5}}$

    $D\to f(x)=e^{\frac{2x}{3}}$
33. (a) $P(0)=100$ means 100% of the words are remembered initially.
    (b) $P(1)\approx 52.5$ means about 52.5% of the words are remembered after 1 week.
    (c) $P(2)\approx 30.1$ means about 30.1% of the words are remembered after 2 weeks.
    (d) $P(7)\approx 10.5$ means about 10.5% of the words are remembered after 7 weeks.
    (e) $P(20)\approx 10.0$ means about 10% of the words are remembered after 20 weeks.
35. (a) A=$17,631.94
    (b) A=$17,831.37
    (c) A=$17,878.15
    (d) A=$17,901.11
37. (a) A=$2462.88
    (b) A=$2465.85
    (c) A=$2467.31
    (d) A=$2467.36
39. $M(1985)\approx 98.2\%$
41. (a) approximately 1424 units
    (b) $149,330
43. (a) $N(0)=2700$ toys
    (b) $N(1)\approx 2395$ toys
    (c) $N(5)\approx 1482$ toys
    (d) $N(12)\approx 640$ toys
45. (a) $A(15)\approx 3.234$ mg
    (b) $A(60)\approx 58.084$ mg
47. (a) $N(0)\approx 1.154$ (1154 people)
    (b) $N(3)\approx 1.521$ (1521 people)
    (c) $N(10)\approx 2.867$ (2867 people)
    (d) $N(26)\approx 10.678$ (10,678 people)
    (e) 45,000 people
53. 7/9
55. 0
57. $\frac{f(7)-f(3)}{7-3}=20$
59. $C'(45)=0.03$ means if the number of sets of golf balls is increased from 45 to 46 thousand, the cost will increase by approximately $0.03.
61. If x increases from 0 to 1, f will increase by approximately 768.
63. (0,0) is a relative maximum, $x=\pm\sqrt{3}$ are vertical asymptotes, y=3 is a horizontal asymptote.

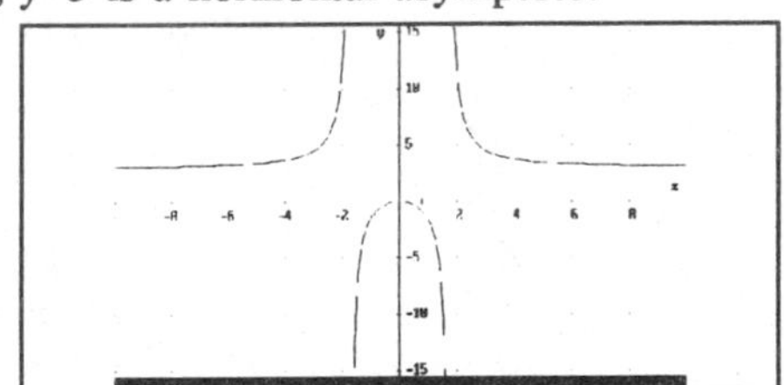

SECTION 5.2

9. $433,319.06
11. 2798 boxes
13. $\ln a=6x$
15. $\ln x=-3k+8$
17. $\ln 8.262=4x-10$
19. $15=e^{k}$
21. $3x+11=e^{2y-4}$
23. $x-1=e^{-0.002}$
25. $x\approx -0.4497$
27. $x\approx 4.2874$
29. $t\approx 2.9997$
31. $x\approx 0.0151$
33. $x=-1$
35. $x\approx 13.012$
37. $h\approx 3.89$ miles

39. t≈79.8 years
41. t≈30 days
43. t≈7.07 minutes
45. t≈27.95 years
47. (a) t≈11.55 years
(b) t≈8.66 years
(c) t≈6.93 years
49. A_0=$13,479.87
51. t≈7.5 years
55. 3
57. 11
59. f′(6)=10 means if x increases from 6 to 7, f will increase by approximately 10.

61. $f'(x)=\frac{56}{(2x+6)^2}$

63. $y'=\frac{-3x^2-4xy^2}{4x^2y+3}$

SECTION 5.3

1. Incorrect
3. Incorrect
5. Incorrect
7. Incorrect
9. Correct

11. $f'(x)=\frac{2}{2x-7}$

13. $f'(x)=\frac{16}{x}$

15. $f'(x)=2x\ln(3x)+x$

17. $f'(x)=\frac{5}{x+4}$

19. $f'(w)=\frac{3[\ln(w+4)]^2}{w+4}$

21. $f'(t)=\frac{3}{2(3t+5)}$

23. $f'(u)=\frac{1}{u\ln u}$

25. $f'(t)=2t\ln(t^2)[\ln(t^2)+2]$

27. $f'(t)=\frac{82.5}{t^{\frac{1}{4}}(40t^{\frac{3}{4}}+15)}$

29. $y'=-\frac{y}{2}(2+\frac{1}{x})$

31. $y'=\frac{3y-9x^3y^5}{12x^4y^4-x}$

33. f′(0)=10 means if x increases from 0 to 1, f will increase by approximately 10.
35. f′(6)=1/6 means if x increases from 6 to 7, f will increase by approximately 1/6.
37. t′(0.1)≈-69.33

39. $\frac{dt}{dN}\Big|_{N=120}\approx-13.66$

41. N′(6)≈0.597 means if the number of complimentary copies is increased from 6000 to 7000, the number of textbooks sold will increase by approximately 597.
45. limit does not exist
47. -1/3

49. $\frac{f(7)-f(1)}{7-1}=15$

51. $f'(x)=-\frac{7}{(x-4)^2}$

53. $y'=\frac{-6xy-2y^2}{3x^2+4xy}$

55. A price of $20 maximizes revenue.
57. t≈7 years

SECTION 5.4

1. Incorrect
3. Incorrect
5. Correct
7. Correct
9. Incorrect

11. $f'(x)=3e^{3x+6}$

13. $f'(x)=6e^{6x}$

15. $f'(x)=10xe^{5x^2+4}$

17. $f'(x)=27e^{3x+1}$

19. $f'(x)=3(5x^2+2+2e^{-3x})$

21. $f'(x)=5xe^{x^2+x}(2x^2+x+2)$

23. $f'(x)=\frac{3(2x-3)e^{\sqrt[4]{x^2-3x}}}{4(x^2-3x)^{\frac{3}{4}}}$

25. $f'(x)=-2e^{-x}\left[\ln(x^2+2x)-\dfrac{2(x+1)}{x^2+2x}\right]$

27. $f'(x)=-2.5e^{-0.05x}$

29. $f'(x)=\dfrac{e^{2x}(e^{x}+2)}{(e^{x}+1)^2}$

31. $y'=-\dfrac{2x^2+ye^{3xy}}{xe^{3xy}}$

33. $y'=-\dfrac{y}{x}$

35. $f''(x)=2e^{x^2+4}(1+2x^2)$

37. $f'(1)\approx21.75$ means if x increases from 1 to 2, f will increase by approximately 21.75.

39. $f'(40)\approx-120.69$ means if x increases from 40 to 41, f will decrease by approximately 120.69.

41. (a) $V'(1)\approx-18{,}875$ sales per day
    (b) $V'(6)\approx-1892$ sales per day

43. $N'(4)\approx382$ more people between days 4 and 5.

45. $A'(10)\approx-0.0144$ grams between years 10 and 11.

51. -1

53. 0

55. $f'(x)=6x$

57. $f''(x)=8$

59. A production level of approximately 74,910 will maximize revenue.

SECTION 5.0

1. Algebraic

3. Exponential

5. Algebraic

7. $f(2)\approx398.107$, growth

9. $f(80)\approx148.780$, growth

11. $f(6)\approx20.086$, decay

13. $f(16)\approx50{,}352.956$, growth

15. $f(18)\approx-51.457$, decay

17. (a) $P(0)=14.7$ pounds per square inch

    (b) $P(1)\approx11.92$ pounds per square inch

    (c) $P(6.5)\approx3.75$ pounds per square inch

19. \$77,679.88

21. $\ln x=-4y$

23. $\ln k=3h-1$

25. $\ln 150=-0.5x+16$

27. $e^{-2}=x+6$

29. $e^{3y+7}=2x-7$

31. $x\approx0.9722$

33. $x\approx10.040$

35. $t=0$

37. $x\approx3.0000$

39. $x\approx4.0001$

41. $x\approx5.6590$

43. $t\approx14$ days

45. $t\approx12.92$ years

47. $f'(x)=-3e^{-3x}$

49. $f'(x)=2xe^{4x}(1+2x)$

51. $f'(x)=\dfrac{2}{x}$

53. $f'(x)=3$

55. $f'(x)=x^2(3\ln x+1)$

57. $f'(x)=(2x+2)e^{x^2+2x}$

59. $f'(x)=12e^{12x+6}$

61. $f'(x)=3e^{e^{3x}+3x}$

63. $f'(x)=\dfrac{5}{2(5x+1)}$

65. $f'(x)=-37.51e^{-0.31x}$

67. $f'(x)=4\left(\dfrac{3}{4x+4}+10e^{4x+4}\right)$

69. $y'=-\dfrac{4+y^2e^{x^2y^2}}{xye^{x^2y^2}}$

71. $y'=\dfrac{e^{x}-ye^{xy}}{xe^{xy}-e^{y}}$

73. $f''(x)=9e^{3x}$
75. f′(1)≈-2.2072 means if x increases from 1 to 2, f will decrease by approximately 2.2072.
77. f′(65)≈ -12.705 means if x increases from 65 to 66, f will decrease by approximately 12.705.

SECTION 6.1

1. N′(130)=8.6 means between days 130 and 131, the number of pants sold increases by approximately 8600. N(130)=98 means that on day 130 a total of 9800 pants are sold.
3. A′(35)=-0.68 means if the attitude coefficient increases from 35 to 36, the amount of work will decrease by approximately 0.68 units. A(35)=4.6 means the amount of work is 4.6 when the attitude coefficient is 35.
5. N′(170)=0.008 means if the speed is increased from 170 to 171 units per minute, the number of failed units will increase by about 0.008. N(170)=2.9 means about 2.9 units fail when the speed is 170 units per minute.
7. R′(60)=-0.25 means if the number of advertisements increase from 60 to 61, the candidate's rating will fall approximately 0.25 points. R(60)=38 means the rating is 38% when 60 advertisements are shown.
9. A′(540)=-0.92 means that 540 minutes after 5 AM, i.e., 2 PM, the number of cars is decreasing by approximately 0.92 per 100 yards. A(540)=7 means at 2 PM there are 7 cars per 100 yards of roadway.

11. $F'(x)=\frac{5x^4}{5}+\frac{15x^2}{3}+\frac{14x}{2}-1$

$= x^4+5x^2+7x-1$
$= f(x)$

13. $F'(x)=3\left(\frac{1}{3}e^{3x}\right)-(-e^{-x})+2x$

$= e^{3x}+e^{-x}+2x$
$= f(x)$

15. Not correctly evaluated

$\int e^{6x}dx=\frac{1}{6}e^{6x}+c$

17. Not correctly evaluated (left off "+c")
19. Correctly evaluated
21. Not correctly evaluated

$\int (e^{3x}-e^{-3x})\,dx=\frac{1}{3}e^{3x}+\frac{1}{3}e^{-3x}+c$

23. Correctly evaluated

25. $\int 6x^5dx=x^6+c$

27. $\int \frac{1}{x^4}dx=-\frac{1}{3x^3}+c$

29. $\int \frac{8}{\sqrt[5]{x^2}}dx=\frac{40x^{\frac{3}{5}}}{3}+c$

31. $\int 6e^x dx=6e^x+c$

33. $\int \left(2x^3+\frac{3}{x^2}+1\right)dx=\frac{x^4}{2}-\frac{3}{x}+x+c$

35. $\int (x^{-3}+x^{-2}-x^{-1})\,dx=-\frac{1}{2x^2}-\frac{1}{x}-\ln|x|+c$

37. $\int (0.03+0.12x^{-\frac{1}{2}})\,dx=0.03x+0.24x^{\frac{1}{2}}+c$

39. $f(x)=x^5-x^4+4x^2+5$

41. $f(x)=3e^x+4\ln|x|+2e$

43. (a) $C(x)=-0.052x^2+82.12x-1800$
(b) C(950)=$29,284.00
(c) C′(950)=-16.68 means if production increases from 950 to 951 units, cost will decrease approximately $16.68. C(950)=$29,284 is the total cost of producing 950 units.
45. (a) A′(5)=0.3475 ppm per year
(b) A(5)=4.0125 ppm
47. (a) S(5)≈8772 units
(b) S′(5)≈1169.6 per year
(c) S(5)+S′(5)≈9942 units
(d) S(6)≈9905.8 units
49. (a) A′(4)=1.2 mg per hour
(b) A(4)=12.7 mg
(c) A′(4)=1.2 means between hours 4 and 5 the amount of glucose increases by approximately 1.2 mg.
51. Between days 5 and 6 the sales volume will decrease by approximately $2997.74. The sales volume on day five is approximately $71,516.83.
59. P′(835)=120.50 means if the number of sales increase from 835 to 836, the profit will increase by approximately $120.50.

61. $f'(x)=(7x+8)^2(2x-5)^3(98x-41)$

63. Power function: $f(x)=x^b$

Exponential function: $f(x)=b^x$

65. $f'(x)=\frac{-8e^{-2x}}{(2-e^{-2x})^2}$

SECTION 6.2

1. $\int(3x-4)^4dx=\frac{(3x-4)^5}{15}+c$

3. $\int\frac{6}{7x+5}dx=\frac{6}{7}\ln|7x+5|+c$

5. $\int 2x\sqrt{x^2+4}dx=\frac{2}{3}(x^2+4)^{\frac{3}{2}}+c$

7. $\int\frac{2}{3x^{\frac{1}{3}}(x^{\frac{2}{3}}+4)^3}dx=-\frac{1}{2(x^{\frac{2}{3}}+4)^2}+c$

9. $\int\frac{e^{4x}}{e^{4x}+2}dx=\frac{1}{4}\ln(e^{4x}+2)+c$

11. $\int x^2e^{5x^3-2}dx=\frac{1}{15}e^{5x^3-2}+c$

13. $\int\frac{12x^2+5}{4x^3+5x-6}dx=\ln|4x^3+5x-6|+c$

15. $\int\frac{(7+\ln x)^{\frac{3}{2}}}{x}dx=\frac{2}{5}(7+\ln x)^{\frac{5}{2}}+c$

17. $\int(2x\ln x)^4(\ln x+1)dx=\frac{1}{10}(2x\ln x)^5+c$

19. $\int\frac{5x}{5x-4}dx=\frac{1}{5}(5x-4)+\frac{4}{5}\ln|5x-4|+c$

21. $\int 4x(2x+5)^5dx=\frac{1}{42}(2x+5)^6(12x-5)+c$

23. $\int\frac{x\ln(x^2+4)}{\ln(x^2+4)}dx=\frac{x^2}{2}+c$

25. $\int\frac{x-10}{(x+2)^3}dx=-\frac{1}{x+2}+\frac{6}{(x+2)^2}+c$

27. $\int e^{-0.02x}dx=-50e^{-0.02x}+c$

29. $\int 45e^{0.15x}dx=300e^{0.15x}+c$

31. $\int(e^{3x}+4e^{-3x})dx=\frac{1}{3}e^{3x}-\frac{4}{3}e^{-3x}+c$

33. $\int(e^{x-2}+\frac{\ln x}{x})dx=e^{x-2}+\frac{(\ln x)^2}{2}+c$

35. $\int x^5(\sqrt[3]{x^3+1})dx=\frac{1}{28}(x^3+1)^{\frac{4}{3}}(4x^3-3)+c$

37. Let u=ax
du=adx
1/a du=dx

$$\int e^{ax}dx=\int e^u\left(\frac{1}{a}\right)du$$
$$=\frac{1}{a}\int e^u du$$
$$=\frac{1}{a}e^u+c$$
$$=\frac{e^{ax}}{a}+c$$

39. (a) $N(t)=300\ln(10t+15)-300\ln 15$
(b) $N(5)\approx 440$ thousand antibodies

41. (a)
$R(t)=395{,}000e^{-0.004t}+380{,}000-395{,}000e^{-0.04}$
(b) R(24)=$359,331
(c) $488

43. (a) $N(t)=21{,}484.375e^{-0.32t}+33{,}516$
(b) $N(10)\approx 34{,}392$ sales
(c) 33,516 sales

45. (a) $P(t)=7\ln(1+e^{18t})+28{,}000-7\ln 2$
(b) $P(10)\approx 29{,}255$
(c) $P'(5)=126$ means between months 5 and 6 the population will increase by approximately 126.

49. 1/3
51. MP(500)=$11

53. $y'=\frac{-14xy}{14x^2+3y}$

57. $f'(x)=6+0.2e^{-0.05x}$

SECTION 6.3

1. $\int xe^xdx=e^x(x-1)+c$

3. $\int xe^{-8x}dx=-\frac{1}{8}e^{-8x}(x+\frac{1}{8})+c$

5. $\int x^3e^{3x}dx=\frac{1}{27}e^{3x}(9x^3-9x^2+6x-2)+c$

7. $\int \ln(x^3)\,dx=3(x\ln x-x)+c$

9. $\int (\ln x)^3dx=x[(\ln x)^3-3(\ln x)^2+6\ln x-6]+c$

11. $\int \frac{1}{x}\ln x\,dx=\frac{(\ln x)^2}{2}+c$

13. $\int x(\ln x)^3dx=\frac{x^2}{8}[4(\ln x)^3-6(\ln x)^2+6\ln x+3]+c$

15. $\int \ln(7x-4)\,dx=\frac{1}{7}(7x-4)[\ln(7x-4)-1]+c$

17. $\int x(1-x)^{\frac{3}{2}}dx=-\frac{2}{35}(1-x)^{\frac{5}{2}}(5x+2)+c$

19. $\int e^{x+3}(2x+1)\,dx=e^{x+3}(2x-1)+c$

21. $\int (2x+1)^2\ln(2x+1)\,dx=\frac{(2x+1)^3}{18}[3\ln(2x+1)-1]+c$

23. $C(x)=\frac{1}{4}[2x^2\ln(x+1)-x^2+2x-2\ln(x+1)]$

25. $R(t)=12,500t-241t^2e^{-\frac{t}{2}}-964te^{-\frac{t}{2}}$

$$-1928e^{-\frac{t}{2}}+14,428$$

27. $N(t)=-1.25te^{-0.8t}-1.5625e^{-0.8t}+c$

31. 0

33. $\frac{f(5)-f(2)}{5-2}=2$

35. $f'(x)=-18(4-3x)^5(2x^3-3)^2(5x^3-4x^2-3)$

37. Revenue is maximized at $95,099 when 39,350 units are produced.

39. $\int \frac{4x}{x^2+3}dx=2\ln|x^2+3|+c$

SECTION 6.4

1. Step 2 -- F(x) should be evaluated at x=6, then at x=2.
3. Step 1 -- Divide by 3, not multiply.
5. $\int_a^b A(t)\,dt$ represents the total accumulation from month *a* to month *b*.
7. $\int_a^b P(t)\,dt$ represents the total population accumulation between years *a* and *b*.
9. $\int_a^b S(t)\,dt$ represents the total number of people who hear the news from days *a* to *b*.
11. $\int_a^b C(x)\,dx$ represents the total cost for producing between *a* and *b* items.
13. $\int_a^b E(t)\,dt$ represents the total evaporation between temperatures *a* and *b*.
15. $\int_a^b U(f)\,df$ represents the total price increase for a change in fuel price from *a* to *b*.
17. $\int_a^b L(t)\,dt$ represents the length of time spent on air travel between years *a* and *b*.
19. $\int_1^3 4x^3dx=80$
21. $\int_1^e \frac{3}{x}dx=3$
23. $\int_0^{15} e^{-0.05}dx=15e^{-0.05}$
25. $\int_4^5 \frac{2}{x-3}dx=2\ln 2$
27. $\int_0^1 (6x+1)e^{3x^2+x}dx=e^4-1$
29. $\int_2^3 \frac{x}{\sqrt{x^2-1}}dx=2\sqrt{2}-\sqrt{3}$
31. $\int_2^5 e^{2x}(e^{2x}-4)\,dx=\frac{1}{4}(e^{20}-8e^{10}-e^8+8e^4)$
33. $\int_{\ln 3}^{\ln 5} 3e^xdx=6$
35. $\int_2^3 \frac{1}{x\ln(3x)}dx=\ln(\ln 9)-\ln(\ln 6)$
37. Approximately $90,952 accumulates
39. Approximately 14.65 sales
41. Machine depreciates approximately $12,650
43. Approximately 3.85 million cubic feet of gas
45. $757.53

49. All the fish die

51. $f'(3)=16$

53. $f'(x)=\dfrac{40(3x-1)^4}{(2x+2)^6}$

55. If the number of copies sold increases from 5000 to 6000, the number of desk copies increases by approximately 15.

57. $\int \dfrac{5x^4+2}{2x^5+4x-4}dx=\dfrac{1}{2}\ln|2x^5+4x-4|+c$

SECTION 6.5

1. $A=\int_a^b f(x)\,dx+\int_b^c f(x)\,dx$

3. $A=\int_a^b f(x)\,dx-\int_a^c f(x)\,dx$

5. $A=63$ sq. units

7. $A=12$ sq. units

9. $A\approx 1.386$ sq. units

11. $A\approx 15{,}319.26$ sq. units

13. $A=9\frac{1}{3}$ sq. units

15. $A\approx 0.332$ sq. units

17. $A\approx 1644.61$ sq. units

19. $A=1125$ sq. units

21. $A=9$ sq. units

23. $A=189\frac{1}{3}$ sq. units

25. \$74,574.59
27. 0.35 gms
29. Approximately 510 deaths per 1000
31. \$176.93
35. Limit does not exist
37. If x increases from 1 to 2, f will increase by approximately 1029.

39. $f'(x)=\dfrac{3}{3x+5}$

41. $\int_1^e x^3\ln x\,dx\approx 10.3$

43. $\int_1^3 \dfrac{x}{x+3}dx\approx 0.784$

SECTION 6.6

1. $\int_0^\infty f(t)\,dt$ represents the total income from the property if the flow continues indefinitely.

3. $\int_{1.5}^\infty E(t)\,dt$ represents the total energy conserved from year 1.5 and on if the rate continues indefinitely.

5. $\int_0^\infty Q(x)\,dx$ represents the total improvement in quality if the rate continues indefinitely.

7. $\int_1^\infty \dfrac{1}{x^4}dx=\dfrac{1}{3}$

9. $\int_e^\infty \dfrac{3}{x}dx$ diverges

11. $\int_{-1}^\infty \dfrac{3}{3x+4}dx$ diverges

13. $\int_0^\infty e^{-4x}dx=\dfrac{1}{4}$

15. $\int_{-\infty}^3 e^{2x}dx=\dfrac{1}{2}e^6$

17. $\int_0^\infty \dfrac{1}{(x+1)^3}dx=\dfrac{1}{2}$

19. $\int_{-\infty}^\infty xe^{-x^2}dx=0$

21. $\int_0^\infty xe^{-x^2}dx=\dfrac{1}{2}$

23. $\int_{50}^\infty \dfrac{6x^2}{(x^3+6)^{\frac{3}{2}}}dx\approx 0.0113$

25. $\int_e^\infty \dfrac{1}{x(\ln x)^2}dx=1$

27. $\int_0^\infty ae^{-ax}dx$ diverges

29. \$2,500,000
31. 60,000 tons of pollutant
33. \$3681
37. 8/3

39. If x increases from 0 to 1, f will increase by approximately 208.
41. The box should be 20 inches by 20 inches by 10 inches to minimize surface area.

43. $\int \frac{5x}{4x^2+1}dx=\frac{5}{8}\ln(4x^2+1)+c$

45. $\int_1^3 \frac{\ln x}{2x}dx \approx 0.302$

SECTION 6.0

1. $P'(14.5)=0.18$ means if the weight is increased from 14.5 to 15.5 ounces, the probability of acceptance will increase by approximately 0.18. $P(14.5)=0.63$ means there is a 63% chance of acceptance if the weight is 14.5 ounces.
3. $E'(6.2)=70$ means if wind speed increases from 6.2 to 7.2 mph, evaporation will increase by approximately 70 gallons per hour. $E(6.2)=1275$ means the evaporation rate is 1275 gallons per hour when the wind speed is 6.2 mph.

5. $F(x)=2\ln x+c$

$$F'(x)=\frac{2}{x}$$
$$=f(x)$$

7. $\int(3x^2-5x+2)dx=x(x^2-\frac{5}{2}x+2)+c$

9. $\int 40e^{-0.02x}dx=-2000e^{-0.02x}+c$

11. $V'(5)\approx 3759$ means between the fifth and sixth day after the campaign ends sales volume will decrease approximately \$3759. $V(5)\approx 67{,}662$ is the sales volume 5 days after the end of the campaign.

13. $\int \frac{x}{x^2+1}dx=\frac{1}{2}\ln(x^2+1)+c$

15. $\int(5x^4+2)e^{x^5+2x}dx=e^{x^5+2x}+c$

17. $\int \frac{5\ln x}{x}dx=\frac{5}{2}(\ln x)^2+c$

19. $\int \frac{3x}{x-4}dx=3(x-4+\ln|x-4|)+c$

21. $\int x^4e^{x^5}dx=\frac{1}{5}e^{x^5}+c$

23. $\int_a^b H(t)\,dt$ represents the total amount of news heard from days a to b.

25. $\int_1^3(3x-1)^3dx=340$

27. $\int_1^3 \frac{2x}{x^2-3}dx=\ln 6-\ln 2$

29. $\int_0^1 x\sqrt{1-x^2}dx=\frac{1}{3}$

31. $\int_0^4 xe^{x^2}dx=\frac{1}{2}(e^{16}-1)$

33. $\int_{\ln 2}^{\ln 5}2e^{x}dx=6$

35. $\int_e^{e^2}\frac{6}{x\ln x^3}dx=2\ln 2$

37. $\int_1^3(6x^2-5)\,dx=42$

39. Approximately 38.06 thousand dollars

41. $\int 3x^3y^2dx=\frac{3}{4}y^2x^4+c$

43. $\int 5xe^{-y}dx=\frac{5}{2}e^{-y}x^2+c$

45. $\int \frac{28xy}{y^2+5}dy=14x\ln(y^2+5)+c$

SECTION 7.1

1. $A=\int_c^d[g(x)-f(x)]\,dx$

3. $A=\int_d^c[k(x)-m(x)]\,dx$

5. $A=\int_a^b[f(x)-h(x)]\,dx+\int_b^c[h(x)-f(x)]\,dx$

7. $A=3\frac{1}{3}$ sq. units

9. $A\approx 6.33$ sq. units

11. $A\approx 4.10$ sq. units

13. $A\approx 11.61$ sq. units

15. $A=0.05$ sq. units

17. $A=20\frac{5}{6}$ sq. units

19. $A=4.5$ sq. units

21. $P\approx 339.91$ million dollars

23. approximately 159,388 insects

25. approximately 3.05 million people

29. 5

31. $y'(3,\frac{7}{2})=\frac{1}{2}$

33. $y'=\frac{xy-8y}{16x+2xy}$

35. $\int xe^{4x}dx=\frac{1}{16}e^{4x}(4x-1)+C$

37. \$3133.86

SECTION 7.2

1. $\int_0^{14}[D(x)-85]\,dx=8500$

   means when the cost is \$85 and demand is D(x), the consumer's surplus is \$8500.

3. $\int_0^{35}[12-S(x)]\,dx=1645.15$

   means when the price is \$12 and the supply is S(x), the producer's surplus is \$1645.15.

7. CS=\$666.67
9. PS=\$11.75
11. CS=\$1000, PS=\$2000
13. CS=\$20.84
15. PS=\$20,040.90
17. PS=\$17,839.56
19. CS=\$392.16, PS=\$587.63
21. CS=\$132,322.02, PS=\$33,790.70
27. f(x) is discontinuous at x=5 since the limit does not exist.

29. $g'(x)=\frac{2}{x}-3$

31. $f'(x)=\frac{25}{(3x-2)(2x+7)}$

33. $\int_0^{\ln 2}xe^{2x}dx=2\ln 2-\frac{3}{4}$

35. $\int_7^{12}(2x-5)\,dx=70$ sq. units

SECTION 7.3

1. Do not buy the machine since the present value is less that the cost of the machine.
3. If the machine costs less than \$151,189.37, buy it. Otherwise, do not.
5. At the end of 10 months the total value of the investment stream is \$3089.23. It would have taken a single \$2914.18 initial investment to achieve the same future value.
7. At the end of 12 years the total value of the investment stream is \$29,192.06. It would have taken a single \$11,868.61 investment to achieve the same future value.
9. At the end of 36 months the total value of the investment stream is \$8396.61. It would have taken a single \$6220.36 initial investment to achieve the same future value.
11. A=\$96,701.79, P=\$37,026.43
13. The cost of the machine is less than the present value (\$372,924.22) of the investment stream. Buy the machine.
15. If the machine costs less than the present value (\$1,715,474.74), buy it. Otherwise, do not.
17. \$228,571.43 would be the required initial investment to match the value of the rental income.
23. -1/4

25. $\frac{f(9)-f(4)}{9-4}=-31$

27. $S'(t)=6t^2(6t-30)^3(7t-15)$

29. Power function: $y=x^b$

    Exponential function: $y=b^x$

31. 70,000 tons

SECTION 7.4

1. $\frac{dy}{dx}=4x$

3. $\frac{dy}{dx}=2x(Ce^{x^2})=2xy$

7. $\frac{dy}{dx}=\frac{y^2}{x}$

5. $\frac{dy}{dx}=Ce^x=y$

9. Separable
11. Not separable
13. Separable
15. Not separable

17. $y=\frac{x^5}{5}+c$

19. y=c

21. $y^2=2x^2+c$

23. $y=\frac{1}{2}e^{x^2}+c$

25. $y=1+x-\ln|1+x|+c$

27. $\ln|\ln y|=\ln|x|+c$

29. $y^3=x^3+64$

31. $y^2=2\ln|x|+16$

33. $y=xe^x-e^x-\frac{x^2}{2}$

35. $\ln(1+y^2)=\ln|x|+\ln 2$

37. $P=\frac{1}{2}kt^2+P_0$

39. 107 words per minute
41. 165 crimes in 1995
43. $P\approx 0.68P_0$
45. C=4.96 ppm
47. Approximately 22.88% vote

49. $a=[\frac{1}{4}(0.2t+8.46)]^4$

51. Approximately 1,292,898 employees
59. $R'(1500)=225$ means if the number of units sold is increased from 1500 to 1501, revenue will increase by approximately $225.

61. $f'(x)=4x(x^2+1)^3(x^2-1)(3x^2-1)$

63. $f'(x)=2e^{2x}(1+2x)$

65. $\int_0^1 3x^2e^{x^3}dx=e-1$

SECTION 7.5

1. Unlimited growth
3. Limited decay
5. Unlimited decay
7. Unlimited decay
9. $P_o=Ce^{k(0)}$, $C=P_o$
11. $P\approx 2800$

13. (a) $T=68-28e^{-0.0131t}$

(b) $T\approx 49°F$

15. (a) $P\approx 650-605e^{-0.105t}$

(b) $P\approx 252$ units

17. (a) $P\approx\frac{2500}{1+2499e^{-1.852t}}$

(b) $P\approx 2410$ people

19. $t\approx$13,412 years
21. $t\approx$20.1 hours
25. -4

27. $f'(x)=8x-1$

29. $f'(x)=-\frac{38}{(6x-7)^2}$

33. $\int_e^{e^2}\frac{5}{x(\ln x)^2}dx=2.5$

SECTION 7.6

1. Approximately 9.52% of all air travellers will be exposed to between 5.00 and 5.50 mrem of radiation.
3. The country will receive between 2.4 and 4.4 inches of rain in February 55.26% of the time.
5. There is a 15.12% chance that any telephone call will last longer than 16 minutes.
7. There is a 0.59% chance that reaction time is more than 0.2 seconds.

9. $\int_2^5 x^2dx=\frac{117}{100}\neq 1$

11. No, since $p(1)=-\frac{1}{16}$

13. No, since $\int_{-2}^2 4x^2dx\neq 1$

15. No, since $\int_1^{(\sqrt[5]{2})}\frac{25}{9}x^4dx\neq 1$

17. $k=\frac{4}{45}$

19. $k=\frac{1}{12}$

21. $k=\frac{1}{2}$

23. $\frac{1}{2}\int_0^1 Xe^{-\frac{X^2}{2}}\,dX \approx 0.197$

25. $\int_1^{10} \frac{1}{9X}\,dX \approx 0.256$

27. $\int_5^{10} 0.04e^{-0.04X}\,dX \approx 0.148$

29. $\int_2^4 \frac{3}{40}(X^2-2X)\,dX = 0.5$

31. (a) $\int_0^3 \frac{6}{1127}(X^2+3X)\,dX \approx 0.120$

(b) $\int_0^1 \frac{6}{1127}(X^2+3X)\,dX \approx 0.0098$

33. (a) $\int_{40}^{50} 0.03e^{-0.03X}\,dX \approx 0.078$

(b) $\int_0^{100} 0.03e^{-0.03X}\,dX \approx 0.950$

35. (a) $\int_5^6 0.14e^{-0.14X}\,dX \approx 0.065$

(b) $\int_0^1 0.14e^{-0.14X}\,dX \approx 0.131$

37. (a) $\int_4^{4.3} \frac{1}{1.4}\,dX \approx 0.214$

(b) $\int_4^{4.5} \frac{1}{1.4}\,dX \approx 0.357$

39. (a) $\int_0^2 \frac{1}{15}\,dX \approx 0.133$

(b) $\int_{10}^{15} \frac{1}{15}\,dX = 0.6$

41. (a) $\int_5^8 0.07e^{-0.07X}\,dX \approx 0.133$

(b) $\int_0^2 0.07e^{-0.07X}\,dX \approx 0.131$

(c) $P(x \geq 10) = 1 - P(0 \leq x \leq 10) \approx 0.497$

43. $P(0.42 \leq Z \leq 0.99) = 0.1761$

45. (a) $P(-1 \leq Z \leq -\frac{1}{2}) = 0.1498$

(b) $P(Z > 0.83) = 0.2033$

47. (a) $P(-0.78 \leq Z \leq -0.67) = 0.0337$

(b) $P(Z < -1.11) = 0.1335$

(c) $P(Z > 2.78) = 0.0027$

51. $\frac{4}{7}$

55. $N'(800)=0.06$ means if the number of homes increases from 800 to 801, the pollution will increase by approximately 0.06 ppm.

57. $f''(x) = \frac{44}{(2x-3)^3}$

59. $\frac{dC}{dt} \approx \$292$

SECTION 7.0

1. $A = 2\frac{2}{3}$ sq. units

3. $A = 93.6$ sq. units

5. $A \approx 3.79$ sq. units

7. $A = 12$ sq. units

9. $A \approx 5.167$ sq. units

11. $A \approx 12.319$ sq. units

13. $CS \approx \$864.74$

15. $PS \approx \$4064.92$

17. $\frac{dy}{dx} = \frac{11}{(x+6)^2}$

19. $\frac{dy}{dx} = C$, but $C = \frac{y+4}{x+4}$, so

$$\frac{dy}{dx} = \frac{y+4}{x+4}$$

21. $y = \frac{1}{3}x^6 + c$

23. $y = \ln(x+c)$

25. $\ln(\ln y) = \frac{1}{2}x^2(\ln x) - \frac{1}{4}x^2 + c$

27. $-\frac{1}{y} = \frac{1}{2}\ln(x^2+4) + c$

29. $y=4x^{\frac{3}{2}}$

31. $y=-1+e^{-e^{-x}+1}$

33. $y=-\ln(-\frac{x^3}{3}-5x^2-25x+1)$

35. P=\$144,080
37. V=\$505,852

39. (a) $V=100-99.6e^{-0.0005025t}$

(b) $V(16)\approx 1.19\%$

41. There is a 1.1% chance that a male age 15-25 will buy the clothes.
43. About 8.5% of the reflectors will last 3 years or less.
45. No, p(X) is not ≥ 0 for all X.

47. $k=\frac{2}{117}$

49. $\int_0^3 2e^{-2X}dX\approx 0.998$

51. $\int_3^{10}\frac{1}{70}(X-3)\,dX=0.35$

53. $\int_{40}^{50}\frac{12}{135}dX\approx 0.889$

55. $P(-1\le Z\le 2)=0.8185$

57. $P(-2.86\le Z\le 2.86)=0.9958$

59. (a) $\int_{10}^{15}\frac{4}{789}(15-X^{\frac{1}{3}})\,dX\approx 0.321$

(b) $\int_0^5\frac{4}{789}(15-X^{\frac{1}{3}})\,dX\approx 0.348$

(c) $1-\int_0^{12}\frac{4}{789}(15-X^{\frac{1}{3}})\,dX\approx 0.192$

61. (a) $\int_0^{15}\frac{1}{35}dX\approx 0.429$

(b) $\int_5^{10}\frac{1}{35}dX\approx 0.143$

(c) $1-\int_0^{25}\frac{1}{35}dX\approx 0.286$

SECTION 8.1

5.

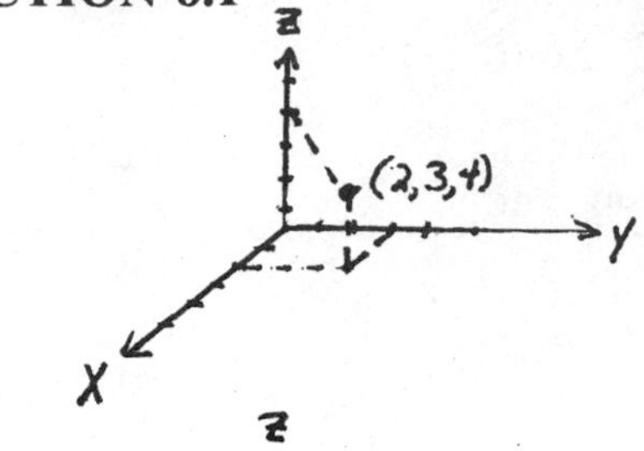

7.

9.

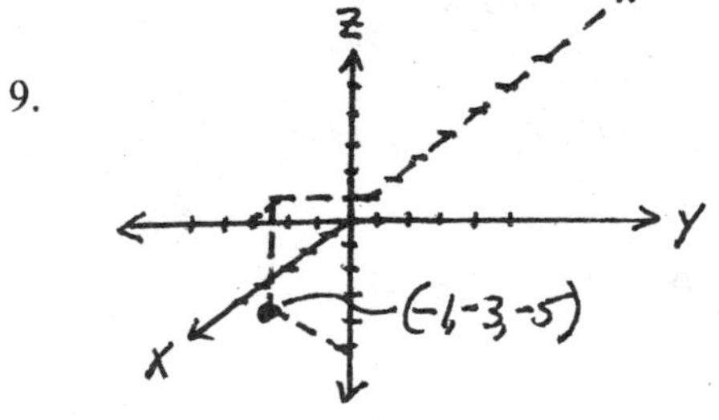

11. f(1,4)=-14
13. f(200,60)≈358.51
15. P(20,27)≈150.14
17. f(5,8)≈2.51
19. P(12,2,20)≈4837.79
21. f(0,12)≈15.62

23. $\frac{f(x+h,y)-f(x,y)}{h}=2x$

25. $\lim_{h\to 0}\frac{f(x,y+h)-f(x,y)}{h}=2y+4x$

27. (a) 42,580 items produced
(b) 141.5 unit increase
(c) 2608.43 unit decrease
(d) $f(3x,3y)=600(3x)^{.4}(3y)^{.6}$

$$=600(3)^{.4}x^{.4}(3)^{.6}y^{.6}$$
$$=600(3)^{(.4+.6)}x^{.4}y^{.6}$$
$$=1800x^{.4}y^{.6}$$
$$=3f(x,y)$$

29. (a) 512.37
(b) 78.22 decrease in depression
35. 7/6
37. Concave down
41. $f'(x)=8e^{8x+4}$

43. $\int \frac{18x}{9x+5}dx = \frac{2}{9}[9x+5-5\ln|9x+5|]+c$

SECTION 8.2

1. $\frac{\partial}{\partial x}P(3,8)=12$ means when x=3 and y=8, if x increases from 3 to 4 the function P will increase by approximately 12. $\frac{\partial}{\partial y}P(5,6)=-9.2$ means when x=5 and y=6, if y increases from 6 to 7 the function P will decrease by approximately 9.2.

3. $\frac{\partial}{\partial x}T(2,3,11)=0.06$ means when x=2, y=3 and z=11, and if x increases from 2 to 3 the function T will increase by approximately 0.06.

   $\frac{\partial}{\partial y}T(3,4,10)=0.008$ means when x=3, y=4 and z=10, if y increases from 4 to 5 the function T will increase by approximately 0.008

   $\frac{\partial}{\partial z}T(5,5,8)=0.4$ means when x=5, y=5 and z=8, if z increases from 8 to 9 the function T will increases by approximately 0.4.

5. $\frac{\partial}{\partial x}P(15,6)>0$ and $\frac{\partial^2}{\partial x^2}P(15,6)<0$ means when \$15,000 is spent on advertising and 6 salespeople are working, the profit is increasing at a decreasing rate.

7. $\frac{\partial}{\partial r}R(10,0.2)<0$ and $\frac{\partial^2}{\partial r^2}R(10,0.2)>0$ means when a blood vessel has length 10 cm and radius 0.2 cm, the resistance is decreasing at a decreasing rate.

9. $C_y(x,y,z)>0$ and $C_{yy}(x,y,z)>0$ means cost is increasing at an increasing rate.

11. $P_x(x,y)>0$ and $P_{xx}(x,y)>0$ means the proportion of the population applying for welfare is increasing at an increasing rate.

13. $\frac{\partial}{\partial w}D(w,d)>0$ and $\frac{\partial^2}{\partial w^2}D(w,d)>0$ means the dose of radiation is increasing at an increasing rate.

15. $\frac{\partial}{\partial y}f(30,100)\approx 0.56$ and

    $\frac{\partial^2}{\partial y^2}f(30,100)\approx -0.001$ means that the number of units produced is increasing at a decreasing rate.

17. $\frac{\partial D}{\partial p_1}>0$ means if the price of the calculator increases, the demand increases. $\frac{\partial D}{\partial p_2}<0$ means if the price of the competitor increases, demand decreases.

19. $\frac{\partial P}{\partial n}>0$ and $\frac{\partial P}{\partial n}<0$

21. (a) $\frac{\partial f}{\partial x}=10x+6y$

    (b) $\frac{\partial f}{\partial y}=6x+24y^2$

    (c) $\frac{\partial}{\partial x}f(2,1)=26$

    (d) $\frac{\partial}{\partial y}f(4,2)=120$

23. (a) $f_x=3x^2+9x^2y^2$

    (b) $f_y=-8y+6x^3y$

    (c) $f_x(0,-1)=0$

    (d) $f_y(-2,1)=-56$

25. (a) $f_x=e^{x+y}$

    (b) $f_y=e^{x+y}$

    (c) $f_x(1,1)=e^2$

    (d) $f_y(2,1)=e^3$

27. (a) $f_x=\frac{4xy^2}{1+2x^2y^2}$

    (b) $f_y=\frac{4x^2y}{1+2x^2y^2}$

    (c) $f_x(0,0)=0$

(d) $f_y(0,0)=0$

29. (a) $f_x=\frac{5x^2-6xy^2-15y}{(5x-3y^2)^2}$

(b) $f_y=\frac{15x+9y^2+6x^2y}{(5x-3y^2)^2}$

(c) $f_x(0,2)=-\frac{5}{24}$

(d) $f_y(0,1)=1$

31. (a) $f_x=2xe^{3y}$

(b) $f_y=3x^2e^{3y}$

(c) $f_x(0,0)=0$

(d) $f_y(0,0)=0$

33. (a) $f_{xx}=4$

(b) $f_{yy}=10$

(c) $f_{xy}=0$

(d) $f_{yx}=0$

35. (a) $f_{xx}=20$

(b) $f_{yy}=10$

(c) $f_{xy}=4$

(d) $f_{yx}=4$

37. (a) $f_{xx}=18$

(b) $f_{yy}=2$

(c) $f_{xy}=-6$

(d) $f_{yx}=-6$

39. (a) $f_{xx}=32ye^{2x}$

(b) $f_{yy}=0$

(c) $f_{xy}=16e^{2x}$

(d) $f_{yx}=16e^{2x}$

41. (a) $f_{xx}=0$

(b) $f_{yy}=36y$

(c) $f_z=-2z$

(d) $f_{zy}=0$

43. $(-\frac{3}{2},-\frac{2}{7})$

45. $(1,\frac{\sqrt{6}}{3}),(1,-\frac{\sqrt{6}}{3}),(-1,\frac{\sqrt{6}}{3}),(-1,-\frac{\sqrt{6}}{3})$

47. (a) 4110 units sold
(b) approximately 320 more units will be sold
(c) approximately 60 fewer units will be sold

49. (a) $IQ(30,22)\approx 136.36$
(b) IQ will decrease by approximately 2.44

55. $\frac{f(9)-f(5)}{9-5}=68$

57. $f''(x)=36x-16$

59. Box should be 6 inches by 6 inches by 3 inches.

61. $A=26\frac{1}{6}$ sq. units

63. $f(1,0,1)=1$

SECTION 8.3

1. (-4,4) is a relative minimum
3. (2,5) is a saddle point
5. (3,5) is a relative minimum
7. (-2,0) is a relative maximum
9. (2,8) is a saddle point
11. (3,7) is a relative maximum
13. $(\frac{48}{19},\frac{61}{19})$ is a saddle point
15. second derivative test is inconclusive at both critical points, (0,1) and (0,-1)
17. (2,6) is a saddle point
19. (0,0) is a saddle point
21. \$33,750 on radio and \$362.50 on newspapers maximizes revenue
23. box should be 4 inches by 4 inches by 4 inches
25. box should be 14 inches by 14 inches by 28 inches
29. limit does not exist
31. If x increases from 2 to 3, f will increase by

approximately 11/64.
33. 2 hours
35. $y=2\ln|x^2-1|+5-2\ln 3$
37. f(2,-3)=30

SECTION 8.4

1. $\lambda=1.3$ means that if c is increased from 80 to 81, the maximum value of f(x,y) will increase by approximately 1.3 units from 85 to 86.3.
3. $\lambda=1.32$ means that if c is increased from 21.66 to 22.66, the maximum value of S(h,r) will increase by approximately 1.32, from 42.98 to 44.3 sq. in.
5. $\lambda=4.09$ means that if c is increased from 35 to 36, the minimum cost will increase approximately $4.09, from $355 to $359.09.
7. relative minimum of 13 at (1,1.5)
9. relative minimum of 0 at (0,6)
   relative maximum of -216 at (6,-6)
11. relative maximum of 24.5 at (-1/2,-1/2)
13. relative minimum of -8 at (-2,-4)
15. relative maximum of approximately 55.68 at (0.952,5.716)
17. Solve the system (see problem 16)

$$aCx^{a-1}y^{1-a}+\lambda p=0 \quad (1)$$

$$(1-a)Cx^{a}y^{-a}+\lambda q=0 \quad (2)$$

$$px+qy-k=0 \quad (3)$$

From (1), $\lambda=\dfrac{-aCx^{a-1}y^{1-a}}{p}$ (4)

From (2), $\lambda=\dfrac{(a-1)Cx^{a}y^{-a}}{q}$ (5)

Then, from (4) and (5),

$$\frac{-aCx^{a-1}y^{1-a}}{p}=\frac{(a-1)Cx^{a}y^{-a}}{q}$$

Multiply both sides by pq and divide by $Ca^{a}y^{1-a}$ to obtain

$$-aqx^{-1}=p(a-1)y^{-1}$$

$$x^{-1}=-\frac{p(a-1)}{aq}y^{-1}$$

$$x=-\frac{aqy}{p(a-1)}$$

Substitute this into (3)

$$p\left(-\frac{aqy}{p(a-1)}\right)+qy-k=0$$

$$-\frac{aq}{a-1}y+qy=k$$

$$\left(-\frac{aq}{a-1}+\frac{q(a-1)}{a-1}\right)y=k$$

$$y=\frac{k(a-1)}{-q}=\frac{k(1-a)}{q}$$

So $x=-\dfrac{aq\left(\dfrac{k(1-a)}{q}\right)}{p(a-1)}=\dfrac{ak}{p}$

The critical value, and therefore the relative maximum, is $\left(\dfrac{ak}{p},\dfrac{k(1-a)}{q}\right)$

19. Production is maximized at approximately 12,928 when 115.2 hours of labor and $837,216 in capital are allocated. Also $|\lambda|\approx|-0.0058|\approx 0.0058$ which means for each additional dollar allocated over $837,216 production will increase by approximately 0.0058 units.
21. $f(x,y)=20x^{\frac{3}{5}}y^{\frac{2}{5}}$
    subject to 50x+300y=50,000
    From problem 17, the critical value is (600,66 2/3), which is a maximum.

Now, $\dfrac{\text{unit cost of labor}}{\text{unit cost of capital}}=\dfrac{50}{300}=\dfrac{1}{6}$

Also, $f_x=12x^{-\frac{2}{5}}y^{\frac{2}{5}}$
$f_y=8x^{\frac{3}{5}}y^{-\frac{3}{5}}$

Then, $\dfrac{f_x}{f_y}=\dfrac{12x^{-\frac{2}{5}}y^{\frac{2}{5}}}{8x^{\frac{3}{5}}y^{-\frac{3}{5}}}=\dfrac{3y}{2x}$

$$\frac{f_x}{f_y}\left(600,66\frac{2}{3}\right)=\frac{3\left(66\frac{2}{3}\right)}{h(600)}=\frac{200}{1200}=\frac{1}{6},$$

which illustrates the law.
23. Cost is minimized at 50 inspections per day at point A and 10 inspections per day at point B.
25. Area is maximized by a square, 9.75 inches by 9.75 inches
27. Maximum occurs at approximately 939,047 at the

point (255,330). $\lambda=-37$ means if the constraint increases from 42,000 to 42,001, then number of people reached will decrease by approximately 37.

29. Use approximately 146 pounds of A and 1854 pounds of B.
33. 0
35. $f'(0)=0$
37. $f^{(4)}(x)=0$
39. $f'(x)=2[\ln(2x)+1]$
41. $\dfrac{e^{12x}(10,368x^5-4320x^4+1440x^3-360x^2+60x-5)}{31,104}$
43. (0,0) is a saddle point
    (3,3) is a relative minimum

SECTION 8.5

1. For df(a,b) to be less than zero, $f_x(a,b)<2f_y(a,b)$ must hold.
3. df(a,b)<0 for all a and b.
5. df(x,y)=16x-6y+14
7. df(3,1)≈297.34
9. df(9,4)=-5/12
11. df(3,1,11)≈$6.25\text{x}10^{-49}$
13. df(150,20)≈0.0022
15. df(1,0,0)=0
17. (a) df(30,90)≈-331.72
    (b) actual change is -332.64
19. (a) dC(14,12)=$148
    (b) actual change is $188
21. (a) df(200,40)≈-0.053 (-53 units per week)
    (b) actual change is -55
25. limit does not exist
27. f'(x)=10x-2
29. f'(x)=$2xe^{2x+7}(1+x)$
31. 643.696 million dollars
33. $\left(\sqrt{\frac{2}{3}},\sqrt{\frac{1}{3}}\right)$ is a relative maximum

    $\left(-\sqrt{\frac{2}{3}},-\sqrt{\frac{1}{3}}\right)$ is a relative minimum

SECTION 8.6

1. $\int 24x^2y^3\,dx=8x^3y^3+g(y)$
3. $\int (x^2y^4-xy^5)\,dx=\frac{x^2y^5}{5}-\frac{xy^6}{6}+g(x)$
5. $\int 9x^2e^{3y}dx=3x^3e^{3y}+g(y)$
7. $\iint 2x\,dx\,dy=x^2y+G(y)+h(x)$
9. $\iint xy\,dy\,dx=\frac{x^2y^2}{4}+G(x)+h(y)$
11. $\iint e^x\,dx\,dy=ye^x+G(y)+h(x)$
13. $\int_0^1\int_1^2 (1+x^2)\,dy\,dx=\frac{4}{3}$
15. $\int_1^3\int_0^2 dy\,dx=4$
17. $\int_1^2\int_0^1 (y-x)\,dx\,dy=1$
19. $\int_1^2\int_2^3 \frac{2y}{x}\,dy\,dx=5\ln 2$
21. $\int_1^2\int_0^y (x+3y)\,dx\,dy=\frac{49}{6}$
23. $\int_0^1\int_0^y e^{y^2}dx\,dy=\frac{1}{2}(e-1)$
25. $\int_0^1\int_0^{\frac{x}{2}} e^{2y-x}dy\,dx=\frac{1}{2e}$
27. $\int_0^3\int_0^{x+2} \frac{y}{x+2}\,dy\,dx=\frac{117}{6}$
29. $\int_0^3\int_y^{2y} (1+3x^2+3y^2)\,dx\,dy=207$
31. $\int_1^3\int_x^{9-x^2} x^2dy\,dx=9.6$
33. $\int_0^1\int_{x^2}^{\sqrt{x}} (1+x^2-y^2)\,dy\,dx=\frac{5}{21}$
35. $\int_0^1\int_y^{y^2} x^2y^2\,dx\,dy=-\frac{1}{54}$
39. 9
41. $f'(x)=\dfrac{4x(3x^2-1)^2(3x^2+23)}{(x^2+5)^2}$

43. $\int 12xe^{3x^2-4}dx=2e^{3x^2-4}+C$

45. \$1183.30

47. (1,-3) is a saddle point

SECTION 8.7

1. $\frac{1}{10-3}\int_3^{10} P(x)\,dx$ represents the average amount of pollution between 3 and 10 miles from the facility.

3. $\frac{1}{24-12}\int_{12}^{24} D(t)\,dt$ represents the average monthly depreciation between months 12 and 24.

5. $\frac{1}{20-0}\int_0^{20} V(t)\,dt$ represents the average value of the piece of art in the first 20 years since its introduction.

7. $\frac{1}{120-60}\int_{60}^{120} L(t)\,dt$ represents the average production level for employees on the job between 60 and 120 days.

9. $\frac{1}{4-0}\int_0^4 (x^2+1)\,dx=\frac{19}{3}$

11. $\frac{1}{16-0}\int_0^{16}\sqrt{x}dx=\frac{8}{3}$

13. $\frac{1}{50-20}\int_{20}^{50} 440e^{-0.04x}dx\approx 115.13$

15. $\frac{1}{5-1}\int_1^5 x\sqrt{x^2-1}dx\approx 9.80$

17. Average value of the house after 4 years is approximately \$221,040.50.
19. Average marginal cost for units 400 to 500 is \$8.
21. Average number of units produced is approximately 4625.5.
23. Average revenue is \$23,550.
25. Average amount of pollution is approximately 9.25 units.
27. Average \$316.67 per week.
29. An average of 34,560 people are reached.
31. An average of 4108 units

35. $\frac{f(8)-f(5)}{8-5}=47$

37. concave up

39. $\int_0^5 \frac{x}{2x+3}dx\approx 1.40$

41. (1,3) is a relative minimum
43. Cost will decrease by approximately \$1509.

SECTION 8.8

1. $\int_c^d\int_{h_1}^{h_2} f(x,y)\,dy\,dx$

3. $\int_0^2\int_0^1 7xy\,dy\,dx=7$ cu. units

5. $\int_1^3\int_0^2 (x-y)\,dy\,dx=4$ cu. units

7. $\int_1^2\int_0^1 (y-x)\,dx\,dy=1$ cu. units

9. $\int_0^2\int_0^{3x} xy^2\,dy\,dx=\frac{288}{5}$ cu. units

11. $\int_0^2\int_x^{3x} xy\,dy\,dx=16$ cu. units

13. $\int_0^1\int_0^x \sqrt{1-x^2}\,dy\,dx=\frac{1}{3}$ cu. units

15. $\int_1^2\int_0^1 (2x+y)\,dy\,dx=3.5$ cu. units

17. $\int_0^4\int_0^2 (6-2y)\,dy\,dx=32$ cu. units

19. $\int_1^e\int_0^1 \ln x\,dy\,dx=1$ cu. unit

21. $\int_0^4\int_0^1 xy\,dx\,dy=4$ cu. units

23. $\int_0^4\int_0^{\sqrt{y}} (x+y)\,dx\,dy=\frac{84}{5}$ cu. units

25. $\int_0^2\int_0^{\sqrt{4-x^2}} x\,dy\,dx=\frac{8}{3}$ cu. units

27. $\int_0^1\int_x^{\sqrt{x}} ye^x\,dy\,dx\approx 0.141$ cu. units

31. 0

33. $S'(t)=3t^4(4t-3)^{-\frac{3}{4}}(7t-5)$

35. $-\frac{1}{4y^4}=\frac{x^3}{3}+C$

37. If 385,000 units of labor and capital is increased by 1 to 385,001 the maximum value of f(x,y) will decrease by approximately 0.41 units.
39. 2

SECTION 8.0

1. R(x) takes into account only newspaper advertising, while R(x,y) includes both newspaper and radio advertising.
3. f(1,3)=17
5. $\lim_{h\to 0}\frac{f(x,y+h)-f(x,y)}{h}=2x$
7. $\frac{\partial}{\partial x}P(4,10)=15$ means if x increases from 4 to 5, P will increase by approximately 15.

   $\frac{\partial}{\partial y}P(10,11)=4$ means if y increases from 11 to 12, P will increase by approximately 4.
9. $P_x(x,y)>0$, $P_y(x,y)>0$, $P_{xx}(x,y)<0$ and $P_{yy}(x,y)<0$ means P is increasing at a decreasing rate.
11. $f_x(2,1)=29,\ f_y(2,2)=-22$
13. $f_{xx}=2+9e^{3x},\ f_{yy}=9e^{3y},\ f_{xy}=f_{yx}=0$
15. (a) Production will increase by approximately 11.25.
    (b) Production will decrease by approximately 53 1/3.
17. $(-\frac{4}{7},\frac{5}{6})$ is a saddle point
19. (-6,1) is a relative minimum
21. (0,0) is a saddle point
    (3,3) is a relative minimum
23. no critical values
25. If the value 9.5 million is increased to 10.5 million, the maximum revenue will decrease by approximately 0.43 million dollars.
27. $(2,\frac{7}{2})$ is a relative maximum
29. (100,100) is a relative maximum
31. Allocate $15,000 for labor and $45,000 for capital to maximize production at approximately 12,432 units.

    $\lambda=-0.207$ means if resource allocation is increased from $60,000 to $60,001 the maximum will decrease by approximately 0.207.
33. $df(1,0)\approx 38.36$
35. (a) df(35,15)=-$905.
    (b) The actual change is -$971.
37. $\int_1^5\int_1^2 2x^2y\,dx\,dy=56$
39. $\int_2^4\int_0^2 (xy+xy^2)\,dy\,dx=28$
41. $\int_0^1\int_0^y (x+y^2)\,dx\,dy=\frac{5}{12}$
43. $\frac{1}{250-150}\int_{150}^{250}C(x)\,dx$ represents the average cost to produce between 150 and 250 units.
45. $\frac{1}{7-3}\int_3^7(-0.3t^2+6t+53)\,dt=75.1°F$
47. $\frac{1}{5}\int_0^5[-26,000(2+0.4t)^{-\frac{1}{2}}+35,000]\,dt\approx 19,769.55$
49. An average of approximately 3727.27 units produced

APPENDIX 1.1

1. 512
3. 128
5. $\frac{1}{xy^2}$
7. $\frac{1}{x}$
9. $x^{17}y^{16}$
11. 1
13. $\sqrt[10]{x^7}\sqrt{y}$
15. $\frac{2a}{4b^2}$
17. $m^{\frac{3}{10}}$
19. $3(\sqrt[4]{x^3})$
21. $\frac{8\sqrt{xy}}{y}$

APPENDIX 1.2

1. $a^2(a-b)$
3. $(x-10)(x+5)$

5. $(x^4+y^4)(x^2+y^2)(x+y)(x-y)$

7. $(x^2+1)(x^4-x^2+1)$

9. $(m-7)(m-6)$

11. $8a(a-2)(a-1)$

13. $(x+5)(x-3)$

15. $(2z+3)(2z+5)$

17. $5(t-2)(t-3)$

19. $(a^3+2)(a-2)(a^2+2a+4)$

21. $(8x-5)(x+2y)$

23. $bc(2a^2+5c)(2a-7c^2)$

APPENDIX 1.3

1. $\frac{-4}{x-3}$

3. $-y-4$

5. $\frac{x-6}{x-1}$

7. -1

9. $\frac{(x+4)(x-2)}{(x+8)(x-6)}$

11. $\frac{3x+6}{5x-2}$

13. $\frac{-24n}{(n+6)(n-6)}$

15. $\frac{k^2+2k-41}{(k+7)(k+2)}$

17. $\frac{(5x+1)(4x^2-x-4)}{(4x+3)(3x-2)(x-1)}$

19. $(y+6)(y-6)$

APPENDIX 1.4

1. $\log_3 81=4$

3. $\log_{64} 4=\frac{1}{3}$

5. $2^2=4$

7. $c^a=2n$

9. $\log_3 3^5=5$

11. 1

13. $\log_3 5x-\log_3 y$

15. $4\log_2 x+\frac{1}{3}\log_2 y-\frac{2}{5}\log_2 z$

17. $\log_2 \frac{xy}{z}$

19. 39

21. $\pm\frac{\sqrt{3}}{15}$

23. $\frac{\log 9}{\log 7}$

25. $7+\frac{\log 10}{\log 4}$

27. $\frac{3+\frac{\log 21}{\log 8}}{4}$

Index

A

B

C

D

E

F

G

H

I

L

M

U

V

X

Special characters